Molecular Biomethods Handbook

Molecular Biomethods Handbook

Edited by

Ralph Rapley

and

John M. Walker

University of Hertfordshire, Hatfield, UK

HUMANA PRESS ✳ TOTOWA, NEW JERSEY

Cover illustration: Adapted from Figure 7 in Chapter 48, "Immunocytochemistry," by Lorette C. Javois.

Cover design by Patricia F. Cleary.

For additional copies, pricing for bulk purchases, and/or information about other Humana titles, contact Humana at the above address or at any of the following numbers: Tel.: 973-256-1699; Fax: 973-256-8341; E-mail: humana@ mindspring.com; Website: http://humanapress.com

Printed in the United States of America. 10 9 8 7 6 5 4 3 2 1

Preface

There have been numerous advances made in many fields throughout the biosciences in recent years, with perhaps the most dramatic being those in our ability to investigate and define cellular processes at the molecular level. These insights have been largely the result of the development and application of powerful new techniques in molecular biology, in particular nucleic acid and protein methodologies.

The purpose of this book is to introduce the reader to a wide-ranging selection of those analytical and preparative techniques that are most frequently used by research workers in the field of molecular biology. Clearly, within the constraints of a single volume, we have had to be selective. However, all of the techniques described are core methods and in daily research use. We have aimed to describe both the theory behind, and the application of, the techniques described. For those who require detailed laboratory protocols, these can be found in the references cited in each chapter and in the laboratory protocol series *Methods in Molecular Biology*™ and *Methods in Molecular Medicine*™ published by Humana Press.

Molecular Biomethods Handbook begins with all the essential core nucleic acid techniques, such as the extraction and separation of nucleic acids, their detection and preliminary characterization, through to the application of gene probes and blotting techniques, each of which are described in separate chapters. The DNA technology theme is then developed to cover more complex areas of characterization, such as gene cloning and library production, mapping, and expression, as well as more applied fields, such as transgenesis, in vitro protein expression, mutagenesis, and DNA profiling. Within a number of larger, more generic chapters, a range of further specific techniques may be found. The latter part of the book similarly focuses on protein and related techniques. In this way we hope to provide the reader with information and background on many of the general concepts and specific techniques used in molecular biology at the present time. Although many chapters may stand alone in their own right, the majority are interlinked. Accordingly, these are cross-referenced in order to provide a coherent context in which a chapter may be read.

Molecular Biomethods Handbook should prove useful to undergraduate students (especially project students), postgraduate researchers, and all research scientists and technicians who wish to understand and use new techniques, but do not yet have the necessary background to set us specific techniques. In addition, it will be useful for all those wishing to update their knowledge of particular techniques. All chapters have been written by well-established research scientists who run their own research programs and who use the methods on a regular basis. In sum, then, our hope is that this book will prove a useful source of information on all the major molecular biotechniques in use today, as well as a valuable text for those already engaged in or just entering the field of molecular biology.

Ralph Rapley
John M. Walker

Contents

Preface v

Contributors xi

 1 Extraction of Total RNA from Tissues and Cultured Cells
    Sandeep Raha, Mingfu Ling, and Frank Merante 1

 2 Isolation of Total Cellular DNA from Tissues and Cultured Cells
    Frank Merante, Sandeep Raha, and Mingfu Ling 9

 3 Gel Electrophoresis of DNA
    Duncan R. Smith 17

 4 S1 Nuclease Analysis of RNA
    Louis Lefebvre and Stéphane Viville 35

 5 Detecting mRNA by Use of the Ribonuclease Protection Assay (RPA)
    Ralf Einspanier and Annette Plath 51

 6 Gene Probes
    Marilena Aquino de Muro 59

 7 Nonradioactive Labeling of DNA
    Thomas P. McCreery and Terrence R. Barrette 73

 8 Southern Blot Analysis
    Paolo A. Sabelli 77

 9 Northern Blot Analysis
    Paolo A. Sabelli 89

10 DNA Sequencing
    Chris Spencer 95

11 Autoradiography and Fluorography
    Bronwen Harvey 109

12 Mobility Shift Assays
    Nigel J. Savery and Stephen J. W. Busby 121

13 cDNA Libraries
    Ian G. Cowell 131

14 Genomic DNA Libraries
    Bimal D. M. Theophilus 145

15 λ as a Cloning Vector
    Rupert Mutzel 153

16 Plasmid-Derived Cloning Vectors
    Craig Winstanley and Ralph Rapley 165

17 M13 and Phagemid-Based Cloning Vectors
    Ralph Rapley 181

18 Retroviral Vectors: *From Laboratory Tools to Molecular Medicines*
    Richard G. Vile, Anna Tuszynski, and Simon Castleden *193*

19 Baculovirus Vectors
    Azeem Ansari and Vincent C. Emery *219*

20 Gene Transfer and Expression in Tissue Culture Cells of Higher Eukaryotes
    M. Alexandra Aitken, Selina Raguz, and Michael Antoniou *235*

21 Plant Transformation
    Andy Prescott, Rob Briddon, and Wendy Harwood *251*

22 Restriction Fragment-Length Ploymorphisms
    Elaine K. Green *271*

23 Genome Mapping
    Jacqueline Boultwood *281*

24 Yeast Artificial Chromosomes
    Angela Flannery and Rakesh Anand *287*

25 Polymerase Chain Reaction
    Ralph Rapley *305*

26 In Vitro Transcription
    Martin J. Tymms *327*

27 In Vitro Translation
    Martin J. Tymms *335*

28 Site-Directed Mutagenesis
    John R. Adair and T. Paul Wallace *347*

29 Transgenic Techniques
    Roberta M. James and Paul Dickinson *361*

30 DNA Profiling: *Theory and Practice*
    Karen M. Sullivan *383*

31 Radiolabeling of Peptides and Proteins
    Arvind C. Patel and Stewart R. Matthewson *401*

32 Protein Electrophoresis
    Paul Richards *413*

33 Free Zone Capillary Electrophoresis
    David J. Begley *425*

34 Protein Blotting: *Principles and Applications*
    Peter R. Shewry and Roger J. Fido *435*

35 Ion-Exchange Chromatography
    David Sheehan and Richard Fitzgerald *445*

36 Size-Exclusion Chromatography
    Paul Cutler *451*

37 Hydrophobic Interaction Chromatography
    Paul A. O'Farrell *461*

38 Affinity Chromatography
    George W. Jack *469*

Contents

39 Reversed-Phase HPLC
   Bill Neville 479
40 Glycoprotein Analysis
   Terry D. Butters 491
41 Protein Sequencing
   Bryan J. Smith and John R. Chapman 503
42 Solid-Phase Peptide Synthesis
   Gregg B. Fields 527
43 Protein Engineering
   Sudhir Paul 547
44 Monoclonal Antibodies
   Christopher Dean and Helmout Modjtahedi 567
45 Phage-Display Libraries
   Julia E. Thompson and Andrew J. Williams 581
46 Enzyme-Linked Immunosorbent Assay (ELISA)
   John R. Crowther 595
47 Epitope Mapping: *Identification of Antibody Binding Sites
   on Protein Antigens*
   Glenn E. Morris 619
48 Immunocytochemistry
   Lorette C. Javois 631
49 Flow Cytometry
   Robert E. Cunningham 653
50 Mass Spectrometry
   John R. Chapman 669
51 The Technique of *In Situ* Hybridization: *Principles and Applications*
   Desiré du Sart and K. H. Andy Choo 697

Index 721

Contributors

JOHN R. ADAIR • *Axis Genetics PLC, Cambridge, UK*

M. ALEXANDRA AITKEN • *Department of Experimental Pathology, UMDS, Guys Hospital, London, UK*

RAKESH ANAND • *Genome Group, Zenecca Pharmaceuticals, Macclesfield, UK*

AZEEM ANSARI • *Department of Virology, Royal Free Hospital School of Medicine, London, UK*

MICHAEL ANTONIOU • *Department of Experimental Pathology, UMDS, Guys Hospital, London, UK*

MARILENA AQUINO DE MURO • *Department of Biological Sciences, Heriot Watt University, Edinburgh, UK*

TERRENCE R. BARRETTE • *ImaRx Pharmaceuticals, Inc., Tucson, AZ*

DAVID J. BEGLEY • *Biomedical Sciences Division, King's College, London, UK*

JACQUELINE BOULTWOOD • *LRF Molecular Hematology Unit, John Radcliffe Hospital, Oxford, UK*

ROB BRIDDON • *John Innes Centre, Norwich Research Park, Norwich, UK*

STEPHEN J. W. BUSBY • *School of Biochemistry, University of Birmingham, Birmingham, UK*

TERRY D. BUTTERS • *Department of Biochemistry, University of Oxford, UK*

SIMON CASTLEDEN • *ICRF, Laboratory of Cancer Gene Therapy, The Rayne Institute, St. Thomas Hospital, London, UK*

JOHN R. CHAPMAN • *Cheshire, UK*

K. H. ANDY CHOO • *Murdoch Institute for Birth Defects, Royal Childrens Hospital, Parkville, Australia*

IAN G. COWELL • *Department of Biochemistry and Genetics, School of Medicine, University of Newcastle, Newcastle upon Tyne, UK*

JOHN R. CROWTHER • *IAEA, Vienna, Austria*

ROBERT E. CUNNINGHAM • *Department of Cellular Pathology, AFIP, Washington, DC*

PAUL CUTLER • *SmithKline Beecham, Welwyn, UK*

CHRISTOPHER DEAN • *Section of Immunology, Institute of Cancer Research, McElwain Laboratories, Sutton, UK*

PAUL DICKINSON • *MRC Human Genetics Unit, Western General Hospital, Edinburgh, UK*

DESIRÉE DU SART • *Murdoch Institute for Birth Defects, Royal Childrens Hospital, Parkville, Australia*

RALF EINSPANIER • *Forschungszentrum, Milch Lebensmittel, Institut Physiologie, Freising, Germany*

VINCENT C. EMERY • *Department of Virology, Royal Free Hospital School of Medicine, London, UK*

ROGER J. FIDO • *IACR, Long Ashton Research Station, Bristol, UK*

GREGG B. FIELDS • *Department of Laboratory Medicine and Pathology, University of Minnesota, Minneapolis, MN*

RICHARD FITZGERALD • *Department of Biochemistry, University College, Cork, Ireland*

ANGELA FLANNERY • *Genome Group, Zenecca Pharmaceuticals, Macclesfield, UK*

ELAINE K. GREEN • *Department of Oncology, Birmingham Childrens Hospital, Birmingham, UK*

BRONWEN HARVEY • *Amersham International, Amersham, UK*

WENDY HARWOOD • *John Innes Centre, Norwich Research Park, Norwich, UK*

GEORGE W. JACK • *CAMR, Salisbury, UK*

ROBERTA M. JAMES • *MRC Human Genetics Unit, Western General Hospital, Edinburgh, UK*

LORETTE C. JAVOIS • *Department of Biology, The Catholic University of America, Washington, DC*

LOUIS LEFEBRVE • *Wellcome/CRC Institute, Cambridge, UK*

MINGFU LING • *Department of Genetics, Hospital for Sick Children, Toronto, Canada*

STEWART R. MATTHEWSON • *Amersham International, Amersham, UK*

THOMAS P. MCCREERY • *ImaRx Pharmaceuticals, Inc., Tucson, AZ*

FRANK MERANTE • *Department of Genetics, Hospital for Sick Children, Toronto, Canada*

HELMOUT MODJTAHEDI • *Section of Immunology, Institute of Cancer Research, McElwain Laboratories, Sutton, UK*

GLENN E. MORRIS • *MRIC, N. E. Wales Institute, Wrexham, UK*

RUPERT MUTZEL • *Fakultät für Biologie, Universität Konstanz, Germany*

BILL NEVILLE • *SmithKline Beecham, The Frythe, Welwyn, UK*

PAUL A. O'FARRELL • *University of Hertfordshire, Hatfield, UK*

ARVIND C. PATEL • *Amersham International, Amersham, UK*

SUDHIR PAUL • *Department of Anesthesiology, University of Nebraska Medical Center, Omaha, NE*

ANNETTE PALTH • *Forschungszentrum, Milch Lebensmittel, Institut Physiologie, Freising, Germany*

ANDY PRESCOTT • *John Innes Centre, Norwich Research Park, Norwich, UK*

SELINA RAGUZ • *Department of Experimental Pathology, UMDS, Guys Hospital, London, UK*

SANDEEP RAHA • *Department of Genetics, Hospital for Sick Children, Toronto, Canada*

RALPH RAPLEY • *University of Hertfordshire, Hatfield, UK*

PAUL RICHARDS • *Leicester University, Leicester, UK*

PAOLO A. SABELLI • *Institute of Molecular and Cell Biology, National University of Singapore, Republic of Singapore*

NIGEL J. SAVERY • *School of Biochemistry, University of Birmingham, Birmingham, UK*

DAVID SHEEHAN • *Department of Biochemistry, University College, Cork, Ireland*

PETER R. SHEWRY • *IACR, Long Ashton Research Station, Bristol, UK*

BRYAN J. SMITH • *Celltech Therapeutics, Slough, UK*

DUNCAN R. SMITH • *Molecular Biology Laboratory, Tan Tock Seng Hospital, Republic of Singapore*

CHRIS SPENCER • *Amersham International, Amersham, UK*

KAREN M. SULLIVAN • *Department of Molecular and Life Sciences, University of Abertay, Dundee, UK*

BIMAL D. M. THEOPHILUS • *Department of Hematology, Birmingham Childrens Hospital, Birmingham, UK*

JULIA E. THOMPSON • *Cambridge Antibody Technology, Inc., Cambridge, UK*

ANNA TUSZYNSKI • *ICRF, Laboratory of Cancer Gene Therapy, The Rayne Institute, St. Thomas Hospital, London, UK*

MARTIN J. TYMMS • *Institute for Reproduction and Development, Monash Medical Center, Clayton, Australia*

RICHARD G. VILE • *ICRF, Laboratory of Cancer Gene Therapy, The Rayne Institute, St. Thomas Hospital, London, UK*

STÉPHANE VIVILLE • *Wellcome/CRC Institute, Cambridge, UK*

T. PAUL WALLACE • *Axis Genetics PLC, Cambridge, UK*

ANDREW J. WILLIAMS • *Cambridge Antibody Technology, Inc., Cambridge, UK*

CRAIG WINSTANLEY • *Department of Biomedical Sciences, Bradford University, Bradford, UK*

1

Extraction of Total RNA from Tissues and Cultured Cells

Sandeep Raha, Mingfu Ling, and Frank Merante

1. Introduction

The isolation of intact, high quality, total cellular RNA is often the starting point for many molecular biological procedures (**Fig. 1**). There are numerous general methods for the isolation of total cellular RNA *(1–9)*. There are also many specialized methods for the isolation of RNA from specific tissues *(8–10)*, various cell types *(11)*, and sub-cellular organelles *(7,12,13)*. In addition, a number of methods describe the simultaneous isolation of RNA and DNA *(14–17)*. Generally, the rationale for any isolation procedure is to solubilize cellular components and simultaneously inactivate intracellular RNases while maintaining biologically active RNA. Therefore, the goal is to acquire purified cellular RNA in an intact form that can be a substrate for further manipulations, such as in vitro translation, RNase protection, reverse transcription, and Northern-blot analysis.

Most isolation procedures combine the use of one or more agents, such as organic solvents (i.e., phenol, chloroform) *(18)*, detergents (i.e., sodium dodecyl sulfate [SDS], *N*-lauryl sarcosyl, Nonidet P 40, sodium deoxycholate) *(12,17–19)*, or chaotropic salts, such as guanidinium isothiocyanate (GITC), trifluoroacetate, or urea *(2,5)*. Often β-mercaptoethanol is added to further assist in protein denaturation *(6)*. In combination, these agents denature proteins, inactivate RNases, and remove lipids, thereby improving yield and the quality of the isolated RNA. Recently a specialized procedure for the isolation of RNA from blood was introduced that permitted the simultaneous lysis of cells and the precipitation of RNA and DNA by the use of a commercially available cationic surfactant (Catrimox-14). The RNA is then extracted using hot formamide and precipitated *(20)*.

Generally, contaminating cellular DNA must be separated from RNA. Many procedures perform this separation by utilizing cesium chloride *(21)* or cesium trifluoroacetate *(22,23)* density gradients. Although effective, the obvious limitations of these methods are the need for an ultracentrifuge, difficulty in handling very small sample sizes, and the number of samples that can be processed at any one time owing to limited rotor space. Another approach is to ethanol-precipitate selectively the DNA from the RNA *(14,15,24)* or RNA from the DNA *(25)*. These procedures capitalize on the large size of genomic DNA (>100 kb) and its ability to be precipitated and removed by

From: *Molecular Biomethods Handbook*
Edited by: R. Rapley and J. M. Walker © Humana Press Inc., Totowa, NJ

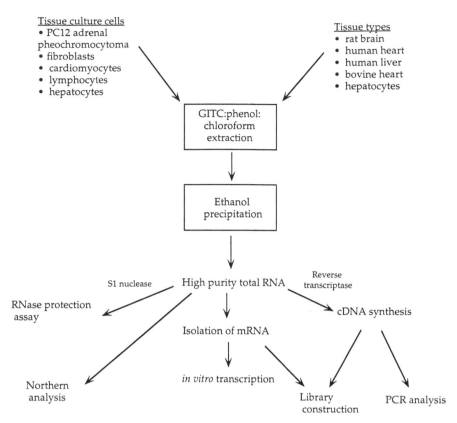

Fig. 1. Schematic diagram of the potential applications of the RNA isolated using the GITC:phenol method.

spooling or centrifugation. Selective precipitation of RNA can also be performed by using LiCl *(21,26)*, but this involves a lengthy precipitation period (for example, 4 h at 4°C).

There are a number of procedures for the isolation of RNA from yeast, bacteria, and plants. These techniques are essentially very similar to those employed for RNA isolation from mammalian cells, except that they use a variety of specialized steps during the cells' lysis procedure. These include the use of glass beads or other grinding agents to disrupt the rigid cell walls present in yeast and plants, as well as lysozyme treatment to target the outer walls of gram-negative bacteria. These methods are highly specialized and vary depending on the type of organism being targeted. These procedures are beyond the scope of this chapter and the reader is referred to one of several reviews for additional information *(27–29)*.

A very efficient RNA isolation procedure was introduced by Chomczynski and Sacchi *(3)*. This method capitalizes on the ability of GITC to denature proteins and inactivate intracellular RNases. The presence of *N*-lauroylsarcosine and β-mercaptoethanol in the mixture also enhances the solubilization properties of the GITC-extraction buffer. In this method, an acidic phenol extraction (at pH < 5.0) selectively retains cellular DNA in the organic phase and aids in the extraction of proteins and lipids. The addition of chloroform further removes lipids and establishes two distinct phases: an organic phase containing the DNA, proteins, and lipids, and an aqueous phase containing the RNA. The advantage of this phase-extraction process is the use of a single tube for extraction, its scalability, and the flexibility to process multiple samples. Because of its usefulness, many modifications to the original method have been described, including precipitation with LiCl *(6,11)*, or 95% ethanol *(25)*. An additional acidic phenol extraction can improve the purity of the isolated RNA *(10)*. In fact, in recent years, several commercial reagents have been made available to the molecular biologist that capitalize on the relative ease of this method. In most cases the reagent is marketed as a single-step RNA isolation solution that contains a mixture of chaotropic agents and phenol. These reagents also contain a proprietary dye that is soluble in the organic phase, which helps to discriminate the organic phase from the aqueous phase during extraction. Such reagents are very useful in isolating RNA from small amounts of tissue or processing only a few plates of cultured cells.

2. General Precautions

The majority of RNA isolation procedures are based on the use of either guanidinium hydrochloride (GHCl) or GITC. These are both extremely toxic substances and should be handled with care.

RNase activity is extremely difficult to inactivate and will survive autoclaving; for this reason it is best to bake glassware at 280°C overnight. Water should be treated with diethylpyrocarbonate (DEPC) (final concentration of 0.1%) overnight and autoclaved *(14,15)*. This treatment is important for the elimination of contaminating RNase activity. All other chemicals should be of molecular biology grade or comparable whenever possible. In addition, it is important to wear gloves during all procedures to prevent contamination of the sample with RNases present on the skin.

3. General Protocols for the Processing of Tissues and Cultured Cells

Isolation of RNA from mammalian sources usually involves the processing of either tissues or cultured cells. In general, isolation of RNA from tissues requires that the tissue be ground thoroughly following freezing in liquid nitrogen prior to cell lysis. To maximize yield, it is important to ensure the complete solubilization of the powdered tissue. If required, the tissue can be gently mixed using a wrist-action shaker for better solubilization. On the other hand, cells growing as a monolayer can be lysed directly on the tissue-culture dish. This results in complete cell lysis and a greater yield of RNA. Regardless of whether tissue or cultured cells are being processed, it is important that the detergent responsible for cell lysis and the chaotropic reagent required for neutralization of RNase activity be present in the same buffer mixture.

The majority of protocols currently employed for RNA isolation exploit either rapid extraction with an organic solvent (either phenol or a phenol:chloroform mixture) or a combination of CsCl-density centrifugation followed by an extraction using solvents. The former phase extraction methods exploit the tendency of most proteins and small DNA fragments (<10 kb) to fractionate into the organic phase at low pH. Larger DNA fragments and some proteins remain at the interphase between the organic and aqueous phases. It is advisable to sacrifice some of the aqueous phase close to the interface in the interest of improved RNA purity. In general, this is only a very small fraction of the total aqueous-phase volume and results in little reduction in RNA yield. If required, a second extraction with chloroform will remove any trace amounts of phenol that may be contaminating the aqueous phase. This in turn improves the purity of the RNA, especially if it is to be used for reverse transcription (RT) and subsequent polymerase chain reaction (PCR).

Despite the rapidity usually associated with the phase-extraction methods, methods employing CsCl density-gradient centrifugation, originally developed by Chirgwin and his colleagues *(1)* remains one route to obtaining high-quality RNA. In fact, we have found that this method remains very useful for isolating large amounts of high-mol-wt RNA, usually from tissues.

4. Quantitation of RNA

A diluted sample should be quantitated by measuring the absorbance (A) at 260 nm. An absorbance of 1 is equivalent to an RNA concentration of 40 µg/mL. Therefore the yield = $A_{260} \times 40 \times$ dilution factor. The purity of the RNA can be assessed in two ways. The first is a determination of the absorbance of the sample at 260 and 280 nm. This is a reflection of the protein contamination in the sample. A ratio of the absorbance at 260/280 nm >1.8 is generally considered good quality RNA. Second, formaldehyde agarose-gel electrophoresis should be performed for the most direct evaluation of RNA quality *(30)*. One should ensure there is little or no ethidium bromide staining near the origin, indicating an absence of DNA contamination. **Figure 2** shows human-liver RNA isolated using a phase-extraction procedure that has been fractionated on a 1.2% agarose gel. One should also ensure that the 28S and 18S ribosomal RNA species are intact. If necessary, add a commercially available RNase inhibitor for long-term storage.

5. Advantages and Disadvantages of Phase-Extraction Methods

One disadvantage of phase extraction procedures is that large tissue samples (>1 g) cannot be processed effectively. For larger sample sizes, it may be more feasible to use a CsCl-gradient procedure *(1)*. Large masses of tissue, even when homogenized, may result in decreased yields. This problem may be circumvented by performing several smaller isolations. The problems with sample size are likely to hamper isolations from particularly fibrous tissues, such as heart muscle. Other tissues, such as liver or brain, which are particularly rich in RNases, may be problematic because lack of rapid homogenization may lead to partial degradation of the RNA.

Methods relying on phase extraction may result in the contamination of the RNA with DNA *(11)*. This may be problematic in PCR procedures in which the cDNA being amplified is the same size as the corresponding genomic product. In such instances, discriminating PCR primers should be used whenever possible, but a few groups have attempted to circumvent this problem by using a modified isolation procedure. Monstein et al. *(31)* subjected the aqueous phase containing the RNA to RNase-free DNase treatment. However, this requires a second phenol/chloroform extraction to repurify the RNA-containing aqueous phase. Others have attempted to shear the DNA, in an attempt to aid its separation into the organic phase, by passing the solubilized fraction through a large-gage syringe needle *(11)*. Nevertheless, it is important to mix well, but gently, during the GITC:phenol:chloroform extraction process to ensure sufficient removal of DNA contaminants.

In considering the advantages of phase extraction, one must first examine the simplicity of this procedure. It is a very rapid, single-step protocol that allows for the

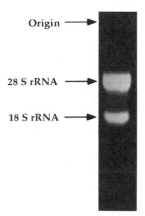

Origin →

28 S rRNA →

18 S rRNA →

Fig. 2. Total cellular RNA from rat liver. RNA was isolated from rat liver using the GITC:phenol method. Twenty-two micrograms of RNA were fractionated on a formaldehyde agarose gel and visualized by ethidium bromide staining. The positions of the 29S and 18S rRNA bands as well as the origin are indicated.

removal of DNA, protein, and lipids in a single extraction. The use of phenol with an acidic buffer assists in the removal of DNA from the aqueous phase *(27)* and maintains the integrity of the RNA during the isolation process *(32)*. Other methods utilize multiple extractions to remove proteins, lipids, and DNA from the sample. A number of procedures also facilitate the removal of DNA by spooling the very large DNA fragments onto glass rods *(14,15)*. Although effective, this process is time-consuming, and one must be careful in the initial steps not to shear these large DNA fragments.

Phase extraction methods are readily scalable. As a result, they are able to accommodate a variety of sample sizes and permit minimal sample handling with minimal reagent preparation. The rapidity of the method may contribute to higher-quality RNA than that obtained using isolation procedures that involve differential precipitation. GITC is a more effective and potent inhibitor of RNases than most of the denaturants used in other methods, and results in the isolation of substantially larger amounts of intact RNA. **Figure 2** shows an ethidium-bromide-stained gel (1.2%) following the separation of total RNA isolated from human liver using a GITC protocol *(3)*. Any good RNA-isolation procedure should yield RNA that exhibits sharp and distinct 28S and 18S rRNA bands (**Fig. 2**).

Phase extraction methods are applicable to the isolation of RNA from a wide variety of tissues and cultured cells. In general, tissue-culture cells provide larger yields, probably owing to more efficient lysis in the initial steps of the procedure and the actively dividing nature of the cells.

The RNA isolated using phase extraction is of high quality, having an A_{260}/A_{280} ratio of at least 1.8, indicating that it is free of contaminating proteins. As a result, the RNA isolated using this procedure is useful in a number of common molecular biological applications.

Such methods provide good quality RNA that is obtainable in a reasonably short period of time (1–2 h). This is substantially more rapid than procedures that employ the use of CsCl gradients *(1)* or differential precipitation of RNA using LiCl *(11)*.

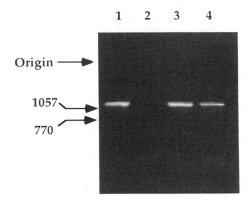

Fig. 3. RT-PCR amplification of cDNA sequences using GITC:phenol-isolated RNA. Ten micrograms of total cellular RNA from human heart (lanes 1 and 3) and human liver (lane 4) were used to synthesize first-strand cDNA. A fraction of each was used to amplify the glyceraldehyde-3-phosphate dehydrogenase (lane 1) or the pyruvate dehydrogenase E1α sequences (lanes 3 and 4). Lane 2 represents a no-DNA control.

An example of the quality of the PCR product obtained using cDNA generated from total RNA obtained by phase extraction is illustrated in **Fig. 3**. This RNA is of sufficient quality to be used for RT and subsequent PCR amplification of specific sequences. In fact, the high quality of the RNA allows us to carry out routinely our PCR reactions using total RNA, rather than purified mRNA. **Figure 3** shows the ethidium-bromide-stained gel following the separation of PCR products amplified from human heart and liver RNA (**Fig. 3**). The isolated RNA was used to generate first-strand cDNA that was subsequently used to amplify a human pyruvate dehydrogenase and a glycerol 3-phosphate dehydrogenase sequence. In both cases, the PCR amplifications yield the desired products, which are approx 1.1 kb in size.

Similarly, RNA isolated via this method can also be used for the construction of cDNA libraries. However, one may consider the isolation of mRNA for the generation of cDNA libraries, especially in cases in which low-copy-number messages are required.

Other applications for this procedure are RNase protection assays or in vitro translation using cell-free translation systems. Therefore, the GITC method of Chomczynski and Sacchi and its various modifications provide high quality RNA required for these procedures in a short period of time.

References

1. Chirgwin, J. M., Przybyla, A. E., MacDonald, R. J., and Rutter, W. J. (1979) Isolation of biologically active ribonucleic acid from sources enriched in ribonuclease. *Biochemistry* **18**, 5394–5399.
2. Auffray, C. and Rougeon, F. (1980) Purification of mouse immunoglobulin heavy-chain messenger RNAs from total myeloma tumor RNA. *Eur. J. Biochem.* **197**, 303–414.
3. Chomczynski, P. and Sacchi, N. (1987) Single step method of RNA extraction by acid guanidinium thiocyanate-phenol-chloroform extraction. *Anal. Biochem.* **162**, 156–159.
4. Emmett, M. and Petrack, B. (1988) Rapid and quantitative preparation of cytoplasmic RNA from small numbers of cells. *Anal. Biochem.* **173**, 93–95.

5. Gough, N. M. (1988) Rapid and quantitative preparation of cytoplasmic RNA from small numbers of cells. *Anal. Biochem.* **173,** 93–95.

6. Meier, R. (1988) A universal and efficient protocol for the isolation of RNA from tissues and cultured cells. *Nucleic Acids Res.* **16,** 2340.

7. Wilkinson, M. (1988) A rapid and convenient method for isolation of nuclear, cytoplasmic and total cellular RNA. *Nucleic Acids. Res.* **16,** 10,934.

8. Nemeth, G. G., Heydemann, A., and Bolander, M. E. (1989) Isolation and analysis of ribonucleic acids from skeletal tissues. *Anal. Biochem.* **183,** 301–304.

9. Tavangar, K., Hoffman, A. R., and Kraemer, F. B. (1990) A micromethod for the isolation of total RNA from adipose tissue. *Anal. Biochem.* **186,** 60–63.

10. Monstein, H.-J., Nylander, A.-G., and Chen, D. (1995) RNA extraction from gastrointestinal tract and pancreas by a modified Chomczynski and Sacchi method. *Biotechniques* **10,** 341–344.

11. Vauti, F. and Siess, W. (1993) Simple method of RNA isolation from human leucocytic cell lines. *Nucleic Acids Res.* **23,** 4852,4853.

12. Wilkinson, M. (1988) RNA isolation: a mini-prep method. *Nucleic Acids Res.* **16,** 10,933.

13. Hatch, C. L. and Bonner, W. M. (1987) Direct analysis of RNA in whole cell and cytoplasmic extracts by gel electrophoresis. *Anal. Biochem.* **162,** 283–290.

14. Raha, S., Merante, F., Proteau, G., and Reed, J. K. (1990) Simultaneous isolation of total cellular RNA and DNA from tissue culture cells using phenol and lithium chloride. *Gene Anal. Tech.* **7,** 173–177.

15. Merante, F., Raha, S., Reed, J. K., and Proteau, G. (1994) The simultaneous isolation of RNA and DNA from tissues and cultured cells. *Methods Mol. Biol.* **32,** 113–120.

16. Kitlinska, J. and Wojcierowski, J. (1995) RNA isolation from solid tumor tissue. *Anal. Biochem.* **228,** 170–172.

17. Coombs, L. M., Pigott, D., Proctor, A., Edymann, M., Denner, J., and Knowles, M. A. (1990) Simultaneous isolation of DNA, RNA and antigenic protein exhibiting kinase activity from small tour samples using guanidine isothiocyanate. *Anal. Biochem.* **188,** 338–343.

18. Ogretmen, B. and Saffa, A. R. (1995) Mini-preparation of total RNA for RT-PCR from cultured human cells. *Biotechniques* **19,** 374–376.

19. Chattopadhyay, N., Kher, R., and Godbole, M. (1993) Inexpensive SDS/phenol method from RNA extraction from tissues. *Biotechniques* **15,** 26.

20. Dhale, C. E. and MacFarlane, D. E. (1993) Isolation of RNA from cells in culture using Catrimox-14 cationic surfactant. *Biotechniques* **15,** 1102–1105.

21. Chan, V.T.-W., Flemming, K. A., and McGee, J. O'D. (1988) Simultaneous extraction from clinical biopsies of high-molecular-weight DNA and RNA: comparative characterization by biotinylated and ^{32}P-labeled probes on Southern and Northern blots. *Anal. Biochem.* **168,** 16–26.

22. Zarlenga, D. S. and Gamble, H. R. (1987) Simultaneous isolation of preparative amounts of RNA and DNA from *Trichinella spiralis* by cesium tribluoroacetate isopycnic centrifugation. *Anal. Biochem.* **162,** 569–574.

23. Mirkes, P. E. (1985) Simultaneous banding of rat embryo DNA, RNA and protein in cesium trifluoroacetate gradients. *Anal. Biochem.* **148,** 376–383.

24. Karlinsey, J., Stamatoyannopoulos, G., and Enver, T. (1989) Simultaneous purification of DNA and RNA from small numbers of eukaryotic cells. *Anal. Biochem.* **180,** 303–306.

25. Siebert, P. D. and Chenchik, A. (1993) Modified acid guanidium thiocyanate-phenol-chloroform RNA extraction method which greatly reduces DNA contamination. *Nucleic Acids Res.* **23,** 2019,2020.

26. Krieg, P., Amtmann, E., and Sauer, G. (1983) The simultaneous extraction of high-molecular-weight DNA and RNA from solid tumors. *Anal. Biochem.* **134,** 288–294.

27. Kingston, R. E., Chomczynski, P., and Sacchi, N. (1988) Single-step RNA isolation from cultured cells or tissues, in *Current Protocols in Molecular Biology* (Ausubel, S. M., Brent, R., Kingston, R. E., Moore, D. D., Seidman, J. G., Smith, J. A., Struhl, K., eds.), Green, Brooklyn, NY, pp. 4.2.4–4.2.8.

28. Kormanec, J. and Farkasovsky, M. (1994) Isolation of total RNA from yeast and bacteria and detection of rRNA in Northern blots. *Biotechniques* **71,** 838–842.

29. Bugos, R. C., Chiang, V. L., Zhang, X. H., Campbell, E. R., Polida, G. K., and Campbell, W. H. (1995) RNA isolation from plant tissues recalcitrant to extraction in guanidine. *Biotechniques* **19,** 734–737.

30. Fourney, R. M., Miyakoshi, J., Day, R. S., and Paterson, M. C. (1988). Northern blotting: efficient RNA staining and transfer. *Focus* **10,** 5–7.

31. Monstein, H., Nylander, G., and Chen, D. (1995) RNA extraction from gastrointestinal tract and pancreas by a modified Chomczynski and Sacchi method. *Biotech.* **19,** 340–343.

32. Noonberg, S. B., Scott, G. K., and Benz, C. C. (1995) Effect of pH on RNA degradation during guanidinium extraction. *Biotechniques* **19,** 731–733.

2

Isolation of Total Cellular DNA
from Tissues and Cultured Cells

Frank Merante, Sandeep Raha, and Mingfu Ling

1. Introduction

The isolation of high-quality cellular DNA is often a starting point for a variety of molecular-biology techniques. These include Southern-blot analysis, PCR amplification, and genomic-library construction (**Fig. 1**). It is often necessary to use the DNA from a single preparation for a number of different applications. Some applications require higher quality DNA than do other applications, such as genomic library construction vs routine polymerase chain reaction (PCR) amplification. Therefore, it is advantageous to employ an isolation procedure that provides high quality DNA. Generally, quality is indicated by the absence of contaminating RNA, proteins, lipids, and other cellular constituents that may interfere with restriction enzymes, ligases, and thermostable DNA polymerases. More importantly, the preparation should be free of contaminating DNA nucleases, which can nick and degrade high-mol-wt DNA. The large size of mammalian genomic DNA also requires that the isolation method be gentle enough to minimize mechanical shear stress, which would fragment the large genomic DNA during the course of purification. In addition, a good DNA isolation method should be able to accommodate a wide variety of tissues and cell types. Many methods have been published that have been optimized for a specific application *(1–3)*, whereas other methods allow the simultaneous isolation of DNA and RNA from the same sample *(4,5)*. Commonly used procedures employ a buffer containing one or several detergents; for example, SDS *(6)*, NP-40, or Triton X-100 *(7)*. These detergents lyse cells and assist in the removal of proteins from the DNA *(8)*. More thorough deproteinization is achieved by the use of proteinase K in the lysis buffer. This enzyme is active in the presence of SDS and remains functional at elevated temperatures (56–65°C) *(3,9)*. Under these conditions denatured or partially denatured proteins are more easily digested by proteinase K. In contrast, most other enzymes (e.g., DNases) are denatured under these conditions. Hence, the combination of proteinase K, the buffer components, and the elevated temperature are important for the inactivation of endogenous nucleases, in the disruption of cellular integrity and in the release of DNA from the nuclear compartment. This enzyme is so effective in the deproteinization of a crude preparation that relatively pure DNA can simply be obtained by a single precipitation

From: *Molecular Biomethods Handbook*
Edited by: R. Rapley and J. M. Walker © Humana Press Inc., Totowa, NJ

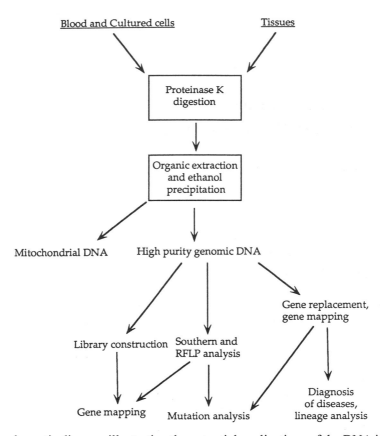

Fig. 1. A schematic diagram illustrating the potential applications of the DNA isolated using the proteinase K:phenol method.

with isopropanol *(10)* following proteinase K treatment. Alternatively, contaminating RNA and proteins can be removed by a selective LiCl precipitation step *(3)*.

Residual protein and contaminating lipids can be effectively removed by extraction with phenol and chloroform *(4,5,8,11)*. When phenol is used for this purpose it is important to pre-equilibrate the organic phase with a buffer of pH ≥ 8.0 to ensure the effective partitioning of the DNA into the aqueous phase *(8)*. Otherwise, the bulk of the DNA will be retained in the organic phase and in the interface. Once cellular proteins and lipids are removed, contaminating RNA can easily be degraded by treating the sample with DNase-free RNase, followed by a chloroform extraction to remove the RNase.

Only high grade, glass distilled phenol should be use for the extraction of nucleic acids to avoid sample degradation by organic peroxides present in aged phenol. Phenol should be discarded if it exhibits an orange or brown discoloration, which indicates significant oxidation. The shelf life of phenol can be extended by storage at 4°C, by the addition of β-mercaptoethanol to a final concentration of 0.2%, and by the addition of hydroxyquinoline to a final concentration of 0.1% *(8,12)*. Hydroxyquinoline also imparts a light orange color to the phenol, enabling easier phase recognition during extraction. Any darkening indicates excessive oxidation and the phenol should not be used. Once equilibrated, phenol should be protected from light and stored refrigerated in small aliquots.

Total cellular DNA can be effectively precipitated from the aqueous phase by the addition of either ethanol (two volumes) or isopropanol (one volume). Throughout the extraction procedure it is recommended that wide bore, disposable, plastic transfer pipets be used to minimize shearing.

In many procedures nuclei are first isolated prior to the release of genomic DNA *(13)*. In these methods, cells are lysed in an osmotically balanced buffer containing Triton X-100 to maintain the integrity of the nuclei. A simple centrifugation step then separates the nuclei away from the bulk of the cellular membranes and other subcellular organelles. Deproteinization is further achieved by proteinase K digestion and the DNA is recovered by ethanol precipitation *(13)*. Although useful, this procedure is less amenable with some cultured cells (e.g., fibroblasts) because they produce a large degree of extracellular matrix. Additionally, nonnuclear DNA (i.e., mitochondrial DNA) is lost during the nuclear isolation step.

Alternative methods avoid the use of organic extractions by employing an overnight proteinase K digestion step followed by the addition of a saturated NaCl solution to precipitate proteins *(14)*. Following a brief centrifugation, the DNA in the supernatant is recovered by ethanol precipitation. Although effective for easily lysed white blood cells, this method may not provide high yields of DNA from cells that produce a rich extracellular matrix (e.g., fibroblasts).

A one-step extraction procedure *(15)* using guanidinium-HCl was introduced that capitalized on its strong chaotropic properties. Although we have found that this method is rapid and simple, the isolated DNA is often difficult to resuspend. A second adaptation uses guanidinium isothiocynate along with a heparin treatment step to dissociate nuclear proteins from DNA *(16)*. Generally, heparin should be avoided, or subsequently removed, if the sample is to be used for PCR because it is a potent inhibitor of Taq DNA polymerase *(17)*.

An additional method embeds the nuclei in agarose *(18)* in order to protect the DNA from mechanical shearing during isolation. In this procedure proteinase K digestion in the presence of SDS is used to remove proteins and lipids from the embedded DNA. However, the embedding can only be performed with suspension cells and is not suitable for adherent cells or tissues *(18)*. Other methods dialyse the DNA against polyethylene glycol *(2)* to minimize shearing. In general, dialysis requires numerous buffer changes, making this procedure laborious and time consuming, thereby limiting the number of samples that can be accommodated. Other options to improve the purity of the isolated DNA involve the use of column chromatography *(6)*, but as with agarose embedding and polyethylene glycol dialysis, the processing of multiple samples is cumbersome.

DNases are dependent on Mg^{2+} and Ca^{2+} for activity *(19)*. EDTA (at least 2 mM) included in the DNA extraction buffer chelates these cations and thereby prevents the degradation and random nicking of high-mol-wt DNA by DNases *(7)*. Washing the cells with phosphate-buffered saline (PBS) without Mg^{2+} and Ca^{2+} minimizes the activities of DNases for the same reason. The procedure presented in this chapter is based on that of Blin and Stafford *(20)* and has incorporated improvements from the literature and from our own experience.

2. Materials and Solutions

All reagents are precooled or kept at 4°C before use.

1. Proteinase K stock: 20 mg/mL in water at –20°C.
2. 10% SDS. Care should be taken in handling SDS.
3. Phenol saturated with TE (pH 8.0)/chloroform (1:1).
4. Chloroform /isoamyl alcohol (24:1).
5. DNase-free RNase stock (1 mg/mL). It is important to treat the commercial RNase by boiling in 10 mM Tris-HCl (pH 7.5)/15 mM NaCl for 15 min and allowing it to cool slowly. This inactivates DNase activities.
6. Extraction solution (ES): 0.1 M EDTA, 0.2 M NaCl, 0.05 M Tris-HCl (pH 8.0), 0.5% SDS, 50 mg/mL RNase (added immediately prior to use from a frozen stock).

3. Procedure

3.1. Tissues

1. Fresh tissue (0.1–1 g) should be rinsed with cold PBS and immediately frozen in liquid nitrogen. Fresh, or previously frozen, tissues are homogenized in liquid nitrogen. The mortar and pestle should be precooled with liquid nitrogen. A Waring blender (precooled), instead, can also be used to homogenize the tissue sample. Gloves and eye protection should be worn while grinding. Also, all biological materials should be treated as potentially hazardous. If possible, **step 1** should be performed in a fume-hood. Grinding in liquid nitrogen serves two purposes: First, nucleases, which result from the disruption of cells during the grinding process, are inactivated temporarily at this temperature. In addition, tissues are generally much easier to grind at this temperature. The grinding may be aided by initially hammering the tissue with a pestle into small pieces. More liquid nitrogen should be added if the tissue appears to be thawing. After the tissue is ground into a fine powder, 10 vol of extraction solution (ES) are added to 1 vol of tissue to completely immerse the tissue. One gram of tissue can be regarded as being approximately equal to 1 mL. It should be noted that relatively large amount of RNase is used in the extraction solution to overcome the partial inhibition from SDS and EDTA present in the ES buffer.
2. Slowly add proteinase K (final concentration of 100 mg/mL) to the above cell suspension while gently mixing with a glass rod. Incubate this solution at 55°C for a minimum of 2–3 h with occasional manual or mechanical gentle mixing. It is important to stress that mixing should be gentle to minimize DNA breakage. If necessary, another aliquot of proteinase K can be added in order to obtain a relatively clarified and viscous solution. Incubation for periods as long as 16 h does not cause damage to DNA *(21)* and is recommended.
3. An equal volume of phenol (buffer saturated)/chloroform is added to the cell lysate. The extraction should be performed in a fumehood. The two phases are mixed gently to form an emulsion. Polystyrene tubes should be avoided because they will be dissolved by organic solvents. Centrifuge at 1500g for 5 min to separate the two phases. The aqueous (top) phase is transferred to a new tube using a wide bore transfer pipet. Two additional extractions, each with an equal volume of chloroform/isoamyl alcohol, are performed. The role of isoamyl alcohol is to prevent foaming that sometimes occurs during the organic extraction. In most cases, however, it can be omitted. Further extractions may improve the purity of the DNA, but at the expense of DNA yield and size.
4. Add 2 vol of ethanol and sodium acetate (NaAc) to a final concentration of 0.3 M (from a 3 M stock, pH 4.8) to the aqueous phase and mix gently. At this point there are two ways to recover the DNA. A sealed Pasteur pipet with its tip bent into a "U" shape can be used to spool the DNA precipitate. Alternatively, centrifugation at 10,000g for 5 min can be

used to collect the DNA. The DNA precipitate still on the Pasteur pipet, or as a pellet, should be washed with 70% ethanol to decrease residual salt and briefly dried under vacuum or air-dried at 37°C to evaporate the ethanol. Overdrying should be avoided since dehydrated DNA is very difficult to resuspend. The DNA should now appear relatively clear (transparent but not white).

5. The ethanol-free DNA is dissolved in TE (pH 8.0) at a concentration of 1 mg/mL. The dissolution sometimes may take up to 16 h. If necessary, this process can be accelerated by gentle mechanical motion. A small amount of this DNA is diluted in water to measure the absorbance at 260 and 280 nm. Both the DNA purity and the DNA concentration are then estimated from the measurements. The concentration (mg/mL) can be calculated as:

$$A_{260} \times 50 \times l \text{ (light path in cm)} \times \text{dilution factor}$$

The purity can be estimated from the ratio of A_{260}/A_{280}. A ratio of 1.8–2.0 suggests minimal protein contamination. Typically, 100–200 mg DNA can be isolated from one plate of cells, 15 mL of blood, or 0.1 g of tissue. Generally, the yield of DNA is variable because of different tissues or cell types. The size as well as the concentration of DNA can also be estimated by agarose gel electrophoresis. The purified DNA is best stored at 4°C. Freezing and thawing cycles may result in the degradation of DNA; therefore, they should be avoided. If necessary, aliquot the DNA before freezing for long-term storage.

3.2. Blood

Ten milliliters of blood is collected in a Vacutainer containing 100 mL of 15% EDTA. The sample is then centrifuged at 1500g for 15 min. The plasma (supernatant) is discarded and the cell pellet is resuspended in 2 mL of ES. Alternatively, white blood cells can be isolated by using Ficoll-Paque (Pharmacia) according to the manufacturer's instructions. Proceed with **steps 2–5**.

3.3. Adherent Cells

In order to minimize contamination with polysaccharides, which are not effectively removed by proteolytic digestion, extractions, or ethanol precipitation, the cells (or animals in the case of whole tissue) should be starved overnight prior to sample preparation. Wash cells with 5 mL of PBS and add 2 mL of ES to a plate containing approx 1×10^7 to 1×10^8 cells. Scrape cells with a rubber policeman and remove the suspension to a fresh tube. Rinse the plate with an additional 1 mL of ES to rinse the plate and combine. Alternatively, cells can be trypsinized with 1 mL 0.1% trypsin + 1 mM EDTA, centrifuged and resuspended in the ES. Proceed with **steps 2–5**.

3.4. Cells in Suspension

Approximately 1×10^7 to 1×10^8 cells are pelleted at 1500g for 5 min in a Sorval or equivalent centrifuge and resuspended in 2 mL of ES. Proceed with **steps 2–5**.

4. Notes and Comments

DNA obtained using this method can be used for most applications (**Fig. 1**). The DNA is sufficiently intact and pure for use in Southern analysis (**Fig. 2**), including restriction fragment length polymorphism (RFLP) analysis. Laird et al. *(10)* have suggested that the phenol/chloroform and chloroform extractions may be avoided and satisfactory Southern blots would still result. Although significant labor is saved, the prepared DNA may not be readily digested with all restriction enzymes. For analyzing

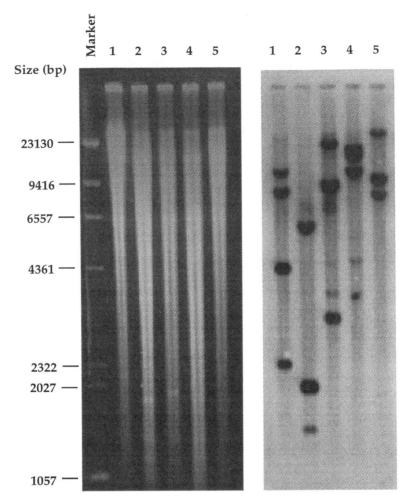

Fig. 2. Southern blot analysis of isolated DNA. Total genomic DNA from human brain was digested with various restriction endonucleases and fractionated by agarose gel electrophoresis. Each lane contains 10 mg of genomic DNA digested with *Bam*HI, *Eco*RI, *Hin*dIII, *Pst*I, and *Sac*I (lanes 1–5, respectively). **(A)** Ethidium bromide stained agarose gel of digested DNA. l-*Hin*dIII/fX174-*Hin*cII digested DNA was used as a size marker. **(B)** The corresponding autoradiogram of the Southern blot hybridized with a human ^{32}P-labeled cytochrome c oxidase subunit VIa cDNA probe.

a large number of samples by PCR, other more rapid DNA preparation methods can be used *(22)*. In the case of blood samples, heparin should be avoided because it inhibits Taq polymerase. If necessary, heparin can be removed by using heparinase II *(16)*. A quick DNA preparation method with a 1 h digestion and without extraction has also been reported *(23)*, showing that PCR products were successfully amplified from such DNA. When a long PCR product is required, the DNA quality (size and purity) becomes very important *(24)*. Usually, genomic library construction is the most demanding application using genomic DNA. In order to completely represent all genes and to facilitate cloning, both the integrity and purity of the DNA are essential. DNAs obtained using the method described here allows amplification of 24-kb products *(24)* and is sufficient for cloning in L-based vectors but may not be suitable for cosmid cloning *(19)*. Noninvasive methods, such as agarose-embedding *(18)*, have been recommended for the amplification of PCR products up to 30 kb *(24)*. The advantages and disadvantages of the method described here are briefly summarized below.

4.1. Advantages

1. High degree of DNA purity. This method uses phenol/chloroform extraction to effectively remove proteins and lipids.
2. High degree of DNA integrity. The size of DNA obtained is around 150 kb, which is sufficient for most applications and is larger than that of DNA obtained from guanidinium-based methods.
3. Shorter labor time. Time required for this method is shorter than that of agarose embedding or PEG dialysis methods.
4. Reasonable DNA yield. The DNA yield of the method described here is comparable to that obtained guanidinium based methods.
5. Flexibility. The scale can be adjusted according to the need.

4.2. Disadvantages

1. In comparison to guanidinium based methods, a relatively long procedure time is required for this method.
2. Organic solvents are necessary.

References

1. Graham, D. D. (1978) The isolation of high molecular weight DNA from whole organisms of large tissue masses. *Anal. Biochem.* **85,** 609–613.
2. Longmire, J. L., Albright, K. L., Meincke, L. J., and Hildebrand, C. E. (1987) A rapid an simple method for the isolation of high molecular weight cellular and chromosome-specific DNA in solution without the use of organic solvents. *Nucleic Acids Res.* **15,** 859.
3. Reymond, C. D. (1987) A rapid method for the preparation of multiple samples of eukaryotic DNA. *Nucleic Acids. Res.* **15,** 8118.
4. Merante, F., Raha, S., Reed, J. K., and Proteau, G. (1994) The simultaneous isolation of RNA and DNA from tissues and cultured cells, in *Methods in Molecular Biology, vol. 31: Protocols for Gene Analysis* (Harwood, A. J., ed.), Humana, Totowa, NJ, p. 113–120.
5. Raha, S., Merante, F., Proteau, G., and Reed, J. K. (1990) Simultaneous isolation of total cellular RNA and DNA from tissue culture cells using phenol and lithium chloride. *Gene Anal. Tech.* **7,** 173–177.
6. Potter, A. W., Hanham, A. F., and Nestmann, E. R. (1985) A rapid method for the extraction and purification of DNA from human leukocytes. *Cancer Lett.* **26,** 335–341.
7. Lahiri, D. K. and Schnebel, B. (1993) DNA isolation by a rapid method from human blood samples: effects of $MgCl_2$, EDTA, storage time, and temperature on DNA yield and quality. *Biochem. Genet.* **(31)7/8,** 321–328.
8. Wallace, D. M. (1987) Large and small scale phenol extractions. *Meth. Enzymol.* **152,** 33–41.
9. Jeanpierre, M. (1987) A rapid method for the purification of DNA from blood. *Nucleic Acids Res.* **(15)22,** 9611.
10. Laird, P. W., Zijderveld, A., Linders, K., Rudnicki, M. A., Jaenisch, R., and Berns, A. (1991) Simplified mammalian DNA isolation procedure. *Nucleic Acids Res.* **19,** 42–93.
11. Albarino, C. G. and Romanowski, V. (1994) Phenol extraction revisited: a rapid method for the isolation and preservation of human genomic DNA from whole blood. *Mol. Cell. Probes* **8,** 423–427.

12. Manitatis, T., Fritsch, E. R., and Sambrook, J. (1982) *Molecular Cloning: A Laboratory Manual.* Cold Spring Harbor Laboratory, Cold Spring Harbor, NY.

13. Grimberg, J., Nawoschik, S., Belluscio, L., McKee, R., Turck, A., and Eisenberg, A. (1989) A simple and efficient non-organic procedure for the isolation of genomic DNA from blood. *Nucleic Acids Res.* **17,** 83–90.

14. Miller, S. A., Dykes, D. D., and Polesky, H. F. (1988) A simple salting out procedure for extracting DNA from human nucleated cells. *Nucleic Acids. Res.* **16,** 12–15.

15. Botwell, D. D. L. (1987) Rapid isolation of eukaryotic DNA. *Anal. Biochem.* **162,** 463–465.

16. Koller, C. A. and Kohli, V. (1993) Purification of genomic DNA using heparin to remove nuclear proteins. *Nucleic Acids Res.* **21,** 29–52.

17. Beutler, E., Gelbart, T., and Kuhl, W. (1990) Interference of heparin with the polymerase chain reaction. *Biotechniques* **9,** 166.

18. Konat, G., Gantt, G., Laszkiewicz, I., and Hogan, E. L. (1990) Rapid isolation of genomic DNA from animal tissues. *Exp. Cell. Res.* **190,** 294–296.

19. Sambrook, J., Fritsch, E. F., and Maniatis, T. (1989) *Molecular Cloning: A Laboratory Manual.* Cold Spring Harbor Laboratory, Cold Spring Harbor, NY.

20. Blin, N. and Stafford, D. N. (1976) A general method for isolation of high molecular weight DNA from eukaryotes. *Nucleic Acids Res.* **3,** 2303–2308.

21. Sharma, R. C., Murphy, A. J. M., DeWald, M. G., and Schimke, R. T. (1993) A rapid procedure for isolation of RNA free genomic DNA from mammalian cells. *Biotechniques* **14,** 176,177.

22. Goldenberger, D., Perschill, I., Ritzler, M., and Altwegg, M. (1995) A simple "universal" DNA extraction procedure using SDS and proteinase K is compatible with direct PCR amplification. *PCR Methods Appl.* **4,** 368–371.

23. Simard, L. R., Gingras, F., and Labuda, D. (1991) Direct analysis of amniotic fluid cells by multiplex PCR provides rapid prenatal diagnosis for Duchenne muscular dystrophy. *Nucleic Acids Res.* **19,** 2501.

24. Cheng, S., Chen, Y., Monforte, J. A., Higuchi, R., and Van Houten, B. (1995) Template integrity is essential for PCR amplification of 20- to 30 kb sequences from genomic DNA. *PCR Meth. Appl.* **4,** 294–298.

3

Gel Electrophoresis of DNA

Duncan R. Smith

1. Introduction

The ability to separate and visualize IDNA strands from as small as 5 base pairs (bp) to as large as 5,000,000 bp forms a fundamental cornerstone of today's techniques of molecular biology. The wide size range of DNA molecules that can be handled effectively derives from the application of three essentially similar—and to a large extent overlapping—techniques of gel electrophoresis, namely polyacrylamide-gel electrophoresis (PAGE), agarose-gel electrophoresis, and pulse-field gel electrophoresis (**Table 1**). In each case, DNA molecules are moved through a gel matrix by the application of an electric field. The gel matrix consists of pores through which the DNA molecule must pass. In both polyacrylamide- and agarose-gel electrophoresis, a voltage applied at the ends of the gel produces an electic field with a strength determined by both the length of the gel and the potential difference at the ends. Owing to the presence of negatively charged phosphate groups along the backbone of the DNA molecule, the DNA chain will migrate toward the anode at the application of an electric field. Because the charge to mass ratio of DNA molecules is constant, the rate of migration in the absence of the gel would also be constant. However, in the gel matrix, it is frictional drag through the gel that essentially governs the rate of migration: Larger molecules move more slowly because of greater frictional drag and because they worm their way through the pores of the gel less efficiently than smaller molecules. In the following sections, the background to and applications of these techniques will be more fully explored.

2. Agarose Gel Electrophoresis

2.1. The Gel Matrix

Agarose is a linear polymer that is extracted from seaweed and is sold as a white powder. The powder is melted in buffer and allowed to cool, whereby the agarose forms a gel by hydrogen bonding. The hardened matrix contains pores, the size of which depends on the concentration of agarose. The concentration of agarose is referred to as a percentage of agarose to a volume of buffer (w/v), and agarose gels are normally in the range of 0–3%. The resolving range of the gel is determined by the concentration of the agarose.

From: *Molecular Biomethods Handbook*
Edited by: R. Rapley and J. M. Walker © Humana Press Inc., Totowa, NJ

Table 1
Resolution of DNA Gels

Gel matrix	%	Range of separation	Comments
Polyacrylamide	20	5–100 bp	Xylene cyanol migrates at around 50 bp
	15	20–150 bp	
	12	50–200 bp	
	8	60–400 bp	Standard concentration for DNA sequencing gels (denaturing)
	6	100–600 bp	
	3.5	1 kb–2 kb	Xylene cyanol migrates at around 450 bp
Agarose	3	0.1–1 kb	Can separate small fragments differing from each other by a small amount. Must be poured rapidly
	2	0.2–1.5 kb	Do not allow to cool to 50°C before pouring
	1.5	0.3–3 kb	As for 0.8%, bromophenol blue runs at about 500 bp
	1	0.5–5 kb	As for 0.8%
	0.8	1–71 kb	General purpose gel; separation not greatly affected by choice of running buffer; bromophenol blue runs at about 1 kb
	0.6	3–10kb	Gel mechanically weak but with care can use low melting-point agarose
	0.4	5–30 kb	As above
	0.3	5–40 kb	Gel very weak mechanically; separation in 20–40 kb range improved by using high ionic-strength buffers; use only high melting-point agarose
Pulse field		up to 2000 kb	FIGE (Field-Inversion Gel Electrophoresis)
		up to 5000 kb	CHEF (Contour-clamped homogenous-field electrophoresis)

2.2. Equipment

Many different apparatus arrangements have been devised to run agarose gels; for example, they can be run horizontally or vertically, and the current can be conducted by wicks or the buffer solution. However, today the "submarine" gel system is almost universally used. In this method, the agarose gel is formed on a supporting plate, and the plate is then submerged into a tank containing a suitable electrophoresis buffer. Wells are preformed in the agarose gel with the aid of a "comb" that is inserted into the cooling agarose before the agarose has gelled. Into these wells are loaded the sample to be analyzed, which has been mixed with a dense solution (a loading buffer) to ensure that the sample sinks into the wells.

Apparatus for submarine gel electrophoresis consists of four main parts: a power supply (capable of at least 100 V and currents of up to 100 mA), an electrophoresis tank, a casting plate, and a well-forming comb. Apparatus for agarose gel electrophoresis is available from many commercial suppliers, but tends to be fairly expensive.

Alternatively, apparatus can be "home-made" with access to a few sheets of Perspex and some minor electrical fittings. The construction of such apparatus is outside the scope of this chapter but can be found in Sambrook et al. *(1)*, Sealey and Southern *(2)*, or Boffey *(3)*.

2.3. Sample Loading

DNA samples to be analyzed by agarose gel electrophoresis are normally mixed with a loading buffer and placed in the preformed wells with the aid of either a glass capillary or, more commonly, a plastic pipet tip. The tip should be rinsed thoroughly between successive samples, or preferably discarded after one use. Loading buffers are usually made as a 3–5× concentrated stock and serve several roles. They normally have three main constituents:

1. High-density solution: Loading buffers normally contain one solution of high density, so when mixed with a DNA sample to be analyzed give the solution sufficient density to sink to the bottom of the gel well. Commonly used high density solutions include 50% glycerol, 30% Ficoll (Pharmacia Fine Chemicals), or 40% sucrose.
2. Tracking dyes: Loading buffers also commonly contain one or two tracking dyes (such as xylene cyanol or bromophenol blue), which migrate in an electric field in the same direction as the DNA. The migration of the tracking dyes is somewhat dependent on the concentration of the gel, but in a standard 0.7–0.8% gel, xylene cyanol comigrates with DNA of about 500 bp and bromophenol blue comigrates with DNA of about 4–5 kbp, enabling a rough idea of the migration of the DNA to be estimated without stopping the gel. Some workers use Orange G, which comigrates with the smaller DNA sizes, and so approximately marks the leading edge of the gel front. All dyes tend to quench the fluorescence of ethidium bromide, and so can obscure the presence of DNA bands.
3. Chelating agent: The final constituent in loading buffers tends to be a chelating agent, normally ethylenediaminetetra-acetic add (EDTA). This complexes with divalent cations (such as Mg^{2+}) and will stop all enzymatic reactions. For this reason, loading buffers can be used as a "stop" solution to terminate enzymatic reactions, such as restriction digests.

Agarose does not have an unlimited capacity for DNA, and the limit depends on both the size range of the DNA and the complexity of the size distribution. For a sample of DNA yielding only one or a few bands, such as a plasmid digest or a polymerase chain reaction (PCR) reaction, the upper limit is approx 10 $\mu g/cm^2$ of pocket face. Where the sample contains a continuous distribution of DNA sizes (such as would result from a digest of genomic DNA), the upper limit can be as high as 50 $\mu g/cm^2$ of pocket face. Beyond these limits the gel becomes overloaded and the DNA tends to run as a smear.

2.4. Migration of DNA

The migration of DNA through a specific gel is modified by a number of parameters, including the conformation of the DNA molecules, the buffer composition, the concentration of agarose, and the applied voltage.

2.4.1. Agarose Concentration

The concentration of agarose plays a role in determining the resolving range of a particular gel (**Table 1**). A standard gel would contain approx 0.7–0.8% agarose, and would resolve bands in the range 500–15 kb. A reduction in the percentage to 0.2% would resolve bands in the range of 5–40 kb, and a concentration of 2–3% would be useful in resolving DNA fragments in the range 100–1000 base pairs.

2.4.2. Buffer Composition

Several commonly used buffers exist (**Table 2**) that have slightly different effects on DNA mobility. In general high ionic-strength buffers (i.e., Loenings E buffer) produce better analytical gels but are not recommended for preparative use. Low ionic-strength buffers (glycine buffer, Tris-borate EDTA buffer, and Tris-acetate buffer) are preferred for use in preparative gel electrophoresis, although resolution tends to be slightly poorer.

2.4.3. DNA Conformation

Whereas linear duplex DNA migrates in a manner that is inversely proportional to $\log_{10}$ of the DNA length *(4)*, this is not true for supercoiled or nicked-circle plasmid DNA *(5)*. Supercoiled plasmids have a smaller cross-section than their DNA size would indicate and so pass through the pores more easily than would be expected. Hence, they migrate faster then the comparative linear DNA molecule of the same molecular weight. Conversely, nicked-circle DNA migrates more slowly than would be expected from its apparent molecular weight. These migration patterns are altered by the inclusion of ethidium bromide in the gel, which alters the mobility of all forms of DNA.

2.4.4. Applied Voltage

The actual voltage to be applied across the gel is a balance between a number of factors. The higher the voltage applied across the gel, the quicker the bands will migrate. However, the application of a voltage across the gel results in a heating effect in the gel. In some cases, this heating effect can lead to distortion of the bands, and in rare cases when a high voltage is applied over a period of time can melt the gel! This, however, is an extreme situation and although the gel is run at room temperature normally the buffering liquid provides sufficient heat dissipation. Another problem at high voltages is buffer breakdown. This can be partially compensated for by circulating the buffer between the two chambers of the gel tank. Normally, however, the resolution of the bands is more important than speed of separation, so a voltage can be chosen that does not give the aforementioned problems. With large DNA bands, optimal resolution can be achieved by using a relatively low voltage for a longer time. With smaller bands, however, diffusion becomes more of a problem, so low voltages for a long period of time can result in diffuse bands. Diffusion can be kept down with higher volts per centimeter of gel, but this will tend to decrease resolution between bands. Therefore, an optimal balance must be found between the desired resolution between bands and the diffuseness of the bands themselves.

Similarly, the length of the gel can play a role. The longer the gel, the further the bands migrate and the further the separation between bands, but again, the greater the diffusion within a band.

2.5. Visualization of DNA

The intercalating dye ethidium bromide (3,8-diamino-6-ethyl-5-phenyl-phenanthridium bromide) is commonly included in both the gel and in the running buffer

Table 2
Commonly Used DNA Electrophoresis Buffers

Buffer	Description	Solution
Leonings E	High ionic-strength buffer for agarose gels, but not recommended for preparative agarose gel electrorophoresis.	For 5 L of 10X: 218 g Tris base, 234 g $NaH_2PO_4 \cdot 2H_2O$ and 18.8 g $Na_2EDTA \cdot 2H_2O$
Glycine	Low ionic-strength buffer for agarose-gel electrophoresis. Very good for preparative gels, but can also be used for analytical gels.	For 2 L 10X: 300 g of glycine, 300 mL of 1 M NaOH (or 12 g pellets), and 80 mL 0.5 M EDTA, pH 8.0
TBE (Tris-Borate EDTA)	Low ionic-strength buffer, can be used for both preparative and analytical agarose gels Most commonly used buffer for polyacrylamide gels. Occasionally used at 0.5X TEE for sequencing qels.	For 5 L 10X: 545 g Tris, 278 g of boric acid, and 46.5 g of EDTA
TAE (Tris-Acetate)	Good for analytical agarose gels and preparative agarose gels when the DNA will be purified by glass beads. Sometimes used in pulse-field electrophoresis, but can lead to gel heating.	For 1 L 50X: 242 g Tris base, 57.1 mL glacial acetic acid, and 100 mL of 0.5 M EDTA, pH 8.0
GTBE buffer	Commonly used buffer for pulse-field gel electrophoresis. Essentially 0.5X TBE with 100 mM glycine added. Alternatively, 0.5X TBE can be used.	For 1 L of 5X: 250 mL of 10X TBE (see above), 250 mL 2 M glycine

(6). The dye intercalates between the stacked bases of duplex DNA, and when illuminated with UV light (260–360 nm) fluoresces orange-red. The nonintercalated dye does not fluoresce. Hence, DNA bands in an agarose gel are seen as orange-red bands when viewed under UV light. This provides a means of following the migration of DNA down the length of the gel so the electrophoretic run can be monitored. This is especially convenient when the gel is poured on UV-transparent plastic rather than glass, because the gel can be removed from the tank and viewed under UV illumination without removing the gel from the supporting plate. However, it must be noted that ethidium bromide promotes the damage of DNA when viewed under UV light (photonicking). For this reason the viewing of gels with ethidium bromide present should be as brief as possible, and less damage to the DNA occurs if the gel is viewed under 300 nm than at 254 nm. The presence of ethidium bromide also causes alterations in the mobility of DNA molecules (especially supercoiled plasmid DNA). For this reason, some workers prefer to perform the electrophoresis minus ethidium bromide in the gel and buffer and stain the gel in a 0.5 µg/mL solution at the end of the electrophoretic run. If the electrophoretic run is performed in the presence of ethidium bromide, it should be noted that ethidium ions migrate in the opposite direction to the DNA in the electric field; hence, if the dye is not included in the running buffer the dye will leach from the gel. Also note that ethidium bromide is both carcinogenic and mutagenic and therefore must be handled with extreme caution and disposed of in accordance with local regulations. It should also be remembered that UV light itself is harmful at the intensities normally used to view gels, and safety goggles (or a full-face mask) should be worn. Some UV transilluminators can give painful burns on the face, forearms, and wrists of workers who spend long periods of time with gels being illuminated by UV light.

A currently used alternative to the use of ethidium bromide and UV transillumination is silver-stain detection (essentially the same stain used to detect nanogram quantities of protein). Although it has been reported to be sensitive to nanogram quantities of DNA, it can only be used at the end of the electrophoretic run (and so does not allow monitoring at intervals during the electrophoretic run) and is not suitable where the recovery of the DNA from the gel is required.

More recently, FMC BioProducts (Rockland, ME) have marketed SYBR® green nucleic acid stain, which they claim is 25–100 times more sensitive than ethidium bromide as well as being less mutagenic. SYBR® green can also be used to stain for single-stranded DNA (which stains poorly with ethidium bromide). Best results are obtained by staining after the electrophoretic run, which may limit utility somewhat. We are still evaluating this stain in my laboratory, but the lower mutagenicity is attractive to some staff, despite the higher cost than ethidium bromide.

2.6. DNA Size Markers

Although bromophenol blue and xylene cyanol give a rough estimate of the migration of DNA through a gel, they do not tell the exact size of the product's run on the gel. Although it is possible to produce migration curves based on the molecular weight and the mobility of the bands, this technique is seldom used now. The quickest and most convenient method of calculating DNA sizes (whether of PCR products or digest products) is by direct comparison with DNA standards. These are loaded onto the gel in a spare pocket at the same time as the samples are loaded. DNA size markers consist of a number of bands of known sizes (calculated through DNA sequencing), and can either be produced in the laboratory (by digesting λ DNA or plasmid DNA) or purchased from commercial companies (although these tend to be expensive). Currently, several different size markers cover the whole range of DNA sizes in relatively short increments. Because these markers are also DNA they will fluoresce with ethidium bromide and will be recorded on the photographic record.

3. Applications of Agarose Gel Electrophoresis

Agarose-gel electrophoresis tends to have two main application, either analytical or preparative.

3.1. Analytical-Gel Electrophoresis

Analytical agarose gels are most commonly run to check the status of a DNA sample, either with respect to yield or to determine the completeness of a restriction digest or PCR reaction. Optimum resolution on a standard gel is obtained at about 10 V/cm, whereas fragments smaller than 1 kb are normally resolved better at higher V/cm. A standard analytical gel can easily detect a single DNA band of 100 ng, and in some cases can detect bands present at as low as 10 ng. Care must be taken not to overload the pocket face of the gel, and too great an amount of DNA loaded will result in smearing of the bands. Samples should not contain excessive salt because this will lead to "necking" of the bands. After electrophoresis, it is normal to view the gel on a UV light transilluminator and record the data via either a photodocumentation system or more commonly a Polaroid camera with an appropriate red filter.

3.2. Analytical-Gel Electrophoresis–Southern Blotting

Southern blotting *(7)* is used to identify the presence of a sequence of interest from a complex digest of either cloned DNA or genomic DNA. Southern blotting has many applications, from gene mapping to the identification of transgenic animals. DNA is initially cleaved with a restriction enzyme to produce fragments. These fragments are then fractionated according to size on an agarose gel.

The DNA is then partially cleaved by depurination (to facilitate transfer of larger DNA fragments) and alkali denatured by sequential soaking of the gel in solutions containing HCl and NaOH, respectively. The denatured DNA fragments are then transferred to a solid matrix or filter (either a nitrocellulose or a nylon membrane) by either capillary transfer or vacuum blotting for subsequent hybridization to a specific labeled probe.

3.3. Preparative-Gel Electrophoresis

Preparative-gel electrophoresis is used to purify a single DNA fragment of interest from a complex mixture of digest products, or to purify a single product after PCR. In each case, the mixture of DNAs is separated by size on an appropriately resolving agarose gel. The gel can be either standard agarose or one of the modified low-gelling agarose powders that can be melted at approx 70°C. When the band of interest has been resolved sufficiently, the DNA can be extracted from the gel matrix. There are a large number of methods of purifyng DNA from gels. The following are two of the more common ones.

1. A trench is cut into the gel immediately in front of the band of interest. The trench is filled with glycerol and the gel briefly re-electrophoresed to move the DNA from the gel into the glycerol solution (this can be monitored by UV). The glycerol-DNA solution is then extracted via pipet.

2. After electrophoresis, a slit is cut into the gel immediately in front of the band of interest into which is inserted a piece of NA-45 paper. The gel is then re-electrophoresed and the DNA band of interest will migrate onto the paper. The paper is then rinsed and DNA eluted by heating the paper to 70°C. DNA can then be phenol-extracted and precipitated.

The following methods involve physically removing the gel slice containing the band of interest from the surrounding gel matrix with a razor blade.

1. If low-gelling agarose has been used, the gel is diluted 1:1 with buffer, salt added to 0.5 *M*, and the slice melted at 70°C. When liquid, the solution is extracted with phenol/chloroform and precipitated. Note that DNA prepared by this method is not suitable for all subsequent procedures because of the presence of inhibitors in the agarose. It is best then to pass the DNA fragment down an exclusion column.

2. The gel slice of interest is placed on a bed of siliconized glass wool in a small (0.5 mL) microcentrifuge tube that has a hole punched in the bottom. This tube is then placed inside a larger microcentrifuge tube (1.5 mL) and subjected to centrifugation at high speed for 10–15 min. A solution of DNA collects at the bottom of the large tube and can be further purified. Some workers freeze the gel slice before centrifugation to improve yield *(8)*.

3. The gel slice of interest is dissolved in NaI, a chaotropic salt, that at concentrations of around 4 *M* is able to solubilize agarose. Glass beads are then added, which in this concentration of NaI efficiently bind the released DNA fragments. RNA, proteins, and other impurities do not bind to the glass beads. Following a few washing and pelleting cycles the purified DNA is eluted from the glass into a low salt buffer. However, this method, although both quick and efficient, has a narrow optimal range. The recovery of small (500–800 bp) DNA fragments is not very efficient and large fragments (>15 kb) can bind to different glass beads and the strands can be broken during the wash cycles *(9)*.

4. Polyacrylamide-Gel Electrophoresis

4.1. The Gel Matrix

Acrylamide is a synthetic monomer that can be polymerized into long chains by free radicals, usually supplied by ammonium persulfate and stabilized by N,N,N',N'-tetramethylethylene-diamine (TEMED). When N,N^l-methylenebisacrylamide is included in the polymerization mix, the polymer chains become crosslinked, forming a porous gel. Acrylamide has been identified as a potent neurotoxin in its unpolymerized form and should be handled with extreme caution. It is also recommended that polymerized gels be handled with similar caution.

The length of the chains, and hence the size of the pores, is determined by the concentration of acrylamide in the reaction and is normally between 3.5 and 20%, and 1 molecule of crosslinker is included for every 29 monomers of acrylamide. Hence, a 10% gel would contain 10% (w/v) of acrylamide plus bisacrylamide, and the higher the percentage, the smaller the pore size. Acrylamide solutions degrade over time, so stock acrylamide solutions should be kept at 4°C in the dark. Unused stock solutions should be discarded after a few months. Ammonium persulfate solution should either be made freshly each time, or kept for not more than 1 wk at 4°C. The rate of gel polymerization depends on a number of factors, such as the ambient temperature and amount of initiator and catalyst present. In general, complete polymerization should occur in about 1 h. Prepared gels can be kept overnight, although care must be taken not to let the wells dry out. This can be prevented by placing a few tissues soaked in buffer over the wells after polymerization is complete and wrapping the top in ding film. It is usual to remove the comb only immediately prior to loading.

In contrast to agarose-gel electrophoresis, polyacrylamide-gel electrophoresis is routinely used to analyze both single-stranded and double-stranded DNA, and polyacrylamide gels can be either denaturing or nondenaturing. Electrophoretic separation in both systems is again by DNA size, with the exception that single-stranded DNA molecules electrophoresed in a nondenaturing polyacrylamide will show a sequence-specific mobility owing to secondary structure formation via intramolecule base-pairing, a fact now extensively used in point-mutation analysis *(10,11)*. When a denaturant, such as urea or formamide, is included in the polymerization mix, intramolecule base-pairing is prevented and electrophoretic separation of single-stranded DNA is again according to molecule size.

4.2. Equipment

Equipment for polyacrylamide gel electrophoresis is more cumbersome and somewhat less robust than that for agarose-gel electrophoresis, and it is also more expensive. Typically apparatus consists of two glass plates that are held apart from each other by a spacer of defined thickness. The glass plates are often 20 × 40 cm, and spacers from 0.2 to 0.8 mm. Different commercial manufacturers have different techniques, but all involve inserting the polymerization mix between the two plates and waiting for polymerization to oocur. A comb is either inserted into the polymerization mix, or, for some applications, such as DNA sequencing, the mix is allowed to polymerize to give a flat top. With polyacrylamide gel electrophoresis, the two buffer reservoirs are kept apart, and current is conducted solely through the gel. A power pack capable of 3000 V and 100 mA is required. Polyacrylamide gels are usually run vertically.

4.3. Sample Loading

Polyacrylamide gels have a greater capacity for DNA than do agarose gels, so a greater amount of DNA can be analyzed, and approx 2–5 μg of DNA can be loaded onto a 1 mm × 2 cm lane, depending on the fragment size (the smaller the DNA fragment, the more that can be loaded). Samples to be analyzed are mixed with loading buffer in a similar manner to agarose gels. If the gel is a nondenaturing polyacrylamide gel, intended for the electrophoretic separation of double-stranded DNA fragments, the loading buffers described for agarose-gel electrophoresis are satisfactory. Again, the buffer serves the dual role of providing density to allow the solution to sink to the bottom of the well, as well as containing one or two tracking dyes to enable the electrophoresis to be monitored.

In a denaturing polyacrylamide gel, the two DNA strands must be denatured and kept denatured prior to loading. For this reason, the sample to be analyzed is mixed with a loading buffer consisting of 96% formamide as well as one or two tracking dyes. Samples are then heat-denatured prior to loading on the polyacrylamide gel. Prior to loading, the comb is removed from the gel and the pockets rinsed copiously with buffer. One important point with loading a polyacrylamide gel, whether denaturing or nondenaturing, is to ensure that the gel is not loaded symmetrically, because the orientation of the gel is often lost with subsequent steps leading to visualization of the bands.

4.4. Migration of DNA Bands

Electrophoretic separation of DNA bands in polyacrylamide is similar to that in agarose gels, in that migration of DNA is toward the anode, and discrimination by size is based on the pore size of the gel. Polyacrylamide gels can resolve bands between 1000 and 2000 bp (3.5% acrylamide w/v) and 6–100 bp (20.0% acrylamide w/v).

4.5. Visualization of DNA

Polyacrylamide gels for DNA analysis are routinely between 0.2 and 0.8 mm thick and 40–50 cm long, and as such they can present difficulties in handling. For this reason, gels are usually supported on either one of the glass plates or are transferred to a filter-paper backing. When double-stranded DNA is analyzed, bands can be visualized by ethidium bromide (either by performing the electrophoresis with ethidium bromide in the gel and the buffer or by staining after electrophoresis) and UV light transillumination in the same way as for agarose gels. However, illuminating the gel with UV can present problems owing to the gel support. In general, visualization of DNA in polyacrylamide gels is undertaken indirectly, by radioactively labeling the DNA and visualizing the resultant bands on autoradiography or X-ray film (although some works prefer nonradioactive labeling methods, such as biotin or digoxigenin [12]). In certain protocols, such as DNA sequencing, radioactive labeling is almost universally used. The preparation of the gel to enable radioactive visualization of bands depends somewhat on the nature of the isotope employed to label the DNA. When ^{32}P is used as the labeling isotope, the gel can be covered with a layer of ding film or Saran Wrap and exposed directly to the film (in the dark, of course). When less energetic isotopes, such as ^{35}S or ^{33}P, are used as the labeling isotope, the thickness of the gel will almost completely absorb the β particles. For this reason gels that have employed weak β emitters as the labeling isotope are dried prior to autoradiography. If the gel is a denaturing gel, the denaturant (usually urea) must be removed prior to drying, which can be achieved by soaking the gel for 10–15 min in a mixture containing acetic acid and ethanol (10% of each). The gel can then be transferred to a filter paper backing and placed in a commercial vacuum drier, which heats the gel to approx 80°C while applying a moderate vacuum. The polyacrylamide gel will then be dried to a glaze on the filter-paper backing and can be exposed to the film.

4.6. DNA Size Markers

As with agarose-gel electrophoresis, the tracking dyes included in the loading buffer can give an approximate indication of the migration of the DNA. In a 5% polyacrylamide nondenaturing gel, xylene cyanol migrates at about 250 bp and bromophenol blue at about 65 bp. In a 20% polyacrylamide gel, the corresponding figures are approx 50 and 10 bp, respectively. In a 5% denaturing gel, xylene cyanol would migrate at about 130 bases and bromophenol blue at about 35 bases, whereas the corresponding figures in a 20% denaturing polyacrylamide gel would be 30 and 10 bases.

For polyacrylamide gels, radioactive size markers can be produced relatively easily. Any commercially purchased or home-produced DNA size markers for agarose-gel electrophoresis can be end-labeled. This involves first dephosphorylating the marker with either bacterial alkaline phosphatase or calf intestinal alkaline phosphatase under conditions recommended by the manufacturer. The DNA can then be purified by phenol extraction and ethanol precipitation. A small aliquot can then be removed and labeled with polynucleotide kinase and $\gamma[^{32}P]$-ATP (or ^{33}P). When mixed with loading buffer, this labeled marker will last for several weeks, and one batch of dephosphorylated marker should provide enough material for several months. The size markers will then appear on the final autoradiograph and provide a permanent record. On a nondenaturing gel, the marker will run as an exact size marker. On a denaturing gel, the bands will tend to split into the separate strands but will still enable the experimenter to work out approximate sizes.

5. Applications of Polyacrylamide-Gel Electrophoresis

5.1. Applications of Nondenaturing Polyacrylamide Gels

5.1.1. Analytical

Nondenaturing polyacrylamide gels are used in much the same way as agarose gels and can be used to confirm restriction digests, especially if the digest products are expected to be in the range 20–500 bp, or below the limit of resolution of agarose gels. Increasingly, polyacrylamide gels are being used in mutation studies to detect the presence of point mutations *(10,11)*. When two DNA strands are denatured and then electrophoresed on a nondenaturing polyacrylamide gel, the two complementary strands will adopt different conformations owing to intramolecular pairing and will therefore migrate differently through the gel matrix. If a proportion of the DNA contains a mutation (or a polymorphism), these two strands will have yet different conformations and so will show up as two extra bands after electrophoresis. The conformations are temperature-dependent, so it is necessary to control the temperature of the gel carefully. Some manufacturers (e.g., StrataTherm Temperature Controller; Stratagene, La Jolla, CA) now produce electrophoresis apparatus that can be temperature-controlled.

5.1.2. Preparative

Nondenaturing polyacrylamide gels can be used preparatively to purify small double-stranded fragments. The band of interest can either be identified by ethidium bromide staining or by autoradiography. If autoradiography is used, the band required can be cut from the X-ray film, then the film reorientated onto the gel and the film used as a template. In each case, the band is excised from the polyacrylamide gel with the aid of a scalpel. Because the gel matrix is formed by crosslinking, chaotropic agents will not solubilize the gel, so recovery is essentially by diffusion–elusion with the gel slice being placed in an elution buffer and being shaken gently for 2 h to overnight, depending on the size of the fragments to be recovered. Small fragments of <250 bp can be eluted in 2–3 h, whereas fragments of up to 1000 bp require overnight shaking at 37°C. Alternatively, good results have been achieved by replacing the bis-acrylamide in the polymerization mix with bisacrylcystamine (BAC). In this case, the gel is held together by disulfide crosslinkages, and the gel slice can easily be solubilized in mercaptoethanol. The DNA then needs to be purified from the mercaptoethanol solution *(13)*.

5.2. Denaturing Polyacrylamide Gels

The majority of analytical-denaturing polyacrylamide gels run today are DNA-sequencing gels. Although some authors produce sequencing ladders by the chemical-cleavage method of Maxam-Gilbert *(14)*, the majority produce sequencing ladders by the chain-terminating method of Sanger *(15)*. For each DNA sequence of interest, four parallel reactions are undertaken. In each reaction, a series of bands are produced, each of which corresponds to a particular DNA base. In this way, by reading up the gel, the sequence of the DNA can be read. Invariably, the reactions are radioactively labeled and the sequencing bands are visualized by autoradiography. Sequencing absolutely depends on the ability to resolve bands that differ from each other by 1 base, and sequencing can routinely achieve runs of 400–500 readable nucleotides.

Several modifications of standard denaturing gels have been employed to increase the utility of sequencing gels. Some workers routinely employ a lower ionic-strength buffer, such as 0.5X TBE. To increase the amount of information generated by a single run, some workers have employed either wedge gels, where the bottom of the gel is physically thicker, or gradient gels, in which the bottom of the gel is of a higher percentage polyacrylamide than the top of the gel. Both of these methods have the result of compressing the lower bands, and enabling a greater amount of sequence to be determined on a single gel. Gradient gels can be tricky to pour manually, and it is up to the individual worker to determine if the greater information is worth the extra effort. The same result can be achieved by running the samples twice, once for a short time to obtain information nearest the start site of sequencing and once for a long time to obtain information furthest away from the sequencing primer (this can be achieved on one gel by loading the long-run samples first, electrophoresing for a few hours, and then loading the short-run samples in adjacent wells and re-electrophoresing the gel for a few hours more). Care must be taken not to over run the longrun, otherwise the overlap region between the two runs can be lost.

Denaturing polyacrylamide gels are employed wherever an accurate size of a single-stranded DNA is required. Other uses include S1-nuclease mapping to determine transcription-start sites and intron/exon boundaries, RNase H mapping to quantitate RNA levels, and primer-extension assays. In each of these methods, it is a DNA product of a specific length that is being analyzed on the gel, although this reflects an RNA status.

5.3. Denaturing-Gradient Gel Electrophoresis

This form of electrophoresis employs the use of a polyacrylamide gel with a gradient of denaturant, and is used extensively in mutation analysis. The system is very sensitive and can detect single mutations in several hundreds of base pairs of DNA. In this method, DNA samples are electrophoresed through a polyacrylamide gel that contains a linear gradient of denaturant. When the double-stranded DNA enters a region where the concentration of denaturant is equal to the bonding forces holding domains of the two strands together, local melting will occur and the DNA will form a branched structure that has reduced mobility compared to the native duplex. Where a mutation or polymorphism is present, this will change the domain melting conditions, and so extra bands will be detected. For a fuller treatment of this application of polyacrylamide-gel electrophoresis, readers are directed to Myers et al. *(16)*.

6. Pulse-Field Gel Electrophoresis

Pulse-field gel electrophoresis is the least widely used of all of the DNA electrophoretic techniques, and tends to be used in only specialized circumstances. Although the least widely used of the three techniques, it is the most difficult technically, and requires good practical skills to obtain high resolution results. This section will therefore be confined to a basic overview of the technique, and those wishing to obtain more detailed experimental methodology are referred to some of the excellent articles elsewhere (*see also* **refs. *3–5***).

Although some authors report the separation of DNA molecules of up to 750 kb by agarose-gel electrophoresis in 0.1–0.2% gels *(17,18)*, this is generally considered to be the maximum upper limit, and generally standard agarose gels poorly resolve DNA molecules above about 25 kb in length. In practice, this upper limit is dictated by the pore size of the gel. The agarose gel itself is composed of pores of varying sizes, and small molecules can pass easily through most pores. The mobility of the molecule is therefore related to the percentage of pores through which it can pass, a process called sieving. The larger the pores in the gel, the larger the DNA that can be sieved. DNA molecules above a certain size can no longer be sieved by the gel, and so pass through the gel at a constant rate, the so-called limiting mobility. This limit is reached when the radius of gyration of the molecule exceeds the pore size of the gel. In this case, the DNA molecules pass through the matrix end on, a process known as "reptation." At this point the molecules are no longer separated according to size. This limiting factor was overcome by Schwartz and Cantor in 1984 *(19)*, who showed that it was possible to resolve molecules up to two megabase pairs in length by alternating the direction of the electric field across the gel. This results in the DNA molecules being forced to alter their orientation to the electric field, before they can again migrate forward in their reptation tube. The larger the molecule, the longer the time required for reorientation to the new electric field. The fields applied to each other must be at least 90° perpendicular to each other, but can be much higher. The highest angle between the two fields that can be achieved is 180°, which is employed in field-inversion gel electrophoresis. The electric fields are switched in a pulse fashion (hence pulse field), and molecules whose reorientation time are less than the pulse time will therefore be resolved according to their size. The original method of Schwartz and Cantor used a nonuniform electric field, so the bands tended to migrate toward the edges of the gels, resulting in some loss of resolution. Today, pulse-field gel electrophoresis apparatus tends to use uniform fields, resulting in straight lanes, and there are two commonly used methods of pulse-field electrophoresis, namely field-inversion gel electrophoresis (FIGE; **ref. *20***), which can separate molecules between 10 and 200 kb, and contour-clamped homogenous electric field (CHEF; **ref. *21***), which can resolve molecules up 5000 kb.

6.1. The Gel Matrix

The gel matrix used in pulse-field electrophoresis is standard DNA agarose, although some manufacturers produce pulse-field grade agarose (e.g., SeaKem FastLane; FMC Bioproducts), and approximately similar results can be obtained with either 1% pulse-field grade agarose or 0.8% standard agarose, although the former will tend to have a greater mechanical strength.

6.2. Equipment

The simplest form of pulse-field gel electrophoresis could be achieved by using a standard agarose gel box and gel, and simply rotating the gel periodically. In practice this would be difficult, because runs can take up to 100 h. For this reason, specialist equipment is invariably used. The easiest method is to use a standard agarose-gel electrophoresis gel tank, coupled with a power source capable of switching the direction of the current for varying times (field-inversion gel electrophoresis). Contour-clamped homogenous electric field electrophoresis requires more specialized apparatus, because the gel is surrounded by a set of electrodes that combine to give a uniform electric field. The current is switched between these electrodes. In contrast to FIGE, a voltage divider is also required, as well as a power pack capable of pulsed electric current.

6.3. Sample Loading

Sample loading is perhaps the most criticial step of this form of electrophoresis. Normal DNA preparation involves shearing forces that can reduce the maximum size of DNA to approx 50 kb, well below the range of DNA capable of being analyzed on these gel systems. For this reason the cells or sample to be analyzed is normally encapsulated in an agarose block or plug, fitted to the size of the preformed well pockets. All of the sample preparation, such as lysing the cells, digesting proteins, and undertaking restriction enzyme digests, are undertaken with the sample embedded in the block. The block is then lowered into the preformed pocket prior to electrophoresis.

6.4. Migration

As described in **Subheading 6.**, the migration of DNA in pulse-field gel electrophoresis is more complex than in either agarose-gel electrophoresis or polyacrylamide-gel electrophoresis. The DNA becomes trapped in its reptation tube and migrates end on toward whichever electrode is currently the anode. In field-inversion gel electrophoresis, two electrodes are constantly switched between anode and cathode. The period of time that an electrode is acting as either the anode or cathode is termed the "pulse time," and the pulse times have to be adjusted to obtain resolution over the DNA range of interest. To enable an overall direction of migration toward the bottom of the gel, the forward pulse time needs to be longer than the reverse pulse time; otherwise there would be no net migration of the DNA down the length of the gel. In field-inversion gel electrophoresis, a single pulse time will result in a very narrow range of separation, and for this reason the electrophoretic run normally cycles through a range of pulse times. In CHEF electrophoresis, the migration of DNA is actually a complex zigzag run, generated by alternating switching between a number of electrodes, which results in an overall direction of migration toward the bottom of the gel. Again, resolution of samples over a broad range may be improved by varying pulse times.

6.5. Visualization

Visualization of the DNA bands can be achieved with the use of ethidium bromide, as for standard agarose gels. Typically pulse field gels are run without ethidium bromide in the gel and the running buffer, because the presence of ethidium can alter the migration of the bands, and the gel can be stained after electrophoresis by immersing into an ethidium bromide solution. Alternatively, samples can be transferred to nylon membranes and subjected to Southern analysis *(7)*, although this requires careful depurination and cleavage of the DNA strands to ensure efficient transfer of the DNA to the membrane.

6.6. DNA Size Markers

DNA size markers for pulse-field electrophoresis can either be purchased from commercial companies (e.g., Boehringer Mannheim; Life Technologies) or produced in the laboratory. Commonly used markers include the chromosomes of strains of yeast (such as *Saccharomyces cerevisiae* or *Schizosaccharromyces pombe*) or concatamers of λ DNA.

6.7. Applications

Pulse-field gel electrophoresis is able to resolve DNA up to 5000 kb, and as such is extensively used in gene mapping, genome analysis, and cloning (especially for YAC cloning). Applications of this technique have resulted in physical maps of large regions of mammalian chromosomes *(22)* and the entire *Escherichia coli* chromosome.

7. Conclusions

This chapter has attempted to provide an overview of the basic methods of DNA gel electrophoresis as employed in the majority of laboratories today. A number of highly specialized modifications of gel electrophoresis have been developed, but these are not in use in the majority of laboratories or have become less common owing to the advent of newer technologies. Examples of the former would include annular preparative gel electrophoresis *(2)* and compound systems for two-dimensional gel electrophoretic separation of restriction fragments *(23)*. Examples of the latter would include the many types of denaturing agarose gel electrophoresis for separating DNA strands using either mercuric hydroxide or sodium hydroxide *(2)*. Wth the exception of these more specialized applications of gel electrophoresis, it is remarkable that the combination of only three techniques, agarose-gel electrophoresis, polyacrylamide-gel electophoresis, and pulse-field gel electrophorcsis, give today's researcher the ability to analyze DNA from the size of a few base pairs to nearly 1% of the human genome.

References

1. Sambrook, J., Fritsch, E. F., and Maniatis, T. (1989) *Molecular Cloning: A Laboratory Manual,* 2nd ed. Cold Spring Harbor Laboratory, Cold Spring Harbor, NY.
2. Sealey, P. G. and Southern, E. M. (1982) Electrophoresis of DNA, in *Gel Electrophoresis of Nucleic Acids: A Practical Approach* (Rickwood, D. and Hames B. D., eds.), IRL Oxford, UK, pp. 39–76.
3. Boffey, S. A. (1984) Agarose gel electrophoresis of DNA, in *Methods in Molecular Biology, vol. 2: Nucleic Acids* (Walker, J. M., ed.), Humana, Clifton, NJ, pp. 43–50.
4. Helling, R. B., Goodman, H. M., and Boyer, H. W. (1974) Analysis of R.*Eco*R1 fragments of DNA from lambdoid bacteriophages and other viruses by agarose gel electrophoresis. *J. Virol.* **14**, 1235–1244.
5. Johnson, P. H. and Grossman, L. I. (1977) Electrophoresis of DNA in agarose gels. Optimising separations of conformational isomers of double- and single-stranded DNAs. *Biochemistry* **15**, 4217–4224.
6. Sharp, P. A., Sugden, B., and Saunders, J. (1973) Detection of two restriction endonuclease activities in Haemophilus parainfluenzae using analytical agarose-ethidium bromide electrophoresis. *Biochemistry* **12**, 3055–3063.

7. Southern, E. (1975) Detection of specific sequences among DNA fragments separated by gel electrophoresis. *J. Mol. Biol.* **98,** 503–517.

8. Thuring, R. W., Sanders, J. B., and Borst, P. A. (1975) Freeze-squeeze method for recovering long DNA from agarose gels. *Anal. Biochem.* **66,** 213–220.

9. Vogelstein, B. and Gillespie, D. (1979) Preparative and analytical purification of DNA from agarose. *Proc. Natl. Acad. Sci. USA* **76,** 615–619.

10. Orita, M., Iwahana, H., Hayashi, K., and Sekiya, T. (1989) Detection of polymorphisms of human DNA by gel electrophoresis as single strand conformation polymorphisms. *Proc. Natl. Acad. Sci. USA* **86,** 2766–2770.

11. Orita, M., Suzuki, Y., Sekilya, T., and Hayashi, K. (1989) Rapid and sensitive detection of point mutations and DNA polymorphisms using the polymerase chain reaction. *Genomics* **5,** 874–879.

12. Kessler, C. (1992) *Non-Radioactive Labeling and Detection of Biomolecules* (Kessler, C., ed.), Springer-Verlag, Berlin.

13. Hansen, J. N. (1981) Use of solubilizable acrylamide disulphide gels for isolation of DNA fragments suitable for sequence analysis. *Anal. Biochem.* **116,** 146–151.

14. Maxam, A. and Gilbert, W. (1980) Sequencing end labeled DNA with base-specific chemical cleavages, in *Methods in Enzymology* (Grossman, L. and Moldave, K., eds.), Academic, New York and London, pp. 499–560.

15. Sanger, F., Nidden, S., and Coulson, A R. (1977) DNA sequencing with chain terminating inhibitors. *Proc. Natl. Acad. Sci. USA* **74,** 5463–5467.

16. Myers, R., Sheffield, V. C., and Cox, R. (1988) Detection of single base changes in DNA: ribonuclease cleavage and denaturing gradient gel electrophoresis, in *Genome Analysis: A Practical Approach* (Davies, K. E., ed.), IRL, Oxford, UK, pp. 98–139.

17. Fangman, W. L. (1978) Separation of very large DNA molecules by gel electrophoresis. *Nucleic Acids Res.* **5,** 653–657.

18. Serwer, P. (1980) Electrophoresis of duplex deoxyribonucleic acid in multiple-concentration agarose gels: fractionation of molecules with molecular weights between 2×10^6 and 110×10^6. *Biochemistry* **19,** 3001–3004.

19. Schwartz, D. C. and Cantor, C. R. (1984) Separation of yeast chromosome-sized DNAs by pulse held gradient electrophoresis. *Cell* **37,** 67–75.

20. Carle, G. F., Frank, M., and Olson, M. V. (1986) Electrophoretic separation of large DNA molecules by periodic inversion of the electric field. *Science* **232,** 65–68.

21. Chu, G., Vollrath, D., and Davis, R. W. (1986) Separation of large DNA molecules by contour clamped homogenous electric fields. *Science* **234,** 1582–1585.

22. Lawrance, S. K., Smith, C. L., Srivastava, R., Cantor, C. R., and Weissman, S. H. (1987) Megabase-scale mapping of the HLA gene complex by pulse field gel electrophoresis. *Science* **235,** 1387–1390.

23. De Wacher, R. and Fiers, W. (1988) Two-dimensional gel electrophoresis of nucleic acids, in *Gel Electrophoresis of Nucleic Acids: A Practical Approach* (Rickwood, D. and Hames, B. D., eds.), IRL, Oxford, UK. pp. 77–116.

Further Reading

1. Ausubel, F. M., Brent, R., Kingston, R. E., Moore, D. D., Seidman, J. G., Smith, J. A., and Struhl, K., eds. *Current Protocols in Molecular Biology Vol. 1.* John Wiley, Brooklyn, NY. *Heavily weighted toward practical protocols, of which several alternatives are usually presented, but also some readable theoretical sections.*

2. Osterman, L. A. Electrophoresis of nucleic acids, in *Methods of Protein and Nucleic Acids Research, vol. 1. Electrophoresis. Isoelectric Focusing. Ultracentrifugation.*

Springer-Verlag, Berlin, pp. 102–151. *A fairly technical analysis of electrophoresis of DNA and various applications.*

3. Rickwood, D. and Hames, B. D., eds. (1988) *Gel Electrophoresis of Nucleic Acids: A Practical Approach,* IRL, Oxford, UK. *Contains six chapters detailing various aspects of electrophoresis of nucleic acids. A good mix of theory and practical tips.*

4. Sambrook, J., Fritsch, E. F., and Maniatis, T. *Molecular Cloning: A Laboratory Manual, 2nd ed.* Cold Spring Harbor Laboratory, Cold Spring Harbor, NY. *Comprehensive section on electrophoresis containing a mix of theory and practice of DNA electrophoresis.*

5. Smith, C. L., Klco, S. R., and Cantor, C. R. (1988) Pulse-field electrophoresis and the technology of large DNA molecules, in *Genome Analysis: A Practical Approach* (Davies, K. E., ed.), IRL, Oxford, UK, pp. 41–72. *Excellent chapter on pulse field gel electrophoresis, especially dealing with preparation and loading of samples. Several examples of what can go wrong (and why) during pulse field experiments.*

4

S1 Nuclease Analysis of RNA

Louis Lefebvre and Stéphane Viville

1. Introduction

The analysis of RNA molecules by S1-nuclease digestion constitutes an important tool for molecular studies of cellular processes, not only because of the historical significance of its contribution to messenger ribonucleic acids (mRNA) structure analyses, but also for its simplicity and wide range of applications. Following a brief introduction to important aspects of RNA studies, this chapter presents a detailed description of the theory and principles underlying the technique of RNA mapping by S1-nuclease digestion, as well as a discussion of the most relevant applications of this molecular technique.

1.1. Why Study RNA Molecules?

Molecular biology teachers, and indeed several textbooks, often rely on the analogy between the genome of a given organism—that is, the complete DNA information contained in its chromosomes—and a library to illustrate the concept of genetic information. According to this view, the code necessary for the propagation and maintenance of any life form can be cataloged and analyzed from the multiple volumes of the library. This analogy is essentially correct as long as its implications are limited to the informational aspect of genetic processes. Indeed, the elucidation of the three-dimensional structure of DNA and of the genetic code supports this strictly linear representation of the genetic content of a cell. Where the analogy fails, however, is in properly conveying our inability to understand how a cell works, even if its entire genome has been sequenced. Although we have made remarkable progress in understanding the structure, alphabet, and grammar of the genetic material, we still have much to learn about how this information, through the processes of transcription and translation (and the corresponding syntheses of RNA molecules and proteins) mediates its diverse and complex cellular functions. The study of the synthesis, structure, and function of RNA molecules thus represents a major contemporary field of research with significant contributions to most, if not all, biomedical sciences.

Although molecular biologists have revealed the rather simple, repetitive, linear, and quite stable nature of DNA, researchers studying RNAs have demonstrated that these molecules, albeit closely related in chemical terms to DNA, represent a much

From: *Molecular Biomethods Handbook*
Edited by: R. Rapley and J. M. Walker © Humana Press Inc., Totowa, NJ

greater experimental challenge. RNA is more labile than DNA and its purification from biological sources is greatly complicated by the presence of ubiquitous ribonucleases. Although the meticulous prevention of RNA degradation during isolation often yields satisfactory results, other parameters can also influence the recovery of specific RNA molecules. First, the abundance of certain RNA transcripts varies among biological sources, tissues, and developmental stages; the synthesis of RNA is often repressed or induced by specific growth conditions or responds to cell-cycle cues. Second, RNAs form a remarkably complex and heterogeneous family of molecules that participate in a broad range of cellular functions. The development of tools and approaches allowing the analysis of a specific RNA among this diverse population of chemically similar compounds has been crucial not only for the evolution of RNA studies, but indeed for the emergence of molecular biology as a powerful approach to probe biomedical phenomena.

1.2. Specificity in RNA Analysis

For simplicity, we will restrict ourselves here to a discussion of the study of mRNA transcripts, although several considerations and applications will also be relevant to the analyses of other types of RNAs. A fundamental question intrinsic to the study of mRNA molecules emerges from the complexity of the mRNA pool found in any living cell at a given time: How does one limit the experimental analysis of RNA to a single or a few specific mRNAs? In order to understand how this experimental problem is solved, some knowledge of the relationship between DNA and mRNA molecules is required.

In biochemical terms, RNA molecules are polymers of ribonucleoside monophosphate units, synthesized by RNA polymerase enzymes. The properties of these enzymes define important structural characteristics of RNA molecules:

1. The RNA polymerases require a DNA template that provides the actual nucleic-acid sequence information to be transcribed into RNA;
2. The nucleotide precursors are incorporated such that the nascent RNA transcript is polymerized from its 5' terminus toward its 3' end; and
3. During polymerization, the nucleoside monophosphate added at the free 3'-hydroxyl group is selected on the basis of its complementarity to the corresponding residue on the template DNA strand. This complementarity is in turn dictated by the specificity of base-pairing (G-C, A-T, or A-U in RNA).

An important consequence of the transcription process is that the nucleic-acid information contained in the resulting RNA molecule is actually present in the DNA sequences from which it was transcribed. **Figure 1** represents this simple relationship for a typical eukaryotic mRNA. In this diagram, the top DNA strand (5' to 3') contains the same genetic information as that present in the RNA synthesized, except that the pyrimidine base thymine is replaced by uracil in RNA. As a consequence, this strand is known as the coding or sense strand; it is usually the DNA strand represented in published gene sequences and from which polypeptide sequences coded by open reading frames can be deduced by virtual translation using the genetic code. The DNA strand that is utilized as a template by RNA polymerase II is the bottom strand in **Fig. 1** (3' to 5'). It is referred to as the template or antisense strand. This DNA strand is complementary to the RNA transcripts derived from it and can hybridize with them in appropriate conditions: It constitutes a specific probe for its transcriptional products. This property of nucleic acids is central to the analysis of RNA by S1-nuclease digestion and to most

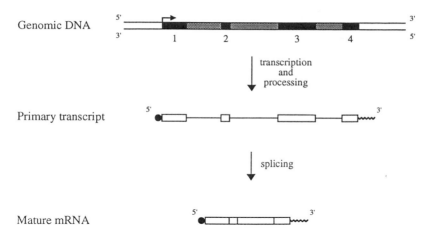

Fig. 1. Schematic representation of the synthesis of a typical eukaryotic mRNA molecule from genomic DNA. The dsDNA is shown by double horizontal lines and the transcription start site is indicated by the arrow. On the DNA, exons are in black and numbered 1–4; introns are shaded. The primary and mature transcripts are drawn to highlight the exon–intron structure (rectangles and line, respectively) as well as the position of the 5' and 3' termini, with their posttranscriptional modifications (5' cap and 3' poly[A] tail).

techniques developed to study RNA molecules because it provides the specificity required to study almost any given RNA from a complex mixture of RNAs.

2. S1 Nuclease Mapping

In the technique of RNA analysis by S1-nuclease digestion (also called S1-nuclease protection assay or simply S1 mapping), hybridization between RNA and antisense DNA is exploited to study qualitative and quantitative aspects of RNA synthesis. S1 mapping *(1–4)* consists primarily of three experimental steps (diagrammed in **Fig. 2**):

1. Formation of RNA:DNA hybrids in solution;
2. Enzymatic degradation of unprotected nucleic acids by the single-strand-specific S1 nuclease; and
3. Analysis of the protected fragments.

The rationale and theory behind each of these steps are discussed in **Subheadings 2.1.–2.3.** Variations to this core protocol are given in **Subheading 2.** under specific applications of the technique.

2.1. Hybridization: Choice of Probe and Experimental Conditions

As introduced in the previous section, the concept of hybridization—the ability of complementary nucleic-acid molecules to form antiparallel duplexes—plays a central role in most techniques used in RNA analysis. Although a detailed discussion of the theory of nucleic-acid hybridization is beyond the scope of this chapter *(2,5,6)*, the successful application of S1 mapping will largely depend on the choice of appropriate hybridization conditions for a given application.

The formation of a double-stranded nucleic acid (or duplex) is a complex thermodynamic process. In solution, the state of nucleic-acid strands is dictated by the equilib-

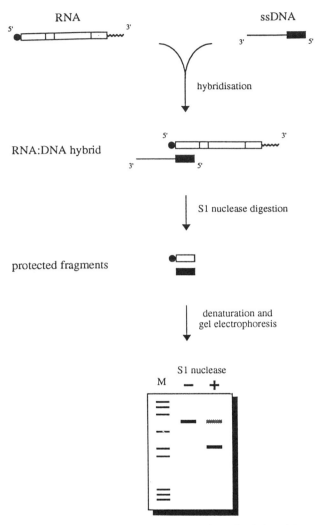

Fig. 2. Analysis of RNA by S1 nuclease mapping. The principle of S1 nuclease analysis is illustrated here by the mapping of the 5' end of an mRNA molecule. A single-stranded antisense probe is first hybridized to its target RNA. The free molecules as well as the RNA:DNA hybrids are digested by S1 nuclease, which degrades all single-stranded nucleic acids. The protected fragment—those that formed double-stranded structures—are purified, denatured, and analyzed by gel electrophoresis. The size of the labeled DNA fragment obtained after digestion (+) gives the distance in nucleotides between the 5' terminus of the probe and the 5' end of the RNA analyzed. The undigested probe (−) is included as a control and a guide for the position of background signal caused by incomplete digestion or reannealing in the case of dsDNA probes.

rium between the dissociation of single-stranded species (melting or denaturation), which is promoted by the electrostatic repulsion between the negatively charged phosphodiester backbones and their association (base-pairing or hybridization), through the formation of hydrogen bonds between complementary base pairs.

Although several different parameters can influence this equilibrium, the temperature of a system is the most readily modified variable experimentally and does not require any changes in the chemical composition of the hybridization solution. Temperature is often used to represent the relative thermodynamic stability of a duplex in a given set of conditions: The melting temperature, or T_m, is the temperature at which 50% of the duplexes are denatured and therefore present as single-stranded molecules in the reaction mixture. The T_m is an experimentally determined parameter, but equations have been derived that allow the calculation of estimated T_m values. These equations express the T_m as a function of intrinsic factors (length of probe, G+C content, degree of mismatch), as well as extrinsic variables (pH, ionic strength, presence organic solvents), which all influence duplex formation *(2)*. Whereas temperatures above the T_m promote strand separation, temperatures below the T_m shift the equilibrium toward reannealing. Finally, even though base-pairing between complementary nucleic acids does not discriminate on the basis of the nature of the phosphodiester backbone, such that DNA:DNA, RNA:DNA, and RNA:RNA hybrids can be obtained in solution, RNA molecules form more stable duplexes. Consequently, the relative thermodynamic stability of double-stranded molecules increases as follows: DNA:DNA < RNA:DNA < RNA:RNA. As will be described below, this experimental observation was crucial to the development of S1 mapping.

In the first step of an S1-mapping experiment, a nucleic-acid probe (DNA or RNA; labeled or not) is added to an RNA preparation of a given cell type and reaction conditions are adjusted to promote the formation of hybrid molecules. The specific hybridization conditions used, as well as the expected results, will depend on the type of probe utilized. In the next section, two different parameters are considered: the nature of the probe (double- or single-stranded, DNA or RNA); and the choice of label (uniform labeling, end labeling).

2.1.1. Double-Stranded Probes

In its original description, S1 mapping of mRNA transcripts was performed using uniformly labeled dsDNA probes *(7)*. Although such probes are relatively easy to obtain and purify from genomic clones, the hybridization reaction is complicated by the presence of the DNA-sense strand, which competes against the specific RNA for duplex formation with the labeled antisense strand of the probe. The successful application of this approach therefore relies on the observation that in certain conditions (80% formamide, 0.4 *M* NaCl) the T_m of DNA:DNA hybrids is 5–10°C lower than that of RNA:DNA hybrids *(6,8)*. By selecting a temperature 2–4°C above the T_m of the DNA:DNA hybrids, it is therefore possible to favor simultaneously the denaturation of the dsDNA probe and the formation of specific RNA:DNA hybrids. Because T_m calculations only give estimates of the true melting point of a duplex, several different temperatures are usually tested in trial experiments to define the best conditions for each probe. Note that organic solvents, such as formamide, act by interfering with the formation of hydrogen bonds between paired bases, thus decreasing the stability of the duplexes and consequently the temperature required for melting.

Temperatures below the T_m of a specific DNA probe lead to reannealing of some dsDNA probes. As a consequence, the concentration of ssDNA-antisense probe available is decreased (which could affect the quantification of RNA; *see* **Subheading 3.3.**). More importantly, a significant proportion of labeled probe is protected from S1 digestion, thus generating a high background signal for the full-length probe. When qualitative aspects of RNA structures are being analyzed, the possible interference caused by this background signal can be minimized by the judicious selection of a probe, including a long region of noncoding or intronic DNA sequences. Because this portion of the probe does not hybridize with the target RNA, it remains unprotected from S1 digestion in RNA:DNA hybrids. As a result, the protected fragments resulting from RNA:DNA and DNA:DNA hybrids show an important size difference and can be resolved easily by gel electrophoresis (**Fig. 2**). Alternatively, the background signal can be eliminated by using an appropriate end-labeled probe instead of a uniformly labeled dsDNA fragment. For example, a dsDNA probe could be obtained by digestion of a genomic clone with a restriction endonuclease generating a single-stranded 5' overhang, followed by labeling of the 5' end with polynucleotide kinase in the presence of $[\gamma\text{-}^{32}P]ATP$. If the labeled 5' end of the antisense strand is complementary to the mRNA (like in **Fig. 2**), the S1-protected fragment resulting from the RNA:DNA hybrids is labeled. But if the dsDNA probe reanneals, the 5' overhangs that carry the label are degraded, the remaining dsDNA protected fragments are not labeled, and no background signal is detected *(9)*.

As the temperature increases away from the T_m of the RNA:DNA hybrid, it is expected that the intensity of the signal corresponding to the RNA-specific protected band decreases. The temperature at which the signal-to-background ratio is optimum can therefore be determined for any probe.

2.1.2. Single-Stranded Probes

Most of the problems experienced with dsDNA probes can be avoided by using antisense-ssDNA probes, which are now the favored probes. Because the sense strand is not present in the probe itself, hybridization with the specific RNA target is a straightforward annealing reaction. As in the case of DNA:DNA duplex formation, the maximum annealing rate is seen in high salt (0.5–1.0 *M* NaCl), at temperatures more than 20°C below the T_m. Hybridizations can also be performed in the presence of formamide (50–80%) and low salt (<0.4 *M* NaCl). Because formamide decreases the T_m of the RNA:DNA hybrid, lower temperatures can be used, thus minimizing the degradation of RNA caused by high temperatures.

Single-stranded DNA molecules were first obtained by electrophoretic strand separation procedures. These approaches were rather time-consuming and the purity of the final preparation was quite variable. New technologies now offer several possibilities for creating single-stranded nucleic acids such that the preparation of strand-specific probes is now a routine molecular biology technique. For example, specific DNA strands can be synthesized in vitro by primer extension reactions on plasmids containing a genomic clone of the gene under study *(10)*. In this approach, a gene-specific primer or a "universal" primer hybridizing next to the cloning site is annealed to a single-stranded plasmid, or to a linearized double-stranded plasmid, and used to prime the synthesis of a DNA strand by DNA polymerase I in the presence of dNTPs. If a single-stranded template is used, the dsDNA molecules formed after polymerization are linearized to define the 3' end of the probe. The strand-specific probe is finally recovered after electrophoresis in denaturing conditions.

Another approach that can be considered in order to achieve even greater signal-to-noise ratios involves the use of antisense RNA probes *(11)*. Since RNA:RNA hybrids are more stable than similar duplexes involving DNA strands, more stringent hybridization conditions can be used with an RNA probe. These probes are synthesized by in vitro transcription with purified bacteriophage RNA polymerases (for example, SP6 or T7 polymerases) in the presence of rNTPs. Plasmid vectors containing the specific SP6 and T7 promoters flanking a cloning site have been designed for the synthesis of sense and antisense RNA probes *(2,3)*.

It is important to note that, by definition, strand-specific probes used in RNA hybridization techniques must represent the template strand in order to anneal with the RNA. They are commonly referred to as antisense probes. The preparation of antisense probes by primer extension thus requires a sense (coding) ssDNA template. Obviously the choice of the specific template strand to use in an application will depend on the orientation of the transcriptional unit relative to the genomic clone available. A direct implication of these considerations is that ssDNA probes can actually be used to determine the polarity of a transcriptional unit with regard to a genomic map of the locus. On the other hand, if the purpose of an experiment is solely to establish whether a specific genomic fragment contains transcribed sequences (without any knowledge of the nature or orientation of the transcript), a dsDNA probe would be the first choice because both sense and antisense strands would be present in the hybridization mixture.

In considering the preparation of gene-specific probes, the techniques presented thus far assume that at least part of the gene under study is accessible as a genomic or a cDNA clone. If such a clone is not available for a given application, two different approaches can generate the necessary probe, provided some sequence information for that gene is already known. First, specific genomic or cDNA fragments can be obtained by the polymerase chain reaction (PCR). These PCR products can be cloned and treated as discussed above to generate the desired probes. Alternatively, they can also act directly as dsDNA probes, or as templates for the synthesis of a strand-specific probe by primer extension. If a synthetic bacteriophage promoter is also included in one of the PCR primers, the products obtained can serve directly as templates for the synthesis of an RNA probe by in vitro transcription *(12)*. Second, oligonucleotides of more than 50 nucleotides can be synthesized on automated synthesizers and used as antisense probes for S1 analysis. Although currently available synthesizers will usually achieve high coupling rates and yield pure preparations of full-length oligonucleotides for short probes (10- to 30-mers), longer probes should be gel-purified before performing S1 analysis. This purification step is particularly important if the probe is designed for the purpose of mapping the 5' boundary of an exon (or the 5' terminus of the mRNA, in the case of exon 1). Since oligonucleotide probes are usually 5'-end-labeled by a kinase reaction in the presence of $[\gamma-{}^{32}P]ATP$, the contamination by shorter species (for example, n-1, n-2, which are shorter at the 5' end) would produce smaller protected fragments wrongly interpreted as indicating heterogeneity in the boundary or terminus under study. As mentioned previously for dsDNA probes, these oligonucleotides should also contain sequences (usually at their 3' end) that are not present in the RNA analyzed to generate a small protected fragment easily distinguished from full-length, undigested probes.

2.1.3. Labeling

The type of labeling used in a particular S1 mapping experiment will depend largely on the specific question being asked and the current information available concerning the structure of the RNA studied. Two different labeling procedures can be envisaged: uniform labeling or end labeling. Because each of these approaches has its own advantages and limitations, the success of an S1 mapping project will require careful selection of the appropriate experimental design.

Uniformly labeled antisense probes can be readily obtained by adding one or several radioactive nucleotides during in vitro synthesis of the probe. The main advantage of this approach is to yield probes of very high specific activity that may be desirable if the mRNA studied is present at low levels in the sample. Because labeled nucleotides are randomly incorporated throughout the length of the probe, all the protected fragments generated after S1 nuclease digestion will be visible on the autoradiogram *(7)*. This strategy can be used to identify genomic fragments containing expressed sequences and to determine the number of exons covered by these fragments. It is a particularly useful approach in initial mapping experiments when the exon–intron structure of the gene is not known, but has few applications for two main reasons: The interpretation of the results can be complicated by the complexity of the patterns obtained, especially if the gene contains several small exons, and the protected fragments obtained cannot be directly assigned to specific regions of the probe used unless the relative positions of exons within the probe have previously been determined. Using this labeling method, the ordering and mapping of exons on genomic clones can be established only if several overlapping genomic fragments are successively used as probes in S1 mapping experiments *(7)*.

An important variation to the original S1 mapping protocol introduced the use of end-labeled probes *(9)*. Since these probes bear a single terminal label, they usually generate a unique protected fragment readily mapped on genomic clones. Furthermore, the size of the protected band, as expressed in nucleotides, provides a direct measurement of the distance between the labeled terminus of the probe and the site of S1 digestion. As previously mentioned in **Subheading 2.1.1.**, the successful application of end-labeled probes depends on the ability of the labeled terminus to be protected from S1 nuclease digestion by base pairing with the complementary RNA. In most situations, this implies that the terminal radioactive nucleotide must fall within an exon (for example, *see* **Fig. 4A, B**). If the labeled terminus does not hybridize with the RNA, it will remain sensitive to S1 digestion and the protected fragment formed by the rest of the probe (if any) will not be labeled. Therefore, the main disadvantages of end-labeled probes over uniformly labeled ones are that some knowledge of the exon–intron structure of the gene is required and that only fragments ending within an exon can be used. Exceptions to theses rules are discussed in **Subheading 3.**

Terminally labeled probes carrying a single radioactive phosphate group at their 5' end are easily obtained by a kinasing reaction catalyzed by the bacteriophage T4 polynucleotide kinase in the presence of ATP labeled at its terminal phosphate ($[\gamma\text{-}^{32}P]ATP$). Even though the kinase will perform an exchange reaction in the presence of a 5'-phosphorylated substrate (such as a DNA fragment obtained by digestion with a restriction enzyme), the forward reaction on a 5'-OH substrate is generally more efficient and yields probes with greater specific activities. 5'-Hydroxylated molecules can be generated by treatment of a 5'-phosphorylated substrate with a phosphatase *(3)*. Antisense probes synthesized in vitro by primer extension are usually 5'-end labeled by kinasing the oligonucleotide primer before the polymerization reaction *(10)*.

The synthesis of 3'-end-labeled probes is slightly less convenient. Such labeling usually involves the incorporation of additional radioactive nucleotides at the 3'-OH terminus of a probe. These extra nucleotides must base-pair with the target RNA in order to be protected from S1 digestion. Therefore, protocols involving the addition of labeled dNTPs by the template-independent enzyme terminal transferase are usually unsuitable for the synthesis of S1 mapping probes. For most applications, the synthesis of 3'-end-labeled probes involves the incorporation of at least one [α-^{32}P]dNTP by the template-dependent enzyme DNA polymerase I on a substrate that is generated by a restriction endonuclease cutting within an exon and leaving a 5' overhang. This approach can be used for strand-specific labeling of a dsDNA fragment or of a single-stranded probe synthesized in vitro by primer extension.

2.2. S1 Nuclease Digestion

Throughout the previous sections, we frequently referred to nucleic acid molecules as being protected or unprotected from nuclease digestion. Indeed, the development of the technique of RNA analysis by nuclease digestion was made possible by the availability of the highly specific enzyme S1 nuclease *(7)*. S1 nuclease (EC 3.1.30.1) is an endonuclease-purified from *Aspergillus oryzae (13,14)*. It is active on both RNA and DNA and thus possesses endoribonuclease and endodeoxyribonuclease activities. From an enzymatic perspective, it acts as a hydrolase on phosphoric monoester bonds and degrades nucleic acids to produce 5'-mononucleotides. Furthermore, S1 nuclease is highly specific for single-stranded nucleic acids: RNA and DNA hybrids are degraded at rates more than a thousand times slower than the single-stranded species. The phosphodiester bonds of base-paired nucleic acids are thus said to be protected from S1 digestion. Although the enzyme will, at high concentration, nick dsDNA molecules, this activity is minimized at high ionic strength (usually at least 0.2 M NaCl) *(15)*. S1 nuclease is a zinc-dependent enzyme and exhibits optimal activity at acid pH (pH 4.0–4.5). Suboptimal conditions (higher pH) can also be used to restrict the nicking of dsDNA molecules *(16)*.

Figure 3 summarizes some important enzymatic activities of S1 nuclease. Since it degrades both ssDNA and ssRNA, S1 nuclease can trim protruding single-stranded ends in dsDNA or RNA:DNA hybrids, leaving intact the shorter base-paired fragments (**Fig. 3A, B**). Its endonuclease activity can also recognize and hydrolyse phosphodiester bonds at mismatched base-pairs (**Fig. 3C**), a property that has been exploited to develop a procedure to map mutations in dsDNA molecules *(17)*. Of more relevance to the analysis of mRNA, a similar endonuclease digestion can degrade internal single-stranded loops in RNA:DNA hybrids, such as those formed by intronic sequences on hybridization with mature mRNA (**Fig. 3D**).

The treatment of RNA:DNA hybrids by S1 nuclease certainly represents the most critical step in the analysis of RNA structure by S1 mapping. Indeed, most of the artifacts encountered using this technique are related to the activities of S1 nuclease. For example, regions of double-stranded nucleic acid rich in A and T residues (AT-rich regions) are thermodynamically less stable than other regions containing evenly distributed C and G residues. Consequently, AT-rich domains are thought to be more prone to "breathing," which refers to a dynamic equilibrium between a double-stranded state and its constituent single-stranded species. Because this momentary melting

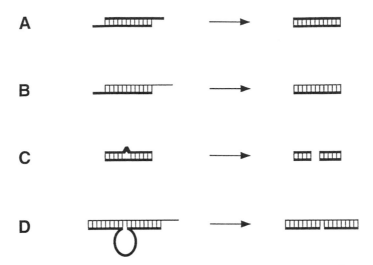

Fig. 3. S1-nuclease activities. **(A)** Digestion of protruding single-stranded ends from DNA:DNA hybrids. **(B)** Digestion of unhybridized regions of RNA:DNA hybrids. **(C)** Formation of double-strand breaks by endonuclease activity at mismatch base pair. **(D)** Digestion of single-stranded loops in RNA:DNA hybrids. At high temperature (45°C) the RNA bond facing the nick is also cleaved.

exposes single-stranded templates for S1 nuclease digestion, the presence of AT-rich regions in otherwise perfectly base-paired RNA:DNA hybrids can lead to undesirable cleavage *(18)*. Similarly, the ends of nucleic acid duplexes are sensitive to trimming by S1 nuclease, particularly at high-enzyme concentration. Therefore, in several mapping experiments, the protected fragments obtained appear as a group of bands varying in size from 2 to 6 nucleotides *(1)*. Other structures, such as trimolecular hybrids *(19)*, certain mismatches in RNA:DNA duplexes *(7)*, phosphodiester bonds facing RNA loops *(20)*, or simply the 5' cap structure of mRNA molecules *(9)*, all show some degree of resistance to S1 nuclease digestion.

Taken together, these cases of hypersensitivity or resistance to S1 nuclease tend to complicate the interpretation of the results and limit the applications of the technique. However, most artifacts can be avoided by covering a range of digestion conditions allowing the identification of reproducible patterns of protected fragments for a specific mapping experiment. Variations in the concentration of S1 nuclease utilized as well as in the temperature of digestion are most commonly used. Other approaches, such as the addition of RNase H *(20)* or confirmation of the mapping results by another technique (such as primer extension, cDNA sequencing, or Northern-blot analysis), can also be considered to further validate the results obtained by S1 digestion.

2.3. Analysis of Protected Fragments

The final step in an S1 mapping experiment usually involves purification of the nucleic acids, denaturation of the double-stranded hybrids resistant to the nuclease activity, separation of the labeled fragments by gel electrophoresis, and their detection by autoradiography (*see* **Figs. 2** and **5**). For most applications, this analysis is rather straightforward and follows well-established molecular biology techniques *(3)*. When precision is desirable in a mapping experiment, electrophoresis on a denaturing poly-acrylamide/urea gel is the method of choice since it offers the highest resolution. However, in order to take full advantage of this gel system, the labeled probe should be appropriately selected to yield protected fragments smaller than 300 nucleotides. Additional accuracy can also be gained if a DNA sequencing reaction performed on the probe fragment is used as a size marker and run next to the digested samples. Such a sequencing ladder can be obtained by chemical Maxam-Gilbert reactions performed directly on the labeled probe *(9,10)*. In applications where a single-stranded probe is synthesized in vitro by extension of a 5'-labeled oligonucleotide, the same primer can be used for dideox sequencing to generate easily the sequencing ladder.

3. Applications

Contrary to other techniques frequently used in RNA analysis, such as primer extension and Northern blot, S1 mapping offers a wide range of applications. As a consequence, important variations exist in the possible experimental design, most notably in the choice and preparation of the DNA probe. The most common applications of S1 nuclease analysis of RNA are discussed below. **Figures 4** and **5** can be referred to for schematic representations of specific RNA:DNA hybrids and the expected band patterns on analysis of the protected fragments, respectively.

3.1. Mapping of 5' and 3' Termini

An important application of S1 analysis is the mapping of the 5' and 3' termini of an RNA molecule. The 5' end of an mRNA defines the first nucleotide from the genomic sequence utilized as a template for RNA synthesis and therefore locates the position of the transcription start site. Several eukaryotic mRNA show considerable heterogeneity in the specific location of their 5' end, for example, through the usage of alternative promoter elements with different developmental and tissue specificities. S1 digestion can be used not only to map the positions of these transcription start sites, but also to quantify the relative abundance of each variant in different samples. As shown in **Fig. 4A**, the 5' terminus of an mRNA can be mapped with a 5'-end-labeled genomic probe that spans the transcription start site. However, if the exon–intron structure of the gene studied is not known, it is preferable to use a uniformly labeled probe since a 5' label is protected from digestion only if the labeled end maps within an exon. In an ingenious variation to this approach, the genomic probe is linearized and 5'-labeled within the first intron (**Fig. 4F**) in order specifically to detect, map, and quantify the unspliced precursor or primary transcript *(9)*. Knowledge of the gene structure is also important to avoid the possibility that the labeled terminus maps within the second exon, downstream of a small exon 1. Since in such a case the S1 analysis would map the beginning of exon 2 instead of that of the mRNA molecule, it is often advisable to confirm the position of the transcription start site(s) through an independent technique, such as primer extension.

The 3' end of an mRNA is generated by a more complex mechanism that is not linked to transcription *per se*, in that transcription terminates at a site downstream of the 3' terminus of the mature mRNA. The 3' end is created by further processing of the longer transcriptional product and defines the site of polyadenylation of the mRNA. S1 nuclease digestion in the presence of a genomic probe including part of the last exon

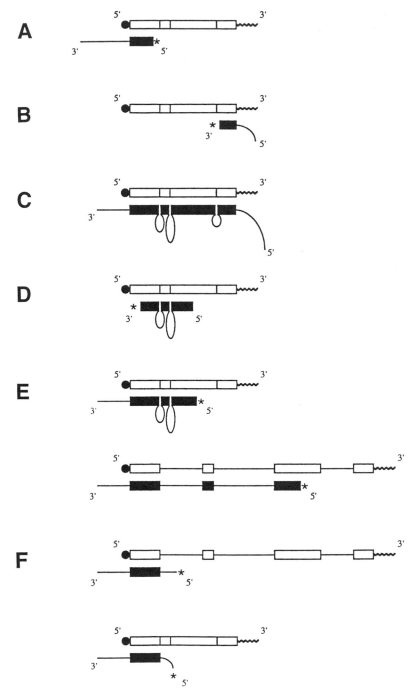

Fig. 4. Schematic representation of RNA:DNA hybrids formed for specific applications of S1 nuclease mapping. *See text* for discussion. Where specified, the position of the label on the ssDNA probe is shown by an asterisk. Features of the RNA molecules are represented as described in **Fig. 1**.

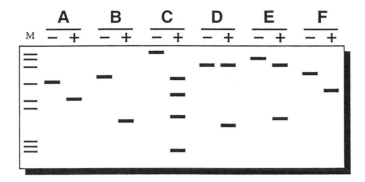

Fig. 5. Analysis of S1-protected fragments by gel electrophoresis. For each RNA:DNA hybrid presented in **Fig. 4**, the expected size of the labeled ssDNA fragment(s) before (–) and after (+) S1-nuclease digestion is shown on a schematic representation of an X-ray film obtained after gel electrophoresis of the purified fragments and autoradiography.

can be used to map the position of the last nucleotide retained on the mRNA (5' of the poly[A] tail). Such a probe can be uniformly labeled or 3'-end-labeled, as shown in **Fig. 4B**. Similar to considerations mentioned for the mapping of 5' termini, the labeled end must fall within the last exon in order to map the expected terminus. The position of the 3' end, as determined by S1 mapping, can be confirmed by sequencing cDNA clones obtained by oligo-dT priming.

3.2. Mapping of Exon–Intron Structure

Historically, S1 analysis played an important role in the characterization of gene structure since it provided the first enzymatic assay to probe the exon–intron organization of transcribed sequences *(7)*. The fact that all exons can base pair with the corresponding mRNA implies that, in theory at least, the number and size of all exons can be determined in a single S1 mapping experiment performed with a large uniformly labeled genomic clone spanning the entire transcriptional unit (**Fig. 4C**). In practice, the applications of such a simple approach are rather limited owing to

1. The complex patterns it generates on mRNAs with more than 4 or 5 exons;
2. The greater chance of obtaining digestion artifacts; and
3. The impossibility to order the exons identified.

Ideally, end-labeled probes linearized within exons can be obtained to map a single splice junction (**Fig. 4D, E**). Here again, this procedure is applicable only to simple transcriptional units of <4 or 5 exons. In most cases, the method of choice available now is the direct comparison between the sequence data obtained from a cDNA and a genomic clone. Nevertheless, the mapping of splice junctions by S1 nuclease digestion can be exploited to detect and quantify a primary transcript [*(9)* and **Fig. 4E**], and more importantly, to study alternative processing events, such as read-through of an exon splice site *(21)*, or alternative splicing *(22)*.

3.3. Quantification of Specific RNAs

In addition to the strictly qualitative aspects of S1 nuclease analysis presented above, this technique is also widely used to quantify specific RNA molecules or to determine their relative abundance in different samples *(2)*. Although the quantification of RNA by S1-nuclease protection assay involves essentially the same steps as the mapping experiments described in the previous sections, certain considerations are critical in order to develop a reliable assay. Whereas in a mapping analysis features of an mRNA are inferred from the size of the protected bands obtained, the quantification of an mRNA is based on the intensity of the signal detected for a given band. Therefore, even if any kind of probe could be used to quantify RNA, it is advisable to select a probe that yields a simple, easily quantified banding pattern, preferably a single band. As mentioned earlier, it is also important to include a significant portion of unprotected sequences in the probe in order to avoid comigration of protected and undigested fragments, which would prevent reliable quantification of the RNA.

As far as the quantitative aspect of the assay is concerned, the technique relies on the assumption that all target RNA molecules present in the sample are hybridized to the probe. In practice, this implies that the probe must always be present in molar excess over the target RNA. An easy way to assess if the hybridization of the RNA is complete and test the reliability of the assay is to perform a series of calibration experiments in which the amount of RNA added is varied, while the probe concentration is kept constant *(18,23)*. For a given amount of probe, it is thus possible to determine the range of RNA concentrations over which a linear relationship exists between the quantity of target RNA present and the intensity of the signal obtained. Even though it is possible to estimate the absolute molar amount of the target RNA from calculations based on the specific activity of the probe and the number of counts present in the protected band *(1)*, S1 mapping is usually used to quantify relative variations of RNA between different samples. In this regard, S1 mapping also offers the possibility to include a secondary probe for a control RNA, which should be present at the same level in all samples studied. Because both the target and the control RNAs can be detected in a single S1-mapping experiment, the ratios of intensities obtained for the corresponding protected bands can be calculated to estimate the relative amounts of the target RNA in the different samples.

Abbreviations

DNA, deoxyribonucleic acid; dNTP, deoxyribonucleotide triphosphate (dATP, dCTP, dGTP, and dTTP); ds, double-stranded nucleic acid, DNA or RNA; mRNA, messenger RNA; PCR, polymerase chain reaction; RNA, ribonucleic acid; rNTP, ribonucleotide triphosphate; ss, single-stranded nucleic acid, DNA or RNA; T_m, melting temperature.

References

1. Berk, A. J. (1989) Characterization of RNA molecules by S1 nuclease analysis. *Methods Enzymol.* **180,** pp. 334–347.
2. Farrell, R. E. (1993) *RNA Methodologies. A Laboratory Guide for Isolation and Characterization.* Academic, San Diego.
3. Sambrook, J., Fritsch, E. F., and Maniatis, T. (1989) *Molecular Cloning: A Laboratory Manual.* (2nd ed.) Cold Spring Harbor Laboratory, Cold Spring Harbor, NY.
4. Williams, J. G. and Mason, P. J. (1985) Hybridisation in the analysis of RNA, in *Nucleic Acid Hybridisation—A Practical Approach* (Hames, B. D. and Higgins, S. J., eds.), IRL, Oxford, UK, pp. 139–160.

5. Britten, R. J. and Davidson, E. H. (1985) Hybridisation strategy, in *Nucleic Acid Hybridisation—A Practical Approach* (Hames, B. D. and Higgins, S. J., eds.), IRL, Oxford, UK, pp. 3–16.

6. Casey, J. and Davidson, N. (1977) Rates of formation and thermal stabilities of RNA:DNA and DNA:DNA duplexes at high concentrations of formamide. *Nucleic Acids Res.* **4**, 1539–1552.

7. Berk, A. J. and Sharp, P. A. (1977) Sizing and mapping of early adenovirus mRNAs by gel electrophoresis of S1 endonuclease-digested hybrids. *Cell* **12**, 721–732.

8. Dean, M. (1987) Determining the hybridization temperature for S1 nuclease mapping. *Nucleic Acids Res.* **15**, 6754.

9. Weaver, R. F. and Weissmann, C. (1979) Mapping of RNA by a modification of the Berk-Sharp procedure: the 5' termini of 15 S β-globin mRNA precursor and mature 10 S β-globin mRNA have identical map coordinates. *Nucleic Acids Res.* **7**, 1175–1193.

10. Viville, S. and Mantovani, R. (1994) S1 mapping using single-stranded DNA probes, in *Methods in Molecular Biology, vol. 31: Protocols for Gene Analysis* (Harwood, A. J., ed.), Humana, Totowa, NJ, pp. 299–305.

11. Quarless, S. A. and Heinrich, G. (1986) The use of complementary RNA and S1 nuclease for the detection of low abundance transcripts. *BioTechniques* **4**, 434–438.

12. Dranginis, A. M. (1990) Binding of yeast α1 and α2 as a heterodimer to the operator DNA of a haploid-specific gene. *Nature* **347**, 682–685.

13. Ando, T. (1966) A nuclease specific for heat-denatured DNA isolated from a product of *Aspergillus oryzae. Biochim. Biophys. Acta* **114**, 158–168.

14. Vogt, V. M. (1973) Purification and further properties of single-strand-specific nuclease from *Aspergillus oryzae. Eur. J. Biochem.* **33**, 192–200.

15. Vogt, V. M. (1980) Purification and properties of S1 nuclease from *Aspergillus. Methods Enzymol.* **65**, 248–255.

16. Weigand, R. C., Godson, G. N., and Radding, C. M. (1975) Specificity of the S1 nuclease from *Aspergillus oryzae. J. Biol. Chem.* **250**, 8848–8855.

17. Shenk, T. E., Rhodes, C., Rigby, P. W. J., and Berg, P. (1975) Biochemical method for mapping mutational alterations in DNA with S1 nuclease: the location of deletions and temperature-sensitive mutations in simian virus 40. *Proc. Natl. Acad. Sci. USA* **72**, 989–993.

18. Miller, K. G. and Sollner-Webb, B. (1981) Transcription of mouse rRNA genes by RNA polymerase I: in vitro and in vivo initiation and processing sites. *Cell* **27**, 165–174.

19. Lopata, M. A., Sollner-Webb, B., and Cleveland, D. W. (1985) Surprising S1-resistant trimolecular hybrids: potential complication in interpretation of S1 mapping analyses. *Mol. Cell. Biol.* **5**, 2842–2846.

20. Sisodia, S. S., Cleveland, D. W., and Sollner-Webb, B. (1987) A combination of RNase H and S1 nuclease circumvents an artefact inherent to conventional S1 analysis of RNA splicing. *Nucleic Acids Res.* **15**, 1995–2011.

21. Scott, G. K., Robles, R., Park, J. W., et al. (1993) A truncated intracellular HER2/neu receptor produced by alternative RNA processing affects growth of human carcinoma cells. *Mol. Cell. Biol.* **13**, 2247–2257.

22. Diebold, R. J., Koch, W. J., Ellinor, P. T., et al. (1992) Mutually exclusive exon splicing of the cardiac calcium channel alpha 1 subunit gene generates developmentally regulated isoforms in the rat heart. *Proc. Natl. Acad. Sci. USA* **89**, 1497–1501.

23. Viville, S., Jongeneel, V., Koch, W., Mantovani, R., Benoist, C., and Mathis, D. (1991) The Eα promoter: a linker-scanning analysis. *J. Immunol.* **146**, 3211–3217.

5

Detecting mRNA by Use
of the Ribonuclease Protection Assay (RPA)

Ralf Einspanier and Annette Plath

1. Introduction

The characterization of gene expression patterns within cells and tissues is a very desirable and highly informative way of providing insights into cell development, maintenance, and cell–cell interactions. Several widely used methods, such as Northern or dot blot, *in situ* hybridization, reverse transcription-polymerase chain reaction (RT- PCR), and ribonuclease protection assay (RPA), have been developed to analyze changing cellular distribution and concentration of mRNAs. The fundamental principle to detect mRNA expression is the specific hybridization of a complementary DNA or RNA probe with its cellular homolog. The relative or absolute quantitation of such mRNAs is complex and each of the techniques has to be judged on its technical limitations and the information it provides. This chapter provides a short introduction to the RPA together with the typical advantages, new nonradioactive approaches, and more simplified gel systems based on original results from our laboratory.

One major advantage of RPA relies on the solution hybridization procedure enabling the detection of very low abundant mRNA molecules, whereas Northern blot and *in situ* hybridization use support-bound RNAs. The term solution hybridization may be used to describe both RPA and S1 nuclease analysis (*see* Chapter 4), which both rely on degrading single-stranded nucleic acids; for the RPA, single-strand specific ribonucleases A and T1 are used. S1 analysis has been mainly used to study the mRNA structure by mapping of exon–intron regions, 5' and 3' ends, or mutational analysis *(1)*. However, RPA has been further developed to enable quantitation study of mRNA. It has been found that by using the selective ribonucleases (A and T1) it is possible to detect relatively low mRNA *(2–4)*. In this chapter not all possible applications or developments of RPA can be dealt with but we have tried to give a proprehensive survey of this important molecular biology technique.

2. Description of the RNase Protection Assay

In principle, cellular mRNA is hybridized with an artificial single-stranded complementary RNA, which has to be specifically labeled, usually with ^{32}P. After the formation of double-stranded complexes, all single-stranded RNA molecules are eliminated

From: *Molecular Biomethods Handbook*
Edited by: R. Rapley and J. M. Walker © Humana Press Inc., Totowa, NJ

A

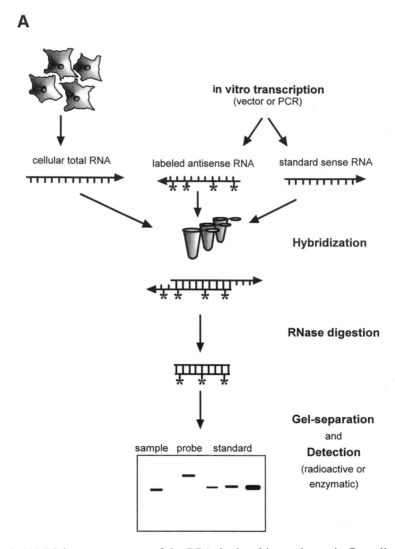

Fig. 1. **(A)** Major components of the RPA depicted in a schematic flow diagram.

by selective ribonuclease digestion. The resulting specifically protected duplex structures are detected through their label, depending on size and concentration.

The following essential steps have to be considered before performing a RPA:

1. Preparation of a gene-specific, labeled, single-strand complementary RNA;
2. Selective hybridization of template mRNA and labeled complementary RNA;
3. Digestion of only unbound single-stranded RNA; and
4. Separation and detection of protected double-stranded fragments.

Figure 1A shows the course of events in detail. Assuming a normally abundant mRNA, generally 10 µg of total cellular RNA is introduced as the template into RPA. Adequate precautions are necessary to provide an RNase-free environment during extraction and handling of RNA. Alternatively, a very quick assay can be performed omitting the RNA isolation by using just crude cell lysates *(5)*. In principle, the mRNA

B

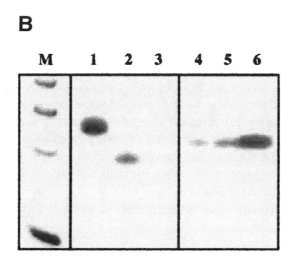

Fig. 1. *(continued)* **(B)** Original autoradiography presenting the untreated probe (1), the results of a TGFα-specific RPA done with bovine mammary total RNA (2), and the background control (3). Additionally, increasing amounts of sense TGFα-RNA-standard (4–6: 0.25, 0.5, 5 pg) were separated on 15% PAGE; exposure 4 d. RNA-ladder (M): 155, 280, 400, 530.

(cDNA) or gene of interest has to be characterized. This means that the whole or partial nucleotide sequence must be available. A labeled complementary RNA is generated from a linear DNA template by in vitro transcription using an appropriate promoter (T3, T7, SP6) and RNA polymerase. This is normally undertaken by subcloning the sequence of interest into a circular transcription vector or alternatively by PCR amplification of the desired DNA fragment using modified primers with attached promoter sequences *(6)*. Following in vitro synthesis of specifically labeled antisense RNA, the yield and quality should be determined, usually by gel electrophoresis. Simultaneously, the necessary purification of the probe is carried out *(7,8)*. For standardized purifications with an excellent recovery we usually prefer gel chromatography using quick spin columns to reduce the troublesome exposure to radioactivity *(9)*. Unlabeled standard sense RNA is produced using the opposite reading frame. Subsequently, the standard-sense RNA is quantitated by gel electrophoresis with an internal DNA marker ladder. If higher yields are available UV-spectroscopy (OD$_{260}$) may be used, alternatively, this may be determined using an incorporated tracer (e.g., ^{3}H).

The labeled antisense RNA in excess is hybridized with known amounts of total cellular RNA and standard-sense RNA, respectively. After several hours the formation of hybrids occurs in the hybridization solution between the labeled probe and cellular RNA or dilutions of standard-sense RNA (for a detailed discussion regarding hybridization conditions *see* Chapter 6). Subsequently, all single-stranded RNA molecules are eliminated using a selected mixture of single-strand-specific RNases (RNase A and T1).

It should be emphasized that in general RPA requires optimization with respect to the conditions of hybridization and RNase digestion for optimal results. The RNases in the reaction mixture are subsequently inactivated and the protected double-stranded molecules, which represent homologous mRNAs compared to the probe, are separated by denaturing polyacrylamide gel electrophoresis (PAGE). Normally this is undertaken on long vertical DNA-sequencing gel systems *(4)*; alternatively, rapid horizontal PAGE systems are now available that reduce gel-running and exposure time, buffer-volumes, and isotopic contamination *(9,11)*. The new horizontal gel systems can be either cast manually or obtained commercially.

The protected ^{32}P-labeled RNA is detected by exposure on X-ray film for several days or up to 1 or 2 wk depending on the required signal-to-noise ratio. Furthermore analysis is also achieved by direct radioactive measurement of separated fragments or by densitometry of the exposed X-ray film. Following calculation of nonspecific signals and standard sense RNA samples the resulting data may be directly correlated to the level of specific gene expression within the cells or tissue of interest. An example of an original RPA is shown in **Fig. 1B**.

Obviously, two different types of information can be obtained from such an autoradiograph:

1. The size of each protected sample may be deduced from the standard marker or ladder.
2. The intensity of the bands directly correspond with the absolute concentration of the specific cellular mRNA

Different computerized systems can be helpful to create relevant statistical data from RPA-derived raw signals. It must be emphasized that the resulting size differences between untreated probe, and protected fragments from samples and those from the standards rely on a system-specific design of antisense probe and standard-sense RNA. A small variable fragment of vector sequence is an obligatory part of the sense as well antisense transcript. Such artificial elements are removed during RNase treatment and each protected fragment appears shorter when compared to the labeled probe.

3. Application and Detection Limits

The RNase protection assay is mainly undertaken for sensitive and reliable mRNA detection and quantitation. This technique is most favorable if low mRNA levels are expected, where limited cell material is present or where mRNA would be too long for Northern analysis, or at worst, when partially degraded total RNA is available. Depending on the individual protocols the sensitivity of the procedure can be increased by adding up to 60 µg total RNA or using polyA+RNA *(7,11)*. A mean probe size between 100 and 500 nucleotides generally appears to be optimum *(8)*. A remaining limitation is the quality of the introduced labeled antisense RNA probe: Careful preparation and quality control is absolutely necessary. Realistic estimation of RNA concentrations is only possible if the absolute mass of a specific mRNA can be analyzed within heterogeneous sample mixtures. Introducing artificial standard RNA into RPA is a method of choice to approach this objective. Such a standard should first have similar properties (sequence and length) to the authentic mRNA to allow the same hybridization conditions and treatment method *(13)*. However, this is not possible for most investigations and one must accept certain limits when claiming a precise quantitation. These general problems are found to more or less the same for all known mRNA quantitation assays. An example for the individual resolution of RT-PCR, RPA, and Northern blot technique in detecting different expression levels of the same gene is shown in **Fig. 2**.

Recently, detection limits of different RPAs have reported to be routinely expected between 0.1 and 1 pg/sample *(9,13,14)*. A further advantage of RPA is the highly spe-

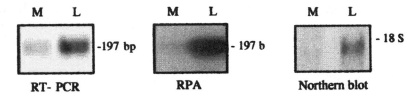

Fig. 2. Detection of TGFα-specific mRNA in bovine mammary (M) and liver (L) total RNA by use of RT-PCR (330 ng RNA), RPA (10 µg RNA), and Northern blot (10 µg RNA).

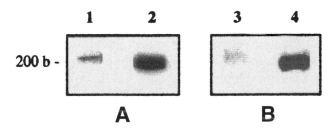

Fig. 3. Comparision of radioactive versus nonradioactive RPA. Detection of mRNA specific for IGF1 (1,3), GAPDH (2,4) in total bovine liver RNA. **(A)** ^{32}P-labeled probes, 10 µg total RNA, separated on 15% PAGE, exposure 12 d. **(B)** Digoxygenin-labeled probes, 15 µg total RNA, separated on 6% PAGE, exposure 1 h.

cific detection of mismatches: false positive hybridization artifacts known from Northern analysis do not generally occur. Therefore, RPA is a tool for distinguishing closely related mRNA species *(15)*.

4. Nonradioactive Approach

Recently, some new nonradioactive approaches are reported to circumvent the use of isotopes (^{32}P, ^{33}P) which are widely used as probe-label for RPA *(11,14)*. Two most frequently used marker molecules, biotin and digoxygenin, have been successfully introduced into the RPA *(9,10,16)* leading to acceptable results. Such nonradioactive labeled probes are relatively stable and can be stored for longer periods. The obligatory blotting and detection steps increase the handling time, sometimes causing a reduced signal strength. A low crosslinked polyacrylamide gel (6%) must be used for this purpose; otherwise, the blotting and resulting signal efficiency is dramatically lowered. In **Fig. 3**, a direct comparison between two RPAs using ^{32}P- and digoxygenin-labeled probes is depicted. Obviously, one can see the detection of abundant mRNA concentrations. The detection limit of 0.5 pg mRNA/10 µg total RNA is apparent. If a greater sensitivity is required, this can be archived by introducing up to a 10-fold increase in the total amount of RNA.

5. Conclusion and Evaluation

The measurement of mRNA levels to illustrate the patterns of gene expression in cells is an innovative approach for biological and medical research. In comparison with Northern analysis the development of the RNase protection assays represents a favorable method of providing highly sensitive, rapid, robust quantitative measurements.

Even partially degraded cellular RNA is not a severe problem with the RPA. On the other hand, the feature of absolute size determination of target RNAs is compromised slightly using RPA. Only RT-PCR can detect extremely rare transcripts down to single molecules, but this often requires severe working precautions because of the problems of contamination and a laborious absolute quantitation process *(17)*. Performing expression analysis by RPA has became easier because of the commercial availability of optimized system components. One great advantage of the RPA-system is the multiple sampling to reduce interassay variations: more than 40 single RNA samples can be processed in parallel during one assay.

In summary, RPA is one of the recommended techniques for detecting and measuring low abundant mRNA levels in large sample scales. Formerly, RNase protection assays appeared to be difficult to set, requiring expert operations. Recent technological improvements of RPA now allow easy performance for every researcher. As a leading technique in this field, future development of the RPA is therefore recommended. New nonradioactive detection systems will make it feasible to use fluorescent labels or ELISA techniques leading to a more automated analysis of gene expression.

References

1. Aranda, M. A., Farile, A., Garcia-Arenal, F., and Malpica, J. M. (1995) Experimental evaluation of the ribonuclease protection assay method for the assessment of genetic heterogeneity in populations of RNA viruses. *Arch. Virol.* **140,** 1373–1383.
2. Sisodia, S. S., Cleveland, D. W., and Sollner-Webb, B. (1987) A combination of RNase H and Si nuclease circumvents an artefact inherent to conventional S1 analysis of RNA splicing. *Nucleic Acids Res.* **15,** 1995–2011.
3. Friedberg, T., Grassow, M. A., and Oesch, F. (1990) Selective detection of mRNA forms encoding the major phenobarbital inducible cytochrome P450 and other members of the P45011B family by the RNase A protection assay. *Arch. Biochem. Biophys.* **279,** 167–172.
4. Sambrook, J., Fritsch, E. F., and Maniatis, T. (1989) *Molecular Cloning: A Laboratory Manual,* 2nd ed. Cold Spring Harbor Laboratory, Cold Spring Harbor, NY.
5. Haines, D. S. and Gillespie, D. H. (1992) RNA abundance measured by a lysate RNase protection assay. *Biotechniques* **12,** 736–741.
6. Yang, H. and Melera, P. W. (1992) Application of the polymerase chain reaction to the ribonuclease protection assay. *Biotechniques* **13,** 922–927.
7. Tymms, M. J. (1995) Quantitative measurement of mRNA using the RNase protection assay, in *Methods in Molecular Biology, 37: In Vitro Transcription and Translation Protocols* (Tymms, M. J., ed.), Humana, Totowa, NJ, pp. 32–46.
8. Ma, Y. J., Dissen, G. A., Rage. F., and Ojeda, S. R. (1996) RNase protection assay. *Methods: A Companion to Methods in Enzymology* **10,** 273–278.
9. Plath, A., Peters, F., and Einspanier, R. (1996) Detection and quantitation of specific mRNAs by ribonuclease protection assay using denaturing horizontal polyacrylamide gel electrophoresis: a radioactive and nonradioactive approach. *Electrophoresis* **17,** 471,472.
10. Turnbow, M. A. and Garner, C. W. (1993) Ribonuclease protection assay: use of biotinylated probes for the detection of two messenger RNAs. *Biotechniques* **15,** 267–270.
11. Belin, D. (1996) The RNase protection assay, in *Methods in Molecular Biology, vol. 58: Basic DNA and RNA Protocols* (Harwood, A., ed.), Humana, Totowa, NJ, pp. 131–136.
12. Pape, M. E., Melchior, G. W., and Marotti, K. R. (1991) mRNA quantitation by a simple and sensitive RNase protection assay. *Genet. Anal. Tech. Appl.* **8,** 206–213.

13. Ludlam, W. H., Zang, Z., McCarson, K., Krause, J. E., Spray, D. C., and Kessler, J. A. (1994) mRNAs encoding muscarinic and substance P receptors in cultured sympathic neurons are differentially regulated by LIF or CNTF. *Dev. Biol.* **164,** 528–539.

14. Meisinger, C. and Grothe, C. (1996) A sensitive RNase protection assay using ^{33}P labeled antisense riboprobes. *Mol. Biotechnol.* **5,** 289–291.

15. Genovese, C., Brufsky, A., Shapiro, J., and Rowe, D. (1989) Detection of mutations in human type I collagen mRNA in osteogenesis imperfecta by indirect RNase protection. *J. Biol. Chem.* **264,** 9632–9637.

16. Wundrack, I. and Dooley, S. (1992) Nonradioactive ribonuclease protection analysis using digoxygenin labelling and chemiluminescent detection. *Electrophoresis* **13,** 637,638.

17. Souaze, E., Ntodou-Thome, A., Tran, C. Y., Rostene, W., and Forgez, P. (1996) Quantitative RT-PCR: limits and accuracy. *Biotechniques* **21,** 280–285.

6

Gene Probes

Marilena Aquino de Muro

1. Introduction

The use of nucleic acid or gene probes as cloning and diagnostic tools has proven to be a powerful technique for isolation and manipulation of genes, as well as for rapid detection and identification of bacteria and other microorganisms, with wide applications in molecular biology, medical, and environmental sciences. By definition, a probe is a DNA molecule with a strong affinity for a specific target, and the hybrid (probe-target combination) can be revealed when an appropriate detection system is used. The nature of interactions between a probe and complementary nucleic acids is primarily through H-bonding, at thousands of sites in some cases, depending on the length of the hybrid. Hydrophobic interactions contribute little to the specificity, although a reduction in stability can be observed in organic solvents.

This chapter covers the main aspects of probe design, labeling and detection, target format, and hybridization conditions, but does not provide details of reagents, buffers, or protocols, which can be found in the references (*see* **Table 1**).

2. Probe Design

The probe design depends on whether a gene probe or an oligonucleotide probe is desired.

2.1. Gene Probes

Gene probes are generally longer than 500 bases and comprise all or most of a target gene. They can be generated in two ways. Cloned probes are normally used when a specific clone is available or when the DNA sequence is unknown and must be cloned first in order to be mapped and sequenced. It is usual to cut the gene with restriction enzymes and excise it from an agarose gel, although if the vector has no homology it may not be necessary.

The polymerase chain reaction (PCR) is a powerful procedure for making gene probes because it is possible to amplify and label (at the same time), long stretches of DNA using chromosomal or plasmid DNA as template as well as labeled nucleotides included in the extension step (*see* **Subheadings 2.2.** and **3.2.3.**). Having the whole sequence of a gene, which can easily be obtained from databases (GenBank, EMBL),

From: *Molecular Biomethods Handbook*
Edited by: R. Rapley and J. M. Walker © Humana Press Inc., Totowa, NJ

Table 1
Examples of The Applications of Gene Probes
or Oligonucleotide Probes in Medical, Environmental, and Food Research

Application	Reference	Comments
Medical		
Detection of multidrug resistance genes in patients with acute leukemia	*11*	Gene probes
Evaluation of alterations in the tumor supressor genes P16 and P15 in human bladder tumors	*12*	Gene probes
Differentiation between new- and old-world *Leishmania* groups	*15*	Tubulin gene probes
Detection of malignant cells in the bone marrow and peripheral blood of patients with multiple myeloma	*20*	IgH gene oligonucleotide probe
Identification of nonmotile *Bacteroidaceae* from dental plaque	*27*	DNA and RNA gene probes
Diagnosis of sexually transmitted bacterial infections (*Neisseria gonorrhoeae* and *Chlamydia trachomatis*)	*28*	Gene probes
Isolation and identification of *Campylobacter* species	*29*	Gene probes
Human-brain tissue *in situ* hybridization to identify and analyze specific mRNA transcripts changes related to neurological diseases (Alzheimer's, Huntington's)	*31*	Oligonucleotide and RNA probes
Localize and quantitate gonadotropin-releasing hormone messenger RNA under different physiological conditions	*32*	Oligonucleotide probes
Environmental		
Evaluation of the genetic diversity and evolutionary relationships among the symbiotic forms of the cyanobacterium *Anabaena*	*13*	Nitrogen fixation gene probes
Detection of novel marine methanotrophus after methane enrichment	*14*	16S rRNA oligo probe
Differentiation, phylogenetic analysis, and identification of novel mosquitocidal strains of *Bacillus sphaericus*	*24–25*	16S rRNA gene and oligonucleotide probe, toxin gene probes
Detection of enteric adenoviruses 40 and 41 in raw and treated waters	*16*	Gene probes
Detection of *E. coli* and *Enterococcus faecalis* in illuminated freshwater and seawater	*17*	*E. coli*-specific *vidA* gene probe, 23S rRNA *Enterococcus* oligonucleotide probe
Identification of aquatic *Burkholderia (Pseudomonas) cepacia*	*21*	16S and 23S rRNA gene probes
Detection of polychlorinated biphenyls degrading organisms in soil	*22*	Biphenyl operon gene probes
Detection of *Desulfovibrio* species in oil field production waters	*23*	Hydrogenase gene probes
Detection and identification of 2,4-dichlorophenoxyacetic acid (herbicide) degrading populations in soil	*25*	Gene probes

Table 1 *(continued)*

Application	Reference	Comments
Food		
Characterization of polyubiquitin genes in tissues from control and heat-shocked maize seedlings	*18*	Gene probes
Differentiation between RNA transcripts of cellulase genes during induced avocado fruit abscission and ripening	*19*	Gene probes
Search for enterotoxin LT-II and STB genes among *E. coli* isolated from raw bovine meat samples	*24*	Enterotoxin gene probes
Detection of enterotoxigenic *E. coli* in waste water, river water, and seawater	*26*	Heat-labile gene probe
Detection and identification of food-borne pathogenic microorganisms	*30*	Gene probes

primers can be designed to amplify the whole gene or gene fragments. A considerable amount of time can be saved when the gene of interest is PCR-amplified, because there is no need for restriction-enzyme digestion, electrophoresis, and elution of DNA fragments from vectors. However, if the PCR amplification gives nonspecific bands, it is recommended to gel-purify the specific band that will be used as a probe. Gene probes generally provide greater specificity than oligonucleotides because of their longer sequence and because more detectable groups per probe molecule can be incorporated into them than into oligonucleotide probes *(1)*.

2.2. Oligonucleotide Probes

Oligonucleotide probes are generally targeted to specific sequences within genes. The most common oligonucleotide probes contain 18–30 bases, but current synthesizers allow efficient synthesis of probes containing at least 100 bases. An oligonucleotide probe can match perfectly its target sequence and is sufficiently long to allow the use of hybridization conditions that will prevent the hybridization to other closely related sequences, making it possible to identify and detect DNA with slight differences in sequence within a highly conserved gene, for example.

The selection of oligonucleotide probe sequences can be carried out manually from a known gene sequence using the following guidelines *(1)*:

1. Probe length should be between 18 and 50 bases. Longer probes will result in longer hybridization times and low synthesis yields; shorter probes will lack specificity.
2. Base composition should be 40–60% G-C. Nonspecific hybridization may increase for G-C ratios outside of this range.
3. Be certain that no complementary regions within the probe are present. These may result in the formation of "hairpin" structures, which will inhibit hybridization of the probe.
4. Avoid sequences containing long stretches (more than four) of a single base.
5. Once a sequence meeting the above criteria has been identified, computerized sequence analysis is highly recommended. The probe sequence should be compared with the sequence region or genome from which it was derived, as well as with the reverse complement of the region. If homologies to nontarget regions > 70% or eight or more bases in a row are found, that probe sequence should be re-evaluated.

However, to determine the optimal hybridization conditions the synthesized probe should be hybridized to specific and nonspecific target nucleic acids over a range of hybridization conditions. The aforementioned guidelines are applicable to design forward and reverse primers for amplification of a particular gene of interest to make a gene probe. It is important to bear in mind that in this case it is essential that the 3' end of both forward and reverse primers have no homology with other stretches of the template DNA other than the region required for amplification.

Numerous software packages are available (Lasergene of DNASTAR Inc., USA, GeneJockey II@, Biosoft, UK) that can be used to design a primer for a particular sequence or simply to check if the pair of primers designed manually will perform as expected.

3. Labeling and Detection

3.1. Types of Label

3.1.1. Radioactive Labels

Radioactive probes coupled with autoradiographic detection provide the highest degree of sensitivity and resolution currently available in hybridization assays *(1,2)*. However, considerations of user safety as well as cost and disposal of radioactive waste products limit the applications of radioactive probes.

^{32}P is the most commonly used isotope, and Keller *(1)* lists a few reasons:

1. ^{32}P has the highest specific activity;
2. It emits β-particles of high energy;
3. ^{32}P-labeled nucleotides do not inhibit the activity of DNA-modifying enzymes, because the structure is essentially identical to that of the nonradioactive counterpart.

The disadvantage is its relatively short half-life (14.3 d), so ^{32}P-labeled probes should be used within 1 wk after preparation. The lower energy of ^{35}S plus its longer half-life (87.4 d) make this radioisotope more useful than ^{32}P for the preparation of more stable, less specific probes. These ^{35}S-labeled probes, although less sensitive, provide higher resolution in autoradiography and are especially suitable for *in situ* hybridization procedures. Another advantage of ^{35}S over ^{32}P is that the ^{35}S-labeled nucleotides present little external hazard to the user. The low energy β-particles barely penetrate the upper dead layer of skin and are easily contained by laboratory tubes and vials.

Similarly, ^{3}H-labeled probes have traditionally been used for *in situ* hybridization because the low-energy β-particle emissions result in maximum resolution with low background. It has the longest half-life (12.3 yr).

The use of ^{125}I and ^{131}I has declined since the 1970s with the availability of ^{125}I-labeled nucleoside triphosphates of high specific activity. ^{125}I has lower energies of emission and longer half-life (60 d) than ^{131}I, and is frequently used for *in situ* hybridization.

3.1.2. Nonradioactive Labels

Compared to radioactive labels, the use of nonradioactive labels has several advantages (*see* Chapter 7):

1. Safety;
2. Higher stability of probe;
3. Efficiency of the labeling reaction;

4. Detection *in situ*; and
5. Less time taken to detect signal.

Commonly used nucleotide derivatives for this purpose include digoxigenin-11-UTP, -11-dUTP, and -11-ddUTP, and biotin-11-dUTP or biotin-14-dATP. After hybridization, these are detected by an antibody or avidin, respectively, followed by a color or chemiluminescent reaction catalyzed by alkaline phosphatase or peroxidase linked to the antibody or avidin. Biotin is potentially a poor choice as a detectable group when working with food or bacterial samples since it may contain biotin-binding proteins or interfering amounts of endogenous biotin *(1)*.

The digoxigenin (DIG) system is the most comprehensive, convenient, and effective system for labeling and detection of DNA, RNA, and oligonucleotides. An antidigoxigenin antibody-alkaline phosphatase conjugate is allowed to bind to the hybridized DIG-labeled probe. The signal is then detected with colorimetric or chemiluminescent alkaline phosphatase substrates. If a colorimetric substrate is used, the signal develops directly on the membrane. The signal is detected on an X-ray film (as with ^{32}P or ^{35}S labeled probes) when a chemiluminescent substrate is used.

Boehringer Mannheim has a series of kits for DIG-labeling and detection, and comprehensive detailed guides *(3,4)* with protocols for single-copy gene detection of human genome on Southern blots, detection of unique mRNA species on Northern blots, colony and plaque screening, slot/dot blots, and *in situ* hybridization.

Other fluorescent derivatives of nucleotides can also be incorporated directly into the probe, or conjugated to the antibody or avidin in place of an enzyme. Another system is the horseradish peroxidase (HRP), which covalently crosslinks the marker enzyme HRP to the DNA probe. In the presence of peroxide and peroxidase, chloronaphthol, a chromogenic substrate for HRP, forms a purple insoluble product. HRP also catalyzes the oxidation of luminol, a chemiluminogenic substrate for HRP *(5)*.

The one area in which nonradioactive probes have a clear advantage is *in situ* hybridization. When the probe is detected by fluorescence or color reaction, the signal is at the exact location of the annealed probe, whereas radioactive probes can only be visualized as silver grains in a photographic emulsion some distance away from the actual annealed probe *(6)*.

3.2. Labeling Methods

The majority of radioactive labeling procedures rely on enzymatic incorporation of a nucleotide labeled into the DNA, RNA, or oligonucleotide.

3.2.1. Nick Translation

Nick translation is one method of labeling DNA that uses the enzymes pancreatic DNase I and *Escherichia coli* DNA polymerase I. The nick translation reaction results from the process by which *E. coli* DNA polymerase I adds nucleotides to the 3'-OH created by the nicking activity of DNase I, while the 5' to 3' exonuclease activity simultaneously removes nucleotides from the 5' side of the nick. If labeled precursor nucleotides are present in the reaction, the pre-existing nucleotides are replaced with labeled nucleotides. For radioactive labeling of DNA, the precursor nucleotide is an $[\alpha\text{-}^{32}\text{P}]$dNTP. For nonradioactive labeling procedures, a digoxigenin or a biotin moiety attached to a dNTP analog is used *(2)*.

3.2.2. Random-Primed Labeling

Gene probes, cloned or PCR-amplified, and oligonucleotide probes can be random-primed labeled with radioactive isotopes and nonradioactive labels. Random-primed labeling of DNA fragments (double or single-stranded DNA) was developed by Feinberg and Vogelstein *(7,8)* as an alternative to nick translation to produce uniformly labeled probes. Double-stranded DNA is denatured and annealed with random oligonucleotide primers (6-mers). The oligonucleotides serve as primers for the 5' to 3' polymerase (the Klenow fragment of *E. coli* DNA polymerase I), which synthesizes labeled probes in the presence of a labeled nucleotide precursor.

3.2.3. PCR Labeling

A very robust method for labeling DNA is by using PCR. The gene probe is PCR-amplified using the same set of primers and thermocycling parameters; however, the dNTP mixture should have less dTTP because the labeled DIG-dUTP will also be added to the reaction. When this method is used with $[\alpha\text{-}^{32}\text{P}]\text{dCTP}$, the dNTP mixture has no or low concentration of dCTP. The incorporation of the large molecule of DIG-dUTP along the DNA strands during the PCR cycles makes the fragment run slightly slower through the agarose gel, so a control PCR reaction without DIG should be prepared to check the size of the desired gene. **Figure 1** shows the steps involved in random primed DIG labeling and PCR-DIG labeling. The advantage of PCR-DIG labeling is the incorporation of higher number of DIG moieties along the amplified DNA strands.

3.2.4. Photobiotin Labeling

Photobiotin labeling is a chemical reaction, not an enzymatic one. The materials for photobiotin labeling are more stable than the enzymes needed in nick translation or oligo labeling and are less expensive, and it is a method of choice when large quantities of probe are needed and when very high sensitivities are not needed *(5)*.

3.2.5. End Labeling

End labeling of probes for hybridization is mainly used to label oligonucleotide probes. Three commercial methods for labeling oligonucleotides with digoxigenin have been developed *(4)*:

1. The 3' end labeling of an oligonucleotide 14–100 nucleotides in length with one residue of DIG-11-ddUTP/molecule;
2. The 3' tailing reaction, where terminal transferase adds a mixture of unlabeled nucleotides and DIG-11-dUTP, producing a tail containing multiple-digoxigenin residues;
3. The 5' end labeling in a two-step synthesis with first an aminolinker residue on the 5' end of the oligonucleotide, and then, after purification, a digoxigenin-NHS ester is covalently linked to the free 5'-amino residue.

Oligonucleotides can also be labeled with radioisotopes by transferring the $\gamma\text{-}^{32}\text{P}$ from $[\gamma\text{-}^{32}\text{P}]\text{ATP}$ to the 5' end using the enzyme bacteriophage T4 polynucleotide kinase. If the reaction is carried out efficiently, the specific activity of such probes can be as high as the specific activity of $[\gamma\text{-}^{32}\text{P}]\text{ATP}$ itself *(2)*.

The choice of probe labeling method will depend on *(4)*:

1. Target format: Southern, Northern, slot/dot, or colony blot (*see* **Subheading 4.**);
2. Type of probe: gene or oligonucleotide probe; and

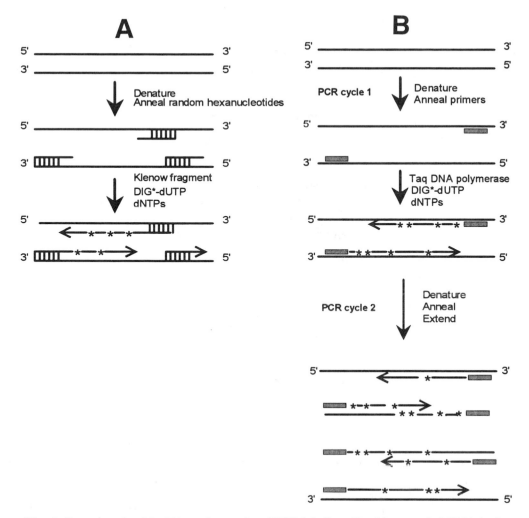

Fig. 1. Steps involved in (A) random-primed DIG labeling: Double-stranded DNA is denatured and annealed with random oligonucleotide primers (6-mers); the oligonucleotides serve as primers for the 5' to 3' Klenow fragment of *E. coli* DNA polymerase I, which synthesizes labeled probes in the presence of DIG-dUTP. (B) PCR-DIG labeling: DIG-dUTP is incorporated during PCR cycles into the DNA strands amplified from DNA target. The asterisk marks represent the digoxigenin molecule incorporated along the DNA strands.

3. Sensitivity required for detection: single-copy gene or detection of PCR-amplified DNA fragments.

For example, 3' and 5' end labeling of oligonucleotides gives good results on slot and colony hybridization in contrast with poor sensitivity when using Southern blotting. **Figure 2** shows an example of the strong hybridization of an end-DIG-labeled 19-mer oligonucleotide to chromosomal DNA from *Bacillus sphaericus*, highlighting the samples that have the same 19-base sequence as the probe in the highly conserved 16S rRNA gene.

4. Target Format

A convenient format for the hybridization of DNA to gene probes or oligonucleotide probes is immobilization of the target nucleic acid (DNA or RNA) onto a solid support,

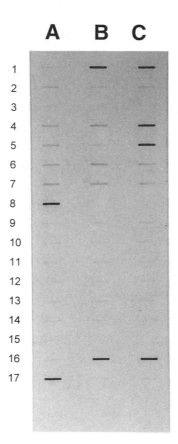

Fig. 2. Slot-blot hybridization (nitrocellulose membrane and Minifold II system, Schleicher and Schuell, Dassel, Germany) of *B. sphaericus* chromosomal DNA with 19-mer oligonucleotide based on the 16S rRNA gene mismatches, end-DIG labeled using digoxigenin-3-*O*-methylcarbonyl-ε-aminocaproic acid-*N*-hydroxy-succinimide ester into the 5'-amino-substituted oligonucleotide. Hybridization was carried out at 32°C for 2.5 h. The color development was carried out at 37°C for 20–30 min. The probe hybridized to the DNA samples in slots B1, C1, C4, C5, A8, B16, C16, and C17.

whereas the probe is free in solution. The solid support can be a nitrocellulose or nylon membrane, latex, magnetic beads, or microtiter plates. Nitrocellulose membranes are very commonly used and produce low background signals; however, they can only be used when colorimetric detection will be performed and no probe stripping and reprobing is planned. For these purposes, positively charged nylon membranes are recommended, and they also ensure optimal signal-to-noise ratio when the DIG system is used. Although nitrocellulose membranes are able to bind large quantities of DNA, they become brittle and gradually release DNA during the hybridization step. Activated cellulose membranes, on the other hand, are more difficult to prepare, but they can be reused many times because the DNA is irreversibly bound *(2)*.

The different immobilization techniques include Southern blots, when whole or digested chromosomal DNA is electrophoresed in an agarose gel, denatured, and blotted onto a membrane (*see* Chapter 8); Northern blots, when the same procedure is used

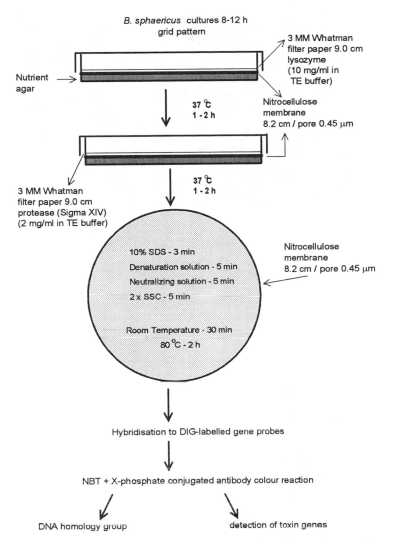

Fig. 3. Colony-blot procedure developed for *B. sphaericus* isolated from various soil and/or mosquito samples.

for RNA (*see* Chapter 9); slot- blots, when whole RNA or denatured DNA is loaded under vacuum into slots onto membranes (similar procedure for dot-blots); colony blots, where colonies are treated with lysozyme on plates and further treatment with protease, and denaturation and neutralization solutions are applied and procedure adjusted according the microorganisms' peculiarities. An example of colony-blot procedure developed for *B. sphaericus*, which can be used for other *Bacillus* species, is shown in **Fig. 3**. The great advantage of colony blotting over slot blotting is that strains with a specific sequence can be rapidly detected from plates, and the DNA-preparation procedure can be then undertaken only for the strains of interest. In a similar way, the slot-blotting procedure has the advantage of quickly highlighting which DNA sample has the gene sequence of interest when a gene probe is hybridized to whole DNA samples. The Southern-blotting procedure, which involves the digest of DNA with restriction

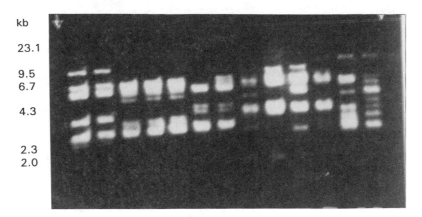

Fig. 4. Autoradiograph of a Southern blot of *Hind*III digests from *B. sphaericus* chromosomal DNA on nitrocellulose hybridized to the ^{32}P-labeled rRNA probe from *E. coli*. Hybridization was carried out at 42°C for 18 h in 40% (v/v) formamide. Autoradiograms on X-Omat film were exposed at −70°C for up 24 h. Fragment sizes are indicated on the left for the *Hind*III-digested fragments of bacteriophage λ DNA.

enzymes and gel electrophoresis, takes a longer preparation time than slot blotting but can provide information on the size and position of the gene, as well as group the samples based on the different patterns when different restriction enzymes are used to digest the samples (**Figs. 4–6**).

5. Hybridization Conditions

Many methods are available to hybridize probes in solution to DNA or RNA immobilized on nitrocellulose membranes. These methods can differ in the solvent and temperature used; the volume of solvent and the length of time of hybridization; method of agitation, when required; the concentration of the labeled probe and its specificity; and the stringency of the washes after the hybridization. However, basically, target nucleic acid immobilized on membranes by the Southern blot, Northern blot, slot- or dot-blot, or colony blot procedures are hybridized in the same way. The membranes are first prehybridized with hybridization buffer minus the probe. Nonspecific DNA-binding sites on the membrane are saturated with carrier DNA and synthetic polymers. The prehybridization buffer is replaced with the hybridization buffer containing the probe and incubated to allow hybridization of the labeled probe to the target nucleic acid. The optimum hybridization temperature is experimentally determined, starting with temperatures 5°C below the melting temperature (T_m). The T_m is defined as the temperature corresponding to the midpoint in transition from helix to random coil, and depends on length, nucleotide composition, and ionic strength for long stretches of nucleic acids. G-C pairs are more stable than A-T pairs because G and C form three H-bonds as opposed to two between A and T. Therefore, double-stranded DNA rich in G and C have higher T_m (more energy required to separate the strands) than A-T-rich DNA. For oligonucleotide probes bound to immobilized DNA the dissociation temperature, T_d, is concentration-dependent. Stahl and Amman *(9)* discuss in detail the empirical formulas used to estimate the T_m and T_d.

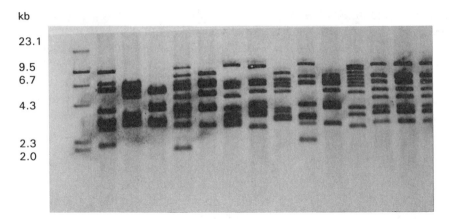

Fig. 5. Southern blot of *Hin*dIII digests from *B. sphaericus* chromosomal DNA on nitrocellulose membrane hybridized to the PCR-DIG-labeled 16S rRNA gene probe from *B. sphaericus* strain 2362. Hybridization was carried out at 62°C for 12 h. The color development was carried out at 37°C for 20–30 min. Fragments sizes are indicated on the left for the *Hin*dIII-digested fragments of bacteriophage λ DNA.

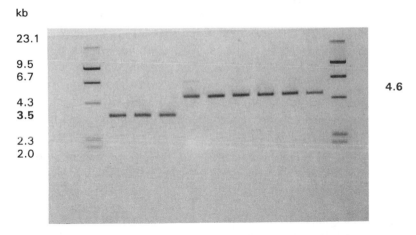

Fig. 6. Southern-blot (nitrocellulose membrane Hybond, Amersham) hybridization of *Hin*dIII-digested chromosomal DNA from *B. sphaericus* strains with the DIG-labeled pBSE-18 probe showing two different sizes of fragment (3.5 or 4.6 kb, in boldface characters on the left and right, respectively) containing the binary-toxin gene. Fragment sizes are indicated on the left and right with DIG-labeled *Hin*dIII-digested fragments of bacteriophage λ DNA.

Following the hybridization, unhybridized probe is removed by a series of washes. The stringency of the washes must be adjusted for the specific probe used. Low-stringency washing conditions (higher salt and lower temperature) increases sensitivity; however, this can give nonspecific hybridization signals and high background. High-stringency washing conditions (lower salt and higher temperature, closer to the hybridization temperature) can reduce background and only the specific signal will remain. The hybridization signal and background can also be affected by probe length, purity, concentration, sequence, and target contamination *(1)*.

In aqueous solution, RNA–RNA hybrids are more stable than RNA–DNA hybrids, which are in turn more stable than DNA–DNA ones. This results in a difference in T_m of approx 10°C between RNA–RNA and DNA–DNA hybrids. Consequently, more stringent conditions should be used with RNA probes *(6)*.

In general, hybridization rate increases with probe concentration. Also, within narrow limits, sensitivity increases with increasing probe concentration. The concentration limit is not determined by any inherent physical property of nucleic-acid probes, but by the type of label and nonspecific binding properties of the immobilization medium involved.

6. Applications

The application of nucleic acid probes has been particularly evident in microbial ecology, where probes can be used to detect unculturable microorganisms, pathogens in the environment, or simply provide rapid identification to species and group levels. Through the development of DNA–DNA, RNA–DNA hybridization procedures, and recombinant DNA methodology, the isolation of species-specific gene sequences is readily achieved *(10,11)*. Gene sequences can be either unique to the particular species or be more highly repeated genes, such as the rRNA genes that contain regions of specificity within them. Because rRNA genes are apparently not subject to horizontal genetic transfer, they represent a given microbial population more directly *(9)*.

Oligonucleotide hybridization probes complementing small ribosomal subunits, large ribosomal subunits, or internal-transcribed spacer regions have now been developed for a wide variety of microorganisms, such as *Actinomyces, Bacteriodes, Borrelia, Clostridium, Campylobacter, Haemophilus, Helicobacter, Lactococcus, Mycoplasma, Neisseria, Proteus, Rickettsia, Vibrio, Streptococcus, Plasmodium, Pneumocystis, Trichomonas, Desulfovibrio,* and *Streptomyces (12)*. These, as well as some uncultivated species, such as marine proteobacteria and thermophilic cyanobacterium, are some of the examples listed by Ward *(12)*.

Table 1 shows just a few examples of recently published literature on applications of gene probes and/or oligonucleotide probes in medical, environmental, and food research.

References

1. Keller, G. H. and Manak, M. M. (1989) *DNA Probes.* Stockton, New York.
2. Sambrook, J., Fritsch, E. M., and Maniatis, T. (1989) M*olecular Cloning: A Laboratory Manual*, 2nd ed., Cold Spring Harbor Laboratory, Cold Spring Harbor, NY.
3. *The DIG System User's Guide for Filter Hybridisation* (1995) Boehringer Mannheim GmbH, D68298 Mannheim, Germany.
4. *Nonradioactive* In Situ *Hybridisation Manual* (1995) Boehringher Mannheim GmbH, D68298 Mannheim, Germany.
5. Karcher, S. J. (1995) *Molecular Biology—A Project Approach.* Academic, San Diego, CA.
6. Alphey, L. and Parry, H. D. (1995) Making nucleic acid probes, in *DNA Cloning 1: Core Techniques* (Glover, D. M. and Hames, B. D., eds.), IRL, Oxford, UK, pp. 121–141.
7. Feinberg, A. P. and Vogelstein, B. (1983) A technique for radiolabeling DNA restriction endonuclease fragments to high specific activity. *Anal. Biochem.* **132,** 6–13.
8. Feinberg, A. P. and Vogelstein, B. (1984) Addendum. *Anal. Biochem.* **137,** 266,267.
9. Stahl, D. A. and Amman, R. (1991) Development and application of nucleic acid probes, in *Nucleic Acid Techniques in Bacterial Systematics* (Stackebrandt, E. and Goodfellow, M., eds.), Wiley, Chichester, UK, pp. 205–244.

10. Brooker, J. D., Lockington, R. A., Attwood, G. T., and Miller, S. (1990) The use of gene and antibody probes in identification and enumeration of rumen bacterial species, in *Gene Probes for Bacteria* (Macario, A. J. L. and Conway de Macario, E., eds.), Academic, San Diego, CA, pp. 390–416.

11. Stahl, D. A. and Kane, M. D. (1992) Methods in microbial identification, tracking and monitoring of function. *Curr. Opin. Biotechnol.* **3,** 244–252.

12. Ward, D. M., Bateson, M. M., Weller, R., and Ruff-Roberts, A. L. (1992) Ribosomal RNA analysis of microorganisms as they occur in nature. *Adv. Microb. Ecol.* **12,** 219–286.

13. Macfarland, A., Dawson, A. A., and Pearson, C. K. (1995) Analysis of MDR1 and MDR3 multidrug-resistance gene expresion and amplification in consecutive samples in patients with acute leukaemia. *Leuk. Lymph.* **19(1–2),** 135–140.

14. Orlow, I., Lacombe, L., Hannon, G. J., Serrano, M., Pellicer, I., Dalbagni, G., Reuter, V. E., Zhang, Z. F., Beach, D., and Cordoncardo, C. (1995) Deletion of the P16 and P15 genes in human bladder-tumours. *J. Natl. Cancer Inst.* **87(20),** 1524–1529.

15. Mendozaleon, A., Havercroft, J. C., and Barker, D. C. (1995) The RFLP analysis of the beta-tubulin gene region in new-world *Leishmania. Parasitology* **111(1),** 1–9.

16. Brown, R., Luo, X. F., Gibson, J., Morley, A., Sykes, P., and Brisco, M. (1995) Idiotypic oligonucleotide probes to detect myeloma cells by messenger-RNA in situ hybridisation. *Br. J. Haematol.* **90(1),** 113–118.

17. Shah, H. N. and Gharbia, S. E. (1994) Progress in the identification of nonmotile *Bacteroidaceae* from dental plaque. *Clin. Infect. Dis.* **18(S4),** 287–292.

18. Goltz, S. P., Donegan, J. J., Yang, H.-L., Pollice, M., Todd, J. A., Molina, M. M., Victor, J., and Keller, N. (1990) The use of nonradioactive DNA probes for rapid diagnosis of sexually transmitted bacterial infections, in *Gene Probes for Bacteria* (Macario, A. J. L. and Conway de Macario, E., eds.), Academic, San Diego, CA, pp. 2–44.

19. Wetherall, B. L. and Johnson, A. M. (1990) Nucleic acid probes for *Campylobacter* species, in *Gene Probes for Bacteria* (Macario, A. J. L. and Conway de Macario, E., eds.), Academic, San Diego, CA, pp. 256–293.

20. Higgins, G. A. and Mah, V. H. (1989) *In situ* hybridisation approaches to human neurological disease, in *Gene Probes* (Conn, P. M., ed.), Academic, San Diego, CA, pp. 183–196.

21. Penschow, J. D., Haralambidis, J., Pownall, S., and Coghlan, J. P. (1989) Location of gene expression in tissue sections by hybridisation histochemistry using oligodeoxyribonucleotide probes, in *Gene Probes* (Conn, P. M., ed.), Academic, San Diego, CA, pp. 222–238.

22. Vancoppenolle, B., Mccouch, S. R., Watanabe, I., Huang, N., and Vanhove, C. (1995) Genetic diversity and phylogeny analysis of Anabaena azzolae based on RFLPs detected in Azolla-Anabaena azollae DNA complexes using NIF gene probes. *Theor. Appl. Genet.* **91(4),** 589–597.

23. Holmes, A. J., Owens, N. J. P., and Murrell, J. C. (1995) Detection of novel marine methanotrophs using phylogenetic and functional gene probes after methane enrichment. *Microbiol. UK* **141(8),** 1947–1955.

24. Aquino de Muro, M., Mitchell, W. J., and Priest, F. G. (1992) Differentiation of mosquito-pathogenic strains of Bacillus sphaericus from non-toxic varieties by ribosomal RNA gene restriction patterns. *J. Gen. Microbiol.* **138,** 1159–1166.

25. Aquino de Muro, M. and Priest, F. G. (1993) Phylogenetic analysis of *Bacillus sphaericus* and development of an oligonucleotide probe specific for mosquito-pathogenic strains. *FEMS Microbiol. Lett.* **112,** 205–210.

26. Aquino de Muro, M. and Priest, F. G. (1994) A colony hybridisation procedure for the identification of mosquitocidal strains of Bacillus sphaericus on isolation plates. *J. Invertebr. Pathol.* **63,** 310–313.

27. Genthe, B., Gericke, M., Bateman, B., Mjoli, N., and Kfir, R. (1995) Detection of enteric adenoviruses in South-African waters using gene probes. *Water Sci. Technol.* **31(5–6),** 345–350.

28. Anderson, S. A., Lewis, G. D., and Pearson, M. N. (1995) Use of gene probes for the detection of quiescent enteric bacteria in marine and fresh-waters. *Water Sci. Technol.* **31(5–6),** 291–298.

29. Leff, L. G., Kernan, R. M., Mcarthur, J. V., and Shimets, L. J. (1995) Identification of aquatic *Burkholderia (Pseudomonas) cepacia* by hybridisation with species-specific ribosomal-RNA gene probes. *Appl. Environ. Microbiol.* **61(4),** 1634–1636.

30. Layton, A. C., Lajoie, C. A., Easter, J. P., Jernigan, R., Sanseverino, J., and Sayler, G. S. (1994) Molecular diagnostics and chemical-analysis for assessing biodegradation of polychlorinated-biphenyl in contaminated soils. *J. Ind. Microbiol.* **13(6),** 392–401.

31. Voordown, G. (1994) From the molecular biology of *Desulfovibrio* to a novel method for defining bacterial communities in oil-field environments. *Fuel Proc. Technol.* **40(2–3),** 331–338.

32. Ka, J. O., Holben, W. E., and Tiedje, J. M. (1994) Use of gene probes to aid in recovery and identification of functionally dominant 2,4-dichlorophenoxyacetic acid-degrading populations in soil. *Appl. Environ. Microbiol.* **60(4),** 1116–1120.

33. Liu, L., Maillet, D. S., Roger, J., Frappier, H., Walden, D. B., and Atkinson, B. G. (1995) Characterization, chromosomal mapping, and expression of different polyubiquitin genes in tissues from control and heat-shocked maize seedlings. *Biochem. Cell Biol.* **73(1–2),** 19–30.

34. Tonutti, P., Cass, L. G., and Chritoffersen, R. E. (1995) The expression of cellulase gene family members during induced avocado fruit abscission and ripening. *Plant Cell Environ.* **18(6),** 709–713.

35. Cerqueira, A. M. F., Tibana, A., Gomes, T. A. T., and Guth, B. E. C. (1994) Search for LT-II and STB DNA-sequences among Escherichia coli isolated from bovine meat products by colony hybridisation. *J. Food Prot.* **57(8),** 734–736.

36. Tamanaishacoori, Z., Jolivetgougeon, A., Pmmepuy, M., Cormier, M., and Colwell, R. R. (1994) Detection of enterotoxigenic *Escherichia coli* in water by polymerase chain reaction amplification and hybridisation. *Can. J. Microbiol.* **40(4),** 243–249.

37. Wernars, K. and Notermans, S. (1990) Gene probes for detection of food-borne pathogens, in *Gene Probes for Bacteria* (Macario, A. J. L. and Conway de Macario, E., eds.), Academic, San Diego, CA, pp. 353–388.

7

Nonradioactive Labeling of DNA

Thomas P. McCreery and Terrence R. Barrette

1. Introduction

Many important advances in molecular biology would not have been possible without the use of radioisotopes. It is relatively simple to substitute a radioactive isotope into a nucleotide to produce a molecule that has the same biological properties as the unlabeled molecule (*see* Chapter 6). These molecules are incorporated into DNA sequences by a variety of protocols. Unfortunately, with the high efficiency and easy incorporation of radionuclides comes a Pandora's Box of difficulties. The short half lives of the most commonly used nucleotides ($^{32}P, ^{33}P, ^{35}S$) necessitate that the material be freshly labeled for optimal efficiency. This alone makes it difficult for scientists in developing nations to use isotopic techniques. The ability of radiation to penetrate human tissue and cause damage requires that all work be done from behind shielding material. Personnel exposure to radiation must be monitored on a regular basis. A problem that is now looming large for the use of radiation in the molecular-biology laboratory is the lack of disposal sites for radioactive waste. Many regulatory agencies will not issue licenses to work with radiation if no disposal site is available.

The problems with radiation have been the impetus for the development of nonradioactive techniques. The methods of nonradioactive labeling depend upon the use of detectable marker molecules that are covalently bound to nucleotides. These markers fall into three classes. One class is the fluorescent dyes, including fluorescein and rhodamine. A second class is the haptens, which are detected by affinity or antibody binding; these include biotin and digoxigenin. A third class involves direct coupling of enzymes to sequences ranging in size from oligonucleotides to 50-kb fragments; enzymes commonly used for this are alkaline phosphatase and horseradish peroxidase (HRP). All of these methods of labeling have a common problem in that the label molecule is relatively large, so the protocols that are used for radioactive labeling have had to be modified to accommodate these new larger labels.

The methods of incorporating digoxigenin, biotin, and the fluorescent dyes will be the most familiar to those who have used isotopic techniques. These labels are available bound to dUTP generally at the 11th position, e.g., digoxigenin-11-dUTP. These labeled nucleotides are incorporated using modifications of the procedures of polymerase chain reaction (PCR) *(1)* and random priming *(2)*. In these techniques, the

From: *Molecular Biomethods Handbook*
Edited by: R. Rapley and J. M. Walker © Humana Press Inc., Totowa, NJ

Table 1
Comparison of a Number of Nonradioactive Labeling and Detection Techniques

Marker molecule	Method of incorporation	Type of hybridization	Enzyme-coupling reaction	Detection method
Digoxigenin	PCR, nick translation, end labeling	Southern, plaque lift, colony lift, *in situ*	Antibody	Colorimetric, chemiluminescence, immunogold, fluorescence
Biotin	"	"	Streptavidin affinity	"
Bromodeoxyuridine				
Horseradish peroxidase	Direct coupling	"	None	Colorimetric, chemiluminescence
Alkaline phosphatase	"	"	"	"
Fluorescein, rhodamine, Texas red	PCR, nick translation, end labeling	*In situ*, DNA binding	None	Fluorescence microscopy, fluorometer

labeling is carried out using a doped nucleotide mix at ratios from 1 to 10% of the labeled nucleotide to its unlabeled counterpart. Biotinylation of DNA sequences can be similarly performed by nick translation with the use of biotinylated calf intestinal alkaline phosphatase *(3)*.

It is a very different matter to directly couple enzymes to a DNA sequence. HRP labeling is carried out using a positively charged HRP-parabenzoquinone-polyethyleneimine complex. Double-stranded DNA is denatured by boiling, then the positively charged labeling reagent is mixed with the negatively charged single-stranded DNA to form a complex. This complex is then crosslinked with glutaraldehyde to form a single-stranded, stable, labeled DNA. This procedure is effective on a wide range of DNA strands from 50 bases to 50 kilobases long *(4)*.

Alkaline phosphatase (AP) is commonly used to label oligonucleotides. An oligonucleotide is constructed with the substitution of an amino-terminated base at either the 5' or the 3' end. This amino terminal is chemically activated and then coupled to the AP enzyme *(5)*. AP labeled oligonucleotide probes are reusable, making them extremely useful for looking for a specific sequence in a large number of hybridizations.

For all of the molecules described in **Table 1** except the fluorescent dyes, detection is carried out through the use of an enzyme and a substrate. It is at this point where the speed advantage of the enzyme-labeled probes is evident. After hybridization, these probes can be detected by adding the substrate and detecting the product either though colorimetric or chemiluminescent methods. The detection protocols are effectively the same for Southern hybridization, colony lifts, fluorescent *in situ* hybridization, and DNA-binding assays.

The first step in the process is preparing the material to be hybridized with the labeled probes. Generally, this process, known as prehybridization, involves incubation in a solution at elevated temperatures. This ensures that the DNA on the surface is available

Table 2
Comparison of the Timing Protocols of Nonradioactive Detection Methods

Method	Hybridization time	Wash time	Exposure time
Radiation	Overnight	1 h	1–2 d
Digoxigenin or biotin chemiluminescence	Overnight	3 h	3 h
Digoxigenin or biotin colorimetric	Overnight	3 h	Instantaneous
Enzyme labeled oligonucleotides	15 min	30 min	3 h
Fluorescent *in situ*	Overnight	15 min	Instantaneous

to hybridize to the probe. While the material to be hybridized is in the prehybridization solution, the probes are denatured by boiling to produce single-stranded probes for hybridization. After prehybridization, the probes are added to a hybridization solution and incubated with the target material.

After hybridization, the material is washed in an sodium dodecyl sulfate (SDS)/SSC solution to remove any unbound material. In many cases, the hybridization solution can be saved and reused. Generally, after this step the material is washed in a buffer to remove the washing solution and to prepare the material for the blocking reagent. It is at this point where protocols for enzyme-labeled probes begin to differ from those using hapten-labeled molecules. For hapten protocols, the target material is incubated in blocking buffer to reduce nonspecific binding of the conjugate. The material is then treated with a solution containing the antibody (antidigoxigenin) or affinity molecule (avidin) conjugated to a reporter enzyme (HRP or AP). A washing step follows to remove the unbound reporter conjugate. The final step is the same for hapten and enzyme-conjugated probes. The material is soaked in the substrate in which the reporter enzyme will act. These molecules include Nitro blue tetrazolium (NBT) and 3,3'-diaminobenzidine (DAB), which are colorimetric substrates for AP and HRP, respectively. Chemiluminescent substrates for these two reporter enzyme are for AP, adamantyl 1,2 dioxetane phosphate (AMPPD), and luminol (ECL) is used for HRP *(6, 7)*.

Colorimetric methods produce an insoluble precipitate that is easily evaluated. The problem with this technique is that, as a rule, the material cannot be stripped and analyzed with a different probe. Chemiluminescence is generally detected by exposing the material to autoradiography film and observing the dark areas on the film produced by the light released from the enzyme-substrate reaction. In fluorescent *in situ* hybridization, the molecule conjugated to the antibody or affinity molecule is fluorescein isothiocyanate (FITC) and detection is carried out using a fluorescence microscope *(8)*.

Another advantage of nonradioactive protocols is speed. As can be seen from **Table 2**, at the most you could do two sets of hybridization a week with radioactive probes. With nonradioactive probes, it is possible to do a hybridization on one day and the next afternoon have the data analyzed and the blots stripped ready for another hybridization. A particularly fast method of analysis is possible when using enzyme-labeled oligos. With this technique, the entire procedure can be done in a day. Colorimetric probes are quick but blots can not be reused.

For a large-scale screening project requiring the labeling a large number of probes, PCR labeling with digoxigenin or biotin is the best technique, because 96 well plate thermal cyclers are widely available and allow the use of this convenient format throughout the study. On the other hand, when one is screening a library for a specific mutant, the speed of an enzyme-labeled oligonucleotide would provide a decided advantage.

There are drawbacks to using nonradioactive methods, although in our opinion these are greatly outweighed by the advantages. The first is that the reagents are fairly sensitive to quality of preparation and cleanliness of glassware; these problems are fairly easily solved by careful handling of reagents and glassware. The second is that the target materials must be handled carefully; the sensitivity of this technique means that scratches in the surface are visible, and fingerprints will show up as dark smudges and can never be removed from the surface.

Hopefully, this chapter has provided an overview of nonradioactive labeling of DNA that will allow you to choose whether these techniques are suitable for your experiments. We feel that the speed, sensitivity, and safety of nonradioactive labeling make it an attractive choice for most molecular biology laboratories.

Acknowledgment

The authors thank Terri New, Michelle Crowell, and Lance Galler for their review and comments on this chapter.

References

1. McCreery, T. and Helentjaris, T. (1994) Production of hybridization probes by the PCR utilizing digoxigenin-modified nucleotides, in *Methods in Molecular Biology, vol. 28: Protocols for Nucleic Acid Analysis by Nonradioactive Probes* (Isaac, P., ed.), Humana, Totowa, NJ, pp. 67–71.
2. McCreery, T. and Helentjaris, T. (1994) Production of hybridization probe with digoxigenin-modified nucleotides by random hexanucleotide priming, in *Methods in Molecular Biology, vol. 28: Protocols for Nucleic Acid Analysis by Nonradioactive Probes* (Isaac, P., ed.), Humana, Totowa, NJ, pp. 73–76.
3. Karp, A. (1994) Labeling of double-stranded DNA probes with biotin, in *Methods in Molecular Biology, vol. 28: Protocols for Nucleic Acid Analysis by Nonradioactive Probes* (Isaac, P., ed.), Humana, Totowa, NJ, pp. 83–87.
4. Durrant, I. and Stone, T. (1994) Preparation of horseradish peroxidase-labeled probes, in *Methods in Molecular Biology, vol. 28: Protocols for Nucleic Acid Analysis by Nonradioactive Probes* (Isaac, P., ed.), Humana, Totowa, NJ, pp. 89–92.
5. *The Guide to Non-Radioactive Products*. (1993) Promega Corporation, Madison WI, pp. 2-2–2-10.
6. McCreery, T. and Helentjaris, T. (1994) Hybridization of digoxigenin-labeled probes to Southern blots and detection by chemiluminescence, in *Methods in Molecular Biology, vol. 28: Protocols for Nucleic Acid Analysis by Nonradioactive Probes* (Isaac, P., ed.), Humana, Totowa, NJ, pp. 107–112.
7. Durrant, I. and Stone, T. (1994) Hybridization of horseradish peroxidase-labeled probes and detection by enhanced chemiluminescence, in *Methods in Molecular Biology, vol. 28: Protocols for Nucleic Acid Analysis by Nonradioactive Probes* (Isaac, P., ed.), Humana, Totowa, NJ, pp. 127–133
8. McQuaid, S., McMahon, J., and Allan, G. (1995) A comparison of digoxigenin and biotin labelled DNA and RNA probes for in situ hybridization. *Biotech. Histochem.* **70,** 147–154.

8

Southern Blot Analysis

Paolo A. Sabelli

1. Introduction

The detection and identification of specific nucleic acid sequences is routine in molecular biology research. The principles that govern molecular annealing (or molecular hybridization) are used in many applications to the study of nucleic acids. Under suitable conditions, two single nucleic-acid chains form a hybrid molecule to an extent that is largely dependent on the degree of their nucleotide complementarity. The formation of such a duplex molecule occurs mainly through hydrogen bonding between guanosine and cytosine bases, and between adenosine and thymidine bases. Other base compositions are not complementary because of steric reasons and are, therefore, incompatible. The composition and distribution of bases differ in distinct nucleic-acid molecules resulting in different hybridization properties; hence, hydrogen bonding (or the molecular hybridization between complementary bases), when coupled to suitable labeling and detection methods, provides a valuable tool to identify identical or related sequences to a reference nucleic-acid sequence (probe).

Among the various procedures that exploit molecular hybridization for the analysis of nucleic acids, perhaps the most frequently used techniques consist of the hybridization of a labeled nucleic acid probe to a target nucleic acid immobilized on a solid support, typically a nitrocellulose or nylon membrane. Examples of these techniques include dot or slot hybridization, and bacterial colony and phage-plaque hybridization for the screening of gene libraries. Besides information on the relatedness of target and probe molecules, filter hybridization, when coupled to gel-fractionation of nucleic acids, can provide additional important information on the size of the target molecules *(1)*. Such information can, in turn, help to solve a variety of molecular biology problems.

Characteristic examples of these techniques are Southern and northern blotting hybridization for the analysis of DNA and RNA, respectively. In this chapter, the principles, important practical aspects, and examples of applications of the Southern-blotting technique will be described, whereas northern blotting is dealt with in Chapter 9 in this volume.

In Southern-blotting hybridization *(1)*, DNA is separated according to size by gel-electrophoresis (typically in a horizontal agarose gel), in-gel denatured into single-stranded molecules by treatment with alkali, neutralized, and transferred (blotted) to a

From: *Molecular Biomethods Handbook*
Edited by: R. Rapley and J. M. Walker © Humana Press Inc., Totowa, NJ

hybridization membrane by capillarity using a high salt concentration buffer. DNA is then irreversibly bound to the membrane either by heat treatment or UV crosslinking. Thus, single-stranded target DNA molecules are available on the filter for hybridization with a labeled single-stranded DNA probe. After removal of the nonspecific hybridization, which would give an unacceptably high background signal, specific interactions between target and probe molecules are detected through a range of procedures, depending on whether the probe has been radiolabeled or generated by nonradioactive methods. Since its introduction in 1975 *(1)*, Southern blotting has become a routine technique in the analysis of gene organization, the identification and cloning of specific sequences, the study of mutants, the characterization of genotypes by restriction fragment length polymorphisms (RFLPs) (*see* Chapter 22), the identification of sequences amplified by the polymerase chain reaction (PCR), the genetic fingerprinting in both medical and forensic analyses, the diagnosis of genetic disorders and cancer, the detection of food-borne microorganisms, and the monitoring of the environment. In this chapter, Southern blotting hybridization will be described, with emphasis on important theoretical and practical aspects. For further details, a comprehensive overview of Southern blotting theory and practice is available in the literature *(2–8)*. Finally, some examples of the applications of this technique to molecular-biology studies based on research carried out in plants will be given.

2. Principles and Practice

A schematic diagram of the Southern blotting procedure is shown in **Fig. 1**. The main steps involved are as follows:

1. DNA digestion and electrophoresis: The DNA containing the target sequence(s) is digested with one or more restriction enzymes. In the examples shown in **Fig. 1** and following figures, genomic DNA is digested to give a wide range of fragments of different length. These are separated according to size by electrophoresis in an agarose gel. In the case of genomic DNA, this results in a smear on the gel.
2. DNA blotting: After digestion, the DNA strands are separated by alkali treatment, neutralized, and transferred (blotted) to a hybridization membrane by a variety of methods, typically by passive diffusion of a highly concentrated saline buffer through the gel and the membrane on which the DNA is trapped and immobilized. Thus, a replica of the electrophoretic pattern is obtained on the hybridization membrane.
3. Hybridization and post hybridization: The membrane is incubated with a single-stranded nucleic-acid probe labeled with ^{32}P or with various nonisotopic methods. The probe sequence is complementary to the target DNA sequence(s) on the membrane and the hybridization conditions are designed to maximize hybridization between target and probe sequences. After removal of the probe, which is not specifically bound to the target DNA, the hybridization signal is detected either by autoradiography (in the case of radioactive probes) or by a range of nonradioactive-detection procedures.

A more detailed discussion on the theoretical and practical aspects of Southern blotting is given in **Subheadings 2.1.–2.3.**

2.1. DNA Digestion and Electrophoresis

The sensitivity limit of Southern blotting, using radioactive probes, and allowing a 1–3 d exposure time, is around 0.5–1.0 pg DNA. Thus, the amount of DNA to be processed depends on the abundance of the target sequence within the sample. If the

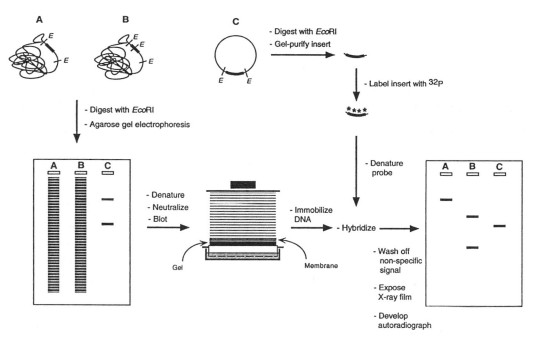

Fig. 1. Schematic diagram of the Southern blotting technique. Target DNA, genomic DNA from samples **(A)** and **(B)** in this example, is digested with one (*Eco*RI) or more restriction cnzymes and size-separated by electrophoresis on an agarose gel. The relevant restriction sites and the complementary sequence to the probe (thick bar) are indicated. Plasmid DNA containing the insert DNA probe **(C)** is also digested and electrophoresed as a positive control. After denaturation and neutralization, single-stranded DNA is transferred onto a hybridization membrane by capillarity and immobilized. For the preparation of the probe, insert DNA (thick bar in C) is purified from the vector by digestion and recovery from an agarose gel, then labeled with ^{32}P and denatured. The membrane-bound and probe DNAs are hybridized, the nonspecific signal removed, and the membrane used to impress an X-ray film. After developing the film, complementary DNA fragments to the probe are revealed as autoradiographic bands. In the figure, the principle underlying the restriction fragment length polymorphism is also described. The DNAs extracted from two individuals (A and B) differ for the presence of one additional *Eco*RI site within the target sequence in the B sample. Such difference in the genome is revealed by a different hybridization pattern. This principle is exploited in many molecular biology applications.

sequence complexity of the sample is high (e.g., genomic DNA), a relatively large amount of DNA should be digested (typically 2–10 µg). For example, 10 µg of genomic human DNA are normally used to detect a single-copy target sequence, although lower quantities may also be analyzed. On the other hand, if the sample consists of homogeneous sequences homologous to the probe (e.g., cloned DNA into plasmid or phage vectors), much smaller quantities of DNA are needed (as little as a few picograms of DNA, although usually several nanograms of DNA are used). The duration of the digestion can vary depending on the amount of enzyme used per mass unit of DNA. Generally, 2–4 h are sufficient for cloned DNA, whereas overnight digestions are routine for genomic DNA.

A photograph of the ethidium bromide-stained gel aligned to a fluorescent ruler prior to blotting is important when estimating the sizes of the hybridizing bands with those of the size-marker DNA. Bacteriophage λ DNA digested with *Hin*dIII or a 1-kb ladder DNA (Gibco, Gaithersburg, MD) are convenient markers for most applications. Alternative methods for obtaining size-reference marks on the membrane are available and their choice depends on individual preferences. These methods include blotting the size-marker DNA on the membrane and relying on the occasional hybridization obtained with any given probe, as well as the use of ^{32}P- or ^{35}S-labeled size markers.

2.2. Blotting

After gel electrophoresis and before blotting, it is a relatively common practice to include a depurination treatment (usually by incubating the gel with a diluted HCl solution), which nicks the DNA and facilitates the transfer of large fragments. However, this step is often difficult to control and may result in the loss of smaller fragments, and it may be omitted if opportune compensations are made in evaluating autoradiographs. On the other hand, depurination is necessary for Southern analysis of very large DNA fragments, e.g., following pulsed-field gel electrophoresis (PFGE).

A range of commercial nitrocellulose and nylon hybridization membranes is available *(8)*. Nylon membranes are preferred because they are more resistant than nitrocellulose membranes, thus allowing several rounds of hybridization. In addition, nylon membranes bind short fragments (<500 nt) more efficiently than nitrocellulose, and the blotted DNA can be covalently bound by a brief UV irradiation, whereas nitrocellulose requires baking at 80°C in a vacuum oven for 2 h. Positively charged nylon membranes are also available and are particularly suited for alkali-blotting because DNA binds covalently to the membrane under these conditions without any treatment.

A direct relationship between salt concentration and the efficiency of capillary transfer has been observed, especially for large fragments *(1)*. Thus, high-salt concentrations should be used for genomic DNA, whereas low-salt concentrations may be used for short sequences (e.g., plasmids and PCR fragments).

Although in the examples given in this chapter an improved version of the blotting procedure by upward capillarity transfer, originally developed by Southern, was used *(6,7)*, a wide range of alternative protocols has been reported in the literature. These include the use of alkaline-blotting buffers *(9)*, positive or negative pressure *(10)*, electroblotting *(11,12)*, and downward blotting *(13,14)*. Such methods usually result in shorter blotting times, and their use should certainly be given consideration for projects that require the analysis of many samples (e.g., RFLP mapping; *see* Chapter 22). In the case of blotting from polyacrylamide gels, the small pore size, which prevents the passive diffusion of DNA, indicates that electroblotting is the method of choice.

2.3. Hybridization and Posthybridization

The prehybridization and hybridization conditions are designed to maximize specific hybridization of the probe to target sequences while minimizing nonspecific binding to both membrane and nontarget DNA. A wide range of hybridization buffers has been used in Southern hybridization experiments *(2–6,8,15)*. Generally, hybridization buffers are characterized by a high-ionic strength, which is an important factor for hybrid stability (*see* **Eqs. 1** and **2**). Several types of blocking agents can be used to suppress nonspecific binding, thereby limiting the background signal. Denhardt's reagent *(16)* and nonfat dried milk *(17)* are among the most commonly used blocking agents. These are usually used in combination with detergents (such as sodium dodecyl sulfate [SDS]) to prevent binding of the probe to the membrane. Denatured, fragmented DNA of high complexity (such as salmon sperm or calf thymus DNA) is also used to block nonspecific interaction between the membrane-bound DNA and probe. The use of an inert high-mol-wt polymer, such as polyethylene glycol (PEG) 6000 or dextran sulfate, results in a much higher actual-probe concentration improving considerably the hybridization efficiency while allowing the operator to work with low nominal concentrations of probe in relatively large volumes of hybridization buffer (resulting in saving of probe and a simplification of the practical aspects of the procedure) *(18)*.

A large number of factors affects hybridization experiments; these can be grouped in those affecting the rate of hybridization and those affecting the stability of duplex nucleic-acid molecules *(2–8)*. In practical terms, only the factors that affect the thermal stability of hybrid nucleic acid molecules play a decisive role in most filter-hybridization experiments. The melting temperature (T_m) of nucleic-acid duplexes indicates the temperature at which the duplex is 50% denatured under the given conditions, and it represents a direct measure of the stability of the nucleic-acid hybrid. The T_m of a particular DNA duplex may be conveniently estimated using **Eq. 1** *(19)*:

$$T_m \; (°C) = 81.5°C + 16.6 \; (\log_{10}M) + 0.41 \; (\%G + C) - 0.72 \; (\% f) - 500/n \qquad (1)$$

whereas **Eq. 2** *(20)* may be used for RNA–DNA hybrids:

$$T_m \; (°C) = 79.8°C + 18.5 \; (\log_{10}M) + 0.58 \; (\%G + C) - 11.8$$
$$(\%G + C)^2 - 0.56 \; (\% f) - 820/n \qquad (2)$$

where M is the molarity of monovalent cations (typically Na^+), (% G + C) is the percentage of guanosine and cytosine nucleotides, (% f) is the percentage of formamide, and n is the length of the hybrid (in bases). These equations indicate that the stability of nucleic-acid duplexes is directly related to the salt concentration, the proportion of triple hydrogen bond-forming nucleotides, and the hybrid length. The T_m is also inversely related to the concentration of double helix-destabilizing agents, such as formamide. Although the use of formamide allows hybridization to be carried out at conveniently low temperatures, even when the probe is very long and contains a high proportion of G and C bases, formamide is toxic and its use does not represent an advantage in most applications. **Equations 1** and **2** are valid for 100% complementary sequences. However, the presence of mismatches (frequently encountered in the study of gene families or in the identification of heterologous sequences to the probe) greatly affects the duplex's stability.

Although difficult to quantify, because the effect of mismatch distribution is also very important, on average each 1% mismatch results in a 1°C decrease in the T_m value *(21)*.

The stringency used in hybridization experiments is an inverse measure of the Criterion value *(C)*, which is given by **Eq. 3**:

$$C \; (°C) = T_m - T_i \qquad (3)$$

where T_i is the experimental temperature. When T_i is low, C is high and the stringency is low, and vice versa. Target sequences identical or closely related to the probe can be identified under high-stringency conditions (e.g., 5°C < C <10°C), whereas low-stringency conditions are appropriate for less complementary sequences.

Usually, hybridization and the initial-washing steps are carried out under low-stringency conditions (e.g., $T_m -30°C$), and the stringency is increased progressively in the subsequent washing steps. This allows distantly related sequences to the probe to be initially detected. Such sequences may be subsequently removed under increased stringency washes. The stringency can be conveniently changed by varying the salt concentration in the hybridization/washing buffers or by modifying the temperature, or a combination of both.

Equations 1 and **2** are accurate only for duplexes longer than 100–200 nucleotides. The use of oligonucleotide probes requires several adjustments. Oligonucleotide probes are frequently used for the screening of cDNA or genomic libraries and northern analysis, and they are also occasionally used in Southern hybridization experiments. Short-hybrid molecules are highly unstable (even if 100% complementary), especially during the washing steps because the probe concentration in the washing buffer is virtually zero. Thus, washing steps are usually kept very short (e.g., 2–5 min).

In addition, the stability of oligonucleotide duplexes is greatly affected by the nucleotide composition and by mismatches. The T_m of oligonucleotide probes shorter than 50 nucleotides is usually calculated using **Eq. 4**:

$$T_m (°C) = 4 (G + C) + 2 (A + T) \tag{4}$$

where A, C, G, and T are the numbers of the corresponding nucleotides. However, often the experimental conditions to obtain reliable hybridization with oligonucleotide probes (especially in the analysis of large genomes) must be determined on an empirical basis. More complicated hybridization experiments using degenerate oligonucleotide sequences require further adjustments *(2,22,23)*.

The majority of filter-hybridization experiments is carried out using radioactive probes labeled with [32]P (although other isotopes, such as ^{35}S, may also be used). These result in the highest sensitivity available, which is often the most important experimental requirement. However, the handling and disposal of radioactive isotopes requires particular training, equipment, and disposal procedures. In addition, because radioactive probes are subjected to decay, they are only useful for a relatively short time. Whatever the method used to radiolabel nucleic acid probes, Southern hybridization requires a minimum probe-specific activity of 10^9 dpm/µg, although an activity of 10^8 dpm/µg may be considered acceptable in applications requiring low sensitivity (e.g., when the abundance of the target DNA in the hybrid is not a limiting factor).

Nonradioactive methods are becoming increasingly common although their sensitivity is not comparable to that of radioactive procedures *(5)*. However, nonradioactive techniques are safe for the operator, produce probes that can be stored for a long time before use, and result in a fast detection of the hybridization signal, thus being particularly suited for the analysis of a large number of samples requiring a relatively low sensitivity.

3. Applications

3.1. Estimation of the Number of Copies of a DNA Sequence in the Genome

Information about the number of genomic DNA sequences isolated by gene-cloning techniques or introduced into organisms, either by conventional breeding or through recombinant DNA technology, is important in the study of animals and plants and for

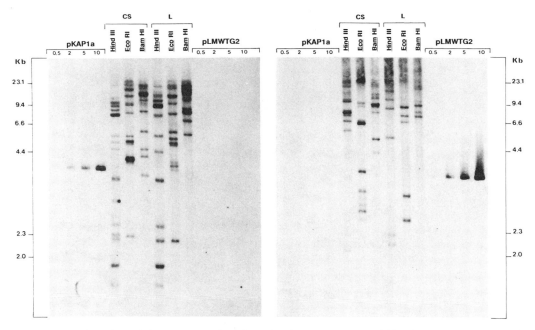

Fig. 2. Estimation of the number of copies of a given sequence within the genome. Genomic DNAs from bread wheat cv. Chinese Spring (CS) and durum wheat cv. Langdon (L) were digested with *Hind*III, *Eco*RI, and *Bam*HI, gel-fractionated, blotted, and hybridized first with a cDNA probe for γ-gliadin genes *(pKAP1a)* **(left)**, then with a LMW-glutenin-specific probe *(pLMWTG2)* **(right)**. Reference amounts of linearized plasmids containing the probes corresponding to 0.5, 2, 5, and 10 copies per haploid bread wheat genome were included as indicated. Hybridizing fragments were estimated to correspond to 1–15 copies of sequences related to the probes. Under the chosen conditions, the two probes did not crosshybridize. Hybridization was at 65°C. Washing was in 0.5X SSC, 0.1% (w/v) SDS at 65°C. Exposure was for 1 d. Reproduced from **ref.** *24* with the permission of Springer-Verlag GmbH & Co. KG.

supporting breeding programs and the generation of transgenics. Such information can be obtained by Southern hybridization. Genomic DNA is digested with one or more endonucleases and gel-separated together with a dilution series of the cloned DNA sequence under analysis, which has also been used to prepare the probe. Following hybridization, the number of copies of the cloned DNA sequence (or related sequences) in the genome is estimated by comparing the intensities of the hybridization signals of the genomic fragments and the reference DNA. An example is given in **Fig. 2**, in which the number of copies of two unrelated sequences encoding gluten proteins (γ-gliadins and LMW-glutenin) are estimated in both hexaploid bread and tetraploid durum wheats *(24)*. A range of fragments is obtained, each containing 1–15 copies of related sequences to the probe. This experiment also indicates that, under the chosen conditions, the two probes do not crosshybridize and, therefore can be effectively used to identify distinct classes of genomic sequences. DNA concentrations should be carefully evaluated, sample spillages avoided during gel loading, and only the cloned DNA sequences should be used as probes because contaminating vector sequences may crosshybridize with genomic DNA. The number of copies of a given DNA sequence is conventionally based on the haploid complement of the genome (1C value). Compilations of such

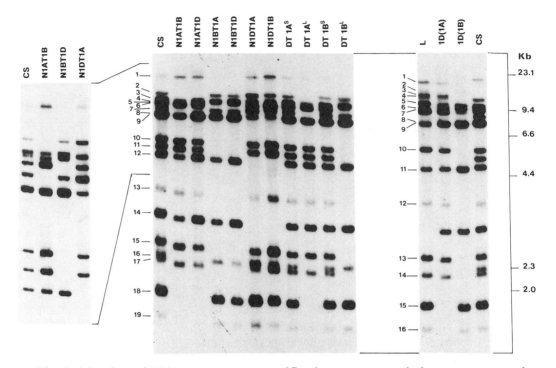

Fig. 3. Mapping of DNA sequences to specific chromosomes and chromosome arms by Southern analysis of aneuploid genotypes. A γ-gliadin probe *(pKAP1a)* was hybridized to *Hin*dIII-digested DNAs from euploid bread wheat (cv. Chinese Spring, CS), and durum wheat (cv. Langdon, L), nullisomic-tetrasomic and ditelosomic lines of Chinese Spring **(center)**, and disomic substitution lines of Langdon that have chromosomes 1A or 1B substituted by 1D chromosomes from Chinese Spring **(right)**. Comparisons of the hybridization patterns allowed most fragments to be assigned to the homeologous group 1 chromosomes and to specific Chinese Spring chromosome arms in the case of Chinese Spring. A higher-band resolution was obtained with a long gel separation of about 50 h **(left)**. Hybridization was at 65°C. Washing was in 0.5X SSC, 0.1% (w/v) SDS at 65°C. Exposure was for 3 d. Reproduced from **ref.** *24* with the permission of Springer-Verlag GmbH & Co. KG.

values for many species are available in the literature *(25,26)*. As in the example given in **Fig. 2**, many genes belong to relatively large gene families and variation in the degree of sequence similarity and the presence of repetitive motifs can complicate considerably the evaluation of the individual copy numbers.

3.2. Mapping of Sequences to Specific Chromosomes

Southern mapping of DNA sequences to specific restriction fragments represents the prerequisite for the construction of genetic maps based on RFLPs, improving breeding programs, and genetic fingerprinting. When Southern hybridization is coupled to the analysis of genetic stocks differing in chromosome arrangements, DNA sequences may also be mapped to specific chromosomes or chromosome regions. In the example given in **Fig. 3**, related sequences to the probe *pKAP1a* are mapped to specific group 1 chromosomes in both bread and durum wheats *(24)*. For bread wheat (a hexaploid species containing three homeologous genomes, designated A, B, and D), comparison of the hybridization pattern

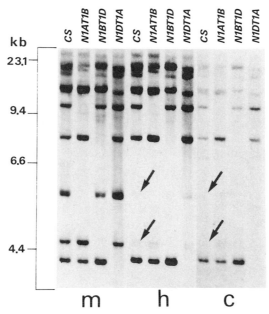

Fig. 4. Identification of a subset of sequences within a multigene family, characterized by a low degree of homology with the probe used. DNAs from euploid (CS) and aneuploid nulli-somic-tetrasomic lines of bread wheat cv. Chinese Spring were digested with *Bam*HI and hybridized to *pKAP1a*, a γ-gliadin cDNA, then washed under conditions of moderate (m) and high (h) stringency. High-stringency conditions were obtained by decreasing the salt concentration in the washing buffer from 0.5X to 0.2X SSC. Fragments that hybridized under moderate stringency conditions, but failed to hybridize under high-stringency conditions, are indicated by arrows in (h). The same fragments also failed to hybridize when the 3' nonrepetitive part of the probe was used under moderate-stringency conditions [arrows in (c)], suggesting that they encode ω-gliadins, which consists almost entirely of repetitive sequences. Hybridization was at 65°C. Washing was in 0.5X or 0.2X SSC, 0.1% (w/v) SDS at 65°C. Exposure was for 2 d. Reproduced from **ref. 6** with the permission of Academic Press Ltd, London.

of the euploid DNA with those of lines that lack one pair of homeologous group 1 chromosomes but possess four copies of another group 1 chromosome (e.g., Nulli 1A, Tetra 1B) allows the mapping of the restriction fragments to specific chromosomes. Similar comparisons with ditelosomic lines lacking the short or long arms of chromosomes 1A or 1B allow the mapping to specific chromosome arms. For example, Chinese Spring fragments 1, 6, 10, 11, 13, 15, and 17 are located on the short arm of chromosome 1B. For durum wheat (a tetraploid species with two homeologous genomes, A and B), comparison of euploid DNA with those of substitution lines, in which the chromosome pair 1A or 1B of Langdon has been substituted by a pair of 1D chromosomes from Chinese Spring, allows the mapping of Langdon fragments to the 1A or 1B chromosomes.

3.3. Analysis of Multigene Families

Members of multigene families often share variable degrees of sequence similarity. By using different stringency conditions or different probe sequences it is possible to discriminate among related but distinct family members. As shown in **Fig. 4**, DNAs

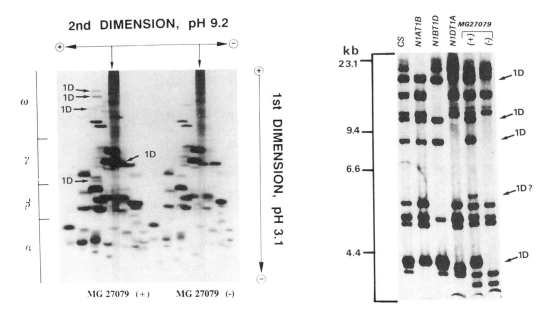

Fig. 5. Characterization of a mutant form of bread wheat with a deletion at the *Gli-D1* locus. **(Left)** Two-dimensional (two-pH) gel electrophoresis of gliadins of bread wheat accession MG27079. 'Normal' (+) and mutant (−) forms were separated on the same gel. The *1D*-encoded components, absent in the mutant, are indicated. **(Right)** Southern-blot analysis of *Eco*RI-digested DNAs from cv. Chinese Spring (CS), the group 1 nulli-tetrasomic aneuploids and accession MG27079 ('+' and '−' forms). The DNAs were probed with *pKAP1a*, a γ-gliadin cDNA, which also hybridizes to ω-gliadin genes. Most fragments missing in the mutant correspond to *1D* fragments of Chinese Spring and to the missing proteins, indicating that substantial deletions at the *Gli-1D* locus were responsible for the mutant phenotype. Reproduced from **ref. 6** with the permission of Academic Press Ltd., London, and based on **ref. 27**.

from euploid and aneuploid Chinese Spring lines were hybridized to *pKAP1a*, a γ-gliadin cDNA whose repetitive 5'-end is similar to ω-gliadin genes. Analysis under conditions of moderate (m) and high (h) stringency conditions revealed different hybridization patterns with several fragments unable to hybridize to the probe under high stringency conditions (arrows). The use of a 3'-end nonrepetitive *pKAP1a* probe specific for γ-gliadin genes (c) gave a similar pattern as in (h), suggesting that the arrowed bands contain ω-gliadin genes whose repetitive sequences hybridize with the 5' repetitive region of *pKAP1a* under moderate, but not under high-stringency, conditions *(24)*.

3.4. Analysis of Mutants

The analysis of mutants can provide important information on the organization and regulation of genes in the genome. To be detectable by Southern analysis, the mutation should affect the hybridization of the duplex molecule(s) or result in a size polymorphism (*see also* **Fig. 1**). Gel-electrophoresis analyses of gluten proteins from wheat seeds have resulted in the identification of a wide range of mutants lacking specific proteins. In the example given in **Fig. 5**, a mutant form of accession MG27079 of bread wheat lacks γ- and ω-gliadins encoded by genes on chromosome 1D. Southern analysis revealed that several restriction fragments containing γ- and ω-gliadin genes were indeed missing from the genome of the mutant form. Thus, substantial deletions at the locus encoding these proteins, and not repression of gene expression, is responsible for the mutant phenotype *(27)*.

References

1. Southern, E. M. (1975) Detection of specific sequences among DNA fragments separated by gel electrophoresis. *J. Mol. Biol.* **98**, 503–517.
2. Sambrook, J., Fritsch, E. F., and Maniatis, T. (1989) *Molecular Cloning. A Laboratory Manual.* Cold Spring Harbor Laboratory, Cold Spring Harbor, NY.
3. Meinkoth, J. and Wahl, G. (1984) Hybridization of nucleic acids immobilized on solid supports. *Anal. Biochem.* **138**, 267–284.
4. Hames, B. D. and Higgins, S. J. (1985) *Nucleic Acid Hybridization. A Practical Approach.* IRL, Oxford, UK.
5. Dyson, N. J. (1991) Immobilization of nucleic acids and hybridization analysis, in *Essential Molecular Biology. A Practical Approach*, vol. II (Brown, T. A., ed.), IRL, Oxford, UK, pp. 111–156.
6. Sabelli, P. A. and Shewry, P. R. (1993) Nucleic acid blotting and hybridization, in *Methods in Plant Biochemistry* (Bryant, J., ed.), Academic, London, UK, pp. 79–100.
7. Sabelli, P. A. and Shewry, P. R. (1995) Gene characterization by Southern analysis, in *Methods in Molecular Biology, vol. 49: Plant Molecular Biology Protocols* (Jones, E., ed.), Humana, Totowa, NJ, pp. 161–180.
8. Brown, T. A. (1993) Hybridization analysis of DNA blots, in *Current Protocols in Molecular Biology* (Ausubel, F. M., Brent, R., Kingston, R. E., Moore, D. D., Seidman, J. G., Smith, J. A., and Struhl, K., eds.), Wiley, New York.
9. Reed, K. C. and Mann, D. A. (1985) Rapid transfer of DNA from agarose gels to nylon membranes. *Nucleic Acids Res.* **13**, 7207–7221.
10. Olszewska, E. and Jones, K. (1988) Vacuum blotting enhances nucleic-acid transfer. *Trends Genet.* **4**, 92–94.
11. Smith, M. R., Devine, C. S., Cohn, S. M., and Lieberman, M. W. (1984) Quantitative electrophoretic transfer of DNA from polyacrylamide or agarose gels to nitrocellulose. *Anal. Biochem.* **137**, 120–124.
12. Ishihara, H. and Shikita, M. (1990) Electroblotting of double-stranded DNA for hybridization experiments: DNA transfer is complete within 10 minutes after pulsed-field gel electrophoresis. *Anal. Biochem.* **184**, 207–212.
13. Lichtenstein, A. V., Moiseev, V. L., and Zaboikin, M. M. (1990) A procedure for DNA and RNA transfer to membrane filters avoiding weight-induced gel flattening. *Anal. Biochem.* **191**, 187–191.
14. Chomczynski, P. (1992) One-hour downward alkaline capillary transfer for blotting of DNA or RNA. *Anal. Biochem.* **201**, 134–139.
15. Boulnois, G. J. (1987) *University of Leicester Gene Cloning and Analysis: A Laboratory Guide.* Blackwell Scientific, Oxford, UK, pp. 47–60.
16. Denhardt, D. (1966) A membrane filter technique for the detection of complementary DNA. *Biochem. Biophys. Res. Commun.* **23**, 641–646.
17. Johnson, D. A., Gautsch, J. W., Sportsman, J. R., and Elder, J. H. (1984) Improved technique utilizing nonfat dry milk for analysis of proteins and nucleic acids transferred to nitrocellulose. *Gene Anal. Techn.* **1**, 3–8.
18. Amasino, R. M. (1986) Acceleration of nucleic acid hybridization rate by polyethylene glycol. *Anal. Biochem.* **152**, 304–307.
19. Schildkraut, C. and Lifson, S. (1965) Dependence of the melting temperature of DNA on salt concentration. *Biopolymers* **3**, 195–208.

9

Northern Blot Analysis

Paolo A. Sabelli

1. Introduction

The analysis of RNA is central to a wide range of molecular-biology studies, because it is often important to obtain information about the expression of genes in living organisms. Filter hybridization of size-separated RNAs with a labeled nucleic-acid probe, a technique known as northern-blot analysis, is designed to address the study of RNA sequences *(1)*. This procedure, which derives from Southern-blot analysis (*see* Chapter 8 in this volume), is based on the ability of complementary single-stranded nucleic acids to form hybrid molecules *(2)*. In brief, RNA is separated according to size on an agarose gel under denaturing conditions, transferred (blotted) and irreversibly bound to a nylon or nitrocellulose membrane, and hybridized with a ^{32}P-labeled nucleic-acid probe. After washing off the unbound and nonspecifically bound probe, hybridizing RNA molecules are revealed as autoradiographic bands on a X-ray film *(3,4)*.

Northern blotting is an important procedure for the study of gene expression, because the RNA present in a given cell or tissue type roughly represents the portion of the genome that is expressed in that cell or tissue. What is detected by northern hybridization is the steady-state level of accumulation of a given RNA (usually mRNA) sequence in the sample under study. Northern blot analysis, therefore, allows one to relate mRNA levels to the physiological and morphological properties of living organisms. Several factors control mRNA abundance:

1. Gene transcription;
2. mRNA processing and transport; and
3. mRNA stability.

Of these, gene transcription is the most important determinant of mRNA expression. In addition to the aforementioned factors, translational and posttranslational controls play additional roles in the overall regulation of gene expression, which ultimately results in a specific level of protein being synthesized. Thus, caution should be used in interpreting northern-hybridization data since several factors, in addition to mRNA levels, concur to phenotypic expression. Although techniques that study the localization of mRNAs at the single-cell level (i.e., *in situ* hybridization; *see* Chapter 51) or more sensitive methods for the analysis of rare transcripts (i.e., reverse transcriptase-

From: *Molecular Biomethods Handbook*
Edited by: R. Rapley and J. M. Walker © Humana Press Inc., Totowa, NJ

mediated polymerase chain reaction [RT-PCR] and RNase protection assay) have recently become common, northern hybridization remains a widely used procedure for the analysis of gene expression. For example, this technique has been extensively used for the characterization of cloned cDNAs and genes, the study of tissue/organ specificity of gene expression, the analysis of the activity of endogenous and exogenous genes in transgenics, and the study of gene expression in relation to development and biotic and abiotic factors.

The principles that govern Southern hybridization are also suitable to a large extent for the analysis of RNA and, therefore, the reader is referred to Chapter 8 in this volume. In the present chapter, discussion will be limited to the modifications required for northern analysis together with some examples of the applications of this technique *(5,6)*.

2. Principles and Practice

RNA is extremely susceptible to degradation by RNases. Solutions and plasticware should be treated with the RNase inhibitor diethylpyrocarbonate (DEPC) *(7)*, and glassware should be baked at 200°C for several hours. DEPC is a suspect carcinogen and care should be taken in handling the hybridization buffer (preferably in a fume hood).

2.1. Gel Electrophoresis

RNA molecules are single nucleic-acid chains that tend to form secondary structures by intramolecular nucleotide pairing and intermolecular interactions. This interferes with reliable size-separations unless denaturing conditions are used. The most commonly used method is based on heat-denaturation of the RNA samples in the presence of formaldehyde prior to loading of the gel, followed by electrophoresis on an agarose-formaldehyde gel *(8)*. Alternative denaturing methods include the glyoxal/DMSO *(9)* and the methylmercuric chloride *(10)* procedures, but these methods are generally more complicated and unsafe for the operator.

The sensitivity limit of Northern hybridization is 1–5 pg of RNA target molecule. Because RNA samples are highly complex mixtures of different sequences, the amount of RNA to be analyzed depends on the abundance of the target molecule within the sample. Cellular RNA comprises ribosomal RNA (rRNA) (approx 80%), transfer RNA (tRNA) and small cytoplasmic RNA (scRNA) (approx 18%), and messenger RNA (mRNA) (approx 2%). For the detection of very abundant RNA species, such as rRNA, 1–2 µg of cellular RNA is usually sufficient. In the case of abundant mRNAs (>0.1% of the mRNA population), 10–20 µg of total cellular RNA should be used. For the detection of rare mRNAs (<0.1% of the total mRNAs), 1–3 µg of purified mRNA should be analyzed. It should be noted that in a typical high-eukaryotic cell there are about 10,000–30,000 different mRNA species, of which as many as one-third are rare mRNAs (with about 5–20 copies per cell each). Thus, high sensitivity is the most important requirement in northern hybridization experiments for the study of many mRNAs.

rRNA bands (28S and 18S) are often used as size-markers (corresponding to 4.7 and 1.9 kb in mammals, respectively). However, more accurate size-estimates for hybridizing bands are obtained with a higher number of reference sizes as in commercial RNA ladders. DNA cannot be used as size-marker because it migrates on denaturing gels more slowly than RNA of equivalent size *(11)*.

There are several alternative procedures for the staining of electrophoresed RNA. The gel may be treated with a high concentration of ethidium bromide for a short time or with a low concentration for a longer time *(3)*; ethidium bromide may be added to the sample before electrophoresis *(12)*, or the blotted RNA stained on the membrane using methylene blue *(13)*. Regardless of the method used, staining of RNA results in an approx 50% reduction in hybridization sensitivity and, therefore, it could be omitted for the detection of rare transcripts. A replica gel could be stained to check the integrity and the relative amount of each RNA sample before proceeding to the hybridization step (*see* **Fig. 2**).

2.2. Blotting and Hybridization

RNAs are short, single-stranded molecules, and thus such steps as fragmentation by depurination, strand-separation by alkali, and neutralization are not required prior to blotting. Probe preparation, prehybridization, and hybridization are performed as in Southern blotting (*see* Chapter 8), but the temperature of the hybridization and washing steps could be slightly increased owing to the higher stability of DNA–RNA duplexes rather than DNA–DNA duplexes. RNA–RNA hybrids are even more stable and, in the case of RNA probes, the hybridization and washing temperature could be further increased. Although nonradioactive labeling and detection methods are available (*see* Chapter 7), they suffer from a general low sensitivity and are rarely used in northern blot applications. In order to perform multiple rounds of hybridization on the same membrane, stripping off of the probe cannot be carried out with alkali as this results in RNA hydrolysis. As an alternative procedure, incubation in a very-low-concentration saline buffer at 65°C can be used *(5,6)*.

3. Applications

Northern blot analysis is used as a wide spread technique in molecular biology studies. Valuable information on the size and occurrence/abundance of transcripts in samples of various origin can be obtained using this method. Such information may provide important clues on gene structure and regulation. Northern blotting is also a valuable tool for the characterization of large numbers of DNA sequences isolated in genome random-sequencing projects. In fact, the presence of an RNA band on a northern blot using such sequences as probes is a good indication of their ability to encode proteins. Many biological problems have benefited from the analysis of RNA by northern blotting, including gene characterization and regulation, analysis of development, and responsiveness to internal and external stimuli.

An example of gene characterization is given in **Fig. 1**. Two cDNA clones (designated 42 and 49) have been isolated from a maize root apex by differential screening of PCR-generated cell population-specific cDNA libraries *(14,15)*. The presence of one intense band in separate northern hybridizations with root tip RNA using the cDNAs as probes indicated, in addition to the approximate transcript sizes, that the corresponding genes are expressed at relatively high levels in the root apex. The example given in **Fig. 2** shows the tissue/organ specific regulation of histone H4 genes in maize. Total-cellular RNA was extracted from different tissues/organs. Similar amounts were loaded on a replica gel, which was stained with ethidium bromide but not blotted (**Fig. 2**). The two RNA ribosomal bands are present in all samples and additional chloroplast ribosomal bands are visible in the samples derived from green tissues (leaf 1–4 and ligule). A similar gel was hybridized with a histone H4 probe (gift of Dr. C. Gigot, Strasburg). A

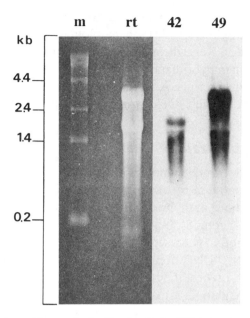

Fig. 1. Example of northern-blot analysis. Total cellular RNA from maize root tips (10 μg/lane) (rt) was gel-fractionated, blotted, and hybridized with cDNA probes 42 and 49 which have been isolated by differential screening of PCR-amplified root tip cDNA libraries *(14,15)*. On the left part of the figure, root tip RNA and the RNA size-marker (0.24–9.5 kb ladder from Gibco) (m) are visualized after electrophoresis by staining with ethidium bromide. The two ribosomal bands are evident in the RNA sample. Hybridization was at 65°C, washing was with 0.5X SSC, 0.1% (w/v) SDS at 65°C. Exposure was overnight. Reproduced with the permission of Wiley Ltd. from **ref.** *15*.

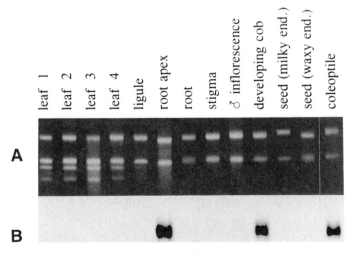

Fig. 2. Northern analysis of gene expression in different plant tissues and organs. Total RNA was extracted from different maize tissues and organs as indicated, gel-separated and stained with ethidium bromide **(A)**. A similar gel, was blotted and hybridized with a histone H4 probe **(B)**. Histone H4 transcripts are accumulated in the root apex, developing cob, and coleoptile. (A), "leaf 1", "2", "3" and "4" indicate different regions of the leaf blade from the top to the bottom, respectively. Two samples of seed RNA, at the milky and waxy endosperm developmental stages, were also analyzed.

band was obtained from the root apex, developing cob and coleoptile samples. Histone H4 genes are transcriptionally regulated in a cell cycle-dependent fashion and the presence of transcripts in specific samples is an indication of the spatial and developmental confinement of cell proliferation in higher plants. Although a large proportion of genes presides to vital functions and are expressed equally in every cell type ("housekeeping" genes), many genes are differentially expressed in distinct tissues or organs.

References

1. Alwine, J. C., Kemp, D. J., and Stark, G. R. (1977) Method for detection of specific RNAs in agarose gels by transfer to diazobenzyloxymethyl-paper and hybridization with DNA probes. *Proc. Natl. Acad. Sci. USA* **74,** 5350–5354.
2. Denhardt, D. (1966) A membrane filter technique for the detection of complementary DNA. *Biochem. Biophys. Res. Commun.* **23,** 641–646.
3. Sambrook, J., Fritsch, E. F., and Maniatis, T. (1989) *Molecular Cloning. A Laboratory Manual.* Cold Spring Harbor Laboratory, Cold Spring Harbor, NY.
4. Brown, T. A. (1993) Analysis of RNA by northern and slot blot hybridization, in *Current Protocols in Molecular Biology* (Ausubel, F. M., Brent, R., Kingston, R. E., Moore, D. D., Seidman, J. G., Smith, J. A., and Struhl, K., eds.), Wiley, New York.
5. Sabelli, P. A. and Shewry, P. R. (1993) Nucleic acid blotting and hybridization, in *Methods in Plant Biochemistry* (Bryant, J., ed.), Academic, London, UK, pp. 79–100.
6. Sabelli, P. A. and Shewry, P. R. (1995) Northern analysis and nucleic acid probes, in *Methods in Molecular Biology, vol. 49: Plant Molecular Biology Protocols* (Jones, E., ed.), Humana, Totowa, NJ, pp. 161–180.
7. Kumar, A. and Lindberg, U. (1972) Characterization of messenger ribonucleoprotein and messenger RNA from KB cells. *Proc. Natl. Acad. Sci. USA* **69,** 681–685.
8. Lehrach, H., Diamond, D., Wozney, J. M., and Boedtker, H. (1977) RNA molecular weight determinations by gel electrophoresis under denaturing conditions: a critical reexamination. *Biochemistry* **16,** 4743–4751.
9. Thomas, P. S. (1980) Hybridization of denatured RNA and small DNA fragments transferred to nitrocellulose. *Proc. Natl. Acad. Sci. USA* **77,** 5201–5205.
10. Bailey, J. M. and Davidson, N. (1976) Methylmercury as a reversible denaturing agent for agarose gel electrophoresis. *Anal. Biochem.* **70,** 75–85.
11. Wicks, R. J. (1986) RNA molecular weight determination by agarose gel electrophoresis using formaldehyde as denaturant: comparison of RNA and DNA molecular weight markers. *Int. J. Biochem.* **18,** 277,278.
12. Fourney, R. M., Miyakoshi, J., Day, R. S., III, and Paterson, M. C. (1988) Northern blotting: efficient RNA staining and transfer. *Bethesda Res. Lab. Focus* **10,** 5,6.
13. Wilkinson, M., Doskow, J., and Lindsey, S. (1991) RNA blots: staining procedures and optimization of conditions. *Nucleic Acids Res.* **19,** 679.
14. Sabelli, P. A., Burgess, S. R., Carbajosa, J. V., Parker, J. S., Halford, N. G., Shewry, P. R., and Barlow, P. W. (1993) Molecular characterization of cell populations in the maize root apex, in *Molecular and Cell Biology of the Plant Cell Cycle* (Ormrod, J. C. and Francis, D., eds.), Kluwer, Dordrecht, The Netherlands, pp. 97–109.
15. Sabelli, P. A. (1996) Gene isolation by differential screening, in *Plant Gene Isolation* (Foster, G. D. and Twell, D., eds.), Wiley, Chichester, UK, pp. 125–155.

10

DNA Sequencing

Chris Spencer

1. Introduction

There are many reasons why it may be desirable to obtain the sequence of a piece of DNA. At the highest level, there are currently many genome-sequencing projects underway, the object of which is to obtain the complete sequence of the DNA of a particular organism; human *(1)*, yeast *(2)* and *Caenorhabditis elegans (3)* are some examples. The hope is that by obtaining the complete sequence, we will gain a much better understanding of how gene expression is controlled and hence a greater understanding of the genetic basis of many diseases *(4)*. Alongside the genome projects, many workers are focusing in on particular genes and at the method of action and control. Eventually, with the introduction of simpler automated systems for sequencing, it may be possible to perform diagnosis of genetic disease by sequencing. All of the areas where sequence information is required mean that sequencing is one of the most important technologies currently available to the molecular biologist.

1.1. What Is a Sequence?

Deoxyribonucleic acid (DNA) consists of a linear arrangement of the four deoxynucleoside bases; adenine (A), cytosine (C), guanine, (G), and thymine (T). The bases may occur in any order. In its normal form, DNA consists of two antiparallel strands wound around each other in a double helix. Although the bases in any one strand can occur in any order, there are specific rules for which bases can exist opposite each other in the double strand. This is determined by the molecular forces that hold the strands together, and the relationship is as follows: adenine (A) pairs with thymine (T), cytosine (C) pairs with guanine (G).

This relationship is very important in DNA sequencing because it means that by sequencing one strand, you will know the sequence of the other strand. For example, if you had the sequence: A C T G C A G T you would know that the sequence of the other strand would be: T G A C G T C A. Each individual strand runs in a particular direction according to the chemical structure at each end of the strand. There is a phosphate (PO_4-) at one end, referred to as the 5' end, and a hydroxyl (OH–) group at the other, referred to as the 3' end.

From: *Molecular Biomethods Handbook*
Edited by: R. Rapley and J. M. Walker © Humana Press Inc., Totowa, NJ

By convention, DNA sequences are written from the 5' end to the 3' end so the sequence above would be written (remembering as well that the strands are anti parallel):

1.2. DNA Replication

The replication of DNA in vivo is a complicated subject that goes far beyond the scope of this elementary introduction. However, a level of understanding of the underlying principles is necessary to appreciate how the mechanisms of DNA replication can be applied in vitro to generate a DNA sequence.

A number of different components must be brought together to bring about DNA replication. These are detailed.

1. The template, the section of DNA to be replicated.
2. The bases themselves, the building blocks of the DNA. These are provided as deoxynucleoside triphosphates (dNTPs) and there is one for each of the four bases found in DNA; dATP, dCTP, dGTP, and TTP.
3. The primer, a short piece of DNA or RNA that is complementary to a specific section of the template strand.
4. The DNA polymerase, which extends the primer from the 3' end by adding on dNTPs. Many different types of DNA polymerase used in sequencing will be covered in detail in **Subheading 4.6.**

The way in which these elements are combined in DNA replication are shown in **Fig. 1**. The next section will consider how this process has been modified in vitro for DNA sequencing.

2. DNA Sequencing Chemistry

DNA sequencing is basically a modified version of DNA replication performed under controlled conditions in vitro. Sequencing differs from normal DNA replication in the inclusion of dideoxynucleotides (ddNTPs) in the reaction. Dideoxynucleotides differ from normal deoxynucleotides in that they lack the 3'-OH group necessary for the DNA polymerase to extend the growing chain. When a dideoxynucleotide is incorporated into DNA, the synthesis is terminated (**Fig. 2**). This method is therefore known as chain termination sequencing *(5)*.

A ddNTP corresponds to each dNTP; ddATP, ddCTP, ddGTP, and ddTTP. In a set of four reactions, each containing all four dNTPs but only one of the ddNTPs, the products of such reactions will be four nested sets of fragments. Each set of fragments will start in the same place (at the primer) but will be terminated according to the ddNTP included in the reaction; in the reaction with ddATP, each will be terminated wherever there is a T in the template strand, in the ddCTP reaction wherever there is a G, in the ddGTP reaction wherever there is a C and in the ddTTP reaction wherever there is an A.

If the ratios of dNTPs and ddNTPs are suitably balanced, then all possible terminated fragments within a defined range (usually from the primer up to around 500–

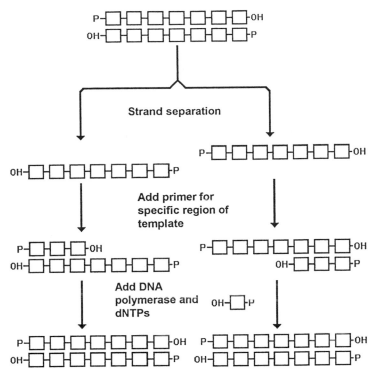

Fig. 1. A simplified mechanism for DNA replication. First, the template DNA strands are separated to allow access for the primers. This allows the DNA polymerase to add dNTPs to the 3'-end of the primer, creating the newly synthesized strand. The products are two new double-stranded molecules identical to the original, each containing one newly synthesized strand and one original strand.

1000 bases) will be produced. The reaction products can be detected by incorporation of a suitable label during the sequencing reaction. This will be covered in more detail in **Subheading 6.**

3. The Sequencing Process

The sequencing process can be split into five sections as shown:

These will each be considered in detail in the following sections.

3.1. Template Preparation

Before a piece of DNA can be sequenced, it is first necessary to obtain it in a format that can be used in the sequencing chemistry. This usually involves cloning the DNA of interest into a suitable vector, although, as will be seen later, the introduction of the polymerase chain reaction (PCR) *(6)* has had a substantial impact on this area.

It is beyond the scope of this text to consider cloning vectors and strategies *(7–9)* and these are dealt with elsewhere in this volume; however, vectors used in DNA

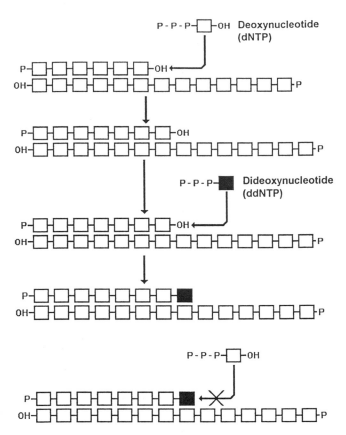

Fig. 2. Chain termination sequencing of DNA. DNA polymerase catalyzes the addition of deoxynucleotides (dNTPs) onto the 3'-end of the primer. The enzyme can incorporate dideoxynucleotides (ddNTPs) as normal into the growing chain. However, ddNTPs lack the 3'-OH group necessary for further extension of the chain so it is no longer possible for the enzyme to add more dNTPs. The chain is therefore terminated.

sequencing. These fall into two main categories: single-stranded and double-stranded. As mentioned in **Subheading 1.1.**, the normal form of DNA is as a double-stranded helix. However, certain viruses and bacteriophages (viruses that grow in bacteria) have DNA that exists in single-stranded form. The most commonly encountered in molecular biology is a bacteriophage called M13 *(10)*. This is a simple virus that replicates in the bacterium *Escherichia coli* and can be easily grown in the laboratory. The simple structure of the virus particle means that the DNA can be easily extracted.

The other class of vector is double-stranded. These are usually autonomously replicating pieces of DNA called plasmids. Again, these can be easily propagated in *E. coli* in the laboratory. Although the natural form of this vector is double-stranded, there are various ways of rendering the DNA single-stranded for sequencing (*see* Chapter 17).

Both types of vector are commonly used as sequencing templates. Some of the features of each are considered in **Table 1**.

On balance, more sequencing is performed using plasmid vectors than single-stranded vectors. This is partly related to the characteristics of the vector and partly to the techniques that are used for sequencing, which will be considered in **Subheading 4.**

Table 1
Characteristics of Single- and Double-Stranded Vectors for Sequencing Applications

Feature	M13 templates (single-stranded)	Plasmid templates (double-stranded)
Quality of sequencing results	Excellent	Very good
Ease of cloning procedures	Moderate (usually requires a subcloning step)	Easy
Stability	Average (prone to spontaneous mutation)	Good
Insert size	Up to 1000 bases	Up to 5000 bases
Other uses for template	Few	Many

The methods for extracting M13 and plasmid DNA *(11)* are now very well charac-terized and most template preparation is undertaken using commercial kits that rapidly produce DNA of very high quality.

3.2. The Use of PCR Products as Sequencing Templates

It is now increasingly common to use the products of a PCR as a sequencing template (*see* Chapter 25). This has a number of attractions; there is frequently no need to go through the cloning stage (although often the PCR products are generated from a library of clones) *(12)*, and no need to go through the bacterial cell culture and template-extrac-tion stages. The template can be generated very rapidly, saving time and materials. The disadvantages are that the size of PCR product that can be reliably produced is still some-what limited (around 1000–2000 bases) and that there is still some need for purification.

The reason for this is that the components of a PCR are very similar to the compo-nents of a sequencing reaction. If the product is not purified, the leftover primers, dNTPs, and enzyme from the PCR can perturb the carefully controlled sequencing-reaction conditions.

4. Sequencing Reactions

The choice of enzyme is often the key factor in the sequencing reactions. At present, there are two classes of DNA polymerase that are commonly used in DNA sequencing. All use the same basic principles for generating the sequence ladder (extension from a known primer followed by chain termination) but the way in which they are applied are somewhat different. They can be broadly categorized as nonthermostable and thermo-stable DNA polymerases, and each will be considered in detail in the following sections.

4.1. Nonthermostable DNA Polymerases

A nonthermostable DNA polymerase will lose all activity at temperatures in excess of 65°C. For many years, these were the only enzymes that were available to the DNA sequencer. Two types of nonthermostable DNA polymerase have traditionally been used in DNA sequencing.

The first enzyme to be used was the large fragment of *E. coli* DNA polymerase I *(13)*, also known as the Klenow fragment. This enzyme worked well in DNA sequenc-ing although its properties were such that interpretation of the data generated from the

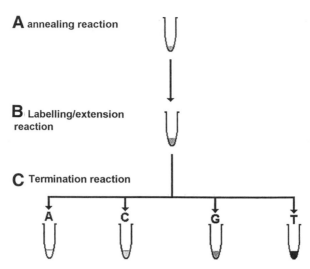

Fig. 3. Practical usage of T7 DNA polymerase in radioactive sequencing. In the first step **(A)**, the template DNA (which will either be single-stranded, or else be double-stranded DNA that has been denatured to render it single-stranded) and primer are mixed and the reaction is heated and cooled to allow annealing of the primer to the template. After this reaction has cooled, the enzyme, labeled dATP, and limiting concentrations of dCTP, dGTP, and TTP are added **(B)**. This reaction is then incubated briefly to allow limited extension of the primer and incorporation of the radioactive label. The reaction is then divided into four **(C)** for addition to the termination mixes, one for each dideoxynucleotide mixture. After a further incubation, loading dye is added in preparation for running on the sequencing gel.

sequencing reaction was not entirely straightforward. This meant that it was very rapidly replaced by enzymes, such as T7 DNA polymerase *(14)* that showed improved qualities for DNA sequencing. Today, very little sequencing is performed with Klenow.

Improvements in gene cloning techniques meant that it was possible to not only clone DNA polymerases into *E. coli* but also to use in vitro mutagenesis to better understand and to therefore modify the properties of the enzymes. The first enzyme to be treated in this way for sequencing was T7 DNA polymerase *(15)*, commercially available under the trademark Sequenase from Amersham. This enzyme was first cloned in 1986 and since then has undergone a number of changes to produce an enzyme with optimal properties for DNA sequencing.

4.2. Practical Steps in Sequenase Sequencing

The most common use of Sequenase DNA polymerase is in radioactive sequencing. This uses the mechanism shown in **Fig. 3**.

4.3. Performance in Sequencing

Sequenase (and other T7 polymerases) has many properties that make it ideal for DNA sequencing. Such properties are:

1. High processivity; the enzyme can copy many hundreds of bases very rapidly.
2. Low exonuclease activity; most DNA polymerases have an associated exonuclease activity. This can be considered to be the opposite of the polymerase in that it will remove bases from

the growing chain (this activity is used for proofreading in vivo). When used in sequencing, it has the effect of removing the dideoxynucleotide from the chain and so producing a background or ghost band. For sequencing, an enzyme with low or no exonuclease is desirable.

3. Low discrimination against dideoxynucleotides; although ddNTPs are chemically similar to dNTPs, they are not identical and so the enzyme can distinguish between the two. This leads to some degree of discrimination, which in turn can lead to uneven band intensities.

4. Even band intensities; it is desirable when visualizing and assigning the sequence for all of the bands to be of equal intensity. This makes sequence assignment easier and more accurate. This is particularly important when the sequence assignment is being done automatically, as in fluorescent sequencing or when using an imaging system with an autoradiograph.

The idea of using genetic manipulation to tailor the properties of enzymes has also been applied to produce the latest generation of thermostable enzymes. This will be covered in **Subheading 4.6.**

4.4. Thermostable DNA Polymerases

The other class of DNA polymerases that are commonly used in sequencing are thermostable; that is, they retain their activity at high temperature. It was the widespread availability of thermostable polymerases that really led to the enormous success of the PCR.

4.5. Practical Steps in Cycle Sequencing

Sequencing with a thermostable polymerase utilizes a process known as cycle sequencing *(16)*. The enzyme, primer, and termination mixes are added to the DNA at the start of the reaction. The reaction is then heated to denature the DNA. After this, it is cooled to allow annealing of the primer. The enzyme then extends from the primer to produce the chain-terminated fragments.

Repeated cycles of heating and cooling are then used, each cycle producing a set of termination products. Typically 20–30 cycles are done leading to an overall amplification of the amount of sequencing reaction product.

This means that cycle sequencing is ideal for applications where the amount of input DNA is limiting, the detection sensitivity for sequence visualization is relatively low; for example, with fluorescent sequencing as compared to radioactivity.

4.6. Examples of Enzymes Used in Cycle Sequencing

Many different types of thermostable enzyme have been used in cycle sequencing applications. One of the most common has traditionally been *Thermus aquaticus* (Taq) polymerase *(17)*. More recently, advances in the understanding of the mechanisms involved in nucleotide recognition have led to a new generation of enzymes specifically engineered for sequencing.

One problem with native thermostable enzymes has been that their ability to recognize nucleotide-base analogs (such as dideoxynucleotides) was not very good. This led to high background and uneven band intensities when compared to T7 DNA polymerase.

Tabor and Richardson *(18)* looked at the nucleotide-binding site of T7 DNA polymerase (which has excellent nucleotide analog recognition) and compared it to the homologous site on a thermostable DNA polymerase. They found that there was only one amino-acid difference between the two sites. When this amino-acid (phenylalanine to tyrosine) change is made on the active site of a thermostable enzyme, the resulting

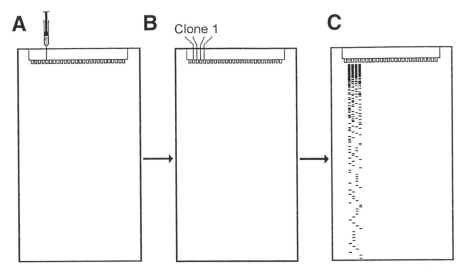

Fig. 4. The sequencing gel is cast between two glass plates and run vertically. Samples are loaded onto the top of the gel (**A**) and then an electric field is applied (**B**). Under the influence of this field, the fragments separate according to size (**C**).

enzyme behaves very like T7 polymerase with respect to the nucleotide-recognition properties while still retaining the advantages of a thermostable polymerase.

This resulted in a thermostable enzyme that gave very even band intensities and very low backgrounds commercially available under the trademark Thermo Sequenase from Amersham. This has had major implications, particularly in fluorescent sequencing where the evenness of the bands enables the software to call the sequence more accurately.

5. Sequence Separation

In all cases, whichever template type, sequencing enzyme, or labeling method is used, the samples are analyzed on a similar type of gel. The type of gel that is used for the analysis of sequencing reactions is usually 4–6% polyacrylamide containing a high concentration of a denaturant, such as urea, that helps keep the strands of DNA (template and sequenced strand) apart. This type of gel is capable of resolving pieces of DNA differing by only one base-pair in length.

The gels are thin (typically 0.2–0.4 mm) to aid resolution and are run vertically between two glass plates. The length of gel varies somewhat, although as a general rule, longer gels will give better resolution and therefore longer sequence. This, however, has to be balanced against the increased difficulty of preparation. Typically, a sequencing gel will be 40–60 cm long and 25–40 cm wide.

The principle behind running a sequencing gel is illustrated in **Fig. 4**. There is no specific order that the individual tracks should be loaded; ACGT and GATC are two popular configurations.

Under the influence of an electric field, the shortest fragments of DNA will migrate the fastest. The separation of the gel is not linear so the fragments at the bottom of the gel are farther apart than the fragments at the top of the gel. It is usually this feature that limits the amount of data that can be collected from a sequencing reaction. When

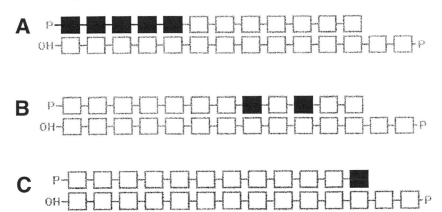

Fig. 5. Alternative labeling positions used in sequencing. **(A)** Labeled primer; the label is usually incorporated on the 5'-end of the primer before the start of the sequencing reaction. **(B)** Labeled nucleotide; one of the nucleotides, usually dATP, is labeled and is incorporated by the enzyme during the sequencing reaction. **(C)** Labeled dideoxynucleotide; in this case, all four labeled ddNTPs must be available, one for each termination reaction.

sequence visualization is considered, we will see how automated DNA-sequencing instruments address this issue to maximize the amount of data that can be collected.

Alternative-gel matrices are now being introduced to offer greater resolution and increased safety over polyacrylamide, which is highly toxic as a monomer. Modified acrylamide-based polymers are now commonly used in commercial preparations, such as Long Ranger from FMC Bioproducts. The preparation and processing of gels is one of the most time-consuming parts of the sequencing process, so much effort is being put into non gel-based methods of sequencing, such as sequencing by hybridization *(19)* or capillary electrophoresis *(20)*.

Even the best gels will only give approx 400–700 bases of reliable data. Using long gels, some labs have claimed 800–1000 bases, although this is the exception rather than the rule.

6. Sequence Visualization

In order to visualize the sequence, it is first necessary to incorporate a detectable label into the sequenced strand. There are three alternative positions in which the label may be incorporated during the reaction. These are illustrated in **Fig. 5**.

There are three common types of labeling used in sequencing; radioactive, four-color fluorescent, and one-color fluorescent. Usually the type of fluorescent labeling is determined by the instrumentation system used.

Although in theory any of the different labeling positions may be used with either radioactivity or fluorescence, in practice there are fairly clear preferences. These are illustrated in **Table 2**.

7. Radioactive Labeling

Traditionally, the sequencing ladder was labeled by the incorporation of a radioactively labeled dNTP (usually dATP) into the extended DNA chain. The DNA was then simply visualized after electrophoresis by drying the gel and then exposing it overnight to a film sensitive to the radioactivity (**Fig. 6**).

Table 2
Common Examples of Label Types and the Usual Position in Which They Are Used

Type of label	Examples	Labeled primer	Labeled dNTP	Labeled ddNTP
Radioactive	^{35}S ^{32}P ^{33}P	Yes (usually cycle sequencing)	Yes	No
Four-color fluorescent	TAMRA ROX JOE FAM	Yes	No	Yes
One-color fluorescent	Fluorescein Cy5 Texas Red	Yes	Yes	No

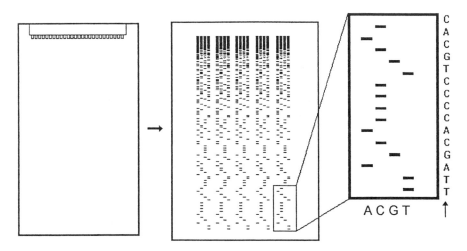

Fig. 6. Radioactive sequence visualization. The samples are loaded on the gel and run (usually for 2.5–5 h). The gel is then transferred to paper and dried down before being exposed to X-ray film (usually overnight). The film is then developed to show the pattern of bands. The sequence is read from the bottom of the gel (smallest fragments) to the top (largest fragments) by comparing the relative positions of the bands in the A, C, G, and T tracks.

The most commonly used isotope for radioactive sequencing is ^{35}S. This offers safe handling (owing to low energy β-emission), reasonable half-life (87.4 d), and reasonable exposure times (typically overnight, 16 h). Other isotopes that are used include ^{32}P, although this is not particularly popular owing to the high energy β-emissions and the short half-life (14.3 d). It does, however, offer the highest available sensitivity and so is still used in some applications. Labeled-primer applications will often use ^{32}P because ^{35}S does not give sufficient sensitivity in this particular application.

More recently, ^{33}P has found applications in DNA sequencing. The properties of ^{33}P (half-life 25.4 d) fall somewhere between those of ^{35}S and ^{32}P; ^{33}P may be used both in labeled primer and labeled-dNTP applications.

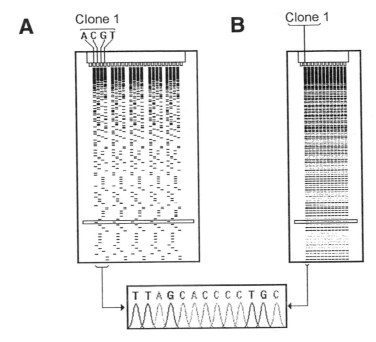

Fig. 7. One-color **(A)** and four-color **(B)** fluorescent sequencing. In both cases, the sequence is detected as the bands migrate past a fixed point in the gel. The output from the fluorescent sequencer is usually represented in a chromatogram format.

7.1. Fluorescent Labeling

The majority of sequencing world-wide is still done using radioactive labeling although the trend is moving toward automated-sequencing systems based on fluorescence. Before moving on to discuss the different fluorescent-sequencing options, it is worth clarifying that such systems automate the sequence separation, visualization, and data-analysis stages, not the template preparation and sequencing reactions.

Fluorescent sequencing relies on exciting a fluorescent moiety on the sequenced-DNA strand so that it emits light that can be detected by an appropriate optical system *(21)*. The excitation is usually performed by shining a laser at a fixed point through the gel. As the bands of DNA pass by the laser, the fluorescent label is excited and the light emitted is detected (**Fig. 7**).

One of the features of fluorescent sequencing is therefore that the sequence information is collected "on line," while the gel is running, rather than after completion of electrophoresis. This has one major advantage over radioactive sequencing. Because the separation of the bands should be constant at a fixed point of the gel, it is possible to collect more data before the resolution of the gel breaks down. Gels can therefore be run for longer (typically overnight). On average, a fluorescent sequencer will yield up to twice as much data for a given length of gel than a radioactive sequence.

The major disadvantage with fluorescent sequencing is the sensitivity. With radioactive sequencing, if the signal is weak, the film can simply be exposed for a longer time in order to detect the sequence. On a fluorescent sequencer, the only chance of detecting the band is as it passes the detector. If the intensity is not strong enough, there is no way that the information can be retrieved. The sensitivity issue has had a major implication for the choice of enzyme and so most fluorescent applications use cycle sequencing in order to maximize the signal levels.

7.2. Four-Color Fluorescence

Four-color fluorescence (for example, as supplied by the Applied Biosystems Division of the Perkin-Elmer corporation) uses a different concept from radioactive or one-color fluorescent sequencing. In radioactive sequencing, each set of terminations (which corresponds to one of the four ddNTPs) is labeled in the same way. This means that when the reaction products are separated on a gel, each termination reaction must be run in a separate track. The sequence is the read by comparing the bands in each track against each other.

With four-color fluorescence, by comparison, a different fluorescent label is used in each termination reaction. There is, therefore, a correlation between the color of the dye and the ddNTP that was used in the termination. For example, all fragments ending in A would be labeled with dye 1, all C's with dye 2, and so on. What this means is that all of the termination-reaction products can be pooled and separated in a single track on the gel. When the data is collected, there will be bands at each position in the ladder, but the color of the band will determine which ddNTP that particular fragment is terminated with (**Fig. 7B**).

This mode of visualization has a number of potential advantages. For instance, four times as many samples can be loaded per gel as with other detection methods (typically up to 36 samples per gel) and variations across the gel, which might cause difficulties in assigning the sequence across four tracks, will be eliminated.

There are also some disadvantages. The presence of the dye labels on the DNA causes mobility differences that are different for each of the four dyes. Also, the dyes are not of equal intensities, which means that some termination reactions have a higher signal than others. However, both of these factors can be corrected by software analysis.

The four-color fluorescence mode of detection has also had implications for the way in which the sequencing chemistry is performed. The association that needs to be made in the sequencing reaction is between the dideoxynucleotide used to terminate the chain and the fluorescent dye. If the dideoxynucleotide is actually labeled with the dye, that association is already made and whenever the dideoxynucleotide is incorporated, the dye will also be incorporated. This means that the reaction can be performed in a single tube rather than as four separate reactions. This considerably simplifies the practical steps required and the dye terminator *(22)* method of sequencing is now extremely popular.

7.3. Single-Color Fluorescence

Alternative automated systems are available from other manufacturers, such as Pharmacia, Licor, and Amersham, which use a single fluorescent label. In this case, similar to radioactivity, it is necessary to run each termination reaction in a separate track. Single-color fluorescence has some advantages over the four-color system in that access to the raw data is generally much better, which can be useful when trying to resolve sequencing ambiguities. However, since four times the number of tracks are required when compared to four-color fluorescence, the throughput of such instruments is generally lower.

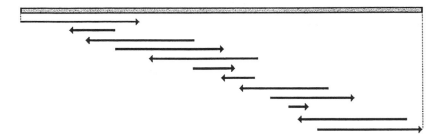

Fig. 8. The complete sequence of the DNA of interest is usually assembled from a number of individual sequences derived from a number of sequencing experiments. This contig assembly is usually done by computer program.

8. Sequence Assignment

At the end of any sequencing experiment is the assignment of the bases. The way in which the sequence is assigned will depend on the method that was used for visualization. The end-point of a radioactive gel is an autoradiograph, which has to be read to obtain the data. Although relatively straightforward, this is an extremely tedious process. Various scanning and image-analysis programs are now available that can aid the reading of a radioactive gel, although most have not yet gained widespread acceptance and most radioactive sequencing gels are read manually. Typically, an expert reader could obtain 250–300 bases from an average length gel at >98.5% accuracy.

Automated fluorescent sequencers not only perform the sequence separation and visualization but will also perform the sequence assignment. The software algorithms for calling the bases are now extremely good and the accuracy of sequence obtainable is as good if not better than an expert reader of an autoradiograph. Typically, 450–600 bases would be available at >98.5% accuracy.

8.1. Sequence Assembly

Depending on the type of sequencing project, it may be necessary to join together a number of sequences in order to generate the complete sequence of interest (**Fig. 8**). This is a process known as contig assembly because the overlapping contiguous sequences are joined together.

This process may mean that each part of the sequence is actually sequenced more than once and should have been sequenced on both strands. This redundancy leads to more accurate sequence because each base in the consensus sequence will be represented more than once. It is not unusual in large sequencing projects to have 8- to 10-fold average redundancy.

References

1. Guyer, M. S. and Collins, F. S. (1995) How is the Human Genome Project doing, and what have we learned so far? *Proc. Natl. Acad. Sci. USA* **92,** 10,841–10,848.
2. Vassarotti, A. and Goffeau, A. (1992) Sequencing the yeast genome: the European effort. *Trends Biotechnol.* **10,** 15–18.
3. Waterston, R. and Sulston, J. (1995) The genome of Caenorhabditis elegans. *Proc. Natl. Acad. Sci. USA* **92,** 10,836–10,840.

4. Green, E. D. and Waterston, R. H. (1991) The human genome project: prospects and implications for clinical medicine *JAMA* **266,** 1966–1975.
5. Sanger, F., et al. (1977) DNA sequencing with chain terminating inhibitors. *Proc. Natl. Acad. Sci. USA* **74,** 5463–5467.
6. Mullis, K. et al. (1986) Specific enzymatic amplification of DNA *in vitro*: the polymerase chain reaction. *Cold Spring Harbor Symp Quant Biol* **51(1),** 263–273.
7. Davison, A. J. (1991) Experience in shotgun sequencing a 134 kilobase pair DNA molecule. *DNA Sequenc.* **1,** 389–394.
8. Henikoff, S. (1990) Ordered deletions for DNA sequencing and in vitro mutagenesis by polymerase extension and exonuclease III gapping of circular templates. *Nucleic Acids Res.* **18,** 2961–2966.
9. Fulton, L. L. et al. (1995) Large-scale complementary DNA sequencing methods. *Methods Cell Biol.* **48,** 571–582.
10. Messing, J. (1983) New M13 vectors for cloning. *Methods Enzymol. Recombinant DNA Techniques* **101 (part C)** 20–78.
11. Birnboim, H. C. and Doly, J. (1979) A rapid alkaline extraction procedure for screening recombinant DNA. *Nucleic Acids Res.* **7,** 1513–1523.
12. Runnebaum, I. B. et al. (1991) Vector PCR. *Biotechniques* **11,** 446–448.
13. Klenow, H. and Henningsen, I. (1970) Selective elimination of the exonuclease activity of the deoxyribonucleic acid polymerase from *Escherichia coli* B by limited proteolysis. *Proc. Natl. Acad. Sci. USA* **65,** 168–175.
14. Tabor, S, and Richardson, C. C. (1987) DNA sequencing analysis with a modified T7 DNA polymerase. *Proc. Natl. Acad. Sci. USA* **84,** 4767–4771.
15. Tabor, S, and Richardson, C. C. (1989) Selective inactivation of the exonuclease activity of bacteriophage T7 DNA polymerase by *in vitro* mutagenesis. *J. Biol. Chem.* **264,** 6447–6458.
16. Murray, V. (1989) Improved double stranded DNA sequencing using the linear polymerase chain reaction *Nucleic Acids Res.* **17,** 8889.
17. Engelke, D. R., et al. (1990) Purification of *Thermus Aquaticus* DNA polymerase expressed in *Escherichia coli. Anal. Biochem.* **191,** 396–400.
18. Tabor, S. and Richardson, C. C. (1995) A single residue in DNA polymerases of the *Escherichia coli* DNA polymerase I family is critical for distinguishing between deoxy- and dideoxyribonucleotides. *Proc. Natl. Acad. Sci. USA* **92,** 6339–6343.
19. Drmanac, R. et al. (1993) DNA sequence determination by hybridisation: a strategy for efficient large scale sequencing. *Science* **260,** 1649–1652.
20. Luckey, J. A. et al. (1993) High speed DNA sequencing by capillary gel electrophoresis. *Methods Enzymol.* **218,** 154–173.
21. Smith, L. M. et al. (1986) Fluorescence detection in automated DNA sequence analysis. *Nature* **321,** 674–679.
22. Prober, J. M. et al. (1987) A system for rapid DNA sequencing with fluorescent chain terminating dideoxynucleotides. *Science* **238,** 336–341.

11

Autoradiography and Fluorography

Bronwen Harvey

1. Introduction

Autoradiography is the localized recording of a radiolabel within a solid specimen. The term does not apply to any single technique. There are many ways of producing an image on a photographic emulsion. The quality and accuracy of the image depends on many factors, including the autoradiographic technique, the isotope employed, and the photographic medium used to record the image. The best results are obtained in the shortest possible time with the correct emulsion and exposure time. There are constraints that can restrict resolution as well as sensitivity. Some of these may be overcome by converting the emitted radiation to light, either by enclosing the autoradiography film between the radioactive sample and a fluorescent-intensifying screen or by incorporating a scintillator into the sample (fluorography).

The major use of autoradiography is at the macroscopic level, where the sample is placed in contact with an X-ray film, for example, hybridization filters. The process can also occur at the microscopic level. Here there is a requirement for a very close contact between the sample or specimen and the photographic emulsion. In this case, the emulsion is applied as a liquid directly to the specimen. After drying, the emulsion is exposed for a period of time and then developed. Typically these reagents are referred to as nuclear emulsion. Such a process allows correlation of the biochemical process at the cellular and subcellular level. The major application areas are receptor localization *(1)*, radio immunochemistry *(2)*, and *in situ* hybridization *(3,4)*.

2. Characteristics of the Different Types of Radiation

The radiations most commonly emitted by radionuclides are alpha (α)-particles, beta(β)-particles, and gamma (γ) emissions.

2.1. α-Particles

α-Particles are positively charged helium nuclei (two protons and two neutrons). They are capable of producing many ionizations throughout their path length, which is short. The resolution of the resulting image produced by these particles is excellent. However, their use in biological work is limited.

From: *Molecular Biomethods Handbook*
Edited by: R. Rapley and J. M. Walker © Humana Press Inc., Totowa, NJ

Table 1
Commonly Used Isotopes in Life Science Applications

Isotopes	Emission	Energy (MeV [max])	Half-life
^{3}H	Weak β	0.019	12.4 yr
^{14}C	Weak β	0.156	5730 yr
^{35}S	Weak β	0.167	87.4 d
^{33}P	Medium β	0.249	25.4 d
^{32}P	Strong β	1.709	14.3 d
^{125}I	γ ray	0.035	60 d
	X rays	0.027	
	Auger electrons	0.030	

2.2. β-Particles

A β-particle has both mass and charge equal in magnitude to an electron. These particles have a longer path length, and fewer ionizations, which decrease in energy with distance compared to α-particles. The path itself is a random one; the β-particle can be easily deflected from its course.

The particles are absorbed by the matter they penetrate. Absorption is dependent on the thickness of the absorbing material. Unlike α-particles, β-particles have many different energies, even those produced from the same isotope. The energy distribution of β-radiation, particularly at the upper end, is important in determining the resolution of any resultant image. A wide variety of radionuclides emitting β-particles of varying energies are available to the Life Science researcher (**Table 1**).

2.3. γ-Emissions and X-Rays

A γ ray is a discrete quantity of energy, without mass or charge, that is propagated as a wave. X-rays, usually produced through electron bombardment of a metal target in a vacuum, have similar properties to γ-rays, but generally have lower energy. Some isotopes, for example ^{125}I, produce X-rays as a result of a K electron absorption by the atomic nucleus, which is replaced by an electron from the next outward shell. In elements with low atomic numbers (<50) the resulting emission is probably absorbed by the electron shell of the same atom. This leads to the formation of Auger electrons, which have an energy similar to tritium. Similar events occur in those isotopes that decay by isomeric transitions.

3. The Photographic Process

Radioactive isotopes were discovered by their ability to blacken photographic film. Capturing the energy released during radioactive decay in a photographic emulsion is still a convenient method of detection, providing an accurate representation of the distribution of a radiolabel within a sample. Photographic emulsions have a clear matrix of gelatin, in which are embedded crystals of silver halide, usually a bromide *(5)*. These crystals, the grain of an emulsion, form a regular lattice structure. The interface between the gelatin and the silver halide crystal is called a fault. The sensitivity of an emulsion is affected by the size and concentration of these crystals. The larger and greater the number, the more sensitive an emulsion.

The resolution of an image depends on the path length of the radioactive emission. This is influenced by the energy and type of emission and the size of the silver halide crystals in the emulsion. The smaller a grain, the greater the resolution. In order to maximize sensitivity, the concentration of silver halide is high.

Low-energy emitting isotopes, such as ^3H, give excellent resolution. In contrast, high-energy emitters give poor resolution. In some cases the use of intensifying screens lower resolution even further. Sensitivity and resolution are mutually exclusive. It is therefore important to select the correct isotopes for the application. Whatever is required, close contact between the sample and the emulsion while keeping the sample as thin as possible produces the best results.

There is a fundamental difference between the response of a photographic emulsion to light and its response to ionizing radiation. In the case of light there is a delayed response. When a photon of light enters the silver halide crystal, there is a high probability that it will give up its energy to one of the orbital electrons in the lattice. Once the electron has acquired sufficient energy, it will leave its orbit. A silver ion accepts the electron to become an atom of metallic silver at a fault. This silver ion constitutes a latent or hidden image, which is then converted to a visible image by the process of "development." To produce a developable image, each silver halide crystal will require the energy from approx 5 photons, each one producing an atom of metallic silver within the crystal *(6)*. By comparison, when a β-particle passes through a photographic emulsion, energy is lost in a series of interactions with the orbital electrons of the silver halide crystal *(5)*, all of which result in a formation of metallic silver, again forming a latent image.

The energy from a single β-particle is sufficient to render each crystal it hits fully developable and many crystals may be converted by a single particle. As a result, the blackening of a film in this type of autoradiography (direct) is directly proportional to the amount of radiation. This allows accurate quantification work to be carried out. During the development process, reducing agents present in the developer solution complete the reduction of the silver halide crystal in the latent image to metallic silver. This is a self-catalytic process, proceeding faster in crystals where reduction has started.

Development is therefore an amplification process. Once silver ions in the crystal are converted to metallic silver, development stops in this crystal. The process does not proceed at the same rate in all crystals, and the process can be stopped at any time. In order to achieve the best results, it is important to stop the process before reduction of crystals not containing a latent image. This will produce a result with an excellent signal-to-noise ratio.

To achieve good reproducible results, development conditions—temperature, time, and agitation—should be optimized and standardized. Manufacturer's recommendations for use and reagent replacement should be followed. This is particularly important in manual processing.

The "fixing" stage of the film's development dissolves any nonmetallic silver halide without affecting the metallic silver. It also renders the emulsion transparent. The emulsion at this stage is soft and easily damaged. The inclusion of a hardener in the fixative prevents this. Most commercial fixers already contain hardeners.

Once fixation is complete, the film should be thoroughly washed in running water. Washing is important, because this will effect the stability of the photographic image. A 5-min wash at room temperature is usually sufficient for films to be evaluated within 1 yr.

4. Autoradiography Products

By far the most common way in which photographic emulsions are used in autoradiography is by coating them onto a thin plastic support, producing an "X-ray film." In principle there are two types of X-ray films used in autoradiography; the direct type and the screen type. Direct films are relatively insensitive to light and are used in direct autoradiography. The film may be sided with the highly sensitive emulsion (high silver halide content) present on only one side of the support. Double-coated film has the advantage of producing images of high resolution. The emulsion may or may not be protected by an antiscratch layer. Films producing the best results with low-energy emitters do not have such a protective layer, because they prevent the weak β-particle reaching the emulsion. Such films are suitable for use with ^{3}H, a weak β-particle emitter, and ^{125}I, which produces Auger electrons during decay. The absence of a protective layer means that care is required in the handling of such films to avoid damage to the emulsion and the creation of artifacts.

Screen-type films are designed to respond to light, usually light in the blue and ultraviolet range of the spectrum. Such films are, therefore, ideal for use with indirect autoradiographic (fluorography) techniques, which involve the use of scintillants. Scintillants convert ionizing radiation to light. Such films are also able to give good results in direct autoradiography. Most films of this type are double-sided and usually have antiscratch layers. The disadvantage of this type of film structure is the reduced resolution compared to single-coated film. These films are particularly suitable for use with high-energy emitters and intensifying screens. Medium-energy, β-emitters, for example, ^{35}S and ^{14}C, may be used with this film. However, most of the radiation will be absorbed by the emulsion layer closest to the sample. Therefore, the second layer does not contribute significantly to the image.

5. Direct Autoradiography

The usefulness of direct autoradiography, where the sample is in direct contact with the photographic emulsion, is limited by the efficiency of the energy transfer between the radioactive isotope and the film. However, direct autoradiography will produce images with good resolution, and is therefore preferred when resolution is more important than sensitivity, for example, in DNA sequencing. Direct autoradiography is suitable for use with the isotopes ^{33}P, ^{35}S, ^{14}C, and ^{3}H, provided the β-particles emitted by these radiolabels are not absorbed by the sample. The thicker the sample, the more absorption. The lower the energy of the emissions, the greater the likelihood that the β-particle fails to reach the film. The problem is most acute with the very weak emissions of ^{3}H. Highly energetic β-particles, like those from ^{32}P or X rays emitted by ^{125}I or ^{131}I, can pass through both sample and film so that only a small proportion of their energy is transferred to produce a latent image. These problems of absorption with a sample and inefficient absorption by the film can be overcome by converting the emission to light.

6. Indirect Autoradiography

This is the technique in which emitted energy is converted to light by means of a scintillator. A scintillator, which can be an inorganic or organic molecule, emits photons of blue or ultraviolet light when excited by β-particles or X-rays. The process is achieved using fluorography or intensifying screens. Fluorography overcomes the prob-

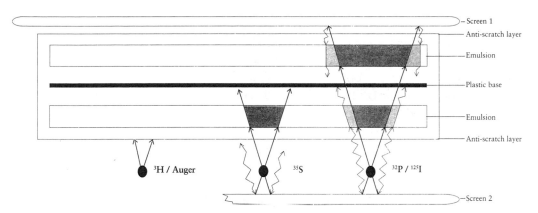

—→ *Radioactive emissions*
⌇⌇⌇ *Light emissions*

Fig. 1. Use of intensifying screens: a compromise between sensitivity gain and resolution loss. Reproduced from Guide to Autoradiography. S544/5385/92/11. Amersham International plc, Amersham Place, Little Chalfont, Buckinghamshire, UK HP7 9NA.

lem of detecting weak β-particles, which are absorbed by the sample before reaching the photographic emulsion, because the sample itself is impregnated with the scintillator. The ultraviolet light produced by scintillator-impregnated samples is captured on the film, provided the sample is translucent and colorless, because it is capable of traveling further than the original β-particle.

Placing a screen incorporating a scintillator behind the film allows highly energetic β-particles or X-rays, which pass straight through the film without transferring their energy, to be converted to light. The resulting photons produce a photographic image, which is superimposed on the autoradiographic image. A diagrammatic representation of the process is shown in **Fig. 1**.

The advantage of converting β-particles/X-rays to light is that the sensitivity of detection is increased. This increase may be as much as 1000-fold with weak emitters, such as ^3H, although the gain is not as great with the isotopes, such as ^{32}P and ^{125}I, which have more penetrative emissions (**Table 3**). Resolution, the distance between the point of maximum exposure and the point at which the number of developed grains is halved *(14)*, may be significantly reduced.

It has already been described that approx 5 photons of light are required for the conversion of several silver ions to atoms and subsequent reduction of the halide. Whereas two or more conversions within the silver halide crystal are stable, one is not. There is a rapid reversion back to the ionic form. This means that a double hit is required within the half-life of the excited form, in order to achieve stabilization. Such a double hit is extremely unlikely in situations where the light is low. Photographic emulsions are therefore disproportionately insensitive to very low intensities of light. However, the half-life of the excited form may be significantly increased at low temperature. The low temperature appears to have little effect on light production by the scintillator at this temperature *(7)*. Investigations suggest that the optimal exposure temperature is around −78°C, efficiency being much lower at −20°C. Despite improving the response to low-intensity light, the relationship between the amount of radioactivity and blackening of the film is still not linear. Preflashing can help to achieve full linearity.

7. Fluorography

In fluorography weaker emitters are brought into contact with the scintillant by impregnation of the sample itself. The procedure used depends on the nature of the sample, **Table 2** summarizes commonly used protocols. The early procedures, usually involving 2,5-diphenyl-oxazole (PPO) and dimethyl sulfoxide (DMSO), produced good results, but are less used today because of safety issues. Several less hazardous, faster aqueous systems are currently available from various suppliers, for example, DuPont/NEN Enlightening system and Amersham's Amplify reagent.

8. Intensifying Screens

A variety of intensifying screens are available using a dense range of inorganic scintillants. Most common is calcium tungstate and lanthanum oxysulfide. The most efficient are those such as calcium tungstate, that produce a blue light. Examples are DuPont/NEN Cronex and Amersham Hyperscreens.

Lanthanum oxysulfide screens produce green light and have been found to be less satisfactory for radioisotope detection *(8)*. It is therefore important to select a film suitable for the wavelength of light emitted by the intensifying screen.

Enhancement of sensitivity is achieved by the use of an intensifying screen placed behind the X-ray film. A second film placed on the far side of the sample can produce a further enhancement of up to twofold. Such a system will reduce exposure time by up to fourfold (**Table 3**).

9. Preflashing

The problem of underrepresentation of low light intensities can be offset to some extent by exposure of fluorographs and intensifying screen at −70°C. The use of preflashing can overcome this effect in all circumstances involving low-intensity light, including chemiluminescence detection, where the use of low temperatures would significantly effect the light producing enzymatic reaction *(9,10)*. Preflashing by increasing sensitivity for the low signals gives a linear film response to light, allowing accurate quantification if required.

The film is pre-exposed to an instantaneous flash of light prior to the sample's exposure to the film. This flash of light allows for at least two conversions of silver ions to atoms within the silver halide crystal. As a result, the reversible stage of latent image formation is overcome. Obviously, the conditions for preflashing require careful optimization (*see* **Fig. 2**). It is important to achieve the minimum stable conditions possible. Preflashing the film, so that its absorbance at 450 nm after development is increased by 0.1–0.2 absorbance units compared to the control, effectively introduces a stable pair of silver atoms per crystal. Preflashing above 0.2 U will disproportionately increase the film's response to low levels of light. The effect of preflashing is greater for longer exposures. The flash of light must be short—1 ms is sufficient—and the light source should be of appropriate intensity and wavelength. These conditions are provided by most photographic flash units.

Table 2
Polyacrylamide Gel Fluorography Protocols

Method[a]	PPO in DMSO (11)	PPO in glacial acetic acid (12)	Sodium salicylate (13)	Amplify[b]
Fixing advisable	Yes	No	Yes	Yes
Recommended fixer/time	Acetic acid: 2 vol Methanol: 9 vol Water: 9 vol 1 h	—	Acetic acid: 2 vol Methanol: 9 vol Water: 9 vol 30 min	7% (v/v) acetic acid 30 min
Presoak required	2X 30 min in DMSO	1X 5 min in glacial acetic acid	1X 30 min in water	None
Scintillant cocktail	4X gel volume of 22.2% (w/v) PPO in DMSO	4X gel volume of 20% (w/v) PPO in glacial acetic acid	10X gel volume of 1 M sodium salicylate pH 5.0–7.0	4X gel volume of Amplify, as supplied
Time to impregnate	3 h	1.5 h	0.5 h	0.3–0.5 h
Rinses	1 h in 20X gel volume of water	1X 30 min in water	None	None
Drying	Under vacuum, 1 h	Under vacuum at 70°C 3–5 h	Under vacuum at 80°C 1 h	Under vacuum at 70–80°C 1 h

[a]Abbreviations: DMSO, dimethyl sulfoxide; PPO, 2,5-diphenyloxazole.
[b]Amplify is a trademark of Amersham International plc.
The information is appropriate for gels up to 1 mm thick. Thicker gels may require longer incubation.

Table 3
Sensitivity Gains Possible

Isotope	^{3}H		^{14}C/^{35}S		^{32}P		^{125}I	
Sample	Direct polyacrylamide gel	Hybridization filter	Direct polyacrylamide gel	Hybridization filter	Direct polyacrylamide gel	Hybridization filter	Direct polyacrylamide gel	Hybridization filter
Direct autoradiography	1×10^7	1×10^7	8×10^3	8×10^3	7×10^2	7×10^2	2.1×10	2×10^{10}
Fluorography 1 or 2	1×10^4	5×10^3	5.3×10^2	—	—	—	—	—
Intensifying screen	—	—	—	—	7×10^1	5×10^1	1.3×10^2	1.3×10^2
Gain	1000	100	15	10	10	14	16	16
Sample preparation advice	Dry sample. Do not wrap. Direct contact with film required.	Direct contact with film required.	Dry sample. Do not wrap. Direct contact with film required.	Direct contact with film required.	Dry sample for best resolution. For convenience wet samples are suitable. Can be wrapped.		Dry sample for best resolution. For convenience wet samples are suitable. Can be wrapped.	

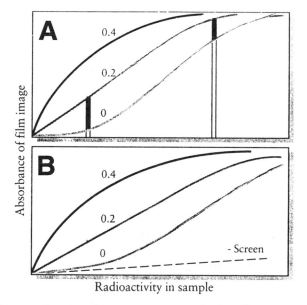

Fig. 2. Preflashing requirement for preexposure to obtain a linear response to film to light from **(A)** a ³H fluorography or **(B)** ³²P with intensifying screen. Preflashing absorbance of 0, 0.2, and 0.4 U above the absorbance of unflashed film are shown. The vertical bars in (A) illustrate how the effect of preflashing is easily under estimated when it is assessed only with large amounts of radioactivity. Reproduced from review booklet number 23. Efficient detection of biomolecules by autoradiography, fluorography or chemiluminescence, by R. Laskey. Amersham International plc, Amersham Place, Little Chalfont, Buckinghamshire, UK HP7 9NA.

10. Appendix: Working Safely with Radioactive Materials

This advice does not replace any instructions or training in your establishment, local rules, or advice from your Radiation Protection Adviser. The safety requirements for many of the less toxic nuclides, for example, ³H, ¹⁴C, and ³⁵S, may be less complex and restrictive than the Regulations or Codes of Practice appear to indicate. This does not mean that these materials may be treated casually.

In general, compounds labeled with low energy β emitters may be handled safely in the small quantities found in most research and teaching laboratories with only modest precautions that are little more than good practice as found in any properly conducted chemical laboratory. When handling high energy β emitters, such as ³²P or γ-labeled compounds, further precautions are necessary. Although radiation protection can be a complex subject, it is possible to simplify it to 10 basic things you should do—the golden rules.

10.1. Rules

1. Understand the nature of the hazard and get practical training. Never work with unprotected cuts or breaks in the skin, particularly on the hands or forearms. Never use any mouth-operated equipment in any area where unsealed radioactive material is used. Always store compounds under the conditions recommended. Label all containers, clearly indicating nuclide, compound, specific activity, total activity, date, and name of user. Containers should be properly sealed.

2. Plan ahead to minimize time spent handling radioactivity. Do a dummy run without radioactivity to check your procedures. The shorter the time the smaller the dose.

3. Distance yourself appropriately from sources of radiation. Doubling the distance from the source quarters the radiation dose (inverse square law).

4. Use appropriate shielding for the radiation. One centimeter of β perspex will stop all β, but beware "Bremsstrahlung" from high energy β emitters. Use suitable thickness of lead for X and γ emitters.

5. Contain radioactive materials in defined work areas. Always keep active and inactive work separate, preferably by maintaining rooms to be used solely for radioactive work. Always work over a spill tray and work in a ventilated enclosure (except with small [a few tens of MBq] quantities of ^{3}H, ^{35}, or ^{14}C compounds in a nonvolatile form in solution).

6. Wear appropriate protective clothing and dosimeter, for example, laboratory overalls, safety glasses, and surgical gloves. However, beware of static charge on gloves when handling fine powders. Local rules will define what dosimeters should be worn; for example, body film badge, thermoluminescent extremity dosimeter for work with high energy β emitters, and so forth.

7. Monitor the work area frequently for contamination control. In the event of a spill, follow the prepared contingency plan:
   a. Verbally warn all people in the vicinity.
   b. Restrict unnecessary movement into and through the area.
   c. Report the spill to the radiation protection supervisor/adviser.
   d. Treat contaminated personnel first.
   e. Follow clean up protocol.

8. Follow the local rules and safe ways of working. Do not eat, drink, smoke, or apply cosmetics in an area where unsealed radioactive substances are handled. Use paper handkerchiefs and dispose of them appropriately. Never pipet radioactive solutions by mouth. Always work carefully and tidily.

9. Minimize accumulation of waste and dispose of it by appropriate routes. Use the minimum quantity of radioactivity needed for the investigation. Disposal of all radioactive waste is subject to statutory control. Be aware of the requirements and use only authorized routes of disposal.

10. After completion of work, monitor yourself, wash, and monitor again. Never forget to do this. Report to the local supervisor if contamination is found.

Extracts from Safe and Secure. A guide to working safely with radiolabelled compounds. RPN 1704. Amersham International plc, Amersham Place, Little Chalfont, Buckinghamshire, England HP7 9NA.

References

1. Clark, C. R. and Hall, M. D. (1986) Hormone receptor autoradiography: recent developments. *Trends in Biochem. Sci.* **11**, 195–199.
2. Pickel, V. M. and Beauchet, A. (1984) *Immunolabelling for Electron Microscopy* (Polak, J. M. and Varndell, I. M., eds.), Elsever, Oxford, UK.
3. McCafferty, J., Cresswell, L., Alldus, C., Terenghi, G., and Fallon R. A. (1989) A shortened protocol for *in situ* hybridization to mRNA using radiolabelled RNA probes. *Technique* **1**, 171–182.
4. Terenghi, G. and Fallon, R. A. (1990) Techniques and applications of *in situ* hybridization. *Current Topics in Pathology, vol. 28, Pathology of the Nucleus* (Underwood, J. C. E., ed.), Springer-Verlag, Berlin, Germany, p. 290–337.
5. Rogers, A. W. (1979) *Techniques of Autoradiography, Third Edition.* Elsevier, Amsterdam.

6. Laskey, R. A. (1990) Radioisotope detection using X-ray film. *Radioisotopes in Biology, A Practical Approach* (Slater, R. J., ed)., IRL, Oxford, UK, p. 87–107.

7. Prydz, S., et al. (1970) Fast radiochromatographic detection of tritium with "liquid" scintillators at low temperatures. *Anal. Chem.* **42,** 156–161.

8. Laskey, R. A. and Mills, A. D. (1977) Enhanced autoradiographic detection of ^{32}P and ^{125}I using intensifying screens and hypersensitized film. *FEBS Lett.* **82,** 314–316.

9. Whitehead, T. P., Thorpe, G. H. G., Carter, C. J. N., Groucutt, C., and Kricka, L. J. (1983) Enhanced luminescence procedure for sensitive determination of peroxidase-labelled conjugates in immunoassay. *Nature* **305,** 158–159.

10. Schaap, A. P., Akhavan-Tafti, H., and Romano, L. J. (1989) Chemi-luminescent substrates for alkaline-phosphatase—application to ultrasensitive enzyme-linked immunoassays and DNA probes. *Clin. Chem.* **35,** 1863,1864.

11. Bonner, W. M. and Laskey, R. A. (1974) Film detection method for tritium-labeled proteins and nucleic acids in polyacrylamide gels. *Eur. J. Biochem.* **46,** 83–88.

12. Skinner, M. K. and Griswold, M. D. (1983) Fluorographic detection of radioactivity in polyacrylamide gels with 2,5-diphenyloxazole in acetic acid and its comparison with existing procedures. *Biochem. J.* **209,** 281–284.

13. Chamberlain, J. P. (1979) Fluorographic detection of radioactivity in polyacrylamide gels with water-soluble fluor, sodium-salicylate. *Anal. Biochem.* **98,** 132–135.

14. Doniach, I. and Pelc, S. R. (1950) Autoradiograph technique. *Brit. J. Radiol.* **23,** 184–192.

12

Mobility Shift Assays

Nigel J. Savery and Stephen J. W. Busby

1. Introduction

The electrophoretic mobility shift assay (EMSA, also known as the gelshift, gel retardation, or bandshift assay) is a technique that is widely used to study interactions between proteins and nucleic acids. The assay is simple to perform, because it requires little specialized equipment and only small quantities of the protein to be studied, yet it can yield detailed information about the binding specificity of a protein, the kinetics of the binding reaction, and the structural consequences of the protein–DNA interaction. Unlike other techniques, such as nitrocellulose filter binding assays, EMSA can also be used to gain information about the stoichiometry of protein–DNA complexes and to study interactions between proteins within a multiprotein-DNA complex.

A straightforward EMSA consists of three steps (**Fig. 1**):

1. Binding: A sample of a DNA-binding protein is mixed with a radiolabeled DNA probe, and the system is allowed to equilibrate.
2. Electrophoresis: The equilibrated mixture of unbound DNA molecules and protein–DNA complexes is subjected to electrophoresis through a nondenaturing polyacrylamide gel. The binding of a protein to a DNA molecule results in a complex with a greater mass, different specific charge, and sometimes different end-to-end length than the unbound DNA molecule, and therefore the protein–DNA complex and the unbound DNA will move through the gel at different rates. Generally, the binding of a protein decreases the electrophoretic mobility of the DNA, and so the protein–DNA complex will be retarded, although there are examples of cases in which protein–DNA complexes migrate more rapidly than the corresponding free DNA.
3. Detection: Following electrophoresis the gel is analyzed using autoradiography or a phosphor screen. In a simple system, where a single molecule of protein binds to each DNA molecule, two radiolabeled bands will be detected, the more retarded corresponding to the protein–DNA complex. EMSA can also be performed using unlabeled DNA, in which case the bound and unbound DNA are detected by staining the gel with ethidium bromide after electrophoresis. This procedure is less sensitive than the use of labeled DNA, however, and precludes some of the more advanced applications of EMSA, which rely on the ability to distinguish between labeled and unlabeled DNA species.

From: *Molecular Biomethods Handbook*
Edited by: R. Rapley and J. M. Walker © Humana Press Inc., Totowa, NJ

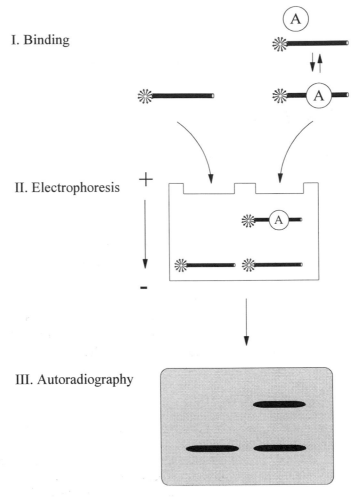

Fig. 1. Principles of EMSA. Binding of a DNA binding protein A to a radiolabeled DNA fragment (**I**) causes it to migrate more slowly than free DNA during electrophoresis (**II**). The two bands containing radiolabeled DNA are detected by autoradiography (**III**).

Generally speaking, the larger the mass of a protein, the greater its effect on the relative mobility of the DNA probe. Conversely, the greater the mass of the target DNA, the more difficult it is to resolve the free species from the protein–DNA complex. This places an effective upper limit on the length of the DNA probe of about 300 bp. The only restraint defining a lower limit on probe size is that all the sequence determinants required for protein binding should be present, and EMSA can be performed satisfactorily with probes that are <20 bp long. Most protein–DNA complexes are remarkably stable during electrophoresis, and several possible mechanisms for the apparent stabilizing effect of the gel matrix have been proposed (*see* **ref. *1***). In the short time interval between each sample being loaded into a well and the complexes actually entering the gel matrix, however, this stabilizing effect will not be in operation and complexes that dissociate are unlikely to reassociate. EMSA can therefore only be used to study complexes that have a lifetime greater than the 1 or 2 min it takes for the sample to enter the gel.

2. Applications

EMSA was initially developed to study interactions involving RNA *(2,3)*, but the technique really rose to prominence when it was applied to the study of DNA-binding proteins involved in the regulation of transcription *(4,5)*. The assay has continued to be widely utilized in the study of transcription factors, and many of the examples cited in this chapter are taken from that field. It should not be forgotten, however, that nucleic acid–protein interactions also play a key role in many other cellular functions, including DNA replication, recombination, and repair, and EMSA is equally applicable to the study of these systems.

2.1. Determination of Binding Specificity

DNA-binding proteins often exhibit a strong sequence specificity, binding far more tightly to a particular target sequence than to "nonspecific" DNA. EMSA is a useful tool in the elucidation of these binding preferences. The initial step, to demonstrate that binding is indeed specific, is simple: When a sequence-specific DNA binding protein is mixed with a labeled fragment of DNA carrying its target sequence, it will bind to that fragment and produce a retarded band on a polyacrylamide gel. If a large excess of unlabeled DNA that also carries the target sequence is included in the binding mixture, the protein will bind to the unlabeled DNA with equal affinity, and there will be less protein available to bind to the labeled probe. The retarded band will therefore decrease in intensity, possibly to such an extent that it becomes undetectable. In contrast, if the unlabeled DNA, which is added in excess to the binding mixture, does not contain the target sequence of the protein in question, the production of the retarded band will be largely unaffected. Taking this procedure one step further, a number of DNA fragments whose sequences vary only slightly from that of a postulated target sequence are used as unlabeled competitor DNA. Some will compete more successfully for the protein than others, enabling the optimal target sequence to be determined. Alternatively, EMSA can easily be adapted to measure the binding affinity of a particular protein for different DNA sequences *(6,7)*.

More specific information about the individual bases involved in the protein–DNA interaction can be obtained from "interference" assays, in which the labeled DNA fragment is treated with a modifying agent before being mixed with the protein. One commonly used modifying agent is dimethyl sulfate (DMS), which methylates N-7 of guanine and, to a lesser extent, N-3 of adenine *(8)*. The modification is performed in conditions such that, on average, each individual DNA molecule is altered only once. A set of DNA molecules carrying modifications at different positions is thus generated, and some of these modifications will occur at bases that are essential for the interaction with the DNA-binding protein. Fragments carrying such critical modifications will remain unbound when the DNA is incubated with the protein, and EMSA is used to separate the bound and unbound molecules. The unbound molecules are extracted from the gel, cleaved at the site of modification (methylated bases can be detected by piperidine cleavage), and analyzed by denaturing gel electrophoresis. Comparison of the ladder of cleavage products with a suitable sequence ladder reveals the position of the bases at which modification interfered with the binding of the protein, and that are therefore likely to be in close proximity to the protein when it is bound *(9)*.

A technique called SELEX (Systematic Evolution of Ligands by EXponential enrichment *[10]*) can be used determine the target sequence of a protein in the absence of any pre-existing knowledge. In this procedure, a degenerate oligonucleotide is used to generate a random library of target DNA molecules. This heterogeneous mixture of DNA is incubated with a limiting concentration of purified protein and EMSA is used to separate those molecules that are bound by the protein from those that are not. The retarded band is extracted from the gel, amplified by PCR, and then subjected to a further EMSA-selection step, in which a lower concentration of the protein is used to encourage greater discrimination between higher and lower affinity binders. After several cycles of selection and amplification the sequence of the "successful" molecules is determined, and a consensus sequence for the protein's target site can be deduced *(11)*.

2.2. Detection of DNA Binding Proteins in Crude Cell Extracts

As well as the study of purified DNA-binding proteins, EMSA can be performed satisfactorily with mixtures of proteins, or even with crude cell extracts, provided that precautions are taken to prevent degradation of the DNA probe by nucleases present in the sample. EMSA is therefore often used to check for the presence of previously identified DNA-binding proteins, in particular cell extracts (for example, to follow the expression of a particular transcription factor over time *[12]*, or to monitor the purification of a DNA binding protein), to identify novel DNA-binding proteins that are able to bind to particular DNA sequences (for example, sequences that have been shown by mutation to be important for the regulation of expression of a particular gene and that are therefore postulated to be the binding site for an unidentified transcription factor *[13]*), or to rapidly test the ability of mutated proteins to bind to a given target sequence (for example, when attempting to define the DNA-binding domain of a protein). Experiments with cell extracts follow the same basic procedures as those with purified proteins, although, because there are likely to be many DNA-binding proteins in a mixture as complex as a cell extract, it is important that specific binding is confirmed by performing the experiment in the presence of excess unlabeled specific and nonspecific competitor DNA.

2.3. Determination of Kinetic Parameters

When carefully performed, EMSA can be more than just a qualitative tool, and can be used to obtain quantitative information about the equilibrium and rate constants of protein–DNA interactions *(14,15)*. Equilibrium constants can be determined by incubating a fixed concentration of target DNA with a range of concentrations of proteins, and then measuring the ratio of bound to unbound DNA at each concentration (**Fig. 2**) *(6,7)*. These assays are performed under conditions of large protein excess, so that the amount of protein bound to the DNA always represents only a small fraction of the total protein present. The concentration of free protein in each reaction can then justifiably be approximated to the total (known) protein concentration and the equilibrium constant is calculated as the reciprocal of the protein concentration at which 50% of the DNA is present in the retarded band.

Rate constants can be determined by removing samples from binding reaction mixes at timed intervals and then applying them to running gels. The later samples are thus subjected to a shorter period of electrophoresis than are the early samples, and as a consequence, the pattern of bands produced is curved. Once the samples have entered the gel, any further association or dissociation is halted or greatly slowed, but, as already

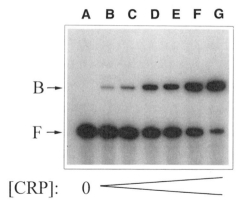

Fig. 2. Use of EMSA to measure equilibrium binding constants. The figure shows an autoradiograph of an EMSA experiment in which a fixed concentration of target DNA was incubated with a range of concentrations of the *E. coli* cAMP receptor protein (CRP). Lane A contained no CRP. Lanes B–G contained increasing concentrations of CRP. As the CRP concentration increased, a greater proportion of the DNA was bound by the protein, and the intensity of the retarded band owing to bound DNA (B) increased as the intensity of the free DNA band (F) decreased.

discussed, EMSA cannot be used to study reactions that occur over a time scale that is short compared with the time taken for the sample to enter the gel after loading. Association-rate constants can be calculated by measuring the ratio of bound to unbound DNA in samples removed at intervals after the binding reaction has been initiated by mixing the protein and DNA samples. Dissociation-rate constants are measured by mixing preformed protein–DNA complexes with a large excess of unlabeled target DNA, and, again, removing samples at timed intervals. As the labeled complex dissociates, the protein reassociates with the unlabeled DNA and thus the proportion of labeled DNA in the retarded (bound) band decreases with time *(16)*.

2.4. DNA Bending—The Cyclic Permutation Assay

Although many DNA binding proteins (for example, the *lac* repressor protein) do not bend DNA when they bind to it, others grossly distort the DNA when binding. For example, the *Escherichia coli* cyclic AMP receptor protein (CRP) causes a bend of >90° when it binds to its 22 bp recognition site, and the TATA box binding protein (TBP) of eukaryotic RNA polymerase II bends its target DNA by 80°. In both of these cases the bends, which have now been characterized by X-ray crystallography of the protein–DNA complex, were previously detected using EMSA *(17,18)*. A great deal of information regarding the location, angle, and direction of bends introduced into DNA can be obtained by a variation of EMSA called the cyclic-permutation assay *(18)* (**Fig. 3**). This technique relies on the observation that bent DNA migrates through polyacrylamide gels more slowly than does straight DNA of the same length. The greater the angle of the bend the more slowly the fragment migrates, and the nearer the bend is to the center of the DNA the greater the extent of the retardation. In a cyclic permutation assay the protein binding site DNA is cloned into a vector between two directly repeated restriction-site polylinkers *(19)*. Digestion with a number of different enzymes thus yields a set of DNA fragments that are all the same number of base pairs long but that

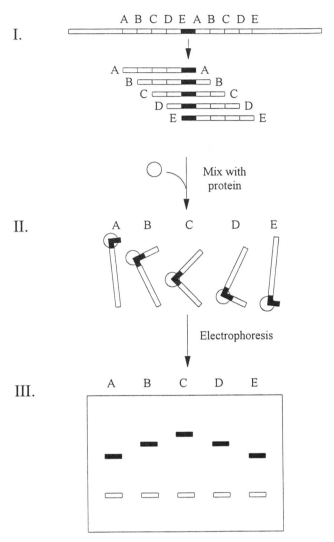

Fig. 3. Cyclic permutation assays. **(I)** The target sequence (shaded box) is cloned between directly repeated polylinkers (restriction sites A–E). Digestion with individual enzymes yields fragments of identical length, but that carry the target sequence at different positions. **(II)** A DNA binding protein (open circle), which bends the DNA when it binds the target, will have different effects on the end-to-end distance at each of the different fragments. **(III)** The protein–DNA complex will exhibit differing mobilities in a polyacrylamide gel, retardation being greatest when the bend is in the middle of the fragment.

carry the protein-binding site at different positions with respect to the end of the fragment. The different fragments are mixed individually with the protein of interest and subjected to gel electrophoresis. If the protein does not bend the DNA, then all of the protein–DNA complexes will be retarded to the same extent. If, however, the protein kinks the DNA on binding, then the relative mobility of the protein–DNA complex will be further decreased. The relative mobility of the complex will be at its lowest with fragments in which the site of the bend (as opposed to the site of protein binding) is at the center of the restriction fragment. Thus, by comparing the relative mobility of a number of different protein–restriction fragment complexes, the approximate position of the bend can be determined. Furthermore, by comparing the bend-induced retardation with calibration curves constructed using bends of a known angle (for example, that induced by CRP binding, or the bend formed by phased runs of "A"s), an estimation of the angle of bending can be made. A final twist on the cyclic-permutation theme allows the direction of the induced bend to be determined. The binding site of interest is cloned adjacent to a well-characterized region of bent DNA, and the distance between the two is increased stepwise by the insertion of small fragments of DNA. Two adjacent bends in the same direction as one another will have an additive effect, further decreasing the relative mobility of the complex, whereas two bends that are in opposing directions will tend to cancel out one another's effect. The relative mobility of the fragments carrying the two curves will therefore vary as the spacing is altered, with a periodicity of one helical turn, allowing the direction of the curve to be deduced *(21)*.

2.5. Supershifts—Multiprotein Complexes and Antibodies

In many instances, more than one molecule of a DNA-binding protein binds to a particular region of DNA. Because each additional protein binding to a protein–DNA complex alters its electrophoretic mobility and results in the production of an additional retarded band, EMSA can yield direct information about the stoichiometry of such complexes *(4)*. It is also common for DNA-binding proteins, particularly those with regulatory roles, to interact with other proteins once they have bound to their target sites (for example, many transcription factors function by making direct contacts with RNA polymerase). These other proteins may themselves have a DNA-binding function and bind to a site adjacent to the primary site, or the interaction may consist entirely of interprotein contacts. Again, the "supershift" caused by the binding of the second protein can be used to measure stoichiometry, and quantitative information concerning the cooperativity of binding and the stability of the resulting complexes can be obtained from experiments similar to those used when only a single protein binds to the probe (Fig. 4). Finally, EMSA is sufficiently sensitive for the binding of a monoclonal antibody to cause a "supershift." The assay can therefore confirm the presence of a particular protein in a complex, because the mobility of the antibody–protein–DNA ternary complex formed in the presence of a specific antibody will be less than that of the protein–DNA complex formed in a control reaction containing no antibody *(22)* (unless the antibody binds to a domain of the protein important for DNA binding, in which case the presence of the antibody will abolish complex formation).

3. Conclusions

The electrophoretic mobility shift assay has many diverse applications, principally owing to the ease with which it can be performed and the simple requirements regarding sample quantity and purity. Often EMSA is a profitable first technique when initiating a study of protein–nucleic acid interactions, the information it provides being supplemented later by more advanced footprinting techniques. EMSA has been applied to many different systems: it is impossible to cover them all in a brief chapter. The interested reader is recommended to consult one of several excellent and comprehensive reviews covering the subject *(14,15,23)*.

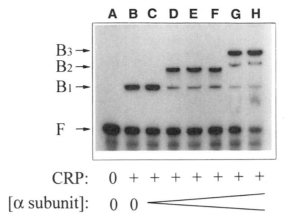

Fig. 4. Supershifts. The figure shows an autoradiograph of an EMSA experiment in which purified α subunits of *E. coli* RNA polymerase were added to preformed CRP-DNA complexes. Lane a contained no CRP, and only the free DNA band is observed (F). Lanes B–H contained a fixed concentration of CRP, which gave rise to a retarded band (B1). Addition of increasing concentrations of α subunits (lanes c–h) resulted in the appearance of first one (B2) and then a second (B3) "supershifted" band, representing complexes containing both CRP and purified α subunits.

References

1. Fried, M. G. and Liu, G. (1994) Molecular sequestration stabilizes CAP-DNA complexes during polyacrylamide gel electrophoresis. *Nucleic Acids Res.* **22,** 5054–5059.
2. Dahlberg, A. E., Dingman, C. W., and Peacock, A. C. (1969) Electrophoretic characterization of bacterial polyribosomes in agarose-acrylamide composite gels. *J. Mol. Biol.* **41,** 139–147.
3. Spassky, A., Busby, S. J. W., Danchin, A., and Buc, H. (1979) On the binding of tRNA to *E. coli* RNA polymerase. *Eur. J. Biochem.* **99,** 187–201.
4. Fried, M. and Crothers, D. M. (1981) Equilibria and kinetics of lac repressor-operator interactions by polyacrylamide gel electrophoresis. *Nucleic Acids Res.* **9,** 6505–6525.
5. Garner, M. M. and Revzin, A. (1981) A gel electrophoresis method for quantifying the binding of proteins to specific DNA regions: application to components of the *Escherichia coli* lactose operon regulatory system. *Nucleic Acids Res.* **9,** 3047–3600.
6. Kolb, A., Busby, S., Herbert, M., Kotlarz, D., and Buc, H. (1983) Comparison of the binding sites for the *Escherichia coli* cAMP receptor protein at the lactose and galactose promoters. *EMBO J.* **2,** 217–222.
7. Kolb, A., Spassky, A., Chapon, C., Blazy, B., and Buc, H. (1983) On the different binding affinities of CRP at the *lac, gal* and *malT* promoter regions. *Nucleic Acids Res.* **11,** 7833–7852.
8. Shaw, P. E. and Stewart, A. F. (1994) Identification of protein–DNA contacts with dimethyl sulfate, in *Methods in Molecular Biology, vol. 30: DNA–Protein Interactions: Principles and Protocols* (Kneale, G. G., ed.), Humana, Totowa, NJ, pp. 79–87.
9. Minchin, S. and Busby, S. (1993) Location of close contacts between *Escherichia coli* RNA polymerase and guanine residues at promoters with or without consensus -35 region sequences. *Biochem. J.* **289,** 771–775.
10. Tuerk, C. and Gold, L. (1990) Systematic evolution of ligands by exponential enrichment: RNA ligands to bacteriophage T4 DNA polymerase. *Science* **249,** 505–510.

11. He, Y., Stockley, P. G., and Gold, L. (1996) In vitro evolution of the DNA binding sites of the *Escherichia coli* methionine repressor, Met J. *J. Mol. Biol.* **255,** 55–66.

12. Adler, V. and Kraft, A. S. (1995) Regulation of AP-3 enhancer activity during hematopoietic differentiation. *J. Cell. Physiol.* **164,** 26–34.

13. Read, M. L., Smith, S. B., and Docherty, K. (1995) The insulin enhancer binding site 2 (IEB2; FAR) box of the insulin gene regulatory region binds at least three factors that can be distinguished by their DNA binding characteristics. *Biochem. J.* **309,** 231–236.

14. Fried, M. G. (1989) Measurement of protein–DNA interaction parameters by electrophoresis mobility shift assay. *Electrophoresis* **10,** 366–376.

15. Gerstle, J. T. and Fried, M. G. (1993) Measurement of binding kinetics using the gel electrophoresis mobility shift assay. *Electrophoresis* **14,** 725–731.

16. Gaston, K., Kolb, A., and Busby, S. (1989) Binding of the *Escherichia coli* cyclic AMP receptor protein to DNA fragments containing consensus nucleotide sequences. *Biochem. J.* **261,** 649–653.

17. Horikoshi, M., Bertuccioli, C., Takada, R., Wang, J., Yamamoto, T., and Roeder, R. G. (1992) Transcription factor TFIID induces DNA bending upon binding to the TATA element. *Proc. Natl. Acad. Sci. USA* **89,** 1060–1064.

18. Wu, H.-M. and Crothers, D. M. (1984) The locus of sequence-directed and protein-induced DNA bending. *Nature* **308,** 509–513.

19. Kim, J., Zwieb, C., Wu, C., and Adhya, S. (1989) Bending of DNA by gene-regulatory proteins: construction and use of a DNA bending vector. *Gene* **85,** 15–23.

20. Thompson, J. F. and Landy, A. (1988) Empirical estimation of protein-induced bending angles: applications to λ site-specific recombination complexes. *Nucleic Acids Res.* **16,** 9687–9705.

21. Zinkel, S. S. and Crothers, D. M. (1987) DNA bend direction by phase sensitive detection. *Nature* **328,** 171–181.

22. Kristie, T. M. and Roizman, B. (1986) α4, the major regulatory protein of herpes simplex virus type 1, is stably and specifically associated with promoter-regulatory domains of α genes and of selected other viral genes. *Proc. Natl. Acad. Sci. USA* **83,** 3218–3222.

23. Lane, D., Prentki, P., and Chandler, M. (1992) Use of gel retardation to analyze protein–nucleic acid interactions. *Microbiol. Rev.* **56,** 509–528.

13

cDNA Libraries

Ian G. Cowell

1. Introduction

Messenger RNA (mRNA) is a complex mixture of easily degraded molecules representing the processed (spliced, polyadenylated, and capped) transcripts from genes that are active in a given tissue or cell type. In the construction of a cDNA library, RNA-dependent DNA polymerase (reverse transcriptase; RT) is used to convert mRNA into double-stranded cDNA molecules suitable for insertion into vectors (plasmids or bacteriophage) that can be propagated in prokaryotes. Because cloning the vectors and hence cDNA libraries can be propagated indefinitely, usually in the bacterial host *Escherichia coli*, this results in an effectively permanent and inexhaustible library or bank of cDNA clones representing the original mRNA population. The real utility of this, however, is the ability to select or screen for specific cDNAs and thus isolate a cDNA copy of a particular mRNA in a pure state.

Enzymatic synthesis of cDNA and the construction of cDNA libraries are outlined in **Subheading 2.** and in the following sections the applications of cDNA library methodology and means of isolating specific cDNA clones are described in more detail.

2. Synthesis of cDNA

The synthesis of double-stranded cDNA occurs through a series of enzymatic steps starting with the enzymatic copying of mRNA into cDNA by RT. Although many variations exist, a typical cDNA synthesis protocol is outlined in **Fig. 1.** The first and probably the most critical step is the copying of polyadenylated mRNA into "first strand" cDNA by RT, to create an mRNA/DNA-hybrid molecule. Like other polymerases, RNA-dependent RT adds nucleotides to the 3'-end of a primer annealed to a nucleic-acid template (RNA or DNA). In most cases, the primer for first strand cDNA synthesis is oligo dT$_{12-18}$, which conveniently anneals to the poly A tail found at the 3' end of most eukaryotic mRNAs (**Fig. 1, step 1**). Following completion of first strand cDNA synthesis, the mRNA strand in each hybrid molecule is replaced by DNA by the combined action of ribonuclease H (RNase H) and *E. coli* DNA polymerase I *(1)*. The role of RNase H is to degrade the mRNA strand, leaving RNA fragments to act as primers for synthesis of the second strand by DNA polymerase I (**Fig. 1, step 3**). The resulting second strand DNA fragments are then joined by the action of T4 DNA ligase. These

From: *Molecular Biomethods Handbook*
Edited by: R. Rapley and J. M. Walker © Humana Press Inc., Totowa, NJ

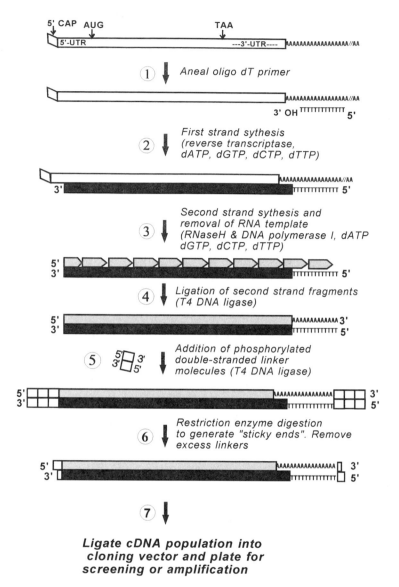

Fig. 1. Scheme for the synthesis of double-stranded cDNA from polyadenylated mRNA.

steps yield double-stranded cDNA with blunt or flush ends. To efficiently ligate this cDNA into the appropriate cloning vector, it is necessary to add cohesive or "sticky" ends to the cDNA molecules. This can be done by ligating linker molecules to the cDNA (*see* **Fig. 1, steps 5** and **6** and later in this section) or by homopolymeric tailing. In the latter method, calf-thymus terminal transferase is used to catalyze the addition of a series of molecules of a single nucleotide to the 3'-hydroxyl terminal of each strand of the cDNA molecules. The complementary nucleotide is added to the vector DNA. Tailed cDNA and vector molecules are then annealed and transformed into host *E. coli* cells. Host-cell enzymes fill and covalently join annealed cDNA/vector molecules. Homopolymeric tailing is suitable for plasmid vectors, but not for the creation of bacteriophage libraries (*see* **Subheading 3.**) and so the addition of linkers is now a more popular route. Linker molecules consist of double-stranded oligonucleotide molecules containing an internal restriction-enzyme site. Linkers are ligated to both ends of the cDNA molecules, which are then digested to completion with the appropriate restriction enzyme (**Fig. 1, steps 5** and **6**), generating recessed ends at both ends of each cDNA. For enzymes such as *Eco*RI, digestion of the cDNA at any internal restriction-enzyme sites can be prevented by treatment of the cDNA with the appropriate DNA methylase prior to addition of linkers. *Eco*RI methylase modifies any internal *Eco*RI-restriction sites, rendering them resistant to cleavage. Following digestion, linkered cDNA molecules are separated from free-linker molecules on the basis of size and are ready for ligation to vector DNA, which has been digested with the same restriction enzyme.

In an interesting variation of this method, different restriction-enzyme sites are added to the two ends of the cDNA allowing the cDNA molecules to be cloned directionally. In place of oligo dT, an oligonucleotide, such as 5'-GAGAGAGACTCGAG TTTTTTTTTTTTTTTTT-3' *(2)*, is used to prime first strand cDNA synthesis, the CTCGAG sequence is the site for cleavage by a restriction enzyme, in this case *Xho*I. cDNA synthesis proceeds otherwise as in **Fig. 1** and double-stranded linkers containing the site for a second restriction enzyme such as *Eco*RI are added as before. *Eco*RI linkers become ligated to both ends of the cDNA molecules, but digestion with *Eco*RI and *Xho*I leaves an *Eco*RI "sticky end" at the end of the DNA molecule corresponding to the mRNA 5' terminus and an *Xho*I "sticky end" at the other. The resulting cDNA molecules can then be ligated directionally into vector DNA digested with *Eco*RI and *Xho*I.

In another variation, oligo dT is replaced with a pool of random primers that anneal at any point in the mRNA sequence, thus producing first strand products that are truncated at the 3'-end. Second-strand synthesis and subsequent steps are carried out as before. This approach is useful in producing cDNAs from very long mRNAs or from mRNAs lacking poly A tails.

Synthesis of cDNA typically requires microgram quantities of polyadenylated RNA. However it may not always be practical or even possible to obtain this much RNA if the starting material from which mRNA is to be prepared is very limited. To solve this problem, the polymerase chain reaction (PCR) has been adapted to allow construction of cDNA libraries from very small quantities of polyadenylated mRNA *(3,4)*.

3. Bacteriophage vs Plasmid Vectors

At this point it is worth pointing out that there are two types of vector for creating cDNA libraries: plasmids and bacteriophage vectors (*see* Chapters 15 and 16). Plasmids are double-stranded circular DNA molecules usually approx 3–10 kb in size that are replicated separately from the chromosomal DNA in the host cell. Among the crucial features of a plasmid cloning vector are: an antibiotic-selection marker such as β-lactamase that confers resistance to ampicillin on bacterial cells harboring the plasmid, an origin of replication, and one or more unique restriction-enzyme sites for insertion of foreign DNAs. Plasmid libraries are used for some specialized purposes, but they suffer from several disadvantages compared to bacteriophage libraries (it is more difficult to make large plasmid libraries and plasmid libraries are more difficult to store and replicate than those created in bacteriophage vectors). Plasmid vectors are usually designed for propagation in the bacterial host *E. coli*. However, specialized plasmids

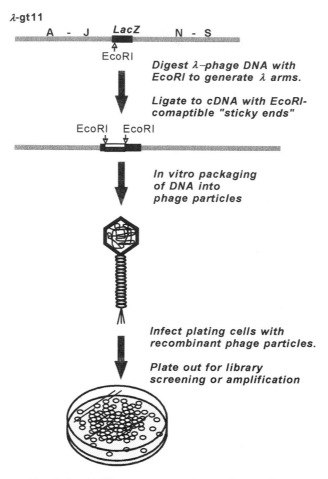

Fig. 2. λgt11 library construction and screening.

have been constructed that can replicate in *E. coli* and yeast or even mammalian cells and may contain promoter sequences to allow expression of cDNAs in those cells (*see* the yeast two-hybrid discussed in **Subheading 11.** below for example).

The wild-type bacteriophage λ genome is a double-stranded DNA molecule with a length of approx 50 kb. It is carried in nucleoprotein bacteriophage particles (*see* **Fig. 2**) as a linear molecule with single-stranded complementary termini (cos ends). A detailed description of the life cycle of bacteriophage λ can be found elsewhere *(5,6)* but briefly, infectious phage particles are absorbed onto the outer membrane of host *E. coli* cells through affinity for the product of the bacterial *lamB* gene. Upon phage entry into the host cell, one of two pathways is followed: lytic growth, where the phage DNA is replicated many fold, phage proteins are synthesized, progeny phage are assembled, and finally, the host cell is lysed, releasing many infectious phage particles, or lysogenic growth where the phage DNA becomes incorporated into the host genome by site-specific recombination. The choice between lysogeny and lytic growth is determined by an intricate balance between viral and host genes in the infected cell, but significantly (*see* **Subheading 4.**), maintenance of the lysogenic state depends on the phage *cI* repressor gene. The lytic cycle is employed in most cDNA-library applications. Phage particles mixed with host *E. coli* cells and spread over agar plates multiply to form plaques of lysed cells in a "lawn" of host cells (*see* **Subheading 5.** for more details).

Many alterations to the wild-type λ genome have been introduced to generate the current generation of λ-cloning vectors, but one of the most important is based on the finding that the central one third of the phage genome (between the genes designated *J* and *N*) is not required for lytic growth and can be replaced with foreign DNA. Although it would not be possible to catalog here all of the different λ-phage cDNA cloning vectors that have been used, three of the most commonly used λ-phage vectors for cDNA library construction are described in **Subheadings 4., 4.1.,** and **4.2.**

4. λgt10 *(7)*

cDNA molecules can be inserted into a unique *Eco*RI site in the phage-repressor gene *cI*. This vector is suitable for cDNA inserts of up to 7 kb and has the advantage that recombinants (phage containing an insert) have a disrupted λ-repressor gene. Plating on a host strain with a high frequency of lysogeny *(hfl)* mutation results in suppression of plaque formation from nonrecombinant (*cI*⁺) phage, but allows normal lytic growth and plaque formation of *cI*⁻ recombinants *(7)*.

4.1. *λgt11* (7–9)

The site for insertion of cDNA in this vector is an *Eco*RI site in the *lacZ* gene that was artificially engineered into the viral genome. The *lacZ* gene, which is bacterial in origin, encodes β-galactosidase and cDNAs cloned into the gene have the potential to be expressed as fusion proteins with β-galactosidase. This allows cDNA libraries constructed in λ-gt11 to be screened with antibody probes for production of the desired fusion proteins in specific recombinant clones (*see also* **Subheading 10.**). Production of recombinant fusion protein, which may be toxic to the host cell, is initially suppressed by the use of a host strain expressing high levels of the *lac* operon repressor. Once plaques have formed, protein synthesis is induced with isopropyl-β-D-thiogalactopyranoside (IPTG) a synthetic activator of the *lac* operon. λgt11 can accept cDNA inserts of up to 7 kb and libraries made in this vector can also be screened by hybridization with DNA probes.

4.2. *λZAP and λZAPII* (2,6,10)

Like λgt11, λZAP is an expression vector. cDNAs of up to 10 kb can be cloned into a multicloning site (a sequence containing a series of unique restriction enzyme sites) located in a portion of the *lacZ* gene, which itself resides within a plasmid vector (pBluescript) embedded within the phage genome. As for λgt11, foreign cDNAs have the potential to be expressed as fusion proteins with β-galactosidase, allowing screening with antibody probes for fusion-protein production by specific recombinant clones. Once individual cDNA clones have been identified and isolated, it is usually necessary to subclone cDNA inserts from the recombinant phage into a plasmid vector for further analysis. But this often time-consuming step has been obviated in λZAP vectors because the embedded plasmid containing the cloned cDNA can be excised automatically from the phage genome in the presence of f1 or M13-helper phage *(6,10)*. The plasmid vector pBluescript is designed to facilitate DNA sequencing of cDNA inserts, in vitro synthesis of RNA transcripts, expression of fusion proteins, and other manipulations. A further use of λZAP and other similar phage strains exhibiting auto-excision properties is the ability to generate a plasmid library directly from a phage library (*see* **ref.** *11* for an example).

5. Construction and Propagation of λ-Phage cDNA Libraries

The λ-phage genome is a double-stranded linear-DNA molecule. Digestion of λ genomic DNA with a restriction enzyme to allow insertion of cDNA results in two DNA fragments termed λ-arms. cDNA produced as shown in **Fig. 1** can be ligated to the λ arms to produce recombinant-phage DNA molecules. In the example shown in **Fig. 2**, cDNA generated with terminal *Eco*RI linkers is ligated into the *Eco*RI cloning site of λgt11. Phage DNA must then be packaged in vitro to generate infective-phage particles. Efficient packaging is crucial to the production of good cDNA libraries and relies on mixtures of extracts from cells infected with λ-phage strains that are defective in phage-particle assembly. Luckily for those engaged in library production, tested packaging extracts are available commercially.

To propagate λ-phage libraries, phage particles are mixed with an excess of host *E. coli* cells for infection to occur. Cells are then mixed with warm molten dilute agar (top agar) and poured over solid agar plates. Uninfected cells grow as a lawn in the top agar, whereas phage multiply in infected cells and produce plaques in the lawn a few millimeters in diameter. Each plaque corresponds to an original phage particle. It is usual to amplify freshly made cDNA libraries by collecting phage particles released from the original plated library in an overlayed buffered magnesium sulfate solution to make a library stock. In this way, each original phage leads to the production of may thousands of progeny. The stock may have a concentration of phage of more than 10^{10} plaque-forming units (pfu) per milliliter, can be stored for extended periods, and contains sufficient phage particles for many platings. However, amplification has the potential to introduce biases in the abundance of different cDNAs if some recombinant clones replicate better than others, and for this reason only one round of amplification is recommended.

6. Propagation of Plasmid Libraries

Having constructed a cDNA library in a plasmid vector, it is necessary to introduce the recombinant plasmid population into suitable host cells in a process known as transformation. This process uses either chemical means to alter the properties of the host-cell walls, allowing entry of recombinant molecules (chemical transformation) *(12)* or uses an intense electric field to achieve the same thing (electroporation). Following transformation, cells are spread over agar plates containing an antibiotic such as ampicillin to which a plasmid-borne gene confers resistance. Cells that have taken up a plasmid molecule and as a result contain an antibiotic-resistance gene form colonies on the agar plates, whereas cells lacking a plasmid molecule (the vast majority) are prevented from multiplying by the antibiotic. Amplification of a plasmid library is similar to that of a λ library as described in **Subheading 5**. The entire cDNA library is transformed into a suitable *E. coli* host strain and plated out on agar plates containing antibiotic to select for plasmid uptake. Once colonies of plasmid-harboring *E. coli* appear, cells are collected and plasmid DNA is prepared. Again, this can result in a large increase in the quantity of library DNA. Alternatively, the transformed cells themselves can be scraped from the agar plates and stored for future use.

$$N = \frac{\ln(1-P)}{\ln(1-1/n)}$$

Where N = the number of clones required to achieve a
probability (P) of a given low abundance cDNA being
represented.

$1/n$ = the fraction of the mRNA population that is
represented by the given mRNA species

If $1/n$ = 1/100,000 and P = 0.99, then $N \cong 500,000$

Fig. 3. Probability calculation for a given mRNA being represented in a cDNA library.

7. The Importance of Library Size

A typical vertebrate cell contains between 10,000 and 30,000 different mRNA species. These mRNAs will not all be expressed at the same steady-state level. Some mRNAs will be abundant owing to active transcription and/or long half-life and other messenger RNAs will be present at only a few copies per cell. When constructing a cDNA library, the minimum number of clones that are required to achieve a given probability that a low-abundance cDNA will be present in the library can be calculated using the simple formula given in **Fig. 3** *(6)*. In practice, a figure of 10^6 independent clones can be used as a guideline for the minimum size or complexity of cDNA library that is useful for cloning medium to low abundance cDNAs. However, this figure must be increased for expression libraries where the cDNA must be in the correct reading frame and orientation in the vector DNA for a fusion protein to be expressed. The rarity of a particular mRNA my be made worse by the fact that it is often necessary to obtain mRNA from a tissue composed of several or many cell types, only some of which contain the mRNA of interest. Several methods of mRNA enrichment have been devised that reduce the number of cDNA clones that are needed to obtain a particular cDNA. The simplest way to do this is to start with a tissue or cell line where the mRNA of interest is expressed more abundantly, perhaps as judged by the abundance of the corresponding protein. However, this may not always be possible. Two of the original methods for enriching mRNA populations are size fractionation of mRNA and immunological purification of polysomes. The former is self-explanatory and the second uses antibodies to purify polysomes that are synthesizing the polypeptide of interest. This is a powerful technique, but cannot always be applied and has a number of drawbacks *(6)*. A different approach to enrichment uses subtractive-hybridization techniques to remove mRNAs that are common to two different cell types or tissues, leaving just those found in one of them. (Examples of this can be found in **refs. *13,14***).

8. Uses of cDNA Libraries

As stated at the beginning of this chapter, the objective of cDNA cloning is usually to isolate a cDNA copy of a particular mRNA in a pure state. It is frequently desired for example to isolate a cDNA corresponding to a particular protein that has previously been purified and characterized. Several routes can be taken to accomplish this goal. Short stretches of amino-acid sequence obtained from a purified protein can be used to derive the corresponding mRNA sequence by "reverse translation" (with some ambiguity owing to the degenerate nature of the genetic code) and hence to design oligonucleotide probes for cDNA library screening by hybridization (*see* **Subheading 9.**) or to generate a longer probe using the PCR to amplify the segment between two oligonucleotide primers. Alternatively, antibodies raised against a purified protein can be used to screen a cDNA expression library (λgt11 or λZAP for example) for recombinant phage expressing the protein of interest. In a variation to this approach, it is sometimes possible to identify clones expressing a given protein by virtue of its affinity for a specific ligand. This method has been used with considerable success to clone transcription factors from expression libraries using as a probe a double-stranded DNA segment containing DNA sequences recognized by the factor of interest *(15–17)*. Molecular biologists are also frequently interested in families of related genes in one organism or shared between several or many organisms and the objective of many cDNA cloning projects is to isolate cDNAs related to a given DNA probe sequence by DNA hybridization.

A second frequent aim of cDNA cloning is the identification of proteins that physically interact with a given known protein. λgt11 and λZAP libraries, for example, can be screened with radioactively labeled proteins to identify clones expressing proteins capable of interacting with the probe protein in vitro *(18)*. Other more specialized methods, such as the yeast two-hybrid system (*see* **ref. *19*** and **Subheading 11.**) have also been devised to identify protein-interaction partners.

A different approach is embodied in the method of complementation cloning. In this case, it is a genetic phenotype endowed by a cDNA clone that leads to its isolation. Usually, the expressed cDNA replaces or complements the function of a mutated gene often using yeast as a host system. For example, the mammalian-transcription factor CP1 which is a homolog of the yeast *hap2* gene was cloned by complementation of a yeast *hap2* mutant *(20)*. *hap2* mutants are viable, but suffer from a respiratory deficiency. When *hap2* mutant yeast were transformed with a plasmid-expression library, transformants containing a clone capable of functional replacing *hap2* were selected by their ability to use a nonfermentable carbon source such as lactose.

Another possible objective of a cDNA cloning project is identification of genes expressed differentially between two tissues or cell types or whose expression is modulated by agents such as drugs or hormones. Several approaches exist to solve this problem, but recent advances have made it easier to select mRNA species present in only one of two RNA populations by depleting mRNAs common to both using sophisticated hybridization techniques *(14)*. The resulting subtracted mRNA, once converted to cDNA, can be used to screen standard cDNA libraries or to make subtracted cDNA libraries of differentially expressed clones.

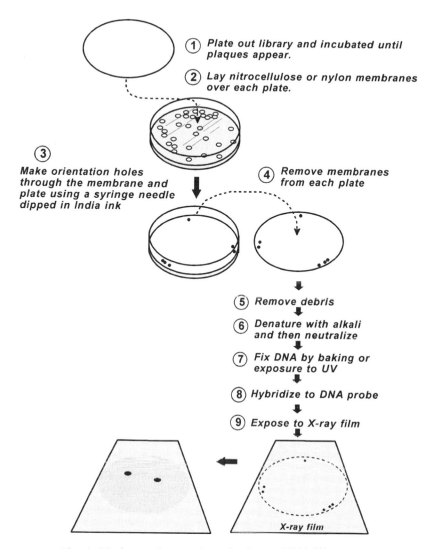

① Plate out library and incubated until plaques appear.

② Lay nitrocellulose or nylon membranes over each plate.

③ Make orientation holes through the membrane and plate using a syringe needle dipped in India ink

④ Remove membranes from each plate

⑤ Remove debris

⑥ Denature with alkali and then neutralize

⑦ Fix DNA by baking or exposure to UV

⑧ Hybridize to DNA probe

⑨ Expose to X-ray film

X-ray film

Fig. 4. Plating and screening a λ-phage cDNA library.

9. Screening Libraries by Hybridization to a DNA Probe

Probably the most common screening method is the detection of specific cDNA clones by hybridization to a DNA probe *(6,21)*. Probes may be short oligonucleotides or longer cDNA or genomic DNA fragments, but in each case, the basis of the screening protocol is as depicted in **Fig. 4**. For a bacteriophage library, the library is plated on a number of agar plates (typically 10 150-mm diameter plates), and when plaques appear replicas are made by laying a nitrocellulose or nylon filter on each plate. Filters and agar plates are marked for later orientation. The filters, which have now adsorbed-phage particles, are removed and phage DNA is denatured with alkali to generate single strands for later hybridization with radiolabeled probe *(6,22)*. After neutralization, phage DNA is fixed to the filters by baking or exposure to UV radiation. The filters are subsequently incubated in a solution containing single-stranded ^{32}P-labeled probe DNA. The probe hybridizes only to phage DNA containing the cDNA of interest and this is detected by autoradiography (**Fig. 4 , step 9**). Dark spots on the autoradiographs are aligned with the plaques on the agar plates and phage from the positive plaque can be isolated. Plasmid libraries can be screened in a similar way by replica plating and lysing bacterial colonies *in situ* with alkali *(6)*.

10. Screening Expression Libraries with Antibody or Other Probes *(8,16)*

Cloning vectors such as λ-gt11 or λZAP are designed to allow immunological screening. The methodology for this follows **steps 1–3** of **Fig. 4**, but the nitrocellulose filters are first soaked in the inducer IPTG, which switches on expression of the β-galactosidase fusion proteins (*see* **Subheadings 4.1.** and **4.2.**). Filters are left in contact with the agar to induce fusion protein expression and for protein to bind to the filter. Once filters are removed, the replica filters are treated with antibody for the detection of a specific fusion protein. Once the position of positively interacting plaques is identified, those plaques can be picked and phage isolated.

Cloning sequence-specific DNA binding proteins, such as transcription factors with ^{32}P-labeled DNA binding site probes, follows the same rationale as immunological screening. Filter lifts are incubated in a solution containing the radiolabeled probe, washed, and subjected to autoradiography as indicated in the bottom part of **Fig. 4**. Recombinant clones expressing a fusion protein that binds to the labeled probe appear as black spots on the autorad.

Filter lifts from the expression library may be probed in essentially the same way with radiolabeled protein probes *(18)* or other ligands to identify clones encoding binding proteins for those ligands.

11. The Yeast Two-Hybrid System

A frequent goal in molecular biology or biochemistry is the identification of proteins that interact with another known protein. Various biochemical approaches exist to do this, but the two-hybrid system *(19,23,24)* is a powerful method that simultaneously identifies interacting proteins in vivo and provides their corresponding cDNA clones. Typically, a plasmid construct is prepared that allows the "bait" protein for which interacting proteins are sought to be expressed in yeast as a fusion protein, with the DNA-binding domain of the yeast transcription factor GAL4. This fusion protein can bind promoter DNA, but does not function as a transcription factor as it lacks a necessary activation domain. Yeast cells expressing this fusion protein are transformed with a plasmid cDNA library in which the cDNA inserts have the potential to be expressed as fusion proteins with the transcriptional-activation domain of the GAL4 transcription factor. If the protein encoded by a cDNA insert binds to the "bait" protein, a functional transcription factor is formed (*see* **Fig. 5**) and this is detected by the activation of a GAL4-induced reporter gene. The specialized yeast strains employed for two-hybrid screening contain a chromosomal *GAL4*-activatable *HIS3* reporter gene, which allows the cells to grow on media lacking histidine only if they contain *GAL4* activity (*see* **Fig. 5**). Thus, when the transformed yeast are plated on media lacking histidine, only cells containing a plasmid encoding an activation domain-fusion protein that can interact with the bait protein can grow. In practice, a significant number of

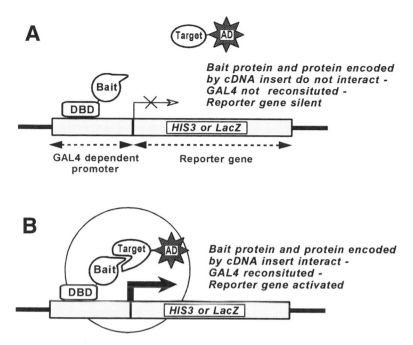

Fig. 5. The principle behind the yeast two-hybrid system. **(A)** In the majority of plated cells the plasmid-encoded GAL4 DNA binding domain (DBD) and activation domain (AD) fusion proteins fail to interact with each other, the *HIS3* reporter gene is silent and cells fail to grow on plates lacking histidine. **(B)** Some cDNA clones will encode a protein that can bind the bait protein. Interaction of the two fusion proteins reconstitutes an active transcription factor. The *HIS3* reporter gene is activated and cells form a colony on plates lacking histidine.

false positives also occur and the yeast strains used usually contain a second *GAL4*-activated reporter gene *(lacZ)* whose product can easily be assayed enzymatically to eliminate some of the spurious positives (*see* **Fig. 5**). This method and the strains and plasmid vectors that are used are discussed in more detail elsewhere *(19,25)*. Activation-domain plasmids are isolated from yeast positive for both reporters for further analysis.

12. What Next?

Once clones have been identified and purified the aims of the cloning project define what happens next. cDNAs are usually excised from λ phage clones and subcloned into a convenient plasmid vector (*see* **Subheading 4.2.** on in vivo excision from λZAP vectors) to facilitate analyses such as DNA sequencing or expression of the peptide sequence encoded by the cDNA. The nucleotide sequence of a newly isolated cDNA clone can be compared to databases of cDNA or gene sequences using PC-based or on-line software.

References

1. Gable, U. and Hoffman, B. J. (1983) A simple and very efficient method for generating cDNA libraries. *Gene* **25,** 263.
2. Snead, M. A., Alting-Mees, M. A., and Short, J. M. (1997) cDNA library construction for the lambda ZAP-based vectors, in *Methods in Molecular Biology, vol. 69: cDNA Library Protocols* (Cowell, I. G. and Austin, C. A., eds.), Humana, Totowa, NJ, pp. 39–51.

3. Lambert, K. N. and Williamson, V. M. (1993) cDNA library construction from small amounts of mRNA using paramagnetic beads and PCR. *Nucleic Acids Res.* **21**, 775,776.

4. Rothstein, J. L., Johson, D., Jessee, J., Skowronski, J., DeLoia, J. A., Solter, D., and Knowles, B. B. (1993) Construction of primary and subtracted cDNA libraries from early embryos. *Methods. Enzymol.* **225**, 587–610.

5. Ptashne, M. (1986) *A Genetic Switch: Gene Control and Phage* λ. Cell Press and Blackwell Scientific Publishing, Cambridge, UK, pp. 1–128.

6. Sambrook, J., Frisch, E. F., and Maniatis, T. (1989) *Molecular Cloning: A Laboratory Manual.* Cold Spring Harbor Laboratory, Cold Spring Harbor, NY.

7. Huynn, T. V., Young, R. W., and Davis, R. W. (1988) Construction and screening of cDNA libraries in lambda gt11 and lambda gt10, in *Cloning: A Practical Approach,* Vol I. (Glover, D. M., ed.), IRL, Oxford, UK, pp. 49–78.

8. Young, R. A. and Davis, R. W. (1983) Efficient isolation of genes by using antibody probes. *Proc. Natl. Acad. Sci. USA* **80**, 1194–1199.

9. Young, R. A. and Davis, R. W. (1983) Yeast RNA polymerase II gene: isolation with antibody probes. *Science* **222**, 778.

10. Short, J. M., Fernandez, J. M., Sorge, J. A., and Huse, W. D. (1988) Lambda ZAP: a bacteriophage lambda expression vector with *in vivo* excision properties. *Nucleic Acids Res.* **16**, 7583.

11. Durfee, T., Becherer, K., Chen, P. L., Yeh, S. H., Yang, Y., Kilburn, A. E., Lee, W. H., and Elledge, S. J. (1993) The retinoblastoma protein associates with the protein phosphatase type 1 catalytic subunit. *Genes Dev.* **7**, 555–569.

12. Hanahan, D., Jessee, J., and Bloom, F. R. (1991) Plasmid transformation in Escherichia coli and other bacteria. *Methods. Enzymol.* **204**, 63–113.

13. Swaroop, A., Xu, J., Agarwal, N., and Weissman, S. M. (1991) A simple and efficient cDNA library procedure: isolation of human retin-specific cDNA clones. *Nucleic Acids Res.* **19**, 1954.

14. Aasheim, H. C., Deggerdal, A., Smeland, E. B., and Hornes, E. (1994) A simple subtraction method for the isolation of cell-specific genes using magnetic monodisperse polymer particles. *BioTechniques* **16**, 716–721.

15. Singh, H., LeBowitz, J. H., Baldwin, A. S., and Sharp, P. A. (1988) Molecular cloning of an enhancer binding protein: isolation by screening of an expression library with a recognition site DNA. *Cell* **52**, 415–423.

16. Cowell, I. G. and Hurst, H. C. (1993) Cloning transcription factors from a cDNA expression library, in *Transcription Factors: A Practical Approach* (Latchman, D. S., ed.), IRL, Oxford, UK, pp. 105–123.

17. Cowell, I. G. and Hurst, H. C. (1994) Cloning DNA binding proteins from cDNA expression libraries using oligonucleotide binding site probes, in *Methods in Molecular Biology, vol. 31: Protocols for Gene Analysis* (Harwood, A. J., ed.), Humana, Totowa, NJ, pp. 363–370.

18. Takayama, S., Sato, T., Krajewski, S., Kochel, K., Irie, S., Millan, J. A., and Reed, J. C. (1995) Cloning and functional analysis of BAG-1: a novel Bcl-2 binding protein with anti-cell death activity. *Cell* **80**, 279–284.

19. Bartel, P. L., Chien, C. T., Sternglanz, R., and Fields, S. (1993) Using the two hybrid system to detect protein-protein interactions, in *Cellular Interactions in Development: A Practical Approach* (Hartley, D. A., ed.), Oxford University Press, Oxford, UK, pp. 153–179.

20. Becker, D. M., Fikes, J. D., and Guarente, L. (1991) A cDNA encoding a human CCAAT-binding protein cloned by functional complementation in yeast. *Proc. Natl. Acad. Sci. USA* **88**, 1968–1972.

21. Austin, C. A. (1997) Article title, in *Methods in Molecular Biology, vol. 69: cDNA Library Protocols* (Cowell, I. G. and Austin, C. A., eds.), Humana, Totowa, NJ, pp. 147–154.

22. Benton, W. D. and Davis, R. W. (1977) Screening lambda gt recombinant clones by hybridization to single plaques *in situ. Science* **196,** 180.

23. Fields, S. and Song, O. (1989) A novel genetic system to detect protein-protein interactions. *Nature* **340,** 245,246.

24. Chien, C. T., Bartel, P. L., Sternglanz, R., and Fields, S. (1991) The two-hybrid system: a method to identify and clone genes for proteins that interact with a protein of interest. *Proc. Natl. Acad. Sci. USA* **88,** 9578–9582.

25. Cowell, I. G. (1997) Yeast two-hybrid library screening, in *Methods in Molecular Biology, vol. 69: cDNA Library Protocols* (Cowell, I. G. and Austin, C. A., eds.), Humana, Totowa, NJ, pp. 185–202.

14

Genomic DNA Libraries

Bimal D. M. Theophilus

1. Theory and Techniques of Library Construction

Genomic DNA libraries are a collection of DNA fragments that together represent the entire (or nearly entire) genome of the individual from which the DNA was derived. These fragments are contained within self-replicating vectors that enable them to be maintained and propagated within the cells of microorganisms, such as *Escherichia coli* or *Saccharomyces cerevisiae* (yeast). Such libraries may be stored as a permanent source of representative sequences of a particular organism. Individual laboratories may obtain such a library from another laboratory or a commercial source as an alternative to constructing their own.

Once obtained, libraries are used predominantly either for screening by one of a variety of methods to isolate a particular sequence(s) of interest, or to determine the location and order of sequences in the genome from which the library is constructed ("physical mapping").

Construction of a genomic DNA library comprises four main stages (**Fig. 1**): DNA digestion, ligation into a vector, packaging of the recombinant clones into a virus protein particle to mediate host-cell entry, or pre-treatment of host cells to render them competent, and introduction of the construct into a host cell.

2. DNA Digestion

DNA is first extracted from an easily accessible source, such as blood cells. The DNA is usually digested with a restriction enzyme that has a 4-bp recognition sequence, such as MboI, which recognizes 5'-GATC-3'; this would be expected to occur every few hundred bases in any random DNA sequence. A partial digest is carried out by using a limited amount of enzyme and incubating for a short time period. This produces a series of overlapping fragments cleaved at only a proportion of the potential cleavage sites. The average fragment size is controlled to produce fragments of average size appropriate for the particular cloning vector to be used (*see* **Subheading 3.**). The DNA is then fractionated and only fragments within the desired size range selected.

In addition to genomic libraries derived from a complete genome, subgenomic libraries may also be constructed that represent individual chromosomes or regions of chromosomes. This is possible by use of flow cytometry to isolate individual human chromosomes *(1)* or chromosome microdissection *(2)*.

From: *Molecular Biomethods Handbook*
Edited by: R. Rapley and J. M. Walker © Humana Press Inc., Totowa, NJ

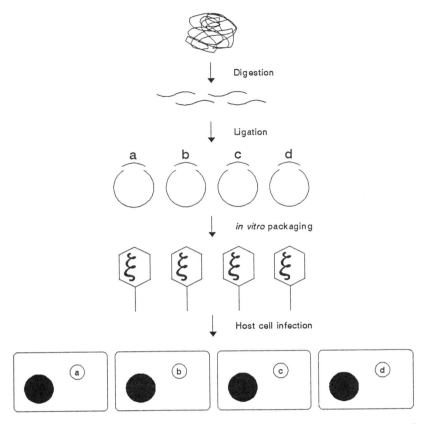

Fig. 1. Construction of a genomic DNA library. Stages in the construction of a genomic DNA library in a virus-based vector. a, b, c, and d represent different but overlapping segments of sequence from a genome, which are incorporated into separate vectors and represented as discrete clones in the final library. *See text* for details and vector-specific variations.

3. Cloning DNA Fragments

DNA fragments resulting from restriction enzyme digestion are covalently ligated in vitro into a cloning vector using the enzyme DNA ligase. Prior to ligation, the vector is digested at a single site, normally with the same enzyme used to digest the source DNA, to produce "cohesive ends," which are complementary to those of the fragments to be inserted. Sometimes the vector and source DNA are digested with a combination of two enzymes to prevent each of the molecular species from self-ligation. Recircularization of the vector may be further minimized by treatment with alkaline phosphatase to remove the 5' phosphate groups from the vector ends.

A large number of cloning vectors have been developed for the construction of DNA libraries, the choice of which depends on the size range of fragments to be cloned, and the subsequent applications.

3.1. Plasmids

Plasmids are covalently closed, circular-supercoiled molecules of a few kilobases in size (*see* Chapter 16). Although plasmids occur naturally, mainly in bacteria, those employed in the construction of DNA libraries are usually purposely designed synthetic constructs. These cloning vectors contain restriction sites for which a given enzyme may have only a single unique site on a plasmid into which foreign DNA may be inserted. The sites for many different enzymes may be clustered in a "multiple-cloning site" or "polylinker."

Plasmids additionally contain one or more "marker" genes, such as genes conferring antibiotic resistance, which enable the growth of only plasmid-containing host cells, and genes for an enzyme, such as β-galactosidase, which enable insert-containing recombinant molecules to be distinguished from nonrecombinants (*see* **Subheading 5.**).

The disadvantages of plasmids as cloning vectors are that their capacity is limited to small fragments of only a few kilobases in size and they transform host cells only at a relatively low frequency.

3.2. Virus-Based Cloning Vectors

Bacteriophages are viruses that infect bacteria. Their genomes may be circular or linear double-stranded DNA, and are contained within a protein coat that mediates their entry into a host cell.

Bacteriophage λ is one of the most popular virus-based cloning vectors used in the construction of libraries. Its genome comprises a 50-kb linear double-stranded DNA molecule that infects *E. coli* within which it then adopts a double-stranded circular conformation. During replication in the "lytic" cycle, concatamers of λ genomes are produced that are subsequently cleaved at specific sequences (COS sites) to produce λ monomers that are packaged into preformed protein heads. As an alternative to the lytic cycle, λ may enter a "lysogenic" state during which it stably integrates into the *E. coli* host chromosome. The genes required for lysogenic function are clustered into a functional group in the center of the phage, and have been removed in the construction of a wide range of λ-derived cloning vectors that can accept up to 20 kb of insert DNA (*see* Chapter 15). λ vectors are normally obtained as left and right "arms" to which insert DNA is ligated. The resulting recombinants are then packaged into a protein coat using an in vitro packaging system, which enables high efficiency infection of a host cell.

For a typical mammalian genome of 3×10^9 bp, approx 7×10^5 λ recombinants of an average size of 20 kb will ensure with 99% probability that any given DNA sequence will be present in the library.

Single-stranded bacteriophage vectors, such as M13, are also often used for library construction. They have circular genomes that replicate initially as a double-stranded molecule following bacterial infection, and subsequently produce single-stranded genomes for further cycles of infection (*see* Chapter 17).

3.3. Cosmids

Cosmids are plasmid vectors that contain the *cos* sites of λ *(3)*. Cosmids can therefore be packaged into λ particles at high efficiency. Following infection of a host cell, the linear-recombinant DNA circularizes at the *cos* sites and is then propagated as a large plasmid. Cosmids, like plasmids, are relatively easy to manipulate, but can accept genomic DNA inserts of ~40–45 kb in size. Although cosmids can accept larger fragments than phage λ, rearrangements of the inserted DNA may occur.

3.4. Yeast Artificial Chromosomes (YACs)

Yeast artificial chromosomes (YACs) are linear cloning vectors based on natural yeast chromosome structure *(4)*. They may be cleaved into two fragments or chromosome arms, which may accept very large fragments of up to 500–1000 kb. YAC constructs may then be taken up into yeast cells that have had their cell walls removed (spheroplasts) where they are maintained as stable, autonomous chromosomes.

YAC libraries have been assembled and employed for mapping large regions of human DNA. Because of their size, pulsed field gel electrophoresis (PFGE) is usually necessary for their analysis *(5)*.

The disadvantages of YACs include the relatively low cloning efficiency of some systems, the occurrence of chimeric clones (transformant cells containing two noncontiguous pieces of DNA), insert instability, and difficulty in their manipulation relative to bacterial systems.

3.5. P1-Derived Artificial Chromosomes (PACs)

Many of the disadvantages of YAC as well as cosmid cloning may be overcome by bacteriophage P1-based vectors *(6,7)*. P1-plasmid vectors are able to package exogenous DNA inserts of 100–300 kb into the P1-protein coat in vitro, followed by adsorption and injection into bacterial cells.

3.6. Bacterial Artificial Chromosomes (BACs)

The bacterial artificial chromosome system is based on the F-plasmid factor, which can replicate in *E. coli* with inserts of >300 kb *(8)*. BAC recombinants are introduced into bacterial cells by electroporation, in which membrane permeability is increased by the application of a high voltage to the cells. BACs are useful for library construction, mapping, and genome analysis. They have a high cloning efficiency, and inserts are easily manipulated and remain stable.

4. Introduction of Foreign DNA into Cells

Once the genomic DNA has been fragmented and ligated into a vector, the resulting clones can be isolated by introduction into cells, where they are amplified as a part of cell growth and division. This also enables individual clones to be screened to detect and analyze inserted sequences. Although some transformation systems may be prone to cotransformation of more than one molecular species (*see* **Subheading 3.**), most cells normally take up only a single recombinant molecule, which is then amplified to produce large numbers of identical clones within a single colony or plaque.

Plasmid vectors require that the host cell be made "competent" to accept vector constructs by direct transformation with the ligated DNA. This involves treating host cells with divalent cations, such as Ca^{2+} or Mg^{2+}. Alternatively, a higher efficiency of transformation may be achieved by electroporation.

Bacteriophage and cosmid vectors are introduced into host cells by in vitro packaging into a protein coat, which mediates adsorption and injection of the enclosed genetic material. Pretreatment of the recipient cell is therefore not necessary.

In Gram-positive bacteria, and also in plants, the outer cell wall may be removed to leave protoplast, which is able to take up DNA and then is regenerated in appropriate media.

Animal cells will accept DNA that has been first precipitated with Ca_2, or by making the membrane permeable with divalent cations or high-mol-weight polymers, such as diethylaminoethyl (DEAE) dextran or polyethylene glycol (PEG). DNA may also be introduced into animal cells by electroporation or mediated by liposomes. Microinjection has been use to introduce DNA into cell nuclei of *Drosophila melanogaster, Xenopus laevis*, and mice.

5. Screening Libraries

DNA cloned into plasmids or YACs produces colonies when the transformed cultures are spread onto an agar plate of growth medium and incubated under appropriate conditions. Bacteriophages lyse the cells they infect and so produce "plaques:" circular (2–3 mm) zones of clearing on a background bacterial lawn.

Vectors, such as those described above, normally contain a genetic marker that enables selective growth of only those host cells that have taken up the vector sequence. Frequently this involves a gene for antibiotic resistance and growth of transformed cells on media containing the corresponding antibiotic.

Often, vectors contain additional genes to discriminate transformed cells containing inserts of foreign DNA from those that simply contain recircularized vector. For example, vectors may encode the β-galactosidase gene, which catalyzes generation of a blue product from a colorless substrate (Xgal), resulting in the growth of blue colonies. Insertion of a DNA fragment inactivates the gene, resulting in production of white colonies (*see* Chapter 16 for a complete description).

Another method enabling selection of transformants containing inserts involves vectors carrying a gene for a suppressor tRNA, which results in the incorporation of an amino acid at the site of a termination codon. When transformed into a host cell that contains a mutation to a termination codon in a gene with an easily scored phenotype (e.g., colored colonies), the suppressor gene restores expression of the host-gene phenotype.

A clone containing the particular sequence of interest may be identified by colony hybridization. A small amount of the transformant colony or plaque is transferred onto a nitrocellulose or nylon filter, which has been overlaid onto the agar plate. The DNA is denatured and fixed onto the filter by baking, and then hybridized in a buffer containing a radioactively labeled probe comprising sequence that is complementary to a part of the sequence to be identified. For example, this may be a synthetic oligonucleotide probe, one derived from partial genomic DNA, cDNA, or protein sequence, or a PCR product. Sometimes the probe may be based on a sequence derived from a homologous gene of another species and hybridized at low stringency with the expectation of sufficient sequence conservation to enable cross-hybridization. Excess probe is washed away and the filter exposed to X-ray film. By orientating the developed film with respect to the original agar plate, it is then possible to match actual colonies with corresponding positively hybridizing colonies on the X-ray film.

Recently a format of gridded arrays has been used as an alternative method of containing and screening libraries. In this approach, the bacterial or yeast colonies are individually picked and spotted in arrays onto membranes or into the wells of microtiter dishes. Each clone can be identified by its individual grid coordinate. In comparison to the traditional methods, this approach lends itself more easily to automation, the distribution of libraries, and the collation of data between different laboratories.

If the desired clone expresses a particular protein or protein fragment, then by constructing the library with an expression vector that contains transcription and translation signals it may be possible to screen directly for the expressed protein. For example, if sufficient corresponding protein is available, specific antibodies may be generated, labeled, and used to detect antigenic determinants on the expressed protein from lysed colonies. If the required insert encodes a biological activity, then screening may be achieved by an assay for that activity, or by direct selection, e.g., selection on incomplete growth media for a biosynthetic enzyme necessary for growth. However, expression-based approaches are generally more applicable to cDNA libraries than genomic libraries (*see* Chapter 15).

6. Applications of Genomic Libraries

Genomic DNA libraries have found extensive application in the physical mapping of DNA and in the identification of disease genes or other DNA sequences of interest for further analysis.

The production of clones with different but overlapping inserts has the advantage that the resulting library may be used in a "chromosome walk." This procedure enables progression from a starting point on a chromosome (e.g., a disease-associated marker sequence) to a nearby unidentified locus (e.g., the gene responsible for the disease itself) by carrying out repeated cycles of clone isolation, characterization, and hybridization using the clone as a probe to identify further clones within the library containing adjacent sequence from the original source DNA (*see* Chapter 23). Chromosome walking is usually carried out with cosmid, λ, or YAC recombinant libraries.

Chromosome walking also enables the construction of a "clone contig"–an ordered series of overlapping DNA clones that together represent a contiguous region of genomic DNA. Clone-contig construction often forms an essential part of the analysis of DNA sequences generated during strategies for the physical mapping of DNA and the identification of disease genes by positional cloning. Clone contigs have been constructed using a variety of vectors, including λ, P1, PACs, BACs, and YACs. Although chromosome walking using whole clones as probes is problematic with mammalian YAC clones because of the large amount of repetitive DNA they contain, it is possible to perform a chromosome walk if the probes are based on short-end fragments recovered from individual YACs. Alternatively, clones may be ordered into a contig by clone fingerprinting, in which overlapping inserts are identified by the pattern of repetitive sequences in their inserts *(9)*. YAC contigs have been particularly useful in the Human Genome Project (*see* Chapter 24) because of the large insert size that they are able to accommodate

The identification of the cystic fibrosis gene illustrates the use of genomic DNA libraries for gene identification by positional cloning, utilizing only knowledge of the gene's subchromosomal localization *(10–12)*. In addition to chromosome walking, this strategy involved the construction of a special type of DNA library to isolate sequences by chromosome jumping *(13,14*; *see* Chapter 23). Chromosome-jumping libraries contain inserts that comprise two sequences, ligated together, that may have been located hundreds of kilobases from each other in the original genome. This is achieved by partial digestion of genomic DNA to produce large fragments. The DNA is diluted to a very low concentration and then ligated. This results in intramolecular ligation, whereby each molecule is more likely to self ligate (circularize) by ligating its two free ends together, rather then being ligated to a separate molecule. The ligation is carried out in the presence of a marker gene, such as a suppressor tRNA gene, which enables selection for constructs containing the marker gene (*see* **Subheading 5.**). Digestion is then carried out with an enzyme that lacks sites in the marker gene followed by cloning into a λ vector to produce a library. Screening such a library with one of the two original sequences enables isolation of adjacent sequence that was originally located hundreds of kilobases away. Carrying out multiple rounds of clone isolation and hybridization enables progression along a chromosome in relatively large "jumps," which speeds up the mapping process.

Genomic DNA libraries are also utilized in the identification of disease genes by functional cloning. In this approach, information about the function of the gene is exploited to isolate the desired gene from a library. An oligonucleotide (or a mixture of "degenerate" oligonucleotides that correspond to the various codon permutations) whose sequence is based on partial amino-acid sequence is used as a probe to isolate a cDNA clone by screening a cDNA library (*see* Chapter 13). The cDNA clone may then be used to screen a genomic library to isolate genomic DNA clones and enable characterization of the full genomic sequence.

This approach has been used to identify the hemophilia A (factor VIII) gene *(15)*. Oligonucleotide probes based on amino-acid sequence of porcine factor VIII protein were initially used to isolate a porcine factor VIII genomic clone, which was then used as a probe against human DNA libraries to identify the human gene.

7. Summary

Genomic DNA libraries have been utilized both for analyzing individual genes and for the physical mapping of large regions of unknown sequence. The latter function has found particular application in the Human Genome Project, which has in turn acted as a catalyst for the development of improved cloning vectors and methods of library construction, further facilitating the efficient analysis of characterized and unknown sequences in the genomes of humans and other species.

Suggested Readings

Old, R. W. and Primrose, S. B. (1994) *Principles of Gene Manipulation: An Introduction to Genetic Engineering*, 5th ed. Blackwell Scientific Publications, Oxford, UK.

Sambrook, J. Fritsch, E. F., and Maniatis, T. (1989) *Molecular Cloning: A Laboratory Manual*, 2nd ed. Cold Spring Harbor Laboratory, Cold Spring Harbor, NY.

Strachan, T. and Read, A. P. (1996) *Human Molecular Genetics*. Bios Scientific Publishers, Oxford, UK.

References

1. Bartholdi, M., Meyne, J., Albright, K., Luedemann, M., Campbell, E., Chritton, D., Deaven, L. L., and Scott Cram, L. (1987) Chromosome sorting by flow cytometry. *Meth. Enzymol* **151,** 252–267.

2. Edstrom, J. E., Kaiser, R., and Rohme, D. (1987) Microcloning of mammalian metaphase chromosomes. *Meth. Enzymol.* **151,** 503–516.

3. Collins, J. and Hohn, B. (1978) A type of plasmid gene-cloning vector that is packageable *in vitro* in bacteriophage λ heads. *Proc. Natl. Acad. Sci. USA* **75,** 4242–4246.

4. Burke, D. T., Carle, G. F., and Olsen, M. V. (1987) Cloning of large segments of exogenous DNA into yeast by means of artificial chromosome vectors. *Science* **236,** 806–812.

5. Schwartz, D. C. and Cantor, C. R. (1984) Separation of yeast chromosomal sized DNAs by pulsed field gradient gel electrophoresis. *Cell* **37,** 67–75.

6. Sternberg, N. (1990) Bacteriophage P1 cloning system for the isolation, amplification and recovery of DNA fragments as large as 100 kilobase pairs. *Proc. Natl. Acad. Sci. USA* **87,** 103–107.

7. Ioannou, P. A., Amemiya, C. T., Games, J., Kroisel, P. M., Shizuya, S., Chen, C., Batzer, M. A., and de Jong, P. J. (1994) A new bacteriophage P1-derived vector for the propagation of large human DNA fragments. *Nat. Genet.* **6,** 84–89.

8. Shizuya, H. Birren, B., Kim, U.-J., Mancino, V., Slepak, T., Tachliri, Y., and Simon, M. (1992) Cloning and stable maintenance of 300-kilobase-pair fragments of human DNA in *Escherichia coli* using an F-factor-based vector. *Proc. Natl. Acad. Sci. USA* **89,** 8794–8797.

9. Bellanne-Chantelot, C., Lacroix, B., Ougen, P., Billault, A., Beaufils, S., Bertrand, S., Georges, I., Gilbert, F., Gros, I., Lucotte, G., Susini, L., Codani, J.-J., Gesnouin, P., Pook, S., Vaysseix, G., Lu-Kuo, J., Ried, T., Ward, D., Chumakov, I., Le Paslier, D., Barillot, E., and Cohen, D. (1992) Mapping the whole human genome by fingerprinting yeast artificial chromosomes. *Cell* **70,** 1059–1068.

10. Rommens, J. M., Iannuzzi, M. C., Kerem, B.-S., Drumm, M. L., Melmer, G., Dean, M., Rozmahel, R., Cole, J. L., Kennedy, D., Hidaka, N., Zsiga, M., Buchwald, M., Riordan, J. R., Tsui, L.-C., and Collins, F. S. (1989) Identification of the cystic fibrosis gene: chromosome walking and jumping. *Science* **245,** 1059–1065.

11. Riordan, J. R., Rommens, J. M., Kerem, B.-S., Alon, N., Rozmahel, R., Zbyszko, G., Zielenski, J., Lok, S., Plavsik, N., Chou, J.-L., Drumm, M. L., Iannuzzi, M. C., Collins, F. S., and Tsui, L.-C. (1989) Identification of the cystic fibrosis gene: cloning and characterization of complementary DNA. *Science* **245,** 1066–1073.

12. Kerem, B.-S., Rommens, J. M., Buchanan, J. A., Markiewicz, D., Cox, T. K., Chakravarti, A., Buchwald, M., and Tsui, L.-C. (1989) Identification of the cystic fibrosis gene: genetic analysis. *Science* **245,** 1073–1080.

13. Collins, F. S. and Weissman, S. M. (1984) Directional cloning of DNA fragments at a large distance from an initial probe: a circularization method. *Proc. Natl. Acad. Sci. USA* **81,** 6812–6816.

14. Poustka, A. and Lehrach, H. (1986) Jumping libraries and linking libraries: the next generation of molecular tools in mammalian genetics. *Trends Genet.* **2,** 174–179.

15. Gitschier, J., Wood, W. I., Goralka, T. M., Wion, K. L., Chen, E. Y., Eaton, D. H., Vehar, G. A., Capon, D. J., and Lawn, R. M. (1984) Characterization of the human factor VIII gene. *Nature* **312,** 326–330.

15

λ *as a Cloning Vector*

Rupert Mutzel

1. Introduction

Genes can be identified and cloned by complementation of mutants, hybridization of nucleic acids, screening for their products, or systematic sequencing of genetic entities. The choice of the strategy to be applied for isolating a gene of interest has to be taken in a space of parameters including, for example, the availability of mutants and the stringency and specificity of selection procedures, the homology of nucleic-acid probes, the possibility to unambiguously identify an encoded function, the abundancy of the desired gene or its product with respect to the background (in other words, the size of the haystack in which to search for the needle), but also the effort and the amount of money that have to be spent to accomplish the task. Evaluation of these parameters will then decide on the tools required, e.g., DNA probes, antibodies, or functional assays, and help to determine the vector/host system and type of library to be used.

The phage λ/*Escherichia coli* system has been one of the work horses in the field of gene cloning, and its applications reach from straightforward complementation of *E. coli* mutants to the systematic determination of coding DNA sequences. It is the aim of this chapter to briefly introduce some of the derivatives of λ as cloning vectors and to summarize examples for applications. There is now a wide variety of λ vectors for almost any possible application, and examples for most of them are legion; therefore, this review can only cover selected examples.

Since their first use as cloning vehicles more than 20 yr ago *(1)*, λ vectors and their bacterial hosts have been constantly improved, and now comprise a broad family of general or specialized vectors with features such as multiple-cloning sites, automatic subcloning of cDNA inserts for further manipulation, the capacity to accommodate large DNA fragments, or the possibility to produce recombinant proteins.

Although it would at first seem paradoxical to use a large, approx 50 kb vector as a cloning vehicle, λ phages became a very successful, versatile, and robust tool in molecular biology. Backed by a profound knowledge of the biology of the phage and its host, the in vivo and in vitro engineering of restriction sites *(2,3)* and the development of reliable in vitro packaging techniques *(4)* allowed the advantage of both in vitro cloning technology and the very high efficiency with which the phage can infect

From: *Molecular Biomethods Handbook*
Edited by: R. Rapley and J. M. Walker © Humana Press Inc., Totowa, NJ

its host. The inconvenience that preparing either genomic or cDNA libraries in λ is still more demanding than preparation of plasmid libraries is overruled by several advantages of the phage system. First, even with the most sophisticated electroporation techniques, the efficiency of plasmid transformation (10^8/μg of DNA) is still well below the efficiency of infection with λ phages (up to 10^{12}/μg of DNA), even though a λ phage carries 10- or 20-fold more nonrecombinant DNA than a typical plasmid vector. Second, phage libraries can be stored for many years without too high a drop in their titer (for example, we are still using with good success a λgt11 library prepared in 1985 *[5]*). Many λ vectors can be propagated both as phage particles and as lysogens, allowing to switch between a single-copy state during growth of the bacterial host and a highly amplified state during lytic growth. An additional important advantage for screening experiments is the high density at which phage plaques can be plated as compared to bacterial colonies. Whereas colonies will become confluent at densities above 10,000 per standard (9 cm) Petri dish, the same surface can accommodate 100,000 phage plaques, allowing screening of highly complex libraries, but also making the percentage of recombinant clones in a λ library less critical; in practice, this greatly reduces the amount of material and work that have to be invested in a screening experiment—screening 10 or 100 plates makes a huge difference.

A definite advance, in the early 1980s, was the construction of the λgt series of vectors, with λgt10 and λgt11 as the most widely used examples *(6)*. Cloning of DNA fragments in λgt10 takes advantage of the facts that a significant part of the λ genome is dispensable for propagation of the phage, and that insertion of foreign DNA into the *cI* repressor gene abolishes the phage's ability to become lysogenic. Thus, large DNA fragments can be inserted and packaged into infectious-phage particles, and recombinants can be selected by plating a library on an *E. coli hfl* host, where recombinant phages form clear plaques, whereas most wild-type λgt10 will enter the host genome as lysogens and become titrated out during an amplification step. Extensions of this strategy are λEMBL and λCharon vectors, which take advantage of "stuffer" fragments inserted into the vector for propagation of the nonrecombinant phage. During cloning of DNA fragments, these stuffers are replaced by the cloned DNA. Constructs lacking the stuffer or a cloned DNA fragment will be less efficiently packaged and are therefore counterselected. The use of stuffers consisting of amplified short DNA fragments containing one or several restriction sites makes removal of these small fragments very easy. Similarly, the Spi phenotype can been used to enrich for vectors carrying inserted DNA *(7)*.

In λgt11 the insertion of an *E. coli lacZ* gene with a unique *Eco*RI cloning site near its 3' end allows the expression of coding DNA fragments (if inserted in the proper orientation and reading frame) as fusion proteins with an N terminal part of β-galactosidase, opening an avenue for isolating DNA clones by their encoded function. Together with additional useful mutations (like an amber mutation in the *S* gene that allows phage products to be accumulated in nonlysed cell "carcasses") and a range of suitable host strains, λgt11 and its relatives have become a widely used system not only for the cloning of genes, but also for the expression of foreign proteins in *E. coli*. λgt10 and λgt11 may be considered the paradigms of the modern family of λ vectors, which now span a wide range of vehicles designed for a variety of special purposes like clon-

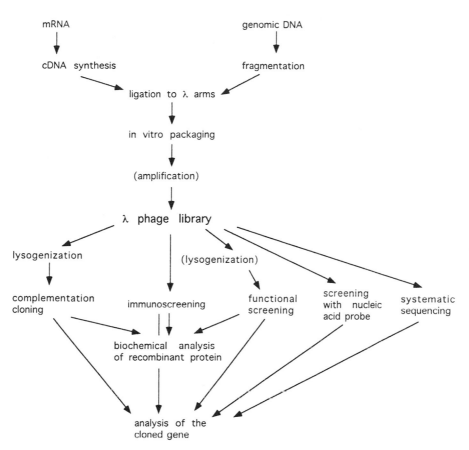

Fig. 1. Strategies for cloning and analysis of genes with λ vectors. mRNA (usually selected from total RNA by binding to an oligo dT matrix) is reverse-transcribed into cDNA *(35)*, genomic DNA is restriction-cut or mechanically trimmed into fragments of a size that can be accommodated by the chosen vector. After rendering the ends compatible to the restriction site for cloning, the fragments are ligated to λ arms and packaged in phage particles in vitro. The original library may be amplified and nonrecombinant phages (e.g., in λgt10 libraries) counterselected by infection of a suitable host (however, extensive amplification may result in counterselection of phages harboring large fragments or genes whose expression can be deleterious to the host). An original or amplified phage library can then be transformed into a bacterial library by lysogenization (**22**; note the possibility to eliminate parental cells either by selecting for an antibiotic resistance encoded by the phage, or by challenging the lysogens with a λ *cI* mutant that will be lytic for cells that do not contain a lysogen and therefore do not express a functional λ repressor *[28]*). Details for the various techniques of isolating genes from λ libraries are described in the text.

ing of DNA fragments as large as 20 kb (λ EMBL and λCharon vectors) to rapid, automatic subcloning of DNA inserts (e.g., λZAP *[8]*) and to λYES-type shuttle vectors that allow to complement mutations of bacteria and yeast cells with a single cDNA library **Subheading 3.** (*See* **ref. 4** for a concise overview on λ vectors and suitable *E. coli* hosts). Applications of the use of λ vectors will be discussed later. **Figure 1** gives a schematic overview of these applications.

2. Systematic Sequencing of cDNAs

Systematic sequencing of expressed sequence tags (ESTs) from several eukaryotic organisms is currently under way (*see* **ref.** *9* for a list of examples); the vector of choice for preparing and analyzing the highly complex cDNA libraries required is λZAP, which combines the high efficiency of cDNA cloning and propagation in λ vectors with the convenience with which cDNA inserts in the bluescript-plasmid vector can be automatically subcloned and sequenced. The most impressive example for the power of this approach is provided by Venter and his coworkers *(9)*, who have assembled a database of 83 million nucleotides of human cDNA sequence, generated from more than 300 cDNA libraries prepared from 37 distinct tissues and defining nearly 90,000 unique sequences. This approach is much more than just a shortcut to systematic sequencing of entire genomes, because it provides information not only about the number and nature of sequences present in a genome, but also about their expression pattern, tissue specificity and the abundancy of individual transcripts in a given cell type at a specific developmental stage.

3. Complementation of *E. coli* and Yeast Mutants

A number of examples of cloning of genes by complementation of *E. coli* and yeast mutants have been published. For *E. coli* mutants, the general strategy consists of introducing either the *hflA* mutation (which allows for efficient lysogenization of λ prophages) into the mutant of interest, or introducing the mutation of interest into a *hflA* strain, e.g., Y1089 *(6)*, suitable for further analysis of recombinant phages. Bacteria are then lysogenized with the λ-phage library, and complementing phages identified by plating the lysogen library on selective medium. Alternatively, λ vectors that allow in vivo excision of expression plasmids (e.g., λZAP) can be converted to plasmid-expression libraries *(10)*. Using these strategies, a cDNA for human holocarboxylase synthetase has been cloned by complementation of an *E. coli* biotin auxotroph *(11)* and cDNAs for enzymes involved in purine synthesis have been cloned from different organisms *(10,12)*. The approach of cloning GMP synthetase by complementation of the *E. coli guaA* mutation has been extended by Kessin's group to cGMP phosphodiesterase by supplying a *guaA* mutant lysogenized with a λgt11 cDNA expression library with cGMP, allowing *guaA* mutants to grow if they can convert cGMP to 5' GMP *(13)*.

λYES, a shuttle vector for use in complementation cloning both in *E. coli* and *Saccharomyces cerevisiae*, has been described by Elledge et al. *(14)*; using cDNA libraries prepared in this vector, it was possible to clone the human *cdc28* homolog by complementation of a *cdc28* mutation of *S. cerevisiae*, and enzymes involved in biosynthesis of amino acids in *Arabidopsis thaliana* by complementation of the corresponding *E. coli* mutations *(14)*.

The bottleneck in this type of cloning experiment is the need for very tight selection: based on the observation that a serine/threonine protein phosphatase from yeast could complement the *E. coli serB* mutation *(15)* (serine biosynthesis in this mutant is blocked at the level of serine phosphate *[16]*), we have tried to clone serine/threonine phosphatases from a *Dictyostelium* λgt11 expression library lysogenized in *E. coli* Y1089 *serB*. About 80 lysogens that grew on minimal medium without serine were identified in a total of 10^5 recombinant lysogens. However, none of the phages supported growth when isolated and relysogenized in Y1089 *serB*, suggesting that the mutation could be readily by-passed by chromosomal mutations of *E. coli* (Q. Husain, M. Ehrmann, and R. M., unpublished observations). Moreover, any corruption of a λ library with bacterial DNA will become apparent if the abundancy of complementing cDNA clones is low *(13)*. Such a corruption would probably not be seen in a screening experiment with a DNA or antibody probe both for the low abundancy of contaminations and their lower degree of homology to the probe or for lower antigenic crossreactivity.

Systematic sequencing of entire λ libraries and complementation cloning may be considered the extremes of how genes can be cloned from λ libraries: there is either no search for any specific property of a gene, or this property is so specific that it can be selected for by complementation of a mutation in a heterologous host. In any other case, the gene of interest has to be searched for by screening a highly complex population of recombinant phages. There are two basic strategies for this, screening plaques formed during lytic growth of the phage and screening lysogens, that is, bacterial cells which harbor recombinant λ prophages; and the gene of interest can be screened for either by homology with a nucleic-acid probe, or by any function of its encoded protein, e.g., binding of antibodies, its enzyme activity, or binding of a ligand.

A typical screening experiment consists of infecting a suitable bacterial host with the λ library, plating on an agar surface at the highest acceptable density, and assaying the phage plaques or lysogen colonies after transfer to a filter support for the presence of the desired gene or its product; the original agar plate is saved as a source of viable phage or lysogens, which are isolated after identification of positive signals by autoradiography or with a chromogenic or chemiluminescent reaction, and rescreened at increasingly lower densities until every single clone will yield a positive signal.

4. Screening with Nucleic Acid Probes

Isolating cDNAs from λ phage libraries with labeled nucleic-acid probes is a standard technique in molecular biology *(4)*. Here, the high stability of cloned DNA fragments in λ vectors, the possibility to chose a λ derivative that can accommodate large pieces of foreign DNA, and the high density at which phage plaques can be plated and assayed, come into play. Originally DNA probes used for screening by hybridization were labeled with ^{32}P either by nick-translation or end-labeling; with the advent of sensitive nonradioactive labeling and detection techniques the use of stable, nonradioactive probes that can be used over many months has proven advantageous. In addition to avoiding the inconvenience of manipulating radioactivity, "second generation" chemiluminescence-detection techniques demand for very short exposure times, allowing performance of a screening experiment in <2 d.

DNA probes can include anything from perfectly matching sequences to synthetic oligonucleotides designed by reverse translation of a piece of amino-acid sequence, to highly degenerated oligonucleotides derived from amino-acid sequence comparison of members of a protein family from phylogenetically distant organisms. The window for successful screening with short degenerated oligonucleotides is usually narrow, and the conditions that allow detection of complementary sequences with acceptable background hybridization to bacterial and λ DNA have to be largely determined by trial and error. PCR-based strategies may allow to speed up the procedure, either by amplifying first a specific DNA fragment from mRNA or DNA of the organism in question and using this as a probe for screening the λ library, or by directly amplifying such a fragment from recombinant DNA present in the library and rescreening for complete clones; it is also possible to amplify specific probes from a λ library by using a specific primer and part of the known λ DNA sequence flanking the cloning site as a general primer *(17)*.

4.1. Screening for Encoded Proteins

λ vectors have proven most useful for cloning and analyzing genes by assaying for a property of their encoded proteins, i.e., immunological crossreaction, binding of low-molecular weight, nucleic-acid or protein ligands, or even enzymatic reactions. For discussion of these approaches, we will focus on λgt11, which has been widely used for this purpose.

A prerequisite for expression of cDNAs in a phage vector is the presence of regulatory elements that allow transcription and translation of cloned DNA fragments in *E. coli* cells. In λgt11 cDNA fragments are cloned into a unique *Eco*RI restriction site located near the 3' end of a truncated *E. coli* lactose operon (*lacP/O-lacZ*), allowing inducible, high-level transcription of *lacZ*-cDNA encoded fusion mRNAs that can be translated into fusion proteins if the cDNA coding strand is properly oriented with respect to *lac* transcription, and in phase with the β-galactosidase coding reading frame. Full-length cDNAs including 5' noncoding sequence can also be obtained by screening for their protein products: eukaryotic 5' noncoding sequences can either harbor Shine-Delgarno like ribosome-binding sites (in this case translation is initiated on a fusion mRNA with the *lacZ* mRNA and the expression of the encoded protein is regulated by the *lac* promoter), or expression of cDNAs inserted in the opposite direction with respect to transcription of the *lac* gene can be driven by the λ *lom* promoter, which is located 3' from the *Eco*RI cloning site and drives transcription of the opposite DNA strand (*18,19*; in this case expression of the cloned DNA is unregulated). Some eukaryotic cDNAs may even harbor a complete set of regulatory elements for transcription and translation initiation *(20)*.

An *E. coli* host strain is chosen that carries a *lacI*[q] allele on a high-copy number plasmid (pCM9), ensuring efficient repression of *lac* transcription in the absence of an inducer and minimizing possible deleterious effects of the expressed foreign protein. Because λgt11 encodes a temperature-sensitive *cI* repressor (*cI*857), it can be lysogenized and stably maintained in a single-copy state as a lysogen at the permissive temperature, and subsequently switched to a "multi-copy vector" by inducing lytic growth of the prophage by a temperature shift. Moreover, an amber mutation in the *S* gene (*S100*amb) renders the phage lysis-defective in a nonsuppressor host, allowing the accumulation of phage-encoded products inside the bacterial-host cells. To maximize stability of *lacZ*-cDNA-encoded fusion proteins, a host strain is recommended *(6)* that carries a deletion in the gene for the *lon* protease.

In a λ library, a nucleic-acid probe will detect any DNA fragment with sufficient homology, irrespective of its orientation, reading frame, and whether it would encode a complete protein or only a small peptide. The number of target molecules will be approximately the same (about 10^6) for every phage plaque on a plate. Probing a library for an expressed protein can not rely on the same set of relatively constant parameters. The fraction of DNA fragments that can be detected is much lower; in λgt11 it is on average one sixth of the cDNAs encoding a particular protein, because they have to be cloned in the proper orientation with respect to transcription from the *lac* promoter (the chance is 50% because there is no selection for the orientation of the insert), and they have to be inserted in frame with the reading frame for *lacZ* in order to be expressed as a C-terminal fusion part to *E. coli* β galactosidase (the chance is one third). The expressed gene product must not be (too) toxic for *E. coli* cells; even though tight repression of the *lac* promoter can be ensured by using a host strain that constitutively overproduces a *lacI^q* gene, a few copies of a highly toxic product may significantly reduce the viability of infected bacteria and hence the yield of phages and phage-encoded proteins. The assay for the protein in question should be highly specific because it has to be detected in a heavy background of *E. coli* and phage proteins, and the assay should be highly sensitive because the amount of protein that can be expressed in a phage plaque is very low (typically about 1 fmol of a *lacZ*-cDNA encoded fusion protein in a single λgt11 phage plaque *[21]*).

4.2. Immunoscreening

Probably owing to the low amounts of cDNA-encoded protein that can be expressed in a phage plaque the original approach for screening λ cDNA expression libraries with antibodies consisted in screening of bacterial cells lysogenized with the recombinant phages *(22)*; however, it was soon realized that most antibodies can be readily detected after binding to their target protein in a phage plaque *(23)*. This again allowed to take advantage of the high density at which phage plaques can be plated, a particularly important factor in this type of screening experiment (*see* **Subheading 4.1.**). The quality of the antibodies is of crucial importance for immunoscreening experiments: polyclonal sera are preferred over monoclonal antibodies because they generally react with a number of different epitopes on the protein, allowing truncated or misfolded proteins to be detected. Crossreaction with *E. coli* or phage proteins can be a serious problem because positive clones have to be detected on a background of nonspecific binding. Nonspecific antibodies against *E. coli* or phage proteins can be removed from a serum either by affinity purification of the antibodies on the authentic antigen (coupled to a gel matrix or transferred to a filter support) or by simply neutralizing them with an excess of a crude extract from isogenic bacteria lysogenized with the nonrecombinant λ vector. However, even when the antibodies recognize apparently common epitopes on *E. coli* proteins and the cDNA-encoded products, it is sometimes possible to identify recombinant phages by the mere overexpression of the epitope; for example, we could isolate cDNA clones for *Dictyostelium* ribosomal proteins by using an antiserum that crossreacted with *E. coli* ribosomal proteins (B. Knoblach and R. Mutzel, unpublished observation).

4.3. Functional Screening

Functional screening of λ libraries became a possibility once it was realized that many *lacZ*-cDNA encoded fusion proteins retain their ligand-binding or even enzymatic functions when expressed in λ lysogens or in phage plaques (e.g., *24–26*). Functional isolation of a gene can be undertaken when specific antibodies or oligonucleotide probes are not available, or heterologous antibodies or DNA probes fail to recognize specifically their counterparts from evolutionary distant organisms. What has been said in **Subheading 4.1.** on the limits of detection applies in an even more stringent manner to this technique. Moreover, if the function depends on a post-translational modification of the protein that cannot be performed by the *E. coli* host, or when several subunits that are encoded by individual genes are required for functionality, screening will not be possible. Background problems will be encountered if *E. coli* cells or the λ vehicle express a similar enzyme.

Although some enzymes may interact strongly enough with their substrates to make interaction cloning feasible *(27)*, functional screening has been mainly applied to "simple" high-affinity ligand-binding proteins in which the binding function is confined to a well defined, stable-binding domain. For example, cDNA clones for the high-affinity cAMP-binding subunit of protein kinase A could be identified by *in situ* binding of ^{32}P-labeled cAMP to λgt11 phage plaques *(21)*, but also by ^{3}H-cAMP binding to the protein expressed in bacteria lysogenized with the same λgt11 library *(28)*. In both cases binding of the labeled nucleotide to the *E. coli* cAMP receptor protein was very low owing to its 1000–10,000-fold lower affinity so that recombinant clones could be readily detected. In a similar experiment, isolation of a cDNA clone for a monomeric GTP-binding protein was reported *(29)*. Here the original library was divided into sublibraries containing each 1000 individual phages; total protein synthesized in each of them was separated by sodium dodecyl sulfate-polyacrylamide gel electrophoresis (SDS-PAGE), transferred to nitrocellulose membranes, probed with labeled GTP, and sublibraries containing phages that encoded a GTP-binding activity were further analyzed by the same procedure.

A classical example for screening a λ library by using a protein ligand is the work of Sikela and Hahn *(30)*, who used iodinated bovine-brain calmodulin to identify clones encoding a rat-brain calmodulin-binding protein in a λgt11 expression library. Repeating their experiment with *Dictyostelium* calmodulin to isolate calmodulin-binding proteins from a *Dictyostelium* λgt11 cDNA library, we failed to detect any positive signals when phage plaques were screened; however, detection of calmodulin-binding proteins such as the catalytic subunit of a calmodulin-dependent protein phosphatase (calcineurin A *[31]*) was possible when the cDNAs were expressed in bacterial colonies lysogenized with the same λgt11 library, demonstrating that the success of this type of screening approach can be critically dependent on the affinity of the target for its ligand, the choice of the labeled probe, and the amount of recombinant protein expressed in an individual clone *(28)*. An elegant extension of this type of interaction cloning has been proposed by Germino et al. *(32)* who constructed plasmids encoding genetic fusions of protein ligands with part of the biotin carboxylase carrier protein, which labels the fusion protein with biotin in vivo. Bacteria transformed with this constructs were then infected with a λ library containing cDNAs fused to a functional *lacZ* gene, and protein complexes formed between the ligand and its binding proteins were immobilized on filter supports containing bound streptavidin, avidin, or antibiotin antibodies, and detected by the β galactosidase activity of the λ-encoded fusion protein.

Finally, labeled nucleic-acid probes have been used in a number of cases to detect DNA-binding proteins with modified "Southwestern" assays *(33)*, taking advantage of the high affinity and specificity of binding of these proteins to their target nucleic acid sequences.

5. Analysis of Positive Clones

Recombinant λ DNA clones obtained from a selection or screening experiment are a very convenient material for further analysis. With the rapidly increasing number of known primary structures that are readily accessible from various databases, determination of part of the sequence of the clone will often be helpful to decide whether the "good" gene has been obtained. Such sequence tags can be obtained very rapidly even from recombinant DNA cloned in λ vectors that do not contain excisable phagemids of the bluescript type by PCR amplification of the cloned insert using primers directed against λ sequences flanking the cloning site and direct sequencing of the amplified fragment. Partial clones isolated by immunoscreening may be used as specific DNA probes for the isolation of full-length cDNA clones or the entire gene in question, but can also be directly used to assess the expression pattern of the gene, e.g., on Northern blots. The λ system also offers direct access to the biochemistry of the encoded protein: a recombinant λ lysogen can serve as a model to establish a purification protocol for the encoded foreign protein (e.g., *see* **ref. *31***). High molecular-weight fusion proteins with β galactosidase can be rapidly purified from SDS polyacrylamide gels and used to raise antibodies against the cDNA-encoded protein (e.g., *see* **ref. *34***).

References

1. Thomas, M., Cameron, J. R., and Davis, R. W. (1974) Viable hybrids of bacteriophage lambda and eukaryotic DNA. *Proc. Natl. Acad. Sci. USA* **71,** 4579–4583.
2. Murray, N. E. and Murray, K. (1974) Manipulation of restriction targets in phage λ to form receptor chromosomes for DNA fragments. *Nature* **251,** 476–478.
3. Rambach, A. and Tiollais, P. (1974) Bacteriophage λ having *EcoRI* endonuclease sites only in the nonessential region of the genome. *Proc. Natl. Acad. Sci. USA* **71,** 3927–3930.
4. Sambrook, J., Fritsch, E. F., and Maniatis, T. (1989) *Molecular Cloning. A Laboratory Manual (Part I)*, Cold Spring Harbor Laboratory, Cold Spring Harbor, NY.
5. Lacombe, M. L., Podgorski, G. J., Franke, J., and Kessin, R. H. (1986) Molecular cloning and developmental expression of the cyclic nucleotide phosphodiesterase gene of *Dictyostelium discoideum. J. Biol. Chem.* **261,** 16,811–16,817.
6. Huynh, T. V., Young, R. A., and Davis, R. W. (1985) Constructing and screening cDNA libraries in λgt10 and λgt11, in *DNA Cloning Vol. I* (Glover, D. M., ed.), IRL, Oxford, UK, pp. 49–78.
7. Karn, J., Brenner, S., Barnett, L., and Cesareni, G. (1980) Novel bacteriophage λ vector. *Proc. Natl. Acad. Sci. USA* **77,** 5172–5176.
8. Short, J. M., Fernandez, J. M., Sorge, J. A., and Huse, W. D. (1988) λ ZAP: a bacteriophage expression vector with *in vivo* excision properties. *Nucleic Acids Res.* **16,** 7583–7600.
9. Adams, M. D., Kerlavage, A. R., Fleischmann, R. D., Fuldner, R. D., Bult, C. J., Lee, N. H., Kirkness, E. F., Weinstock, K. G., Gocayne, J. D., White, O., Sutton, G., Blake, J. A., Brandon, R. C., Chiu, M.-W., Clayton, R. A., Cline, R. T., Cotton, M. D., Earle-Hughes, J., Fine, L. D., FitzGerald, L. M., FitzHugh, W. M., Fritchman, J. L., Geoghagen, N. S. M., Glodek, A., Gnehm, C. L., Liu, L.-I., Marmaros, S. M., Merrick, J. M., Moreno-Palanques, R. F., McDonald, L. A., Nguyen, D. T., Pellegrino, S. M., Phillips, C. A., Rydeer, S. E., Scott, J. L., Saudek, D. M., Shirley, R., Small, K. V., Spriggs, T. A., Utterback, T. R., Weldman, J. F., Li, Y., Barthlow, R., Bednarik, D. P., Cao, L., Cepeda, M. A., Coleman, T. A., Collins, E.-J., Dimke, D., Feng, P., Ferrie, A., Fischer, C., Hastings, G. A., He, W.-W., Hu, J.-S., Hudleston, K. A., Greene, J. M., Gruber, J., Hudson, P., Kim, A., Kozak, D. L., Kunsch, C., Ji, H., Li, H., Meissner, P. S., Olsen, H., Raymond, L., Wei, Y.-F., Wing, J., Xu, C., Yu, G.-L., Ruben, S. M., Dillon, P. J., Fannon, M. R., Rosen, C. A., Haseltine, W. A., Fields, C., Fraser, C. M., and Venter, C. A. (1995) Initial assessment of human gene diversity and expression patterns based upon 83 million nucleotides of cDNA sequence. *Nature* **377(Suppl.),** 3–17.

10. Chen, Z., Dixon, J. E., and Zalkin, H. (1990) Coning of a chicken liver cDNA encoding 5-aminoimdazole ribonucleotide carboxylase and 5-aminoimidazole-4-*N*-succino-carboxamide ribonucleotide synthetase by functional complementation of *Escherichia coli pur* mutants. *Proc. Natl. Acad. Sci. USA* **87,** 3097–3101.

11. Léon-Del-Rio, A., Leclerc, D., Akerman, B., Wakamatsu, N., and Gravel, R. A. (1995) Isolation of a cDNA encoding human holocarboxylase synthetase by functional complementation of a biotin auxotroph of *Escherichia coli. Proc. Natl. Acad. Sci. USA* **92,** 4626–4630.

12. Van Lookeren Campagne, M. M., Franke, J., and Kessin, R. H. (1991) Functional cloning of a *Dictyostelium discoideum* cDNA encoding GMP synthase. *J. Biol. Chem.* **266,** 16,448–16,452.

13. Van Lookeren Campagne, M. M., Villalba-Diaz, F., Meacci, E., Manganiello, V. C., and Kessin, R. H. (1992) Selection of cDNAs for phosphodiesterases that hydrolyze guanosine 3';5'-monophosphate in *Escherichia coli. Sec. Mess. Phosphoprot.* **14,** 127–137.

14. Elledge, S. J., Mulligan, J. T., Ramer, S. W., Spottswood, M., and Davis, R. W. (1991) λYES: a multifunctional cDNA vector for the isolation of genes by complementation of yeast and *Escherichia coli* mutations. *Proc. Natl. Acad. Sci. USA* **88,** 1731–1735.

15. Hoffmann, R., Jung, S., Ehrmann, M., and Hofer, H. W. (1994) The *Saccharomyces cerevisiae* gene PPH3 encodes a protein phosphatase with properties different from PPX, PP1 and PP2A. *Yeast* **10,** 567–578.

16. Roeder, W. and Sommerville, A. (1979) Cloning of the *trpR* gene. *Mol. Gen. Genet.* **176,** 361–368.

17. Gonzalez, D. H. and Chan, R. L. (1993) Screening cDNA libraries using λ sequencing primers and degenerate oligonucleotides. *Trends Genet.* **9,** 231,232.

18. Cirala, S. S. (1986) The nucleotide sequence of the *lac* operon and phage junction in lambda gt11. *Nucleic Acids Res.* **14,** 5935.

19. Mutzel, R., Lacombe, M. L., Simon, M. N., de Gunzburg, J., and Véron, M. (1987) Cloning and cDNA sequence of the regulatory subunit of cAMP-dependent protein kinase from *Dictyostelium discoideum. Proc. Natl. Acad. Sci. USA* **84,** 6–10.

20. Simpson, P. J., Chou, W.-G., Whitbeck, A. A., Young, F. E., and Zain, S. (1987) Evidence for procaryotic transcription and translation control regions in the human factor IX gene. *Gene* **61,** 373–383.

21. Lacombe, M. L., Ladant, D., Mutzel, R., and Véron, M. (1987) Gene isolation by direct in situ cAMP binding. *Gene* **58,** 29–36.

22. Young, R. A. and Davis, R. W. (1983) Efficient isolation of genes by using antibody probes. *Proc. Natl. Acad. Sci. USA* **80,** 1194–1198.

23. Young, R. A. and Davis, R. W. (1983) Yeast RNA polymerase II genes: isolation with antibody probes. *Science* **222,** 778–782.

24. Kaufman, D. L., McGinnis, J. F., Krieger, N. R., and Tobin, A. J. (1986) Brain glutamate decarboxylase cloned in λgt-11: fusion protein produces γ-aminobutyric acid. *Science* **232,** 1138–1140.

25. Bernier, L., Alvarez, F., Norgard, E. M., Raible, D. W., Mentaberry, A., Schembri, J. G., Sabatini, D. D., and Colman, D. R. (1987) Molecular cloning of a 2',3'-cyclic nucleotide 3'-phosphodiesterase: mRNAs with different 5' ends encode the same set of proteins in nervous and lymphoid tissues. *J. Neurosci.* **7,** 2703–2710.

26. Mutzel, R., Simon, M. N., Lacombe, M. L., and Véron, M. (1988) Expression and properties of the regulatory subunit of *Dictyostelium* cAMP-dependent protein kinase encoded by λgt11 cDNA clones. *Biochemistry* **27,** 481–486.

27. Chapline, C., Ramsay, K., Klauck, T., and Jaken, S. (1993) Interaction cloning of protein kinase C substrates. *J. Biol. Chem.* **268,** 6858–6861.

28. Mutzel, R., Bäuerle, A., Jung, S., and Dammann, H. (1990) Prophage lambda libraries for isolating cDNA clones by functional screening. *Gene* **96,** 205–211.

29. Nagano, Y., Matsuno, R., and Sasaki, Y. (1993) A cDNA cloning method for monomeric GTP-binding proteins by ligand blotting. *Anal. Biochem.* **211,** 197–199.

30. Sikela, J. M. and Hahn, W. E. (1987) Screening an expression library with a ligand probe: isolation and sequence of a cDNA corresponding to a brain calmodulin-binding protein. *Proc. Natl. Acad. Sci. USA* **84,** 3038–3042.

31. Dammann, H., Hellstern, S., Husain, Q., and Mutzel, R. (1996) Primary structure, expression and developmental regulation of a calcineurin A homologue from *Dictyostelium discoideum. Eur. J. Biochem.* **238,** 391–399.

32. Germino, F. J., Wang, Z. X., and Weissman, S. M. (1993) Screening for *in vivo* protein-protein interactions. *Proc. Natl. Acad. Sci. USA* **90,** 933–937.

33. Singh, H., LeBowitz, J. H., Baldwin, A. S., Jr., and Sharp, P. A. (1988) Molecular cloning of an enhancer binding protein: isolation by screening of an expression library with a recognition site DNA. *Cell* **52,** 415–423.

34. Sonnemann, J., Bäuerle, A., Winckler, T., and Mutzel, R. (1991) A ribosomal calmodulin-binding protein from *Dictyostelium discoideum. J. Biol. Chem.* **266,** 23,091–23,096.

35. Gubler, U. and Hoffman, B. J. (1983) A simple and very efficient method for generating cDNA libraries. *Gene* **25,** 263–269.

16

Plasmid-Derived Cloning Vectors

Craig Winstanley and Ralph Rapley

1. Introduction

Plasmids, circular molecules of DNA that replicate autonomously in a bacterial host cell, are widely distributed among prokaryotes. They vary in size from a few to several hundred kilobase pairs. Although generally not essential to cellular survival, plasmids can confer a selective advantage on their host by encoding phenotypes, such as resistance to antibiotics, heavy-metal resistance, production of toxins or other virulence factors, bacteriocin production, or the catabolism of aromatic compounds. Naturally occurring plasmids often have characteristics that make them ideally suited for development as cloning vectors. Particularly important in this respect has been the exploitation of the many and varied antibiotic-resistance mechanisms encoded by plasmids (1,2).

Plasmids are thought to play a major role in bacterial gene transfer. The process of conjugation allows plasmids to be transmitted between bacterial strains. Plasmids encoding the ability to transfer are described as conjugative. Conjugation is controlled by a set of transfer or *tra* genes located on the plasmid. Although plasmids may be conjugative or nonconjugative, even nonconjugative plasmids can be transferred by a process called mobilization, so long as they carry the necessary *mob* genes.

Plasmids are grouped on the basis of their ability to coexist in the same bacterial host. Two plasmids that are unable to be propagated stably in the same cell line are assigned to a common incompatibility group. Incompatibility is caused by the sharing of elements involved in the replication of plasmids and can be considered to be a measure of relatedness. Plasmids can also vary in copy number, a measure of how many copies of the plasmid may exist within a cell. Members of the same incompatibility group tend to have similar copy numbers. The host-range of a plasmid can also vary. Although the ability of a plasmid to replicate may often be restricted to a single species or genus, members of some incompatibility groups, such as IncP or IncQ, are capable of being transferred between, and maintained within, a wide range of diverse hosts. Plasmids belonging to the IncP group are conjugative and have a host-range that includes virtually all gram-negative bacteria. The smaller IncQ plasmids have a similar wide host-range and are readily mobilized by IncP plasmids. Such so-called promiscuous plasmids (3) are particularly useful for transferring cloned DNA into many bacterial species. Some plasmids, termed episomes, such as the classical F factor of

From: *Molecular Biomethods Handbook*
Edited by: R. Rapley and J. M. Walker © Humana Press Inc., Totowa, NJ

Escherichia coli, are able to integrate into the host chromosome and can promote the transfer of chromosomal DNA between bacterial cells.

One common feature of plasmids is that they all possess some form of a specific sequence, termed an origin of replication *(ori).* This allows a plasmid to replicate within a host cell independently of host-cell replication. These *ori* sequences and associated regulatory sequences are jointly referred to as a replicon. Those replicons maintained at high copy number are termed "relaxed" plasmids and are generally more useful in recombinant DNA studies than those maintained at low copy numbers per cell, termed "stringent" plasmids.

Plasmids have historically been subgrouped into five main types on the basis of the phenotypic function encoded by the genes they contain. The characteristics of these five plasmid types are indicated as follows:

1. **R plasmids:** Plasmids carrying genes encoding resistance to antibiotics. The plasmids may confer on their host resistance to a single or multiple antibiotics. Plasmid-encoded resistance genes have been identified that are active against many different antibiotics, including ampicillin (a β-lactam), streptomycin (an aminoglycoside), sulfonilamide, chloramphenicol, and tetracycline.
2. **Col plasmids:** Plasmids conferring on their host the ability to produce antibacterial polypeptides, called bacteriocins, that are often lethal to closely related or other bacteria. The Col proteins of *E. coli* are encoded by plasmids, such as ColE1.
3. **F plasmids:** Plasmids containing the F or fertility system required for conjugation.
4. **Degradative or catabolic plasmids:** Plasmids, generally found in pseudomonads and related organisms, allowing a host bacterium to metabolize normally recalcitrant, often complex aromatic chemicals, such as xylenes, naphthalenes, or various pesticides.
5. **Virulence plasmids:** Plasmids that are able to confer pathogenicity on a host bacterium by the production of toxins or other virulence factors.

2. General Purpose Plasmid Cloning Vectors

2.1. Introduction

The phenotypes encoded by some plasmids make them suitable for development as tools in gene cloning procedures, but naturally occurring plasmids are rarely ideal for the procedures of genetic manipulation. It has been necessary to amend the basic plasmid starting material in order to develop the range of versatile vectors now available. Since the early 1970s naturally occurring plasmids have been altered genetically in order to facilitate the insertion of fragments of DNA in cloning procedures.

A number of properties are desirable for a useful plasmid cloning vector. These include a region that allows the introduction into the plasmid of DNA digested with different restriction enzymes. Many vectors have a cassette or polylinker containing multiple restriction enzyme sites, sometimes termed multiple cloning sites (MCS). A useful method of selection and a high replication rate are also desirable, as is low-mol-wt (3–5 kb) to allow insertion of large DNA fragments and ease of purification.

The first plasmids to be used as vectors for introducing foreign DNA into cells were developed from *E. coli* plasmids, such as pSC101. This plasmid contains a gene encoding an antibiotic resistance marker and a single *Eco*RI restriction enzyme site into which foreign DNA could be ligated. The first major commonly used plasmid-derived vector was pBR322, and a number of improved or specialized vectors have subsequently been derived from this plasmid.

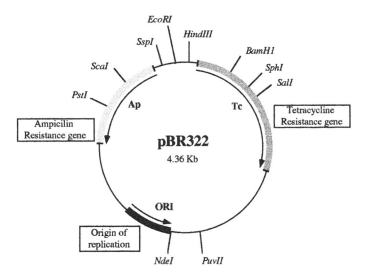

Fig. 1. A map of the cloning vector pBR322.

2.2. Plasmid pBR322

The construction of plasmid pBR322 was carried out by recombination of three plasmids, pSC101, pSF2124, and pMB1, although several others, such as R1 and ColE1, served as intermediaries *(4,5)*. pBR322, for which the complete nucleotide sequence is known *(6)*, contains over 20 unique restriction enzyme recognition sites of which 12 occur in the ampicillin- and tetracycline-resistance genes (**Fig. 1**). Cloning into these 12 sites makes selection of recombinants much simpler since it results in insertional inactivation of the antibiotic-resistance genes. The copy number of pBR322 is approx 15 per cell. This can be dramatically increased to several thousand by plasmid amplification in the presence of the protein synthesis inhibitor chloramphenicol, which prevents replication of bacterial chromosomal DNA but not plasmid DNA.

2.3. Improved pBR322-Derived Vectors

Many different vectors suited to specific cloning requirements have been derived from pBR322. The vectors pBR324 and pBR325 *(7)* were constructed to allow insertional inactivation of selectable markers (derived from other plasmids or bacteriophages) using restriction sites for commonly employed enzymes, such as *Eco*RI and *Sma*I. A widely used pBR322 derivative, pAT153, was generated by removal of a *Hae*II restriction fragment from pBR322. This increases the copy number of plasmids per host cell by up to three times compared to pBR322. It also increases the amount of plasmid- specific protein subsequently produced.

2.4. pUC Vectors

Histological identification of recombinant clones provides an alternative, more direct strategy to insertional inactivation of antibiotic resistance. The most widely used vectors of this kind carry the region of the *E. coli lac* operon coding for the amino-terminal fragment of β-D-galactosidase (product of the *lacZ* gene). Induction by the gratuitous inducer isopropylthiop-D-galactoside (IPTG) leads to intra-allelic (α) complementa-

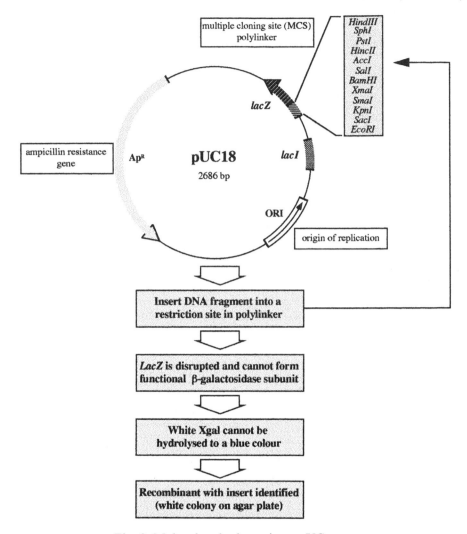

Fig. 2. Molecular cloning using a pUC vector.

tion with a defective form of β-galactosidase encoded by the host. Exposure to IPTG leads to synthesis of both fragments of the enzyme and the formation of blue colonies when the chromogenic substrate 5-bromo-4-chloro-3-indoyl-,8-D-galactoside (X-gal) is also provided in the medium. A series of pUC vectors *(8)* have been generated by inserting the polylinker sequences of phage M13mp vectors (Chapter 17) into the *lacZ* gene without adversely altering function. Consequently, the *lacZ* gene continues to produce β-galactosidase unless foreign DNA is inserted. Insertional inactivation of the amino--terminal fragment of β-galactosidase abolishes the α-complementation and results in colonies of cells harboring pUC recombinants bearing a white phenotype on X-gal-containing media, whereas nonrecombinant colonies remain blue. This blue/white system is used as a basic selection feature of many plasmid and phage vectors (**Fig. 2**). Particular clones can be identified from within a population of recombinant colonies by techniques such as colony hybridization using labeled nucleic acid probes (**Fig. 3**; Chapter 6).

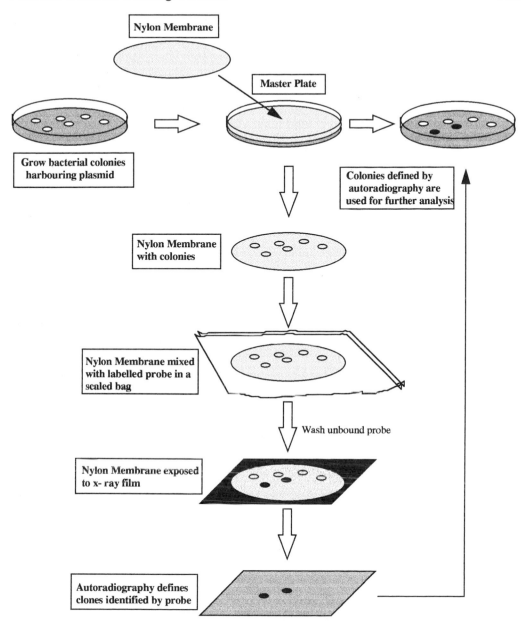

Fig. 3. Screening clones by colony hybridization.

2.5. Direct Selection Vectors

Most plasmid vectors used today have been developed to contain two selectable markers; one used to select transformants and one used to select recombinants (via insertional inactivation). Although this enhances the ability to select recombinants, plasmid vectors have also been constructed to allow direct selection, whereby only transformed host cells harboring a vector containing foreign DNA are able to grow on the selective media. For example, the vector pLX100 has been constructed containing the *E. coli* gene for xylose isomerase *(9)*. Accordingly, *E. coli* cells transformed by pLX100 can only survive in minimal media containing xylose if DNA has been inserted into the sites adjacent to the xylose isomerase gene. Killer vectors, such as the pZErO vectors (Invitrogen, Carlsbad, CA), use positive selection via disruption of lethal genes. Without an insert the vector will express a lethal protein, such as CcdB protein, which causes damage to cellular DNA, leading eventually to cell death. When an insert is ligated into the vector, insertional inactivation of the lethal gene occurs, ensuring that all surviving colonies contain recombinant plasmids.

3. Specialized Plasmid Cloning Vectors
3.1. Low Copy and Runaway Vectors

Most vectors have been specifically engineered to increase plasmid copy number and the amount of cloned protein that can be produced per host cell. There are, however, a few instances where low copy number vectors are desirable. This is particularly important when high-level expression of the encoded protein has a deleterious or lethal effect on the host cell. A series of vectors retaining the two selection markers and multiple cloning sites of pBR322 derivatives, as well as the low copy number characteristic of pSC101 have therefore been developed *(10)*. Although such low copy number vectors allow "lethal" proteins to be cloned, they limit the amount of protein that can be expressed. In order to overcome this, so-called "runaway" vectors have been developed *(11)*. At the maintenance temperature (which depends on the particular vector) the "runaway" plasmid is present at low copy number in the transformed cell. Above this temperature, however, the plasmid undergoes uncontrolled replication, which allows the protein to be overproduced. After a few hours the cell stops growing and becomes nonviable; however, by this time it contains significant amounts of both plasmid and expressed protein.

3.2. Broad-Host-Range Vectors

Early vectors, developed specifically for use in *E. coli,* were limited by their inability to replicate in nonenteric bacterial hosts. The development of cloning vectors suitable for use in a wide range of gram-negative bacteria focused on three naturally occurring plasmids: RSF 1010 (IncQ), RK2 (IncP), and pSa (IncW). All three of these plasmids have the ability to replicate in almost any species of gram-negative bacteria, and carry multiple antibiotic resistance markers, making them attractive candidates for the development of widely applicable vectors *(3)*.

The IncQ plasmid RSF1010 (**Fig. 4**), encoding resistance to streptomycin and sulfonamides, is 8.9 kb in size, making it relatively easy to manipulate genetically. RK2, a 60 kb conjugative plasmid also known as RP1, R68, and R18, encodes resistance to ampicillin, tetracycline, and kanamycin. The 39 kb conjugative plasmid pSa encodes resistance to kanamycin, streptomycin, sulfamides and chloramphenicol.

A considerable range of vectors has been constructed from these naturally occurring broad-host-range plasmids. RSF1010 contains only one restriction site (*Bst*EII) that is useful for cloning and does not cause inactivation of streptomycin-resistance, the only practically useful selective marker. Initial developments therefore concentrated on incorporating segments of DNA specifying alternative selective markers and introducing more flexibility in the choice of restriction site. For example, pKT231 was produced by replacing a *Pst*I fragment in RSF1010 with a *Pst*I fragment derived from the

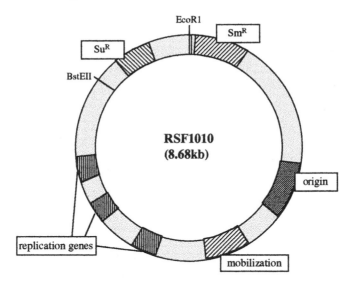

Fig. 4. A map of the broad-host-range plasmid RSF1010.

R6-5 miniplasmid pKT105 *(12)*. The new fragment added kanamycin-resistance and several additional potential cloning sites. A number of such modifications have been carried out to produce a range of vectors derived from RSF1010, some of which contain multiple restriction sites for cloning. All of these plasmids are considerably larger (typically 10–18 kb) than the narrow host–range vectors commonly used in *E. coli.* Because of the complexity of the plasmid maintenance functions, few areas of RSF1010 are dispensable. In contrast to the nonconjugative but immobilizeable RSF1010, both RK2 and pSa are conjugative. The initial development of vectors based on these plasmids involved the deletion of regions not directly participating in maintenance, such as the *tra* genes. Recently, a whole series of RK2-derived cloning vectors ranging in size from 4.8 to 7.1 kb, offering choices in unique cloning sites and antibiotic resistance markers and including expression vectors and vectors susceptible to copy number modification, have been constructed *(13)*.

Some bacteria have developed resistance to most of the commonly used antibiotics, making the choice of vectors for use in cloning experiments extremely limited. For example, the human pathogens *Burkholderia cepacia* and *Burkholderia pseudomallei* often carry resistance to ampicillin, gentamycin, chloramphenicol, kanarmycin or tetracycline. It has been necessary therefore to construct broad host–range vectors encoding trimethoprim resistance specifically for use in these organisms *(14)*.

3.3. Vectors for Gram-Positive Bacteria

Gram-positive bacteria, such as *Bacillus* spp., have a number of interesting metabolic and physiological properties that make them interesting targets for study. Molecular understanding of these properties has necessitated the development of cloning vectors for use in gram-positive hosts. Because of the relative rarity of plasmids encoding antibiotic resistance markers in *Bacillus subtilis,* small *Staphylococcus aureus* plasmids, such as the 2.9 kb plasmid pC194 *(15)*, which encodes chloramphenicol-resistance, have been used for the development of *Bacillus* spp. cloning vectors. Although such plasmids could be improved as vectors by the addition of other antibiotic-resistance markers, their use in *Bacillus* spp. is limited because of structural and segregational instability. This has led to the development of more stable *Bacillus sp.* vectors based on indigenous cryptic plasmids, such as pTA1060 *(16)*, which can be modified to carry antibiotic-resistance genes.

3.4. Shuttle Vectors

Because of the difficulties associated with cloning in *B. subtilis,* hybrid plasmids capable of replication in *E. coli* and *B. subtilis* were developed. The original approach taken to the development of such shuttle vectors was to fuse an *E. coli* plasmid, such as pBR322, with a *B. subtilis* plasmid, such as pCl94. *E. coli* can then be used as an efficient host for the propagation and identification of clones and for the extraction of plasmid clones that can be introduced subsequently into *B. subtilis* by transformation. There are disadvantages to this approach. Many gram-positive bacterial genes, such as those involved in sporulation or competence, are difficult to clone in *E. coli*. The procedures are also time-consuming and do not readily allow selection by complementation.

A number of shuttle vectors also have been developed between *E. coli* and eukaryotic cells. Eukaryotic events, such as protein transport, posttranslational modification, and RNA splicing, cannot readily be studied in *E. coli*. By using shuttle vectors, DNA manipulation can be carried out in *E. coli* prior to the transfer of recombinants into yeasts, filamentous fungi *(17)*, plants (with *Agrobacterium tumafaciens* as an intermediate; *18*; Chapter 21), or mammalian hosts. Plasmids for use in mammalian hosts will be discussed in **Subheading 3.6.**

Many of the *E. coli*-yeast shuttle vectors have been constructed using a naturally occuring yeast plasmid (the 2 µm plasmid; *19)*. Although this plasmid is found in many strains of *Saccharomyces cerevisiae* in 50–100 copies per haploid cell, it has no known function. Chimeric plasmids capable of replication in both yeast and *E. coli* were constructed from the 2 µ yeast plasmid, fragments of yeast nuclear DNA, and the *E. coli* vector PMB9 *(20)*. By using this kind of approach, many yeast episomal plasmid (YEp) vectors with high copy number and good yeast transformation frequency have been developed *(21)*. The *Leu2* gene, encoding an enzyme β-isopropylmalate dehydrogenase; involved in the conversion of pyruvic acid to leucine) has been cloned into the 2 µ plasmid to enable selection of transformed yeast cells. If a yeast host unable to synthesize leucine is used, only those cells containing the plasmid will grow on minimal media lacking leucine.

The inability of the 2 µ yeast plasmid to propagate efficiently in genera other than *Saccharomyces* has limited it's use. The plasmid pKD1 has been developed as a vector for the introduction and propagation of sequences in *Kluyreromyces lactis (22)*. Indeed, the broad host range linear killer plasmids of *K. lactis* offer the possibility of vectors with far wider applicability than offered by those based on the 2 µ yeast plasmid *(23)*.

3.5. PCR Product Cloning Vectors

The development of the polymerase chain reaction (PCR; Chapter 25) led to demand for strategies for cloning amplified products. A common early approach involved appending additional nucleotides onto PCR primers to introduce suitable restriction sites. Following purification of PCR products, restriction digestion could then be used to generate suitable overhangs for ligation into a standard cloning vector. This approach involves the additional expense of generating longer oligonucleotide primers and can be hampered by the inefficient digestion that occasionally occurs at the ends of linearized molecules. An alternative strategy involves blunt-ended ligation, which is also often inefficient. These problems can be overcome by the use of vectors, such as pCR2001 *(24)*, which is designed to exploit the terminal transferase activity of *Taq* polymerase, and other enzymes commonly employed in the PCR, resulting in the addition of a single 3' deoxyadenylate residue at each end of an amplified molecule. pCR2001 can be digested to produce complementary single base 3' deoxythymidine residues, enabling ligation that is more efficient than blunt-ended ligation. Commercial kits, including the TA Cloning System (Invitrogen) and pCR-Script (Stratagene, La Jolla, CA), are now widely used to clone PCR products.

3.6. Expression Vectors

In order to study the properties of a protein, it is often desirable to obtain large quantities of the protein by cloning the relevant gene into a vector designed to express that gene at high levels. Synthesis of a functional protein requires that the appropriate gene be transcribed and that the resulting mRNA be efficiently translated. In order for a cloned gene to be transcribed, the presence of a promoter recognized by the host cell RNA polymerase is required. Foreign DNA under the transcriptional control of its own promoter may only be weakly transcribed or, if the promoter is not recognized by the host RNA polymerase, there may be no transcription at all. For example, when present in an *E. coli* host, promoters from other gram-negative bacteria, such as *Pseudomonas* spp., often direct only weak transcription in *E. coli,* whereas *B. subtilis* promoters are usually inactive. Expression vectors, containing a cloning site located downstream of a strong promoter, have therefore been developed. A number of vectors have been designed such that the cloning site is located at the end of a signal sequence. Consequently, the product of interest is fused to an N-terminal signal peptide, thus facilitating purification of the protein. In this type of vector it is essential that the foreign gene dependent on fusion to an *E. coli* gene is inserted in such a way as to maintain the correct reading frame. To overcome this, a series of vectors with cloning sites in all three reading frames, such as the pEX1-3 expression vectors *(25)*, can be constructed making it possible to guarantee an in-frame fusion for virtually any cloned gene.

A number of strong promoters have been employed in the development of *E. coli* expression vectors. Among the most commonly used are the λP_R and λP_L promoters of bacteriophage λ *(26)*. These promoters have the additional advantage that they can be tightly thermoregulated by adding the cI_{857}, element, which encodes a temperature-sensitive repressor. This allows temperature-controlled gene expression whereby genes under the control of the λ promoters are switched off until the temperature is raised to 42°C. Such control can be particularly important when the cloned gene product is lethal or deleterious to the bacterial host.

Promoters from the *E. coli* *trp* and *lac* operons can be controlled effectively by the presence or absence of the appropriate inducer molecule, such as tryptophan or IPTG. A fusion of *trp* and *lac* promoters has been used to construct stronger artificial *tac* promoters, which can be regulated by incorporating the *lac* repressor gene into the expression vector *(27)*.

A number of plasmid vectors have been constructed to carry promoters derived from bacteriophages T3, T7, and/or SP6 adjacent to a polycloning site to enable the transcription in vitro of inserted DNA. Recombinant plasmids are linearized prior to addition of the appropriate DNA-dependent RNA polymerase and ribonucleotide precursors. Often different promoters are positioned adjacent to each side of the cloning site to enable transcription from either end and either side of the inserted DNA. The RNA generated can be translated in cell-free translation systems or used to synthesize probes for hybridization experiments *(28)*.

In order to clone and study DNA fragments involved in transcriptional control, a number of reporter vectors containing promoter-less marker genes have been constructed. A variety of marker genes are available for the development of such vectors. Expression of the marker gene can be detected by a number of approaches, including colorimetric tests, luminescence, and fluorescence (**Table 1**). The constructs can be used to create fusion proteins when elements involved in transcriptional control are cloned.

Although some eukaryotic proteins have been expressed efficiently in prokaryotic hosts, many eukaryotic proteins fold inefficiently or incorrectly when synthesized in bacteria. In addition, the posttranslational modifications often required for active eukaryotic proteins are not performed by bacterial cells and intron-containing genomic sequences cannot be transcribed correctly in prokaryotes. Systems developed for the expression of mammalian proteins in mammalian hosts include some that utilize plasmid-based vectors. These include simple vectors containing no eukaryotic replicon and more complex plasmids that incorporate elements from the genomes of eukaryotic viruses. The latter are designed to increase the copy number of the transfected DNA and the efficiency with which foreign proteins are expressed. Such mammalian expression vectors require both prokaryotic sequences, often derived from pBR322, to facilitate construction and replication in *E. coli,* and eukaryotic sequences, inserted into nonessential regions of the plasmid, and providing the means for expression of eukaryotic genes *(29)*. The eukaryotic expression module must contain a promoter element to drive transcription, and the necessary signals for polyadenylation of the transcript. Enhancer sequences, which can stimulate transcription up to 1000-fold, and introns with functional splice donor and acceptor sites may also be included. The choice of promoter and enhancer sequences is determined largely by the cell type in which the recombinant gene is to be expressed. The isolation of transfected cell lines normally requires the introduction of a selectable marker into the plasmid vector. Markers include genes encoding thymidine kinase, dihydrofolate reductase, or aminoglycoside phosphotransferase (APH). Aminoglycoside antibiotics, such as kanamycin and neomycin, inhibit protein synthesis in both prokaryotic and eukaryotic cells. When fused to eukaryotic transcriptional regulatory elements, obtained from eukaryotic viruses, such as herpes simplex virus (HSV) or simian virus 40 (SV40), APH can be used as a dominant marker to select mammalian cells successfully transfected with recombinant plasmid *(30)*.

Although the most common approach to the cloning of mammalian genes in plasmid vectors involves the construction of a cDNA library (Chapter 13), genomic DNA, in which coding sequences may be interrupted by introns, may also be cloned. In the latter case, the requirement for controlling sequences to enable efficient expression

Table 1
Marker Genes Used in Reporter Vectors and Methods for Detection of Expression

Marker	Enzyme/protein	Source	Detection
lacZ	β-galactosidase	*E. coli*	Chromogenic substrate X-Gal, included in the media, is converted to a blue color.
xylE	Catechol 2,3 dioxygenase	*Pseudomonas putida*	Colonies sprayed with catechol produce the yellow product 2-hydroxymuconic semialdehyde.
GUS	β-glucuronidase	*E. coli*	A range of histochemical (such as X-Gluc), chromogenic, or fluorescent substrates can be used to assay activity.
luxAB	Luciferase	*Vibrio fischeri* or *Vibrio harveyi*	Following the addition of luciferin, luminescence can be detected by eye in a darkened room, by photography, by X-ray film, or with the use of a luminomete.
GFP	Green fluorescent protein	Jelly fish (*Aqueoria victoria*)	Fluorescence, not requiring additional substrates or cofactors, can be detected by fluorescence microscopy.
TFD	2,4-Dichlorophenoxyacetate (TFD) monooxygenase	*Pseudomonas*	Colonies sprayed with phenoxyacetate are converted to phenol. Subsequent spraying with potassium ferrocyanide and 4 aminoantipyrene causes production under alkaline conditions of a red antipyrine dye.
cysG or *cobA*	Uroporphyrinogen III methyltransferase	*E. coli*, *Pseudomonas denitrificans*, or *Propionibacterium freudenreichii*	Fluorescent porphyrinoid compounds accumulate. When illuminated with UV light, cells fluoresce with a bright red color.
inaZ	Ice nucleation	*Pseudomonas syringae*	Ice nucleation is detected using droplet-freezing assays.

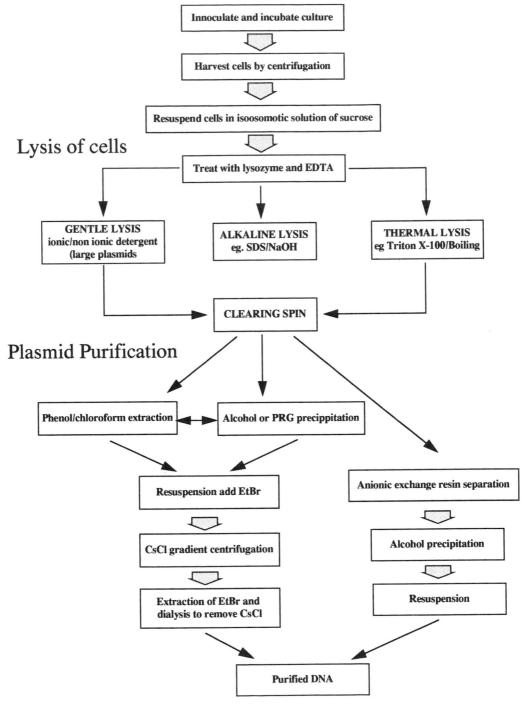

Fig. 5. Common strategies for the isolation of plasmid DNA from *E. coli.*

may be less, although this will depend on the cell type in which the sequence is required to be expressed. A large selection of mammalian expression vectors is available. The best choice depends largely on the individual requirements of the user.

4. Isolation and Purification of Plasmid Vectors

The purification of plasmid DNA from a bacterial culture may be carried out in the laboratory in a number of ways (**Fig. 5**). Initially, a bacterium harboring the plasmid is cultured overnight in liquid media containing the relevant antibiotic and the cells are harvested. In order to release the desired DNA from a bacterial cell, disruption of the wall is necessary. This must be achieved in such a way as to minimize the amount of any contaminating bacterial chromosomal DNA. Gentle lysis of the bacterial wall can be carried out using ionic or anionic detergents, such as SDS or Triton X-100, in conjunction with lysozyme to degrade the peptidoglycan and the chelating agent EDTA. For smaller plasmids somewhat harsher methods for lysis can be employed. These include alkaline lysis, often involving SDS/sodium hydroxide, and thermal lysis, employing a combination of detergent and brief boiling.

The isolation and purification of plasmid DNA relies on exploitation of the covalently closed circular conformation (ccc) of plasmids, an effect of DNA supercoiling, and the consequent differential properties of supercoiled plasmid DNA in comparison to chromosomal nonsupercoiled DNA with respect to denaturation. Such techniques as alkaline and thermal denaturation make use of this variation to enable the separation of plasmid and chromosomal DNA. Under alkaline conditions the bacterial chromosome will denature, whereas the topologically intertwined strands of closed circular plasmid DNA will not. Decreasing the pH to 7.0 will cause aggregation of the chromosomal DNA. A similar effect can be seen during thermal lysis. These methods require a fine balance to be maintained, however, because prolonged heat or alkali treatment will lead to irreversible denaturation of even plasmid DNA.

Once gentle disruption on the cell wall has been achieved, centrifugation is used to produce a cleared lysate containing plasmid DNA. The insoluble aggregated chromosomal DNA should be pelleted along with the cell debris. After centrifugation to produce a lysate, supercoiled plasmid DNA may be separated from any remaining chromosomal DNA by centrifugation through a cesium chloride (CsCl) gradient in the presence of the intercalating agent ethidium bromide. This fluorescent chemical is less able to penetrate supercoiled DNA but may freely bind linear chromosomal DNA. Chromosomal and plasmid DNA will form at different positions through a density gradient of CsCl, and may be isolated following examination of bands using a UV source. Alternatively, plasmid DNA can be purified from lysates by using anion-exchange resins based on the interaction between negatively charged phosphates of the DNA backbone and positively charged DEAE groups on the surface of the resin. DNA is eluted from the column by using a high- salt buffer. Commercial kits, based on this approach, are now widely available for the rapid isolation of plasmid DNA from *E. coli* cells.

References

1. Cohen, S. N. and Miller, C. A. (1969) Multiple molecular species of circular it-factor DNA isolated from *Escherichia coli. Nature* **224,** 1273–1277.
2. Russell, A. D. and Chopra, I. (1996) *Understanding Antibacterial Action and Resistance.* Horwood, London, UK.
3. Thomas, C. M. (1989) *Promiscuous Plasmids of Gram-Negative Bacteria.* Academic, UK.
4. Bolivar, F., Rodriguez, R. L., Betlach, M. C., and Boyer, H. W. (1977) Construction and characterisation of new cloning vehicles. I. Ampicillin resistant derivatives of the plasmid pMB9. *Gene* **2,** 75–93.

5. Bolivar, F., Rodriguez, R. L., Greene, P. J., Betlach, M. C., Heyneker, H. L., Boyer, H. W., Crosa, J. H., and Falkow, S. (1977) Construction and characterisation of new cloning vectors. II. A multipurpose cloning system. *Gene* **2**, 95–113.

6. Sutcliffe, G. (1979) Complete nucleotide sequence of the *Escherichia coli* plasmid pBR322. *Cold Spring Harbor Symp. Quant. Biol.* **43**, 77–90.

7. Bolivar, F. (1978) Construction and characterisation of new cloning vehicles. III. Derivatives of plasmid pBR322 carrying unique *Eco*RI sites for selection of *Eco*RI generated recombinant DNA molecules. *Gene* **4**, 121–136.

8. Vieira, J. and Messing, J. (1982) The pUC plasmids: an M13mp7-derived system for insertion mutagenesis and sequencing with synthetic universal primers. *Gene* **19**, 259–268.

9. Stevis, P. E. and Ho, N. W. Y. (1987) Positive selec‚tion vectors based on xylose ultilisationsupression. *Gene* **55**, 67–74.

10. Stoker, N. G., Fairweather, N. F., and Syratt, B. G. (1982) Versatile low-copy-number plasmid vectors for cloning in *Escherichia coli*. *Gene* **18**, 335–341.

11. Uhlin, B. E., Schweickart, V., and Clark, A. J. (1983) New runaway-replication-plasmid cloning vectors and suppression of runaway replication by novobiocin. *Gene* **22**, 255–265.

12. Bagdasarian, M., Franklin, F. C. H., Lurz, R., Ruckert, B., Elagdasarian, M. M., and Timmis, K. N. (1981) Specific purpose cloning vectors. II. Broad host range, high copy number RSF1010-derived vectors and a host:vector system for gene cloning in *Pseudomonas*. *Gene* **16**, 237–247.

13. Blatny, J. M., Brautaset, T., Winther-Larsen, H. C., Haugan, K. and Valla, S. (1997) Construction and use of a versatile set of broad-host-range cloning and expression vectors based on the RK2 replicon. *Appl. Environ. Microbiol.* **63**, 370–379.

14. DeShazer, D. and Woods, D. E. (1996) Broad-host-range cloning and cassette vectors based on the R388 trimethoprim resistance gene. *BioTechniques* **20**, 762–764.

15. Ehrlich, S. D. (1977) Replication and expression of plasmids from *Staphylococcus aureus* in *Bacillus subtilis*. *Proc. Natl. Acad. Sci. USA* **74**, 1680–1682.

16. Haima, P., Bron, S., and Venema, G. (1987) The effect of restriction on shotgun cloning and plasmid stability in *Bacillus subtilis* Marburg. *Mol. Gen. Genet.* **209**, 335–342.

17. Yelton, M. M., Hamer, J. E., and Timberlake, W. E. (1984) Transformation of *Aspergillus nidulans* by using a *trpC* plasmid. *Proc. Natl. Acad. Sci. USA* **81**, 1470–1474.

18. Leemans, J., Langenakens, J., de Greve, H., Deblaere, R., van Montagu, M., and Schell, J. (1982) Broad host range cloning vectors derived from the W. plasmid Sa. *Gene* **19**, 361–364.

19. Murray, J. A. H. (1987) Bending the rules: the 2μ plasmid of yeast. *Mol. Microbiol.* **1**, 114.

20. Beggs, J. D. (1978) Transformation of yeast by a replicating hybrid plasmid. *Nature* **275**, 104–109.

21. Broach, J. R. (1983) Construction of high copy yeast vectors using 2-micron circle sequences. *Methods Enzymol.* **101**, 307–325.

22. Chen, X. J., Wesolowski-Louvel, M., Tanguy-Rougeau, C., Bianchi, M. M., Fabiani, L., Saliola, M., Falcone, C., Frontali L. and Fukuhara, H. (1988) A gene-cloning system for *Kluyveromyces lactis* and isolation of a chromosomal gc ne required for killer toxin production. *J. Basic Microbiol.* **28**, 211–220.

23. Gunge, N. and Sakaguchi, K. (1981) Intergeneric transfer of deoxyribonucleic acid killer plasmids pGKL1 and pGKL2 from *Kluyveromyces lactis* into *Saccharomyces cerevisiae* by cell fusion. *J. Bacteriol.* **145**, 382–390.

24. Mead, D. A., Kristy Pey, N., Herrnstadt, C., Marcil, R. H., and Smith, L. M. (1991) A universal method for the direct cloning of PCR amplified nucleic acid. *Bio/Technology* **9**, 657–663.

25. Stanley, K. K. and Luzio, J. P. (1984) Construction of a new family of high efficiency bacterial expression vectors: identification of cDNA clones coding for human liver proteins. *EMBO J.* **3,** 1429–1434.

26. Remaut, E., Stanssens, P., and Fiers, W. (1981) Plasmid vectors for high-efficiency expression controlled by the p_L promoter of coliphage lambda. *Gene* **15,** 81–83.

27. Stark, M. J. R. (1987) Multicopy expression vectors carrying the *lac* repressor gene for regulated high-level expression of genes in *Escherichia coli. Gene* **51,** 255–267.

28. Melton, D. A., Krieg, P. A., Rebagliati, M. R., Maniatis, T., Zinn, K., and Green, M. R. (1984) Efficient in vitro synthesis of biologically active RNA and DNA hybridisation probes from plasmids containing bacteriophage SP6 promoter. *Nucleic Acids Res.* **12,** 7035–7056.

29. Lusky, M. and Botshan, M. (1981) Inhibition of SV40 replication in simian cells by specific pBR322 sequences. *Nature* **293,** 79–81.

30. Jiminez, A. and Davies, J. (1980) Expression of a transposable antibiotic resistance element in *Saccharomyces. Nature* **287,** 869–871.

17

M13 and Phagemid-Based Cloning Vectors

Ralph Rapley

1. Single-Stranded DNA Bacteriophage Viruses

Bacteriophage-based vectors, such as λ have been useful in creating gene libraries and in the expression of proteins. However, a number of techniques used in molecular biology require DNA to be in a single-stranded form, such as chain termination sequencing, oligonucleotide-directed mutagenesis, and certain probe synthesis strategies. A small number of naturally occurring viruses produce single-stranded DNA as part of their life cycle; these include M13, f1, and fd, and a number of them have been manipulated and adapted as cloning vectors for the purpose of single-strand DNA production.

2. The Biology of Filamentous Bacteriophage

The filamentous coliphages such as M13 or fd, infect strains of enteric bacteria, such as *Escherichia coli* harboring male specific F pili. They enter the cell by adsorption to this structure through interaction of a coat protein at one end of the phage known as geneIII protein. Following decapsidation and once inside the cell, the phage DNA is converted to a double-stranded "replicative form" or RF DNA (**Fig. 1**). Replication then proceeds rapidly until approx 100 RF molecules are produced within the *E. coli* cell. The simultaneous accumulation of geneII protein then dictates rolling circle replication to begin by binding to the complementary RF strand. A further protein, geneV protein, switches DNA synthesis to the production of single strands and the DNA is assembled and packaged into the capsid at the bacterial periplasm. GeneV protein extrudes the DNA through the bacterial periplasm and is removed. At this point the bacteriophage DNA is encapsulated by the major coat protein, geneVIII protein, of which there are approx 2800 copies with 3–6 copies of the geneIII protein at one end of the particle. The extrusion of the bacteriophage through the bacterial periplasm results in a decreased growth rate of the bacterial cell rather than host cell lysis and is visible on a bacterial lawn as an area of clearing. Approximately 1000 packaged phage particles may be released into the medium in one cell generation.

3. Uses of Single-Stranded DNA Vectors

In addition to producing single-stranded DNA the coliphage vectors have a number of other features that make them attractive as cloning vectors. The bacteriophage DNA

From: *Molecular Biomethods Handbook*
Edited by: R. Rapley and J. M. Walker © Humana Press Inc., Totowa, NJ

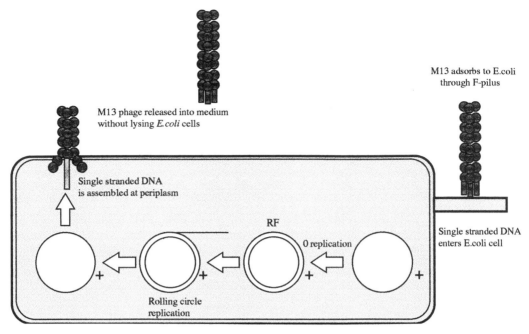

Fig. 1. Life cycle of M13.

is replicated and may be recovered as a double-stranded RF intermediate DNA molecule, which allows a number of regular DNA manipulations to be performed, such as restriction digestion, DNA ligation, and the mapping of DNA fragments. Both forms of the M13 DNA may be used to infect competent bacterial *E. coli* cells, allowing either plaques or infected colonies to be produced. This allows different techniques to be employed in their isolation and purification. RF DNA may be prepared by lysing infected *E. coli* cells and purifying the supercoiled circular phage DNA with the same methods used for plasmid isolation. Intact single-stranded DNA packaged in the phage protein coat in the supernatant may be precipitated with reagents, such as polyethelyene glycol, and the DNA purified with phenol/chloroform. Thus, the bacteriophage may act as a plasmid under certain circumstances and at other times produce DNA in the fashion of a virus.

4. The Construction of M13 Cloning Vectors

The wild-type M13 genome is 6407 bp encoding ten genes essential for its life cycle. However only 507 bp of the genome, termed the intergenic region, is not protein-encoding although this contains the origin of replication (ori). Using a number of manipulations Messing was able to introduce the *lacZ* gene into the intergenic region *(1)*. This encodes the α peptide of the enzyme β galactosidase. This recombinant vector, termed M13mp1, was further enhanced by mutagenesis to create a single restriction site for EcoRT (GAATTC) near the start site of the *lacZ* gene to create M13mp2. Further manipulations enabled a synthetic multiple cloning site or polylinker containing a number of unique restriction sites to be located in the *lacZ* gene without disruption of the reading frame of the gene. This derivative, M13mp7, has been further manipulated by increasing the number of restriction sites in an asymmetric fashion,

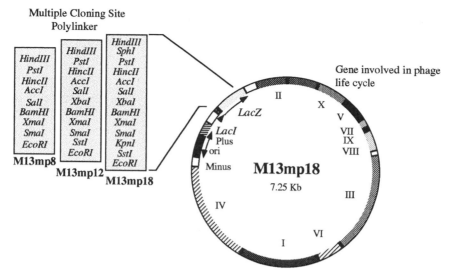

Fig. 2. Fig map of M13 and various derivatives.

resulting in M13mp8, mp12, mp18, and sister vectors that have the same polylinker but in reverse orientation, M13mp9, mp13, and mp19, respectively *(2)*. **Figure 2** indicates the restriction endonuclease sites in a number of the M13mp series.

4.1. Polylinker Design of pUC/M13mp Vectors

The construction of M13 using a polylinker was a significant step in recombinant DNA technology. Not only can a DNA fragment be inserted into, for example, M13mp8 in a defined orientation, termed directional cloning, but it may also be excised and inserted into the same M13 vector in the opposite orientation using a reversed polylinker M13mp9. This allows ease of nucleotide sequencing of an insert in both directions *(3)*. In addition, this convenience is not only confined to M13, because Messing and coworkers also devised a means of having an identical polylinker in the pUC series of plasmids. Thus, M13mp18 and pUC18 have identical polylinkers (**Fig. 3**). This facilitates the direct subcloning of a fragment from a general plasmid cloning vector to a single-stranded phage vector for further manipulations.

4.2. Convenient Features of M13mp Vectors

M13 vectors have a number of useful properties that make them an ideal choice for a cloning and sequencing vector. The presence of the *lacZ* site allows efficient screening to be undertaken based on the technique of blue/white selection. The molecule isopropyl-β-D-thiogalactoside (IPTG) induces the expression of the *lacZ* gene and produces the α peptide of β galactosidase, which is able to complement defective enzyme produced in an appropriate host *E. coli* strain containing lacZΔM15 on the F', such as JM109. A nonrecombinant produces a blue plaque on screening since an intact β galactosidase is able to hydrolyse the colorless chromogenic substance X-gal (5-bromo-4-chloro-3-indolyl-β-D-galactopyranoside) into a blue insoluble product. However, any insert in the *lacZ* gene renders it inactive and therefore recombinants are seen as colorless or white plaques and may be further analyzed.

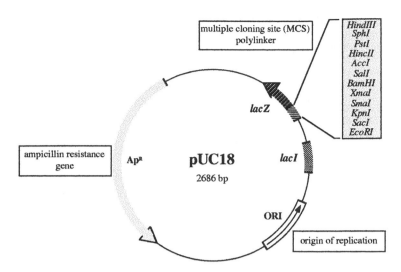

Fig. 3. Elements of pUC18 plasmid.

4.3. Template Production for Nucleotide Sequencing

M13 vectors are the traditional choice for nucleotide sequencing by the chain termination method because of the single-stranded nature of their DNA *(4)*. This sequencing method utilizes a DNA polymerase to synthesize a complementary DNA strand on an existing single strand derived from the M13 vector and insert. Chain extension takes place from a defined oligonucleotide primer, and during the synthesis dideoxynucleotide analogs that lack a 3' hydroxyl group (ddNTPs) are incorporated that terminate chain synthesis at specific points at one of the four bases. In this way multiple fragments of DNA are constructed of sequences differing by one base and ending in a specific chain terminator, ddA, ddT, ddC, or ddG. These fragments are usually visualized as a ladder following autoradiography although a number of automated methods are now generally available. It is possible to sequence double-stranded DNA from plasmid, other bacteriophage vectors, or from the polymerase chain reaction (PCR); however, in general the quality of sequence derived from M13 templates is superior.

A further modification that makes M13 useful in chain termination sequencing is the placement of universal priming sites at –20 or –40 bases from the start of the *lacZ* gene *(5)*. This allows any gene to be sequenced by using one universal primer since annealing of the primer prior to sequencing occurs outside the polylinker and so is M13-specific rather than gene-specific *(6)*. A further, reverse priming site is also located at the opposite end of the polylinker. The advent of the PCR also further reduces the time to screen M13 recombinants since PCR primers designed to anneal to the polylinker or universal priming site and the reverse site allow a recombinant to be distinguished from a nonrecombinant based on the size of the PCR product. Nonrecombinants provide a PCR fragment of approx 125 bp in M13mp18, i.e., the size of the polylinker, whereas a recombinant will be larger. In general, however, clones are usually identified by plaque hybridization and probing with an appropriate gene probe *(7)*. One problem frequently encountered with M13 however is the instability of large inserts. These are accepted by M13, which increases the number of geneVIII proteins and hence the size of the phage particle to compensate. However, under certain circumstances fragments larger than 3 Kb results in loss of the recombinant clone. Furthermore, vectors carrying large DNA inserts tend to be less efficient in host-cell transformation and may be underrepresented in a gene library.

4.4. Template Production for Site Directed Mutagenesis

A further technique that requires single-stranded DNA as a prerequisite is site-directed mutagenesis. This is a very powerful technique for introducing a predefined mutation into a gene *(8)*. This allows the effects of the mutation to be analyzed on the protein it encodes following expression of the gene. The gene is usually cloned into a suitable single-stranded vector although the emergence of mutagenesis techniques based on the PCR have changed this position in the last few years. Complete nucleotide sequence information of the gene is essential to identify a potential region of interest. A short oligonucleotide is then synthesized that is complementary to the region to be mutated but in which is included the nucleotide alterations. Even with the inclusion of extra nucleotides the oligonucleotide will anneal to the single-stranded DNA and act as a primer for a DNA polymerase to synthesize a complementary strand. This reaction is continued until a complete DNA chain is synthesized that also includes the additional sequence mutation. The cloning of this vector then produces multiple copies, of which half contain a sequence with the mutation and half contain the original wild-type sequence (**Fig. 4**). Plaque hybridization using the oligonucleotide as the probe is then used at a stringency that allows only those plaques containing a mutated sequence to be identified.

It is also possible to suppress the growth of the nonmutated wild-type sequence using a number of alternative methods, such as the gapped duplex and primer selection system *(9)*. Many in use at present rely on the phage being grown on a specialized host before the mutations are introduced *(10)*. One in particular involves the incorporation of uracil in the parental strand by growing in hosts containing deficient dUTPase and uracil glycosylase. The mutated strand is produced with dTTP rather than uracil and subsequent isolation is carried out on wild-type hosts that select for the mutant strands. Alternatively, mutant strands can be synthesized in the presence of thionucleotides to give phosphorothioated circular strands *(11)*. These are resistent to digestion by certain restriction enzymes and hence the selection of the mutant strands is permitted.

5. The Production of Phagemid Cloning Vectors

The universal applicability of the M13 system for producing single-stranded DNA has led to further developments that have addressed some of the disadvantages of the vector and incorporated other useful functions of filamentous bacteriophages and plasmid vectors. M13, f1, or fd have all been utilized in the development of phage and plasmid hybrid constructs, termed phagemid vectors. These all contain a single stranded circular genome and replicate in the host *E. coli* following infection at the F pillus. In theory phagemids combine the key features of single-stranded bacteriophage vectors with the small size and insert stability of plasmid vectors. A number of phagemid vectors have been developed and marketed commercially, including Bluescript, pUC118/119, and pGEM.

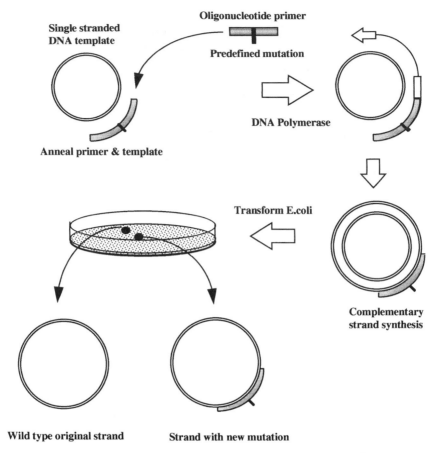

Fig. 4. Basic strategy for M13 site directed mutagenesis.

5.1. Creating Phagemid DNA Cloning Vectors

Plasmids have an advantage over virus-based vectors in that they are relatively easy to manipulate and DNA inserted into the cloning sites is relatively stable. This allows their isolation and analysis to be performed rapidly by conventional techniques, such as alkaline lysis. Plasmids, such as those developed by Messing (pUC), in addition to their *lacZ* selection system and multiple cloning site are also small molecules (2.9 Kb) which facilitate efficient transformation of host bacteria *(6)*. In 1981, experimental evidence provided a means of inserting into a plasmid a region that controls single-stranded phage DNA replication and morphogenesis *(12)*. These are cis-acting elements that are activated on infection of the same host cell with an additional helper phage in a process termed superinfection. Helper phage are introduced to the host cell at a multiplicity of infection (MOI) of 20–40. These phages, such as VCSM13 or R408, provide the necessary gene products for preferential secretion of single-stranded vectors containing intact segments of an M13 intergenic region over the helper phage DNA. Thus, a plasmid containing regions of the control sequences from a phage allows it to act as a phage under the appropriate conditions.

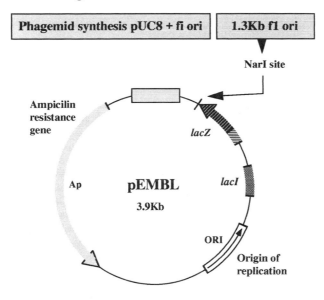

Fig. 5. General map of pEMBL phagemid.

5.2. Phage-Plasmid Hybrids: pEMBL Vectors

One of the first hybrid phage plasmid vectors was pEMBL *(13)*. This vector was constructed by inserting a 1.3 Kb fragment of the f1 phage ori and morphogenesis elements into a unique *Nar*I site (**Fig. 5**). This was undertaken so that both orientations were present, designated (-) or (+). Following superinfection with f1 helper phage the f1 ori is activated, allowing single-stranded DNA to be produced. The phage is assembled into a phage coat extruded through the periplasm and secreted into the culture medium. Without superinfection the phagemid replicates as a pUC type plasmid and in the replicative form (RF) the DNA isolated is double-stranded. This allows further manipulations, such as restriction digestion, ligation, and mapping analysis to be performed. The yields of DNA from the phagemid is usually regarded as superior to those obtained from RF preparations of the M13mp series of vectors.

5.3. Sophisticated λ-f1 Phage Hybrid Vectors: λZapII

λZapII is a commercially produced cloning vector based on a λ insertional vector that has found widespread use in gene cloning and the construction of gene libraries *(14)*. Cloning vectors derived from λ bacteriophage are advantageous since on a molar basis they offer approx 16-fold advantage in cloning efficiency in comparison with the most efficient plasmid cloning vectors. This allows a greater chance of representing a fragment or cDNA when a library is prepared in a λ-derived system. λZapII has all the basic features of such a vector, including six unique cloning sites clustered into a multiple cloning site (MCS), allowing an insert capacity of approx 10 Kb. Furthermore, the MCS is located within a *LacZ* region providing a blue/white screening system based on insertional inactivation and the possibility to express cloned products as fusion proteins with β galactosidase. This is the result of an inducible lac promoter upstream of the *lacZ* gene and allows screening with nucleic acid or antibody probes. Other features

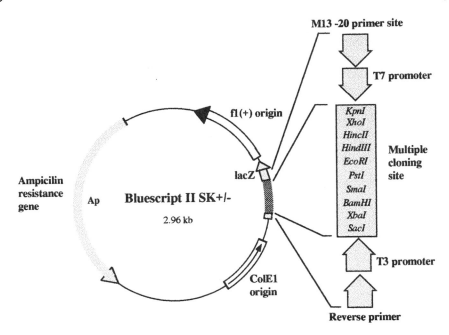

Fig. 6. Elements of Bluescript II SK+/– phagemid.

that make this a desirable cloning vector are the ability to produce RNA transcripts driven from a T7 or a T3 promoter that flank the cloning sites.

One of the most useful features of λZapII is that it has been designed to allow automatic excision in vivo of a small 2.9 kb colony-producing phagemid, Bluescript SK- (**Fig. 6**). This technique is sometimes termed single-stranded DNA rescue and occurs as the result of superinfection of the host cell, analogous to that described for the pEMBL vectors. Initially F' *E. coli* is infected with the recombinant λZapII. Helper phage M13VCS or R408 is then added to the cells, which are grown for an additional period of approx 4 h. The transacting proteins from the helper phage recognize two separate domains (initiator and terminator) within the λZapII vector. These signals are recognized by the phage geneII protein and a new DNA strand is synthesized, displacing the existing strand. The displaced strand is circularized and packaged as a filamentous phage by the helper phage proteins after which the packaged phagemid is secreted from the cell. The *E. coli* and λZapII are then heated for 20 min at 70°C, The Bluescript SK- and helper phage are resistant to this treatment. Bluescript SK- phagemids in the supernatant may be recovered by infecting an F' strain and plating on ampicillin plates, producing bacterial colonies. Thus, the λZapII-Bluescript SK- allows a number of diverse manipulations to be undertaken without the necessity of subcloning that many other strategies use (**Fig. 7**).

The Bluescript vector may be used in its own right as a cloning vector and manipulated as if it were a plasmid. It may, like M13, be used as a vector in nucleotide sequencing and site-directed mutagenesis. It is also possible to produce RNA transcripts that may be used in the production of labeled RNA probes of high specific activity in vitro. It is possible to select which transcript will be produced by supplying the appropriate RNA polymerase, either T3 or T7, prior to linearizing the vector at one

Insert DNA in a unique cloning sites in N terminal lacZ gene

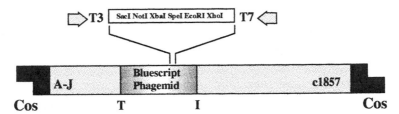

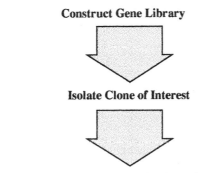

Construct Gene Library

Isolate Clone of Interest

Excise phagemid by co infection with helper phage

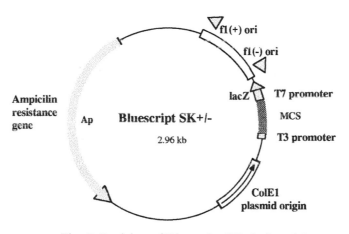

Fig. 7. Excision of Bluescript SK+/– from λ2ap vector.

of the restriction sites, such as *Not*I contained in the multiple cloning site. These "riboprobes" are useful in techniques, such as northern hybridization, RNase mapping, and *in situ* hybridization.

6. Surface Display Phagemid Vectors

As a result of the production of phagemid vectors and as a means of overcoming the problems of screening large numbers of clones generated from genomic libraries, a method for linking the phenotype or expressed protein with the genotype has been developed *(15)*. This is termed phage display since a functional protein is linked to a major coat protein of a coliphage whereas the single-stranded gene encoding the protein is packaged within the virion. There are now a number of different formats for phage display using f1, M13, or phagemid with the protein of interest being linked to major coat protein gIII or gVIII *(16)*. One system exploits the ability to switch among separate, soluble, or linked proteins through the use of an amber codon in the presence of an appropriate growth medium *(17)*.

6.1. Applications of Surface Display Vectors

There are numerous applications for the display of proteins on the surface of bacteriophage virus and commercial organizations have been quick to exploit this technology, e.g., Stratagene SurfZap phagemid display system. One major application is the analysis and production of engineered antibodies, from which the technology was mainly developed. In general, phage-based systems have a number of novel applications in terms of ease of selection rather than screening of antibody fragments *(18)*. Surface expression libraries also compare well with the in vivo antibody repertoire in terms of the potential size and diversity. It also allows a rapid means of selection using antigen–affinity columns, thus obviating the need to screen for antibody binding activity as with recombinatorial libraries by allowing analysis using affinity chromatography. Phage-based cloning methods also offer the advantage of allowing mutagenesis to be performed with relative ease. This may allow the production of antibodies with affinities approaching those derived from the in vivo somatic hypermutation mechanism. It is possible that these types of libraries may provide a route to high-affinity recombinant antibody fragments that are difficult to produce by more conventional means.

Surface display libraries have also been prepared for the selection of ligands, hormones, and other polypeptides in addition to allowing studies on protein–protein or protein–DNA interactions or determining the precise binding domains in these receptor-ligand interactions *(20)*.

In general there have been many rapid and novel developments in the construction of gene libraries using phage molecules. The specialized vectors for producing single-stranded DNA are now commonplace and the combination of phage and plasmids also make phagemids a definite choice when considering a cloning vector. The extension of phagemid development into surface display will also no doubt provide a rapid means of clone characterization, expression, and manipulation.

References

1. Messing, J., Gronenborn, B., Muller-Hill, B., and Hofschneider, P. H. (1977) Filamentous coliphage M13 as a cloning vehicle insertion of a HindII fragment of the lac regulatory region in M13 replicative form *in vitro*. *Proc. Natl. Acad. Sci. USA* **74**, 3642–3646.
2. Yanisch-Peron, C., Vieira, J., and Messing, J. (1985) Improved M13 cloning phage cloning vectors and host strains: nucleotide sequences of the M13mp18 and pUC19 vectors. *Gene* **33**, 103–119.
3. Messing, J. and Vieira, J. (1982) A new pair of M13 vectors for selecting either DNA strand of double digest restriction fragments. *Gene* **19**, 269–276.
4. Sanger, F., Nicklen, S., and Coulsen, A. R. (1977) DNA sequencing with chain terminating inhibitors. *Proc. Natl. Acad. Sci. USA* **74**, 5463–5467.
5. Sanger, F., et al. (1980) Cloning in single-stranded bacteriophage as an aid to rapid DNA sequencing. *J. Mol. Biol.* **143**, 161–178.

6. Vieira, J. and Messing, J. (1982) The pUC plasmids, an M13mp7 derived system for insertion mutagenesis and sequencing with universal primers. *Gene* **19**, 259–268.

7. Benton, W. D. and Davis, R. W. (1977) Screening λgt recombinant clones by hydridization to single plaques *in situ*. *Science* **196**, 180–182.

8. Zoller, M. J. and Smith, M. (1983) Oligonuleotide-directed mutagenesis of DNA fragments cloned into M13 vectors. *Methods Enzymol.* **100**, 468–500.

9. Kramer, W., Drutsa, V., Jansen, H.-W., Kramer, B., Pflugfelder, M., and Fritz, H.-J. (1984) The gapped duplex DNA approach to oligonucleotide-directed mutation construction. *Nucleic Acids Res.* **12**, 9441–9456.

10. Kunkel, T. A. (1985) Rapid and efficient site specific mutagenesis without phenotypic selection. *Proc. Natl. Acad. Sci. USA* **82**, 488–492.

11. Taylor, J. W., Schmidt, W., Cosstick, R., Okruszek, A., and Eckstein, F. (1985) The use of phosphorothioate-modified DNA in restriction enzyme reactions to prepare nicked DNA. *Nucleic Acids Res.* **13**, 8749–8764.

12. Dotto, G. P., Enea, V., and Zinder, N. G. (1981) Functional analysis of bacteriophage f1 intergenic region. *Virology* **114**, 463–473.

13. Dente, L., Cesareni, G., and Cortese, R. (1983) pEMBL a new family of single stranded plasmids. *Nucleic Acids Res.* **11**, 1645–1655.

14. Short, J. M., Fernandez, J. M., Sorge, J. A., and Huse, W. D. (1988) λZAP: a bacteriophage lambda expression vector with *in vivo* excision properties. *Nucleic Acids Res.* **16**, 7583–600.

15. Parmley, S. E. and Smith, P. G. (1988) Antibody selectable fillamentous fd phage vectors: affinity purification of target genes. *Gene* **73**, 305–318.

16. Barbas, C. F., et al. (1991) Assembly of combinatorial libraries on phage surfaces (Phabs): the gene III site. *Proc. Natl. Acad. Sci. USA* **88**, 7978–7982.

17. Winter, G. and Milstein, C. (1991) Man made antibodies. *Nature* **349**, 293–297.

18. Rapley, R. (1995) The biotechnology of antibody engineering. *Mol. Biotechniques* **3(2)**, 139–154.

19 Lowman, H. B., Bass, S. H., Simpson, N., and Wells, J.A. (1991) Selecting high affinity binding proteins by monovalent phage display. *Biochemistry* **30**, 10,832–10,838.

20. Crameri, R. and Suter, M. (1993) Display of biologically active proteins on the surface of filamentous phages: cDNA cloning system for selection of functional gene products linked to the genetic information responsible for their production. *Gene* **137**, 69–75.

18

Retroviral Vectors

From Laboratory Tools to Molecular Medicines

Richard G. Vile, Anna Tuszynski, and Simon Castleden

1. Introduction

The description of the first retroviral packaging cell line in the early 1980s introduced the technology by which a wide variety of different genes could be reliably and efficiently transduced into target cells using retroviral vectors *(1)*. Although the early prototype packaging systems were not optimal, the use of retroviral vectors in the laboratory rapidly became widespread. Recombinant retroviruses can be used to introduce single copies of cloned genes into cultured cells in vitro, allowing investigators to study gene function very precisely in both qualitative and quantitative ways. Subsequently, recombinant retroviruses have been used for in vivo infections providing data on developmental and cell lineage relationships in intact organisms. As the understanding of retrovirology and vectorology has become more refined, the opportunities for the use of retroviral vectors have expanded to answer a variety of more specialized questions both in vitro and in vivo. Most recently, the issues of safety and efficiency of gene transfer have been addressed, and retroviral vectors have become the most frequently used vectors for therapeutic intervention using gene transfer to treat genetic disease (gene therapy). Although much remains to be done before retroviral vectors are optimally useful in this field, the transition from crude experimental tools to potential molecular medicines in just over a decade is a testament to the potential value that retroviral vectors offer to the many different branches of the research community.

2. Retrovirus Structure

A mature retrovirus comprises an inner core (nucleoid) enclosed in a phospholipid envelope (**Fig. 1**). The core comprises a protein capsid with icosahedral symmetry separated from the envelope by matrix protein. The capsid houses two copies of the positive sense viral mRNA genome, including 5'cap and 3'poly(A) structures. The two protein-bound RNA subunits are linked together close to their 5'-ends, at a dimer linkage structure, and there is a tRNA primer hybridized to each strand at a primer binding site, also close to the 5'-end of the genome. Also contained within the capsid are sev-

From: *Molecular Biomethods Handbook*
Edited by: R. Rapley and J. M. Walker © Humana Press Inc., Totowa, NJ

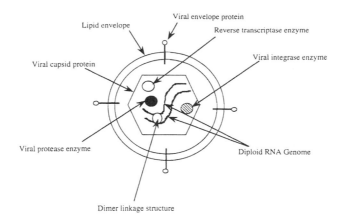

Fig. 1. Anatomy of the retroviral particle. A mature retrovirus comprises an inner core (nucleoid) enclosed in a phospholipid envelope. The capsid houses two copies of the positive sense viral mRNA genome linked together close to their 5'-ends at a dimer linkage structure. Contained within the capsid are several molecules of the virally encoded protease, reverse transcriptase, and integrase enzymes. The envelope is a roughly spherical phospholipid bilayer that derives from the plasma membrane of the virus-producing cell and is studded with closely packed oligomeric membrane spike glycoproteins.

eral molecules of the virally encoded protease, reverse transcriptase, and integrase enzymes. The envelope is a roughly spherical phospholipid bilayer that derives from the plasma membrane of the virus-producing cell and is studded with closely packed oligomeric membrane spike glycoproteins, which appear as surface projections on electron microscopy.

3. Taxonomy

The family of retroviridae consists of three groups, the Spumaviruses (or Foamy viruses), such as the Simian Syncytial Virus and Human Foamy Virus (HFV); the lentiviruses, such as the Human Immunodeficiency Virus types 1 and 2, as well as the Visna virus of sheep; and the Oncoviruses. This latter group can be further subdivided into tour classifications based on morphology, as seen under the electron microscope during viral maturation *(2)*. The group most commonly used to derive retroviral vectors are the C-type oncoviruses, of which Moloney Murine Leukemia (Mo-MLV) is the prototype.

4. Life Cycle

The initial step in retrovirus infection involves the binding of virions to specific cell-surface receptors through the cell recognition domains of their envelope spike glycoproteins **(Fig. 2)**. Following attachment, further events lead to membrane fusion either at the cell surface or in endocytic vesicles, with release of the viral core into the cytoplasm.

Reverse transcription of the single-stranded viral RNA into double-stranded DNA proceeds in the cytoplasm within the viral core, catalyzed by the virally encoded reverse transcriptase enzyme. This enzyme has RNA-dependent DNA polymerase, RNaseH, and DNA-dependent DNA polymerase activities. Reverse transcription

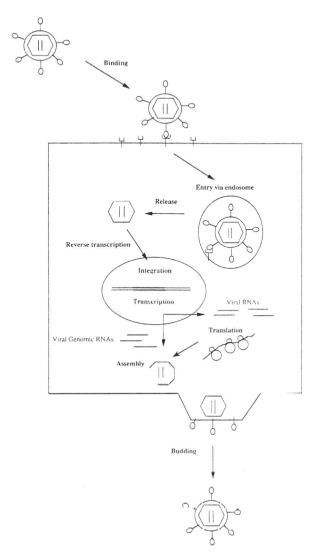

Fig. 2. The retroviral life cycle. The initial steps in retrovirus infection involve the binding of virions to specific cell-surface receptors, membrane fusion either at the cell surface or in endocytic vesicles, and release of the viral core into the cytoplasm Reverse transcription of the single-stranded viral RNA into double-stranded DNN proceeds in the cytoplasm within the viral core, catalyzed by the virally encoded reverse transcriptase enzyme. Reverse transcription generates a linear duplex DNA flanked by two identical viral LTRs, which is integrated into the cellular genome, catalyzed by the virally encoded integrase. Promoter/enhancer elements in the U3 region of the viral LTR drive transcription of the viral genome, and viral proteins are translated from genomic and subgenomic mRNAs. Viral morphogenesis includes packaging of genomic-length RNA molecules by the capsid proteins, and viral particles bud from the infected cell enclosed in the ENV-studded phospholipid bilayer derived from the cell membrane *(4,8)*.

involves two distinct DNA strand transfer reactions *(3)*. In the first, a short nascent segment of minus strand DNA, initiated by the bound tRNA primer, is translocated from the 5'-end of the plus strand viral RNA template to the 3'-end of the second RNA molecule. Plus strand DNA synthesis initiates at a polypurine tract on the growing

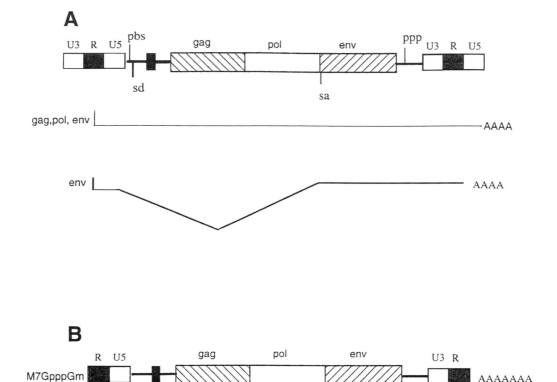

Fig. 3. **(A)** The proviral DNA genome of Mo-MLV as integrated into the host cell genome, showing the full-length genomic RNA transcript (from which the viral *gag* and *pol* genes are translated) and the subgenomic transcript produced by splicing of the full-length transcript, from which *env* is translated. PBS, primer binding site; sd, splice donor site; sa, splice acceptor site; PPP, priming site for synthesis of the plus strand DNA molecule. **(B)** The viral genomic RNA transcript as packaged into the viral particle. The proteins encoded by the *gag*, *pol*, and *env* genes are also shown.

minus strand, primed by an RNA primer generated by RNaseH cleavage of the viral RNA/ DNA hybrid. This generates a short plus strand DNA that includes a primer binding site sequence complementary to the tRNA still bound to the 5'-end of the minus strand DNA template. In the second strand transfer, this short plus strand DNA is transferred to the 3'-end of the same minus strand, allowing the synthesis of a linear duplex DNA flanked by two identical viral LTRs. This complex set of molecular reactions is reviewed in more detail in ref. *(4)*.

5. Transport into the Nucleus

The immediate precursor for integration into the host cell genome is a linear molecule of double-stranded DNA with two bases removed from the 3'-terminus of both strands *(5)*. Retroviral integration is nonhomologous with some degree of sequence selectivity *(6)* and a preference for transcriptionally active host integration sites *(7)*. The virally encoded integrase binds specifically to the termini of linear viral DNA and removes two 3'-terminal nucleotides. The 3'-ends are then covalently attached to the 5'-phosphoryl ends of staggered nicks (4 bp long) in host cell DNA, which may also be generated by the integrase protein. Auxiliary cellular proteins are required for gap repair at the site of insertion.

Promoter/enhancer elements in the U3 region of the viral LTR drive transcription of the viral genome **(Fig. 3)**. For MLVs, the full-length transcript encodes the Pr65^{gag} (core) and Pr180$^{gag-pol}$ (core enzyme) polyproteins, whereas a smaller, spliced transcript encodes the envelope proteins **(Fig. 3A)**. A 10:1 of Pr65^{gag} to Pr180$^{gag-pol}$ is maintained by means of an amber suppressor codon at the end of Pr65^{gag}.

The nascent translation product of the *env* gene enters the endoplasmic reticulum where the signal peptide is cleaved and the protein undergoes core (N-linked) glycosylation, folding, and oligomerization before it is transported to the Golgi apparatus for further processing of N-linked carbohydrate, addition of O-linked carbohydrate, and proteolytic cleavage. The proteolytic cleavage that occurs during transport to the cell surface divides each protomeric subunit in the oligomeric assembly into a small membrane-anchored TM component linked through its extracellular domain to a larger extracellular SU component. The SU/TM spikes are trimers or tetramers and the SU-TM linkage may be relatively unstable, such that SU is variably shed from viruses or infected cells. The viral life cycle is summarized in **Fig. 2** *(4,8)*.

6. Vectors and Packaging Cells

6.1. Principles of Vector Construction

Wild-type retroviruses have been modified to become vehicles for the delivery, stable integration, and expression of cloned genes into a wide variety of cells, both for experimental and, recently, for therapeutic purposes. Although several other viruses have also been used *(9)*, the prototypic retrovirus on which many vectors are derived is the Mo-MLV.

To achieve the aims of carriage and expression of nonviral genes, the vector must be able to behave as a retroviral genome to allow it to pass as a virus from the producer cell line. Hence, its DNA must contain the regions of the wild-type retroviral genome required *in cis* for incorporation in a, retroviral particle. In addition, the vector must contain regulatory signals that lead to the optimization of the expression of the cloned gene once the vector is integrated in the target cell as a provirus. These regulatory signals may or may not be provided by viral DNA sequences.

Although all the viral structural genes can be discarded and replaced by heterologous coding sequences, certain essential sequence elements must be retained within the vector **(Fig. 4)**. These include:

1. The packaging sequence that ensures the encapsidation of the vector RNA into virions (Psi in murine vectors or E in avian systems). More recent vectors retain an extended Psi sequence, which incorporates the start of the *gag* gene (Psi$^+$), but with the AUG start codon of the viral gene mutated *(10,11)*.
2. The tRNA binding site that is necessary to prime reverse transcription of the RNA form of the vector (in the virion) into DNA within the target cell (–PBS).

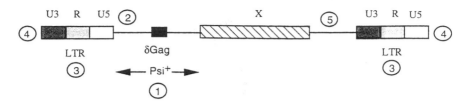

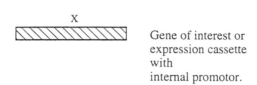

Gene of interest or
expression cassette
with
internal promotor.

Fig. 4. The essential cis-acting components of a retroviral vector. To achieve the aims of
carriage and expression of nonviral genes, the vector must be able to behave as a retroviral genome
to allow it to pass as a virus from the producer cell line. All the viral structural genes can be
discarded and replaced by heterologous coding sequences (X), but certain essential sequence
elements must be retained within the vector, including (1) the packaging sequence, (2) the tRNA
binding site (PBS–), (3) sequences in the LTRs that permit the "jumping" of the reverse transcriptase between RNA strands during DNA synthesis, (4) specific sequences near the ends of the
LTRs for integration of the vector DNA, and (5) the sequences adjoining the 3'LTR that serve as
the priming site for synthesis of the plus strand DNA molecule (+PBS).

3. Sequences in the LTRs that permit the "jumping" of the reverse transcriptase between
   RNA strands during DNA synthesis.
4. Specific sequences near the ends of the LTRs that are necessary for the integration of the
   vector DNA into the host cell chromosome in the ordered and reproducible manner characteristic of retroviruses *(12)*.
5. The sequences adjoining the 3'LTR that serve as the priming site for synthesis of the plus
   strand DNA molecule (+PBS).

Since vector genomes do not require that the viral structural genes *gag*, *pow*, and *env*
be retained, nonviral genes can be cloned into the space vacated by their removal. However, the overall length of the construct cannot exceed 10 kbp, since packaging efficiency declines with increasing length of RNA. This places a restriction on the choice
of genes—and accompanying expression regulatory signals—that can be incorporated
into retroviral vectors *(see* **Subheading 8.3.***)* to an operational limit of about 8 kbp of
added sequence.

Once the vector RNA has been passed as virus into a target cell, it will be reverse-
transcribed *(3,4)*, transported into the nucleus, and integrated into the host cell genome.
Murine oncoviruses require active cell division for this process to be completed *(13)*,
but this is not the case for all retroviruses *(14) (see* **Subheading 10.1.***)*. Following
integration, the LTR can be used to direct transcription of the gene cloned into the
construct. The transcription signals in the LTR reside in the U3 and R regions, and
include recognition sequences for RNA polymerase II and a strong, and largely cell

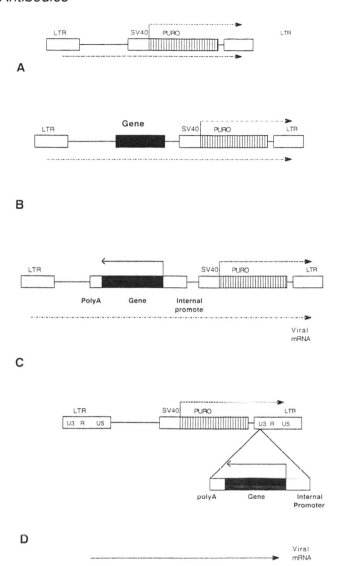

Fig. 5. A selection of different retroviral vector designs. **(A)** pBabe Puro in which a selectable marker gene is expressed from an internal SV40 early promoter *(11)*, allowing simple cell marking. **(B)** pBabe Puro with an additional gene cloned under the control of the viral LTR. This type of vector allows gene transfer and selection clef infected clones. **(C)** Elaboration of B allows the gene of interest to be expressed from its own internal promoter, which might be used for cell-type-specific gene expression *(16)*. **(D)** Expression cassettes cloned into the 3'LTR become duplicated and appear in the integrated provirus in both 3' and 5'-LTRs giving so-called double-expression vectors, which can in theory produce twice as many transcripts of the gene of interest *(17)*. Other types of vector design are described in **refs. 9,20**.

nonspecific, enhancer **(Fig. 4)**. In addition, a polyadenylation site exists in the R region, which is recognized by the processing RNA polymerase II only in the 3'LTR.

The earliest vectors used the viral LTR to direct expression of the inserted gene **(Fig. 5B)** but since then, a wide variety of different vectors have been constructed **(Fig. 5)**.

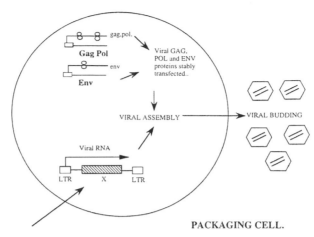

Fig. 6. A "third-generation" retroviral packaging cell line *(30,31)*. Expression cassettes with minimal homology to retroviral sequences outside of the coding sequences are stably transfected into a cell line, such as NIH3T3 fibroblasts. These cassettes constitutively express high levels of the viral structural and catalytic proteins, GAG, POL, and ENV. When a retroviral vector is transferred into the packaging cell, genomic-length RNA transcripts are transcribed from the LTR, and these transcripts can be packaged into viral particles because they contain the Psi packaging signal.

These differ in the number of genes they can express (as many as three) and the promoter/enhancer elements used to express them (ranging from the viral LTR to tissue-specific internal promoters) *(15)*, and the position and orientation in which the genes/expression cassettes have been inserted into the vector backbone *(16)* (including placement of cassettes within the 3'LTR to generate double-copy vectors *(17)* **(Fig. 5)**. For a comprehensive discussion of the varied range of vector types, the reader is referred to more detailed reviews *(9,18–20)*.

6.2. Retroviral Packaging Cell Lines

Retroviral vectors retain only those cis-acting sequences from the original viral genome that are sufficient to identify them as "retroviral." In order to package these vectors into infectious virions, the other viral functions, deleted from the vector genome, must be provided *in trans* in cells that, therefore, are known as packaging cells. Such packaging cells stably express the viral *gag*, *pol* and *env* genes from plasmids that cannot themselves be packaged by their own encoded proteins, because they lack the essential Psi packaging sequence. However, when a vector genome is transfected into such a packaging cell, the viral proteins recognize and package the vector RNA genome into viral particles that are released into the culture supernatant **(Fig. 6)**.

Retrovirus titer is limited by the rate at which infectious particles bud from the producer cells into the culture supernatant and the rate at which they lose their infectivity (i.e., their stability). Production of infectious particles comprises at least three independent concentration-driven processes:

1. Assembly of nucleocapsids from *gag* and *pol* gene products and their subsequent budding from the producer cell surface;
2. Incorporation of packageable RNA transcripts into nucleocapsids; and
3. Incorporation of envelope spike glycoproteins into budding virions.

In addition, loss of infectivity for MLV-based vectors is significant during continued incubation at 37°C (t1/2 ~ 8 h), after each freeze–thaw cycle (~50%) and after virus concentration by centrifugation. This poor stability has been attributed to dissociation of gp70SU envelope protein from the particle surface.

Packaging cell lines have been constructed using both avian and murine retroviral structural proteins. They are reviewed in **ref. 21** and discussed in more detail below.

7. Retroviral Vectors as Laboratory Tools

Investigators wishing to study the biological effects of expression of a particular gene in a specified in vitro system have used retroviral-mediated gene transfer widely. The availability of (relatively) high-titer stocks of well-characterized recombinant retroviruses enables defined genetic sequences to be transferred into target cells of choice both stably and efficiently. The inclusion of a selectable marker in the vector allows long-term selection for clones containing the gene *(22)*, and the efficiency of viral transfer makes this the method of choice over other physical transfer techniques, such as calcium phosphate transfection. In theory following integration, the provirus is passed to all daughter cells heritably and stably, and expression of the transferred gene can be stable for many cell generations (although inactivation of gene expression has been reported as discussed below). The only other genetic information transferred to the cell line is the viral sequences of the vector, and it is, therefore, possible to create infected cell lines that differ from the parental line only in expression of the gene of interest. It must be remembered, however, that integration of the vector might occur within an important locus in the target cell and that the cell is often under selective pressure, so that the infection event itself may alter the cell properties in ways unrelated to the transferred gene(s). With the advent of vectors encoding more than one gene or by multiple infections with vectors encoding different genes (and selectable markers), the interaction of different introduced genes can be studied within a defined system, although the number of variable confounding factors that might distort the results increases correspondingly.

Retroviral vectors have subsequently been used for experiments other than simple transfer of genes into cultured cells. For example, infection of embryos with vectors encoding marker genes has allowed developmental pathways to be followed and cell lineage relationships to be established *(23)*. In addition, more intricate designs of vectors have been developed to identify cellular promoters *(24)*, and retroviral insertional mutagenesis has been exploited to clone cellular genes involved in various metabolic pathways *(25)*. However, with developments in the design of packaging cells that have made them progressively safer and more efficient *(9)*, the most exciting prospective use of retroviral vectors in the coming years has become their potential as therapeutic agents for gene transfer either directly into patients or into patients' cells in human gene therapy protocols *(26)*.

8. Retroviral Vectors in Gene Therapy

8.1. Safety

Irrespective of the type of gene therapy for which retroviral vectors are being used, they must be safe following transfer into patients. Therefore, vector stocks must include only recombinant virus, with no contamination by replication competent: helper virus. In addition, the recombinant virus should be well characterized and, ideally, homogeneously pure with no mutant forms present.

1. Contamination with replication-competent helper virus: For applications in gene therapy, it is essential that stocks of recombinant virus released from the packaging cells do not contain contaminating replication-competent virus. Release of so-called helper virus has been detected from some of the early producer cell lines, making them unsuitable for the production of vectors for human gene therapy. Early packaging cells contained retroviral genomes deleted only in the Psi region *(1)* and, because of a leakiness of the deletion, a small proportion (0.1%) of genomes transcribed from these modified helper plasmids can be packaged into budding virions. Recombination between a co-packaged vector genome (Psi + ve), and this deleted vector (Psi − ve) can lead to rescue of the Psi deletion into a genome encoding all the viral structural genes and the Psi sequence—generating essentially a genome that can replicate independently in the target cells.

   Second- and third-"generation" packaging cell lines have been developed in which more than a single recombination event is required to generate wild-type helper genomes within the virion. Thus, deletion of the 3'-sequences, as well as Psi, from the helper plasmid expressing *gag, pol*, and *env* removed the polypurine tract required for reverse transcription and the 3'LTR to generate the PA317 cell line *(27)* (second generation), which has been used to produce recombinant virus used in human clinical trials (e.g., *28*). No patients who have been exposed to such stocks have been shown to have received replication competent virus, and toxicity studies in primates have been very encouraging *(29)*. The third generation of packaging cells has effectively solved the problem of helper virus generation by separating the *gag, pol* transcription unit from the *env* transcription unit on different plasmids *(30,31)* (**Fig. 6**) and by using packaging constructs that have minimal areas of homology with the vector genome *(32–34)*. In these ways, even if the packaging constructs are copackaged with a vector genome, multiple recombination events would be required to rescue an intact genome *(35)*. However, it is still possible that such cell lines continue to package endogenous viral genomes, which may then be transferred into human patients during gene therapy protocols *(36)*. To avoid such risks, new packaging cells will have to be developed based on primate cell lines already characterized for release of such endogenous viruses.

   It also appears that the margin for error associated with inadvertent infection with replicating virus is relatively wide. In experiments that tested the dangers of accidental contamination, 3 of 10 severely immunosuppressed monkeys treated with recombinant vector stocks contaminated with wild-type virus developed T-cell lymphomas, which all contained replication competent proviral DNA *(37)*. However, the levels of contaminating wild-type virus required to produce disease were orders of magnitude above those that can routinely be detected in sensitive in vitro assays for replicating virus *(38)*. Other studies in which monkeys were also exposed to high levels of replicating MLV (in excess of 10^9 ffu) showed no signs of malignancy even after 7 yr of followup *(39,40)*, suggesting that wild-type MLV is not an acute pathogen for primates. Therefore, although the presence of such contamination must continually be assayed, the risk of transferring replicating virus to a patient should be very small.

A second consideration for the use of retroviral vectors in gene therapy is the possibility that the vector will cause neoplastic transformation of some of the patient's cells by integrating within, or near, critical cellular genes *(41)*. Retroviral-mediated cell transformation has been well documented in vitro in animal systems and in humans, so intentional exposure of patients to retroviral vectors must be carried out with caution *(41)*. However, theoretical models of the chances of causing insertional mutagenesis by a retroviral vector suggest that the overall risk is very low *(42)*. In addition, further improvement in vector design will reduce these risks still further by the development of high-titer self-inactivating vectors in which the strong viral enhancers in the LTRs are rendered transcriptionally silent in the infected cells *(43)*. Nonetheless, there remains the risk of insertional mutagenesis from physical disruption of a critical gene (such as a tumor-suppressor gene) or transcriptional activation of a proto-oncogene by any internal promoters within the vector. These risks can, in turn, be further restricted by using promoter/enhancer elements, which will only be active in the target cell type (*see* **Subheading 8.4.**).

The murine retroviruses that are currently used in clinical trials are rapidly inactivated by human complement *(44,45)*. Although this is a drawback with regard to the efficiency with which such viruses can infect target cells in vivo, it can also be considered advantageous in preventing the spread of viruses through the body in the unlikely event of introduction of replication competent virus into a patient. Several groups are now developing packaging cell lines to produce complement-insensitive virus to increase the in vivo titers attainable for gene delivery *(45)*.

2. Purity of recombinant viral stocks: Ideally, the nature and properties of the recombinant vectors administered to a patient should be well characterized. This can be achieved to high levels of confidence at the level of preparation of the plasmid DNA that is transfected into the packaging line, but is far more difficult to control and monitor thereafter. A characteristic feature of the retroviral reverse transcriptase is that it is highly error-prone *(46,47)*. Therefore, following infection of target cells, it is probable that the proviral DNA generated from the RNA genomes of the virions will differ in a significant proportion of the cells from the intended sequences. In many cases, the mutations introduced by reverse transcriptase will not affect the sequence of the inserted gene, will be silent, or may even destroy its function. However, in some circumstances, such error-prone RT-induced mutation may be deleterious to the target cell by creating a new coding sequence. For example, recombinant retroviruses may be used to deliver correct copies of tumor-suppressor genes, such as *p53*, to tumor cells in order to reverse the malignant phenotype *(48)*. However, a fraction of the infected cells may instead receive a provirus encoding a mutated *p53* gene, as a result of fidelity errors of reverse transcription. If the mutation converts the normal *p53* into a transforming mutant of *p53*, the infected cells may become transformed themselves. In the absence of engineered reverse transcriptase enzymes, which have a higher intrinsic fidelity, the risk of such events cannot currently be controlled, but should be considered in the use of gene therapy strategies where the introduction of mutated "therapeutic" gene may be damaging.

Other criteria for the purity of the recombinant retroviral stocks include screening for mycoplasma, bacteria, and endogenous viruses released by the producer cells. Although these are stringently tested for, their presence is a continual cause for concern.

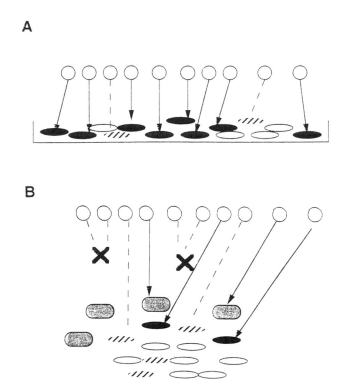

Fig. 7. The difference between real and effective retroviral titer. **(A)** In vitro, dividing target cells are easily accessible to virus particles in the supernatant, and a large percentage of cells can be infected. **(B)** In vivo, the target cells are obstructed by surrounding host tissue, and a significant proportion may not be dividing when the virus is present; in addition, many virions will be inactivated by host immune cells and complement. Therefore, the proportion of cells that can be productively infected by the same amount of virus is greatly reduced. Key: ⬬, Target cell productively infected by a retroviral particle; ⬭, uninfected target cell; ⬭, quiescent target cell (uninfectable); ○, retroviral particle; ⬬, surrounding host cell; **X**, virus inactivation (e.g., human complement or immune cell); ⟶, productive infection; – – – –, abortive infection.

8.2. Retroviral Titer

Typically, retroviral producer clones can generate virus at levels of 10^6–10^7 infectious particles/mL of tissue-culture supernatant. This figure will vary between lines, vectors, and indicator cell lines, but represents the order of magnitude that retroviral-mediated gene transfer is currently likely to achieve. However, although these figures represent "real" viral titers, as judged under optimal in vitro conditions, the "effective" viral titers that such stocks can reach when used for in vivo gene delivery to human patients are likely to be much lower **(Fig. 7)** for a variety of reasons:

1. Complement inactivation: Murine retroviruses, on which current vectors for use in gene therapy are based, are rapidly inactivated by human complement. Wherefore, the half life of viruses in the culture dish is considerably longer than that in the circulation. In addition, retroviral particles are themselves relatively unstable *(see above)*. An area of active research

is the development of viruses with modified envelope proteins which are no longer sensitive to human complement inactivation, thereby increasing "effective" titers in vivo *(45)*.

2. Murine retroviral vectors only infect actively dividing cells: C-type oncoviruses require mitotically active cells for productive infection to occur (involving reverse transcription, nuclear transport, and integration) *(13,14)*. Hence, target cells must be dividing during the short window of time during which they are exposed to the virus. Protocols in which cells can be removed from the body and stimulated to proliferate (such as lymphocytes for transfer of the ADA gene or tumor cells for cytokine gene transfer *[49]*) are, therefore, more suited to retroviral infection. Some retroviruses, such as HIV, do not appear to be completely restricted to infection of dividing cells *(14)* and vectors based on these viruses may expand the number of situations for which retroviral gene transfer may be used in the future *(see below)*.

3. Penetration of target tissues: Viral stocks titered in vitro are exposed to ordered monolayers of proliferating cells in the absence of inactivating mechanisms. This is unlikely to be the case for in vivo gene delivery. For instance, virus delivered to the site of a tumor deposit may be able to infect the cells around the periphery of the tumor, but may not penetrate more than two or three cell layers deep.

Taken together, it is clear that the "effective" titer of retroviral stocks is often likely to be considerably lower than that calculated in the culture dish **(Fig. 7)**, and for effective gene therapy of many human somatic tissues, these titers are really too low. Efforts are now focusing on the production of higher titers of retroviral vectors while maintaining the safety of the packaging lines. It may prove possible to achieve this by optimizing the physical conditions under which virus is harvested or by modifying the biological properties of the packaging systems.

This latter goal could be achieved by increasing either the copy number of the vector in, or the concentration of structural proteins produced from, each packaging cell, or both. Thus, titers have been increased by coculturing producer cells that release viruses with different tropisms (ping-pong), so that the average copy number of proviral copies per producer cell is raised above one *(50)*. Packaging plasmids have also been constructed incorporating sequences that lead to amplification of the plasmid within the packaging line so that higher levels of retroviral proteins can be produced *(34,51)*. Finally, the ability of envelope proteins of one virus to pseudotype core particles of a different retrovirus *(52)* has been used to generate pseudotype viral stocks to very high relative titers *(53)*. Nonetheless, the inability to produce safe stocks of recombinant retroviruses to titers above 10^7/mL remains a major obstacle to the effective use of retroviral vectors in gene therapy protocols (*see* **Subheading 10.2.**).

8.3. Gene Expression

The principal attraction of the use of retroviruses as vectors is that viral enzymes catalyze the eventual integration of the vector genome into the host cell chromosome of dividing cells. Therefore, genes expressed from within the vector can be expressed from a transcription unit that is, in theory, stable within the infected cell and is also passed onto all of its daughter cells following cell division.

However, to ensure stable and high levels of expression, it is now clear that the choice of promoter to drive expression of the therapeutic gene is also critical. In many cases, the viral LTR has been used to direct transcription of inserted genes, since it is a strong promoter that is expressed in a wide variety of different cell types **(Fig. 5B)**. Unfortunately, although many cells infected with such vectors do indeed show good levels of expression when maintained in vitro, they often appear to lose expression after a few weeks when the cells are returned in vivo. The mechanisms for loss of expression are currently poorly understood, but include shutdown of expression from viral promoters by methylation *(54)*, or by other means even though the cells clearly maintain integrated proviruses within their genomes *(55)*, or by loss of the proviral DNA itself *(56)*. This loss of expression detracts greatly from the proposed use of retroviral vectors for gene therapy of diseases where chronic expression of a therapeutic gene (such as CFTR or ADA) is required to correct a metabolic disorder over the life-time of the patient. However, in other protocols where high-level expression of the therapeutic gene may be required for only relatively short periods, such as for the delivery of cytokine genes to tumor cells *(57,58)*, these factors may be less significant.

To circumvent these problems, vectors are now being developed in which the therapeutic gene is expressed from cellular (Fig. C,D), rather than viral promoters, since it has been suggested that vector transcription is subject to programmed control of cellular gene expression *(55)*. Promoters that are normally constitutively active in the target cells may be less susceptible to transcriptional shutdown than "foreign" promoters appear to be. However, despite encouraging results when housekeeping gene or tissue-specific promoters have been used to express marker genes within a retroviral vector *(59,60)*, placement of cellular promoters within the context of a retroviral vector can lead to interference effects between promoters *(16,61–64)* as well as a blanket methylation shutdown of all vector-associated promoters *(54)*. Such interactions may be decreased in the future by the construction of self-inactivating vectors in which all viral enhancer and promoter elements have been removed in the infected cells, leaving just a cellular promoter that is known to be highly active in the target tissue to express the therapeutic gene *(43)*.

Another drawback to the use of retroviral vectors for gene therapy is that there is a tight restriction on the amount of foreign DNA, up to 8 kbp, which can be inserted into the vector. Therefore, such vectors cannot deliver genes that are large, such as the Duchenne Muscular Dystrophy, in protocols that require gene replacement in affected cells. In addition, protocols that require the codelivery of more than one gene may also be problematic owing to size considerations, especially if internal promoters are required for optimal levels of expression (*see* **Subheading 6.1.**). However, recently, picornavirus sequences, known as Internal Ribosome Entry Sites (IRES), have been used within retroviral vectors to direct translation of LTR-driven polycistronic RNA molecules *(65–67)*. Picornaviruses can express several proteins from a single polycistronic mRNA, because these IRES can direct cap-independent translation initiation at internal initiation codons; however, normal cellular mRNAs lacking IRES are incapable of translating a second gene from the same mRNA. Therefore, inclusion of poliovirus or encephalomyocarditids IRES sequences within retroviral transcription units now makes it possible to express several genes from the same promoter *(68)*, although retroviral vectors are still restricted to a total of 8 kbp of inserted DNA. This approach is particularly attractive for expression of multiple, small therapeutic genes, such as the cytokines, in which combination therapies may be more effective than single-expression constructs.

In summary, vectors of the future may contain multiple cDNAs driven by the viral LTR, or inserted into vector backbones in which viral transcriptional elements have been deleted and cellular promoters are used to direct high-level, stable expression.

8.4. Targeted Gene Therapy Using Retroviral Vectors

1. Cell-surface targeting: It is currently not possible to target infection of distinct cell types by retroviral vectors at the level of cell-surface binding, unless the homogeneous population of target cells is physically separate (for example, by cell explant from the patient) *(63)*. Therefore, use of retroviral vectors for in vivo gene therapist has been limited because of the risk of infecting bystander cell types. In vivo targeting has been achieved in some cases by exploiting the inability of retroviruses to infect nondividing cells; therefore, retroviruses have been, used to deliver cytotoxic genes to (dividing) tumor cells in the brain where the surrounding neuronal tissue is largely quiescent *(69)* *(see below)*.

   Lack of tight cell targeting may not be a major factor in gene therapy protocols in which expression of the therapeutic gene is unlikely to be toxic to normal cells—such as for the delivery of extra copies of a tumor-suppressor gene *(70)* or a corrective gene, such as the CFTR gene *(71)*. However, expression of cytotoxic genes and possibly of immunomodulatory genes for cancer gene therapy in vivo requires a much higher stringency of delivery *(72)*.

   In principle, retroviral vectors provide excellent opportunities for the surface targeting of the gene delivery vehicle because of the well-characterized packaging systems that already exist *(30)*, the cloning of relevant retroviral receptors *(73)*, and the ability to modify retroviral structural genes at appropriate domains relevant for receptor binding and entry *(74–76)*. Several groups have already demonstrated that the tropism of retroviral vectors can be altered by the incorporation of foreign or hybrid proteins into the envelope of the virus *(63,77–81)* or by formation of pseudotypes, which assume the tissue tropism of the pseudotyping envelope protein *(82)*. Although this technology is not yet well enough advanced for the production of high-titer, envelope-modified retroviral stocks, there is hope that the identification of target cell-specific molecules, such as tumor antigens *(83)*, may eventually allow surface targeted delivery of retroviral vectors in vivo.

2. Transcriptional targeting: A second level of targeting can be achieved by the incorporation of transcriptional control elements into the retroviral vector that restrict expression of the inserted gene to the target cells *(84)*. If such vectors can be developed, relatively promiscuous gene delivery could be tolerated (although the risks associated with retroviral infection still remain). Many retroviral vectors have been described in which tissue-specific transcriptional regulatory elements (promoters/enhancers/locus control regions) *(85)* have been used to drive different genes *(16,86–90)* **(Fig. 5C,D)**. In many cases, correct control of expression was maintained from the tissue-specific promoter, although there have also been reports or interference between different regulatory elements within such vectors leading to partial loss of cell type-specific expression, depending on the precise design of the vectored *(16,62,88)*. Future refinements of vector design may be able to overcome such problems by the identification of compact regulatory motifs that confer high-level tissue-specific expression onto heterologous genes *(85)* and by the development of vectors in which the number of interacting promoter elements is reduced to a minimum *(61)*.

3. Targeting retroviral integration: The safety of retroviral vectors is currently compromised because of the possibilities associated with insertional mutagenesis leading to inactivation of a crucial cellular gene or activation of a cellular proto-oncogene *(42)*. This arises because the chromosomal site of retroviral integration is essentially random. If this integration event could be directed to a known, transcriptionally active chromosomal loca-

tion, which is not close to important cellular loci, some of the safety concerns relating to use of these vectors would be alleviated. Although molecular understanding of the mechanisms of the viral integrase proteins is increasing *(5)*, it is still not possible to engineer targeted retroviral integration to specific sites. Whether the emerging technology on site-specific, homologous recombination and gene targeting with plasmid vectors can be applied to manipulation of retroviral vector sequences and integrase proteins used in packaging cells remains to be seen.

9. The Use of Retroviral Vectors in Current Clinical Trials

Retroviral vectors have properties that can be seen as either advantages, or simultaneously as disadvantages, to their use in clinical trials of gene therapy depending on the disease to be treated, the rationale of the therapeutic approach being tried, and the ethical/safety issues involved. For example, the ability of retroviral vectors to integrate is often quoted as a benefit because it leads, at least in theory, to long-term stable expression of the therapeutic gene; for replacement of missing genes, such as in ADA deficiency, or even potentially cystic fibrosis, this is clearly desirable *(91)*. In contrast, the integrating ability of retroviruses, compared to the episomal maintenance of adenovirus vectors, for example, is potentially dangerous, since it might lead to insertional mutagenesis and oncogenic transformation of the patient's cells. Therefore, each situation requires careful consideration of the desired goals compared to the properties that individual vectors offer.

9.1. Ex Vivo *Gene Transfer by Retroviral Vectors*

Retroviral vectors are currently the vector of choice for those clinical protocols that permit the explant of patient cells, their genetic modification in vitro, followed by their return in vivo. The problems of relatively low titers of retroviruses in vivo are not so important when gene transfer is accomplished in the culture dish and cells can be stimulated to divide to increase the rate of productive infection in a short time *(49)*. In addition, the transduced cells can, in theory, be suitably tested *(92)* or treated (irradiated) *(58)* to decrease the chance of their becoming transformed by the virus itself. However, problems associated with low levels of prolonged gene expression from retroviral vectors in vivo remain (*see* **Subheading 8.3.**).

Retroviral vectors are currently being used in the treatment/study of several different diseases by gene therapy using *ex vivo* gene transfer *(91)*. These include the transfer of:

1. (Marker and) cytokine genes to tumor-infiltrating cells for use in adoptive immunotherapy of cancer;
2. The adenosine deaminase (ADA) gene to autologous T-lymphocytes for treatment of severe combined immunodeficiency;
3. Marker genes to investigate marrow reconstitution and the mechanism of relapse following chemotherapy;
4. Cytokine genes to autologous or allogeneic tumor cells or fibroblasts prior to their use as in vivo vaccines;
5. The low density lipoprotein (LDL) receptor to hepatocytes for treatment of familial hypercholesterolemia;
6. The HSVtk cytotoxic gene to autologous tumor cells for treatment of ovarian cancer metastases;

7. The gene for glucocerebrosidase into autologous CD34+ cells for treatment of Gaucher disease;
8. Marker/suicide therapeutic vectors into T-cells of patients infected with HIV;
9. The multi drug resistance (MDR) gene to bone marrow cells for chemoprotective therapy in cancer patients; and
10. The factor IX gene into autologous skin fibroblasts to treat hemophilia B.

9.2. In Vivo Gene Transfer by Retroviral Vectors

The problems of low "effective" titers in vivo **(Fig. 7)**, the lack of precise cell targeting, and the requirement for cell division for infection has restricted the number of protocols that use retroviral vectors as the vector of choice for direct in viva gene delivery. For retroviral delivery to be effective in vivo, the target cell population must be localized and accessible to viral injection. Infection of widespread, disseminated cells is unlikely because of the low titers and short serum half-life of murine vectors in the circulation.

Retroviruses have only been approved for direct in vivo delivery in two cases, both for the treatment of cancer. In the first, retroviruses are used to deliver a cytotoxic gene (the Herpes Simplex thymidine kinase, HSVtk, gene) to glioma or metastatic cells localized in the brain. The number of tumor cells that can be killed by this approach is significantly greater than the number of cells that actually become infected because of a "bystander" killing effect when the drug ganciclovir is administered to the patients *(93)*. Retroviral producer cells are also injected stereotactically directly into the tumor, so localized injection, coupled with continual virus production and a powerful bystander killing all combine to increase the effective titer of the virus at the site of disease. In addition, infection of unaffected neuronal tissue is kept to a minimum because these cells are not dividing, unlike the tumor cells *(69,94)*. In these ways, the problems associated with targeting and titer are decreased.

In the second clinical protocol, retroviral vectors will be administered intratracheally to deliver either antisense constructs against activated Ki-*ras* oncogenes or functional *p53* tumor-suppressor genes to small-cell lung cancer cells *(95,96)*. A bystander killing effect of uninfected tumor cells has also been reported for this approach, although its origin is unclear. It remains to be seen if retroviral vectors, aimed at correction of a single genetic defect in multimutated human tumor cells, will be sufficiently efficient or effective to give clinical benefits in human patients *(72,97)*.

Other clinical protocols that necessitate in vivo gene delivery have largely used other vector systems for gene transfer. For instance, replacement of the cystic fibrosis gene into airway epithelial cells is currently being tried using liposome-mediated plasmid delivery or adenoviral vectors rather than retroviruses *(98–100)*. The issues of safety (oncogenic transformation), efficiency of delivery (low titers), and targeting of nondividing epithelial cells have mediated against the use of retroviral vectors in these cases. For the future, it seems likely that retroviruses cells may be useful for the delivery of genes where only a proportion of target cells need to be transduced to elicit a therapeutic response, such as for transfer of immunomodulatory genes to tumor cells *in situ* *(101)*. Implantation of retroviral producer cells may also be important in such cases both to increase effective viral titers and to augment antitumor immune responses *(102)*.

10. Retroviral Vectors—Future Directions

10.1. Novel Vectors

It is clear that there are several drawbacks to the use of murine retroviral vectors in human gene therapy protocols, including low titers, complement sensitivity, the risks associated with insertional mutagenesis, and the inability to infect nondividing cells. Certain of these problems might be solved by the development of retroviral vectors based on different viruses. Thus, the ability of other viruses, such as the vesicular stomatitis virus (VSV), to pseudotype the Mo-MLV core particle to high titer (~10^9/mL) has recently been reported *(53)*. Similarly, classes of retroviruses other than the C-type oncoviruses have not yet been fully exploited for use as vectors, even though they share essentially the same life cycle.

For instance, the HFV (a member of the Spumaviridae class of retroviruses) can infect humans, but as yet, is not clearly associated with any disease and does not appear to be so sensitive to inactivation by human complement *(103)*. Although HFV is a "complex" retrovirus (it contains accessory viral genes involved in transactivation of its LTR), several groups are currently trying to develop vectors based on this genome, which may, therefore, retain higher titers in vivo. Similarly, the D-type oncoviruses, such as Mason Pfizer Monkey virus (MPMV), are not known to be linked to human disease and might provide another source of retroviral vectors *(104)*. The lentiviruses, of which the HIV types 1 and 2 are members, are also potential candidates for vectors for gene therapy, especially of AIDS *(105)*. Since the genetic controls of HIV gene expression are so complex (involving the regulated expression of a series of viral transactivator proteins not found in the C-type oncoviruses), it is unlikely that HIV vectors would be used for gene therapy protocols in which simple on/off high-level expression of therapeutic genes is required (such as expression of the ADA or cystic fibrosis genes). However, paradoxically, the complexity of the expression of genes from the HIV LTR might be used against the virus; one potential strategy is to infect CD4+ T-cells—the primary target of HIV infection—with HIV-based vectors encoding suicide genes. Expression of the suicide gene would require HIV transactivator proteins to be supplied by wild-type infection of the T-cells. Administration of the appropriate prodrug could then be used to destroy HIV-infected T-cells before they can release virus to infect more lymphocytes *(106)*. This and other similar approaches may be useful in controlling HIV infection *(107)*.

HIV may also be of value in overcoming the inability of C-type virus vectors to infect quiescent cells. Recent evidence suggests that HIV can productively infect nondividing cells *(14)*, unlike murine retroviruses *(13)*. This property may be owing, at least in part, to the presence of a nuclear transport signal that allows transport of the reverse-transcribed provirus to the nucleus even in the absence of mitosis *(108)*. It has not set been shown if incorporation of this sequence into conventional vectors can allow similar division-independent infection to occur, but it is clear that lessons will be learned from a wide variety of different retroviruses in construction of the vectors of the future.

10.2. Replication-Competent Retroviral Vectors

The overriding drawback to the use of retroviral vectors in their current form is their low titers. Although efforts are being made to construct more efficient packaging cell lines *(34)*, more highly packageable vectors *(9,11)*, and novel retroviral vectors *(104,107)*/pseudotypes *(53)*, it seems unlikely that retroviral stocks will be developed that have titers high enough to be realistically useful for gene delivery to most target populations in vivo. Therefore, it has been suggested that replication-competent (retro)viruses may eventually be developed which can overcome the disadvantage of low titers inherent to current retroviral vectors *(109,110)*. Only a small inoculum of replication-competent retroviruses carrying the appropriate therapeutic gene(s) would be required to establish a spreading infection of the target tissue, leading potentially to infection of 100% of the target cells. This would overcome both the low titers of current recombinant stocks as well as the problems associated with poor penetration of "one-hit" viruses into target tissues beyond the first few cell layers. The use of replication-competent viruses would, however, only be acceptable if stringent genetic controls could be engineered into the virus to limit its spread only to the target tissue type and not beyond into uninvolved cells. This may eventually become possible by:

1. The inclusion of tissue-specific promoter elements into the provirus (to restrict its replication to cells of one particular type);
2. Use of a hybrid envelope protein to restrict viral infection to target cells expressing a defined, target cell-specific ligand;
3. Inclusion of a suicide gene into the vector, either as part of the therapeutic strategy or as an accessory safety feature, which can be activated by administration of a drug to kill all infected cells and terminate the treatment.

Clearly, much research remains to be done before the use of such replicating retro- or other viral vectors can be sanctioned in patients, but their use for certain gene therapy situations is conceptually attractive *(72,110–114)*.

11. Conclusions

The ease with which retroviral vectors can be constructed and used, combined with the high confidence limits that recombinant virus stocks can be safely generated to high titers, has meant that they have now become widely used in the laboratory. They are routinely employed for simple gene transfer studies, as well as for more complex marking and experimental procedures both in vitro and in vivo. More recently, retroviral technology has become so well characterized that retroviral vectors have already begun to be used as therapeutic agents in the treatment of patients. The efficacy of such trials remains to be seen, but happily, so far there have been no reports of adverse effects related to the exposure of patients or explanted patients' cells to recombinant retroviruses.

Retroviral vectors have a variety of properties that can be proposed as either strengths or weaknesses, depending on the individual disease that is being targeted. Currently, investigators must fit the clinical situation they wish to treat to the vector systems that are available. However, the hope is that future trials of gene therapy will be able to call on vectors that have been custom-designed to fit specific clinical situations, rather than vice versa. The understanding of and technology associated with retroviral vectors are currently among the most advanced of all vector systems; it seems then that retroviral vectors are well placed to move into an era of "designer vectorology," in which specific properties can be incorporated into viral genomes and packaging systems to produce hybrid vectors with beneficial features suited to specific tasks. For retroviruses, these properties will include selective surface and transcriptional targeting, increased titers, reduced potential for insertional mutagenesis and inadvertent generation of replication-competent viruses, and more flexible expression capabilities.

Beyond even this phase of vector construction, it can be envisaged that the most powerful delivery systems of the future will be those that cross the boundaries of individual viruses and amalgamate the most desirable properties of different vectors to generate hybrid delivery vehicles, which can be synthesized according to individual clinical requirements *(115)*.

References

1. Mann, R. M., Mulligan, R. C., and Baltimore, D. (1983) Construction of a retrovirus packaging mutant and its use to produce helper free defective retrovirus. *Cell* **33,** 149–153.
2. Weiss, R. A., Teich, N., Varmus, H. E., and Coffin, J. (1982) 1985) *Molecular Biology of RNA Tumor Viruses,* vols. 1 and 2. Cold Spring Harbor Laboratory Press, Cold Spring Harbor, NY.
3. Panganiban, A. T. and Fiore, D. (1988) Ordered interstrand and intrastrand DNA transfer during reverse transcription. *Science* **241,** 1064–1069.
4. Vile, R. G. (1991) The retroviral tile cycle and the molecular construction of retrovirus vectors, in *Practical Molecular Virology: Viral Vectors for Gene Expression* (Collins, M. K. L., ed.), Humana Press, Totowa, NJ, pp. 1–16.
5. Whitcomb, J. M. and Hughes, S. H. (1992) Retroviral reverse transcription and integration: progress and problems. *Annu. Rev. Cell Biol.* **8,** 275–306.
6. Shih, C. C., Stoye, J. P., and Coffin, J. M. (1988) Highly preferred targets for retrovirus integration. *Cell* **53,** 531–537.
7. Mooslehner, K., Karls, U., and Harbors, K. (1990) Retroviral integration sites in transgenic mice frequently map in the vicinity of transcribed DNA regions. *J. Virol.* **64,** 3056–3058.
8. Varmus, H. E. (1988) Retroviruses. *Science* **240,** 1427–1435.
9. Miller, A. D. (1992) Retroviral vectors. *Curr. Top. Microbiol. Immunol.* **158,** 1–24.
10. Bender, M. A., Palmer, T. D., Gelinas, R. E., and Miller, A. D. (1987) Evidence that the packaging sequence of Moloney murine leukemia virus extends into the gag region. *J. Virol.* **61,** 1639–1646.
11. Morgenstern, J. P. and Land, H. (1990) Advanced mammalian gene transfer: high titre retroviral vectors with multiple drug selection markers and a complementary helper-free packaging cell line. *Nucleic Acids Res.* **18,** 3587–3596.
12. Panganiban, A. T. and Varmus, H. M. (1983) The terminal nucleotides of retrovirus DNA are required for integration but not virus production. *Nature* **306,** 155–160.
13. Miller, D. G., Adam, M. A., and Miller, A. D. (1990) Gene transfer by retrovirus vectors occurs only in cells that are actively replicating at the time of infection. *Mol. Cell. Biol.* **10,** 4239–4242.
14. Lewis, P. F. and Emerman, M. (1993) Passage through mitosis is required for oncoretroviruses but not for the human immunodeficiency virus. *J. Virol.* **68,** 510–516.
15. Overell, R. W., Weisser, K. E., and Cosman, D. (1988) Stably transmitted triple-promoter retroviral vectors and their use in transformation of primary mammalian cells. *Mol. Cell. Biol.* **8,** 1803–1808.
16. Vile, R. G., Miller, N., Chernajovsky, Y., and Hart, I. R. (1994) A comparison of the properties of different retroviral vectors containing the murine tyrosinase promoter to achieve transcriptionally targeted expression of the HSVtk or IL-2 genes. *Gene Ther.* **1,** 307–316.
17. Hantzopoulos, P. A., Sullenger, B. A., Ungers, G., and Gilboa, E. (1989) Improved gene expression upon transfer of the adenosine deaminase minigene outside the transcriptional unit of a retroviral vector. *Proc. Natl. Acad. Sci. USA* **86,** 3519–3523.

18. Mulligan, R. C. (1993) The basic science of gene therapy. *Science* **260,** 926–932.

19. Stoker, A., ed. (1993) Retroviral vectors, in *Molecular Virology: A Practical Approach.* IRL, Oxford, pp. 172–197.

20. Vile, R. G. and Russell, S. J. (1995) Retroviruses as vectors. *Br. Med. Bull.* **51,** 12–30.

21. Miller, A. D. (1990) Retrovirus packaging cells. *Hum. Gene Ther.* **1,** 5–14.

22. Vile, R. G. (1991) Selectable markers for eukaryotic cells, in *Practical Molecular Virology: Viral Vectors for Gene Expression* (Collins, M. K. L., ed.), Humana, Totowa, NJ, pp. 49–60.

23. Price, J., Turner, D., and Cepko, C. (1987) Lineage analysis in the vertebrate nervous system by retrovirus-mediated gene transfer. *Proc. Natl. Acad. Sci. USA* **84,** 156–160.

24. Von Melchner, H. and Ruley, H. E. (1989) Identification of cellular promoters by using a retrovirus promoter trap. *J. Virol.* **63,** 3227–3233.

25. Lu, S. J., Man, S., Bani, M. R., Adachi, D., Hawley, R. G., Kerbel, R. S., and Ben-David, Y. (1995) Retroviral insertional mutagenesis as a strategy for the identification of genes associated with cis-diaminedichloroplatinum(II) resistance. *Cancer Res.* **55,** 1139–1145.

26. Miller, A. D. (1992) Human gene therapy comes of age. *Nature* **357,** 455–460.

27. Miller, A. D. and Buttimore, C. (1986) Redesign of retrovirus packaging cell line to avoid recombination leading to helper virus formation. *Mol. Cell. Biol.* **6,** 2895–2902.

28. Rosenberg, S. A., Aebersold, P., and Cornetta, K. (1990) Gene transfer into humans-immunotherapy of patients with advanced melanoma using tumor-infiltrating lymphocytes modified by retroviral gene transduction. *N. Engl. J. Med.* **323,** 570–578.

29. Ram, Z., et al. (15193) Toxicity studies of retroviral-mediated gene transfer for the treatment of brain tumors. *J. Neurosurg.* **79,** 400–407.

30. Danos, O. and Mulligan, R. C. (1988) Safe and efficient generation of recombinant retroviruses with amphotropic and ecotropic host range. *Proc. Natl. Acad. Sci. USA* **85,** 6460–6464.

31. Markowitz, D., Goff, S., and Bank, A. (1988) A safe packaging line for gene transfer: separating viral genes on two different plasmids. *J. Virol.* **62,** 1120–1124.

32. Bosselman, R. A., Hsu, R. Y., Bruszewski, J., Hu, S., Martin, F., and Nicolson, M. (1987) Replication defective chimeric helper proviruses and factors affecting generation of competent virus: expression of Moloney murine leukemia virus structural genes via the metallothionein promoter. *Mol. Cell. Biol.* **7,** 1797–1806.

33. Dougherty, J. P., Wisniewski, R., Yang, S., Rhode, B. W., and Temin, H. M. (1989) New retrovirus helper cells with almost no nucleotide homology to retroviral vectors. *J. Virol.* **63,** 3209–3212.

34. Takahara, Y., Hamada, K., and Housman, D. E. (1992) A new retrovirus packaging cell for gene transfer constructed from amplified long terminal repeat-free chimeric proviral genes. *J. Virol.* **66,** 3725–3732.

35. Muenchau, D. D., Freeman, S. M., Cornatta, K., Zwiebel, J. A., and Anderson, W. F. (1990) Analysis of retroviral packaging lines for generation of replication-competent virus. *Virology* **176,** 262–265.

36. Scadden, D. T., Fullers B., and Cunningham, J. M. (1990) Human cells infected with retrovirus vectors acquire an endogenous murine provirus. *J. Virol.* **64,** 424–427.

37. Donahue, R. E., Kessler, S. W., Bodine, D., McDonagh, K., and Dunbar, C. (1992) Helper virus induced T cell lymphoma in non human primates after retroviral mediated gene transfer. *J. Exp. Med.* **176,** 1125–1135.

38. Anderson, W. F. (1993) What about those monkeys that got T-cell lymphoma? *Hum. Gene Ther.* **4,** 1,2.

39. Cornetta, K., et al. (1990) Amphotropic murine leukaemia retrovirus is not an acute pathogen for primates. *Hum. Gene Ther.* **1,** 15–30.

40. Cornetta, K., et al. (1991) No retroviremia or pathology in long-term follow up of monkeys exposed to a murine amphotropic retrovirus. *Hum. Gene Ther.* **2,** 215–219.

41. Cornetta, K. (1992) Safety aspects of gene therapy. *Br. J. Haematol.* **80,** 421–426.

42. Moolten, F. L. and Cupples, L. A. (1992) A model for predicting the risk of cancer consequent to retroviral gene therapy. *Hum. Gene Ther.* **3,** 479–486.

43. Yee, J. K., Moores, J. C., Jolly, D. J., Wolff, J. A., Respress, J. G., and Friedmann, T. (1987) Gene expression from transcriptionally disabled retroviral vectors. *Proc. Natl. Acad. Sci. USA* **84,** 5197–5201.

44. Welsh, R. M., Cooper, N. R., Jensen, F. C., and Oldstone, M. B. A. (1975) Human serum lyses RNA tumour viruses. *Nature* **257,** 612–614.

45. Takeuchi, Y., Cosset, F. L., Lachmann, P. J., Okada, H., Weiss, R. A., and Collins, M. K. L. (1994) Type C retrovirus inactivation by human complement is determined by both the viral genome and the producer cell. *J. Virol.* **68,** 8001–8007.

46. Varela-Echavarria, A., Prorock, C. M., Ron, Y., and Dougherty, J. P. (1993) High rate of genetic rearrangement during replication of a Moloney Murine Leukemia virus-based vector. *J. Virol.* **67,** 6357–6364.

47. Dougherty, J. P. and Temin, H. M. (1986) High mutation rate of a spleen necrosis virus-based retrovirus vector. *Mol. Cell. Biol.* **6,** 4387–4395.

48. Fujiwara, T., Cai, D. W., Georges, R. N., Mukhopadhyay, T., Grimm, E. A., and Roth, J. A. (1994) Therapeutic effects of a retroviral wild-type p53 expression vector in an orthotopic lung cancer model. *J. Natl. Cancer Inst.* **86,** 1458–1462.

49. Jaffee, E. M., Dranoff, G., Cohen, L. K., Hauda, K. M., Clift, S., Marshall, F. F., Mulligan, R. C., and Pardoll, D. M. (1993) High efficiency gene transfer into primary human tumor explants without cell selection. *Cancer Res.* **53,** 2221–2226.

50. Bestwick, R. K., Kozak, S. L., and Kabat, D. (1988) Overcoming interference to retroviral super infection results in amplified expression and transmission of cloned genes. *Proc. Natl. Acad. Sci. USA* **85,** 5404–5408.

51. Landau, N. R. and Littman, D. R. (1992) Packaging system for rapid production of murine leukemia virus vectors with variable tropism. *J. Virol.* **66,** 5110–5113.

52. Weiss, R. A. (1993) Cellular receptors and viral glycoproteins involved in retrovirus entry, in *The Retroviridae,* vol. 2 (Levy, J., ed.), Plenum, New York, pp. 1–108.

53. Burns, J. C., Friedmann, T., Driever, W., Burrascano, M., and Yee, J. K. (1993) Vesicular stomatitis virus G glycoprotein pseudotyped retroviral vectors: concentration to very high titer and efficient gene transfer into mammalian and nonmammalian cells. *Proc. Natl. Acad. Sci. USA* **90,** 8033–8037.

54. Richards, C. A. and Huber, B. E. (1993) Generation of a transgenic model for retrovirus-mediated gene therapy for hepatocellular carcinoma is thwarted by the lack of transgene expression. *Hum. Gene Ther.* **4,** 143–150.

55. Palmer, T. D., Rosman, G. J., Osborne, W. R. A., and Miller, A. D. (1991) Genetically modified skin fibroblasts persist long after transplantation but gradually inactivate introduced genes. *Proc. Natl. Acad. Sci. USA* **88,** 1330–1334.

56. Russell, S. J., Eccles, S. A., Flemming, C. L., Johnson, C. A., and Collins, M. K. L. (1991) Decreased tumorigenicity of a transplantable rat sarcoma following transfer and expression of an IL-2 cDNA. *Int. J. Cancer* **47,** 244–251.

57. Gansbacher, B., Houghton, A., Livingston, P., Minasian, L., Rosenthal, F., Gilboa, E., Golde, D., Oettgen, H., Steffans, T., Yang, S. Y., and Wong, G. (1992) Clinical Protocol: a pilot study of immunization with HLA-A2 matched allogeneic melanoma cells that secrete interleukin-2 in patients with metastatic melanoma. *Hum. Gene Ther.* **3,** 677–690.

58. Dranoff, G., et al. (1993) Vaccination with irradiated tumor cells engineered to secrete murine granulocyte macrophage colony stimulating factor stimulates potent, specific, and long lasting anti-tumor immunity. *Proc. Natl. Acad. Sci. USA* **90**, 3539–3543.

59. Petropoulos, C. J., Payne, W., Salter, D. W., and Hughes, S. H. (1992) Using avian retroviral vectors for gene transfer. *J. Virol.* **66**, 3391–3397.

60. Scharfmann, R., Axelrod, J. H., and Verma, I. M. (1991) Long term in vivo expression of retrovirus mediated gene transfer in mouse fibroblast implants. *Proc. Natl. Acad. Sci. USA* **88**, 4626–4530.

61. Li, M., Hantzopoulos, M. L. P. A., Banerjee, D., Schweitzer, B. I., Gilboa, E., and Bertino, J. R. (1992) Comparison of the expression of a mutant dihydrofolate reductase gene under control of different internal promoters in retroviral vectors. *Hum. Gene Ther.* **3**, 381–390.

62. Emerman, M. and Temin, H. M. 1984) Genes with promoters in retroviral vectors can be independently suppressed by an epigenic mechanism. *Cell* **39**, 459–467.

63. Salmons, B. and Gunzburg, W. H. (1993) Targeting of retroviral vectors for gene therapy. *Hum. Gene Ther.* **4**, 129–141.

64. Soriano, P., Friedrich, G., and Lawinger, P. (1991) Promoter interactions in retrovirus vectors introduced into fibroblasts and embryonic stem cells. *J. Virol.* 2314–2313.

65. Morgan, R. A., Couture, L., Elroy-Stein, O., Ragheb, J., Moss, B., and French Anderson, W. (1992) Retroviral vectors containing putative internal ribosome entry sites: development of a polycistronic gene transfer system and applications to human gene therapy. *Nucleic Acids Res.* **20**, 1293–1299.

66. Adam, M. A., Ramesh, N., Dusty Miller, A., and Osborne, W. R. A. (1991) Internal initiation of translation in retroviral vectors carrying picornavirus 5' nontranslated regions. *J. Virol.* **65**, 4985–4990.

67. Koo, H., Brown, A. M. C., Kaufman, R. J., Prorock, C. M., Ron, Y., and Dougherty, J. P. (1992) A spleen necrosis virus-based retroviral vector which expresses two genes from a dicistronic mRNA. *Virology* **186**, 669–675.

68. Boris-Lawrie, K. A. and Temin, H. M. (1993) Recent advances in retrovirus vector technology. *Curr. Opin. Gene. Dev.* **3**, 102–109.

69. Oldfield, E. H., et al. (1993) Clinical Protocol: gene therapy for the treatment of brain tumors using intra-tumoral transduction with the thymidine kinase gene and intravenous ganciclovir. *Hum. Gene Ther.* **4**, 39–69.

70. Friedmann, T. (1992) Gene therapy of cancer through restoration of tumor-suppressor functions? *Cancer* **70(Suppl.)**, 1810–1817.

71. Collins, F. S. (1992) Cystic fibrosis: molecular biology and therapeutic implications. *Science* **256**, 774–779.

72. Vile, R. G. and Russell, S. J. (1994) Gene transfer technologies for the gene therapy of cancer. *Gene Ther.* **1**, 88–98.

73. Miller, D. G., Edwards, R. H., and Miller, A. D. (1994) Cloning of the cellular receptor for amphotropic murine retroviruses reveals homology to that for gibbon ape leukemia virus. *Proc. Natl. Acad. Sci. USA* **91**, 78–82.

74. Battini, J. L., Heard, J. M., and Danos, O. (1992) Receptor choice determinants in the envelope glycoproteins of amphotropic, xenotropic and polytropic murine leukemia viruses. *J. Virol.* **66**, 1468–1475.

75. Morgan, R. A., Nussbaum, O., Muenchau, D. D., Shu, L., Couture, L., and French Anderson, W. (1993) Analysis of the functional and host range-determining regions of the murine ecotropic and amphotropic retrovirus envelope proteins. *J. Virol.* **67**, 4712–4721.

76. Ott, D. and Rein, A. (1992) Basis for receptor specificity of nonecotropic murine leukemis virus surface glycoprotein gp70SU. *J. Virol.* **66,** 4632–4638.

77. Dong, J., Roth, M. G., and Hunter, E. (1992) A chimeric avian retrovirus containing the influenza virus hemagglutinin gene has an expanded host range. *J. Virol.* **66,** 7374–7382.

78. Roux, P., Jeanteur, P., and Piechaczyk, M. (1989) A versatile and potentially general approach to the targeting of specific cell types by retroviruses: application to the infection of human cells by means of major histocompatibility complex class I and class II antigens by mouse ecotropic murine leukemia virus-derived viruses. *Proc. Natl. Acad. Sci. USA* **86,** 9079–9083.

79. Russell, S. J., Hawkins, R. E., and Winter, G. (1993) Retroviral vectors displaying functional antibody fragments. *Nucleic Acids Res.* **21,** 1081–1085.

80. Young, J. A. T., Bates, P., Willert, K., and Varmous, H. E. (1990) Efficient incorporation of human CD4 protein into Avian Leukosis Virus particles. *Science* **250,** 1421.

81. Etienne-Julan, M., Roux, P., Carillo, S., Jeanteur, P., and Piechaczyk, M. (1992) The efficiency of cell targeting by recombinant retroviruses depends on the nature of the receptor and the composition of the artificial cell-virus linker. *J. General Virol.* **73,** 3251–3255.

82. Cosset, F. L., Ronfort, C., Molina, R. M., Flamant, F., Drynda, A., Benchaibi, M., Valesia, S., Nigon, V. M., and Verdier, G. (1992) Packaging cells for avian leukosis virus-based vectors with various host ranges. *J. Virol.* **66,** 5671–5676.

83. Van der Bruggen, P. and Van den Eynde, B. (1992) Molecular definition of tumor antigens recognized by T-lymphocytes. *Curr. Opin. Immunol.* **4,** 608–612.

84. Vile, R. G. (1994) Tumour specific gene expression. *Semin. Cancer Biol.* **5,** 429–436.

85. Dillon, N. (1993) Regulating gene expression in gene therapy. *Trends Biotechnol.* **11,** 167–173.

86. Dzierzak, E. A., Papayannopoulou, T., and Mulligan, R. C. (1988) Lineage specific expression of a human β-globin gene in murine bone marrow transplant recipients reconstituted with retrovirus-transduced stem cells. *Nature* **331,** 3541.

87. Hatzoglou, M., Bosch, F., Park, E. A., and Hanson, R. W. (1990) Hepatic gene transfer in animals using retroviruses containing the promoter from the gene for phosphoenolpyruvate carboxykinase. *J. Biol. Chem.* **265,** 17,285–17,293.

88. Hatzoglou, M., Bosch, F., Park, E. A., and Hanson, R. W. (1991) Hormonal control of interacting promoters introduced into cells by retroviruses. *J. Biolog. Chem.* **266,** 8416–8425.

89. Harris, J. D., Gutierrez, A. A., Hurst, H. C., Sikora, K., and Lemoine, N. R. (1994) Gene therapy for carcinoma using tumour-specific prodrug activation. *Gene Ther.* **1,** 170–175.

90. Huber, B. E., Richards, C. A., and Krenitsky, T. A. (1991) Retroviral-mediated gene therapy for the treatment of hepatocellular carcinoma: An innovative approach for cancer therapy. *Proc. Natl. Acad. Sci. USA* **88,** 8039–8043.

91. Anderson, W. F. (1992) Human Gene therapy. *Science* **256,** 808–813.

92. Rosenberg, S. A., French Anderson, W., Blaese, M., Hwu, P., Yannelli, J. R., Yang, J. C., Topalian, S., Schwartzentruber, D. J., Weber, J. S., Ettinghausen, S. E., Parkinson, D. N., and White, D. E. (1993) The development of gene therapy for the treatment of cancer. *Ann. Surg.* **218,** 455–464.

93. Culver, K. W., Ram, Z., Wallbridge, S., Ishii, H., Oldfield, E. H., and Blaese, R. M. (1992) In vivo gene transfer with retroviral vector-producer cells for treatment of experimental brain tumors. *Science* **256,** 1550–1552.

94. Ram, Z., Culver, K. W., Walbridge, S., Blaese, R. M., and Oldfield, E. H. (1993) In situ retroviral mediated gene transfer for the treatment of brain tumours in rats. *Cancer Res.* **53,** 83–88.

95. Georges, R. N., Mukhopadhyay, T., Zhang, Y., Yen, N., and Roth, J. A. (1993) Prevention of orthotopic human lung cancer growth by intratracheal instillation of a retroviral antisense K-ras construct. *Cancer Res.* **53,** 1743–1746.

96. Fujiwara, T., Grimm, E. A., Mukhopadhyay, T., Cai, D. E., Owen-Schaub, L. B., and Roth, J. A. (1993) A retroviral wild-type p53 expression vector penetrates human lung cancer spheroids and inhibits growth by inducing apoptosis. *Cancer Res.* **53,** 4129–4133.

97. Whartenby, K. A., Abboud, C. N., Marrogi, A. J., Ramesh, R., and Freeman, S. M. (1995) The biology of cancer gene therapy. *Lab. Invest.* **72,** 131–145.

98. Alton, E. W. F. W, Middleton, P. G., Caplen, N. J., Smith, S. N., Steel, D. M., Munkonge, F. M., Jeffery, P, K., Geddes, D. M., Hart, S. L., Williamson, R., and Fasold, K. I. (1993) Non-invasive liposome-mediated gene delivery can correct the ion transport detect in cystic fibrosis mutant mice. *Nature Gene.* **5,** 135–142.

99. Hyde, S. C., Gill, D. R., Higgins, C. F., Trezise, A. E. O., MacVinish, L. J., Cuthbert, A. W., Ratclif, R., Evans, M. J., and Colledge, W. H. (1993) Correction of the ion transport defect in cystic fibrosis transgenic mice by gene therapy. *Nature* **362,** 250–255.

100. Zabner, J., Couture, L. A., Gregory, R. J., Graham, S. M., Smith, A. E., and Welsh, M. J. (1993) Adenovirus-mediated gene transfer transiently corrects the chloride transport defect in nasal epithelia of patients with cystic fibrosis. *Cell* **75,** 207–216.

101. Nabel, G. J., Chang, A., Nabel, E. G., Plautz, G., Fox, B. A., Huang, L., and Shu, S. (1992) Clinical Protocol: immunotherapy of malignancy by in vivo gene transfer into tumors. *Hum. Gene Ther.* **3,** 399–410.

102. Kim, T. S., Russell, S. J., Collins, M. K. L., and Cohen, E. P. (1992) Immunity to B16 melanoma in mice immunized with IL-2 secreting allogeneic mouse fibroblasts expressing melanoma-associated antigens. *Int. J. Cancer* **51,** 283–289.

103. Flugel, R. M. (1991) Spumaviruses: a group of complex retroviruses. *J. Acquired Immune Defic. Syndromes* **4,** 739–750.

104. Vile, R. G., Ali, M., Hunter, E., and McClure, M. O. (1992) Identification of a generalised packaging sequence for D-type retroviruses and generation of a D-type retroviral vector. *Virology* **189,** 786–791.

105. Gilboa, E. and Smith, C. (1994) Gene therapy for infectious diseases: the AIDS model. *Trends Gene.* **10,** 109–114.

106. Brady, H. J. M., Miles, C. G., Pennington, D. J., and Dzierzak, E. A. (1994) Specific ablation of human immunodeficiency virus Tat-expressing cells by conditionally toxic retroviruses. *Proc. Natl. Acad. Sci. USA* **91,** 365–369.

107. Buchschacher, G. L. and Panganiban, A. T. (1992) Human immunodeficiency virus vectors for inducible expression of foreign genes. *J. Virol.* **66,** 2731–2739.

108. Bukrinsky, M. I., Haggerty, S., Dempsey, M. P., Sharova, N., Adzhubel, A., Spitz, L., Lewis, P., Goldfarb, D., Emerman, M., and Stevenson, M. (1993) A nuclear localisation signal within HIV-1 matrix protein that governs infection of non-dividing cells. *Nature* **365,** 666–669.

109. Russell, S. J. (1993) Gene therapy for cancer. *Cancer J.* **6,** 21–25.

110. Russell, S. J. (1994) Replicating vectors for cancer therapy: a question of strategy. *Semin. Cancer Biol.* **5,** 437–443.

111. Dupressoir, T., Vanacker, J. M., Cornelis, J. J., Duponchel, N., and Rommelaere, J. (1989) Inhibition by parvovirus H-1 of the formation of tumors in nude mice and colonies in vitro by transformed human mammary epithelial cells. *Cancer Res.* **49,** 3203–3208.

112. Asada, T. (1974) Treatment of human cancer with mumps virus. *Cancer* **34,** 1907–1928.

113. Southam, C. M. (1960) Present status of oncolytic virus studies. *Trans NY Acad. Sci.* **22,** 657–673.

114. Takamiya, Y., Short, M. P., Ezzeddine, Z. D., Moolten, F. L., Breakefield, X. O., and Martuza, R. L. (1992) Gene therapy of malignant brain tumors: a rat glioma line bearing the Herpes Simplex Virus Type 1-thymidine kinase gene and wild type retrovirus kilts other tumor cells. *J. Neurosci. Res.* **33,** 493–503.

115. Miller, N. and Vile R. G. (1995) Targeted vectors for gene therapy. *FASEB J.* **9,** 190–199.

19

Baculovirus Vectors

Azeem Ansari and Vincent C. Emery

1. Introduction

Recombinant baculovirus expression systems are now firmly established as one of the tools for producing foreign proteins in high yield within a eukaryotic environment. The evolution of both single- and multiple-transfer vectors and different forms of baculoviral DNA has greatly streamlined the isolation and purification of a recombinant virus carrying the foreign gene of interest. This chapter will summarize the wealth of transfer vectors and baculovirus DNAs available to the researcher to generate both single- and multiple-expressing recombinant baculoviruses and highlights the specific areas in which we feel that recombinant protein expressed via the baculovirus system has enabled significant advances to be made.

2. Baculoviruses as Expression Systems

The prototype virus of the subgroup of multiple nuclear polyhedrosis viruses (MNPV) is autographica californica MNPV (AcMNPV), which naturally infects the alfalfa looper. AcMNPV and other members of the group including *Bombyx mori* (Silkworm) and *Orgyla pseudosugata* MNPVs are characterized by the formation of large intranuclear occlusions in infected cells known as polyhedra derived virus (PDV), which consist largely of a single protein of molecular weight of approx 29 kDa known as polyhedrin. When an infected insect dies, thousands of PDVs are released as the host decomposes. The PDVs serve to protect the encapsulated virions from external factors, such as heat and desiccation, which would otherwise destroy the unprotected virions.

When insect larvae feed on contaminated vegetation, the PDVs dissolve in the insect midgut, releasing the virus particles and thus starting another round of infection. Secondary infection disseminates the virus within the insect larvae via the extracellular form (ECV) of AcMNPV. Although the polyhedrin protein is the most abundant protein in infected insect cells (30–50% of the total-cell protein) it is dispensable under insect-cell culture conditions and hence has allowed the design of an expression system based on the replacement of the polyhedrin open reading frame (ORF) by a foreign ORF. In theory, a foreign ORF under the control of the polyhedrin promoter should be expressed to the same level as native polyhedrin. However, expression levels can vary depending on the inherent nature of the protein in question and the effective translat-

From: *Molecular Biomethods Handbook*
Edited by: R. Rapley and J. M. Walker © Humana Press Inc., Totowa, NJ

ability of the mRNA encoding the protein. In addition, the p10 protein has also been shown to be dispensable for virus growth in cell culture and the promoter has been similarly exploited for foreign gene expression. Three other promoters used in a similar fashion are those for the 39k protein, (a delayed early gene) and the basic protein, which forms part of the virus particle and also the glycoprotein 64 promoter. The latter two are late genes.

3. Generating a Recombinant Baculovirus

There are many excellent manuals available that give a detailed description of the experimental steps involved in generating a recombinant baculovirus (1–3). In this chapter, we merely precis the salient points. The steps involved are

1. Cloning the gene of interest into an appropriate transfer vector,
2. Cotransfection of the recombinant transfer vector and a suitable baculovirus DNA into cultured insect cells,
3. Isolation of a single recombinant virus by plaque assay,
4. Analysis at the DNA level for successful recombination,
5. Amplification of the virus stock for large-scale protein production, and
6. Purification of the recombinant protein for further studies.

The first step to construct a recombinant virus involves the insertion of the foreign gene of interest into a suitable transfer vector. All of the transfer vectors available have *Escherichia coli* origins of replication and an antibiotic resistance gene therefore insertion of the foreign gene can be carried out using standard plasmid methodologies. The original vector design to produce a single-expression cassette can be traced back to the laboratory of Professor M. Summers (University of Texas A & M, USA) (4). The first high-level expression vector was produced by Dr. Y. Matsuura working with Professor D. Bishop at the Institute of Virology and Environmental Microbiology in Oxford (5). Concurrent and subsequent work by Possee and colleagues and Roy and colleagues at the Institute of Virology and Environmental Microbiology, among others, has allowed the generation of a broad spectrum of single- and multiple-expression system (6–12). Advances in vector design have resulted in the availability of a wide variety of transfer vectors (**Table 1**). The list in **Table 1** is by no means exhaustive because as technologies advance we are sure the numbers of transfer vectors will continue to rise. The transfer vectors can be broadly categorised into three groups depending largely on whether or not the researcher wishes to express the protein of interest as a fused or nonfused product or perform single- or multiple-expression studies. Expressing protein as a fusion product can facilitate isciation and purification of the recombinant protein. For example, proteins fused to a tag containing six histidine residues can be purified rapidly using one step nickel-agarose column chromatography. Proteins fused with secretory signals can be purified away from culture media (13). If the insect cells are adapted to serum-free media, then the later process is made substantially easier. The transfer vectors also differ in the viral promoter used to drive expression of the foreign gene. The most widely used are the very late promoters, p10 and polybedrin (11). These have a distinct advantage over the other promoters available, the 39k, basic protein, and glycoprotein 64 (7,14), in that during the very late phase of baculovirus replication (48 h postinfection), the vast majority of protein synthesis, both viral and cellular, is shut down. It is only the p10 and polyhedrin promoters that remain active,

Table 1
Baculovirus Transfer Vectors

Plasmid	Promoter	Flanking region	Comments
Polyhedrin single promoter			
pAcYM1	Polyhedrin	Polyhedrin	Full-length leader sequence of polyhedrin including A of ATG; original high level expression transfer vector; ATG must be provided by insert: single cloning site.
pVL941	Polyhedrin	Polyhedrin	Full-length leader sequence of polyhedrin including residues up to +36; polyhednn ATG mutated to ATT; ATG must be provided by insert; product is nonfused; limited multiple cloning site (mcs).
pVL1392/1393	Polyhedrin	Polyhedrin	Derivative of the above; much improved mcs; pVL1392/1393 are identical except the mcs is in the opposite orientation.
pAcSG1	Polyhedrin	Polyhedrin	Derivative of pVL941; smaller in size therefore can accomodate larger inserts.
pAcSGHisNT	Polyhedrin	Polyhedrin	Derivative of pAcSG1; protein expressed as a histidine tagged fusion.
pAcJP1	39k protein	Polyhedrin	Full-length 39k protein promoter; ATG must be provided by insert; product is nonfused; single cloning site.
pAcMP2/3	Basic protein	Polyhedrin	Derivative of the pVL vectors; basic protein promoter is used instead of polyhedrin.
pAc360	Polyhedrin	Polyhedrin	Full-length leader sequence of polyhedrin including residues up to +36; protein expressed as a polyhedrin N terminus fusion: single cloning site.
pBlueBacHis	Polyhedrin	Polyhedrin	Protein expressed as a histiaine tagged Xpress™ (antibody epitope) fusion; limited mcs; recombinant plaques appear blue on staining with X-Gal.
pBlue*Bac*III	Polyhedrin	Polyhedrin	ATG must be provided by insert; product is nonfused: limited multiple cloning site (mcs); recombinant plaques appear blue on staining with X-Gal.

(continued)

Table 1 *(continued)*

Plasmid	Promoter	Flanking region	Comments
pBlueBac4	Polyhedrin	Polyhedrin	Derivative of pBlueBacIII; much improved mcs; smaller in size therefore can accomodate larger inserts.
pBlueBacHis2	Polyhedrin	Polyhedrin	Derivative of pBlueBacHis; smaller in size therefore can accommodate larger inserts much improved MCS.
pMelBac	Polyhedrin	Polyhedrin	Protein expressed as a honybee melitin secretion signal fusion; smaller in size therefore can accomodate larger inserts; recombinant plaques appear blue on staining with X-Gal.
pAcATM3	GP64	Polyhedrin	Full-length GP64 promoter; ATG must be provided by insert: product is nonfused: single cloning site.
pAcATM2	GP64	Polyhedrin	Full-length GP64 promoter; protein expressed as a GP64 secretion signal fusion.
Polyhedrin multiple promoter			
pAcVC3	Two polyhedrin	Polyhedrin	Two full-length polyhedrin promoters in tandem but in opposite orientation; limited cloning sites; ATG must be provided by inserts.
pAcUW51	One p10 one polyhedrin	Polyhedrin	Full-length p10 promoter; *Eco*RI and *Bgl*II cloning site; full-length polyhedrin promoter downstream and in tandem to the polyhedrin promoter but in opposite orientation; single *Bam*HI cloning site; ATG must be provided by inserts.
pAcAB3	Two p10 one polyhedrin	Polyhedrin	Two full-length p10 promoters; *Sma*I and *Bam*HI cloning site downstream of first; upstream an inverted full-length polyhedrin promoter, *Xba*I and *Stu*I sites, second p10 upstream of polyhedrin promoter in tandem; single *Bgl*II site; ATG must be provided by inserts.

(continued)

Table 1 (continued)

Plasmid	Promoter	Flanking region	Comments
p2 Bac	One p10, one polyhedrin	Polyhedrin	Full-length p10 promoter; full-length polyhedrin promoter downstream and in tandem to the polyhedrin promoter but in opposite orientation; improved mcs; ATG must be provided by inserts.
p10 Single promoter			
pAcUW1	p10	p10	Full-length p10 promoter; ATG must be provided by insert; small in size therefore can accommodate larger inserts; single cloning site.
pAcUW21	p10	Polyhedrin	Full-length p10 promoter; ATG must be provided by insert; single cloning site.
p10 Multiple promoter			
pAcUW41	One p10, one polyhedrin	p10	Full-length p10 promoter; single *Bgl*II cloning site; fold length polyhedrin promoter downstream and in tandem to the polyhedrin promoter; single *Bam*HI cloning site; ATG must be provided by inserts.
pAcUW42/43	One p10, one polyhedrin	p10	Derivative of pAcUW41; improved mcs for the p10 promoter.

All transfer vectors have *E. coli* origins of replication therefore cloning of a gene of interest can be cloned using standard methodologies. Each transfer vector has its own individual properties which should be considered in the context of the expression required. A brief summary of the characteristics of the transfer vectors is included as an *aide memoire*.

thus serving as a "natural" enrichment for the protein of interest. In addition to single-promoter transfer vectors, there are multiple-promoter transfer vectors where two or more ORFs can be inserted *(6,9–12)* (*see* **Table 1**). The resulting recombinant virus will express multiple proteins and are particularly useful when the researcher wishes to study protein–protein interactions or multiprotein complexes. Once the cloning has been confirmed by restriction-enzyme analysis and DNA sequencing, the researcher can proceed and cotransfect the recombinant-transfer vector with a baculovirus DNA most suitable for the chosen transfer vector (**Fig. 1**).

Recombinant baculoviruses are almost exclusively derived by homologous recombination between the transfer vector and viral DNA, resulting in the insertion of the foreign ORF into the baculovirus DNA. However, novel AcNMPV DNAs such as AcOmega have been described that allow the direct cloning of genes into the

Baculovirus DNA

Wild-Type DNA (AcMNPV)

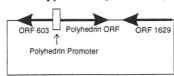

(i)   AcMNPV infection results in the production of occlusion bodies, of which polyhedrin is the major constituent.
(ii)   On recombination the polyhedrin gene is replaced by the protein of interest. Recombinant viruses form occlusion body-negative plaques and can be visually identified.
(iii)   Recombination frequencies are very low, usually in the ratio of 1000:1, nonrecombinants to recombinants.
(iv)   Wild-type DNA is becoming less common.

Baculogold™ (Pharmigen)

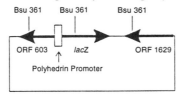

(i)   *LacZ* has replaced wild-type polyhedrin.
(ii)   The Bsu361 sites allow linearization and lethal deletion of the part of the essential open reading frame (ORF) 1629.
(iii)   Baculogold can only be used with transfer vectors carrying ORF 1629.
(iv)   Recombination frequencies of nearly 100% can be achieved.

AcRP23.lacZ

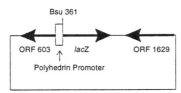

(i)   *LacZ* has replaced wild-type polyhedrin.
(ii)   A single Bsu361 has been added adjacent to the polyhedrin promoter, which allows linearization.
(iii)   AcRP23.lacZ can only be used with polyhedrin-based transfer vectors.
(iv)   Recombination efficiencies of 30% can be achieved.
(v)   Recombinants will appear as white plaques on staining with X-gal; nonrecombinants will be blue.

AcUW1.lacZ

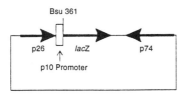

(i)   *LacZ* has replaced the viral p10 gene.
(ii)   A single Bsu361 has been added adjacent to the p10 promoter, which allows linearization.
(iii)   AcUW1.*lacZ* can only be used with p10-based transfer vectors.
(iv)   Recombination efficiencies of 30% can be achieved.
(v)   Recombinants will appear as white plaques on staining with X-gal; nonrecombinants will be blue.
(vi)   Since the polyhedrin locus is not affected during recombination, recombinant viruses will be occlusion-body positive.

Bac 'N' Blu™(Invitrogen)

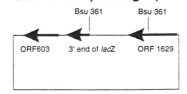

(i)   *LacZ* has replaced wild-type polyhedrin.
(ii)   The Bsu361 sites allow the linearization and deletion of part of the essential ORF 1629.
(iii)   When Bac 'N' Blu is used with transfer vectors carrying the β-galactosidase gene, recombinant plaques are blue.
(iv)   Recombination frequencies of 90–95% can be achieved.

Fig. 1. A schematic representation of the baculovirus DNAs available for cotransfection with a transfer vector containing the gene of interest to produce a recombinant baculovirus. Wild-type DNA, Baculogold™ DNA, AcRP23.lacZ and Bac 'N' Blue™ DNA can only be used with polyhedrin locus-based transfer vectors whereas AcUW1 lacZ can only be used with p10 locus-based transfer vectors. Baculogold™, AcRP23.lacZ, AcUW1.lac Z and Bac-'N'-Blue™ are available predigested from commercial sources.

baculovirus genome *(15)*. Historically, wild-type baculovirus DNA (AcMNPV) was used for cotransfection, which resulted in an extremely low-recombination frequency (0.1%) and necessitated the screening of large numbers of plaques to identify ones with a polyhedrin negative (recombinant) phenotype. Later it was realized that linearization of the viral DNA improved recombination frequencies such that up to 30% of progeny virus could be recombinant *(16)*. With the advent of newer baculovirus DNAs (**Fig. 1**), the time taken to produce a recombinant baculovirus has decreased and generally become less demanding. For example the AcRP23.lacZ and AcUW1.lacZ viruses express the bacterial β-galactosidase gene and upon recombination, the β-galactosidase gene is replaced by the foreign ORF. Therefore, screening for recombinant virus is facilitated because nonrecombinants will appear as blue plaques and recombinants will appear as white plaques upon staining a plaque assay with X-gal. Both AcRP23.lacZ and AcUW1.lacZ are available in linearized form from commercial sources. Baculogoid™ (Pharmigen) is a lethal deletion mutant of wild-type DNA where part of the essential ORF 1629 is deleted. When this DNA is cotransfected in conjunction with a polyhedrin locus based transfer vector, the lethal deletion is repaired upon homologous recombination. Therefore, replication-competent virus will only be generated by successful recombination. Bac 'N' Blu™ (Invitrogen) has further facilitated the screening for recombinant virus. This carries the same deletion as the Baculogold™ DNA but also carries a mutant form of the β-galactosidase gene. When Bac 'N' Blu™ is cotransfected with a polyhedrin locus based transfer vector carrying the β-galactosidase gene, recombination not only results in viable virus but also repairs the β-galactosidase gene thereby allowing recombinant virus plaques to be identified by virtue of their blue color upon staining with X-gal. These deletion mutants have increased recombination frequencies to nearly 100%. For this reason, we strongly recommend the use of the deletion mutants for cotransfection, doing so will certainly decrease the time taken for the isolation of a pure recombinant virus.

Various methods have been used for cotransfection such as electroporation, calcium phosphate precipitation, lipofectin, and Insectin™ (Invitrogen). Cotransfection efficiencies with electroporation and calcium phosphate can be variable whereas generally, in the authors' laboratory, liposome-mediated transfer methods are the most efficient. The cationic liposomes form complexes with the DNA and facilitate interaction with the negatively charged insect-cell membrane. The complexes are taken up by the cell, thus transporting the liposome containing transfer vector DNA and AcNMPV DNA into the cell where recombination occurs. Because illegitimate recombination events may occur, plaque purification of a number of recombinant virus dewed from a transfection is optimal.

4. Properties of Proteins Expressed with Recombinant Baculoviruses

Using the baculovirus system for the expression of eukaryotic genes has many clear advantages. These include the ability of the insect cell to perform the majority of post-translational modification required to produce biologically active protein. However, in some cases, these modifications are not 100% efficient, if the very late promoters are used, possibly owing to alterations in the cellular-modification pathways in the terminal phases of viral replication. This can be overcome by the use of early promoters (*see* **Table 1**), but protein yields may be compromised. Phosphorylation has been shown to occur at authentic sites for baculovirus expressed SV40 T-antigen when compared to protein purified from a native source *(17)*. Although serine residues were shown to be underphosphorylated, it did not affect the biological properties of the protein *(17)*. The insect cell has been shown to carry out two types of fatty acid acylation, palmitoylation and myristylation. Addition of palmitic acid to SV40 T-antigen has been reported *(18)*. *N*-myristylation of human-immunodeficiency virus type 1 (HIV-1) gag allowed the formation of membrane-enveloped virus-like particles to bud from the plasma membrane, whereas nonmyristylated Gag assembled intracellularly *(19)*. There has been much debate about whether or not the insect cell has the necessary machinery required to form the complex-carbohydrate side chains seen on higher eukaryotic proteins. Evidence suggested that the hepatitis-B surface antigen is not processed to the complex-oligosaccharide type as seen on the plasma derived hepatitis-B surface antigen, but remained as a N-linked high mannose form *(20)* characteristic of the insect cell. However, Davidson et al. *(21)* showed that recombinant human-plasminogen glycosylation was of the complex carbohydrate form and found the processing to be time-dependent. Between 60 and 96 h post-infection, essentially all the protein expressed was of the complex-carbohydrate class, suggesting the machinery to assemble N-linked complex oligosaccharides are encoded by the insect cell and can be utilized under certain conditions.

The accumulation of baculovirus encoded proteins is facilitated by the presence of antiapoptotic proteins encoded within the genome *(22)*.

5. Use of Recombinant Baculoviruses to Investigate Virus Structure and Assembly

The ability to produce single- and multiple-gene products within the same cell has enabled the structure and subtleties of virus assembly to be determined. The structure of many virus capsids have been determined through the availability of correctly folded, three-dimensionally (3D) formed products produced within the insect cell. For example, adenovirus-penton fibers have been expressed and shown to be morphologically identical to those extracted from cultures of wild-type viruses *(23)*. Similarly, genes encoding the hepatitis-B core protein *(24)*, the rotavirus-capsid protein *(25)*, the HIV-1 gag precursor *(26)*, and bluetongue-virus (BTV) capsid proteins *(10,27)* have all been shown to self-assemble into empty capsids with the correct conformation. In the case of human parvovirus B19, the capsid protein VP2 was shown to self-assemble into empty capsids, which were then subjected to X-ray diffraction to allow the 3D structure of the capsid to be determined *(28)*. In addition, cryoelectron microscopic studies coupled with high resolution reconstruction and computer image processing have been used to determine the structure of parvoviruses to a resolution of 2.2 nm *(29)*. These studies have revealed the structural uniqueness of the parvovirus capsid with respect to its lack of prominent spikes on the threefold chiral axis, which are present in the related canine parvoviruses. Similar structural studies have been carried out for the Norwalk virus *(30)* and the human-papillomavirus capsid composed of the L1 protein *(31)*. The latter protein has been shown to induce protective immunity in animal models *(32,33)*.

The use of multiple-expression systems has enabled the production of double layered rotavirus particles *(34)* and BTV particles *(27)*. These virus-like particles maintain the structural and functional characteristics of native particles and are being utilized in structural studies and in subunit vaccine studies. In the case of BTV, >50 μg of the capsid protein VP2 was effective at protecting all sheep against challenge with BTV. Addition of 20 μg/sheep of VP5 at the same time as 50 μg of VP2 also resulted in complete protection. Interestingly, inclusion of four core proteins did not enhance the neutralizing antibody response *(35)*. Homologous and partial heterologous protection could also be demonstrated *(36)*. More recently, these particles have been subjected to X-ray diffraction and high-resolution electron microscopy (EM) studies to reveal the structural characteristics of the BTV particle *(37–41)*. In addition, the ability to use the tubule structures associated with nonstructural protein I and the inherent nature of the capsid structure as peptide display vehicles has been demonstrated *(42–44)*.

The ability to produce complex structures by either multiple infections or through reconstitution of solubilized protein products has been specifically used to further our understanding of herpesvirus capsid assembly *(45–50)*. These elegant studies of Homa and colleagues have allowed the delineation of the proteins absolutely required for production of immature and mature herpesvirus capsids. In the case of herpes simplex virus (HSV) type 1, the use of computer-aided cryoelectron microscopy has enabled subtle differences in capsids lacking one or more components of the mature capsid to be identified. Clearly, these studies are providing new insight into the molecular architecture of virus-capsid structure and in the case of HSV, reconstitution of solubilized baculovirus-expressed protein was sufficient to produce capsids in vitro. This observation highlights the thermodynamic stability of the formed capsid vs that of the separate proteins comprising the capsid. In addition, such in vitro production of mature capsids will allow the molecular dissection of residues required for protein–protein interaction and hence, formation of the mature herpesvirus capsid.

Our understanding of the molecular architecture of the HIV virion has been significantly advanced by the availability of recombinant baculovirus produced gag particles and proteolytically processed moieties, such as p17 *(26,51–54)*. The group of Jones and colleagues have shown that gag-particle formation does not require viral-RNA encapsidation and via high-resolution EM the morphology of the gag protein could be classified into particulate spherical tubular shells or nonparticulate two-dimensional flat, curved, or convoluted sheets. These structural features have allowed an assessment of functional characteristics of assembly to be mapped via mutagenic procedures. The culmination of this work has been the crystal structure of the simian immunodeficiency virus (SIV) matrix antigen, which revealed that the p17 protein dictates the overall structure of the gag shell *(54)*. Thus, the SIV p17 moiety exists as a trimer which can adopt a hexameric configuration. The 3D structure of the HIV p24 moiety is required before a complete picture of HIV-particle assembly can be formulated.

6. Functional Analysis of Cytochrome P450 Enzymes

Cytochrome P450-dependent enzymes are central in the metabolism of toxic substances and in the biosynthesis of biologically important macromolecules. In order to facilitate physicochemical studies of these enzymes and for the design of inhibitors, large quantities of functional protein are required. A number of cytochrome-P450 enzymes have now been produced in insect cells using recombinant baculoviruses *(55–62)*. In most studies, addition of hemin (as a source of iron) to the culture medium results in correctly folded and functional enzyme. For example, human-placental aromatase, which is the cytochrome P450 component of the P450/NADPH cytochrome

P450 reductase-enzyme complex responsible for the production of estrogens from androgens, has been expressed in high levels by recombinant baculoviruses *(55,56)*.

The recombinant protein had a specific P450 content similar to enzyme purified from human-placental microsomes, whereas the spectral properties of the insect cell-produced aromatase were also identical to native material *(55)*. Because oestrogens play a role in breast cancer and other hormone sensitive cancers an understanding of the structure-function relationships within the complex may aid the design and evaluation of mechanism-based aromatase inhibitors. In order to circumvent the problems associated with purification of membrane-bound proteins, a truncated form of the human-placental aromatase lacking the first 41 amino acids and the hydrophobic membrane-spanning region has been produced *(56)*. Inclusion of a His-6 tag at the N-terminus resulted in a fully functional protein that could be purified to homogeneity by nickel-agarose chromatography. In this study, only one-third of the expressed protein was functional, soluble, and in the native conformation. Nevertheless expression at 22 μg/100 mm Petri dish of fully functional material that could be maintained in low concentrations of detergent (0.003% NP-40) is a significant advance in producing protein for structural studies.

The high level synthesis of proteins within the baculovirus system is aptly demonstrated by the expression of human 17β hydroxysteroid dehydrogenase, another enzyme involved in estrogen biosynthesis *(59)*. At 60 h postinfection, recombinant protein accounted for approx 5% of total soluble proteins and could be purified with an overall yield of 70%. This level of expression is far greater than can be achieved by transient expression in Hela cells or in any human tissue. In addition, the protein had a higher specific activity than that purified from other sources, possibly owing to the rapid-purification method utilized.

7. Production of Immunologically Important Molecules

Recombinant baculoviruses have been generated that express the heavy and light chains of monoclonal antibodies *(63)*. Structural and functional studies of the molecules indicated that the two chains correctly assemble to form the heterodimer and that both polyclonal and monoclonal idiotypes were present. The expressed antibodies have been demonstrated to interact with other components of the immune system for example, heterologous sources of complement and to participate in antibody-dependent cell-mediated cytotoxicity.

Despite the availability of a 3D structure for human-leukocyte antigen (HLA) class I for many years *(64)* the structure of the HLA class II molecule proved more difficult to solve. However, production of the HLA class II heterodimer in a soluble form using recombinant baculoviruses was instrumental in allowing the resolution of the three-dimensional structure of this molecule at 2.75Å by X-ray diffraction *(65,66)*. As a consequence of these data the topography of the peptide binding site has been elucidated. For an influenza-hemagglutinin peptide bound to HLA DRI, the peptide side chains were accommodated by pockets in the HLA structure, whereas hydrogen bonding between conserved HLA residues and the peptide residues provide a universal mode of peptide binding. These findings show that although the architecture of the peptide-binding site of HLA class I and II molecules are similar, peptide binding is very different between the two molecules.

The assembly of the ternary complex of class I HLA and its peptide requires the limited proteolysis of intracellular antigens followed by translocation of the peptide into the lumen of the endoplasmic reticulum. The proteins that perform peptide transportation are the transporter associated with antigen presentation (TAP) 1 and TAP 2 gene products *(67)*. To date, two human viruses, HSV and human papillomaviruses have been shown to downregulate HLA-class I cell surface display by interfering with TAP1/2 activity *(68,69)*. The availability of recombinant TAP1/2 should facilitate the study of the molecular requirements for peptide transportation and the mechanism by which the viral proteins in HSV and human papillomaviruses interfere with this process. To this end, the transporter has been successfully expressed in a dual-recombinant baculovirus and has been shown to bind and translocate peptides in the presence of ATP into the microsomes of Sf9 insect cells *(70)*. Furthermore, the availability of large quantities of the Tap1/2 complex should facilitate physicochemical studies.

8. Concluding Remarks

In this chapter, we have attempted to summarize the major advances made in the technology available to the researcher to produce recombinant baculoviruses. The availability of a multiplicity of vectors and various forms of AcNMPV DNA places the baculovirus expression system at the forefront of eukaryotic expression options. The ability of the insect cell to perform the modifications encountered in higher eukaryotes has enabled workers to produce large quantities of recombinant protein, that is conformationally and immunologically intact. Multiple-expression systems, which allow a variety of combinations of recombinant proteins to be produced, place the system second to none in the recombinant multiple-expression vector arena. These systems have greatly facilitated the production of proteins for physicochemical studies and are providing large quantities of recombinant proteins hitherto difficult to purify from natural sources. We expect that the insight provided by investigations on proteins produced in this system into basic biochemical processes and in the evaluation of compounds that modulate their activities are likely to allow major advances in many areas in the near future.

References

1. O'Reilly, D., Miller, L. K., and Luckow, V. A. (1992) *Baculovirus Expression Vectors: A Laboratory Manual.* Freeman, New York.
2. King, L. A. and Possee, R. D. (1991) *The Baculovirus Expression System: A Laboratory Guide.* Chapman and Hall, New York.
3. Summers, M. D. and Smith, G. E. (1987) Manual of methods for baculovirus vectors and insect cell culture procedures. *Texas Agri. Exp. Stat. Bull.* No. 1555.
4. Smith, G. E., Summers, M. D., and Fraser, M. J. (1983) Production of human beta interferon in insect cells infected with a baculovirus expression vector. *Mol. Cell. Biol.* **3,** 2156–2165.
5. Matsuura, Y., Possee, R. D., Overton, H. A., and Bishop, D. H. (1987) Baculovirus expression vectors, the requirements for high level expression of proteins, including glycoproteins. *J. Gen. Virol.* **68,** 1233–1250.
6. Emery, V. C. and Bishop, D. H. (1987) The development of multiple expression vectors for hish level synthesis of eukaryotic proteins, expression of LCMV-N and AcNFV polyhedrin protein by a recombinant baculovirus. *Prot. Engineer.* **1,** 359–366.
7. Hill-Perkins, M. S. and Possee, R. D. (1990) A baculovirus expression vector derived from the basic protein promoter of *Autographa californica* nuclear polyhedrosis virus. *J. Gen. Virol.* **71,** 971–976.

8. French, T. J., Marshall, J. J., and Roy, P. (1990) Assembly of double-shelled, viruslike particles of bluetongue virus by the simultaneous expression of four structural proteins. *J. Virol.* **64,** 5695–5700.

9. Belyaev, A. S. and Roy, P. (1993) Development of baculovirus triple and quadruple expression vectors, co-expression of three or four bluetongue virus proteins and the synthesis of biuetongue virus-like particles in insect ceils. *Nucleic Acids Res.* **21,** 1219–1223.

10. French, T. J. and Roy, P. (1990) Synthesis of bluetongue virus (BTV) corelike particles by a recombinant baculovirus expressing the two major structural core proteins of BTV. *J. Virol.* **64,** 1530–1536.

11. Wang, X. Z., Ooi, B. G., and Miller, L. K. (1991) Baculovirus vectors for multiple gene expression. *Gene* **100,** 131–137.

12. Belysev, A. S., Hails, R. S., and Roy, P. (1995) High-level expression of five foreign genes by a single recombinant baculovirus. *Gene* **156,** 229–233.

13. Sarvari, M., Csikos, G., Sass, M., Gal, P., Schumaker, V. N., and Zavodszky, P. (1990) Ecdysteroids increase the yield of recombinant protein produced in baculovirus insect cell expression system. *Biochem. Biophys. Res. Comm.* **167,** 1154–1161.

14. Bishop, D. H. L. (1992) Baculovirus expression vectors. *Sem. Virol.* **3,** 253–264.

15. Ernst, W. J., Grabherr, R. M., and Katinger, H. W. D. (1994) Direct cloning into the *Autographa californica* nuclear polyhedrosis virus for generation of recombinant baculoviruses. *Nucleic Acids Res.* **22,** 2855–2856.

16. Kitts, P. A., Ayres, M. D., and Possee, R. D. (1990) Linearization of baculovirus DNA enhances the recovery of recombinant virus expression vectors. *Nucleic Acids Res.* **18,** 5667–5672.

17. Hoss, A., Moarefi, I., Scheidtmann, K. H., et al. (1990) Altered phosphorylation pattern of simian virus 40 T antigen expressed in insect cells by using a baculovirus vector. *J. Virol.* **64,** 4799–4807.

18. Lanford, R. E. (1988) Expression of simian virus 40 T antigen in insect cells using a baculovirus expression vector. *Virology* **167,** 72–81.

19. Chazal, N., Carriere, C., Gay, B., and Boulanger, P. (1994) Phenotypic characterization of insertion mutants of the human immunodeficiency virus type 1 Gag precursor expressed in recombinant baculovirus-infected cells. *J. Virol.* **68,** 111–122.

20. Lanford, R. E., Luckow, V., Kennedy, R. C., Dreesman, G. R., Notvall, L., and Summers, M. D. (1989) Expression and characterization of hepatitis B virus surface antigen polypeptides in insect cells with a baculovirus expression system. *J. Virol.* **63,** 1549–1557.

21. Davidson, D. J. and Casteliino, F. J. (1991) Asparagine-linked oligosaccharide processing in lepidopteran insect cells. Temporal dependence of the nature of the oligosaccharides assembled on asparagine-289 of recombinant human plasminogen produced in baculovirus vector infected Spodoptera frugiperda (IPLB-SF-21AE) cells. *Biochemistry* **30,** 6165–6174.

22. Clem, R. J. and Miller, L. K. (1994) Control of programmed cell death by the baculovirus genes p35 and lap. *Mol. Cell. Biol.* **14,** 5212–5222.

23. Novelli, A. and Boulanger, P. A. (1991) Deletion analysis of functional domains in baculovirus- expressed adenovirus type 2 fiber. *Virology* **185,** 365–376.

24. Hilditch, C. M., Rogers, L. J., and Bishop, D. H. (1990) Physicochemical analysis of the hepatitis B virus core antigen produced by a baculovirus expression vector. *J. Gen. Virol.* **71,** 2755–2759.

25. Labbe, M., Charpilienne, A., Crawford, S. E., Estes, M. K., and Cohen, J. (1991) Expression of rotavirus VP2 produces empty corelike particles. *J. Virol.* **65,** 2946–2952.

26. Gheysen, D., Jacobs, E., de Foresta, F., et al. (1989) Assembly and release of HIV-1 precursor Pr55gag virus-like particles from recombinant baculovirus-infected insect cells. *Cell* **59,** 103–112.

27. Loudon, P. T. and Roy, P. (1991) Assembly of five bluetongue virus proteins expressed by recombinant baculoviruses, inclusion of the largest protein VP1 in the core and virus-like proteins. *Virology* **180,** 798–802.

28. Agbandje, M., McKenna, R., Rossmann, M. G., Kajigaya, S., and Young, N. S. (1991) Preliminary X-ray crystallographic investigation of human parvovirus B19. *Virology* **184,** 170–174.

29. Agbandie, M., Kajigaya, S., Mckennia, R., Young, N. S., and Rossmann, M. G. (1994) The structure of human parvovirus B19 at 8 A resolution. *Virology* **203,** 106–115.

30. Prasad, B. V., Rothnagel, R., Jiang, X., and Estes, M. K. (1994) Three-dimensional structure of baculovirus-expressed Norwalk virus capsids. *J. Virol.* **68,** 5117–5125.

31. Kimbauer, R., Booy, F., Cheng, N., Lowy, D. R., and Schiller, J. T. (1992) Papillomavirus L1 major capsid protein self-assembles into virus-like particles that are highly immunogenic. *Proc. Natl. Acad. Sci. USA* **89,** 12,180–12,184.

32. Breitburd, F., Kimbauer, R., Hubbert, N. L., et al. (1995) Immunization with viruslike particles from cottontail rabbit papillomavirus (CRPV) can protect against experimental CRPV infection. *J. Virol.* **69,** 3959–3963.

33. Suzich, J. A., Ghim, S. J., Palmer-Hill, F. J., et al. (1995) Systemic immunization with papillomavirus L1 protein completely prevents the development of viral mucosal papillomas. *Proc. Natl. Acad. Sci. USA* **92,** 11,553–11,557.

34. Sabara, M., Parker, M., Aha, P., Cosco, C., Gibbons, E., Parsons, S., and Babiuk, L. A. (1991) Assembly of double-shelled rotaviruslike particles by simultaneous expression of recombinant VP6 and VP7 proteins. *J. Virol.* **65,** 6994–6997.

35. Roy, P., Urakawa, T., Van Dijk, A. A., and Erasmus, B. J. (1990) Recombinant virus vaccine for bluetongue disease in sheep. *J. Virol.* **64,** 1998–2003.

36. Roy, P., Bishop, D. H., LeBlois, H., and Erasmus, B. J. (1994) Long-lasting protection of sheep against bluetongue challenge after vaccination with virus-like particles, evidence for homologous and partial heterologous protection. *Vaccine* **12,** 805–811.

37. Grimes, J., Basak, A. K., Roy, P., and Stuart, D. (1995) The crystal structure of bluetongue virus VP7. *Nature* **373,** 167–170.

38. Hewat, E. A., Booth, T. F., and Roy, P. (1992) Structure of bluetongue virus particles by cryoelectron microscopy. *J. Struct. Biol.* **109,** 61–69.

39. Hewat, E. A., Booth, T. F., Wade, R. H., and Roy, P. (1992) 3-D reconstruction of bluetongue virus tubules using cryoelectron microscopy. *J. Struct. Biol.* **108,** 35–48.

40. Hewat, E. A., Booth, T. F., and Roy, P. (1994) Structure of correctly self-assembled bluetongue virus-like particles. *J. Struct. Biol.* **112,** 183–191.

41. Roy, P. (1993) Dissecting the assembly of orbiviruses. *Trend. Microbiol.* **1,** 299–305.

42. Belyaev, A. S. and Roy, P. (1992) Presentation of hepatitis B virus preS2 epitope on bluetongue virus core-like particles. *Virology* **190,** 840–844.

43. Monastyrskaya, K., Gould, E. A., and Roy, P. (1995) Characterization and modification of the carboxy-terminal sequences of bluetongue virus type 10 NS1 protein in relation to tubule formation and location of an antigenic epitope in the vicinity of the carboxy terminus of the protein. *J. Virol.* **69,** 2831–2841.

44. Mikhailov, M., Monastyrskaya, K., Bakker, T., and Roy, P. (1996) A new form of particulate single and multiple immunogen delivery system based on recombinant bluetongue virus-derived tubules. *Virology* **217,** 323–331.

45. Thomsen, D. R., Newcomb, W. W., Brown, J. C., and Homa, F. L. (1995) Assembly of the herpes simplex virus capsid, requirement for the carboxyl-terminal twenty-five amino acids of the proteins encoded by the UL26 and UL26.5 genes. *J. Virol.* **69,** 3690–3703.

46. Newcomb, W. W., Homa, F. L., Thomsen, D. R., Ye, Z., and Brown, J. C. (1994) Cell-free assembly of the herpes simplex virus capsid. *J. Virol.* **68,** 6059–6063.

47. Thomsen, D. R., Roof, L. L., and Homa, F. L. (1994) Assembly of herpes simplex virus (HSV) intermediate capsids in insect cells infected with recombinant baculoviruses expressing HSV capsid proteins. *J. Virol.* **68,** 2442–2457.

48. Tengelsen, L. A., Pederson, N. E., Shaver, P. R., Wathen, M. W., and Homa, F. L. (1993) Herpes simplex virus type 1 DNA cleavage and encapsidation require the product of the UL28 gene, isolation and characterization of two UL28 deletion mutants. *J. Virol.* **67,** 3470–3480.

49. Trus, B. L, Homa, F. L., Bcoy, F. P., et al. (1995) Herpes simplex virus capsids assembled in insect cells infected with recombinant baculoviruses, structural authenticity and localization of VP26. *J. Virol.* **69,** 7362–7366.

50. Jowett, J. B., Hockley, D. J., Nermut, M. V., and Jones, I. M. (1992) Distinct signals in human immunodeficiency virus type 1 Pr55 necessary for RNA binding and particle formation. *J. Gen. Virol.* **73,** 3079–3086.

51. Morikawa, Y., Kishi, T., Zhang, W. H., Nermut, M. V., Hockley, D. J., and Jones, I. M. (1995) A molecular determinant of human immunodeficiency virus particle assembly located in matrix antigen p17. *J. Virol.* **69,** 4519–4523.

52. Hockley, D. J., Nermut, M. V., Grief, C., Jowett, J. B., and Jones, I. M. (1994) Comparative morphology of Gag protein structures produced by mutants of the gag gene of human immunodeficiency virus type 1. *J. Gen. Virol.* **75,** 2985–2997.

53. Nermut, M. V., Hockley, D. J., Jowett, J. B., Jones, I. M., Garreau, M., and Thomas, D. (1994) Fullerene-like organization of HIV gag-protein shell in virus-like particles produced by recombinant baculovirus. *Virology* **198,** 288–296.

54. Rao, Z., Belyaev, A. S., Fry, E., Roy, P., Jones, I. M., and Stuart, D. I. (1995) Crystal structure of SIV matrix antigen and implications Ior virus assembly. *Nature* **378,** 743–747.

55. Sigle, R. O., Titus, M. A., Harada, N., and Nelson, S. D. (1994) Baculovirus mediated high level expression of human placental aromatase (CYP1 9A1). *Biochem. Biophy. Res. Comm.* **201,** 634–700.

56. Amarneh, B. and Simpson, E. R. (1995) Expression of a recombinant derivative of human aromatase P450 in insect cells utilizing the baculovirus vector system. *Mol. Cell. Endocrin.* **109,** R1–R5.

57. Patten, C. J. and Koch, P. (1995) Baculovirus expression of human P450 2E1 and cytochrome b5, spectral and catalytic properties and effect of b5 on the stoichiometry of P450 2E1-catalyzed reactions. *Arch. Biochem. Biophys.* **317,** 504–513.

58. Zeldin, D. C., DuBois, R. N., Falck, J. R., and Capdeviia, J. H. (1995) Molecular cloning, expression and characterization of an endogenous human cytochrome P450 arachidonic acid epcxygenase isoform. *Arch. Biochem. Biophys.* **322,** 76–86.

59. Nash, W. E., Mercer, R. W., Blanco, G., Stickler, R. C., Mason, J. I., and Thomas, J. L. (1994) Over-expression of Human type 1 (placental) 3B-hydroxy-5-ene-steroid dehydrogenase/isomerase in insect cells infected with recombinant baculovirus. *J. Ster. Biochem. Mol. Biol.* **50,** 235–240.

60. Breton, R., Yang, F., Jin, J. Z., Li, B., Labrie, F., and Lin, S. X. (1994) Human 17B-Hydroxysteroid dehydrogenase, overproduction using a baculovirus expression system and characterization. *J. Ster. Biochem. Mol. Biol.* **50,** 275–282.

61. Kraus, P. F. X. and Kutchan, T. M. (1995) Molecular cloning and heterologous expression of a cDNA encoding berbamunine synthase, a C-O phenol-coupling cytochrome P450 from the higher plant Berberis stolonifera. *Proc. Natl. Acad. Sci. USA* **92,** 2071–2075.

62. Koley, A. P., Buters, J. T. M., Robinson, R. C., Markowitz, A., and Friedman, F. K. (1995) CO binding kinetics of human cytochrome P450 3A4. *J. Biol. Chem.* **270,** 5014–5018.

63. Hasemann, C. A. and Capra, J. D. High-level production of a functional immunoglobulin heterodimer in a baculovirus expression system. *Proc. Natl. Acad. Sci. USA* **87,** 3942–3946.

64. Bjorkman, P. J., Saper, M. A., Samraoui, B., et al. (1987) Structure of the human class I histocompatibility antigen, HLA-A2. *Nature* **329,** 506–512.

65. Brown, J. H., Jardetzky, T. S., Gorga, J. C., et al. (1993) Three-dimensional structure of the human class II histocompatibility antigen HLA-DR1. *Nature* **364,** 33–39.

66. Stern, L. J., Brown, J. H., Jardetzky, T. S., et al. (1994) Crystal structure of the human class 11 MHC protein HLA-DR1 complexed with an influenza virus peptide. *Nature* **368,** 215–221.

67. Heemels, M. T. and Ploegh, H. Generation, translocation and presentation of MHC class l-restricted peptides. *Ann. Rev. Biochem.* **64,** 463–491.

68. Fruh, K., Ahn, K., Djaballah, H., et al. (1995) A viral inhibitor of peptide transporters for antigen presentation *Nature* **375,** 415–418.

69. Cromme, F. V., McCance, D. J., Straight, S. W., et al. (1994) Abstract No. 49 from the 13th International Papillomavirus Conference, Amsterdam. Expression of HPV 16-E5 protein in keratinocytes in vitro leads to post-transcriptional loss of MHC- I and TAP-I expression.

70. Meyer, T. H., van Endert, P. M., Uebel, S., Ehring, B., and Tampe, R. (1994) Functional expression and purification of the ABC transporter complex associated with antigen processing (TAP) in insect cells. *FEBS Lett.* **351,** 443–447.

20

Gene Transfer and Expression in Tissue Culture Cells of Higher Eukaryotes

M. Alexandra Aitken, Selina Raguz, and Michael Antoniou

1. Introduction

The transfer or "transfection" of genetic material into tissue-culture cell lines and the subsequent analysis of its RNA and protein products, has become a key and central technology in contemporary cell biology, molecular biology, and protein chemistry.

Transfection of genes provides a functional end-point assay for experiments addressing a number of issues including:

1. The mapping of transcriptional control elements (promoters, enhancers).
2. The biological function of a protein product within a whole-cell context.
3. The production of model systems to investigate cell-biological processes.
4. The production of large quantities of a protein product for biochemical and structure/ functional studies.
5. The production of large quantities of a protein product for clinical use.
6. Model systems for testing new therapeutic strategies, including gene therapy.

The aim of this chapter is to act as a guide to newcomers in the field offering advice as to the selection and optimization of a gene transfer protocol to meet particular needs, illustrating this from our own experience. We will begin with a brief review of the most commonly used transfection methods and possible future developments.

2. Transfection Methods

Over the years many methods of transfection have been described. The wide range of techniques that have been developed reflects a general problem that still plagues the field, namely, that the efficiency of transfection by any given method varies greatly between different cell types. Therefore, although procedures have evolved toward being more universally applicable, it is still possible that one of the older (and more inexpensive) procedures may be the most suitable for a given cell type.

The first transfection method to be described was that involving the coprecipitation of DNA with calcium phosphate *(1)*. A solution of DNA and calcium chloride is mixed gently with a sodium phosphate buffer to form a fine precipitate of calcium phosphate in which the nucleic acid has been trapped. This is then layered onto the cells to be

From: *Molecular Biomethods Handbook*
Edited by: R. Rapley and J. M. Walker © Humana Press Inc., Totowa, NJ

transfected *(2)*. Uptake of the DNA/CaPO$_4$ coprecipitate is thought to be by phagocytosis. This very simple procedure is still widely used despite being highly variable in its efficiency. However, with some cell types (e.g., HeLa, L-cells) this method can produce transfection efficiencies as high as several percent, which is sufficient for most studies.

Since the advent of the DNA/CaPO$_4$ procedure, a number of other chemical-based methods have been described employing reagents such as diethylaminoethyl (DEAE)-dextran *(3,4)* and polybrene *(5–8)*. All share the common feature of producing charged complexes that can readily adsorb to the cell surface with subsequent internalization by phagocytosis or endocytotic mechanisms. It is often possible to augment DNA uptake by careful treatment with substances that perturb plasma-membrane integrity, such as glycerol *(9)* or dimethyl sulfoxide (DMSO) *(5)*. The method employing polybrene and DMSO treatment is described in detail in **Subheading 6.2.**

Among the most popular transfection methods currently in use, and that shows a good general efficiency, employs cationic liposomes ("lipofection"; *10*). The cationic lipids possess a positively charged head group *(11)* and under aqueous conditions form bilayer liposomes with a highly positively charged exterior. These cationic liposomes are therefore able to strongly associate with the negatively charged phosphate backbone of DNA (and RNA). The resulting complex is in turn able to adsorb efficiently to the cell surface with subsequent internalization either by fusion to the plasma membrane or endocytosis. Endosomal-disrupting agents, such as those used in receptor-mediated endocytosis (RME)-gene delivery, may also improve the efficiency of lipofection. Lipofection reagents consisting of various cocktails of cationic and neutral lipids, have been commercially available for several years (e.g., Lipofectin, Lipofectamine from Gibco-BRL) and their use is detailed in **Subheading 6.1.** The one main disadvantage of lipofection is that the cationic-liposome preparations are very expensive compared to more classical, chemical-based transfection reagents such as DEAE-dextran or polybrene, which may preclude their routine use in many laboratories. However, because many cell types are refractory to these older methods, lipofection offers a good general alternative.

A related and very inexpensive cationic reagent that is also worth considering is polyethylenimine (PEI; *12*). PEI has a complex, three-dimensional structure with the highest possible positive-charge density of any readily available reagent. It is therefore able to associate with DNA and the cell surface by a similar mechanism to that of cationic lipids. Transfection efficiencies comparable to those obtained by lipofection can be achieved with PEI for some cell types *(12)*.

Transfection by "electroporation" is another method in widespread use today that has a high success rate with many different cell types. Many excellent reviews covering this topic have been written over the years to which we refer the interested reader *(13–15)*. This type of gene transfer evolved from observations that eukaryotic-cell membranes could be induced to fuse by the application of a high electrical discharge. In this protocol, a DNA/cell suspension is subjected to a single pulse of a high-current electric field (multiple-pulse discharges do not improve DNA transfer, *see* **ref. *16***). This electrical discharge causes a polarization of the plasma-cell membrane with the transient generation of pores through which the DNA can diffuse into the cytoplasm. One published result suggests that the movement of DNA into the cell is by electrophoresis with DNA passing end on through the field *(17)*. The inclusion of polycations, such as DEAE-dextran, in the electroporation mixture may also enhance DNA uptake *(18)*. Once these pores have resealed, normal cell function can resume. This physical-based gene transfer method has the potential to be used on most cells as it is independent of biological properties, such as tolerance to toxic effects of chemically mediated DNA delivery or phagocytotic/endocytotic capability of the cells. The use of this method is described in greater detail in **Subheading 6.3.**

Most, if not all, of the methods described so far are on the whole very idiosyncratic in regard to the cell lines they will transfect. However, procedures that employ very inexpensive, readily available reagents and that are very simple to carry out (e.g., CaPO$_4$/DNA coprecipitation, DEAE-dextran, polybrene/DMSO), are always worth trying for the cell line of choice before resorting to more complex and more costly methods (e.g., lipofection) that may also require expensive and sophisticated equipment (e.g., electroporation).

A method that holds a great deal of promise for the future delivers DNA to cells by subverting the natural physiological process of receptor-mediated endocytosis (RME; *19*). In its simplest form, this protocol consists of a ligand or antibody directed to a cell-surface receptor conjugated to polylysine. The cationic nature of polylysine allows it to bind DNA to produce a tightly condensed structure. Many of the ligand or antibody molecules remain free to bind to their cell-surface targets. Upon binding to the cell surface, the complex is internalized by endocytosis where it enters the endosomal pathway. Under normal circumstances, the DNA/ligand complex would be degraded as the endosome matures to the lysosomal compartment. Therefore, components that allow for "endosomal escape" must also be used as part of or during the transfection procedure.

Lysosomaltropic agents, such as chloroquine *(20)*, have been used in this context but are in general of limited value owing to their highly variable efficacy with different cell types. A far more effective and universal endosomal escape can be achieved by incorporating into the DNA/ligand complex a virus or viral components that have evolved an inherent endosomal-membrane disrupting capability. The inclusion of inactivated adenovirus as part of the DNA/ligand complex results in 100% transfection efficiencies *(21,22)*. Alternatively, synthetic peptides corresponding to regions of the influenza hemagglutinin molecule can also be used *(23,24)*. A major drawback, which currently limits the application of this technique, is the lack of ready-to-use commercially available reagents. Although the production of the ligand/antibody-polylysine conjugates requires only common reagents employing simple-protein chemistry, it nevertheless needs specialist skills and equipment, which therefore prevents its routine use in most laboratories. If in the future components and or reagents become available to target, for example, ubiquitously expressed cell-surface receptor(s), then this technology stands to revolutionize the way in which transfection of tissue-culture cells is performed on a routine basis.

3. Reporter Genes

A number of genes that code for easily assayed enzymes can be used as "reporters" to assess the efficiency of transfection. The gene of choice is under the control of strong ubiquitous promoter/enhancer combinations, such as those from the immediate early genes of human CMV *(25)*, the RSV LTR *(26)*, and the human β-actin gene *(27)*, for general use. Assays are performed either with a whole-cell protein extract, which is readily prepared in a universal lysis buffer (Reporter Lysis Buffer, Promega), or direct visualization *in situ.*

The first reporter-gene system of this type to be described is that employing the chloramphenicol acetyltransferase (CAT) gene found in the *E. coli* transposon Tn9 *(28)*. The usual substrate for this assay is ^{14}C-chloramphenicol. The resulting acetylated products are resolved and quantified by thin-layer chromatography *(28)* or by direct-scintillation counting *(29)*. A nonradioactive substrate allowing quantitation by fluorimetry has also been developed *(30)*. Assays involving CAT are therefore generally easy to perform and do not necessarily require complex specialist equipment.

The highest sensitivity is afforded by firefly luciferase *(31)*. This enzyme, which produces bioluminescence in the firefly, catalyzes the oxidation of D(–)luciferin in the presence of ATP, Mg^{2+} and O_2 to give oxyluciferin, CO_2 and a photon. The reaction in vitro is extremely rapid, reaching maximum levels of emitted light within 3–5 s after which there is an equally rapid decay to 10% of the peak value. Therefore, in order to take full advantage of the sensitivity of the assay, a luminometer has to be employed that automatically mixes sample and substrate and measures light emission over a 10-s period. A scintillation counter set in single photon-counting mode may also be used, but this only allows quantitation over the plateau value of 10% of the peak. The need for an expensive instrument such as a luminometer obviously negates a great deal of the convenience of this system.

The most versatile reporter-gene system is that based on the *E. coli lacZ* gene, which codes for the glycoside hydrolase β-D-galactosidase (β-gal; *32*). A variety of substrates are available for this enzyme, allowing quantitation by spectrometry *(33)*, fluorimetry *(32)*, and by fluorescence-activated cell sorting (FACS) analysis *(34)*. However, the most common use of β-gal is in combination with X-gal from which it produces an insoluble blue product clearly visible under low-power light microscopy (*see* **Subheading 6.4.**). This permits the direct staining of transfected cells *in situ* (*see* **Figs. 1** and **2**) and therefore the easy assessment of the transfection efficiency on a per-cell basis.

Recently, green fluorescent protein (GFP) from the jellyfish *Aequorea victoria (35)* has also been developed as a reporter-gene system for the direct observation of transfected cells. GFP fluoresces directly upon illumination with UV or blue light *(36)* and therefore permits the visualization of transfected cells by conventional fluorescence microscopy *(37)*. GFP-expression vectors are now commercially available (CLONTECH, Palo Alto, CA).

4. Generation of Stable Transfected Cell Lines

Upon transfection, a given DNA molecule has two possible fates. In the majority of cases, the transfected plasmid is retained episomally and is subsequently lost and/or degraded as the cells divide. This gives rise to "transient" gene expression from the

---

Fig. 1. *(opposite page; top)* Histochemical staining for β-gal activity in untransfected cells and cells transfected by electroporation with RSVβgal (*see* **Subheading 6.3.**).

Fig. 2. *(opposite page; bottom)* Histochemical staining of HL-60 cells for β-gal activity (*see* **Subheading 6.4.**) stably transfected with β-actin-gal/neo (*see* **Subheading 6.5.**).

UNTRANSFECTED   **TRANSFECTED**

A
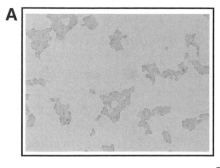 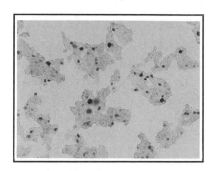

K562

B
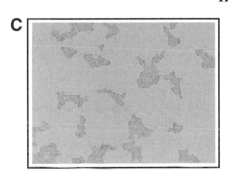 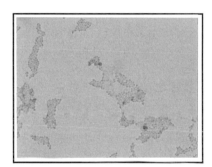

HL-60

C
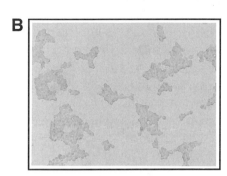 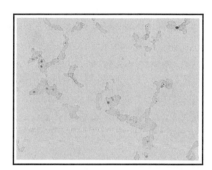

HL-60 EOS

Fig. 1.

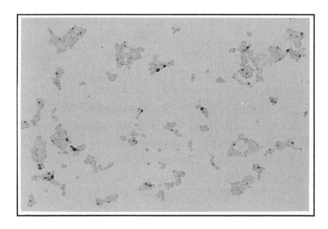

transfected DNA, which is normally maximal between 24 and 72 h post-transfection. Such a transient transfection assay is used to assess and optimize the transfection efficiency of the desired cell type (*see* **Subheading 6.**) as well as the mapping of promoter/ enhancer elements with measurement of reporter-gene expression over this short time scale. Alternatively, transfected DNA molecules may integrate by recombination into the host-cell genome, thereby becoming a stable, heritable genetic component. Many of the applications of gene transfer into tissue-culture cells (*see* **Subheading 1.**) require long-term gene expression that can only be afforded by stable, transfected cell lines. Unfortunately, the frequency of stable integration of transfected DNA using the common, standard methods described here, is normally very low ($1:10^3$–10^6). Therefore, in order to establish permanent stable, transfected cell lines, it is necessary to select for these from the bulk of the culture that is either only transiently or nontransfected. This is achieved by including a gene that confers resistance to a toxic drug as part of the transfected DNA. Only those cells that stably integrate and express the resistance gene will survive and grow upon prolonged exposure to the drug.

The first selection systems were based on cells carrying a mutation in genes encoding enzymes involved in purine/pyrimidine salvage pathways, namely, hypoxanthine-guanine phosphoribosyl transferase (HPRT), adenine phosphoribosyl transferase (APRT) and thymidine kinase (TK) *(38–40)*. These enzymes are normally nonessential for the growth of tissue-culture cells. However, the inhibition of the purine/pyrimidine biosynthetic pathways renders the cells dependent on the salvage enzymes for growth. Therefore, for example, a TK⁻ cell line treated with aminopterin will survive if stably transfected with the TK gene and grown in the presence of thymidine. The use of these genes obviously relies on the availability of mutant-cell lines deficient in one of these enzymes. Indeed, it is often necessary to first generate the desired mutant-cell line in order to subsequently exploit this selection system in transfection experiments *(40)*. This requirement clearly limits the general application of this approach.

The advent of positive drug-selection genes has allowed the stable transfection of all cell types and is now almost invariably the system of choice. The gene/drug combinations that are most commonly used are listed in **Table 1**. Except for dihydrofolate reductase (DHFR), all genes confer resistance to drugs that inhibit various aspects of protein synthesis. DHFR overcomes the effects of methotrexate, which blocks DNA synthesis. The minimum concentration of drug at which cell death occurs varies between different cell types depending on the growth (and hence metabolic/turnover) rate of the cells and uptake efficiency of the drug. It is therefore necessary to assess empirically the amount of drug to be used for a given cell line (*see* **Subheading 6.5.**). As these inhibitors are expensive reagents, this is a worthwhile and cost-effective exercise. The optimal concentration of geneticin/G418 and hygromycin B is normally between 200–1000 µg/mL and results in cell death in 1–2 wk. In contrast, the effective concentration for puromycin is 1–4 µg/mL, with cell death complete in 2–3 d. The use of puromycin as a selection system, therefore results in major savings of time and money compared to either G418 or hygromycin B and is the system of choice in our laboratory.

The main use of the DHFR-methotrexate system is in the generation of Chinese hamster ovary (CHO) cell lines in industry for the large-scale production of proteins for clinical applications. The gradual increase of methotrexate in the culture medium

Table 1
Dominant Selection Markers Gene/Drug Combinations
for the Selection of Stable Transfected Cell Lines (*see* Subheading 4.)

Resistance gene	Drug	Reference
Aminoglycoside phosphotransferase (AGPT; neo[r])	Neomycin geneticin sulfate (G418)	*36*
Hygromycin phosphotransferase (hyg[r])	Hygromycin B	*37*
Puromycin *N*-acetyl transferase (puro[r])	Puromycin	*38*
Histidinol dehydrogenase (*S. typhimurium hisD*)	Histidinol	*39*
Dihydrofolate reductase (DHFR)	Methotrexate	*40*

results in the selection of cells that have undergone amplification of the integrated transgene with the consequent rise in the level of DHFR gene expression, allowing survival at higher drug concentrations *(45)*. This, in turn, normally results in a concomitant increase in the production of the desired protein product from the gene linked in *cis* to that for DHFR.

5. General Rules

There are a few general rules that one should bear in mind when designing a transfection experiment.

1. Cells must be in log-phase growth on the day of transfection. This normally requires that cultures be set up 1–2 d prior to commencement of the experiment.
2. Transient-transfection assays are most effective with supercoiled plasmid-DNA preparations.
3. The efficiency of stable transfection is enhanced by using linearized molecules. The linearizing of the transfected DNA at a pre-determined restriction enzyme site fulfils a number of functions. Prior to integration, a supercoiled molecule would have to initially break to form a linear molecule to generate free ends that can recombine into the host-cell genome. This breaking of the plasmid occurs at random at any point on the molecule, including within the drug resistance or expression genes under study. Linearization at a predetermined site on the plasmid leaves the genes of interest intact, while at the same time creates a molecule that can more efficiently integrate into the genome. In addition, as integration can frequently result in the deletion of some DNA from the ends of the linear molecule, the linearization of the plasmid at a site distant from the drug selection and expression genes helps to preserve their integrity.
4. Stable transfection, as expected, is more efficient when the drug selection gene and the expression gene being investigated are on the same linear-plasmid molecule. The cotransfection of two linear plasmids, with each harboring one gene of interest, is also feasible but less efficient. This is particularly evident when electroporation is used as the transfection method where relatively few molecules are taken up by any given cell. If cotransfection is the only option, we recommend an expression gene to drug-selection plasmid-molar ratio of at least 3:1.

6. Optimization of Transfection Conditions
for Human-Granulocytic Cell Lines

We will now describe the approach we took to optimize the transfection conditions for two human-granulocytic cell lines, namely HL-60 (*see* **ref.** *46*) and HL-60 EOS *(47)*. HL-60 was isolated from a patient with acute promyelocytic leukemia and can be induced to undergo differentiation into various cell types of the myeloid hematopoietic lineage, such as granulocytes (neutrophils, eosinophils), monocytes, and macrophages *(46)*. HL-60 EOS is a subclone of HL-60 selected for constitutive eosinophilic character *(47)*. Transfection experiments using these two cell lines therefore provides the opportunity to study granulocyte-specific gene expression.

We conducted transient-transfection assays comparing three methods; lipofection, electroporation, and polybrene plus DMSO shock. The most efficient, optimized procedure was then used to generate stable transfected cells. The related human-myelogenous leukemia cell line K562 *(48)*, which is known to transfect relatively well by these methods, was included in these experiments as a positive control.

All these cell lines grow in suspension and were routinely cultured in 75 cm^2 flasks with 25 mL medium (Dulbecco's Modified Eagle's medium [DMEM] plus 10% fetal calf serum; FCS). The cells were kept in a 37°C, 5% CO_2 humidified incubator, and all experimental manipulations were carried out under these conditions. Subculture every 3–4 d maintained the cells in log phase growth.

6.1. Lipofection

Lipofection *(10)* was performed with Lipofectin and Lipofectamine (Gibco-BRL). Lipofectamine reagent is a 3:1 (w/w) liposome formulation of the polycationic lipid 2,3-dioleyloxy-*N*[2(sperminecarboxamido)ethyl]-*N*,*N*-dimethyl-1-propanaminium trifluoroacetate (DOPSA) and the neutral lipid dioleoyl phosphatidylethanolamine (DOPE) in water. This is a product that yields up to 90% positively transfected cells and has a 2–30-fold higher activity than the monocationic lipid compound Lipofectin. This greater efficiency is possibly caused by the DOPSA spermine head group *(49)*. The positively charged DOPSA (or in the case of Lipofectin DOTMA *N* [1-(2,3-dioleyloxy) propyl]-*N*,*N*,*N*-trimethylammonium chloride) forms lipid bilayers that rapidly and spontaneously interact with polyanionic DNA to form liposome/polynucleotide complexes *(50)* that capture 100% of the polynucleotide, which is 5 to >100-fold more effective than either the calcium phosphate or DEAE-dextran techniques *(10)*. These complexes are taken up onto the anionic surfaces of the tissue-culture cells and the fusogenic behavior of the DOPSA/DOTMA results in the functional intracellular delivery of the DNA *(51)*.

Lipofection is a very straightforward procedure. The DNA and cationic-liposome preparation are separately taken up in low-serum medium as polyvalent, negatively charged serum components are inhibitory on the action of DOPSA/DOTMA *(50)*. These are then mixed, allowed to stand to permit the formation of DNA/liposome complexes, and the mixture then applied to the cells and incubated for a pre-determined time (usually 5–6 h). Extra medium is then added to return to normal serum concentrations. The main parameters that are important to optimize for this procedure are cell-plating density, DNA:lipid ratio and exposure time to the DNA:lipid complexes, as toxicity can be a limiting factor *(49)*. The efficiency and reproducibility is unaffected by pH values between 6.0 and 8.0, by fluctuations in the level of DNA used or by a low level of contaminants.

A variation on this basic procedure, which is applicable to adherent cell lines, is what is known as "spin lipofection." After addition of the DNA:lipid mixture to cells in six-well plates, centrifugation in a microtiter plate carrier is then carried out for 10 min. The procedure can then be repeated to enhance DNA uptake still further.

6.1.1. Experimental Procedure

Reagents:

1. Lipofectamine (Gibco-BRL).
2. Lipofectin (Gibco-BRL).
3. Optimem 1 reduced serum medium (Gibco-BRL).
4. Growth medium with 20% FCS.

Lipofectamine or Lipofectin (2–100 μL) and the plasmid (2–20 μg) were diluted in Optimem 1 in separate microcentrifuge tubes. These were then mixed and incubated at room temperature for 15–30 min to allow the DNA/liposome complexes to form. The cells (3×10^6/treatment) were washed once with Optimem 1, the suspension of DNA/liposome complexes added and the mixture incubated for 5–72 h at 37°C. After this period complete medium containing 20% FCS was added to the cultures to double the incubation volume (the total serum content is therefore normal at 10%) and returned to the tissue-culture incubator.

6.2. Polybrene-Mediated Delivery

Polybrene (1,5-dimethyl-1,5-undecamethylene polymethobromide) was chosen over other available polycations such as DEAE-dextran *(52)* or poly-L-ornithine *(53)* as these are reported to be toxic at low concentrations and even short-exposure times. Polybrene is free of such problems and can be used for prolonged incubations *(54)*. These are necessary as it is a somewhat smaller polycation and therefore takes a longer period to facilitate DNA adsorption to cells *(5)*. Polybrene has also a better record at reproducibility of transfection frequencies than the calcium phosphate precipitation procedure *(7)*.

Polybrene is generally known for its ability to enhance the infectivity of retroviruses by serving as an electrostatic bridge between the negative viral particles and the anionic components of the recipient-cell membrane *(5)*. However, it was subsequently demonstrated to promote uptake of naked DNA *(8)*. Its use involves a two-stage protocol where first, during a prolonged incubation, the polybrene promotes the binding of the DNA to target cells. This is followed by exposure to DMSO, which permeabilizes the cells and allows uptake of the DNA/polycation complexes.

The transient-expression frequencies obtained are not high (0.01–1.0%) but in the absence of another successful technique this is an easy and convenient method. Optimization of the concentration of polybrene, plasmid DNA, the cell density and the amount and the exposure time to DMSO is required *(6)*. The exposure to DMSO is a crucial factor for the success of this protocol.

6.2.1. Experimental Procedure

Reagents:

1. 1 mg/mL polybrene (Aldrich Chemical Co., Milwaukee, WI) in Hank's balanced salt solution (Gibco-BRL) filter-sterilized, and stored in aliquots at –20°C.
2. Dimethyl sulfoxide (DMSO; Sigma).

A transfection mixture was prepared with additions in the following order to prevent irreversible precipitation.

1. The volume of complete medium needed;
2. The RSVβ-gal plasmid (2–20 μg/mL); and
3. The polybrene (5–100 μg/sample). After each addition, the mixture was briefly vortexed. The cells (3 × 10^6/treatment) were harvested and resuspended in the transfection mixture. The cultures were then returned to the incubator for 20 h. DMSO (15 or 20%) was then added directly in the culture for the desired time (3 or 4.5 min). The samples were then immediately pelleted and the cells washed twice with complete medium to remove all traces of the DMSO, and finally resuspended in complete medium and incubated until assayed for reporter gene activity.

6.3. Electroporation

There are now several commercially available electroporation units. All work on the common principle of capacitance discharge; that is, the required amount of current to be delivered is stored in capacitors and is then pulsed through the cell suspension at a preset potential difference. The exponential decay rate of the current is dictated by the conductance of the DNA/cell suspension, the gap distance between the electrodes within the electroporation chamber and the pre-set potential difference on the apparatus. It was originally thought that a discharge at high voltages (1–3 kV), which gave a current decay rate half-life of a few microseconds, was necessary for efficient transfection *(55)*. However, it has since been demonstrated that much better results are obtained at low voltages, which give a 10–20 millis decay time *(56)*. In our experience, optimal transfection for most cell types occurs at high current (900–1000 μF) discharge and at a potential difference of 200–400 V, which results in a 10–20 millis decay (*see* **Subheading 6.6.**).

The efficiency of transfection increases almost linearly with increasing DNA concentration as might be expected from an uptake mechanism by passive diffusion *(56)*. The transfection efficiency with increasing cell density remains constant. In addition, the preincubation of the DNA/cell mixture with periodic shaking for 10 min and for 5 min postelectroporation before diluting in medium, are important steps in the procedure. It is generally experienced that approx 50% cell death occurs at the optimum-electroporation conditions. An approximation can therefore be obtained by electroporating without plasmid to assess capacitance and voltage settings that give approx 50% cell survival. These data can then be used with 20–100 μg/mL of plasmid to more accurately define the optimal parameters *(14)*.

The choice of electroporation buffer is also crucial as it can effect the electrical resistance and thus effect the decay-time constant. An ideal solution would lead to minimal detrimental effects during the critical period when pores are created in the membrane. Previous experiments have shown success with HEPES buffered DMEM without serum *(14,57)*.

There are conflicting reports with regard to the temperature at which the electroporation should be carried out. Some reports recommend 4°C rather than at room temperature. Some claim that the lower temperature may even prove detrimental *(58)*. The prechilling may protect the cells from any local heating during pulsing, and postchilling may prolong the period of time the plasma-membrane pores remain open for DNA to enter the cells. However, raising the temperature after the pulse delivery speeds up the resealing of the pores and the cell has a better chance of surviving unfavorable conditions *(14)*. We routinely carry out all manipulations at room temperature.

6.3.1. Experimental Procedure

Reagents:

1. Electroshock buffer (ESB): HEPES-buffered DMEM (Hyclone, Northumberland, UK).
2. Phosphate-buffered saline (PBS): 170 mM NaCl, 3.3 mM KCl, 10 mM Na$_2$HPO$_4$, 1.8 mM KH$_2$PO$_4$, pH 7.0.

The cells (1 × 10^7/shock) were harvested and the cell pellet washed two times with PBS. The cells were then resuspended in ESB (0.7 mL). The plasmid (20 µg) was resuspended in ESB (100 µL) and was added to the cells and the mixture allowed to stand with periodic shaking at room temperature for 10 min. The cell/DNA suspension was then transferred to an electroporation cuvet. The cuvet was placed in the electroporation chamber of the Bio-Rad Gene Pulser and a single pulse at 25 or 960 µF, 200–400 V was delivered. The sample was allowed to stand again for 5 min at room temperature. The contents of the cuvet were then diluted into complete medium (10–25 mL) and transferred to the incubator.

6.4. Assay for Reporter Gene Activity

The reporter gene chosen for these experiments was *E. coli LacZ* gene (β-gal). A histochemical staining with X-gal (5-bromo-4-chloro-3-indolyl-β-D-galactoside) was used to determine the β-gal activity *in situ* and therefore assess the transfection efficiency on a per-cell basis *(32)*. The X-gal is hydrolyzed by the expressed β-galactosidase to generate D-galactose and soluble indoxyl molecules, which in turn are oxidized to insoluble indigo, which is clearly visible as a deep blue color under low-power light microscopy. The method generally involves fixing the cells before applying the stain. This comprises sodium phosphate to buffer the staining solution and to provide sodium ions necessary to activate the β-gal enzyme. Magnesium ions are also needed as a cofactor for the enzyme. Potassium ferrocyanide and potassium ferricyanide are required to act as oxidizing agents to increase the rate of conversion of the soluble/diffusible indoxyl molecules to insoluble indigo.

6.4.1. Histochemical Staining for β-Galactosidase

Reagents:

1. Fixative: 2% formal saline (from a 40% solution; BDH, Poole, Dorset, UK), 0.2% glutaraldehyde (from a 25% solution; Sigma), and prepared fresh in PBS immediately prior to use.
2. X-gal stain: 100 mM sodium phosphate, pH 7.3 (77 mM Na$_2$HPO$_4$, 23 mM NaH$_2$PO$_4$), 1.3 mM MgCl$_2$, 3 mM K$_3$Fe(CN)$_6$, 3 mM K$_4$Fe(CN)$_6$, 1 mg/mL X-gal (Gibco-BRL), and stored in the dark at 4°C.
3. Diff-Quik stain I (Baxter Healthcare Ltd., Norfolk, UK).

Cells were pelleted at 1000g for 5 min, washed once with PBS, and resuspended in fixative (2 mL), and incubated for 5 min at 4°C. Cells were then repelleted as before, washed twice with PBS, and resuspended in the X-gal stain (1–2 mL), and incubated overnight at 37°C.

Cells positive for β-galactosidase expression develop a blue color. The percentage of blue cells within the total number of cells present was assayed microscopically by examination with a hemocytometer. Photographs of these cells were obtained from slides. These were generated by fixing an aliquot of the suspension of cells in the X-gal stain to a glass slide by centrifugation in a cytospin centrifuge at $1000g$ for 5 min. It was necessary to counter-stain in order to highlight the untransfected, negative cells that are not otherwise apparent in the photographs. This was done by dipping the slides into Diff-Quik stain I for 15 s and briefly rinsing in water. This counter-stain produces an orange color.

In these experiments, reporter-gene activity was assessed over a 1–6 d period after transfection.

6.5. Generation of Stable Transfected HL-60 Cells

Reagents:

1. Geneticin (G418) sulfate (Gibco-BRL);
2. 500 mg/mL G418 in PBS, filter-sterilized and stored at –20°C.

6.5.1. Toxicity-Dose Response

HL-60 cells (10^7) were incubated with increasing amounts (0–1000 µg/mL) G418 in 200 µg increments. Complete cell death was obtained in 7–10 d with 600 µg/mL and was therefore chosen as the working concentration for subsequent, stable-transfection experiments.

It should be borne in mind that the potency of different batches of G418 can vary quite markedly and a reassessment of the concentration required for selection may have to be carried out with each new stock that is purchased.

6.5.2. Stable Transfection

HL-60 cells (10^7) were electroporated with 50 µg of β-actin-gal/neo plasmid that had been linearized with *Pvu*I using the conditions that had been found to be optimal from transient-transfection assays (*see* **Subheading 6.6.**). The β-actin-gal/neo construct consists of the β-gal gene sequences *(59)* inserted into the human β-actin expression vector pHβ Apr-1 neo *(27)*. G418 (600 µg/mL) was added to the culture medium 24 h post-electroporation. After complete cell death, the drug concentration was reduced by 50%. The resulting pools of stably transfected cells were then stained for β-gal gene expression as described above (**Subheading 6.4.1.** and **Fig. 2**).

6.6. Results

The HL-60 and HL-60 EOS cell lines generally proved difficult to transfect. Lipofection with either Lipofectin or Lipofectamine was unsuccessful at any DNA:lipid ratio and incubation time that was tested. The polybrene/DMSO protocol gave a weak positive result (0.07%) with HL-60 cells using 50 µg polybrene and 20% DMSO shock for 3 min. HL-60 EOS was refractory to this procedure. The highest transfection efficiencies were obtained with electroporation, with 960 µF and 300 V being the optimum conditions for both HL-60 (7.6%) and HL-60 EOS (7%) with maximum numbers of transfected cells observed after 4 and 6 d, respectively (**Fig. 1**). This compared with an efficiency of 11% for K562, 2 d post-electroporation at the same capacitance and voltage settings.

Stable transfected HL-60 cells were also successfully produced (**Fig. 2**) although at a low level of efficiency. The heterogeneity in the level of β-gal expression seen within the pool (**Fig. 2**), reflects clonal variations in chromatin position effects arising from different sites of integration of the transgene.

Acknowledgments

This work was supported by the Medical Research Council, UK.

References

1. Graham, F. L. and van der Eb, A. J. (1973) A new technique for the assay of infectivity of human adenovirus 5 DNA. *Virology* **52**, 456–467.
2. Okayama, H. and Chen, C. (1991) Calcium phosphate mediated gene transfer into established cell lines, in *Methods in Molecular Biology, Vol. 7: Gene Transfer and Expression Protocols* (Murray, E. J., ed.), Humana, Totowa, NJ, pp. 15–21.
3. McCutchan, J. H. and Pagano, J. S. (1968) Enhancement of the infectivity of simian virus 40 deoxyribonucleic acid with diethylaminoethyl-dextran. *J. Natl. Cancer Inst.* **41**, 351–356.
4. Lake, R. A. and Owen, M. J. (1991) Transfection of the Chloramphenicol acetyltransferase gene into Eukaryotic Cells Using Diethyl-Aminoethyl (DEAE)-Dextran, in *Methods in Molecular Biology, Vol. 7: Gene Transfer and Transfection Protocols* (Murray, E. J., ed.), Humana, Totowa, NJ, pp. 23–33.
5. Kawai, S. and Nishizawa, M. (1984) A new procedure for DNA transfection with polycation and dimethyl sulphoxide. *Mol. Cell. Biol.* **4**, 1172–1174.
6. Morgan, T. L., Maher, V. M., and McCormick, J. J. (1980) Optimal parameters for the polybrene induced DNA transfection of diploid human fibroblasts. *In Vitro Cell. Dev. Biol.* **22**, 317–319.
7. Chaney, W. G., Howard, D. R., Pollard, J. W., Sallustio, S., and Stardey, P. (1986) High-frequency transfection of CHO cells using Polybrene. *Somatic Cell. Mol. Genet.* **12**, 237–244.
8. Aubin, R. J., Weinfeld, M., and Peterson, M. C. (1988) Factors influencing efficiency and reproducibility of Polybrene-assisted gene transfer. *Somatic Cell. Mol. Genet.* **14**, 155–167.
9. Luthman, H. and Magnusson, G. (1983) High efficiency polyoma DNA transfection of chloroquine treated cells. *Nucleic Acids Res.* **11**, 1295 1307.
10. Felgner, P. L., Gadek, T. R., Holm, M., Roman, R., Chan, H. W., Wenze, M., Northrop, J. P., Ringold, G. M., and Danielsen, M. (1987) Lipofection: a highly efficient, lipid-mediated DNA-transfection procedure. *Proc. Natl. Acad. Sci. USA* **84**, 7413–7417.
11. Felgner, J. H., Kumar, R., Sridhar, C. N., Wheeler, C. J., Elorder, R., Ramsey, P., Martin, M., and Felgner, P. L. (1994) Enhanced gene delivery and mechanism studies with a novel series of cationic lipid formulations. *J. Biol. Chem.* **269**, 2550–2561.
12. Boussif, O., Lezoualc'h, F., Zanta, M. A., Mergny, M. D., Scherman, D., Demeneix, B., and Behr, J.-P. (1995) A versatile vector for gene and oligonucleotide transfer into cells in culture and *in vivo:* polyethylenimine. *Proc. Natl. Acad. Sci. USA* **92**, 7297–7301.
13. Knutson, J. C. and Yee, D. (1987) Electroporation: parameters affecting transfer of DNA into mammalian cells. *Anal. Biochem.* **164**, 44–52.
14. Andreason, G. L. and Evans, G. A. (1989) Optimisation of electroporation for transfection of mammalian cell lines. *Anal. Biochem.* **180**, 269–275.
15. Chang, D., Chassey, M., Saunders, J., and Sowers, A., eds. (1992) *Guide to Electroporation and Electrofusion.* Academic, New York.

16. Toneguzzo, F., Keating, A., Glynn, S., and McDonald, K. (1988) Electric field-mediated gene transfer: characterisation of DNA transfer and patterns of integration in lymphoid cells. *Nucleic Acid. Res.* **16,** 5515–5532.

17. Winterbourne, D. J., Thimas, S., Hermon-Taylor, J., Hussain, I., and Johnstone, A. P. (1988) Electroshock-mediated transfection of cells. *Biochem. J.* **251,** 427–434.

18. Gauss, G. H. and Lieber, M. (1992) DEAE-dextran enhances electroporation of mammalian cells. *Nucleic Acid Res.* **20,** 6739–6740.

19. Guy, J., Drabek, D., and Antoniou, M. (1995) Delivery of DNA into mammalian cells by receptor-mediated endocytosis and gene therapy. *Mol. Biotechnol.* **3,** 237–248.

20. Cotten, M., Längle-Rouault, F., Kirlappos, H., Wagner, E., Mechtler, K., Zenke, M., Beug, H., and Birnstiel, M. L. (1990) Transferrin-polycation mediated introduction of DNA into human leukemic cells: stimulation by agents that effect the survival of transfected DNA or modulate transferrin receptor levels. *Proc. Natl. Acad. Sci. USA* **87,** 4033–4037.

21. Wagner, E., Zatloukal, K., Cotten, M., Kirlappos, H., Mechtler, K., Curiel, D. T., and Birnstiel, M. L. (1992) Coupling of adenovirus to transferrin-polylysine/DNA complexes greatly enhances receptor-mediated gene delivery and expression of transfected genes. *Proc. Natl. Acad. Sci. USA* **89,** 6099–6103.

22. Cotten, M., Wagner, E., Zatloukal, K., and Birnstiel, M. L. (1993) Chicken adenovirus (CELO virus) particles augment receptor-mediated DNA delivery to mammalian cells and yield exceptional levels of stable transformants. *J. Virol.* **67,** 3777–3785.

23. Wagner, E., Plank, C., Zatloukal, K., Cotten, M., and Birnstiel, M. L. (1992) Influenza virus hemagglutinin HA-2 N-terminal fusogenic peptides augment gene transfer by transferrin-polylysine-DNA complexes: toward a synthetic virus-like gene-transfer vehicle. *Proc. Natl. Acad. Sci USA* **89,** 7934–7938.

24. Midoux, P., Mendes, C., Legrand, A., Raimond, J., Mayer, R., Monsigny, M., and Roche, A. C. (1993) Specific gene transfer mediated by lactosylated poly-L-lysine into hepatoma cells. *Nucleic Acids Res.* **21,** 871–878.

25. Boshart, M., Weber, F., Jahn, G., Dorsch-Hasler, K., Fleckenstein, B., and Schaffner, W. (1985) A very strong enhancer is located upstream of an immediate early gene of human cytomegalovirus. *Cell* **41,** 521–530.

26. Gorman, C. M., Merlino, G. T., Willingham, M. C., Pastan, I., and Howard, B. H. (1982) The Rous sarcoma virus long terminal repeat is a strong promoter when introduced into a variety of eukaryotic cells by DNA-mediated transfection. *Proc. Natl. Acad. Sci. USA* **79,** 6777–6781.

27. Gunning, P., Leavitt, J., Muscat, G., Ng, S.-Y., and Kedes, L. (1987) A human β-actin expression vector system directs high-level accumulation of antisense transcripts. *Proc. Natl. Acad. Sci. USA* **84,** 4831–4835.

28. Gorman, C. M., Moffat, L. F., and Howard, B. H. (1982) Recombinant genomes that express chloramphenicol acetyltransferase in mammalian cells. *Mol. Cell. Biol.* **2,** 1044–1051.

29. Sleigh, M. J. (1986) A nonchromatographic assay for expression of the chloramphenicol acetyltransferase gene in eukaryotic cells. *Anal. Biochem.* **156,** 251–256.

30. Young, S. L., Barbera, L., Kaynard, A. H., Haugland, R. P., Kang, H. C., Brinkley, M., and Melner, M. H. (1991) A non-radioactive assay for transfected chloramphenicol acetyltransferase activity using fluorescent substrates. *Anal. Biochem.* **197,** 401–407.

31. de Wet, J. R., Wood, K. V., DeLuca, M., Helinski, D. R., and Subramani, S. (1987) Firefly luciferase gene: structure and expression in mammalian cells. *Mol. Cell. Biol.* **7,** 725–737.

32. MacGregor, G. R., Nolan, G. P., Fiering, S., Roederer, M., and Herzenberg, L. A. (1991) Use of *E. coli lacZ* (β-Galactosidase) as a reporter gene, in *Methods in Molecular Biology, Vol. 7: Gene Transfer and Expression Protocols* (Murray, E. J., ed.), Humana, Totowa, NJ, pp. 217–235.

33. Eustice, D. C., Feldman, P. A., Colberg-Poley, A. M., Buckery, R. M., and Neubauer, R. H. (1991) A sensitive method for the detection of β-galactosidase in transfected cells. *Biotechniques* **11,** 739–743.

34. Nolan, G. P., Fiering, S., Nicolas, J.-F., and Herzenberg, L. A. (1988) Fluorescence-activated cell analysis and sorting of viable mammalian cells based on β-D-galactosidase activity after transduction of E. *coli lacZ. Proc. Natl. Acad. Sci. USA* **85,** 2603–2607.

35. Prasher, D. C., Eckenrode, V. K., Ward, W. W., Prendergast, F. G., and Cormier, M. J. (1992) Primary structure of the Aequorea victoria green fluorescent protein. *Gene* **111,** 229–233.

36. Heim, R. and Tsien, R. (1996) Engineering green fluorescent protein for improved brightness, longer wavelengths and fluorescence resonance energy transfer. *Curr. Biol.* **6,** 178–182.

37. Chalfie, M., Tu, Y., Euskirchen, G., Ward, W. W., and Prasher, D. C. (1994) Green fluorescent protein as a marker for gene expression. *Science* **263,** 802–805.

38. Harris, M. (1982) Induction of thymidine kinase in enzyme deficient Chinese hamster cells. *Cell* **29,** 483–492.

39. Jones, G. E. and Sargent, P. A. (1974) Mutants of cultured cells deficient in adenine phosphoribosyl transferase. *Cell* **2,** 43–54.

40. Mouth, M. and Harwood, J. (1991) Selection of cells defective in pyrimidine (TK) and purine (APRT- and HPRT) salvage, in *Methods in Molecular Biology, Vol. 7: Gene Transfer and Expression Protocols* (Murray, E. J., ed.), Humana, Totowa, NJ, pp. 257–267.

41. Colbere-Garapin, F., Horodniceanu, F., Kourilsky, P., and Garapin, A.-C. (1981) A new dominant hybrid selection marker for higher eukaryotic cells. *J. Mol. Biol.* **150,** 1–14.

42. Santerre, R. F., Allen, N. E., Hobbs, Jr., J. N., Rao, R. N., and Schmidt, R. J. (1984) Expression of prokaryotic genes for hygromycin B and G418 resistance as dominant selection markers in mouse L-cells. *Gene* **30,** 147–156.

43. Vara, J. A., Portela, A., Ortin, J., and Jimenez, A. (1986) Expression in mammalian cells of a gene from *Streptomyces alboniger* conferring puromycin resistance. *Nucleic Acids Res.* **14,** 4617–4624.

44. Hartman, S. C. and Mulligan, R. C. (1988) Two dominant-acting selectable markers for gene transfer studies in mammalian cell. *Proc. Natl. Acad Sci. USA* **85,** 8047–8051.

45. Stark, G. R. and Wahl, G. M. (1984) Gene amplification. *Ann. Rev. Biochem.* **53,** 447–491.

46. Collins, S. J. (1987) The HL-60 promyelotic leukaemia cell line: proliferation, differentiation, and cellular oncogene expression. *Blood* **70,** 1233–1244.

47. Tomonaga, M., Gasson, J. C., Quan, S. G., and Golde, D. W. (1986) Establishment of eosinophilic sublines from human promyelotic leukaemia (HL-60) cells: demonstration of multipotentiality and single-lineage commitment of HL-60 stem cells. *Blood* **67,** 1433–1441.

48. Lozio, C. B. and Lozzio, B. B. (1975) Human chronic myelogenous leukaemia cell line with positive Philadelphia chromosome. *Blood* **45,** 321–334.

49. Hawley-Nelson, P., Ciccarone, C., Gebeyehn, G., and Jessee, J. (1993) Lipofectamine reagent: a new, higher efficiency polycationic liposome transfection reagent. *Focus* **15,** 73–79.

50. Felgner, P. L. and Ringold, G. M. (1989) Cationic liposome-mediated transfection. *Nature* **337,** 387,388.

51. Duzgunes, N., Goldstein, J. A., Friend, D. S., and Felgner, P. L. (1989) Fusion of liposomes containing a novel cationic lipid N [1-(2,3-dioleyloxy) propyl]-N,N,N-trimethyl-ammonium: induction by multivalent anions and asymmetric fusion with acidic phospholipid vesicles. *Biochem.* **28,** 9179–9184.

52. Pangano, J. S., McCutchan, J. H., and Vaheri, A. (1967) Factors influencing the enhancement of poliovirus ribonucleic acid by diethylaminoethyl-dextran. *J. Virol.* **1,** 891–897.

53. Bond, C. V. and Wold, B. (1987) Poly-L-ornithine-mediated transformation of mammalian cells. *Mol. Cell. Biol.* **7,** 2286–2293.

54. Lasfargues, E. Y., Vaidya, A. B., Lasfargues, J. C., and Moore, D. H. (1976) *In Vitro* susceptibility of mink lung cells to the mouse mammary tumour virus. *J. Natl. Cancer Inst.* **57,** 447–449.

55. Boggs, S. S., Gregg, R. G., Borenstein, N., and Smithies, O. (1986) Efficient transformation and single-site, single-copy insertion of DNA can be obtained in mouse erythroleukemia cells transformed by electroporation. *Exp. Haematol.* **14,** 988–994.

56. Chu, G., Hayakawa, H., and Berg, P. (1987) Electroporation for the efficient transfection of mammalian cells with DNA. *Nucleic Acids Res.* **15,** 1311–1326.

57. Michel, M. R., Eligzoli, M., Koblet, H., and Kempf, C. (1988) Diffusion loading conditions determine recovery of protein synthesis in electroporated P3X63Ag8 cells. *Experimentia* **44,** 199–203.

58. Potter, H. (1988) Electroporation in biology: methods, applications and instrumentation. *Anal Biochem.* **174,** 361–373.

59. Kalnins A., Otto K., and Muller-Hill, B. (1983) Sequence of the *lacZ* gene of *Escherichia coli*. *EMBO J.* **2,** 593–597.

21

Plant Transformation

Andy Prescott, Rob Briddon, and Wendy Harwood

1. Introduction

Plant transformation involves the insertion of "foreign" genetic material into plant cells and the regeneration of transgenic plants. Over the past 15 yr or so, a variety of methods for plant genetic engineering have been described; these fall into two broad categories: *Agrobacterium*-based and direct gene transfer methodologies. The choice of transformation method depends on a number of variables, principally the plant species to be transformed, the purpose of the experiment and the availability of the necessary equipment. The vast majority of plant transformation techniques require the use of explants with high regeneration capacities as starting material; the choice of explant often determines the suitability of particular methods. This chapter describes the different plant-transformation methods available and discusses the factors that must be taken into account in planning this type of experiment. In addition, the range of applications for which plant transformation can be utilized are listed.

2. Agrobacterium-Based Plant Transformation

Two species of *Agrobacterium*, *A. tumefaciens*, and *A. rhizogenes* have been widely utilized in plant transformation experiments. Both are Gram-negative, soil-inhabiting plant pathogens. The interactions between *Agrobacterium* spp. and their plant hosts have been reviewed many times; the reader is directed to the following articles (*1–4b*) for more detail and specific references.

A. tumefaciens causes crown gall disease in susceptible dicotyledonous hosts; this disease is characterized by the formation of neoplastic tumors at wound sites infected by the bacterium. *A. rhizogenes* causes a similar disease that results in the proliferation of roots with pronounced root hairs at the wound site. The rationale behind utilizing these plant pathogens as vectors for plant transformation is based on the observation that the causal agent for the proliferation of tissue at the inoculated wound site is a segment of DNA that is transferred from the bacteria to the plant cell. All virulent *Agrobacterium* strains contain a large (approx 200 kb) plasmid known as either a Ti- (tumor-inducing) or Ri- (root-inducing) plasmid. Heat-cured and avirulent *Agrobacterium* strains do not contain these plasmids and are incapable of interacting with plant cells. Only a small (approx 20 kb) portion of the Ti- or Ri-plasmid, known as the

From: *Molecular Biomethods Handbook*
Edited by: R. Rapley and J. M. Walker © Humana Press Inc., Totowa, NJ

T-DNA (transferred DNA), is transported into the plant cell where it is integrated into the plant genomic DNA. *Agrobacterium*-based transformation methods utilize the ability of these bacteria to transfer discrete DNA segments into the plant genome, while removing the deleterious effects of the pathogen.

2.1. The Interaction of Wild-Type Agrobacterium Strains and Plant Cells

The process of infection of wound sites and T-DNA transfer by wild-type *Agrobacterium* strains can be divided into a number of stages (**Fig. 1**), the majority of which are also common to the interaction of modified ("disarmed") *Agrobacterium* strains with explants during plant transformation experiments.

2.1.1. Attraction to the Wound Site

Agrobacterium spp. are motile bacteria that are attracted to wounds of susceptible plants by chemotaxis towards wound exudates. Attractants include sugars, amino acids, and a variety of phenolic compounds including acetosyringone and coniferyl alcohol.

2.1.2. Binding to Plant Cells

The chromosomal genes, *chvA*, *chvB*, and *exoC*, of *Agrobacterium* spp. are required for the synthesis and export of β-1,2-glucans, which are implicated in the binding of the bacteria to the plant cell surface. Other mutations affecting the ability of the bacterium to attach to plant cells include those at the *att* loci, which lack at least one of the outer membrane proteins normally produced by *Agrobacterium*, and the *cel* mutants that are defective in cellulose biosynthesis. Human vitronectin and anti-vitronectin antibodies inhibit the binding of *Agrobacterium* to plant cells, indicating the possible involvement of a plant vitronectin-like receptor for bacterial attachment.

2.1.3. Activation of Virulence (vir) Genes

The genes required for a successful infection by *Agrobacterium* spp. are located on both the chromosomal genome and the Ti/Ri plasmids. The plasmid-encoded virulence genes occur as a regulon consisting of at least 7 operons, the precise number of which varies between different strains of *Agrobacterium*. Four of the plasmid-encoded *vir* operons (*virA*, *virB*, *virD*, and *virG*) are essential for the infection of any plant host. The products of the remaining operons are involved in the determination of host range and/or the degree of virulence. The *vir* genes are located outside of the T-DNA region and are not transferred to the plant cell (**Fig. 2**).

The activation of *vir* gene expression principally occurs via a two-component regulatory system comprised of the VirA and VirG proteins. The *virA* and *virG* genes are constitutively expressed in the bacteria, whereas expression of other *vir* genes is inducible. *VirG* gene expression is also enhanced by phenolic signals, phosphate starvation, extremes of pH, and other stresses. VirA is a transmembrane histidine protein kinase that can auto-phosphorylate and also phosphorylate the transcription factor, VirG. VirA proteins accumulate as homodimers in the inner membrane of the bacterium *(5)*. The current model for *vir* gene activation is as follows:

1. Phenolic substances released from the wound site interact either with VirA directly *(6)* or possibly via other periplasmically-located binding proteins *(7)* to induce auto-phosphorylation of the C-terminal cytoplasmic domain of VirA. The monosaccharides and acidic pH

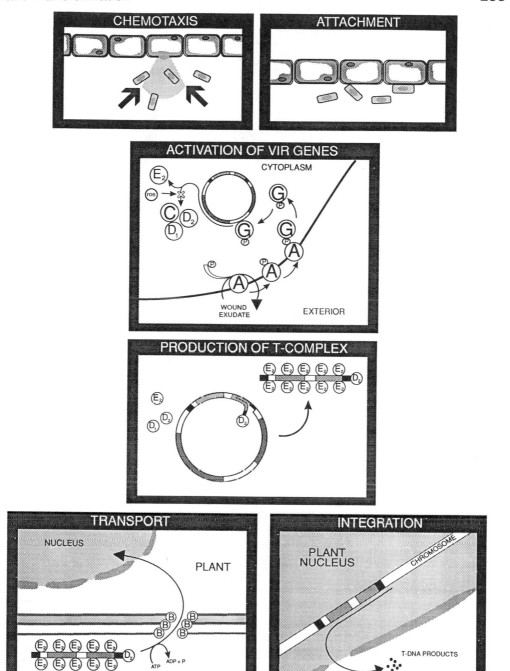

Fig. 1. A diagram showing the different stages in plant infection and transformation by *Agrobacterium*. (A, B, C, D1, D2, E2, G – *vir* gene products; P – inorganic phosphate).

of the wound exudate act synergistically with the phenolic signals to enhance *vir* gene activation. Monosaccharides bind to the chromosomally encoded ChvE sugar-binding protein, which interacts with the periplasmic domain of VirA to enhance induction.

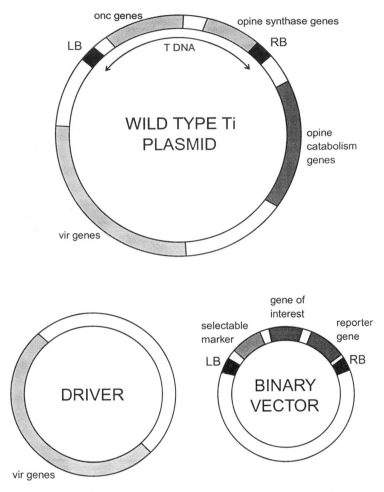

Fig. 2. This figure highlights the differences between a wild-type *Agrobacterium tumefaciens* Ti plasmid, a disarmed driver plasmid and a binary vector and demonstrates how the *Agrobacterium* system has been modified for the purposes of plant transformation. (LB, left T-DNA border; RB, right T-DNA border; onc, oncogenic; *vir*, virulence)

2. After activation of VirA by phosphorylation, a phosphate group is transferred from VirA to VirG. The VirG protein binds to a set of sequences (*vir*-boxes) found in the promoter region of each *vir* operon and induces expression of these operons. VirG can function as a DNA-binding protein in both phosphorylated and unphosphorylated forms, but the phosphorylated form appears to have a higher affinity for DNA-binding, possibly because of the formation of oligomers of VirG *(8)*.

3. The expression of the *virC* and *virD* operons is controlled by a chromosomally encoded repressor, *ros*, in addition to *VirG*.

2.1.4. Production of T-Strands

After *vir* gene induction, the next stage of the infection process entails the production of a T-DNA intermediate, known as the T-strand. The T-DNA region of Ti-plasmids is bounded by a pair of 25 bp imperfect repeats known as the T-DNA borders. The presence of a right border is essential for T-strand synthesis, which occurs in a unidirectional fashion. T-strand production is mediated by two products of the *virD* operon, VirD1 and VirD2, which cause one DNA strand to be nicked between the third and fourth basepairs of each T-DNA border, releasing the single-stranded T-strand. The role of VirD1 in this process is unclear, but VirD2 has been shown to catalyze site-specific DNA nicking in vitro *(9)*. A molecule of VirD2 remains covalently complexed to the 5' end of the T-strand during transport between the bacterial and plant cells. The T-strand is also coated by multiple VirE2 proteins, which may protect the DNA from exonuclease and endonuclease digestion and also promote the unfolding of the DNA to form a long (3600 nm), thin (2 nm) T-complex.

2.1.5. Transport of the T-Complex into the Plant Nucleus

Transport of the T-complex between the bacterial and plant cells involves the products of the *virB* operon. The *virB* operon encodes 11 genes, the majority of which are located in either the periplasm, inner, or outer membrane of the bacterium. The transport of the T-complex has two parallels in bacteriology: conjugation and the export of pertussis toxin from *Bordella pertussis* (reviewed in **ref. 10**). It is thought that the majority of the VirB products interact to form a membrane-associated export structure. Translocation of the T-complex requires energy that is probably supplied from ATP hydrolysis catalyzed by VirB4 and/or VirB11.

The T-complex passes into the nucleus of the recipient plant cell via nuclear pores. Both the VirD2 molecule (which is bound to the 5' end of the T-strand) and the VirE2 proteins bound along the length of the T-strand contain potential nuclear localization targeting motifs. It is possible that the nuclear-localization signal of VirD2 is responsible for the initial interaction of the T-complex with the nuclear pore, whereas that of the VirE proteins aids in the continuation of the uptake process.

2.1.6. Integration of T-DNA into the Plant Genome

The precise method of T-DNA integration into the plant DNA is unknown, but it appears to be a similar process to illegitimate recombination *(11)*. T-DNA integration appears to occur at random sites within the plant genome. Analysis has shown that the number of integrated T-DNA copies may vary between 1 and >10 and that integration of several copies at a single locus is not uncommon. Sequencing the junctions between the plant chromosomal DNA and the T-DNA has revealed that integration is usually accompanied by deletions and rearrangements of the plant DNA. Within the T-DNA, the right border junction tends to show less alteration than that of the left border.

2.1.7. Expression of T-DNA Genes

Following infection by a wild-type *Agrobacterium*, T-DNA transfer results in the expression of the genes situated between the two T-DNA borders. These genes are unusual in that despite their bacterial origin, their promoters resemble those of eukaryotes, thus allowing their expression within a plant cell. The transferred genes fall into two types: those involved in the manipulation of plant growth regulator levels within the plant cell (oncogenes) and those involved in the synthesis of opines.

The oncogenes differ between *A. tumefaciens* and *A. rhizogenes* but each results in the perturbation of normal plant growth regulator levels, resulting in the production of the characteristic phenotypes for infection, tumors or hairy roots. Opines are modified amino acids or sugars that are secreted by transformed plant cells. Opines act as transcriptional regulators, inducing the expression of Ti-plasmid-encoded genes responsible for their catabolism. Thus opines are a source of carbon and nitrogen for the invading *Agrobacterium* strain. In addition, some opines also induce the transcription of conjugal transfer genes, allowing the transfer of the Ti-plasmid capable of utilizing the opines to other *Agrobacterium* strains in the rhizosphere.

2.2. The Production of Disarmed Agrobacterium Strains and Binary Vectors

After the discovery that *Agrobacterium* infection results in the transfer of DNA from the bacterium to plant cells, methods were devised allowing the insertion of foreign genes into the T-DNA. Whilst transgenic plants can be regenerated from hairy roots or tumorigenic callus in some cases, the resulting plants usually have abnormal phenotypes because of their disturbed phytohormone balance. In addition, transgenic plants derived from wild-type *Agrobacterium* infections often exhibit infertility. Thus as well as adding genes to the T-DNA, it is also necessary to remove the native oncogenes in order to facilitate plant regeneration. There are two ways of producing a "disarmed" *Agrobacterium*; the first of these involves engineering a wild-type Ti-plasmid directly by replacing the T-DNA with an engineered T-DNA by a double-recombination event *(12)*. This method is difficult as it requires working directly with a large (*c.* 200 kb) plasmid. The alternative binary vector strategy *(13)* is used in the vast majority of *Agrobacterium*-based plant transformation experiments. In this case, the engineered T-DNA is present on a separate small plasmid (binary vector) which can be constructed in *E. coli* and subsequently transferred to a disarmed *Agrobacterium* strain. The disarmed *Agrobacterium* strain already contains a mutated Ti-plasmid (driver) from which the native T-DNA has been deleted. The driver plasmid supplies the *vir* functions necessary for the transfer of the engineered T-DNA (**Fig. 2**).

The disarmed *Agrobacterium* strain containing the engineered T-DNA is used to infect explants of plant material in tissue culture, which are subsequently regenerated to give transgenic plants.

2.3. Alternative Methods for Agrobacterium-Based Plant Transformation

In addition to using *Agrobacterium* to directly infect plant explants in tissue culture, a number of other methods for transforming plant tissue using *Agrobacterium* have been reported. These include using electroporation to transfer a binary vector from *Agrobacterium* into wheat callus cells *(14)*; using *Agrobacterium* in conjunction with microprojectile bombardment to enhance wounding *(15)*; imbibing germinating seeds with *Agrobacterium (16)*; vacuum infiltration of plant shoots with *Agrobacterium (17)* and *in planta* inoculation of severed plant stems with *Agrobacterium (18)*.

3. Direct Gene Transfer

A number of important crop plants, in particular the cereals, are not generally amenable to *Agrobacterium*-mediated transformation. Wheat and rice, the two most important world crops, fall into this group. It should however be noted that there has been considerable recent interest in using *Agrobacterium* vectors to transform monocots, in particular rice, and there have been reports of the successful production of transgenic plants *(19,19a)*. For most monocotyledonous plants, alternative gene-transfer techniques are preferred and there is a range of such techniques available. An important requirement for most of these techniques is an efficient plant-regeneration system from a plant cell or explant that is suitable for the particular DNA-delivery method.

3.1. Target Tissues and Regeneration Systems

3.1.1. Suspension Cultures

Embryogenic cell-suspension cultures have provided suitable target tissue for transformation procedures. They are usually initiated from embryogenic callus derived from immature embryos or anther cultures. The first report of fertile-transgenic maize resulted from the transformation of an embryogenic suspension culture *(20)*. One disadvantage of suspension cultures is that they are time-consuming to initiate and often only retain their embryogenic potential for a few months at most. It is now possible to cryopreserve suspension cultures *(20)*, which means that their useful life can be considerably extended.

3.1.2. Protoplasts

Much of the early transformation work with plants that were not amenable to *Agrobacterium* focused on the use of protoplasts, that is, plant cells from which the cell wall has been enzymically removed, as transformation targets. Indeed the first success in producing transgenic cereal plants was the result of introducing DNA to regenerable protoplasts of rice *(21)*. Cereal protoplasts are usually isolated from embryogenic suspension cultures, so this target tissue is subject to the same limitations as the suspensions described in **Subheading 3.1.1.** There are additional problems associated with the regeneration of fertile plants from cereal protoplasts. The time from initiation of the suspension to regeneration of plants from suspension-derived protoplasts can be considerable. In general, the longer the tissue-culture phase leading to the production of a regenerated plant, the greater are the chances of problems of somoclonal variation and reduced fertility in the regenerant. In species other than the cereals, for example, tobacco, *Arabidopsis*, and soybean, protoplasts can be isolated directly from leaves, roots, or immature embryos. Such protoplasts are not subject to the limitations of those isolated from suspension cultures and can be regenerated to fertile plants following DNA delivery.

3.1.3. Immature Embryos

Immature embryos are perhaps the most important target tissues for transformation of crop plants that are not susceptible to *Agrobacterium*. Recently there have been notable successes in the transformation of the important cereals (rice, wheat, and barley) using immature embryos *(22–24)*. Regenerated plants are often derived from somatic embryos that originate from the scutellum of the embryo. For this reason, the scutellum alone is sometimes used as a transformation target. In wheat and barley, immature embryos are usually isolated about 14 d after anthesis when they are 1–2 mm in diameter. This target tissue is dependent only on the availability of good-quality donor plants at the correct developmental stage.

3.1.4. Microspores

The regeneration of plants from anther or microspore culture offers important advantages over other target-tissue regeneration systems. Microspores provide single-cell transformation targets, thus avoiding possible problems with chimeric regenerants, but without the disadvantages of the protoplast single-cell system. Microspores of some species, for example, barley, can be highly regenerable *(25)* and regenerated plants are usually haploid or doubled haploids. Transformed plants derived from microspores are then homozygous for the transgene, which is a considerable advantage to breeders in the generation of pure lines. Barley has recently been transformed using isolated microspores as the target tissue *(26)*. The use of microspores as transformation targets is reviewed by Harwood et al. *(27)*.

3.1.5. Other Target Tissues

Mature embryos have been used in transformation studies because they provide a readily available source of target material. In general, however, regeneration frequencies are less than from immature embryos. Embryogenic callus derived from mature or immature embryos has also been successfully used as a transformation target *(28)*. In certain species other explants such as hypocotyl sections have also been used for transformation *(29)*.

3.2. DNA Delivery Methods

Successful methods for the introduction of DNA to the target tissues described in **Subheading 3.1.** must first ensure that the DNA is delivered into the plant cells in a suitable form so that there is a chance of it being integrated into the host genome. Second, the DNA delivery method must leave the plant cells with as little damage as possible so that their regeneration potential is not reduced. Most of the techniques that will be described allow the plant cell wall and plasma membrane to be breached in some way so that the DNA can enter. The first method to be considered, particle bombardment, is the most popular alternative to *Agrobacterium* at the present time.

3.2.1. Particle Bombardment

A particle-bombardment system uses a biolistic[R] device or gene gun to shoot heavy metal particles, or microprojectiles, coated with DNA into the target tissue. The development of this method was based on the original observation by Klein et al. *(30)* that DNA could be "shot" into onion epidermal cells, where it was transiently expressed. The method has been described in more detail by Sanford *(31,32)*, and many alternatives to the original particle delivery system have been designed and built. These often vary in the mechanism of propelling the DNA-coated particles into the plant cells, although the principle of the technique remains the same. Christou *(33)* has reviewed the transformation of crop plants using particle bombardment. Here we describe one of the most popular particle-delivery systems and some of the variations on this design.

3.2.1.1. THE PDS-1000/HE PARTICLE GUN

The first particle delivery system to be commercially available was the PDS-1000 device which was based on the gun designed by Sanford *(31)* and marketed by DuPont

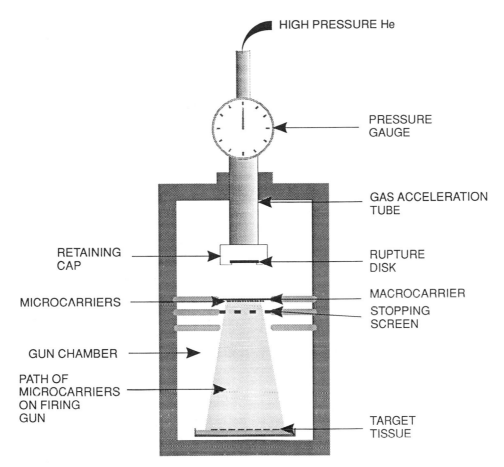

HIGH PRESSURE He

PRESSURE
GAUGE

GAS ACCELERATION
TUBE

RETAINING
CAP

RUPTURE
DISK

MICROCARRIERS

MACROCARRIER

STOPPING
SCREEN

GUN CHAMBER

PATH OF
MICROCARRIERS
ON FIRING
GUN

TARGET
TISSUE

Fig. 3. A simplified diagram of the PDS-1000/He gene gun, demonstrating how the particle bombardment of plant tissues is achieved.

(Wilminston, DE). This has now been updated to the PDS-1000/He gun and this device, which is available from BioRad (Hercules, CA), can be found in many plant-transformation laboratories around the world. Whereas the original PDS-1000 relied on gunpowder to provide the propulsive force, the PDS-1000/He uses the sudden release of a burst of helium gas to propel the DNA coated particles into the plant cells. A simplified schematic diagram of the PDS-1000/He is shown in **Fig. 3.**

The DNA of interest is coated onto gold particles with a diameter of approx 1 μm by a precipitation process involving calcium chloride and spermidine. Details of this procedure are given by Kikkert *(34)*. The gold particles or microcarriers, coated with DNA, are loaded onto a circular carrier disc referred to as the macrocarrier. This is then placed into the launch-assembly unit with the microcarriers facing down. Below the macrocarrier is a stopping screen, which is a wire mesh designed to retain the macrocarrier while allowing the microcarriers to pass through. The target tissue is placed below the launch assembly unit.

To fire the gun, the main gun chamber is evacuated and then helium is allowed to fill the gas-acceleration tube. The helium pressure builds up behind a rupture disk, which bursts at a specific pressure, thus releasing a shock wave of helium that forces the macrocarrier down onto the stopping screen. The microcarriers leave the macrocarrier and continue down the chamber to hit and penetrate the target tissue, thus delivering the DNA. A range of rupture disks are available that burst at different pressures, thus allowing the force with which the gold particles hit the plant cells and the depth penetrated by the particles to be controlled. The operation of the PDS-1000/He is described in detail by Kikkert *(34)*.

The PDS-1000/He gun has been used to successfully transform a number of plants. Using immature embryos as the target tissue, wheat *(23)* and barley *(24,35)* have been transformed. Transformed barley has also been produced using the same gun but with microspores as the target tissue *(26)*. Other uses of the gun are described by Sanford et al. *(36)*, who also cover the optimization of the device for different target tissue.

3.2.1.2. ALTERNATIVE PARTICLE-DELIVERY SYSTEMS

Several alternative particle delivery systems have been developed and one of these, the electric discharge gun (ACCELL™ technology) has given particularly good results in the transformation of rice, cotton, and soybean *(37)*. In this device the DNA- coated gold particles are accelerated by a shock wave produced by an electrical discharge. Advantages of this device over other particle-delivery systems are that it causes minimal damage to the target cells and it is also possible to precisely control the depth to which the particles penetrate. The ACCELL gun is not commercially available.

Another alternative DNA delivery device is the particle inflow gun (PIG). This gun has been built relatively easily and cheaply in a number of laboratories following its development by Finer et al. *(38)*. The gun accelerates DNA-coated tungsten particles directly in a stream of helium rather than coating the particles onto a macrocarrier as in the other devices. High-transformation frequencies have been reported using the PIG, for example, transgenic maize was produced *(39)* using suspension culture cells as target tissue.

The idea of precisely targeting a small area such as a floral meristem with a particle gun led to the development of a biolistic microtargeting device *(40)*. This device has been shown to be effective at delivering DNA to meristems *(41)* but has not yet been widely used.

3.2.2. Electroporation

The technique of electroporation uses electrical pulses to increase the permeability of the cell wall or protoplast membrane to DNA and simply involves mixing the target cells with the DNA before applying the electrical pulse. The technique was originally developed for protoplasts but has subsequently been shown to work with intact plant cells. The first report of fertile transgenic rice used electroporation to introduce DNA into embryogenic protoplasts, which were subsequently regenerated to give plants *(21)*. A similar procedure has now been used for wheat but the transgenic plants regenerated from embryogenic protoplasts failed to set seed *(42)*. This illustrates one of the main problems of techniques for the transformation of monocots that rely on protoplasts as target tissues; that of fertility problems in the regenerants.

Tissue electroporation, which dispenses with the need for protoplasts, was first demonstrated in maize *(43)*. The target tissue in this case was immature zygotic embryos or embryogenic callus which had been wounded prior to electroporation. The authors suggest that this technique should be widely applicable over a range of maize lines and there is now considerable interest in using the method for other monocot crop species as an alternative to particle bombardment.

3.2.3. PEG-Mediated Transformation

Polyethylene glycol (PEG)-mediated transformation is suitable only for plants in which a protoplast regeneration system has been developed. The method is similar to electroporation in that the DNA to be introduced is simply mixed with the protoplasts; in this case, however, the uptake of DNA is stimulated by the addition of PEG rather than by an electrical pulse. PEG treatment has been used to produce transgenic maize *(44)* and has also been used recently to produce fertile transgenic barley *(45)*. These examples demonstrate that protoplast transformation is feasible in the cereals even though problems of fertility in the regenerants are often encountered. PEG-mediated transformation is of particular importance as this method requires no specialist equipment and is also not subject to restrictions of use as is the case for particle bombardment.

3.2.4. Microinjection

The direct physical introduction of DNA to plant cells using micro-needles has also led to the production of transgenic plant material. This method again requires specialist equipment and skilled operators and has largely been superseded by particle-gun techniques. Microinjection does offer the possibility of introducing isolated chromosomes or organelles, as well as nucleic acids, into plant cells, but it is possible to manipulate only a very small number of cells in each experiment. Plant transformation using microinjection techniques is reviewed by Neuhaus and Spangenberg *(46)*.

3.2.5. Silicon Carbide Whisker-Mediated Transformation

A new method, which allows the treatment of a large number of target cells and does not require specialist equipment, is silicon carbide whisker-mediated transformation. In this procedure, first described by Kaeppler et al. *(47)*, the DNA is mixed with the cells to be transformed, for example by vortexing, in the presence of silicon carbide whiskers. The whiskers pierce the cells and thus allow the DNA to enter by a mechanism that is not fully understood. The method is simple, inexpensive, and has been shown to lead to the production of fertile transgenic maize *(48)*. But it is limited because it requires regenerable suspension culture cells as a target tissue and these are often difficult to establish, particularly in the cereals. Certain other target tissues, for example, microspores, may prove amenable to this technique in the future.

3.2.6. Alternative Transformation Methods

There has been considerable interest in using pollen as a transformation vector; this area has been recently reviewed by Harwood et al. *(27)*. Mature pollen grains are attractive vectors to deliver DNA because no in vitro culture steps are needed. The pollen containing the introduced DNA can simply be used for pollination. All of the techniques described for the introduction of DNA to plant cells or explants can also be applied to deliver DNA to pollen. Although there have been reports of successful transformation using mature pollen-based systems, many attempts to repeat this work in different laboratories around the world have failed. Therefore, it must be concluded that at the present time this method is not a reliable technique for the production of transgenic plants.

Several alternative direct gene-transfer techniques have also been reported and will be briefly mentioned. A method of laser-mediated transfer of genes to rice cells has been reported *(49)*. Here a laser microbeam pierced the plant cells to allow access to the DNA. In a similar way ultrasonication has been used to transfer genes to tobacco leaf segments and transgenic plants have been obtained *(50)*. There have also been attempts to introduce DNA to plant protoplasts by fusing them with liposomes containing the DNA of interest *(51)*. However, the method is limited as it requires regenerable protoplasts, and alternative methods for introducing DNA to protoplasts have proved far more popular than this approach.

3.3. Designing a Plant Transformation Experiment

3.3.1. Choice of Explant and Transformation Strategy

The choice of explant material is usually limited to those that are regenerable at high frequency in tissue-culture. Thus, explants may be protoplasts, leaf disks, root or shoot tissue, cotyledons, or somatic embryos. The transformation of some species, such as many members of the Solanaceae, is relatively easy and the choice of explant source or transformation strategy is not so important. For more recalcitrant species, such as members of the Gramineae, the development of a good regenerable system and optimization of the transformation efficiency is usually necessary. In general, *Agrobacterium*-based techniques tend to be used for the plant species that are easy to transform, such as *Arabidopsis*, tobacco, potato, tomato, and petunia; with direct gene transfer preferred for the more difficult species, such as soybean, maize, wheat, barley, and rice *(51a)*. It should be noted that strains of *Agrobacterium* have different host ranges and that this is another factor that must be taken into account in experimental design.

3.3.2. Construction of Plasmid Vectors

The two crucial decisions to be made in designing a transformation vector concern the choice of selectable markers and reporter genes *(51b)* and the type of promoter used to express the gene of interest. Selectable markers are those that can be used in tissue-culture to encourage the growth of transgenic tissue at the expense of nontransformed tissue. Selectable markers include antibiotic-resistance genes, e.g., neomycin phosphotransferase II, which allows selection of transgenic tissue on kanamycin, and herbicide-resistance genes, e.g., phosphoinothricin-*N*-acetyltransferase, which confers resistance to glutosinate and bialaphos. There is some unease at using selectable markers for commercial production of transgenic plants owing to the possibility of accidental transfer of the selectable marker to other species.

Reporter genes can be used in conjunction with, or as alternatives to, selectable markers. Reporter genes are used in assays to determine whether a tissue sample contains transgenic cells. They include β-glucuronidase (GUS; *52*), which can be assayed both biochemically and histochemically, luciferase (LUC; *53*), green fluorescent protein (GFP; *54*) and chloramphenicol acetyltransferase (CAT; *55*). Utilization of CAT as a reporter gene has been largely superseded by the availability of easily detectable visible markers, such as GUS, LUC, and GFP. The choice of whether to use selectable markers and/or reporter genes depends on the nature of the experiment. Some plant species are naturally resistant to some antibiotics or herbicides, whereas others may produce high levels of background activity for reporter-gene assays. Recently, binary vectors have been developed carrying two independent T-DNAs *(55a)*; these allow segregation of selectable markers and genes of interest in the progeny of transgenic plants.

An increasingly important decision for the fine manipulation of biochemical pathways is the type of promoter used to express the "gene of interest." For many purposes use of the 35S promoter of cauliflower mosaic virus, which gives constitutive expression of the transgene, is acceptable. However, a number of promoters giving tissue-specific expression or developmentally specific expression have been discovered, allowing expression of the transgene to be limited in time and space. Alternatively, the gene of interest can be regulated by an inducible promoter *(56,57)*.

3.3.3. Screening Putative Transgenic Plants

Apart from utilizing selectable markers and reporter genes, transgenic material is also usually screened for the presence and expression of the gene of interest by Southern blotting, Northern blotting, or by polymerase chain reaction (PCR) (*see* Chapters 8, 9, and 26, respectively). The transgene is transferred to the progeny by Mendelian inheritance; however, expression of transgenes is not always stable *(58,58a)*.

3.3.4. Licenses and Containment Facilities

Production of transgenic material and use of plant pathogens such as *Agrobacterium* spp. requires the experimenter to obtain the appropriate licenses from the relevant licensing authorities for the country concerned. In addition, attention must be paid to any ethical concerns associated with the planned experiment, particularly in terms of any possible accidental release of any genetically modified plant or bacterium to the environment. The level of containment and the precautionary procedures necessary for the production and growth of transgenic plant material depends on the nature of the experiment and its associated hazards.

3.4. Uses of Plant Transformation Technology

3.4.1. Production of New Crop Varieties with Altered Characteristics

Between 1986 and 1993, 1025 field trials of genetically modified plants were carried out in 32 countries (reviewed in **ref. *59***). Field trials have been carried out in Europe; North, Central and South America; Asia and Africa. The number of trials in developing countries has been increasing steadily over the past 5–6 yr as plant-transformation technology has become more widespread. The first genetically modified plant product for sale to consumers, Flavr Savr tomatoes (Calgene), became available in the United States in the mid 1990s and the first genetically engineered plant products in the United Kingdom were on sale in 1996. Genetically-modified traits include herbicide tolerance, virus resistance, resistance to insect pests, and factors affecting product quality.

3.4.2. Use of Transgenic Plant Organs to Produce Secondary Metabolites

Many plant secondary products are commercially valuable including some pharmaceuticals and food additives. Suspension cultures of plant cells have been used in biofermenters to produce phytochemicals on an industrial scale (reviewed in **ref. *60***). However, undifferentiated plant cells rarely produce large quantities of secondary metabolites and also exhibit both biochemical and genetic instability (reviewed in **refs. *61*** and ***62***). The use of *A. rhizogenes* and *A. tumefaciens* to create transgenic roots or teratomata *(61)* has been investigated, as transgenic organs tend to be more genetically stable and do not require exogenous plant growth regulators. Transgenic organs often produce larger amounts of secondary metabolites than suspension cultures, but they are more problematic to grow on an industrial scale owing to the difficulties in supplying nutrients and oxygen to complex tissues (reviewed in **ref. *62***).

3.4.3. Analysis of Plant Gene Expression and Function

Plant transformation has played an important role in fundamental plant biology as a tool to uncover the function of "unknown" genes. Isolating and cloning plant genes is a relatively straightforward technique, but determining the function of a given gene can be more difficult. Sequencing a gene and comparing the sequence to the sequence databases may give clues to its possible function but do not provide definitive proof. One of the ways of determining the function of a gene is to reintroduce it into plants by genetic engineering. In this way, it is possible to complement mutants *(63)*, or to over-express the gene and determine its function by analysis of the phenotype of the transgenic plant *(64)*. Alternatively, it is possible to silence or inactivate a gene by expressing the transgene in an antisense (reverse) orientation *(65)*.

3.4.4. Agrobacterium-*Mediated Inoculation of Plant Infectious Agents*

Agrobacterium-mediated inoculation (referred to as agroinfection *[66]* or agro-inoculation *[67]*) is an efficient method to introduce plant viruses and viroids into plants. Two approaches to agroinoculation have been employed. In the first, a tandem repeat of the viral genome is placed in the T-DNA. Following transfer to the plant cell a replicating, unit-length viral genome is produced either by homologous recombination or replication/transcription. This method is used extensively for plant viruses with circular DNA genomes including the geminiviruses *(68)*, the caulimoviruses *(69)*, and BADNAviruses *(70)*. The second approach to agroinoculation has been developed for plus-strand RNA viruses. For these viruses, the T-DNA contains a full-length viral cDNA clone inserted downstream of a suitable promoter (usually the cauliflower mosaic virus 35S promoter) and ahead of a suitable polyadenylation signal. The construct is designed to produce a transcript with as few nonviral sequences at both the 5' and 3' ends as possible, because these can reduce infectivity. This method has been used to agroinoculate beet western yellows virus *(71)* and tobacco mosaic virus *(72)* but should be suitable for all positive strand RNA viruses. Both methods have been used to agroinoculate viroids into plants *(73)*.

Although many monocotyledonous plants do not produce tumors when infected with wild-type *Agrobacterium* strains, the use of agroinfection of monocot-infecting geminiviruses has provided evidence of T-DNA transfer to these plants. This technique has provided a useful marker in studying the interaction between *Agrobacterium* and nonhost plants, such as cereals *(74)*. In addition, agroinfection is the only method of introducing some viruses into their monocotyledonous hosts.

3.4.5. Genetic Manipulation of Organelles

Foreign material can be expressed in plant organelles by using normal transformation procedures where the product of the transgene is fused to a signal peptide directing transport of the protein to the appropriate cellular compartment. Direct transformation of chloroplasts is also possible *(75,76)*.

References

1. Winans, S.C. (1992) Two-way chemical signaling in *Agrobacterium*-plant interactions. *Microbiol. Rev.* **56,** 12–31.
2. Zambryski, P. C. (1992) Chronicles from the *Agrobacterium*-plant cell DNA transfer story. *Annu. Rev. Plant Physiol. Plant Mol. Biol.* **43,** 465–490.
3. Hooykaas, P. J. J. and Beijersbergen, A. G. M. (1994) The virulence system of *Agrobacterium tumefaciens. Annu. Rev. Phytopathol.* **32,** 157–179.
4. Zupan, J. R. and Zambryski, P. (1995) Transfer of T-DNA from *Agrobacterium* to the plant cell. *Plant Physiol.* **107,** 1041–1047.
4a. Tinland, B. (1996) The integration of T-DNA into plant genomes. *Trends Plant Sci.* **1,** 178–184.
4b. Zupan, J. and Zambryski, P. (1997) The *Agrobacterium* DNA transfer complex. *Crit. Rev. Plant Sci.* **16,** 279–295.
5. Pan, S. Q., Charles, T., Jin, S., Wu, Z.-L., and Nester, E. W. (1993) Preformed dimeric state of the sensor protein VirA is involved in plant-*Agrobacterium* signal transduction. *Proc. Natl. Acad. Sci. USA* **90,** 9939–9943.
6. Chang, C.-H. and Winans, S. C. (1992) Functional roles assigned to the periplasmic, linker, and receiver domains of the *Agrobacterium tumefaciens* VirA protein. *J. Bacteriol.* **174,** 7033–7039.
7. Lee, K., Dudley, M. W., Hess, K. M., Lynn, D. G., Joerger, R. D., and Binns, A. N. (1992) Mechanism of activation of *Agrobacterium* virulence genes: identification of phenol-binding proteins. *Proc. Natl. Acad. Sci. USA* **89,** 8666–8670.
8. Han, D. C. and Winans, S. C. (1994) A mutation in the receiver domain of the *Agrobacterium tumefaciens* transcriptional regulator VirG increases its affinity for operator DNA. *Mol. Microbiol.* **12,** 23–30.
9. Scheiffele, P., Pansegrau,W., and Lanka, E. (1995) Initiation of *Agrobacterium tumefaciens* T-DNA processing. *J. Biol. Chem.* **270,** 1269–1276.
10. Kado, C. I. (1994) Promiscuous DNA transfer system of *Agrobacterium tumefaciens*: role of the VirB operon in sex pilus assembly and synthesis. *Mol. Microbiol.* **12,** 17–22.
10a. Christie, P. J. (1997) *Agrobacterium tumefaciens* T-complex transport apparatus: a paradigm for a new family of multifunctional transporters in eubacteria. *J. Bacteriol.* **179,** 3085–3094.
11. Gheysen, G., Villarroel, R., and Van Montagu, M. (1991) Illegitimate recombination in plants: a model for T-DNA integration. *Genes Dev.* **5,** 287–297.
12. Zambryski, P., Joos, H., Genetello, C., Leemans, J., Van Montagu, M., and and Schell, J. (1983) Ti plasmid vector for the introduction of DNA into plant cells without alteration of their normal regeneration capacity. *EMBO J.* **2,** 2143–2150.
13. Hoekema, A., Hirsch, P. R., Hooykaas, P. J. J., and Schilperoort, R. A. (1983) A binary plant vector strategy based on the separation of *vir*- and T-region of the *Agrobacterium tumefaciens* Ti plasmid. *Nature* **303,** 179,180.
14. Zaghmout, O. M.-F. and Trolinder, N. L. (1993) Simple and efficient method for directly electroporating *Agrobacterium* plasmid DNA into wheat callus cells. *Nucleic Acids Res.* **21,** 1048.

15. Bidney, D., Scelonge, C., Martich, J., Burrus, M., Sims, L., and and Huffman, G. (1992) Microprojectile bombardment of plant tissues increases transformation frequency by *Agrobacterium tumefaciens*. *Plant Mol. Biol.* **18,** 301–313.

16. Feldmann, K. A. and Marks, M. D. (1987) *Agrobacterium*-mediated transformation of germinating seeds of *Arabidopsis thaliana*: a non-tissue culture approach. *Mol. Gen. Genet.* **208,** 1–9.

17. Bechtold, N., Ellis, J., and Pelletier, G. (1993) *In planta Agrobacterium* mediated gene transfer by infiltration of adult *Arabidopsis thaliana* plants. *C. R. Acad. Sci. Paris, Sciences de la vie/Life Sciences* **316,** 1194–1199.

18. Chang, S. S., Park, S. K., Kim, B. C., Kang, B. J., Kim, D. U., and Nam, H. G. (1994) Stable genetic transformation of *Arabidopsis thaliana* by *Agrobacterium* inoculation *in planta*. *Plant J.* **5,** 551–558.

19. Chan, M. T., Chang, H.H., Ho, S. L., Tong, W. F., and Yu, S. M. (1993) *Agrobacterium*-mediated production of transgenic rice plants expressing a chimeric α-amylase promoter/β-glucuronidase gene. *Plant Mol. Biol.* **22,** 491–506.

19a. Rashid, H., Yokoi, S., Toriyama, K., and Hinate, K. (1996) Transgenic plant production mediated by *Agrobacterium* in *Indica* rice. *Plant Cell Rep.* **15,** 727–730.

20. Gordon-Kamm, W. J., Spencer, T. M., Mangano, M., Adams, T. R., Daines, R. J., Start, W. G., O'Brian, J. V., Chambers, S. A., Adams, W. R., Willetts, N. G., Rice, T. B., Mackey, C. J., Krueger, R. W., Kausch, A. P., and Lemaux, P. G. (1990) Transformation of maize cells and regeneration of fertile transgenic plants. *Plant Cell* **2,** 603–618.

21. Shimamoto, K., Teralda, R., Izawa, T., and Fujimoto, H. (1989) Fertile transgenic rice plants regenerated from transformed protoplasts. *Nature* **338,** 274–276.

22. Christou, P. and Ford, T. (1995) Parameters influencing stable transformation of rice immature embryos and recovery of transgenic plants using electric discharge particle acceleration. *Ann. Bot.* **75,** 407–413.

23. Weeks, J. T., Anderson, O. D., and Blechl, A. E. (1993) Rapid production of multiple independent lines of fertile transgenic wheat *(Triticum aestivum)*. *Plant Physiol.* **102,** 1077–1084.

24. Wan, Y. and Lemaux, P. G. (1994) Generation of large numbers of independently transformed fertile barley plants. *Plant Physiol.* **104,** 37–48.

25. Harwood, W. A., Bean, S. J., Chen, D.-F., Mullineaux, P. M., and Snape, J. W. (1995) Transformation studies in *Hordeum vulgare* using a highly regenerable microspore system. *Euphytica* **85,** 113–118.

26. Jähne, A., Becker, D., Brettschneider, R., and Lörz, H. (1994) Regeneration of transgenic, microspore-derived, fertile barley. *Theor. Appl. Genet.* **89,** 525–533.

27. Harwood, W. A., Chen, D.-F., and Creissen, G. P. (1996) Transformation of pollen and microspores—a review, in In Vitro *Haploid Production in Higher Plants,* vol. 2 (Jain, S. M., Sopory, S. K., and Veilleux, R. E., eds.), Kluwer Academic Publishers, Netherlands, pp. 55–72.

28. Li, L., Qu, R., de Kochko, A., Fauquet, C., and Beachy, R. N. (1993) An improved rice transformation system using the biolistic method. *Plant Cell Rep.* **12,** 250–255.

29. Radke, S. E., Turner, J. C., and Facciotti, D. (1992) Transformation and regeneration of *Brassica rapa* using *Agrobacterium tumefaciens*. *Plant Cell Reports* **11,** 499–505.

30. Klein, T. M., Wolf, E. D., Wu, R., and Sanford, J. C. (1987) High-velocity microprojectiles for delivering nucleic acids into living cells. *Nature* **327,** 70–73.

31. Sanford, J. C. (1988) The Biolistic process. *TIBTECH* **6,** 299–302.

32. Sanford, J. C. (1990) Biolistic plant transformation. *Physiol. Plant.* **79,** 206–209.

33. Christou, P. (1992) Genetic transformation of crop plants using microprojectile bombardment. *Plant J.* **2,** 275–281.

34. Kikkert, J. R. (1993) The Biolistic® PDS-1000/He device. *Plant Cell Tiss. Org. Cult.* **33,** 221–226.

35. Hagio, T., Hirabayashi, T., Machii, H., and Tomotsune, H. (1995) Production of fertile transgenic barley (*Hordeum vulgare* L.) plant using the hygromycin-resistance marker. *Plant Cell Rep.* **14,** 329–334.

36. Sanford, J. C., Smith, F. D., and Russell, J. A. (1993) Optimizing the Biolistic process for different applications. *Methods Enzymol.* **217,** 483–509.

37. McCabe, D. and Christou, P. (1993) Direct DNA transfer using electric discharge particle acceleration (ACCELL™ technology). *Plant Cell Tiss. Org. Cult.* **33,** 227–236.

38. Finer, J. J., Vain, P., Jones, M. W., and McMullen, M. D. (1992) Development of the particle inflow gun for DNA delivery to plant cells. *Plant Cell Rep.* **11,** 323–328.

39. Vain, P., McMullen, M. D. and Finer, J. J. (1993) Osmotic treatment enhances particle bombardment-mediated transient and stable transformation of maize. *Plant Cell Rep.* **12,** 84–88.

40. Sautter, C., Walder, H., Neuhaus-Url, G., Galli, A., Neuhaus, G., and Potrykus, I. (1991) Micro-targeting: high efficiency gene transfer using a novel approach for the acceleration of micro-projectiles. *Bio/technology* **9,** 1080–1085.

41. Leduc, N., Iglesias, V. A., Bilang, R., Gisel, A., Potrykus, I., and Sautter, C. (1994) Gene transfer to inflorescence and flower meristems using ballistic micro-targeting. *Sex. Plant Reprod.* **7,** 135–143.

42. He, D.G., Mouradov, A., Yang, Y. M., Mouradova, E., and Scott, K. J. (1994) Transformation of wheat (*Triticum aestivum* L.) through electroporation of protoplasts. *Plant Cell Rep.* **14,** 192–196.

43. D'Halluin, K., Bonne, E., Bossut, M., De Beuckeleer, M., and Leemans, J. (1992) Transgenic maize plants by tissue electroporation. *Plant Cell* **4,** 1495–1505.

44. Golovkin, M. V., Abraham, M., Morocz, S., Bottka, S., Feher, A., and Dudits, D. (1993) Production of transgenic maize plants by direct DNA uptake into embryogenic protoplasts. *Plant Sci.* **90,** 41–52.

45. Funatsuki, H., Kuroda, H., Kihara, M., Lazzeri, P. A., Muller, E., Lorz, H., and Kishinami, I. (1995) Fertile transgenic barley generated by direct DNA transfer to protoplasts. *Theor. Appl. Genet.* **91,** 707–712.

46. Neuhaus, G. and Spangenberg, G. (1990) Plant transformation by microinjection techniques. *Physiol. Plant.* **79,** 213–217.

47. Kaeppler, H. F., Gu, W., Somers D. A., Rines, H. W., and Cockburn, A. F. (1990) Silicon carbide fiber-mediated DNA delivery into plant cells. *Plant Cell Rep.* **9,** 415–418.

48. Frame, B. R., Drayton, P. R., Bagnell, S. V., Lewnau, C. J., Bullock, W. P., Wilson, H. M., Dunwell, J. M., Tompson, J. A., and Wang, K. (1994) Production of fertile transgenic maize plants by silicon carbide whisker-mediated transformation. *Plant J.* **6,** 941–948.

49. Guo, Y., Liang, H., and Berns, M. W. (1995) Laser-mediated gene transfer in rice. *Physiol. Plant.* **93,** 19–24.

50. Zhang, L-J., Cheng, L-M., Xu, N., Zhao, N-M., Li, C-G., Yuan, J., and Jia S-R. (1991) Efficient transformation of tobacco by ultrasonication. *Bio/technology* **9,** 996,997.

51. Lurquin, P. F. and Rollo, F. (1993) Liposome-mediated delivery of nucleic acids into plant protoplasts. *Methods Enzymol.* **221,** 409–415.

51a. Christou, P. (1996) Transformation technology. *Trends Plant Sci.* **1,** 423–431.

51b. Metz, P. L. J. and Nap, J. P. (1997) A transgene-centred approach to the biosafety of transgenic plants: overview of selection and reporter genes. *Acta Bot. Neerl.* **46,** 25–50.

52. Jefferson, R., Kavanagh, T., and Bevan, M. (1987) GUS fusion: β-glucuronidase as a sensitive and versatile gene fusion marker in higher plants. *EMBO J.* **6,** 3901–3907.

53. Ow, D. W., Wood, K. V., DeLuca, M., De Wet, J. R., Helinski, D. R., and Howell, S. H. (1986) Transient and stable expression of the firefly luciferase gene in plant cells and transgenic plants. *Science* **234,** 856–859.

54. Niedz, R. P., Sussman, M. R., and Satterlee, J. S. (1995) Green fluorescent protein: an *in vivo* reporter of plant gene expression. *Plant Cell Rep.* **14,** 403–406.

55. Herrera-Estrella, L., Depicker, A., Van Montagu, M., and Schell, J. (1983) Expression of chimaeric genes transferred into plant cells using a Ti-plasmid-derived vector. *Nature* **303,** 209–213.

56. Schena, M., Lloyd, A. M., and Davies, R. (1991) A steroid-inducible gene expression system for plant cells. *Proc. Natl. Acad. Sci. USA* **88,** 10,421–10,425.

57. Gatz, C., Kaiser, A., and Wendenburg, R. (1991) Regulation of a modified CAMV 35S promoter by the Tn10-encoded Tet repressor in transgenic tobacco. *Mol. Gen. Genet.* **227,** 229–237.

57a. Komari, T., Hiei, Y., Saito, Y., Murai, N., and Kumashiro, T. (1996) Vectors carrying two separate T-DNAs for cotransformation of higher plants mediated by *Agrobacterium tumefaciens* and segregation of transformants free from selection markers. *Plant J.* **10,** 165–174.

58. Meyer, P., Linn, F., Heidmann, I., Meyer, H., Neidenhof, I., and Saedler, H. (1992) Endogenous and environmental factors influence 35S promoter methylation of a maize A1 gene construct in transgenic petunia and its colour phenotype. *Mol. Gen. Genet.* **231,** 345–352.

58a. Maessen, G. D. G. (1997) Genomic stability and stability of expression in genetically modified plants. *Acta Bot. Neerl.* **46,** 3–24.

59. Ahl Goy, P. and Duesing, J. H. (1995) From pots to plots: genetically modified plants on trial. *Bio/technology* **13,** 454–458.

60. Fujita, Y. (1990) The production of industrial compounds, in *Plant Tissue Culture: Applications and Limitations* (Bhojwani, S. S., ed.), Elsevier, Amsterdam, pp. 259–275.

61. Towers, G. H. N. and Ellis, S. (1993) Secondary metabolism in plant tissue cultures transformed with *Agrobacterium tumefaciens* and *Agrobacterium rhizogenes. ACS Symposium* series **534,** 56–78.

62. Doran, P. M. (1994) Production of chemicals using genetically transformed plant organs. *Ann. NY Acad. Sci.* **745,** 426–441.

63. Giraudat, J., Hauge, B. M., Valon, C., Smalle, J., Parcy, F., and Goodman, H. M. (1992) Isolation of the Arabidopsis *ABI3* gene by positional cloning. *Plant Cell* **4,** 1251–1261.

64. McNellis, T. W., von Arnim, A. G., and Deng, X.-W. (1994) Overexpression of Arabidopsis COP1 results in partial suppression of light-mediated development: evidence for a light-inactivable repressor of photomorphogenesis. *Plant Cell* **6,** 1391–1400.

65. Hamilton, A. J., Lycett, G. W., and Grierson, D. (1990) Antisense gene that inhibits synthesis of the hormone ethylene in transgenic plants. *Nature* **346,** 284–287.

66. Grimsley, N. H., Hohn, B., Hohn, T., and Walden R. (1986) Agroinfection, an alternative route for viral infection of plants by using the Ti plasmid. *Proc. Natl. Acad. Sci. USA* **83,** 3282–3286.

67. Elmer, J. S., Sunter, G., Gardiner, W. E., Brand, L., Browning, C. K., Bisaro, D. M., and Rogers, S. G. (1988) *Agrobacterium*-mediated inoculation of plants with tomato golden mosaic virus DNAs. *Plant Mol. Biol.* **10,** 225–234.

68. Timmermans, M. C. P., Das, O. P., and Messing, J. (1994) Geminiviruses and their uses as extrachromosomal replicons. *Ann. Rev. Plant Physiol. Plant Mol. Biol.* **45,** 79–112.

69. Gal, S. Pisan, B., Hohn, T. Grimsley, N., and Hohn, B. (1992) Agroinfection of transgenic plants leads to viable cauliflower mosaic virus by intermolecular recombination. *Virology* **187,** 525–533.

70. Medberry, S. L., Lockhart, B. E. L., and Olszewski, N. E. (1990) Properties of *Commelina* yellow mottle virus's complete DNA sequence, genomic discontinuities and transcript suggest that it is a pararetrovirus. *Nucleic Acids Res.* **18,** 5505–5513.

71. Leiser, R. M., Ziegler-Graff, V., Reutenauer, A., Herrbach, E., Lemaire, E., Guilley, H., Richards, K., and Jonard, G. (1992) Agroinfection as an alternative to insects for infecting plants with beet western yellows luteovirus. *Proc. Natl. Acad. Sci. USA* **89,** 9136–9140.

72. Turpen, T. H., Turpen, A. M., Weinzettl, N., Kumagai, M. H., and Dawson, W. O. (1993) Transfection of whole plants from wounds inoculated with *Agrobacterium tumefaciens* containing cDNA of tobacco mosaic virus. *J. Virol. Methods* **42,** 227–240.

73. Gardner, R. C., Chonoles, K. R., and Owens, R. A. (1986) Potato spindle tuber viroid infections mediated by the Ti plasmid of *Agrobacterium tumefaciens. Plant Mol. Biol.* **6,** 221–228.

74. Boulton, M. I. and Davies, J. W. (1990) Monopartite geminiviruses: markers for gene transfer to cereals. *Aspects Appl. Biol.* **24,** 79–86.

75. Maliga, P. (1993) Towards plastid transformation in flowering plants. *TIBTECH* **11,** 101–107.

76. Carrer, H. and Maliga, P. (1995) Targeted insertion of foreign genes into the tobacco plastid genome without physical linkage to the selectable marker gene. *Bio/technology* **13,** 791–794.

22

Restriction Fragment-Length Polymorphisms

Elaine K. Green

1. Introduction

Knowledge of eukaryotic-gene structure has grown exponentially over the past two decades, mainly owing to the introduction of new molecular techniques. These have enabled the manipulation and analysis of the DNA molecule and led to the identification of mutations causing single gene disorders. An important factor in the development of such techniques was the discovery of a family of bacterial enzymes, the restriction endonucleases *(1)*.

2. Restriction Endonucleases

Restriction endonucleases are bacterial enzymes that cut double-stranded DNA molecules in a precise and reproducible manner. Three different classes of restriction endonucleases have been recognized, each with a slightly different mode of action. Type II restriction endonucleases are the most important and frequently used cutting enzymes. Such enzymes are named usually by a three-letter abbreviation that identifies their bacterial origin. To distinguish between enzymes with the same origin, Roman numerals are added, for example *Hpa*I is derived from *Haemophilus parainfluenza* and *Eco*RI from *Escherichia coli* RY13 *(2)* (*see* **Table 1**).

Restriction enzymes recognize specific sequences in double-stranded DNA and cleave the DNA, usually within the recognition site, to yield fragments of defined length. The specific recognition sequence of double stranded DNA for the vast majority of type II restriction enzymes is usually between 4 and 6 base pairs in length and generally palindromic; in other words, a sequence of bases is the same on both strands when read in a 5' to 3' direction. Restriction enzymes cut the DNA at every point at which the target sequence occurs. Many restriction endonucleases cut in the middle of the recognition sequence to produce a "blunt end," whereas others cut at different nucleotides usually 2 or 4 bases apart resulting in DNA fragments with short single stranded overhangs at each end known as "sticky or cohesive ends." (*See* **Fig. 1**.) Using restriction endonucleases to digest DNA a physical map of the molecule can be obtained, identifying the sites of cleavage, separated by actual distance along the strand. This is known as a restriction map. Such a map can obtained for any DNA sequence regardless of the presence of mutations or if its function is unknown.

From: *Molecular Biomethods Handbook*
Edited by: R. Rapley and J. M. Walker © Humana Press Inc., Totowa, NJ

Table 1
Recognition Sequences of Frequently Used Restriction Endonucleases[a]

Enzyme	Organism	Recognition sequence	Cut sequence
*Alu*I	*Arthobacter luteus*	AG/CT	–AG CT–
		TC/GA	–TC GA–
*Bam*HI	*Bacillus amyloliquefaciens*	G/GATCC	–G GATCC–
		CCTAG/G	–CCCTAG G–
*Bgl*II	*Bacillus globigii*	A/GATCT	–A GATCC–
		TATAG/A	–TGTAG G–
*Eco*RI	*Escherichia coli*	G/AATTC	–G AATTC–
		CTTAA/G	–CTTAA G–
*Hae*III	*Haemophilus aegyptius*	GG/CC	–GG CC–
		CC/GG	–CC GG–
*Hin*dIII	*Haemophilus influenzae* Rd	A/AGCTT	–A AGCTT–
		TTCGA/A	–T TCGAA–
*Kpn*I	*Klebsiella pneumonia*	GGTAC/C	–GTGGTAC C–
		C/CATGG	–C CATGG–
*Pst*I	*Providencia stuartti*	CTGC/C	–CTGC G–
		G/ACGC	–G ACGC–
*Sma*I	*Serratia marcescens*	CCC/GGG	–CCC GGG–
		GGG/CCC	–GGG CCC–

[a]Both strands of the recognition sequence given in a 5' to 3' direction with the site of cleavage shown by a line, and the resulting digested sequence.

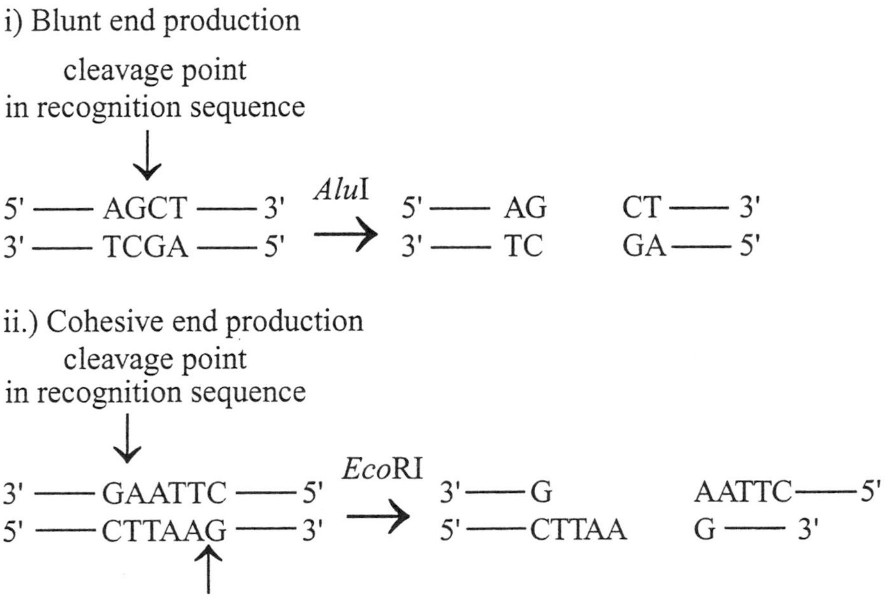

i) Blunt end production

cleavage point
in recognition sequence

ii.) Cohesive end production
cleavage point
in recognition sequence

Fig. 1. Restriction endonucleases generate either blunt-ended or cohesive-ended fragments during cleavage of DNA.

3. Occurrence of RFLPs

Apart from fractionating DNA, restriction endonucleases also provide a valuable approach to analyzing genetic diversity. Throughout the human genome are random-base changes (pathogenic and harmless) both intragenic, within noncoding DNA segments, and extragenic, within coding DNA segments. Such random-base changes may either produce new restriction enzyme sites, or delete pre-existing ones. This variability is inherited in a Mendelian fashion and hence can be followed through generations of families *(3)*. The existence of more than one of the variations or alleles in the genome is called genetic polymorphism, thus any position along a gene at which multiple alleles exist as stable components of the population is by definition polymorphic. Such restriction-site polymorphism may not actually affect the function of the gene—in fact, the majority do not—hence, the phenotype is not altered. The difference in size of the DNA fragments cut with the same restriction endonucleases between individuals is known as restriction fragment-length polymorphism, RFLP (*see* **Fig. 2**). The sickle-cell mutation is an example of a pathogenic mutation where a single nucleotide substitution of A to T at codon 6 in the β globin gene, abolishes a *Mst*II restriction site, as shown in **Fig. 3**.

Although most polymorphisms appear to be randomly distributed throughout a genome, there are certain regions where a particularly high concentration of polymorphisms exist known as hypervariable regions (HVR). Such regions result from DNA sequences repeated in tandem arrays in which the number of repeats at each location may vary from person to person. A region of small individual repeat units of 1–5 nucleotides is known as a short-tandem repeat or microsatellite, whereas larger repeat units of usually 10–60 nucleotides are known as variable number of tandem repeats (VNTRs) (**Fig. 2**).

4. Detection of RFLPs

Initially, RFLPs were typed using Southern blot hybridization and radiolabeled probes *(4)* (*see* Chapters 8 and 6, respectively). This method involves the transfer of DNA fragments that have been cleaved using restriction enzymes and separated by electrophoresis, onto nitrocellulose or nylon paper and subsequently hybridized with a specific radiolabeled DNA probe (a DNA fragment from genome used to identify DNA sequences that are closely related to it in sequence). The position of the fragments containing the gene to be analyzed is then determined by autoradiography. Because the DNA segments obtained following digestion are fractionated by size, any mutations that alter a restriction site, or the presence of large insertions or deletions between two restriction sites, can be typed and result in different restriction-fragment lengths (**Fig. 4**).

Unfortunately, the use of Southern blotting as a technique to detect RFLPs is time consuming, laborious, and requires a large amount of DNA, approx 8–12 µg. However, the advantages of the polymerase chain reaction (PCR) technique overcame these problems to some extent *(5)* (*see* Chapter 26). PCR enables an in vitro amplification of the DNA sequence of interest, using primers designed to regions flanking the polymorphic-restriction site. Subsequent restriction enzyme digestion of the PCR product and size fractionation by gel electrophoresis allows identification of any restriction-site polymorphisms (**Fig. 5**).

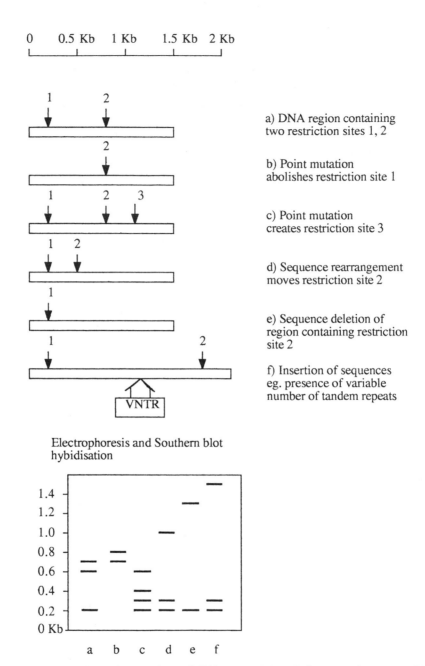

a) DNA region containing two restriction sites 1, 2

b) Point mutation abolishes restriction site 1

c) Point mutation creates restriction site 3

d) Sequence rearrangement moves restriction site 2

e) Sequence deletion of region containing restriction site 2

f) Insertion of sequences eg. presence of variable number of tandem repeats

Fig. 2. Representation of a number of different RFLPs. Point mutations can either create new restriction sites or delete pre-existing ones. The position of a restriction site may also be altered by sequence rearrangement, or by insertion or deletion of sequence-variable number of tandem repeats (VNTRs).

5. Gene Mapping

A complete map of the human genome showing the exact position of all the genes arranged along the chromosomes would provide a invaluable means of studying

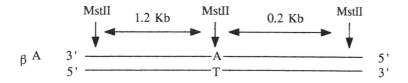

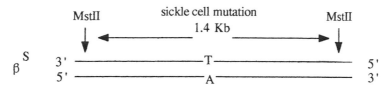

Fig. 3. Sickle-cell mutation causes deletion of an *Mst*II site producing a RFLP. A single point sickle-cell mutation of an A to T deletes a MstII-restriction site occurring in normal B A-globin allele. The flanking *Mst*II sites of the β-globin gene are still present and hence, Msm digestion results in a 1.2 and 0.2 kb *Mst*II RFLP for βA-globin and a 1.4 Kb *Mst*II RFLP for sickle-cell mutation of the βS globin gene.

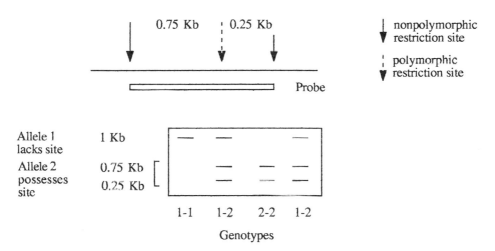

Fig. 4. Detection of RFLPs using Southern blotting and radiolabeled probe.

recombination in humans, as well as aiding gene localization, cloning, and prenatal or presymptomatic diagnosis of individuals at risk of diseases.

Genetic maps have been available for yeast, *Drosphilia*, and the mouse for many decades. Such genetic maps are constructed by crossing different mutants in order to determine whether two gene loci are linked or not. Genes are said to be linked when they are inherited together owing their physical proximity on a single chromosome. In a pedigree with a genetic disease, if a marker occurs both with and without the disease, the marker and disease locus are on separate chromosomes and are not linked. If the marker and disease do not occur together, they are on different members of a homologous pair. However, they are close together on the same chromosome if there is a bias

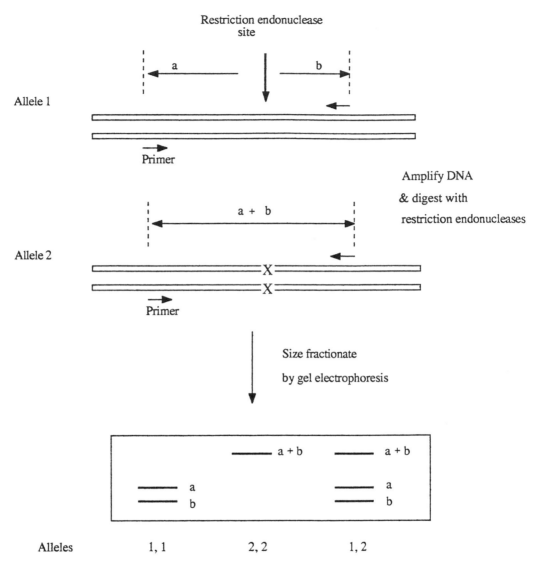

Fig. 5. Restriction endonuclease typing by PCR. Alleles 1 and 2 possess a polymorphism that alters a nucleotide at the recognition site of an restriction endonuclease. Allele 1 possesses the recognition site, but allele 2 lacks the site. Hence, PCR amplification of the DNA segment containing the sites of interest, followed by digestion with the appropriate restriction endonuclease produces either 2 short products as for allele 1 or an undigested fragment as for allele 2.

for the marker and the disease to occur together and are said to be linked. Unfortunately, it is not feasible to construct such a genetic map for humans by the same method because human families in which two diseases are segregating are very rare and if they do exist the number of children may be few or be unsuitable for genetic analysis in some way. Thus, most human genetic maps are based on inherited markers along a chromosome. The markers need not necessarily be related to a disease or genes, as long as they show Mendelian inheritance and are polymorphic, i.e., have at least two forms, known as alleles, that can be distinguished between different members of the family

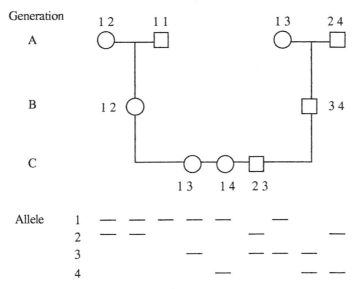

Fig. 6. Restriction site polymorphisms are inherited in a Mendelian fashion. Four alleles for a restriction marker are found in pairwise combinations and segregate independently at each generation. In generation A, three people are heterozygous (1, 2) (1, 3), (2, 4) and one homozygous (1, 1). Their children are heterozygous and hence in generation C the offspring gain either alleles 1 or 2 from one parent, or 3 or 4 from the other.

studies. The presence of different alleles will allow the marker to be followed between generations. Hence, the greater the number of alleles the more easily the inheritance pattern of that particular locus or chromosomal segment can be followed and recombinants with other loci detected (**Fig. 6**). In order to determine the usefulness of a particular marker, two frequently used measurements are the polymorphism informative content (PIC) *(6,7)* and the closely related average heterozygosity in the population.

However, suitable polymorphic markers were not available until in 1980, Botstein et al. *(6)* suggested that it would be possible to construct a complete linkage map of the human genome using common variations in the DNA sequence, conveniently visualized as RFLPs used as genetic markers. Because restriction polymorphisms occur randomly in the genome, some should be present near any particular gene. If one of these markers is tightly linked to the mutant phenotype, comparison of patients suffering with the disease with normal DNA will enable identification of the disease at the genetic level owing to the restriction site that is always present or absent in the patient (**Fig. 6**). Studies of randomly selected RFLPs in human families have found linkage in certain human diseases, for example Duchenne muscular dystrophy *(8)*, Huntington's disease *(9)*, and familial Alzheimer's disease *(10)*.

6. Hypervariable Regions

A governing factor concerning the usefulness of a particular marker system for providing linkage information is the frequency of heterozygosity. Restriction-site polymorphisms caused by single-point mutations have only two alleles and hence the fact that at least half of the individuals in the population are homozygous will make it impossible to follow the transmission of all chromosomes from parent to child.

The fact that the use of RFLPs relies on the presence or absence of a restriction-enzyme site owing to a nucleotide alteration at that particular site can be limiting in that restriction-enzyme sites only represent a small percentage of the human genome that nucleotide variations occur. However, variable number of tandem repeat polymorphisms (VNTRs) are more informative markers. Such hypervariable regions of DNA yield restriction fragments that vary in length widely (10–60 bp) within the population. These regions of DNA have no function attributed to them as yet. The repeat structure of these regions is known as "minisatellite" DNA.

The repeat regions are prone to recombination, which results in the allelic difference in the number of repeated units present at the hypervariable region, HVR, loci, and hence, in length polymorphism. This leads to a high degree of heterozygosity, and thus can be used in genetic-linkage studies *(11)*. The repeats may be detected by digesting genomic DNA with restriction endonucleases that flank the VNTR, Southern blotting and hybridization with a probe to a unique sequence of the locus, and also by PCR amplification technique. However, Southern-blotting results of VNTRs can be difficult to interpret owing to inherited alleles of similar size.

Such VNTRs were first used by Jeffreys et al. *(12)* to distinguish between any two individuals who were not identical twins, by a method known as DNA fingerprinting (*see* Chapter 31). A probe based on a sequence that is repeated in members of a DNA family is used, which can detect, by hybridization to genomic DNA, a large number of loci containing tandem repeats of similar sequence. The restriction fragment pattern revealed by the sum of the HVR loci containing such related sequences, scattered throughout the whole human genome, constitutes a "genetic fingerprint" that is completely unique to any one individual.

7. Highly Informative Multiallelic Markers

Although the detection of HVRs has been useful for the mapping of the genome and for finding individual genes, there are a few points that need to be considered. Owing to the large size of some of the HVRs, they are difficult to amplify and detect by PCR. They tend also not to be widely distributed throughout the genome, clustering in the telomeric region of the chromosome, thereby reducing the usefulness in covering the human genome. However, the identified microsatellites (also known as short-tandem repeat polymorphisms; STRPs) are of greater use for genetic mapping and linkage studies *(13)*. These repeats can be blocks of di-, tri-, or tetra-nucleotides (sometimes more), although the majority are interspersed $(CA)_n$ repeat sequences, which show length polymorphisms. The advantage of these repeats is that they are abundant and dispersed throughout the genome (approximately one every 60 kb), highly informative, and easy to type. Length variables can be detected by restriction digestion of genomic DNA, Southern blotting, and hybridization with probes flanking the repeat sequences, or by PCR. Clearer results can be obtained by PCR when amplifying tri and tetranucleotide repeats giving a single band from each allele. Whereas, studying dinucleotide-repeat sequences tend to be prone to replication slippage during PCR amplification. Slippage occurs when two complementary strands of a double helix do not pair normally; instead, the pairing is altered by staggering of the repeats on the 2 strands resulting in incorrect pairing of the repeats. This leads to an insertion or deletion in the newly synthesized strand, which is seen as a ladder of "stutter band" when gel electrophoresed, making the results difficult to interpret.

8. Summary

The discovery of restriction endonucleases have greatly aided the analysis of the human genome. The existence of sequence variations throughout the genome creates polymorphisms for the site of restriction-enzyme cleavage, known as RFLPs. Such RFLPs can be used as genetic markers providing a means of gene mapping, as well as prenatal and pre-symptomatic diagnosis of individuals.

References

1. Smith, H. O. and Wilcox, K. W. (1970) A restriction enzyme from *Hemophilus influenzae* I. Purification and general properties. *J. Mol. Biol.* **51,** 379–391

2. Smith, H. O. and Nathans, D. (1973) A suggested nomenclature for the bacterial host modification and restriction systems and their enzymes. *J. Mol. Biol.* **81,** 419–423

3. Jeffrey, A. J. (1979) DNA sequence variants in Gγ-, Aγ-,δ-, and β-globin genes of man. *Cell* **18,** 1–10.

4. Southern, E. M. (1975) Detection of specific sequences among DNA fragments separated by gel electrophoresis. *J. Mol. Biol.* **98,** 503–517.

5. Saiki, R. K., Scharf, S., Faloona, F., Mullis, K. B., Horn, G. T., Erlich, H. A., and Arnheim, N. (1985) Enzymatic amplification of b-globin genomic sequences and restriction site analysis for diagnosis of sickle cell anaemia. *Science* **230,** 1350–1354.

6. Botstein, D., White, R. L., Skolnick, M., and Davis, R. W. (1980) Construction of a genetic linkage map in man using restriction fragment length polymorphisms. *J. Hum. Genet.* **32,** 314–331.

7. Willard, H. F., Skolnick, M., Pearson, P., and Mandel, J.-L. (1985) Report of the committee on human gene mapping by recombinant DNA techniques in human gene mapping 8. *Cytogenet. Cell Genet.* **40,** 364–489.

8. Davies, K. E., Pearson, P. L., Harper, P. S., Murray, J. M., O'Brien, T., Sarfarazi, M., and Williamson, R. (1983) Linkage analysis of two cloned sequences flanking the Duchenne muscular dystrophy locus on the short arm of the human X chromosome. *Nucleic Acids Res.* **11,** 2303–2312.

9. Gusella, J. F., Wexler, N. S., Conneally, P. M., et al. (1983) A polymorphic DNA marker genetically linked to Huntington's disease. *Nature* **306,** 234–238.

10. St. George-Hyslop, P. H., Tanzi, R. E., Polinsky, R. J., et al. (1987) The genetic defect causing familial Alzheimer's disease maps on chromosome 21. *Science* **235,** 885–890.

11. Wyman, A. R. and White, R. (1980) A highly polymorphic locus in human DNA. *Proc. Natl. Acad. Sci. USA* **77,** 6754–6758.

12. Jeffreys, A. J., Wilson, V., and Thein, S. L. (1985) Hypervariable 'minisatellite' regions in human DNA. *Nature* **314,** 67–73.

13. Weber, J. L. (1990) Human DNA polymorphisms and methods of analysis. *Curr. Opin. Biotechnol.* **1,** 166–171.

23

Genome Mapping

Jacqueline Boultwood

1. Introduction

The genome maps of a small number of prokaryotes have been completed. In recent years, efforts have been made to map the entire mouse and human genomes; indeed the human-genome project—a global effort to clone, sequence, and map the human genome—represents the most ambitious and challenging task of modern human biology.

Genome mapping may be subdivided into genetic and physical mapping. Early attempts to create maps of the mouse and human genomes were derived primarily from genetic-linkage data. However, as the appropriate technology has become available, the construction of fine physical maps has predominated. In this chapter, each of the techniques employed currently in genome mapping will be reviewed, with most emphasis being given to the most recent and efficient methodologies.

1.1. Linkage Analysis (Genetic Mapping)

Linkage analysis *(1,2)* is based on the Mendelian principle of random assortment of gene pairs when passed from generation to generation. Two genes are completely unlinked when they are localized to different chromosomes. Conversely, two genes are completely linked when there is no recombination between them, i.e., the same alleles are transmitted together, without exception, from generation to generation within a family. A large proportion of gene pairs are considered to be incompletely linked. This occurs when there is a consistent recombination fraction between the two genes, but also a consistent and quantifiable deviation from random assortment. The recombination fraction is approximately proportional to the physical distance separating the two genes, and it is this principle on which gene mapping by linkage analysis is based.

The transmission of alleles of polymorphic genes from informative (heterozygous) individuals within pedigrees and the assessment of recombinants and nonrecombinants forms the basis of simple linkage analysis. In recent years linkage analysis has been much developed. The use of highly informative microsatellite markers, for example, has become widespread and complex computer programs are now routinely used to calculate allele frequencies and penetrances, as well as to estimate recombination fractions via likelihood calculations *(3–6)*. Detailed genetic maps with precise ordering of loci are now thus possible.

From: *Molecular Biomethods Handbook*
Edited by: R. Rapley and J. M. Walker © Humana Press Inc., Totowa, NJ

Although physical mapping is undoubtedly at the forefront of modern genome mapping, genetic mapping remains of great value, particularly where very large distances are involved. Physical maps are frequently integrated into existing genetic maps and vice versa.

1.2. Construction And Use of Cosmids and YACs in Physical Mapping

A partial digest of genomic DNA may be cloned into a yeast vector, which provides selectable markers for the vector arms (Ura and Trp) in addition to a centromere, telomeres, and origin of replication *(7)*. The main advantage conferred by the use of Yeast Artificial Chromosomes (YACs) as vectors for the construction of a physical-genome map is the large size fragment (>1 Mb DNA) that may be cloned *(8)* (*see* Chapter 24). A significant disadvantage to their use, however, is the high frequency (>50% in some libraries) of chimerism.

For most workers involved in positional cloning, the construction of a YAC contig is the method of choice for achieving genomic representation of the region of interest. A YAC contig may be generated either by walking out from a single marker, or by screening a YAC library with a number of known markers mapping to the region of interest. As a large proportion of the human genome is now mapped in some detail with known genes and/or sequence-tagged sites (STSs) the latter method predominates. Many YAC libraries are now fully accessible to the scientific community, including, for example, the ICI and CEPH YAC libraries. These YAC libraries have insert sizes in the 100–500 kb range. However, recent developments have allowed for the cloning of very large fragments of DNA (1–2 Mb) into YAC vectors. The CEPH mega YAC library, for example, represents the most fully characterized of the large insert size YAC libraries currently available *(9)*.

The construction of long-range genomic maps by cloning DNA fragments into cosmid vectors as a first step has been superseded by the advent of YACs. This is primarily because of the larger insert size possible with YAC vectors (>1 Mb) as compared to cosmids (50 kb). Cosmid contigs still, however, have an important intermediary role in genomic mapping *(10–12)*. Individual YACs may be reduced to cosmid contigs, thus generating a fine physical map of a region of interest *(12)*. The individual cosmids may then be manipulated with greater ease and efficiency than YACs in, for example, the generation of STSs and expressed sequence tagged sites (ESTs). More recently, P1-based artificial chromosomes (PACs) and bacterial artificial chromosomes (BACs) have become the vectors of choice for the generation of contigs of large chromosome regions.

1.3. Pulsed Field Gel Electrophoresis in Physical Mapping

DNA fragments greater than approx 25 kb are poorly resolved by standard gel electrophoresis. The technique of pulsed field gel electrophoresis (PFGE) was first developed by Schartz and Cantor *(13)* and allows for the separation of large DNA fragments by periodically alternating the direction of the electric field applied to the gel. This results in a near linearity of separation. PFGE enables the investigator to separate both very large DNA fragments of several megabase pairs (Mb) in length as well as small fragments of between 10 and 20 kb *(13)*. Thus the technique of PFGE fills the niche in genome mapping between standard Southern blotting and techniques such as fluorescence *in situ* hybridization (FISH) and radiation hybrid mapping.

Many advances have been made since the original PFGE equipment was described by Schwartz and Cantor, and the apparatus now used by most investigators is the highly effective CHEF (contour-clamped homogeneous electric field) system *(14)*. Restriction enzymes that cleave DNA infrequently are employed in long-range PFGE analysis of mammalian genomes. These rare cutter enzymes usually contain one or more CpG dinucleotide sequences in their recognition sequence and a number recognize octarmeric sequences, for example, *Not*I. The CpG dinucleotide is under represented in the genome and the cytosine residue may be methylated, thus inhibiting cleavage by methylation-sensitive enzymes.

PFGE was originally applied to the analysis of yeast and protozoan genomes and soon used widely for the long-range physical mapping of eukaryotic genomes. PFGE has been of great benefit in the physical mapping of large regions of the human genome *(15)*. PFGE has, for example, been widely and successfully used in positional cloning approaches; fine physical maps of regions of the human genome harboring disease loci have been invaluable in the identification and cloning of many hereditary-disease genes *(16)*. Single- and double-restriction enzyme digests are necessary for the generation of such maps. The long-range restriction map generated by PFGE of the cystic fibrosis region was, for example, most useful in the isolation of the causative gene *(16)*.

PFGE may be applied to the restriction mapping of YACs and this data then integrated into existing long-range physical maps. PFGE is also of great value in the assembly of YACs into contigs.

1.4. Gene Mapping Using Somatic Cell Hybrids

Human-rodent somatic-cell hybrids (monochromosomal) have been widely used since the 1970s for mapping genes and other DNA markers to individual chromosomes *(17,18)*. The first human–mouse somatic-cell hybrids generated were from a spontaneous fusion between mouse L-cells deficient in thymidine kinase and normal human-embryonic fibroblasts *(19)*. The hybrids were selected in hypoxanthine-aminopterin thymidine (HAT). The mouse-cell line was thymidine-kinase deficient and thus it was necessary for the human–mouse hybrids selected in HAT to retain the human gene for thymidine kinase localized to human chromosome 17 (the other human chromosomes being lost during culture). More recently, however, the development of deletion or radiation-reduced mapping hybrid panels has allowed for precise regional localization of such DNA markers *(20,21)*. Indeed, the radiation-reduced-mapping hybrid panels represent an increasingly important mapping resource for the human-genome mapping project, in particular for the regional localization of STSs and ESTs *(21)*.

Gene assignments depend upon the loss or restriction of specific chromosome fragments in somatic-cell hybrids and whether using Southern-blot hybridization or PCR amplification, a species difference must be determined. Regions of DNA that are far apart on a chromosome are more likely to be separated by radiation-induced fracture and consequently segregate independently in the radiation-hybrid cells than if they are tightly linked together. Information regarding the order and distance of DNA markers is thus obtainable.

1.5. Mapping Using Fluorescence In Situ Hybridization

An important landmark in gene mapping was achieved in the 1970s with the development of the technique of *in situ* hybridization of isotopically labeled DNA probes to metaphase chromosome spreads *(22,23)*. This method allowed genes to be mapped to single chromosomes and in many cases to a single chromosome band. There are a number of disadvantages associated with the use of isotopically labeled probes, however, including for example, poor spatial resolution. Most of these problems were overcome in the late 1980s by the use of probes labeled with fluorochromes *(24,25)*. FISH is fast, efficient, the spatial resolution and sensitivity is high, and results can be obtained from nondividing as well as dividing cells *(26)*. YACs and cosmids are widely used as DNA probes for FISH experiments but small plasmid-cloned sequences (0.5–5 kb) may also be used with some success.

FISH has become an important tool in the construction of physical maps. This technique allows the simultaneous analysis of multiple probes and consequently it is used frequently for ordering and distancing of DNA markers *(25)*. In particular, physical mapping using extended chromatin-fiber preparations allows for the generation of precise long-range physical maps *(26,27)*.

References

1. Ott, J. (1991) *Analysis of Human Genetic Linkage*. Johns Hopkins University Press, Baltimore and London.
2. Terwilliger, J. D. and Ott, J. (1994) *Handbook of Human Genetic Linkage*. Johns Hopkins University Press, Baltimore.
3. Cottingham, R. W., Idury, R. M., and Schaffer, A. A. (1993) Faster sequential genetic linkage computations. *Am. J. Hum. Genet.* **53,** 252–263.
4. Lathrop, G. M., Lalouel, J. M., Julier, C., and Ott, J. (1984) Strategies for multiocus analysis in humans. *Proc. Nat. Acad. Sci. USA* **81,** 3443–3446.
5. Lathrop, G. M., Lalouel, J. M., and White, R. L. (1986) Construction of human genetic linkage maps: likelihood calculations for multilocus analysis. *Genet. Epidemiol.* **3,** 39–52.
6. O'Connell, J. (1995) The Vitesse Algorithm for rapid exact multilocus linkage analysis via genotype set-recording and fuzzy inheritance. *Nature Genet.* **11,** 402–408.
7. Burke, D. T., Carle, G. F., and Olson, M. V. (1987) Cloning of large segments of exogenous DNA into yeast by means of artificial chromosome vectors. *Science* **236,** 806–812.
8. Monaco, A. P. and Larin, Z. (1994) YACs, BACs, PACs and MACs: artificial chromosomes as research tools. *Trends Biotechnol.* **12,** 280–286.
9. Albertsen, H. A., Abderrahim, H., Cann, H. M., Dausset, J., Le Paslier, D., and Cohen, D. (1990) Construction and characterization of a yeast artificial chromosome library containing seven haploid human genome equivalents. *Proc. Nat. Acad. Sci. USA* **87,** 4256–4260.
10. Holland, J., Coffey, A. J., Giannelli, F., and Bentley, D. R. (1993) Vertical intergration of cosmid and YAC resources for the interval mapping on the X-chromosome. *Genomics* **15,** 297–304.
11. Heding, I. J. J. P., Ivens, A. C., Wilson, J., Striven, M., Gregory, S., Hoovers, J. M. N., Mannens, M., Redeker, B., Porteous, D., van Heyningen, V., and Little, P. F. R. (1992) The generation of ordered sets of cosmid DNA clones from human chromosome region 11p. *Genomics* **13,** 89–94.
12. Bellanné-Chantelot, C., Barillot, E., Lacroix, B., Le Paslier, D., and Cohen, D. (1991) A test case for physical mapping of human genome by repetitive sequence fingerprints: construction of a physical map of a 420kb YAC subcloned into cosmids. *Nucleic Acids Res.* **19,** 505–510.
13. Schwartz, D. C., and Cantor, C. R. (1984) Separation of yeast chromosome sized DNAs by pulsed field gradient gel electrophoresis. *Cell* **37,** 67–75.

14. Chu, G., Vollrath, D., and Davis, R. W. (1986) Separation of large DNA molecules by contour-clamped homogeneous electric fields. *Science* **234,** 1582–1585.

15. Dunham, J., Sargent, C. A., Dawkins, R. L., and Campbell, R. D. (1989) An analysis of variation in the long-range genomic organisation of the human major histocompatibility complex class II region by pulsed field gel electrophoresis. *Genomics* **5,** 787–796.

16. Rommens, J. M., Ianuzzi, M. C., Kerem, B., Drumm, M. L., Melmer, G., Dean, M., Rozmahel, R., Cole, J. L., Kennedy, D., Hidaka, N., Zsicia, M., Buchwald, M., Riordan, J. R., Tsui, L., and Collins, F. S. (1989) Identification of the cystic fibrosis gene: chromosome walking and jumping. *Science* **245,** 1059–1065.

17. O'Brien, S. J., Simonson, J. M., and Eichelberger, M. (1982) Genetic analysis of hybrid cells using isozyme markers as markers of chromosome segregation, in *Techniques in Somatic Cell Genetics* (Shaw, J. W., ed.), Plenum, New York.

18. Tunnacliffe, A. and Goodfellow, P. (1984) Analysis of the human cell surface by somatic cell genetics, *Genetic Analysis of the Cell Surface: Receptors and Recognition Series* (Goodfellow P., ed.), Chapman and Hall, London, pp. 57–82.

19. Weiss, M. C. and Green, H. (1967) Human-mouse hybrid cell lines containing partial complements of human chromosomes and functioning human genes. *Proc. Natl. Acad. Sci. USA* **58,** 1104–1111.

20. Walter, M.A., and Goodfellow, P.N. (1993) Radiation Hybrids: irradiation and fusion gene transfer. *Trends Genet.* **9,** 352–356.

21. James, M.R., Richard, C.W. III, Schott, J. J., Yoursy, C., Clark, K., Bell, J., Tersilliger, J. D., Hazan, J., Dubay, C., Viginal, A., et al. (1994). A radiation hybrid map of 506 STS markers spanning human chromosome 11. *Nat. Genet.* **8,** 70–76.

22. Evans, H. J., Buckland, R. A., and Pardu, M. L. (1974) Location of the genes coding for 18S and 28S ribosomal RNA in the human genome. *Chromosoma* **48,** 405–426.

23. Harper, M. E. and Saunders, G. F. (1981) Localization of single copy DNA sequences on G-banded chromosomes by in situ hybridization. *Chromosoma* **83,** 431–439.

24. Bauman, J. G., Wiegant, J., Borst, P., and van Duijn, P. (1980) A new method for fluorescence microscopical localization of specific sequences by in situ hybridization of fluorochrome labelled RNA. *Exp. Cell Res.* **128,** 485–490.

25. Lawrence, J. B., Villnave, C. A., and Singer, R. H. (1988) Sensitive, high-resolution chromatin and chromosome mapping in situ: Presence and orientation of two closely integrated copies of EBV in a lymphoma line. *Cell* **52,** 51–61.

26. Trask, B. J. (1991) Fluorescence in situ hybridization: Applications in cytogenetics and gene mapping. *Trends Genet.* **7,** 149–154.

27. Cai, W., Aburatani, H., Stanton, V. P., Housman, D. E., Wang, Y. K., and Schwartz, D. C. (1995) Ordered restriction endonuclease maps of yeast artificial chromosomes created by optical mapping on surfaces. *Proc. Natl. Acad. Sci. USA* **92,** 5164–5168.

24

Yeast Artificial Chromosomes

Angela Flannery and Rakesh Anand

1. Introduction

Since their introduction in 1987, yeast artificial chromosomes (commonly referred to as YACs) have had a tremendous impact on the field of molecular genetics. The ability of YACs to carry and propagate DNA fragments of up to ~2 million base pairs in size has been instrumental in the construction of the first- and second-generation physical maps of the human genome *(1,2)*. YACs have assisted in the generation of detailed maps of discrete chromosomal regions and significantly speeded up positional cloning projects leading to the identification and cloning of a variety of genes involved in inherited diseases *(3,4)*. YACs have also been used for studies on gene function via complementation of mutations as well as for the transfer of complete genomic gene sequences for the generation of transgenic animals. Widespread use of this broad range of applications has been facilitated by the availability of a number of high-quality genomic YAC libraries via centralized screening centers. This clearly illustrates the scientific advances that can be made when tools and technology are shared in the spirit of cooperation.

In this chapter, we look at the history of the development of YACs, followed by a brief description of some of the key elements of this technology. We then present examples of the applications, with appropriate cross references for further details.

2. Background

Artificial chromosomes of the laboratory yeast *Saccharomyces cerevisiae* were originally developed to study the structure and behavior of eukaryotic chromosomes. *S. cerevisiae* is an attractive organism to study because it can exist in a haploid state, has a relatively small genome of about 13 million base pairs (Megabases, Mb), and undergoes mitotic division or budding. Haploid cells of opposite mating types can also conjugate and fuse to form diploid cells which, under certain conditions, exhibit both mitotic- and meiotic-cell division *(5)*.

Certain functional elements are known to be vital for chromosome stability in eukaryotes. The three major *cis*-acting elements required are replication origins, centromeres, and telomeres. Yeast chromosome autonomous replication sequences (ARS) were identified by Struhl et al. *(6)*. They were found to contain functional origins of

From: *Molecular Biomethods Handbook*
Edited by: R. Rapley and J. M. Walker © Humana Press Inc., Totowa, NJ

replication and to confer on any sequence linked to them the ability to replicate. Two such elements have proved very useful in constructing YACs; ARS1, which is contained in a 837 bp *Eco*RI-*Hin*dIII fragment *(7)* and ARS-H4, which resides on a 374 bp *Sau*3A fragment *(8)*. Functional centromeric (CEN) DNA was isolated from 12 of the 16 yeast chromosomes and a 125-bp DNA fragment was identified that provides an attachment site for microtubules from the mitotic spindle and is both necessary and sufficient for mitotic and meiotic division *(9)*. Yeast telomere (TEL) sequences were isolated by virtue of their high degree of conservation *(10)*. TEL DNA forms the physical ends of chromosomes, allows replication and maintenance of linear molecules, and contains variable numbers of repeated (C_4A_2) units.

These elements were individually cloned and brought together in yeast cells on a linear plasmid vector, creating the first artificial chromosomes in 1983 *(11)*. In the initial experiments, the inability of yeast-linear plasmids to replicate in *E. coli* necessitated the construction of circular-plasmid molecules containing both ARS and CEN fragments, which were first propagated in *E. coli* and then ligated in vitro with TEL sequences prior to transformation of yeast cells.

It quickly became apparent that the stability, replication, and segregation of such chromosomes in the yeast host cells were enhanced as their size was increased from its initial <20 kb to at least 50 kb when phage DNA was inserted between the centromere and telomere elements *(11)*. This immediately raised the possibility of using the artificial chromosomes as vectors for exogenous DNA. Because the original method for constructing YACs was not amenable to large scale cloning work, modifications were made to incorporate all the necessary functions onto a single plasmid that could replicate in *E. coli* and would allow the insertion of exogenous DNA prior to transformation of yeast cells *(12)*.

3. YAC Cloning Vectors

Prior to 1987, the available cloning systems were based mainly on *E. coli* and could only accept relatively small fragments of DNA. The bacterial vectors with the largest capacity were eosmicls, which can accept DNA in the size range 35–45 kb. YACs, with an average capacity of 100–2000 kb, offered a large increase in the size that could be cloned. Since then, new bacterial vectors with larger cloning capacities have been developed, but so far none have reached the capacity of YACs. A listing of vectors and their cloning capacities is shown in **Table 1**.

The YAC vector described in the previous section and originally described by Burke *(12)* was called pYAC2 and further modifications rapidly led to the construction of pYAC4. This has been the most widely used vector in the construction of YAC libraries and a diagram showing the relevant features is shown in **Fig. 1**. Digestion of the 11 kb circular plasmid with the restriction enzyme *Bam*HI results in a linear molecule with telomere (TEL) sequences at either end. The fragment lying between the two *Bam*HI sites containing the his3 gene is a 1.8 kb "stuffer" fragment that is discarded.

The cloning site for insertion of exogenous DNA is an *Eco*RI-restriction site that cleaves the linear molecule into two YAC "arms"; the right arm, a 3.4 kb fragment and the left arm, a 6 kb fragment containing the CEN4 and ARS I elements. The *Eco*RI site lies within the *sup4* gene. The product of this gene is a mutant tRNA, which suppresses the ochre (nonsense) mutation in the *ade2* gene. When a DNA fragment is cloned within

Table 1
Comparative Features of Cloning Vectors

Vector	Year introduced	Host	Structure	Insert size (kb)
Plasmid	1974	*E. coli*	Circular	< 10
Bacteriophage λ	1977	*E. coli*	Linear	6–22
Cosmid	1978	*E. coli*	Circular	35–45
P1	1990	*E. coli*	Circular	70–100
BAC	1992	*E. coli*	Circular	<300
PAC	1994	*E. coli*	Circular	100–300
YAC	1987	*S. cerevisiae*	Linear chromosome	100–2000

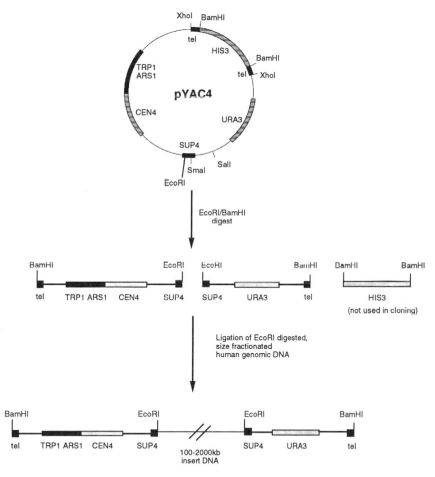

Fig. 1. Schematic drawing of plasmid pYAC4 (not to scale). The cloning site *Eco*RI is within the *Sup4* gene. TRP1 and URA3 are the left and right YAC arm yeast selectable markers respectively. tel are the two telomere sequences and CEN4 provides the centromere function.

the *sup4* gene, the suppression is eliminated and adenine metabolism is interrupted, resulting in the accumulation of a red pre-metabolite (phosphoribosyl-aminoimidazole) giving the recombinant clones a distinctive red color.

Two yeast selectable markers, *trpl* and *ura3* on the left and right arms, respectively, encode enzymes which enable yeast cells carrying the YAC to make tryptophan and uracil *de novo*. The yeast host *S. cerevisiae* AB 1380 (MATaψ+, *ura3, trpl, ade2-1, can1-100, lys2-1, his5*) alone does not have this ability, and growth of this yeast in medium deficient in such metabolites enables positive selection for recombinant molecules containing both YAC arms.

4. YAC Libraries

Following the description of pYAC4, several laboratories set about making representative genomic libraries. However, this turned out to be a challenging task, partly owing to the difficulty in isolating intact high-mol-wt genomic DNA fragments to clone into the YAC vector, but also owing to technical problems in transforming yeast host cells with the very large constructs. Only a relatively small number of quality libraries have been constructed worldwide and the techniques have come to be regarded as fairly specialized. However, most libraries have been made widely available to the research community to screen for YACs of interest, enabling a large number of laboratories to take advantage of the specialist YAC technology. (Details of YAC-library construction are outside the scope of this chapter, but can be found in **refs. *13, 25,*** and ***26***.) Details of the currently available YAC libraries, together with some suppliers, are given in the Appendix to this chapter.

The YAC libraries listed in the Appendix have been constructed from high-mol-wt human-genomic DNA obtained directly from cultured cells or lymphocytes isolated from peripheral blood. An alternative approach is to use flow-sorted chromosomes *(14)* or monochromosomal somatic cell hybrids *(15)* to produce chromosome-specific YAC libraries *(16–20)*. YAC libraries have also been made from the genomic DNA of many nonhuman species, including plants (e.g., *Arabidopsis thaliana, 21*) invertebrates (e.g., *Caenorhabditis elegans, 20*), and mammals (e.g., mouse, *3,23*).

5. Complexity of a Human YAC Library

The size of the human genome is approx 3000 Mb. In order to be reasonably certain that a given library will contain sequences from the entire genome, the following calculation is used:

$$N = \frac{[\ln(1-P)]}{[\ln(1-a/b)]} \tag{1}$$

where N = number of clones required; P = probability (e.g., 95% probability that any given sequence is present); a = average size of the DNA fragment inserted into the vector; and b = total size of the genome.

In practice, it is necessary to have a library complexity of between 3 and 4 times the size of the genome to ensure a 95% probability of obtaining any given sequence. For a cosmid with an average insert size of 30 kb, this would mean screening 300,000 clones, whereas for a YAC with an average insert size of 300 kb, only 30,000 clones need to be screened, saving considerable time and effort. YAC libraries have been constructed with higher representations of the genome, such as 6–10 times coverage *(24–27)*. Such libraries will give a significantly higher probability (~99%) of finding given sequences, but a larger number of positive clones will be identified for each probe used; a situation that can lead to an increase in the amount of work required to analyze clones at each stage of screening.

6. Screening YAC Libraries

YAC libraries can be screened either by PCR or hybridization methods. In both cases, the ordering of libraries in mictotiter plate format has enabled the development of rapid and simple screening strategies. This has been assisted by the introduction of semiautomated robotic systems for replicating clones.

Screening by PCR is the preferred method for isolating specific clones from YAC libraries. This is only possible where an STS is available. This is (normally) a short stretch of DNA sequence flanked by oligonucleotide primers, which gives a unique product when used to amplify genomic DNA. Such sequences are generally obtained from cloned DNA whose approximate chromosomal location is known, and hence they provide a physical marker relative to others in the region. PCR screening is normally performed in stages, starting with DNA pooled from many clones to identify subsets of the library containing positive clones and eventually leading to the identification of individual clones *(28,29)*. One such pooling strategy is shown in **Fig. 2**. The advantage of such a strategy is that it reduces the number of PCR reactions required to identify the individual clone.

Screening by hybridization is normally performed on membranes containing gridded arrays of YAC clones, often at high density *(30,31)*. The recombinant yeast cells are spotted onto nitrocellulose or nylon membranes lying on nutrient agar. The cells are allowed to grow on the membrane and when the colonies have grown sufficiently large, they are lysed *in situ* and the DNA is fixed onto the membranes. Following hybridization with the appropriate radioactively labeled probe and autoradiography, positive signals can be used to identify individual clones whose grid position is known.

7. Characterizing YACs

Once an individual YAC has been identified by screening to contain the marker or probe of interest, it is retrieved from its microtiter plate well by taking a small aliquot of the glycerol stock and streaking it out on an agar plate containing selection media. Following growth, several individual colonies are picked and grown up in liquid media for DNA preparation. This is a precaution, because some YAC clones are inherently more unstable than others, and tend to delete out portions of the DNA during replication. There is also the possibility that during library construction, more than 1 YAC clone is picked into a given microtiter plate well from the original plate. A preliminary analysis of the DNA from individual clones can quickly detect any such events.

The preparation of DNA from the yeast cells harboring the YAC must be done with care to ensure that the DNA is isolated as intact chromosomes. (Detailed procedures can be found in **ref. 29**.) Briefly, yeast cells are harvested by centrifugation and washed to remove any contaminants or inhibitory substances from the media. The cells are then resuspended in an appropriate buffer containing an enzyme (lyticase or novazyme) which digests the tough outer cell wall to generate spheroplasts. The cells are fragile at this stage and must be handled gently. Low melting temperature agarose is added to the solution and the resulting agarose mixture is transferred to plug moulds to solidify. The plugs are incubated in a solution containing lithium dodecyl sulfate and ethylene

Library consisting of 360 microtitre plates, each containing 96 clones
(total 34,560 clones)

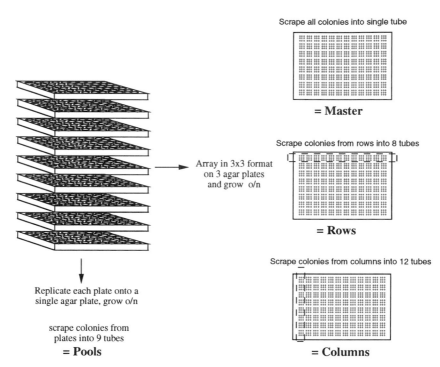

Fig. 2. Pooling strategy for PCR-based screening of YAC library: The library is divided into 40 sets of 9 plates. Each plate is replicated onto selective agar using a replicating tool. When the colonies have become sufficiently large, they are scraped together in water and used to prepare DNA in plugs. The 9 plates in a set are gridded in a 3 × 3 array onto 3 selective agar plates, using an automated gridding robot. When the colonies have become sufficiently large, they are scraped together shown to give Master, Row, and Column pools, and used to prepare DNA in plugs. A single plug can then be melted and used for PCR. The entire library of 34,540 clones can then be screened in two stages. First, 40 PCR reactions of the Masters identifies a set of 9 plates. Second, 9 PCR reactions of the Pools together with 20 PCR reactions of the (8 & 12) Rows and Columns identifies the individual clone.

diamine tetra-acetic acid (EDTA) which liberates the DNA from the spheroplasts. DNA Plugs can be stored at room temperature for many months without degradation of the DNA, which is protected form shearing forces by the agarose matrix.

In order to assess the size and stability of YACs, pulsed-field gel electrophoresis (PFGE) analysis is undertaken. This technique permits the resolution of DNA fragments up to several Mb in size *(32–34)*. Designated apparatus is required to run such gels, the most commonly used being the CHEF (Contour-Clamped Homogeneous Electric-Field) electrophoresis system (Biorad). **Figure 3** shows the separation of chromosomes from the yeast host AB1380 and five YAC clones using the CHEF-DRII system.

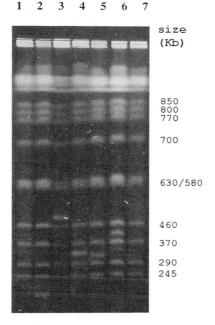

Fig. 3. Yeast chromosomes separated using a CHEF DRII pulsed field gel electrophoresis system. Lanes 1 and 7 *Saccharomyces cerevisiae* host strain. Lanes 2–6 Yeast Artificial Chromosomes present within the host strain are visible as extra chromosome bands of 200 (2), 480 (3), 330 (4), 320 (5), and 400 kb (6).

A commonly used technique for initial assessment of YACs is Fluorescence *In Situ* Hybridization (FISH) onto metaphase chromosomes *(35,36)*. In this technique, the YAC is used as a probe, and is labeled at many sites by incorporation of a reporter molecule (e.g., biotin or digoxigenin). It is then hybridized to chromosomal preparations and visualized by the addition of fluorochromes with affinity to the reporter molecule. The combination of large target size and many labeled sites on the YAC enables its chromosomal position to be visualized using a fluorescent microscope. The technique is also very useful for ascertaining whether a YAC is chimeric. All YAC libraries have some clones comprising DNA from unrelated regions. Such clones have arisen as an artifact during ligation and subsequent transformation, where two or more restriction fragments have become co-cloned into the same vector. These are termed chimaeric YACs and can cause serious problems if encountered during chromosome walking projects. Some libraries have a higher rate of chimerism than others (*see* Appendix). However, such chimeric YACs can be detected by FISH analysis because the YAC insert DNA will hybridize to more than one chromosomal location. This situation will result in signals occurring at more than one site in contrast to a nonchimaeric YAC, which will give a signal at a unique position.

Analysis of YAC insert DNA is rather more difficult than that of DNA cloned into *E. coli* systems because of their large size and the fact that they cannot easily be separated from the rest of the yeast chromosomes. In addition, YACs are replicated in the same way as the yeast host chromosomes, resulting in a single copy of the YAC in each cell. This means that the yield of DNA per cell is quite low. The most useful analytical

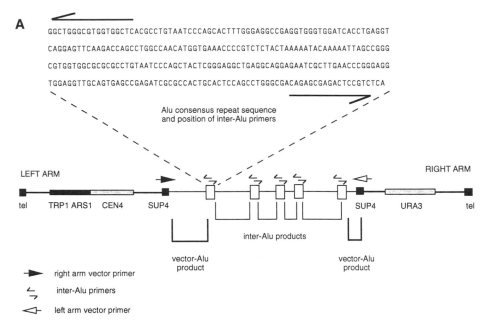

A

GGCTGGGCGTGGTGGCTCACGCCTGTAATCCCAGCACTTTGGGAGGCCGAGGTGGGTGGATCACCTGAGGT

CAGGAGTTCAAGACCAGCCTGGCCAACATGGTGAAACCCCGTCTCTACTAAAAATACAAAAATTAGCCGGG

CGTGGTGGCGCGCGCCTGTAATCCCAGCTACTCGGGAGGCTGAGGCAGGAGAATCGCTTGAACCCGGGAGG

TGGAGGTTGCAGTGAGCCGAGATCGCGCCACTGCACTCCAGCCTGGGCGACAGAGCGAGACTCCGTCTCA

Alu consensus repeat sequence
and position of inter-Alu primers

LEFT ARM   RIGHT ARM

tel   TRP1 ARS1 CEN4   SUP4   SUP4   URA3   tel

inter-Alu products

vector-Alu   vector-Alu
product   product

▶   right arm vector primer

⬱   inter-Alu primers

◁⊢   left arm vector primer

Fig. 4. PCR-based methods for terminal sequence isolation from YACs: **(A)** Alu-VectorPCR; **(B)** Inverse PCR; **(C)** Vectorette PCR.

techniques are therefore those that do not require the YAC to be purified from the rest of the yeast genome, and those that utilize PCR, where the amount of YAC DNA required is not limiting. However, there will occasionally be a requirement for purification.

YACs are generally purified by preparative PFGE in high purity low gelling temperature agarose (such as Seaplaque FMC). The gel is run such that the YAC is separated from the other yeast chromosomes and is preferably resolved within the first 5 cm of the gel. The YAC is then cut out in an agarose slice. It is best not to stain the gel with Ethidium bromide if the YAC DNA is required intact. The position of the YAC can be found by running marker lanes at either end of the gel, cutting these off and staining them and then repositioning them on either side of the gel to give a guide for cutting out the central portion. The DNA is then isolated from the agarose slice by one of several methods, including Agarase (Calbiochem) digestion followed by phenolchloroform extraction and ethanol precipitation or Prep-A-Gene (Biorad) purification. Most of these methods yield a reasonable amount of DNA for further analyses.

8. Chromosome Walking and Physical Mapping with YACs

Where YACs are to be used for chromosome walking, the generation of overlapping contiguous clones ("contigs") is essential. This requires that the YACs be orientated and mapped relative to each other and the outermost ends be identified and isolated for finding further YACs. In practice these two criteria are interdependent, because probes derived from the ends of individual YACs can identify which other YACs in a set overlap the original. In order to extend such contigs, sequences at the ends of the YACs must be isolated for use as probes that can be used to rescreen the YAC library and identify more clones.

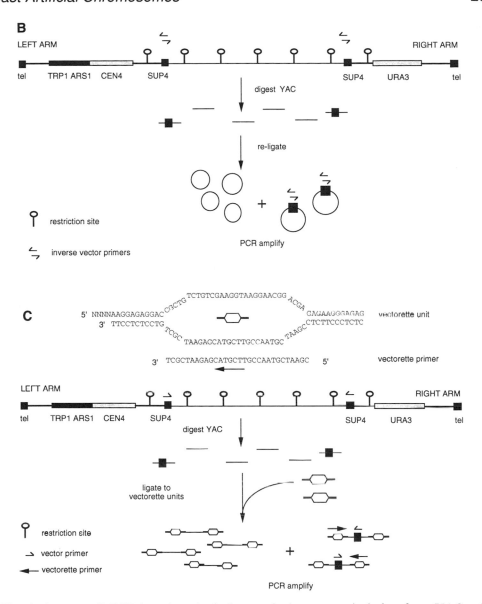

Fig. 4. *(continued)* PCR-based methods for terminal sequence isolation from YACs: **(A)** Alu-VectorPCR; **(B)** Inverse PCR; **(C)** Vectorette PCR.

YAC DNA inserts are difficult to dissect, but the terminal sequences are more amenable to analysis, owing to their position adjacent to the flanking vector DNA whose sequence is known. Many techniques have evolved for this purpose and these are comprehensively reviewed in **ref. *37***. The methods can be broadly divided into three categories; plasmid rescue, subcloning, and PCR-based methods. All have their adherents, but the PCR-based techniques are probably the most popular. These include inverse PCR *(38)*, Alu-vector PCR *(39)*, and Anchor or Vectorette PCR *(40)*. The three techniques are depicted diagrammatically in **Fig. 4**. In inverse PCR, the YAC is digested with a restriction enzyme that cleaves within the vector arm and at an unknown location within the insert DNA. The fragments generated are then ligated such that they circularize and primers which were in inverse orientation in the original vector arm can now be used to amplify the circular template. Alu-vector PCR utilizes the presence in the human genome of a very common 300 bp repeat sequence, the Alu repeat, which occurs on average once every 4 kilobases. Primers which permit inter-Alu amplification are used in combination with vector-specific primers to amplify vector-Alu fragments. This method is only successful if there is an Alu repeat lying relatively close to the insert/vector junction. In vectorette PCR, the YAC is digested in a similar way to the inverse PCR method, following which the fragments are ligated to a small "vectorette units" consisting of a partial duplex of DNA with an internal "bubble" of noncomplementary DNA. One primer specific for the vector and one for the internal portion of the vectorette unit are used to amplify the DNA. Amplification can only occur where both vector sequence and a vectorette unit are present in the template DNA. All of the techniques described above are generally used either to produce probes for hybridization or to generate sequence data from which new PCR primers can be designed.

Further mapping of regions within the YACs may also be necessary, and can be done by digesting the YACs with restriction enzymes, which cut relatively infrequently. The digested material is then subjected to PFGE and Southern blotting, followed by hybridization with specific unique probes from the left and right YAC vector arms. Partial digestion can often assist in building a comprehensive restriction map, although a series of such digests is normally necessary to give a complete picture. Comparative maps can then be built up for YACs known to overlap by screening. Fingerprinting of YACs using the technique of inter-Alu PCR can also be informative because Alu repeats occur frequently in the genome, so that a YAC is likely to contain at least several such sequences. Amplification of YAC DNA with primers complementary to the repeats will produce a pattern of bands specific for a given YAC. Comparison of the banding patterns of overlapping YACs can identify which are common and hence indicate overlapping regions.

Often in the course of positional cloning exercises, YACs that are generated by chromosome walking can be usefully employed in finding additional genetic marker probes or sequence tagged sites (STSs), which can be used for further genetic analysis in family studies. The most consistently useful type of STS for this purpose is the dinucleotide repeat, of which the CA/GT repeat occurs most frequently in the human genome (approximately every 25–100 kb). Such sequences are highly polymorphic in terms of the number of dinucleotide repeats they contain *(41)* and so can be used to distinguish between alleles inherited from the mother and the father. Sequences containing >15 dinucleotide repeats are very likely to be highly informative in genetic analysis, and the information gained can narrow the region of search for a given gene. Several methods exist for the isolation of CA/GT repeats from YACs. Most commonly, YAC DNA is digested with a frequent cutting enzyme, such as *Sau*3AI, and the resulting small fragments subcloned into a complementary restriction site (*Bam*HI) in a plasmid vector. The resulting library of clones is then screened using a $CA_{(n)}$ oligonucleotide probe *(42,43)*. Clones isolated by this method are subjected to sequence analysis, where the number of CA repeats can be counted, and the flanking sequences, which are usually unique, used to design PCR primers to perform genetic analysis.

9. Identifying Expressed Genes in YACs

The final goal in most genome-mapping projects and all positional cloning projects is the identification of genes. When transcribed sequences are thought to occur in the region bounded by a YAC or YAC contig, several methods exist to isolate such sequences.

Often clues to the presence of genes can be gained simply by information from restriction enzyme maps of YACs. The enzymes used to generate such maps cut genomic DNA relatively infrequently. This is because they recognize sequences containing one or more unmethylated CpG dinucleotides. Such sites are underrepresented throughout the genome, but chiefly occur clustered in regions called CpG or HTF (*Hpa*II-tiny-fragment) islands. These HTF islands are often found in or near the RNA polymerase II promoter regions in many housekeeping genes *(44,45)*. They are therefore highly predictive of the presence of genes. If several such enzymes are used in the course of a YAC-mapping experiment, the YAC-restriction map will identify regions where sites are clustered together, leading to the identification of the adjacent gene.

YACs can be used as hybridization probes to identify transcribed sequences from cDNA libraries *(46)*. This requires that the YAC be purified away from other yeast chromosomes so that it can be radiolabeled to a high specific activity. The kinetics of labeling are such that good specific activity is only achieved with smaller YACs, and even then the probes are not optimal. One problem associated with direct screening is the large amount of repetitive DNA in the YAC genomic insert, which can hybridize to similar sequences within the cDNA leading to a large number of nonspecific signals that mask the true signals. To prevent this, the YAC DNA is reassociated with total human DNA to a cot value of >100 *(47)* to block the repetitive sequences. Because of these constraints, direct screening with YAC DNA is difficult to perform successfully and has a limited value.

More recently, some new techniques have evolved for finding cDNAs. Direct selection *(48,49)* is one such method, where cDNA is hybridized to YAC DNA immobilized either on a filter or other support. Nonspecific hybrids are eliminated by washing and the bound material is eluted and PCR amplified. Using this method, very small amounts of cDNA can be enriched via the PCR step, making the technique very sensitive. A second method does not require cDNA, but instead relies on the ability of eukaryotic cells to process RNA transcripts into mature mRNA by the splicing together of exons and elimination of introns. This method, called exon trapping or exon amplification *(50–52)*, overcomes the problems associated with obtaining good cDNA libraries from tissues that express the gene of interest. In exon trapping, YACs are subcloned into a vector (e.g., plasmid pSPL3, Life Technologies Inc.) whose cloning site lies in an intron flanked by functional 5' and 3' splice sites. The entire construct is used to transfect COS-7 cells and an upstream SV40 promoter in the vector enables transcription of the construct. Because COS-7 cells are mammalian cells, the transcript is spliced and, if the subcloned fragment contains an intact exon, this will be present in the mRNA of the cell. The cells are harvested and the RNA isolated, following which PCR primers specific for the vector are used to amplify novel exons. Because exons are relatively small compared to introns, this technique can only produce partial cDNA sequences, but these can then be used as probes to find full-length clones from cDNA libraries.

10. Functional Studies Using YACs

The ability of YACs to harbor complete genes, including flanking control regions, as well as the entire coding region makes them very useful vectors for the transfer of genes into mammalian cells. Intact YACs have been transferred into mammalian cells by a variety of methods. These inclucle fusion of mammalian cells with YAC-bearing yeast spheroplasts, electroporation, calcium phosphate precipitation, lipofection, and microinjection.

The YAC DNA must be integrated into the host genome for the gene(s) to be stably expressed. However, YACs lack a suitable selectable mammalian marker, and early studies relied on the use of YACs bearing genes that encoded enymes involved in metabolic pathways. These were then used to transfect cell lines deficient in the enyme and hence complement the deficiency. The successful transfer and expression of the human hypoxanthine-guanine phosphoribosyltransferase (HPRT) gene and the trifunctional phosphoribosylglycinamide formyltransferase (GART) gene, both of which are involved in purine synthesis, was carried out in this way *(53,54)*.

Subsequently, vectors were developed that enabled the integration into YACs of mammalian selectable markers such as the neomycin-resistance gene *(54–57)*. YACs modified (or "retrofitted") in this way with neomycin markers have been used to generate transgenic mice either by pronuclear injection, or by introduction into embryonic stem (ES) cells *(58)*. In one such study *(59)* a 248 kb YAC-containing the entire human β-globin cluster, including the 5' locus control region (LCR) was transferred into the genome of transgenic mice, enabling the regulation of the gene cluster to be examined. The regulation of the apolipoprotein(a) gene has also been studied in YAC-transgenic mice *(60)* and a mouse model for the peripheral neuropathy Charcot-Marie-Tooth disease type 1 A has been constructed by the introduction of YAC DNA *(61)*.

11. The Post-YAC Era

Despite the success of YACs, there are features of the YAC-cloning system that cause problems. A percentage of clones, especially in human YAC libraries, are chimeric, and this can cause problems in chromosome walking, because the ends of a YAC used for the next round of screening may come from different regions of the genome. This problem can be overcome by performing FISH of metaphase chromosomes with each newly identified YAC to confirm that a signal is seen only in the correct chromosomal region. Other problems associated with YACs are the instability of some clones, and the fact that the YAC clones cannot easily be separated from the yeast chromosomal background. For these reasons, YACs have remained rather specialized tools and their handling has not become routinely integrated into the protocols of general molecular-biology laboratories.

Several alternative cloning systems have been developed in recently that allow the cloning of insert sizes that are intermediate between the original *E. coli* vectors and YACs. These are P1 phage *(62)*, PACs (P1-derived artificial chromosomes, *63*), and BACs (Bacterial artificial chromosomes, *64*). Each system has its inherent advantages and disadvantages, but the main advantages that each of these cloning vectors has over YACs include ease of isolating and purifying insert DNA and easier generation of libraries owing to higher transformation efficiency. Early evidence also suggests that the clones are more stable and less chimeric than YACs.

The successful construction of yeast artificial chromosomes also raised the possibility of making synthetic mammalian artificial chromosomes *(65)*, opening up opportunities for somatic gene therapy. It was recently shown that the transfection of centromeric alpha satellite, telomere, and genomic carrier DNA into human HT1080 cells resulted in the generation of stable microchromosomes *(66)*. This finding should greatly enhance the prospects for both the construction and application of mammalian artificial chromosomes.

YACs have played a key role in genome research. The physical maps of the human genome that provide the framework for the current international Human-Genome Sequencing Project would not exist but for YACs. Their contribution extends even further and is now entering the realms of research into gene regulation and gene therapy. Despite the advent of vectors that may be easier to handle, YACs are likely to continue to play an important role in genome research well into the foreseeable future.

Appendix: YAC Library Availability

St. Louis YAC Library

Brownstein et al. (1989) *Science* **244,** 1348–1351. The library consists of approx 56,000 clones, with an average insert size of just over 200 kb. Data suggests that ~40–60% of the clones are chimeric.

ICI YAC Library

Anand et al. (1990) *Nucleic Acids Res.* **18,** 1951–1955. The library consists of 34,500 clones with an average insert size of 350 kb and a >3.5X coverage of the human genome, and was made from a human lymphoblastoid 48XXXX cell line. Data suggests that ~10% of the clones are chimeric.

CEPH YAC Library

Albertson et al. (1990) *Proc. Natl. Acad. Sci. USA* **87,** 4256–4260. The "mega-YAC" part of this library (plates 613–984) consists of 35,600 clones with an average insert size of 920 kb and a 7–8X coverage of the genome. Data suggests that ~40% of the clones are chimeric.

ICRF YAC Library

Larin et al. (1991) *Proc. Natl. Acad. Sci. USA* **88,** 4123–4127. The library consists of 20,500 clones and combines three separate libraries:

1. 4X (human female lymphoblastoid 48XXXX cell line): 15,700 clones; average insert size 600 kb, 3X coverage of autosomes and 6X coverage of X.
2. 4Y (human male lymphoblastoid cell line 49XYYYY): 2300 clones; insert size 400–500 kb.
3. HD (human female lymphoblastoid, Huntington's disease 46XX): 2500 clones; average insert size 600 kb.

Data suggests that ~30% of the clones are chimeric.

Centers with YAC Library Clone Availability

Noncommercial Suppliers

1. IGBE CNR (Torino, Italy): D. Toniolo, Fax: [39](382) 422286, e-mail: ipvg03%39162 @icil641c:ilea.it

2. Leiden University (Leiden, The Netherlands): Gert-Jan van Ommen & Johan den Dunnen, Fax: [31](71) 276075, e-mail: ddunnen@ruly46.LeidenUniv.nl

3. HGMP Resource Center (Cambridge, UK): Fax: (1223) 494-510

Some YAC library resources are also available from:

1. Fondation Jean Dausset - CEPH (Paris, France): Denis Le Paslier, e-mail: denis@ceph.cephb.fr
2. Whitehead Institute Genome center (Boston, MA) Eric Lander, e-mail: lander@genome.wi.mit.edu
3. ICRF (UK) Mark Ross, e-mail: m_ross@icrf.icnet.uk
4. RIKEN (Japan): Fax: [81](298) 36 9140.
5. Cancer Institute (Tokyo, Japan): Y. Nakamura, Fax: [81](3) 3918 0342.
6. Shanghai Medical University N.2 (China): Z. Chen, Fax: [86](21) 3180 300.

Commercial Suppliers

1. Research Genetics Inc. (Huntsville, AL): Fax: [205] (536)9016, Homepage: http// www.resgen.com
2. Genome Systems Inc. (St. Louis, MO): Fax: [314] (692) 0033, Homepage: http// www.genomesystems.com

References

1. Chumakov, I., Weissenbach, J., and Cohen, D. (1993) A first generation physical map of the human genome. *Nature* **336,** 698–701.
2. Chumakov, I., Rigault, P., LeGall, I., and Cohen, D. (1995) A YAC contig map of the human genome. *Nature* **377(Suppl.),** *Genome Dir.* 175–297.
3. Collins, F. S. (1992) Positional cloning: Lets not call it reverse anymore. *Nat. Genet.* **1,** 3–6.
4. Nelson, D. L. (1995) Positional cloning comes of age. *Curr. Opinion Gen. Dev.* **5,** 298–303.
5. Herskowitz, I. (1988) Life cycle of the budding yeast *Saccharomyces cerevisiae. Microbiol. Rev.* **52,** 536.
6. Struhl, K., Stinchomb, D. T., Scherer, S., and Davis, R. W. (1979) High frequency transformation of yeast: Autonomous replication of hybrid DNA molecules. *Proc. Natl. Acad. Sci. USA* **76,** 1035–1039.
7. Stinchomb, D. T., Struhl, K., and Davis, R. W. (1979) Isolation and characterization of a yeast chromosomal replicator. *Nature* **282,** 39–43.
8. Bouton, A. H. and Smith, M. M. (1986) Fine-structure analysis of the DNA sequence requirement for autonomous replication of Saccharomyces cerevisiae plasmids. *Mol. Cell. Biol.* **6,** 2354–2357.
9. Clarke, L. and Carbon, J. (1980) Isolation of a yeast centromere and construction of functional small circular chromosomes. *Nature* **287,** 504–509.
10. Szostak, J. W. and Blackburn, E. H. (1982) Cloning yeast telomeres on linear plasmid vectors. *Cell* **29,** 245–255.
11. Murray, A. W. and Szostak, J. W. (1983) Construction of artificial chromosomes in yeast. *Nature* **305,** 189–193.
12. Burke, D. T., Carle, G. F., and Olson, M. V. (1987) Cloning of large segments of exogenous DNA into yeast by means of artificial chromosome vectors. *Science* **236,** 806–812.
13. Burke, D. T. and Olson, M. V. (1991) Preparation of clone libraries in yeast artificial chromsosome vectors. *Meth. Enzymol.* **194,** 251–270.
14. Bartholdi, M. F., Meyne, J., Albright, K., Luedemann, M., Campbell, E., Chritton, D., Daven, L., Van Dilla, M., and Cram, L. S. (1987) Chromosome sorting by flow cytometry, in *Methods in Enzymology* vol. 151 (Gottesman M., ed.), Academic, San Diego, CA, pp. 252–267.

15. Abbott, C. and Povey, S. (1994) *Somatic Cell Hybrids: The Basics* (Beynond, R., Brown, T. A., and Howe, C., eds.), Oxford University Press, Cambridge, UK.

16. Abidi, F. E., Wada, M., Little, R. D., and Schlessinger, D. (1990) Yeast artificial chromosomes containing human Xq2-Xq28 DNA: Library construction and representation of probe sequences. *Genomics* **7**, 363–376.

17. Den Dunnen, J. T., Grootscholten, P. M., Dauwerse, J. G., Walker, A. P., Monaco, A. P., Butler, R., Anand, R., Coffey, A J., Bentley, D. R., Steensma, H. Y., and van Ommen, G. J. B. (1992) Reconstructdion of the 2. 4 Mb human DMD-gene by homologous YAC recombination. *Hum. Mol. Genet.* **1**, 19–23.

18. McCormick, M. K., Campbell, E., Deaven, L., and Moyzis, R. K. (1993) Low frequency chimeric yeast artificial chromosome libraries from flow-sorted human chromosomes 16 and 21. *Proc. Natl. Acad. Sci. USA* **90**, 1063–1067.

19. McCormick, M. K., Buckler, A., Bruno, W., Campbell, E., Shera, K., Torney, D., Deaven, L., and Moyzis, R. (1993) Construction and characterisation of a YAC library with a low frequency of chimeric clones from flow-sorted human chromosome 9. *Genomics* **18**, 553–558.

20. Little, R. D., Porta, G., Carle, G. F., Schlessinger, D., and D'Urso, M. (1989) Yeast artificial chromosomes with 200 to 800 kilobase insrts of human DNA containing HLA, Vk, 5S and Xq24-Xq28 sequences. *Proc. Natl. Acad. Sci. USA* **86**, 1598–1602.

21. Edwards, K. J., Thompson, H., Edwards, D., deSaizieul, A., Sparks, C., Thompson, S. A., Greenland, A. J., Eyeres, M., and Schuch, W. (1992) Construction and characterisation of a yeast artificial chromosome library containing three haploid maize genome equivalents. *Plant Mol. Biol.* **19**, 299–308.

22. Coulson, A. R., Waterston, R., Kiff, J., Sulston, J., and Kohara, Y. (1988) Genome linking with yeast artificial chromosomes. *Nature* **335**, 181–186.

23. Chartier, F. L., Keer, J. T., Sutcliffe, M. J., Henriques, D. A., Mileham, P., and Brown, S. D. M. (1992) Construction of a mouse yeast artificial chromosome library in a recombination-deficient strain of yeast. *Nat. Genet.* **1**, 132–136.

24. Brownstein, B. H., Silverman, G. A., Little, R. D., Burke, D. T., Korsmeyer, S. J., Schlessinger, D., and Olson, M. V. (1989) Isolation of single-copy human genes from a library of yeast artificial chromosomes. *Science* **244**, 1348–1351.

25. Anand, R., Riley, J. H., Butler, R., Smith, J. C., and Markham, A. F. (1990) A 3.5 genome equivalent multi access YAC library: construction, characterisation, screening and storage. *Nucleic Acids Res.* **18**, 1951–1956.

26. Larin, Z., Monaco, A. P., and Lehrach, H. (1991) Yeast artificial chromosome libraries containing large inserts from mouse and human DNA. *Proc. Natl. Acad. Sci. USA* **88**, 4123–4127.

27. Albertsen, H. M., Abderrahim, H., Cann, H. M., Dausset, J., LePaslier, D., and Cohen, D. (1990) Construction and characteisation of a yeast artificial chromosome library containing seven haploid human genome equivalents. *Proc. Natl. Acad. Sci. USA* **87**, 4256–4260.

28. Jones, M. H., Khwaja, O. S. A., Briggs, H., Lambson, B., Davey, P. M., Chalmers, J., Zhou, C.-Y., Walker, E. M., Zhang, Y., Todd, C., Ferguson-Smith, M. A., and Affara, N. A. (1994) A set of ninety-seven overlapping yeast artificial chromosome clones spanning the human Y chromosome euchromatin. *Genomics* **24**, 266–275.

29. Anand, R. (1995) Cloning into yeast artificial chromosomes, in *DNA Cloning 3: A Practical Approach* (Glover, D. M. and Hames, B. D., eds.), Oxford University Press, Cambridge, UK, pp. 103–125.

30. Ross, M. T., Hoheisel, J. D., Monaco, A. P., Larin, Z., Zehetner, G., and Lehrach, H. (1992) High-density gidded YAC filters: their potential as genome mapping tools, in *Techniques for the Analysis of Complex Genomes* (Anand, R., ed.), Academic, San Diego, CA, pp. 137–153.

31. Bentley, D. R., Todd, C., Collins, J., Holland, J., Dunham, I., Hassock, S., Bankier, A., and Gianelli, F. (1992) The development and application of automated ridding for efficient screening of yeast and bacterial ordered libraries. *Genomics* **12,** 534–541.

32. Schwartz, D. C. and Cantor, C. R. (1984) Separation of yeast chromosome-sized DNAs by pulsed field gradient gel electrophoresis. *Cell* **37,** 67–75.

33. Lai, E., Birren, B. W., Clark, S. M., Simon, M. I., and Hood, L. (1989) Pulsed field gel electrophoresis. *Biotechniques* **7,** 34–42.

34. Monaco, A. P. (1995) *Pulsed Field Gel Electrophoresis: A Practical Approach* (Rickwood, D. and Hames, B. D., eds.), Oxford University Press, Cambridge, UK.

35. Trask, B. J. (1991) Fluorescence in situ hybridisation: applications in cytogenetics and gene mapping. *Trend. Genet.* **7,** 149–154.

36. Buckle, V. J. and Rack, K. A. (1993) Fluorescent in situ hybridisation, in *Human Genetic Disease Analysis: A Practical Approach* (Davies, K. E., ed.), Oxford University Press, Cambridge, UK, pp. 59–82

37. Silverman, G. A. (1993) Isolating vector-insert junctions from yeast artificial chromosomes. *PCR Meth. Appl.* **3,** 141–150.

38. Silverman, G. A., Ye, R. D., Pollock, K. M., Sadler, J. E., and Korsmeyer, S. J. (1989) Use of yeast artificial chromosome clones for mapping and walking within human chromosome segment 18q21. 3. *Proc. Natl. Acad. Sci. USA* **86,** 7485–7489.

39. Nelson, D. L., Ballabio, A., Victoria, M. F., Pieretti, M., Bies, R. D., Gibbs, R. A., Maley, J. A., Chinault A. C., Webster, T. D., and Caskey, C. T. (1991) Mu-primed polymerase chain reaction for regional assignment of 110 yeast artificial chromosome clones from the human X chromosome: Identification of clones associated with a disease locus. *Proc. Natl. Acad. Sci. USA* **88,** 6157–6161.

40. Riley, J., Butler, R., Ogilvie, D., Finniear, R., Jenner, D., Powell, S., Anand, R., Smith, J. C., and Markham, A. F. (1990) A novel, rapid method for the isolation of terminal sequences from yeast artificial chromosome (YAC) clones. *Nucleic Acids Res.* **18,** 2887–2890.

41. Weber, J. L. (1990) Abundant class of human DNA polymorphisms which can be typed using the polymerase chain reaction. *Genomics* **7,** 524–530.

42. Cornelis, F., Hashimoto, L., Loveridge, J., MacCarthy, A., Buckle, V., Julier, C., and Bell, J. (1992) Identification of a CA repeat at the TCRA locus using yeast artificial chromosomes: A general method for generating highly polymorphic markers at chosen loci. *Genomics* **13,** 820–825.

43. Iizuka, M., Makin, R., Sekiya, T., and Hayashi, K. (1993) Selective isolation of,highly polymorphic (dC-dA)n-(dG-dT)n microsatellites by stringent hybridisation. *GATA* **10,** 2–5.

44. Brown, W. R. A. and Bird, A. P. (1986) Long-range restriction site mapping of mammalian genomic DNA. *Nature* **322,** 477–481.

45. Bird, A. P. (1987) CpG islands as gene markers in the vertebrate nucleus. *Trends Genet.* **3,** 342–346.

46. Elvin, P., Butler, R., and Hedge, P. J. (1992) Transcribed sequences within YACs: HTF island cloning and cDNA library screening, in *Techniques of the Analysis of Complex Genomes* (Anand, R., ed.), Academic, San Diego, CA, pp. 155–171.

47. Sealey, P. G., Whittaker, P. A., and Southern, E. M. (1985) Removal of repeated sequences from hybridisation probes. *Nucleic Acids Res.* **13,** 1905–1932.

48. Parimoo, S., Patanjali, S. R., Shukla, H., Chaplin, D. D., and Weissman, S. M. (1991) cDNA selection: Efficient PCR approach for the selection of cDNAs encoded in large chromosomal DNA fragments. *Proc. Natl. Acad. Sci. USA* **88,** 9623–9627.

49. Lovett, M., Kere, J., and Hinton, L. M. (1991) Direct selection: a method for the isolation of large genomic regions. *Proc. Natl. Acad. Sci. USA* **88,** 9628–9632.

50. Auch, D. and Reth, M. (1990) Exon trap cloning: Using PCR to rapidly detect and clone exons from genomic DNA fragments. *Nucleic Acids Res.* **25,** 6743,6744.

51. Duyk, G. M., Kim, S., Myers, R. M., and Cox, D. R. (1990) Exon trapping: A genetic screen to identify candidate transcribed sequences in cloned mammalian DNA. *Proc. Natl. Acad. Sci. USA* **87,** 8995–8999.

52. Buckler, A. J., Chang, D. D., Graw, S. L., Brook, J. D., Haber, D. A., Sharp, P. A., and Housman, D. E. (1991) Exon amplification: A strategy to isolate mammalian genes based on RNA splicing. *Proc. Natl. Acad. Sci. USA* **88,** 4005–4009.

52. Gnirke, A., Barnes, T. S., Patterson, D., Schild, D., Featherstone, T., and Olson, M. V. (1991) Cloning and in vivo expression of the human GART gene using yeast artifiiciarli chromosomes. *EMBO J.* **10,** 1629–1634.

53. Gnirke, A. and Huxley, C. (1991) Transfer of the human HPRT and GART genes from yeast to mamimalian cells by mricroinjection of YAC DNA. *Som. Cell Mol. Genet.* **17,** 573–580.

54. Pachnis, V., Pevuey, L., Rothstein, R., and Constantim, F. (1990) Transfer of a yeast artificial chromosome carrying human DNA from *Saccharomyces cerevisiae* into mammalian cells. *Proc. Natl. Acad. Sci. USA* **87,** 5109–5113.

55. Pavan, W. J., Hieter, P., and Reeves, R. H. (1990) Modification and transfer into an embryonal carcinoma cell line of a 360-kilobase human-derived yeast artificial chromosome. *Mol. Cell Biol.* **10,** 4163–4169.

56. Srivastava, A. K. and Schlessinger, D. (1991) Vectors for inserting selectable markers in vector arms and human DNA inserts of yeast artificial chromosomes (YACs). *Gene* **103,** 53059.

57. Riley, J. H., Morten, J. E. N., and Anand, R. (1992) Targeted integration of neomycin into yeast artificial chromosomes (YACs) for transfection into mammalian cells. *Nucleic Acids Res.* **20,** 2971–2976.

58. Forget, B. G. (1993) YAC transgenes: bigger is probably better (Review). *Proc. Natl. Acad. Sci. USA* **90,** 7909–7911.

59. Peterson, K. R., Clegg, C. H., Huxley, C., Josephson, B. M., Haugen, H. S., Furukawa, T., and Stamatoyannopoulos, G. (1993) Transgenic mice containing a 248-kb yeast artificial chromosome carrying the human beta-globin locus display proper developmental control of human globin genes. *Proc. Natl. Acad. Sci. USA* **90,** 7593–7597.

60. Frazer, K. A., Narla, G., Zhang, J. L., and Rubin, E. M. (1995) The apolipoprotein(a) gene is regulated by sex hormones, and acute-phase inducers in YAC transgenic. *Nat. Gen.* **9,** 424–431.

61. Huxley, C., Passage, E., Manson, A., Putzu, G., Figarella-Branger, D., Pellissier, J. F., and Fontes, M. (1996) Construction of a mouse model of Charcot-Marie-Tooth disease type 1A by pronuclear injection of human YAC DNA. *Hum. Mol. Genet.* **5,** 563–569.

62. Sternberg, N. (1990) Bacteriophage P1 cloning system for the isolation, amplification, and recovery of DNA fragments as large as 100 kilobase pairs. *Proc. Natl. Acad. Sci. USA* **87,** 103–107.

63. Ioannou, P. A., et al. (1994) A new bacteriophage P1-derived vector for the propagation of large human DNA fragments. *Nat. Genet.* **6,** 84–89.

64. Shizuya, H., et al. (1992)Cloning and stable maintenance of 30-kilobase-pair fragments of human DNA in Eschrerichia cold using an F-factor-based vector. *Proc. Natl. Acad. Sci. USA* **89,** 8794–8797.

65. Huxley, C. (1994) Mammalian artificial chromosomes: a new tool for gene therapy. *Gene Ther.* **1,** 7–12.

66. Harrington, J. J., Van Bokkelen, G., Mays, R. W., Gustashaw, K., and Willard, H. F. (1997) Formation of *de novo* centromeres and construction of first-generation human artificial microchromosomes. *Nature Genet.* **15,** 345–355.

25

Polymerase Chain Reaction

Ralph Rapley

1. Introduction

In general, advances in biological and biochemical research are brought about by improvements and refinement of methods in current use. There are times, however, when a technique is developed that revolutionizes a particular field of research. The polymerase chain reaction, or PCR, developed at the Cetus Corporation in 1985 is one such example. The technique enables large amounts of DNA to be produced from very small amounts of starting material and mimics the basic mechanism of DNA replication and the manner in which it is carried out. The PCR was first described in 1985 and is a technique resulting in near exponential enzymatic amplification of DNA to a level easily detected by conventional methods, such as gel electrophoresis *(1)*.

2. Basic Aspects of the PCR

The PCR is an in vitro DNA amplification method that involves a repeated cycling process of a number of defined stages. The reagents required for the PCR include a DNA polymerase, each of the four nucleotide dNTP building blocks of DNA in equimolar amounts, and a source of template DNA, such as genomic DNA or cDNA containing the target sequence. The technique also demands the availability of two oligonucleotide primers designed to complement DNA sequences flanking the region of interest. The primers may be of variable length but are usually in the region of 15–30 bp and although for some specific purposes may vary or be degenerate in their sequence, they are usually directly complementary. This demands that, in general, sequence information be available for part of the DNA sequence that is to be amplified.

Amplification takes place in repeated cycles made up of three defined stages, termed denaturation, annealing, and extension (*see* **Fig. 1**). In the first stage of denaturation the template DNA is heated in excess of 90°C for at least 60 s to separate the double-stranded DNA and produce two single strands. This is followed by a second stage, known as annealing, where the temperature is reduced to 35–55°C for a time interval of between 60 and 120 s to allow the oligonucleotide primers to bind to their complementary DNA sequences on the single strands produced in the previous stage. In the final

From: *Molecular Biomethods Handbook*
Edited by: R. Rapley and J. M. Walker © Humana Press Inc., Totowa, NJ

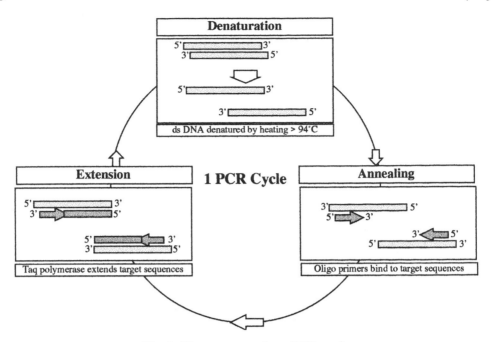

Fig. 1. Three stages of one PCR cycle.

stage of the cycle an enzymatic primer extension reaction is carried out producing complementary copies of the initial single strands from the primers bound to the DNA. This step usually takes place at 72°C for 60–180 s by a DNA polymerase that is able to withstand the high denaturation temperatures. This heat-stable enzyme is the key to the PCR and was initially isolated from the bacterium *Thermus aquaticus* found in hot springs and termed Taq DNA polymerase. The three separate stages are usually repeated between 25 and 40 times. In this way the double-stranded products of the previous cycle become new templates for the next cycle such that in each round the amount of the specific target DNA flanked by the primers essentially doubles (*see* **Fig. 2**). This results in the near exponential accumulation of the specific target DNA sequence of up to a million-fold in 3–4 h. Linear amplification of the initial strands also takes place and contributes to the PCR not being 100% efficient. Because of the repeated nature of the cycles, the PCR may be automated and numerous automatic thermal cyclers have been designed and produced specifically for the PCR.

In general, lengthy template sample preparations are not required for the PCR to work efficiently, and because of the sensitivity of the technique relatively crude DNA samples may be used as templates. For example, in the amplification of DNA from blood samples a simple boiling step is often sufficient to release the DNA, and provided this is diluted to negate the effects of potential inhibitors, such as porphyrins, provides a convenient way of rapidly analyzing large numbers of samples. It is also possible to analyze DNA samples of poor condition by PCR as only relatively short intact sequences are required. Thus, paraffin embedded material, or even ancient samples may be analyzed. This now makes previously difficult retrospective studies quite straightforward to perform *(2)*.

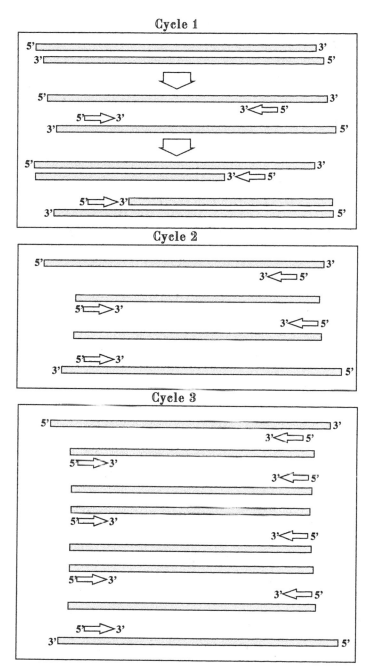

Fig. 2. First few cycles in the PCR.

2.1. Consideration and Design of Primers

Primers used in the PCR are generally designed using information based on existing sequences of close similarities or evolutionary conserved sequences found by searching genetic databases, such as Genbank/EMBL. Amino acid sequence information may also be used to provide deduced nucleotide sequences from which primers may be designed. In general, the primers should have a matched GC content of approx 50% and must not have the potential to form primer–dimer structures or be self complementary, both of which adversely affect the PCR *(3)*.

A modification of the PCR, termed anchored PCR, has been developed for cases in which either the 3' or 5' portion of the sequence under study is not known. It is possible when amplifying cDNA produced from eukaryotic mRNA to make use of the characteristic polyA tail found at the 3' end. In this case, a polyA primer binding to the polydT in the cDNA acts as one primer leaving only one flanking primer to be designed. A similar effect may be obtained by ligating a linker sequence to the end of the cDNA, such as poly dG; a poly dC primer may be used in this case (linker-PCR). Finally, where sequence information is limited it is possible to use gene families and sequences of close similarity to design degenerate primers having either a base analog such as inosine at a particular position within the primer, or by having alternative bases at that position *(4)*. These primers require a degree of optimization of the annealing conditions since they may not directly match the sequence they are designed to bind, but are very useful where sequence information is limited. The choice of annealing temperature is critical in most PCRs since if a temperature is chosen that is very close to the melting temperatures of the primers no mismatches are likely to be tolerated. However, in some cases it may be desirable to lower the annealing temperature to allow the reaction to proceed even if there are one or two mismatches. The design of primers is critical not only for specific amplification but also to allow further post-PCR manipulations through the inclusion of restriction sites or promoter sites in primers. A number of computer programs are currently available to aid in the process of primer design and optimization of annealing temperatures. It is also possible to quantify the PCR in certain circumstances by amplification of an internal standard. This usually consists of the simultaneous amplification from a standard template in addition to the PCR being undertaken. This is especially useful for the determining the extent of viral or bacterial infections or in gene expression studies *(5)*.

2.2. Thermostable DNA Polymerases

During the initial development of the PCR the enzyme used to carry out the extension step was usually the Klenow fragment of DNA polymerase I. However, because this is heat labile, fresh enzyme was required during each cycle since the high denaturation temperatures denatured not only the template DNA but also the enzyme. This made the technique labor-intensive and quite costly. The introduction of thermostable DNA polymerases into the PCR transformed the technique and allowed full automation since only one aliquot of an enzyme needed to be added at the start of the reaction.

2.3. Taq DNA Polymerases

The first and most commonly employed thermostable DNA polymerase was that isolated from a bacterium *T. aquaticus* found in the hot springs of Yellowstone National Park. Taq and its recombinant form Amplitaq® both have relatively high processivity, a 5'–3' exonuclease activity, and a temperature optimum of 72°C *(6)*. The polymerase does lack 3'–5' proofreading exonuclease activity and does appear to contribute to misincorporation of nucleotides. A further derivative of these enzymes is the Stoffel fragment, which has a higher thermostability and is less sensitive to changes in Mg^{2+}

concentration and has no 5'–3' exonuclease activity. The higher thermal stability makes it particularly useful in amplifying GC-rich regions where high or prolonged denaturation temperatures are required.

2.4. Other Thermostable DNA Polymerases

A number of other thermostable DNA polymerases have been discovered and marketed commercially. Those isolated from *Thermococcus litoralis* (Vent™), found in deep ocean floors, are highly thermostable and capable of extending templates in excess of 12 kb pairs. They also have proofreading ability and so have a high degree of fidelity in comparison to Taq DNA polymerases. Further manipulation of Vent has resulted in derivatives having higher thermostability (Deep-Vent™) and a derivative lacking the exonuclease component (exo-). Pfu DNA polymerase isolated from a marine bacterium also has proofreading activity and incorporates radiolabeled nucleotides and analogs efficiently; it is thus particularly useful when producing radiolabeled gene probes or performing such techniques as cycle sequencing. One minor problem with these and another DNA polymerase isolated from *Thermotoga maritima* (UlTma™) is their 3'–5' exonuclease activity, which may cause modification and degradation of primers under initial suboptimal conditions. This may be overcome by using various exonuclease-deficient forms (exo-) of the enzyme.

One further interesting thermostable DNA polymerase is that isolated from *Thermus thermophilus* (Tth), which at 70°C and in the presence of Mn^{2+} is able to carry out reverse transcription reactions *(7)*. Following cDNA synthesis and chelation of Mn^{2+} the polymerase is then able to carry out polymerization of the template. This dual activity of Tth DNA polymerase allows RNA–PCR to be carried out in a single tube and obviates the need for a separate cDNA synthesis reaction.

2.5. Other Components of the PCR

One important consideration in the PCR is the $MgCl_2$ concentration. It is a critical component because it is not only required by the DNA polymerase for efficient activity but also forms a soluble complex with the dNTPs, which is essential for incorporation in the extension step of the PCR cycle. It also affects the specificity of the primer template interaction and denaturation of the double-stranded template by increasing the melting temperature. In general, insufficient Mg^{2+} results in low yields whereas in excess it gives rise to nonspecific products. The optimal Mg^{2+} concentration is usually determined by titration for each PCR and DNA polymerase and usually lies within the range 1–2 mM $MgCl_2$. One further component is the buffer/salt composition, usually 50 mM KCl and 10 mM Tris-HCl; this maintains the pH at 8.3 at room temperature. The pH of the reaction is critical and it is interesting to note that where the reaction is buffered with Tris-HCl significant changes in pH appear to be accompanied with changes in temperature, which may affect the amplification, especially of long fragments *(8)*.

The efficiency of the PCR may also be affected by factors other than the template-primer specificity and the action of the DNA polymerase. These may be present as contaminants from the extraction procedure used to prepare the template DNA. Compounds, such as heparin and porphyrins, appear to be potent inhibitors of the PCR, as are high concentrations of ionic detergents, such as SDS, proteinase K, and traces of phenol, which are all routinely used in nucleic acid extraction techniques. Some compounds are also known to enhance the PCR, although in many cases their exact mode of action is unknown. Denaturants, such as formamide, dimethylsulfoxide (DMSO), polyethylene glycol (PEG), glycerol, and DNA-binding proteins, appear to have an enhancing effect *(9)*. However, this usually has to be determined empirically for each individual amplification reaction.

A further technique that aims to increase the specificity of the PCR has been termed hot-start. This ensures separation of one or more of the important reagents of the PCR such that all reaction components are mixed after denaturation of the DNA template *(10)*. Initially this was carried out by mixing the DNA polymerase with a preheated and denatured reaction mixture. However, it is possible to separate the reaction components physically with a formulated wax bead (AmpliWax) material that melts on raising the temperature, allowing the components to mix, and on cooling solidifies. Once the wax melts it rises above the aqueous PCR reaction mix and causes mixing of the reagents, allowing all reactions to start simultaneously. PCR using the hot-start technique minimizes nonspecific annealing of primers to nontarget DNA sequences and decreases primer oligomerization. It is also useful when using degenerate primers which may misprime and cause nonspecific products. In cases where nonspecific priming occurs it may sometimes be overcome by using a first round PCR with nested primers that amplify outside the area of interest. A second round PCR using an aliquot of the first PCR reaction product together with the specific primers may then allow efficient amplification of the desired region.

3. Confirmation of PCR Products

Confirmation of the correctly sized amplified PCR product is usually carried out by agarose gel electrophoresis. It may also be possible to confirm the PCR product by analysis with restriction endonucleases. This depends on the presence of a suitable restriction site within the amplified sequence. Further confirmation that the amplified product is the expected sequence may be derived from hybridization studies, such as Southern blotting, with a labeled probe that anneals to a position internal to the amplified sequence.

The PCR has found many applications and may be used in the diagnosis of disease states or species identification. It is capable of detecting many sources of DNA with great specificity and therefore can be used to detect viral or bacterial infection. Multiplex PCR may be undertaken to analyze more than one product in the same reaction. This involves adding more than one set of specific oligonucleotide primer pairs to the reaction mix *(11)*. If the primers produce amplified products of different size their detection is greatly facilitated, requiring only simple electrophoresis and staining. This gives a considerable saving in sample DNA, time, and cost. Recently the introduction of fluorogenic probe-based PCR assays (TaqMan™) have allowed direct detection of PCR products within minutes by monitoring the increase in fluorescence of dye-labeled oligonucleotide probe *(12)*.

3.1. Contamination Problems of the PCR

Many of the advantages of PCR will already be apparent. It is an exquisitely sensitive technique, is relatively rapid, and theoretically very simple to perform. Unfortunately, however, the main advantage is also the main disadvantage, i.e., the sensitivity. Since the PCR is capable of amplifying a single copy of DNA, any degree of contamination may result in false or unwanted amplification. This makes the method extremely prone to giving false positive results unless great care is taken when setting up the reactions to avoid any possibility of contamination. The potential sources of contamination not only include positive sample DNA and positive control DNA but also previously amplified material *(12)*. These may be found on laboratory surfaces, pipets, and even in aerosols. It is recommended, therefore, that PCR reactions be set up in a designated location within the laboratory that is physically separated from DNA extraction areas. Dedicated equipment, such as pipets and microcentrifuges, are also desirable. Suitable positive and negative controls are also necessary and should always be implemented.

One useful method of preventing contamination in the PCR, results in the conversion of the PCR product into a nonamplifiable form by UV irradiation. The precise times and conditions of the irradiation depend on the thymidine content, length of DNA, and the energy of the UV radiation. Enzymatic degradation of DNA has also been undertaken with DNase or exonuclease digestion. One further method makes use of the treatment of PCR products with the enzyme uracil *N*-glycosylase following amplification with dUTP substituted for dTTP. This cleaves carryover PCR products that have dUTP incorporated so that they cannot be used as templates in further PCRs *(14)*. This is available commercially as AmpErase™.

4. Analyzing RNA by PCR (RNA–PCR)

A highly useful modification of the original PCR procedure may be used to amplify RNA species *(15)*. This requires the production of complementary DNA (cDNA) to the target RNA sequence. This is undertaken using an appropriate oligonucleotide primer, dNTPs, and the enzyme reverse transcriptase. The resulting DNA may then be used in the PCR directly (*see* **Fig. 3**). It is common when processing many samples to carry out one-tube RNA-PCR where the reverse transcription step is included in the first round of the PCR, provided the reverse transcriptase and the DNA polymerase both work efficiently in the same buffer. The following round generates another strand of DNA complementary to the cDNA that is then used as the template in the PCR.

RNA–PCR has a number of applications but is especially useful for analyzing the transcriptional activity of genes and gene isoforms. This is further enhanced if the PCR primers are designed such that they span intron sequences. This is because although standard PCR would amplify the complete DNA sequence, the RNA–PCR product will be shorter resulting from the splicing of the intron sequence in the RNA message and hence in the cDNA. Many other methods for the analysis of RNA species are detailed elsewhere; however, the PCR has a major advantage over many techniques because of its great specificity and sensitivity in being able to amplify only a single copy of target nucleic acid. This is dictated by the design of the primers and whether they will detect a single target or amplify disparate but related sequences. A useful adaptation to RNA–PCR is mRNA differential display, where short arbitrary primers are used in addition to the polydT primer having modified 3' ends *(16)*. The resulting amplification produces complex patterns of

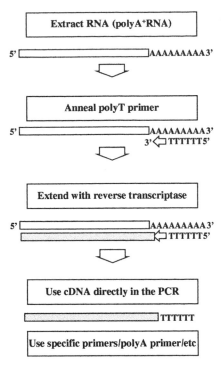

Fig. 3. Representation of RNA–PCR.

bands that are resolved on polyacrylamide gels. This method is useful in comparing the differential expression patterns of genes or subsets of genes during for example, development. Specific PCR products may be removed from the gel and analyzed further (*see* **Fig. 4**).

5. Sequencing PCR Products

One of the main benefits of the PCR has been the production of sufficient quantities of a single species of DNA. This has allowed the means of rapidly deriving nucleotide sequence information without the need for cloning into single-stranded vectors, such as M13. Alternatives for sequencing of PCR products essentially fall into one of two categories where single stranded DNA is generated specifically for sequencing or the direct sequencing of double-stranded PCR product. In general, the latter is more likely to be of greatest immediate significance because of its general applicability and rapidity. Double-stranded sequencing allows the use of the PCR product for other purposes either prior to or subsequent to generation of sequence data. The single-stranded sequencing methods generally require some prior decision regarding sequencing of the product to be made. Assisted by automated methods, sequencing of single-stranded DNA PCR products generated either during thermal cycling or following affinity capture strand separation is significant, particularly in genome mapping and routine clinical diagnosis. Despite template type and protocol differences, in all situations the purity and concentration of PCR-amplified DNA template used remains the most critical factor determining the efficiency and reliability of nucleotide sequencing methods *(17)*.

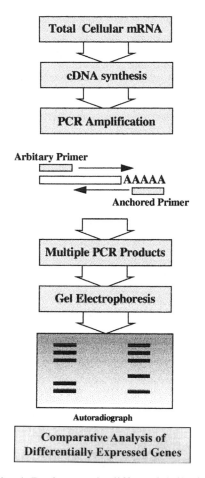

Fig. 4. Basic steps in differential display.

5.1. Direct PCR Sequencing

A number of chain termination methods are available, especially in simplified kit formats that avoid the use of toxic chemicals and are more amenable to automation given the capability of using readily available labeled terminators. All chain termination sequencing reactions demand that the DNA template be single-stranded for the priming reaction to occur. Present methods essentially differ in the timing of the generation of the single-stranded DNA. Most require further manipulations to be undertaken before a single-stranded template is produced of sufficient quality or quantity, a potential drawback in terms of simplicity, reliability, and automation. The PCR can readily be modified to the preferential production of single-stranded product by limiting the availability of one of the pair of oligonucleotide primers; this process is termed asymmetric PCR. Primer ratios of 50:1 to 100:1 are most frequently used to generate single-stranded products. During the first 15–20 cycles double-stranded DNA accumulates in an exponential fashion until limited by primer depletion *(18)*. This continues to act as a template for linear copying of single-stranded DNA primed from the unexhausted oligonucleotide. However, single-stranded products generally require further separation prior to sequencing, which may be somewhat problematic given the range of mobilities that may be observed with the use of certain electrophoresis buffers. Thermostable DNA polymerases may also be used in the sequencing reactions, allowing the reaction to be performed at around 70°C, reducing the risk of target DNA reannealing and the sequencing of templates with significant secondary structure. Modifications of the linear PCR using thermostable DNA polymerase in direct sequencing claims reduced background signals, and increased rapidity and automation. Sequencing reaction conditions must, however, be varied according to template and reaction volumes. Of particular importance is the ddNTP concentration since some polymerases, such as Taq DNA polymerase incorporate ddNTPs less efficiently than dNTPs, although this enzyme has high processivity and more readily allows the incorporation of helix-destabilizing nucleotide analogs, such as 7-deaza-2'-dGTP, permitting increased resolution of band compressions *(19)*.

A number of reagents have also been investigated in the denaturation of the double-stranded product prior to the annealing of a sequencing primer. These, such as DMSO, detergents, single-stranded binding protein, and formamide, act to promote the continued dissociation of the PCR product following the initial denaturation step *(20)*. As with all sequencing reactions the purity of the template is essential and the use of spin columns, gel electrophoretic separation, and centrifuge ultrafiltration methods have also been important in allowing recovery and concentration of the specific PCR products. The application of more sophisticated chromatographic techniques, such as HPLC, is also useful for large throughput reactions.

One currently useful method is cycle sequencing, in which double-stranded PCR products are mixed with a single primer and a thermostable DNA polymerase together with dNTPs and ddNTPs. Linear amplification is then performed for approx 20 cycles. The dideoxy-terminated extension products are then resolved on a sequencing gel. This is a universal approach and provides a long, accurate, readable sequence and has the advantage that it may be applied to unpurified samples *(21)*.

5.2. Sequencing Affinity-Captured PCR Products

Sequencing of affinity-captured products has also been successfully applied in many situations. The use of affinity-capture of single-stranded DNA for sequencing relies on the incorporation of a ligand, such as biotin, in one of the primers, generally at the 5' terminus. Following amplification the double-stranded product is passed through a column or over a matrix containing a ligand-binder, such as streptavidin, to which the biotinylated primer-derived strand binds (*see* **Fig. 5**). This allows subsequent denaturation and elution of the unbiotinylated strand with NaOH. Following this procedure it is possible to use either the eluted, unlabeled strand of DNA for sequencing or, alternatively, the affinity-matrix-bound strand *(22)*.

The affinity-capture approach to single-stranded DNA PCR product separation prior to sequencing allows flexibility and universal applicability in terms of matrices (e.g., cellulose and magnetic microspheres), ligands, and receptor molecules. A potential drawback to this and several other approaches is, however, that the original PCR reaction must not generate any nonspecific amplification products since these will also be captured, resulting in "ghost sequences" overlaying the target sequence generated. Many of these problems may be eliminated by optimization of amplification specificity or the use of a nested sequencing primer.

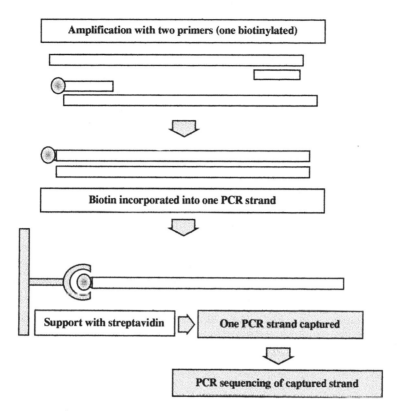

Fig. 5. Representation of the step in affinity capture.

5.3. Sequencing PCR-Derived Transcripts

A relatively simple method for generating single-stranded DNA from double-stranded DNA amplification products was originally developed to facilitate in vitro probe generation. This procedure relies on the addition of a 5' T7 promoter sequence onto one of the primers. Thus, following amplification, T7 DNA polymerase may be used in a second reaction to generate single-stranded DNA. This approach is applicable to both RNA (RAWTS) and genomic DNA (GAWTS) sequencing *(23)*. However, the complexity and expense of synthesizing new oligonucleotides has led to the limited application of this method.

6. Cloning PCR Products

Although PCR has to some extent replaced cloning as a method for the generation of large quantities of a desired DNA fragment, there is in certain circumstances still a requirement for the cloning of PCR amplified DNA *(24)*. For example, certain techniques of gene manipulation, such as gene expression, mutagenesis, and sequencing may be best achieved with the DNA fragment inserted into an appropriate cloning vector. Although PCR-based methodologies are rapidly being developed and evolving, many still remain to be fully optimized and established. In addition to the absence of information on which to base the design of PCR primers, it may be necessary to follow the more traditional approach of constructing a library of clones and screening them for the desired sequence. Screening may be achieved by genetic, functional, or structural analysis. Alternatively, PCR may be used to detect clones containing inserts of expected size by designing primers that anneal to the flanking vector sequence. A further reason for cloning PCR fragments arises from the fact that the latter may comprise a heterogeneous mixture of products. Cloning therefore enables individual products of a PCR reaction to be analyzed in isolation. Many vectors derived from plasmid or phage may be used as host vehicles for cloning products amplified by the PCR. The exact choice of vector depends on the ultimate application of the cloned product. These methods follow closely the cloning of DNA fragments derived from the conventional manipulation of DNA. This manipulation may be achieved through one of two ways, blunt-ended or cohesive-ended cloning.

6.1. Blunt-Ended Cloning of PCR Products

Blunt-ended cloning assumes that products amplified by the PCR are blunt- or flush-ended with no overhanging nucleotides. The vector to be used as the cloning vehicle is digested with a restriction endonuclease that leaves blunt ends, such as *Sma*I. The two molecules are ligated at low temperatures, usually 4°C, in the presence of a DNA ligase. The resulting recombinant construct may then be used to transform a suitable host, such as *Escherichia coli* to produce large amounts of the desired insert for further manipulation. In general, although this is the simplest method to clone a PCR fragment, blunt-ended ligation is an inefficient process, and it also allows multiple tandomly repeated inserts to be cloned and does not allow directional cloning. A further problem lies in the use of certain thermostable DNA polymerases, such as Taq DNA polymerase and Tth DNA polymerase, which tend to give rise to the PCR product having a 3' overhanging A residue and compounds the problem of low ligation efficiency. Methods have been adopted to overcome this problem and usually follow the form of producing flush ends with Klenow DNA polymerase or exonuclease digestion. This problem has, however, been exploited and has led to a useful technique of cloning using dT vectors termed dA:dT cloning. This makes use of the fact that the terminal additions of A residues may be successfully ligated to vectors prepared with T residue overhangs to allow efficient ligation of the PCR product *(25)*.

6.2. Cohesive-Ended Cloning of PCR Products

Cohesive-ended cloning is usually the method of choice when cloning DNA fragments. For PCR fragments this requires that the termini of the PCR products have appropriate restriction sites. It is possible to ligate linkers or adaptors to the ends of the PCR fragments; however, this is usually undertaken by designing oligonucleotide primers for use in the PCR with sequences recognized by specific restriction endonucleases. Since the complementarity of the primers needs to be absolute at the 3' end, the 5' end of the primer is usually the region for the restriction site. This needs to be designed with care since the efficiency of digestion with certain restriction endonuclease decreases if extra nucleotides not involved in recognition are absent at the 5' end. It is also possible by designing different restriction sites in each primer to perform directional cloning. Further methods, such as ligation-independent cloning (LIC), and the generation of half sites have made the efficient cloning of PCR fragments possible.

6.3. Templates Derived from Cloning Vectors

Cloning of PCR products into certain phage vectors, such as M13, via cohesive- or blunt-ended ligation or ligation-free methods allows the production of templates suitable for chain termination sequencing or in vitro mutagenesis and is subject to the same restraints as nonamplified fragments. Often PCR products are required for expression studies and must be cloned into a vector; therefore, it may be preferential to clone into a dual-purpose vector suitable for both sequencing and expression studies. Additional problems inherent with cloning PCR products do, however, decrease the desirability of performing cloning steps prior to sequencing. Furthermore, rigorous screening of many clones may also be required in order to avoid artifacts, such as shuffle clones, misprimed products, or misincorporation errors. Indeed, sequencing of numerous clones to gain consensus sequences was initially deemed imperative in order to take into account possible amplification artifacts, especially the introduction of base substitution errors during amplification.

Misincorporation has largely proven to be a less significant problem with the availability of certain thermostable DNA polymerases with increased fidelity or proofreading activity. There are, however, certain circumstances whereby laborious and time-consuming cloning of PCR products remains advantageous; for example, in order to discriminate between polymorphic or gene-family related products coamplified as a single-molecular-weight DNA species in a single tube using degenerate or mixed allele-specific primers.

7. Site-Directed and PCR Mutagenesis

Site-directed mutagenesis is a very powerful technique for introducing a mutation into a gene and studying its effects on the protein it encodes. There are a number of different methods of site-directed mutagenesis including cassette-mediated mutagenesis, oligonucleotide or primer extension mutagenesis, and more recently mutagenesis methods based on the PCR. Single bases mismatched between the amplification primer and the template become incorporated into the amplified sequence following thermal cycling. The basic PCR mutagenesis system involves the use of two primary PCR reactions to produce two overlapping DNA fragments, both bearing the same mutation in the overlap region. The overlap in sequence allows the fragments to hybridize, after which one of the two possible hybrids is extended to produce a duplex fragment. The other is not a substrate in the reaction (*see* **Fig. 6**). Deletions and insertions may also be created, although the requirements of four primers and three PCR reactions limits the applicability of the technique *(26)*.

7.1. Megaprimer Technique of PCR Mutagenesis

The megaprimer method utilizes three oligonucleotide primers to perform two rounds of PCR where the mutated PCR product (megaprimer) formed in the first reaction is used itself in a further PCR reaction with another oligonucleotide product. Although these are efficient procedures the PCR product would normally have to be cloned into an expression vector in order to produce a mutated protein. This is now being overcome with the advent of in vitro PCR translation systems termed expression PCR (E-PCR). A further problem is that because of the low fidelity of Taq DNA polymerase, it is usually necessary to sequence the entire amplified segment, although,

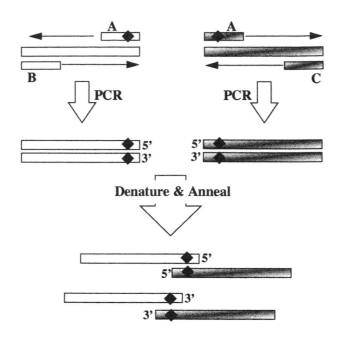

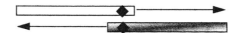

Fig. 6. Standard PCR mutagenesis protocol.

again, the availability of high fidelity thermostable polymerases may obviate this requirement.

7.2. Sticky-Feet Directed PCR Mutagenesis

The modularity of protein structures lends them conveniently to the domain swapping technique of sticky-feet directed mutagenesis. This involves production of PCR primers that anneal to a gene fragment that is to be replaced. PCR of a desired fragment with these primers produces an amplified fragment where part of the primer (sticky feet) may be used to anneal to the gene to be replaced that is cloned into a single-stranded vector. The efficiency of the selection may be enhanced by carrying out the primer extension and polymerization in the presence of thionucleotides and then destroying the parental strand *(28)*.

8. Detecting Polymorphisms by PCR

A number of methods are available to determine the presence of a specific sequence at a specified position in a region of DNA *(29)*. These may be used to detect a disease-causing mutation or for genotyping where determining which of a number of possible alternative sequence polymorphisms are present at a given position in a particular individual. Changes involving one or a few nucleotides sometimes destroy or create a restriction enzyme site, which can be detected by the resulting difference in fragment

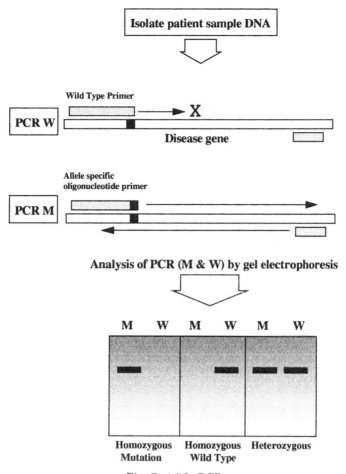

Fig. 7. ASO–PCR.

pattern following digestion with the appropriate restriction enzyme. Differences in restriction fragment patterns between individuals may also be produced by cleavage at restriction sites that flank variable numbers of tandem repeats, or VNTRs. These are repeat sequences whose number at a particular position varies between individuals. Previously, radioactive hybridization methods were required for visualization of fragments, but PCR enables the generation of sufficient specific DNA for observation by ethidium bromide staining of bands on agarose gels.

8.1. Detecting Mutations by PCR

Other methods to detect characterized nucleotide changes are based on PCRs utilizing allele-specific oligonucleotides, or ASOs. These are a pair or series of oligonucleotides, each designed to be a perfect match for one of the alternative sequences that may be present at a particular site. The primer that is the direct complement to the template will give rise to amplification. However, if there is a 3' mismatch no amplification takes place and so may be used as an effective genotyping system (*see* **Fig. 7**). This provides the basis for a commonly used diagnostic technique, the ampli-

fication refractory mutation system system (ARMS) *(30)*. Competitive oligonucleotide priming (COP-PCR) has further simplified these strategies by using differential fluorescent labeling of the primers, obviating the need for an electrophoresis
step. PCR amplification of the target prior to analysis has also overcome the previously encountered problems of high background signals and the requirement for large
amounts of starting material when using traditional methods, such as Southern blotting. Sequence deletions large enough to affect the mobility of a segment of DNA
through a gel may also be detected by performing PCR using primers that flank the
deletion site. This approach has been used to detect mutations in the dystrophin gene
that results in muscular dystrophy. Many regions of the dystrophin gene are susceptible to deletions, many of which may be analyzed simultaneously in a single multiplex PCR reaction.

8.2. Gel-Based PCR Mutation Detection Systems

For mutations involving only one or a few bases, methods are available that
depend on enzymatic or chemical activities, or on the melting properties of duplex
DNA. In all three approaches, heteroduplexes are initially formed in which one
strand derives from the test sequence, and the other from a wild-type normal sequence.
These are then either treated with enzymes, such as RNase A, or chemicals, e.g.,
hydroxylamine or osmium tetroxide followed by piperidine, that preferentially
cleave heteroduplexes at sites of mismatch. They may also be analyzed by denaturing gradient gel electrophoresis (DGGE), in which a mutation containing heteroduplexes is identified by its altered melting properties relative to a perfectly matched
homoduplex on passage through a denaturing gradient gel (*see* **Fig. 8**). Other methods, such as temperature-gradient gel electrophoresis (TGGE) or single-stranded
conformational polymorphism (SSCP), are also widely used mutation-detection
techniques *(31,32)*. The sensitivity of all these methods has been substantially
increased by the use of PCR to amplify both the test DNA and the normal sequence
used to form the heteroduplex. The addition of restriction sites to the PCR primers
also enables subsequent cloning of the fragments in order to characterize the nature
of the detected mutation. Sequencing of the complete region under investigation is
also able to detect all base changes, and although time-consuming at present, rapid
advances in sequencing technologies may make this the future method of choice
for detecting mutations.

8.3. Investigating Unknown DNA Regions by PCR

Frequently it is useful to characterize stretches of DNA flanking a known region of
DNA. A number of techniques have been developed that allow this to take place.
Inverse PCR (I-PCR) uses primers that are divergent or have 3' termini that point away
from each other, as opposed to the standard PCR (*see* **Fig. 9**). The DNA is digested by
restriction enzymes and circularized using low-temperature ligation. PCR is then performed in which the primers are used to amplify the unknown region. This has been
further enhanced by long range I-PCR, in which large DNA fragments have been
amplified and then analyzed using I-PCR *(33)*. An alternative method, termed
vectorette PCR, also allows the characterization of DNA fragments that flank a
known region and has found particular application for amplifying exon/intron bound-

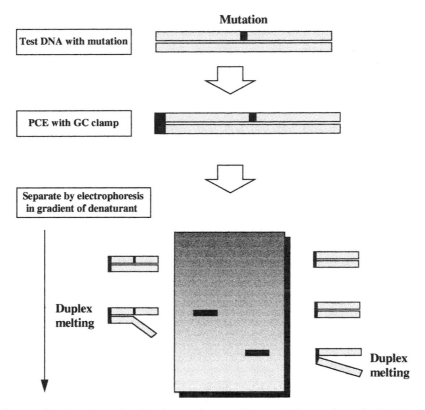

Fig. 8. Mutation detection by denaturing gradient gel electrophoresis (DGGE).

aries and isolating the ends of gene libraries constructed in yeast artificial chromosomes (YACs).

9. Recent PCR Developments

Numerous development and modifications have allowed the adaptation of the PCR to the amplification, detection, and characterization of specific DNA sequences. Recent methods have focused on the amplification of gene regions from single cells (single-cell PCR) identified by microscopy or by fluorescence-activated cell sorting (FACS). This has allowed detailed analysis of cellular development, such as B cell subset differentiation and detection of single-copy virus-infected cells. Primed *in situ* hybridization (PRINS) has also allowed the visualization of amplified DNA fragments on microscope slides. This allows the identification of specific fragments *in situ* using appropriate labeling strategies and counterstaining techniques *(34)*.

10. Alternative Nucleic Acid Amplification Techniques

A number of alternative methodologies have been developed for the amplification of nucleic acids. In general, they rely on the amplification of target sequences although there are exceptions, for example, when the signal is amplified, which makes them suitable for quantitation *(35)*. Primer based techniques are however useful in that they may amplify more than one target in a single reaction. They also tend to fall into two broad categories depending on thermal or nonthermal amplification. One of the most

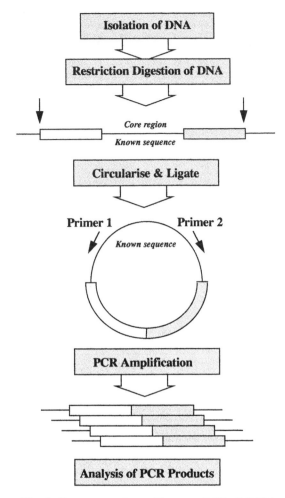

Fig. 9. Representation of inverse-PCR (I-PCR).

common alternatives, used extensively for identification of chlamydia, is the LCR or ligase chain reaction (*see* **Fig. 10**). This is a technique requiring thermal cycling and requires a set of four oligonucleotides complementary to a target sequence containing a mutation. Two pairs of oligos on each sequence bind with a gap of one nucleotide that is sealed by a thermostable DNA ligase. This is then used as further substrate following denaturation. Any sequence containing a mutation will not allow the 3' end of the oligo to bind, so no ligation takes place (*see* **Fig. 9**).

A number of nonthermal cycling techniques have also been devised, such as the Q-β replicase system, 3SR or self-sustained sequence replication, and bDNA or branched DNA amplification. bDNA is an interesting signal amplification method in which the probe for a target sequence includes a special branched nucleotide. If the target is present the probe attaches, grows, and is ultimately detected by a chemiluminescent end point.

There is no doubt that the impact of the PCR on molecular biology has been profound and there are and will be numerous future adaptations to the technique. However, the new techniques of thermal and nonthermal amplification will also no doubt

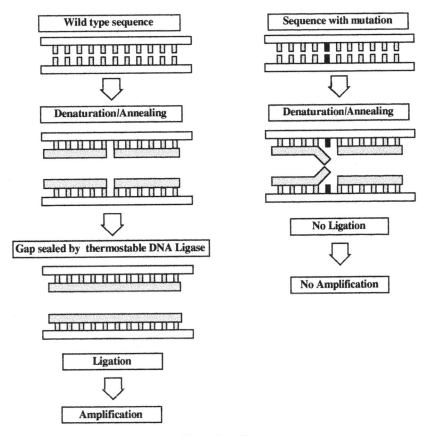

Fig. 10. LCR.

play a large part in the future of biochemical research and routine clinical genetic applications.

References

1. Saiki, R. K., Scharf, S. J., Faloona, F., Mullis, K. B., Horn, G. T., Erlich, H. A., and Arnheim, N. (1985) Enzymatic amplification of β globin genomic sequences and restriction site analysis for diagnosis of sickle cell anemia. *Science* **230,** 1350–1354.
2. Paabo, S. (1989) Ancient DNA extraction characterisation molecular cloning and amplification. *Proc. Natl. Acad. Sci. USA* **86,** 1939–1943.
3. Rychlik, W. (1994) New algorithm for determining primer efficiency in PCR and sequencing. *J. NIH Res.* **6,** 78.
4. Knoth, K., Roberds, S., Poteet, C., and Tamkun, M. (1988) Highly degenerate inosine containing primers specifically amplify rare cDNA using the polymerase chain reaction. *Nucleic Acids Res.* **16,** 932.
5. Reischl, U. and Kochanowski, B. (1995) Quantitative PCR A survey of the present technology. *Mol. Biotechnol.* **3,** 55–71.
6. Eckert, K. A. and Kunkel, T. A. (1990) High fidelity DNA synthesis by the *Thermus aquaticus* DNA polymerase. *Nucleic Acids Res.* **18,** 3739–3744.
7. Myers, T. W. and Gelfand, D. H. (1991) *Biochemistry* **30,** 7661.
8. Cheng, S., Cheng, S-Y., Gravitt, P., and Respess, R. (1994) Long PCR. *Nature* **369,** 684–685.

9. Hung, T., Mak, K., and Fong, K. (1990) A specificity enhancer for the polymerase chain reaction. *Nucleic Acids Res.* **18,** 1666.

10. Chou, Q., Russell, M., Birch, D. E., Raymond, J. and Bloch, W. (1992) Prevention of pre-PCR mispriming and primer dimerization improves low copy number amplification. *Nucleic Acids Res.* **20,** 1717–1723.

11. Chamberlain, J. S., Gibbs, R. A., Ranier, J. E., Nguyen, P. N., and Caskey, C. T. (1988) Deletion screening of the Duchenne muscular dystrophy locus via multiplex DNA amplification. *Nucleic Acids Res.* **16,** 11,141–11,156.

12. Marmo, J., Wyatt, P., Flood, S. J., and McBride, L. (1997) A TaqMan multiplex PCR mesenger RNA assay. *Clin. Chem.*

13. Kwok, S. and Higuchi, R. (1989) Avoiding false positives with PCR. *Nature* **339,** 237–238.

14. Longo, N., Berninger, N. S., and Hartley, J. L. (1990) Use of uracil N-glycolyase to control carry-over contamination in polymerase chain reaction. *Gene* **93,** 125–128.

15. Kawasaki, E. S. (1990) Amplification of RNA, in *PCR Protocols A Guide to Methods & Applications* (Innis, M. A., Gelfand, D., Sninsky, J. J., and White, T. J., eds.), Academic, San Diego, CA, pp. 21–27.

16. Liang, P. and Pardee, A. B. (1992) Differenyial display of eukaryotic messenger RNA by means of the polymerase chain reaction. *Science* **257,** 967–971.

17. Bevan, I. S., Rapley, R., and Walker, M. R. (1992) Sequencing of PCR-amplified DNA. *PCR Methods Appl.* **1 (4),** 222–227.

18. Gyllenstein, U. B. and Erlich, H. A. (1988) Generation of single stranded DNA by the polymerase chain reaction and its application to direct sequencing of the HLA-DQA locus. *Proc. Natl. Acad. Sci. USA* **85,** 7652–7656.

19. McConlogue, L., Brow, M. A. D., and Innis, M. A. (1988) Structure-independent DNA amplification by PCR using 7-deaza-2'deoxyguanosine. *Nucleic Acids Res.* **16,** 9869.

20. Rapley, R. (1996) *Methods in Molecular Biology, vol. 65: PCR Sequencing Protocols.* Humana, Totowa, NJ.

21. Slatko, B. E. (1996) Thermal cycle dideoxy DNA sequencing. *Mol. Biotech.* **6,** 311–322.

22. Mitchell, L. G. and Merril, C. R. (1989) Affinity generation of single stranded DNA for dideoxy sequencing following the polymerase chain reaction. *Anal. Biochem.* **178,** 239–242.

23. Stoflet, E. S., Koeberl, D. D., Sarker, G., and Sommer, S. S. (1988) Genomic amplification with transcript sequencing. *Science* **239,** 491–494.

24. White, B. A., ed. (1996) *Methods in Molecular Biology, vol. 67: PCR Cloning Protocols: From Molecular Cloning to Genetic Engineering.* Humana, Totowa, NJ.

25. Mead, D. A., Pey, N. K., Herrnstadt, C., Marcil, R. A., and Smith, L. M. (1991) A universal method for the direct cloning of PCR amplified nucleic acid. *Bio/Technology* **9,** 657–663.

26. Higuchi, R., Krummel, B., and Saiki, R. K. (1988) A general method of in vitro preparation and specific mutagenesis of DNA fragments: study of protein and DNA interactions. *Nucleic Acids Res.* **16,** 7351–7367.

27. Sarker, G. and Sommer, S. S. (1990) The megaprimer method of site directed mutagenesis. *Biotechniques* **8,** 404–407.

28. Clackson, T. and Winter, G. (1989) Sticky feet directed mutagenesis and its applications to swapping antibody domains. *Nucleic Acids Res.* **17,** 10,163–10,170.

29. Cotton, R. G. H. (1992) Detection of mutations in DNA. *Current Opin. Biotechnol.* **3,** 24–30.

30. Newton, C. R., Graham, A., Heptinstall, L. E., Powell, S. J., Summers, C., Kalsheker, N., Smith, J. C., and Markham, A. F. (1989) Analysis of any point mutation in DNA. The amplification refractory mutation system. *Nucleic Acids Res.* **17,** 2503–2516.

31. Wartell, R. M., Hosseini, S. H., and Moran, C. P. Jr. (1990) Detecting base pair substitutions in DNA fragments by temperature gradient gel electrophoresis. *Nucleic Acids Res.* **18**, 2699–2705.

32. Dockhorn-Dworniczak, B., Dworniczak, B., Brommelkamp, L., Bulles, J., Horst, J., and Bocker, W. (1991) Non-isotopic detection of single strand conformational polymorphism (PCR-SSCP): a rapid and sensitive technique in the diagnosis of phenylketonuria. *Nucleic Acids Res.* **19**, 2500–2505.

33 Ochman, H., Gerber, S. A., and Hartl, D. L. (1988) Genetic applications of an inverse polymerase chain reaction. *Genetics* **120**, 621–625.

34. Gosden, J., ed. (1997) *Methods in Molecular Biology, vol. 71: PRINS and In Situ PCR Protocols.* Humana, Totowa, NJ.

35. Winn-Deen, E. S. (1997) Automation of molecular genetic methods: DNA amplification techniques. *J. Clin. Ligand Assay* **19**, 21–26.

26

In Vitro Transcription

Martin J. Tymms

1. Introduction

The process of transcription in which a DNA template is used to derive RNA copies is a fundamental process in prokaryotic and eukaryotic cells. The recapitulation of these processes in cell-free systems (in vitro transcription) has led to a number of powerful techniques for studying gene structure and regulation. The transcription machinery of prokaryotes is much simpler than that of eukaryotes, with a single RNA-polymerase enzyme composed of a number of polypeptide chains responsible for RNA synthesis. Eukaryotic transcription is more complex, with three types of RNA polymerase involved. RNA-polymerase II holoenzyme, which transcribes genes that are translated into proteins, is a complex involving a large number of proteins in addition to core RNA-polymerase II enzyme. In addition to the holoenzyme, a complex array of proteins is also involved in controlling the rate of initiation of transcription, and the regulation of a gene may involve a suite of transcription factors binding to the promoter and enhancer sequences. The structure of the eukaryotic-transcription machinery and its regulation is being explored using in vitro transcription systems, and a range of different approaches are being used. A large effort has been put into identifying, cloning, and purifying components of the transcriptional complex and the reconstitution of transcription. Other studies use nuclear extracts to study the regulation of specific genes. In addition to nuclear transcription, eukaryotes also have transcriptional machinery in organelles. The transcriptional machinery of chloroplasts and mitochondria is similar to prokaryotic organisms, and in vitro transcription systems have been developed to study these systems *(1,2)*. This chapter will deal with in vitro transcription by eukaryotic polymerase II (class II transcription) and the in vitro generation of RNA for study of eukaryotic transcription.

2. Generation of RNA Using Bacteriophage RNA Polymerases

Bacteriophage-RNA polymerases are very robust, and the availability of purified recombinant proteins allows the easy in vitro generation of large quantities of RNA from cloned cDNA sequences (**Fig. 1**). Eukaryotic in vitro transcription systems yield relatively small quantities of RNA; these systems are used primarily for the study of transcriptional mechanisms. Bacteriophage-RNA polymerases, such as those obtained

From: *Molecular Biomethods Handbook*
Edited by: R. Rapley and J. M. Walker © Humana Press Inc., Totowa, NJ

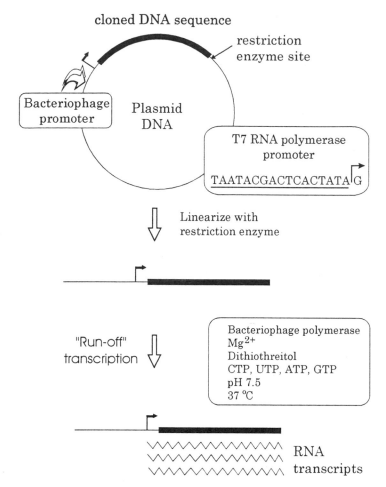

Fig 1. "Run off" in vitro transcription using bacteriophage RNA polymerases and plasmid vectors containing bacteriophage promoters.

from bacteriophages SP6, T7, and T3, are especially suitable for the generation of RNA from DNA sequences cloned downstream of their specific promoters because, first, their promoters are small and easily incorporated into plasmid vectors and second, the polymerases are quite specific for their promoters, which results in very little incorrect initiation from DNA templates *(3)*. Strong termination sequences are not available for these polymerases so that DNA templates are generally linearized with a restriction enzyme 3' to the desired end of the RNA transcript and the polymerase is forced to stop at this point. This process is referred to as "run-off" transcription.

3. Applications

3.1. Synthesis of mRNA for In Vitro Translation

Cloned cDNA sequences can be conveniently transcribed by bacteriophage RNA polymerases and translated in vitro using extracts from rabbit reticulocytes *(4)* or wheat germ *(5)*. In vitro translation is covered in Chapter 28 of this volume.

3.2. Antisense RNA Probes

The detection of complementary DNA or RNA sequences is at the heart of a number of key techniques in molecular biology, such as Southern blots, Northern blots, RNase protections, and *in situ* hybridization. In vitro transcription with bacteriophage-RNA polymerases allows the synthesis of a range of different RNA probes with different detection systems, because radiolabeled or chemically modified nucleotides can be substituted for UTP, CTP, GTP, and ATP in the transcription reaction. Radiolabeled probes can be made using a variety of isotopes, including tritium, carbon-14, phosphorus-33, and phosphorus-32. RNA probes can also be made using modified nucleotides containing biotin or digoxigenin, which allow the use of nonradioactive detection systems (*see* Chapter 7) *(3)*.

3.3. Study of RNA Processing

The primary transcripts from eukaryotic polymerase II transcription are spliced in the nucleus to remove introns. The mature mRNA produced can differ owing to variations in the splicing process, which result in some exons being "skipped." The regulation of this process is of considerable biological interest because alternatively spliced mRNAs can result in proteins with altered properties. The study of eukaryotic RNA splicing in vitro makes use of RNA generated from bacteriophage polymerases. Genomic DNA containing exons and introns is cloned in plasmid vectors containing a bacteriophage-RNA polymerase promoter. The processing of RNA transcripts generated in vitro is then studied using nuclear fractions capable of splicing activity. These studies usually make use of radiolabeled RNA, which allows the sensitive detection of less-abundant splice products *(6)*.

4. In Vitro Class II Transcription
4.1. Reconstituted Systems Using Purified Components

The initiation of mRNA synthesis in eukaryotes involves the recruitment of RNA polymerase II to the TFIID protein complex near the site of transcriptional initiation. TFIID consists of the TATA-binding protein (TBP) and a set of tightly bound subunits, the TBP-associated factors (TAFs). TBP alone can direct basal levels of transcription in vitro, but requires TAF subunits to mediate activator-dependent transcription. TFIID complexes have been assembled from recombinant subunits that support activation from transcription-factor activators *(7)*. The availability of purified components of the RNA-polymerase II transcriptional machinery is leading to the elaboration of transcription mechanisms and their regulation.

4.2. Transcription Using Nuclear Extracts

Nuclear extracts prepared by appropriate methods contain all of the components for transcription *(8)*. The source of the nuclear extract will determine the type and quantity of transcription factors present, which will affect the level of transcription obtained with a particular DNA template. For example, nuclear extracts from the fly *Drosophila (9)* accurately initiate transcription from many human promoters, although they lack a number of transcription factors that are present in nuclear extracts made from the human cell line HeLa. Consequently, human-promoter elements that are tested in *Drosophila* extracts may have lower activities compared with HeLa extracts. Although nuclear extracts are not defined like transcription systems utilizing purified-transcription components, the extracts can be used to study aspects of promoter regulation. The rate of gene transcription is regulated by activators and repressors that bind to the promoter and/or enhancer/silencer sequences. The regulation of transcriptional activators and repressors can be tissue-specific and developmentally programmed. If suitable nuclear extracts can be prepared that contain the necessary transcription factors and suitable DNA templates containing regulator sequences are available, transcription of specific genes can be reconstituted in vitro. In some cases, the relative levels of transcription observed in vitro parallel in vivo activities seen in the tissues and cell lines from which nuclear extracts were made. In many cases, however, faithful in vitro transcription cannot be reconstituted, which reflects transcriptional rates in vivo. This may be caused by a variety of reasons, including poor recovery of active transcription factors and shortcomings in the DNA template. For example, many genes have regulatory sequences at considerable distances from the site of transcriptional initiation and the templates for in vitro transcription may not include these elements. A serious problem with in vitro transcriptions that is not easily resolved is the absence of the structure found in vivo. The DNA in a eukaryotic nucleus is packaged into a nucleosome array and the precise positioning of nucleosomes along the promoter region is critical for appropriate regulation of gene expression *(10)*. Some inducible genes are accessible to transcription factors without chromatin remodeling, but others require remodeling before DNA is accessible *(11)* (also reviewed in **ref. *12***). Nevertheless, nuclear extracts can be made from tissues and cells, which to some degree recapitulate the differential gene expression observed in vivo *(13)*. Attempts have been made to reconstitute chromatin active for polymerase II transcription using crude cellular extracts *(14,15)*. Progress in this area will be important for the wider application of in vitro systems to the study of gene regulation. Some success has been achieved in reconstituting transcriptionally active 5S rDNA gene chromatin in vitro because this class III system is relatively simple and well characterized *(16)*.

In vitro transcription reactions using nuclear extracts can be experimentally manipulated to examine the role of specific transcription factors and the promoter elements to which they bind on transcriptional activity (**Fig. 2**).

Antibodies to transcription factors can be used to dissect the role of specific transcription factors in in vitro transcription driven by nuclear extracts. For example, an antibody that binds to a transcription factor and prevents binding to DNA can be used to assess the role of that factor in the overall transcriptional activity. Alternatively, specific antibodies can be used to remove transcription factors from nuclear extracts prior to the transcription reaction. If antibodies are not available, alternative strategies can be used *(8)*. Since transcription factors bind in a sequence-specific manner, the addition of DNA sequences that will bind a factor results in competition between the transcription template and the added DNA, which usually consists of a molar excess of short double-stranded oligonucleotides 18–30 bp in length. This type of analysis can establish the importance of the promoter element by effectively stopping the protein(s) binding the element from interacting with the transcriptional template.

In vitro transcription systems allow the addition of a wide range of exogenous proteins to study transcriptional regulation. Recombinant-transcription factors or purified

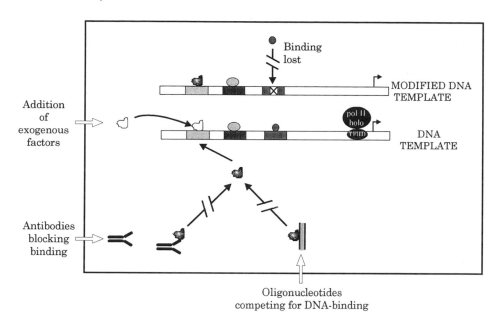

Fig. 2. Experimental manipulation of in vitro transcription using nuclear extracts and plasmid DNA templates.

factors can be added to in vitro transcription reactions. This approach can be useful when the nuclear extracts have been depleted of endogenous transcription factors. For example, a nuclear extract depleted of AP1 transcription factors can be used to study the transcriptional activation of different combinations of recombinant AP1 transcription factors *(17)*. Furthermore, this approach can be used to study the activity of modified proteins and explore structure–function relationships *(18,19)*.

In addition to manipulations of the protein components involved in regulating transcription, the DNA template can also be modified. This can include testing different promoter deletions or making smaller deletions or mutations. This analysis can be used to establish the importance of regulatory sequences to the level of transcription.

5. Quantifying In Vitro Transcription

Application of methodologies appropriate for the type of in vitro transcription performed can be critical for detection of a specific signal. In vitro transcription systems using bacteriophage polymerases generate microgram quantities of RNA, which can be easily visualized on agarose or polyacrylamide gels (**Fig. 3A**). Alternatively, radiolabeled nucleotides can be incorporated into the transcription mix and the radiolabeled RNA product visualized on X-ray film or imaging screens (**Fig. 3A**). Class II (DNA polymerase II) in vitro transcription results in very small quantities of transcript. When class II transcription is reconstituted with highly purified components that do not have contaminating DNA, the addition of radiolabeled nucleotides enables the direct detection of transcripts (**Fig. 3B**). However, when nuclear extracts or components with DNA contamination are present, spurious transcription occurs from contaminating DNA, which can lead to a significant background incorporation that can mask specific transcription products from the added DNA template. In these cases, a number of method-

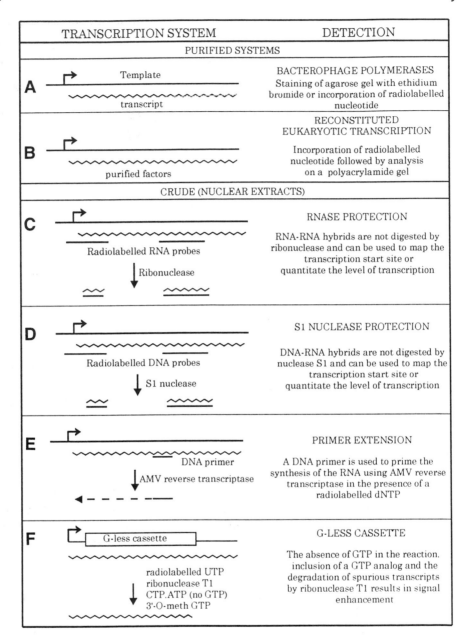

Fig. 3. Detection and quantitation of in vitro transcription.

ologies have been adopted. The transcription reactions can be carried out without radiolabeled nucleotides, and the subsequent detection of specific transcripts carried out in using specific protocols. The RNase protection *(20)* and S1 nuclease-protection protocols *(21)* use radiolabeled antisense RNA and DNA probes, respectively, to detect specific RNA transcripts generated from the transcription template (**Fig. 3C,D**). Furthermore, the use of probes spanning the start of transcription also allows transcriptional start sites to be accurately mapped. Another protocol commonly employed is primer-extension analysis (**Fig. 3E**), in which a transcript-specific DNA primer and reverse transcriptase are used to make a radiolabeled cDNA copy of transcripts up to the site of initiation *(22)*. This technique is the method of first choice for mapping transcriptional-initiation sites. One of the best techniques for detecting specific transcription in crude transcription systems is the G-less cassette method *(13,23)* (**Fig. 3F**). This technique allows the detection of specific transcripts directly in the transcription reaction without a separate "detection step." The key to this ingenious method is reducing background transcription through the use of special plasmid vectors with a "reporter" DNA sequence that does not contain G residues in the noncoding, or mRNA strand. Promoter elements are cloned into these vectors with no G residues between the start of transcription and the G-less cassette (a synthetic DNA sequence that contains no G residues in the noncoding strand). The technique relies on three modifications to normal transcription reactions, which reduces nonspecific transcripts:

1. No GTP is added to the reaction mix: Correctly initiated transcripts from the added plasmid template will not contain any G residues.
2. GTP analog 3'-*O*-methyl GTP is added: This analog, when incorporated in the place of GTP, causes termination of transcription. Because the correctly initiated transcripts do not have G residues, the analog will only terminate spurious transcripts.
3. As the final coup, ribonuclease T1 is added to the reaction: Ribonuclease T1 cleaves RNA adjacent to G nucleotides and will degrade RNA that contains G residues. Because the transcripts from the G-less vector template do not contain G residues, they are not degraded.

The combination of all three of these measures results in clear transcription signals in crude nuclear extracts containing high levels of nuclear DNA contamination.

References

1. Tiller, K. and Link, G. (1995) Plastid in vitro transcription, in *Methods in Molecular Biology, vol. 37: In Vitro Transcription and Translation Protocols* (Tymms, M. J., ed.), Humana, Totowa, NJ, pp. 121–146.
2. Kruse, B., Murdter, N. N., and Attardi, G. (1995) Transcription system using a HeLa cell mitochondrial lysate, in *Methods in Molecular Biology, vol. 37: In Vitro Transcription and Translation Protocols* (Tymms, M. J., ed.), Humana, Totowa, NJ, pp. 179–197.
3. Schenborn, E. T. (1995) Transcription in vitro using bacteriophage RNA polymerases, in *Methods in Molecular Biology, vol. 37: In Vitro Transcription and Translation Protocols* (Tymms, M. J., ed.), Humana, Totowa, NJ, pp. 1–12.
4. Beckler, G. S., Thompson, D., and Van Oosbree, T. (1995) In vitro translation using rabbit reticulocyte lysate, in *Methods in Molecular Biology, vol. 37: In Vitro Transcription and Translation Protocols* (Tymms, M. J., ed.), Humana, Totowa, NJ, pp. 215–232.
5. Van Herwynen, J. F. and Beckler, G. S. (1995) Translation using a wheat-germ extract, in *Methods in Molecular Biology, vol. 37: In Vitro Transcription and Translation Protocols* (Tymms, M. J., ed.), Humana, Totowa, NJ, pp. 245–251.
6. Johansen, S. and Vogt, V. M. (1994) An intron in the nuclear ribosomal DNA of Didymium iridis codes for a group I ribozyme and a novel ribozyme that cooperate in self-splicing. *Cell* **76,** 725–734.
7. Chen, J. L., Attardi, L. D., Verrijzer, C. P., Yokomori, K., and Tjian, R. (1994) Assembly of recombinant TFIID reveals differential coactivator requirements for distinct transcriptional activators. *Cell* **79,** 93–105.
8. Pugh, B. F. (1995) Preparation of HeLa nuclear extracts, in *Methods in Molecular Biology, vol. 37: In Vitro Transcription and Translation Protocols* (Tymms, M. J., ed.), Humana, Totowa, NJ, pp. 349–357.

9. Kadonaga, J. T. (1990) Assembly and disassembly of the *Drosophila* RNA polymerase II complex during transcription. *J. Biol. Chem.* **265,** 2624–2631.

10. Wallrath, L. L., Lu, Q., Granok, H., and Elgin, S. C. R. (1994) Architectural variations of inducible eukaryotic promoters: preset and remodeling chromatin structures. *Bioessays* **16,** 165–170.

11. Archer, T. K., Cordingley, M. G., Wolford, R. G., and Hager, G. L. (1991) Transcription factor access is mediated by accurately positioned nucleosomes on the mouse mammary tumor virus promoter. *Mol. Cell Biol.* **11,** 688–698.

12. Wolffe, A. P. (1994) Transcription: in tune with the histones. *Cell* **77,** 13–16.

13. Bagchi, M. K., Tsai, S. Y., and Tsai, M.-J. (1995) In vitro reconstitution of progesterone-dependent RNA transcription in nuclear extracts of human breast carcinoma cells, in *Methods in Molecular Biology, vol. 37: In Vitro Transcription and Translation Protocols* (Tymms, M. J., ed.), Humana, Totowa, NJ, pp. 107–120.

14. Matsui, T. (1987) Transcription of adenovirus 2 major late and peptide IX genes under conditions of *in vitro* nucleosome assembly. *Mol. Cell. Biol.* **7,** 1401–1408.

15. Workman, J. L. and Roeder, R. G. (1987) Binding of transcription factor TFIID to the major late promoter during *in vitro* nucleosome assembly potentiates subsequent initiation by polymerase II. *Cell* **51,** 613–622.

16. Felts, S. J. (1995) Assembly and transcription of chromatin templates using RNA polymerase III, in *Methods in Molecular Biology, vol. 37: In Vitro Transcription and Translation Protocols* (Tymms, M. J., ed.), Humana, Totowa, NJ, pp. 47–66.

17. Abate, C., Luk, D., and Curran, T. (1991) Transcriptional regulation by Fos and Jun in vitro: interaction among multiple activator and regulatory domains. *Mol. Cell. Biol.* **11,** 3624–3632.

18. Goldberg, Y., Treier, M., Ghysdael, J., and Bohmann, D. (1994) Repression of AP-1-stimulated transcription by c-Ets-1. *J. Biol. Chem.* **269,** 16,566–16,576.

19. Song, C. Z., Tierney, C. J., Loewenstein, P. M., Pusztai, R., Symington, J. S., Tang, Q. Q., Toth, K., Nishikawa, A., Bayley, S. T., and Green, M. (1995) Transcriptional repression by human adenovirus E1A N terminus/conserved domain 1 polypeptides in vivo and in vitro in the absence of protein synthesis. *J. Biol. Chem.* **270,** 23,263–23,267.

20. Tymms, M. J. (1995) Quantitative measurement of mRNA using the RNase protection assay, in *Methods in Molecular Biology, vol. 37: In Vitro Transcription and Translation Protocols* (Tymms, M. J., ed.), Humana, Totowa, NJ, pp. 31–46.

21. Viville, S. and Mantovani, R. (1994) S1 mapping using single-stranded DNA probes, in *Methods in Molecular Biology, vol. 37:Protocols for Gene Analysis* (Harwood, A. J., ed.), Humana, Totowa, NJ, pp. 299–305.

22. Simpson, C. G. and Brown, J. W. (1995) Primer extension assay, in *Methods in Molecular Biology, vol. 49: Plant Gene Transfer and Expression Protocols* (Jones, H., ed.), Totowa, NJ, Humana, pp. 249–256.

23. Sawadogo, M. and Roeder, R. G. (1985) Factors involved in specific transcription by human RNA polymerase II: analysis by a rapid and quantitative in vitro assay. *Proc. Natl. Acad. Sci. USA* **82,** 4394–4398.

27

In Vitro Translation

Martin J. Tymms

1. Introduction

Transcription is a fundamental step in the process of gene expression where the information encoded by messenger RNA (mRNA) is translated into a polypeptide sequence. This process requires ribosomes, an array of translation factors, a supply of transfer-RNAs loaded with amino acids, and an energy supply. Much of the early discovery and elucidation of the fundamental process of translation was carried out using extracts from the prokaryotic bacteria *Escherichia coli*, which are capable of supporting the process of protein synthesis. Extracts from special strains of *E. coli* are still used today for the in vitro translation of prokaryotic genes and to a much lesser extent eukaryotic genes. Eukaryotic genes need to be modified before they can be translated in prokaryotic translation systems as a consequence of fundamental differences in the ribosome translation systems of eukaryotes and prokaryotes (**Fig. 1**). In prokaryotes, transcription and translation are a coupled process with the ribosome recognizing a site just 5' of the initiation AUG codon (the ribosome-binding site, RBS) in the RNA transcript before the process of transcription is complete. Eukaryotic mRNAs need to be modified to include a RBS in order to be translated in a prokaryotic translation system. In eukaryotes transcription takes place in the nucleus, where the primary transcript, which usually contains introns, is processed to yield a mature mRNA that contains a 7-methyl guanosine cap at the 5' end and in most cases a tail of 100–200 adenosine nucleotides at the 3' end. The mature mRNA moves to the cytoplasm, where it is recognized by ribosomes by features in the 5' end of the mRNA that differ from the prokaryotic RBS. The eukaryotic ribosome recognizes the 7-methyl CAP, and sequences around the AUG initiation codon are crucial for correct initiation. The 7-methyl CAP is crucial for efficient translation in vivo, but mRNA without a cap can be translated in eukaryotic cell-free translation systems with a reduced efficiency. Many RNA viruses have overcome the need for a 7-methyl CAP for efficient translation by incorporating sequences in the 5' UTR that allow efficient initiation of translation. When these sequences are attached to other nonviral RNAs the need for capping is removed *(1)*.

From: *Molecular Biomethods Handbook*
Edited by: R. Rapley and J. M. Walker © Humana Press Inc., Totowa, NJ

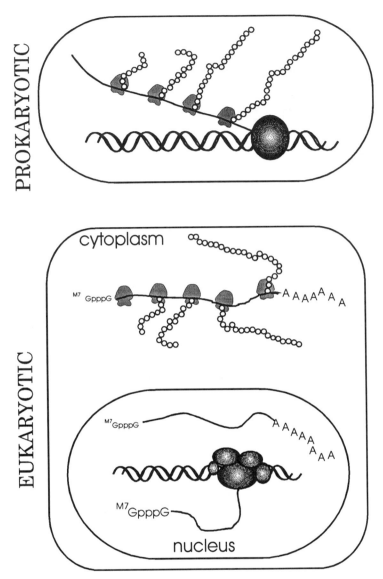

Fig. 1. Schematic representation of translation in eukaryotic and prokaryotic organisms. In prokaryotic organisms transcription and translation are a coupled process. In eukaryotes transcription and translation are independent processes in the nucleus and cytoplasm respectively.

2. Translation Systems

2.1. Coupled Transcription–Translation in Prokaryotic Extracts

E. coli extracts have been used for many years for the study of translation and have been important in the elucidation of basic mechanisms. The *E. coli*-based system has been improved from its inception through the genetic improvement of the *E. coli* strains used to make the extracts *(2)*. Translation is achieved in extracts from the prokaryote *E. coli* as a coupled transcription–translation rather than by a direct translation of mRNA, which is very inefficient. Transcription and translation are achieved by adding a DNA template that contains a cDNA sequence under the control of a strong bacterial promoter. These DNA templates must contain a strong bacterial promoter and a prokary-otic ribosome-binding site prior to the translational initiation site. Although eukaryotic cDNA sequences can be successfully translated in the *E. coli* transcription–translation system, there are problems in addition to the requirement of engineering the cDNA sequence into a plasmid vector with an appropriate 5' end containing a ribosome-bind-ing site. Obtaining full-length proteins from some cDNA sequences can be a problem and correct folding of proteins, which is often necessary for biological activity, is often not achieved in the *E. coli* system. Eukaryotic translation systems, such as rabbit reticu-locyte lysate (RRL), are favored for the translation of eukaryotic mRNA because they generally provide greater quantities of biologically active protein.

2.2. Eukaryotic Translation Systems

The rabbit reticulocyte provides an excellent cell-free translation system for eukary-otic mRNA. The original protocol for preparing rabbit reticulocyte lysate (RRL) described in 1976 has been modified and optimized to provide a very reliable translation system in its commercial form *(3)*, (**Fig. 2**). RRL can carry out a number of posttranslational modifications, including phosphorylation, acetylation, and myristoylation, and, with the addition of microsomal membranes, can also achieve signal peptide cleavage, core glycosylation and insertion into membranes. Other major eukaryotic systems include cell-free systems from wheat-germ (WG) *(4)*, and recently a translation system based on a *Xenopus* extract has been described *(5)*. The *Xenopus* oocyte can be injected with a range of substances and mRNA microinjected into the nucleus is very efficiently translated provided that it is 5' capped *(6)*. The advent of efficient transcription–trans-lation systems optimized for RRL and wheat-germ, which use DNA templates contain-ing bacteriophage promoters, eliminates the separate transcription step needed when cloned sequences are used for in vitro translation studies *(7)*. Bacteriophage promoters are quite small and are easily incorporated into oligonucleotide primers that can be used to generate DNA templates directly from mRNA or genomic DNA (**Fig. 3**).

3. Applications

3.1. Study of Protein Synthesis

In vitro translation systems have been useful for studying mechanisms in protein synthesis and the action of protein synthesis inhibitors. Translation extracts can be fractionated by centrifugation at 100,000*g* into a ribosome pellet and a soluble (S100) fraction that contains translation factors, tRNAs, and other necessary enzymes. This S100 fraction can be further fractionated to purify specific components of the transla-tion system. Protein synthesis can be reconstituted using purified ribosomes and the necessary soluble components. Many of the protein factors in the soluble fraction have been cloned, and recombinant proteins can be used in the reconstitution of in vitro protein synthesis. Many important antibiotics and toxins act by inhibiting protein syn-thesis, and in vitro systems are useful for studying the mechanisms of action of sub-stances that interfere with the translational machinery *(8,9)*.

3.2. Translation of Cellular mRNA and Viral RNA

One of the oldest applications for in vitro translation is for the analysis of cellular mRNA and viral RNA. Viral RNA can be extracted from viral particles and encodes a

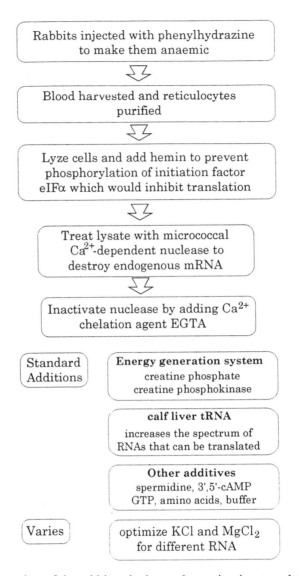

Fig. 2. Preparation of the rabbit reticulocyte lysate in vitro translation system.

small number of genes, including those for the viral coat protein. In vitro translation systems have been used to analyze the proteins of a wide range of RNA viruses (**Table 1**). The genomes of DNA viruses have also been analyzed, but RNA for this analysis is generated in vitro prior to translation. mRNA isolated from cells and tissues can be translated in vitro in the presence of ^{35}S methionine to produce a display of radiolabeled proteins that can be separated on two-dimensional gels *(10,11)*. This type of analysis can be used to identify proteins being differentially expressed in different cellular contexts, and the fractionation of mRNA can lead to the enrichment of these proteins leading to their identification through cDNA cloning. If some of the DNA sequence of a mRNA is already known a technique called hybrid-arrest translation can be used to select out homologous RNA from total mRNA, which can be translated *(12)*.

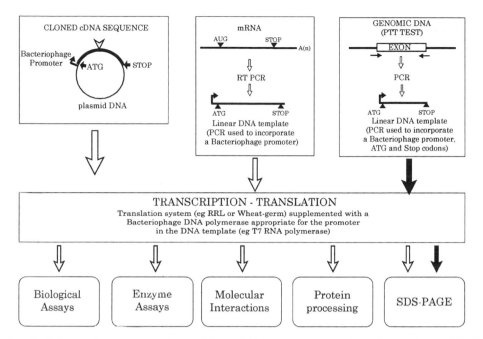

Fig. 3. Schematic representation of how DNA sequences coding for amino acids from genomic DNA, mRNA, or cDNA can be transcribed and translated using in vitro expression systems and studied using a range of techniques.

Table 1
Translation of Viral RNA In Vitro[a]

RNA	System	Reference
Cucumber necrosis virus	RRL, WG	*19*
Hepatitis A virus	RRL	*20*
Rotavirus	RRL	*21*
Alfalfa mosaic virus	RRL, WG	*22*
Foot-and mouth virus	RRL	*23*
Sinbis virus	RRL	*24*
Avian erythroblastosis virus	RRL	*25*
Nepoviruses	RRL	*26,27*

[a]Examples of viral RNAs that have been translated in vitro using rabbit reticulocyte lysate (RRL) and wheat-germ extract (WG).

3.3. Rapid Expression of cDNA Sequences

The study of complex genomes of mammals, in particular human and mouse, has led to the identification and characterization of thousands of new genes. In the majority of cases the starting point for characterization of a newly isolated gene is a full-length cDNA sequence. In vitro translation systems have proven to be an invaluable tool for first verifying the cDNA and second, in many cases, as a starting point for characterization of the protein. In most cases in vitro-translated proteins have properties indistinguishable from native protein, and in many cases this protein is better than that

Table 2
Assays Performed with In Vitro Translated Proteins[a]

Assay	Protein	Reference
Biological assays	Antiviral activity of human interferon-α	*28*
Enzyme assays	Thymidine kinase	*29*
	Ornithine decarboxylase	*30*
	Hepatitis B virus polymerase	*31*
	Dehydroepiandosterone sulfotransferase	*32*
Molecular interactions	Heparin cofactor II binding with thrombin	*33*
	G protein α subunit interactions	*34*
	DNA binding of Myb transcription factor	*35*
	DNA binding of Ets1 transcription factor	*36*
	Use of IA2 tyrosine phosphatase for autoantigen detection	*37*
Protein processing	Mitochondrial import	*38*

[a]Although RRL lysate systems contain high levels of endogenous protein and synthesize only small quantities of proteins in translation reactions, many enzymes and regulatory proteins, such as cytokines and transcription factors, have very sensitive assay systems that require only small amounts of protein. This table gives some examples of the range of assays that have been performed using in vitro synthesized protein.

obtained in prokaryotic expression systems where protein can be proteolytically cleaved, insoluble, or incorrectly folded into a nonnative inactive form. **Table 2** lists examples of the range of assays that have been successfully performed using in vitro-synthesized protein. It must be stressed that the amount of protein synthesized in translation systems is relatively small compared with the level of exogenous protein.

The routine method for determining the amount of programmed protein synthesis is to measure the incorporation of a radiolabeled amino acid and then examine the products of the translation by sodium dodecyl sulfate-polyacrylamide gel electrophoresis (SDS-PAGE) followed by autoradiography or fluorography (*see* Chapters 11 and 32). The amount of background incorporation in commercial quality nuclease-treated rabbit reticulocyte lysates and wheat germ extracts is usually very small, so almost all of the amino acid incorporation results from translation of exogenous mRNA. It is essential to verify the fidelity of translations because some RNA species can be incorrectly initiated and translation conditions (salt concentration, amino acid availability) can affect the yield of full-length protein.

3.4. Screening for Mutations Resulting in Protein Truncation

Mutations in genes that result in synthesis of truncated proteins with defective function can lead to a disease state in humans. A polymerase chain reaction (PCR) amplification technique has been developed for screening for truncation mutations (protein truncation test, PTT) in genetic diseases and cancers *(13)* (*see* **Fig. 3**). PCR can be used to generate a DNA template from either RNA using reverse transcriptase (RT)-PCR or directly from genomic DNA by incorporating a bacterial RNA polymerase promoter, usually the T7 promoter, upstream of the ATG start site. PCR-generated DNA-template can be used to prime in vitro transcription–translation in the presence of ^{35}S-methionine, resulting in the synthesis of polypeptides corresponding to the coding sequences in

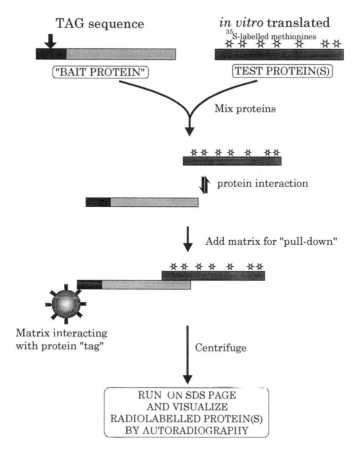

Fig. 4. Schematic representation of how radiolabeled in vitro translated proteins can be used to study protein–protein interactions in the "pull-down" assay.

the DNA template. These radiolabeled polypeptides can then be separated on the basis of size by SDS-PAGE. Templates with mutations that lead to generation of in-frame stop codons will yield smaller polypeptides. This technique can be applied to the analysis of any coding sequence. For example, aberrant rearrangements in the immunoglobulin VL κ gene give rise to truncated product in hybridomas. These truncations can be detected using RT-PCR followed by transcription–translation, with the additional virtue that the correct protein can be identified by its ability to bind the immunizing antigen *(14)*.

3.5. Study of Protein–Protein Interactions

In vitro translation provides a simple method for synthesis of proteins for the study of molecular interactions. Proteins can specifically recognize other proteins, DNA, and RNA, and the specialized immunoglobulins can interact with a huge range of proteins. The ability of in vitro translation to provide proteins incorporating radiolabeled amino acids, while maintaining full biological activity, provides an invaluable tool for many molecular interaction studies. For example, in the study of protein–protein interactions, proteins synthesized in vitro incorporating ^{35}S methionine can be used in a "pull-down assay" (**Fig. 4**). ^{35}S-labeled proteins are mixed with a putative interacting "bait" protein that is synthesized as a fusion protein with a GST, polyhistidine, biotinylation domain, or a short epitope tag sequence to allow the proteins to be recognized by an affinity matrix. Interacting proteins can be separated and analyzed on a polyacrylamide gel.

3.6. Study of Protein–Nucleic Acid Interactions

The study of protein–DNA interactions is a key component in the study of transcriptional regulation of gene expression. One of the central techniques in studying DNA–protein interactions is the electrophoretic mobility shift assay (EMSA), in which nondenaturing polyacrylamide gels are used to show the interaction of DNA-binding proteins with short radiolabeled DNA fragments (*see* Chapter 6). In vitro translations are highly suited to these studies because only small quantities of protein are required, which are easily generated. RRL contains some endogenous DNA-binding proteins that can bind certain DNA sequences, but protocols to remove these proteins from lysates have been devised. RRL contains protein kinases that may effect the activity of some transcription factors *(15)*. Wheat-germ extracts, on the other hand, have low levels of DNA-binding proteins and low levels of posttranslational modification, which makes them useful for the translation of DNA-binding proteins *(16)*.

3.7. Protein Processing Using Microsomal Membranes

Microsomal membranes prepared from dog pancreas give in vitro translation systems the ability to process proteins with signal or secretory signals to their mature form. Microsomes are commercially available and must be isolated free from contaminating membrane fractions, stripped of ribosomes and mRNA to give a preparation suitable for addition to translation systems *(17)*.

Microsomal membranes allow the cleavage of proteins with signal sequences and their correct insertion into membranes; this process can be assayed by a resistance to protease degradation. The presence of microsomal membranes in in vitro translations can often enhance the translation of membrane-associated proteins that are inherently insoluble and precipitate in the absence of membranes.

3.8. Structure–Function Studies

The advent of efficient procedures for site-directed in vitro mutagenesis has led to detailed exploration of protein structure–function relationships (*see* Chapter 29). In the absence of X-ray crystallographic data, amino acid residues critical for protein activity have been determined. An important prerequisite for any study examining protein structure–function relationships is an efficient expression system that on the one hand gives adequate protein yield but also fully native protein.

Expression of mammalian genes in *E. coli* is often very successful, with good protein yields, but some proteins are not correctly folded and as a consequence are not fully biologically active. Structure–function studies can be hampered by poor expression of modified proteins in both yeast and *E. coli* *(18)*. Mammalian expression systems can give high-fidelity protein, but often the protein yield is poor and significant purification may be required. Systematic studies of protein using mutagenesis often require the analysis of a large number of modified proteins. If this process requires recloning into expression vectors, expression of protein, and purification, the analysis of a large number of proteins becomes prohibitive. The RRL translation system can be

Table 3
Structure–Function Studies Using In Vitro Expression in Rabbit Reticulocyte Lysate

Protein	Functional assay	Reference
Bovine lectin	Carbohydrate recognition	*39*
Human antithrombin III	Binding to α thrombin	*40*
Herpes simplex virus thymidine kinase	Kinase activity	*29*
Cowpea mosaic virus protease	Protease activity	*41*
Human estrogen receptor	Estradiol binding	*42*
Poliovirus capsid protein	Myristoylation	*43*
Skeletal muscle actin	Polymerization, amino terminal processing, binding to myosin	*44*
Human interferon-α4	Antiviral, antiproliferative activity	*45*

[a]Examples of proteins on which structure–function studies have been performed using in vitro expression in RRL, and the functional evaluation used on the proteins produced.

used to provide small quantities of protein when primed with synthetic mRNA generated in vitro from DNA templates with bacterial RNA polymerases. Regulatory proteins, such as cytokines, growth factors, and transcription factors, have very sensitive assay systems that require very little protein for extensive characterization.

Structure–function studies have been performed using in vitro expression on a wide range of proteins; examples are given in **Table 3**. The choice of the appropriate in vitro expression system for a particular protein will depend on the posttranslational modifications required and the endogenous activity in the translation system of any assay to be performed. An additional consideration is the conditions under which functional proteins are formed. For example, some proteins must be synthesized under reducing conditions to be active, and wheat-germ extracts tolerate thiols much better than RRL.

References

1. Gallie, D. R., Sleat, D. E., Watts, J. W., Turner, P. C., and Wilson, T. M. (1987) The 5'-leader sequence of tobacco mosaic virus RNA enhances the expression of foreign gene transcripts in vitro and in vivo. *Nucleic Acids Res.* **15,** 3257–3273.
2. Lesley, S. A. (1995) Preparation and use of *E. coli* S-30 extracts, in *Methods in Molecular Biology, vol. 37: In Vitro Transcription and Translation Protocols* (Tymms, M. J., ed.), Humana, Totowa, NJ, pp. 265–278.
3. Beckler, G. S., Thompson, D., and Van Oosbree, T. (1995) In vitro translation using rabbit reticulocyte lysate, in *Methods in Molecular Biology, vol. 37: In Vitro Transcription and Translation Protocols* (Tymms, M. J., ed.), Humana, Totowa, NJ, pp. 215–232.
4. Van Herwynen, J. F. and Beckler, G. S. (1995) Translation using a wheat-germ extract, in *Methods in Molecular Biology, vol. 37: In Vitro Transcription and Translation Protocols* (Tymms, M. J., ed.), Humana, Totowa, NJ, pp. 245–251.
5. Matthews, G. M. and Colman, A. (1995) The *Xenopus* egg extract translation system, in *Methods in Molecular Biology, vol. 37: In Vitro Transcription and Translation Protocols* (Tymms, M. J., ed.), Humana, Totowa, NJ, pp. 199–214.
6. Ceriotti, A. and Colman, A. (1995) mRNA translation in *Xenopus* oocytes, in *Methods in Molecular Biology, vol. 37: In Vitro Transcription and Translation Protocols* (Tymms, M. J., ed.), Humana, Totowa, NJ, pp. 151–178.

7. Craig, D., Howell, M. T., Gibbs, C. L., Hunt, T., and Jackson, R. J. (1992) Plasmid cDNA-directed protein synthesis in a coupled eukaryotic in vitro transcription–translation system. *Nucleic Acids Res.* **20,** 4987–4995.

8. Matias, W. G., Bonini, M., and Creppy, E. E. (1996) Inhibition of protein synthesis in a cell-free system and Vero cells by okadaic acid, a diarrhetic shellfish toxin. *J. Toxicol. Environ. Health* **48,** 309–317.

9. Oka, T., Tsuji, H., Noda, C., Sakai, K., Hong, Y. M., Suzuki, I., Munoz, S., and Natori, Y. (1995) Isolation and characterization of a novel perchloric acid-soluble protein inhibiting cell-free protein synthesis. *J. Biol. Chem.* **270,** 30,060–30,067.

10. Aggeler, J., Frisch, S. M., and Werb, Z. (1984) Collagenase is a major gene product of induced rabbit synovial fibroblasts. *J. Cell Biol.* **98,** 1656–1661.

11. Imai, T., Fujimaki, H., Abe, T., and Befus, D. (1993) In vitro translation of mRNA from rat peritoneal and intestinal mucosal mast cells. *Int. Arch. Allergy Immunol.* **102,** 26–32.

12. Chandler, P. M. (1982) The use of single-stranded phage DNAs in hybrid arrest and release translation. *Anal. Biochem.* **127,** 9–16.

13. Roest, P. A., Roberts, R. G., van der Tuijn, A. C., Heikoop, J. C., van Ommen, G. J., and den Dunnen, J. T. (1993) Protein truncation test (PTT) to rapidly screen the DMD gene for translation terminating mutations. *Neuromuscul. Disord.* **3,** 391–394.

14. Nicholls, P. J., Johnson, V. G., Blanford, M. D., and Andrew, S. M. (1993) An improved method for generating single-chain antibodies from hybridomas. *J. Immunol. Methods* **165,** 81–91.

15. Curran, T., Gordon, M. B., Rubino, K. L., and Sambucetti, L. C. (1987) Isolation and characterization of the c-fos(rat) cDNA and analysis of post-translational modification in vitro. *Oncogene* **2,** 79–84.

16. Mercurio, F., DiDonato, J., Rosette, C., and Karin, M. (1993) p105 and p98 percursor proteins play an active role in NF-κB-mediated signal transduction. *Genes Devel.* **7,** 705–718.

17. Walter, P. and Blobel, G. (1983) Preparation of microsomal membranes for cotranslational protein translocation. *Methods Enzymol.* **96,** 84–93.

18. Tymms, M. J. and McInnes, B. (1988) Efficient *In Vitro* expression of interferon α analogs using SP6 polymerase and rabbit reticulocyte lysate. *Gene Anal. Techn.* **5,** 9–15.

19. Johnston, J. C. and Rochon, D. M. (1990) Translation of cucumber necrosis virus RNA in vitro. *J. Gen. Virol.* **71,** 2233–2241.

20. Brown, E. A., Zajac, A. J., and Lemon, S. M. (1994) In vitro characterization of an internal ribosomal entry site (IRES) present within the 5' nontranslated region of hepatitis A virus RNA: comparison with the IRES of encephalomyocarditis virus. *J. Virol.* **68,** 1066–1074.

21. Clapp, L. L. and Patton, J. T. (1991) Rotavirus morphogenesis: domains in the major inner capsid protein essential for binding to single-shelled particles and for trimerization. *Virology* **180,** 697–708.

22. Gerlinger, P., Mohier, E., Le Meur, M. A., and Hirth, L. (1977) Monocistronic translation of alfalfa mosaic virus RNAs. *Nucleic Acids Res.* **4,** 813–826.

23. Grubman, M. J., Morgan, D. O., Kendall, J., and Baxt, B. (1985) Capsid intermediates assembled in a foot-and-mouth disease virus genome RNA-programmed cell-free translation system and in infected cells. *J. Virol.* **56,** 120–126.

24. Hardy, W. R. and Strauss, J. H. (1989) Processing the nonstructural polyproteins of sindbis virus: nonstructural proteinase is in the C-terminal half of nsP2 and functions both in cis and in trans. *J. Virol.* **63,** 4653–4664.

25. Pawson, T. and Martin, G. S. (1980) Cell-free translation of avian erythroblastosis virus RNA. *J. Virol.* **34,** 280–284.

26. Demangeat, G., Hemmer, O., Fritsch, C., Le Gall, O., and Candresse, T. (1991) In vitro processing of the RNA-2-encoded polyprotein of two nepoviruses: tomato black ring virus and grapevine chrome mosaic virus. *J. Gen. Virol.* **72,** 247–252.

27. Hellen, C. U., Liu, Y. Y., and Cooper, J. I. (1991) Synthesis and proteolytic processing of arabis mosaic nepovirus, cherry leaf roll nepovirus, and strawberry latent ringspot nepovirus proteins in reticulocyte lysate. *Arch. Virol.* **120,** 19–31.

28. Tymms, M. J. (1995) Quantitative measurement of mRNA using the RNase protection assay, in *Methods in Molecular Biology, vol. 37: In Vitro Transcription and Translation Protocols* (Tymms, M. J., ed.), Humana, Totowa, NJ, pp. 31–46.

29. Black, M. E. and Loeb, L. A. (1993) Identification of important residues within the putative nucleoside binding site of HSV-1 thymidine kinase by random sequence selection: analysis of selected mutants in vitro. *Biochemistry* **32,** 11,618–11,626.

30. Tobias, K. E. and Kahana, C. (1993) Intersubunit location of the active site of mammalian ornithine decarboxylase as determined by hybridization of site-directed mutants. *Biochemistry* **32,** 5842–5847.

31. Howe, A. Y., Elliott, J. F., and Tyrrell, D. L. (1992) Duck hepatitis B virus polymerase produced by in vitro transcription and translation possesses DNA polymerase and reverse transcriptase activities. *Biochem. Biophys. Res. Commun* **189,** 1170–1176.

32. Otterness, D. M., Wieben, E. D., Wood, T. C., Watson, W. G., Madden, B. J., McCormick, D. J., and Weinshilboum, R. M. (1992) Human liver dehydroepiandrosterone sulfotransferase: molecular cloning and expression of cDNA. *Mol. Pharmacol.* **41,** 865–872.

33. Sheffield, W. P. and Blajchman, M. A. (1995) Deletion mutagenesis of heparin cofactor II: defining the minimum size of a thrombin inhibiting serpin. *FEBS Lett.* **365,** 189–192.

34. Denker, B. M., Neer, E. J., and Schmidt, C. J. (1992) Mutagenesis of the amino terminus of the alpha subunit of the G protein Go. In vitro characterization of alpha o beta gamma interactions. *J. Biol. Chem.* **267,** 6272–6277.

35. Ramsay, R. G. (1995) DNA-binding studies using in vitro synthesized Myb proteins, in *Methods in Molecular Biology, vol. 37: In Vitro Transcription and Translation Protocols* (Tymms, M. J., ed.), Humana, Totowa, NJ, pp. 369–378.

36. Lim, F., Kraut, N., Frampton, J., and Graf, T. (1992) DNA binding by c-Ets-1, but not v-Ets, is repressed by an intramolecular mechanism. *EMBO J.* **11,** 643–652.

37. Lan, M. S., Wasserfall, C., Maclaren, N. K., and Notkins, A. L. (1996) IA-2, a transmembrane protein of the protein tyrosine phosphatase family, is a major autoantigen in insulin-dependent diabetes mellitus. *Proc. Natl. Acad. Sci. USA* **93,** 6367–6370.

38. Law, R. H. P. and Nagley, P. (1995) Import in isolated yeast mitochondria of radiolabelled proteins synthesized in vitro, in *Methods in Molecular Biology, vol. 37: In Vitro Transcription and Translation Protocols* (Tymms, M. J., ed.), Humana, Totowa, NJ, pp. 293–315.

39. Abbott, W. M. and Feizi, T. (1991) Soluble 14-kDa beta-galactoside-specific bovine lectin. Evidence from mutagenesis and proteolysis that almost the complete polypeptide chain is necessary for integrity of the carbohydrate recognition domain. *J. Biol. Chem.* **266,** 5552–5557.

40. Austin, R. C., Rachubinski, R. A., Fernandez-Rachubinski, F., and Blajchman, M. A. (1990) Expression in a cell-free system of normal and variant forms of human antithrombin III. Ability to bind heparin and react with alpha-thrombin. *Blood* **76,** 1521–1529.

41. Dessens, J. T. and Lomonossoff, G. P. (1991) Mutational analysis of the putative catalytic triad of the cowpea mosaic virus 24K protease. *Virology* **184,** 738–746.

42. Kumar, V., Green, S., Staub, A., and Chambon, P. (1986) Localisation of the oestradiol-binding and putative DNA-binding domains of the human oestrogen receptor. *EMBO J.* **5,** 2231–2236.

43. Marc, D., Drugeon, G., Haenni, A. L., Girard, M., and van der Werf, S. (1989) Role of myristoylation of poliovirus capsid protein VP4 as determined by site-directed mutagenesis of its N-terminal sequence. *EMBO J.* **8,** 2661–2668.

44. Solomon, T. L., Solomon, L. R., Gay, L. S., and Rubenstein, P. A. (1988) Studies on the role of actin's aspartic acid 3 and aspartic acid 11 using oligodeoxynucleotide-directed site-specific mutagenesis. *J. Biol. Chem.* **263,** 19,662–19,669.

45. Waine, G. J., Tymms, M. J., Brandt, E. R., Cheetham, B. F., and Linnane, A. W. (1992) Structure-function study of the region encompassing residues 26–40 of human interferon-α4: identification of residues important for antiviral and antiproliferative activities. *J. Interferon. Res.* **12,** 43–48.

28

Site-Directed Mutagenesis

John R. Adair and T. Paul Wallace

1. Introduction

The ability to create defined mutations and to generate precise fusions of protein domains by modifications at the DNA level is now an accepted and routine technique in molecular biology. This ability has developed over the past 10–15 years and rests on a foundation of experiments on DNA:DNA hybridization that stretches back to the 1960s. The 1993 Nobel lectures *(1,2)* provide a timely, historical survey of the development of key stages in the development of current methods.

In this chapter, we describe a number of approaches to the oligonucleotide-directed manipulation of DNA sequences. We give an overview of the methods, identify some of the main problems, offer the solutions we have found to be of benefit, and indicate our experiences with using these methods. As examples, we describe various reconstructions of the antigen-binding domains of antibodies, to understand structure and function and to produce commercially useful therapeutic and diagnostic reagents.

2. Site-Specific Modification of DNA

Standard methods for oligonucleotide-directed site-specific manipulation of DNA sequences can conveniently be divided into two schools. Both involve annealing of one or more oligonucleotides to a region of (at least temporarily) single-stranded DNA (ssDNA) followed by in vitro DNA-polymerase-directed extension of the oligonucleotide(s). In the first (older) approach the DNA sequence under modification is linked to a replication origin (e.g., in a plasmid, phage, or phagemid), allowing in vivo amplification of the modified and parental genotypes. The desired modification is then identified by any of a number of screening and selection procedures (a summary of the earlier methods is given in **ref. 3**). In the second approach, the desired modification is amplified in vitro by means of the polymerase chain reaction (PCR) and the experimental protocols can be designed to ensure that all of the progeny have the desired mutation. The amplified DNA is then linked to a bacterial-replication origin in a plasmid or phage and cloned by introduction into bacteria (**Fig. 2**).

From: *Molecular Biomethods Handbook*
Edited by: R. Rapley and J. M. Walker © Humana Press Inc., Totowa, NJ

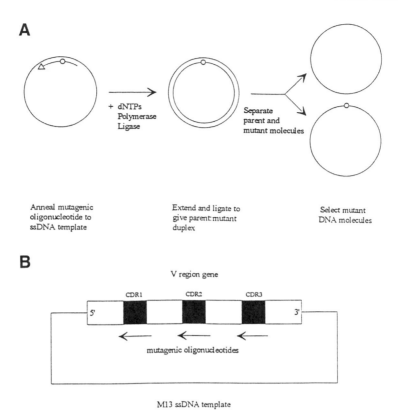

Fig. 1. **(A)** Strategy for non-PCR mutagenesis of ssDNA. **(B)** Simultaneous CDR-replacement mutagenesis.

3. Non-PCR Mutagenesis

The theory behind non-PCR mutagenesis is simple: a synthetic oligonucleotide harboring the required mutation is annealed to circular ssDNA template, the oligonucleotide acts as a primer for in vitro DNA synthesis using DNA polymerase, and in the presence of DNA ligase, circular-mutant daughter molecules are generated that are then isolated from nonmutant template (*see* **Fig. 1A**).

This basic method is flexible and can be adapted for more complex mutagenesis applications or to suit different circumstances. For instance, mutagenic-oligonucleotide primers can be used at the same time on the same template DNA; an example of the application of this method is the mutagenesis of the three amino-acid sequence strings of an antibody heavy- or light-chain variable region during antibody humanization (*see* **Fig. 1B**, and later for a more detailed explanation). It is also possible to mutate regions distant from each other on the same template. The method can be adapted to use double-stranded DNA (dsDNA); although the isolation of mutant molecules may be more complex.

The major advances in the development of the non-PCR mutagenesis technique have come from the availability of DNA polymerases especially suited for the process, and in the development of methods for the efficient isolation of the mutant DNA molecules.

3.1. DNA Polymerases

Escherichia coli DNA polymerase I *(4)* was the first polymerase to be used for mutant-oligonucleotide primer extension in mutagenesis experiments. Along with the 5'-3' polymerase activity, this enzyme has a beneficial 3'-5' "proofreading" exonuclease activity. However *E. coli* DNA polymerase I also possesses a 5'-3' exonuclease activity. This activity can result in the removal of downstream mutant oligonucleotides from the daughter strand when attempting to generate mutant-DNA molecules. Later, a fragment of the polymerase, the "Klenow" fragment, which lacks the 5'-3' exonuclease activity, was developed and became the favored enzyme for mutagenesis *(5,6)*. More recently, polymerases from cloned-bacteriophage genes have become available for use in mutagenesis experiments, e.g., T4-DNA polymerase *(7)* and native T7-DNA polymerase *(8)*. Both of these have efficient proofreading activity and lack the detrimental 5'-3' exonuclease activity.

Processivity is the term used to describe the enzyme's ability to remain associated with a primer and continue to extend the daughter strand. Klenow fragment and T4-DNA polymerase have low processivity, whereas native T7-DNA polymerase exhibits high processivity. The processivity of T4-DNA polymerase can be improved by the inclusion of certain accessory proteins in the polymerase reaction *(9)*. T4- and T7-DNA polymerases are now the preferred polymerases for non-PCR mutagenesis experiments.

3.2. Isolation of Mutant DNA Molecules

A critical step in determining the efficiency of a mutagenesis experiment is the selection and isolation of newly synthesized mutant DNA molecules from nonmutant template. These can conveniently be subdivided into those methods that involve screening the population for the desired mutation and those that seek to enrich the frequency of the mutation within the population. These methods can frequently be used together to efficiently identify the desired mutation(s). **Subheading 3.3.** lists the available methods and highlight their merits for particular applications.

3.3. Screening Methods

3.3.1. Hybridization of Mutant-Specific Oligonucleotides

Oligonucleotides hybridize to DNA with great specificity under the correct conditions. The thermal stability of the annealed oligonucleotides is affected by sequence mismatch. Therefore, the oligonucleotide used to introduce the mutation can often be used to identify the mutation because it will display a higher thermal stability for the mutation compared to the parental sequence. This property can be exploited for the identification of mutants by using the mutagenic oligonucleotide as a labeled probe to screen plaques or colonies of *E. coli* transformed with the mutant-phage DNA or plasmid (reviewed in **ref.** *1*). The labeling can be radioactive, but nonradioactive detection methods are becoming increasingly more common. A mutant can easily be detected among thousands of wild-type sequences.

3.3.2. Direct DNA Sequencing

DNA sequencing of cloned, potential mutant molecules is a direct way of selecting mutants. This method may be time-consuming if the mutagenesis is not efficient. As the mutation is defined, single-track sequencing (e.g., T-tracking) using the nucleotide or one of the affected nucleotides in comparison to the parent sequence can be used as a rapid initial screen. In conjunction with automated DNA-sequencing equipment, screening by sequence can be a routine procedure, because most methods provide the desired mutation at significant (1–50%) yield.

3.4. Enrichment Methods

3.4.1. Phenotypic Selection for Altered Gene Product

If the mutagenesis is designed to alter a biological activity of a particular gene, the gene could be cloned into an appropriate expression vector and mutants selected on the basis of phenotype *(1)*.

3.4.2. Phosphorothioate, Nick/Repair

In this strategy, mutagenic primer extension is performed in the presence of three normal deoxynucleotides, but with the fourth nucleotide substituted by the deoxynucleotide 5'-*O*-(α-thiotriphosphate) analog. The mutant strand is thus synthesized with thiophosphate internucleotide linkages, which are resistant to digestion with some restriction endonucleases. During subsequent digestion with an appropriate restriction enzyme, only the parental template strand is nicked. This strand can then be partially degraded by digestion with an exonuclease and the region repaired using DNA polymerase and DNA ligase. In this way, double-stranded molecules with the required mutation in both strands can be generated *(10)*.

3.4.3. Kunkel Method

For this method, the starting point is ssDNA template isolated from an appropriate *E. coli* strain, such that it contains uracil instead of thymine. Subsequent in vitro primer extension and ligation using the mutagenic oligonucleotide, polymerase, and ligase results in the generation of double-stranded DNA, with uracil in only the parental-template DNA strand. Selection against the uracil-template DNA strand can be achieved by selective digestion of the template using uracil-DNA glycosylase in vitro followed by recovery of the mutant strand by transformation in *E. coli*. Alternatively, the double-stranded DNA molecule can be transformed into *E. coli* TG1, where the endogenous uracil-DNA glycosylase will selectively destroy the template uracil-DNA *(11)*.

3.4.4. Methylation-Directed Selection

In principle, when using the simplest method (*see* **Fig. 1A**) involving oligonucleotide extension and ligation on a ssDNA closed-circular template, the parental sequence and the in vitro-generated mutant sequence should be present in equal proportions after in vivo repair of the mutation. However, this is not usually found. The yield of mutations in the progeny phage will depend on which strand is initially repaired by the *E. coli* mismatch repair system. In vivo mismatch repair is guided by the methylation status of the two DNA strands. Therefore, a parental, in vivo-generated, strand can be distinguished from the in vitro generated, nonmethylated, mutagenic strand by the repair system. The nonmethylated strand is then repaired in vivo using the parental sequence as the template, thus reducing the mutant yield. In the early days of site-directed mutagenesis, it was observed that the yield of mutants found using two different bacteriophage vectors, ΘX174 and M13 differed, with ΘX174 being much more efficient. It was proposed that this was because ΘX174 does not become methylated in vivo at 5'dGATC sequences by the *dam* methylase. ΘX174 does not have any of these motifs in its genome, whereas M13 does (*see*, for example, **ref. *12***). Unfortunately, M13 is the more generally useful of the two vectors, so attempts were made to circumvent the problem. The situation is alleviated to a certain degree because the asymmetric replication process of M13 biases the progeny in favor of the minus (mutagenized) strand. To avoid the methylation problem and enrich further for the desired mutation, the template can be prepared in an *E. coli dam* host so that neither the template nor the in vitro prepared strand are methylated. Mismatch repair then uses either strand as the template. To improve even further the yield of mutants, nonmethylated ssDNA phage can be annealed to linearized methylated phage-replicative form dsDNA, giving a heteroduplex in which only the cloned sequence is present as ssDNA. The mutagenic primer is annealed in this region and the gap is filled and ligated. This method has the advantage that annealing of the mutagenic primer to incorrect locations is suppressed, the gap-filling reaction is more efficient, and mismatch repair is biased toward using the mutated strand as template (*see*, for example, **ref. *13***).

It has been noted that differing mismatches are repaired with differing efficiency by the repair system *(14)* and that the mutator loci *mutL*, *mutS*, and *mutH* are also involved in the methyl-directed mismatch repair. The effects of methyl-directed repair can be avoided by using a *mutL* or *mutS* strain of *E. coli*, such as BMH71-18 *mutL* *(14)* or HB2154 *(15)*.

3.4.5. Comutagenesis of a Selectable Marker on the Vector

In this method, the mutagenesis is deliberately carried out using a vector with a mutation in a selectable marker, for example, the β-lactamase gene, which confers resistance to the penicillin type antibiotics. During annealing of the mutagenic oligonucleotide, a second oligonucleotide is included that is designed to repair the mutation in the selectable marker. Mutant DNA molecules can be identified, following transformation into *E. coli*, by the presence of the selectable marker. The method relies on the cosegregation of the two mutations. This does seem to occur at high frequency, probably because in vivo mismatch repair involves replacing large stretches of DNA sequence around the mutation *(16)*.

3.4.6. Comutagenesis to Abolish a Unique Restriction Site

This method involves the selective removal of a unique restriction site in the vector (but not in the cloned sequence). Phage or plasmid DNA can be used. In a similar fashion to the "co-mutagenesis of selectable-marker" method, a second oligonucleotide is used to eliminate the unique restriction site from the daughter strand. Following this, the product DNA can be selected against by restriction endonuclease digestion of the unique restriction site and transformation in *E. coli* *(17)*.

3.4.7. Selection for Phage Assembly

A number of the M13 vector series (e.g., M13 mp8, mp9) have amber mutations in essential genes I and II. Functional phage can only be assembled when grown in an amber suppressor, *supE*, host, e.g., *E. coli* JM101 or TG1. However, the phage cannot be propagated in a non-*supE* host, e.g., *E. coli* HB2154 *(15)*. Therefore, enrichment for the mutation can be achieved by annealing a second oligonucleotide that repairs the amber mutation, or prehybridizing the amber-phage ssDNA with linearized wild-type phage (e.g., M13, mp18, mp19) to form a heteroduplex that is then used for the annealing/extension/ligation reaction with the mutagenic oligonucleotide. Transformation into a nonsuppressor strain, such as *E. coli* HB2154, then causes selection for the strand with the wild-type sequence in genes I and II and coselection of the desired mutation *(18)*. Two further advantages of this method are that the in vitro extension reaction is limited to the inserted sequence in the vector, increasing the yield of fully completed RF DNA, and the use of HB2154, which is *mutL*, that removing any mismatch repair bias (*see* **Subheading 3.4.4.**).

3.4.8. Selection by Host Restriction

This method takes advantage of the bacterial host-restriction defense mechanism and involves the incorporation into the vector of one or more copies of the recognition sequences for the *EcoK/B* restriction system. This method is particularly useful when consecutive mutations are required in a sequence. To begin the process, the template, which, for example, encodes the *EcoK* site, is generated in an $r^- m^-$ host, e.g., *E. coli* TG1, and then the mutation of interest is incorporated along with a mutation that switches the restriction site from *EcoK* to *EcoB* *(15)* by annealing two oligonucleotides followed by in vitro extension and ligation. The resultant in vitro-generated progeny are transformed into an $r_K^+ m_K^+$ host, e.g., *E. coli* JM101. Progeny that derive from the template strand will be restricted, whereas progeny deriving from the mutagenic strand will not. When coupled with mismatch-repair deficiency, this gives a strong selection for the mutagenic strand.

A second round of mutagenesis can then be done on the template by switching the restriction sequence back to the *EcoK* and selecting on $r_B^+ m_B^+$ cells, e.g., *E. coli* AC2522. Recently the system has been improved by using tandem *EcoK* or *EcoB* sites to improve the stringency of the restriction *(19)*.

4. PCR-Based Methods

PCR *(20)* is a method that enables amplification of DNA (*see* Chapter 25). The method involves annealing two primers, one upstream and the other downstream of the region of DNA to be amplified. The primers are designed to bind to opposite strands of the DNA molecule so that extension using a polymerase results in DNA synthesis in the direction of the other primer. One cycle of PCR involves annealing the primers, extension of the primers using a thermostable DNA polymerase, and then melting of the newly synthesized DNA duplex to allow access for further primer molecules to bind, and the start of another cycle of PCR. The use of a thermostable polymerase enables repeated PCR cycles without the need to introduce fresh polymerase at each extension step. The ability to perform repeated cycles enables the amplification of a region of DNA from even a very small amount of DNA; in theory, only one molecule of starting material is required *(21–24)*.

The basis of PCR mutagenesis methods is that at least one of the oligonucleotides employed in the amplification process has one or more sequence mismatches to the template, so that on amplification a mutation is fixed in the progeny DNA. In early methods, two oligonucleotides were employed in a single-step process (**Fig. 2**). The amplified fragment was cloned by blunt-end cloning, by the incorporation of restriction sites into one or both of the oligonucleotides, by using restriction sites internal to the priming sites for subsequent cloning, or a combination of these options.

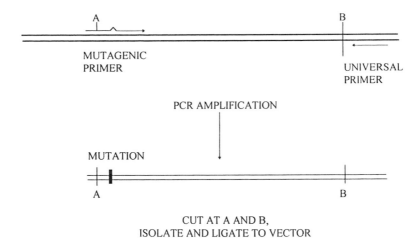

MUTAGENIC
PRIMER

UNIVERSAL
PRIMER

PCR AMPLIFICATION

MUTATION

CUT AT A AND B,
ISOLATE AND LIGATE TO VECTOR

Fig. 2. Original PCR mutagenesis method.

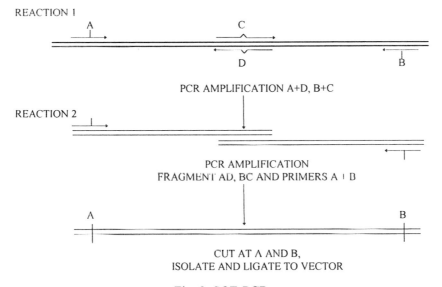

REACTION 1

PCR AMPLIFICATION A+D, B+C

REACTION 2

PCR AMPLIFICATION
FRAGMENT AD, BC AND PRIMERS A + B

CUT AT A AND B,
ISOLATE AND LIGATE TO VECTOR

Fig. 3. SOE-PCR.

The problem of generating mutants distal from a convenient restriction site has been approached in several ways involving two or three primers (e.g., megaprimer methods; *25*, but more general methods are available if one is prepared to make two oligonucleotides for each mutation (not a difficulty these days) and have two universal primers that anneal in the DNA sequence flanking the region of interest and outside of the cloning points. (These universal primers are also useful sequencing primers). Overlap extension PCR mutagenesis, or the Single Overlap Extension-PCR (SOE-PCR) method *(26–28)*, is of general use not only for point mutations but also for sequence deletion, as well as domain construction and gene fusions. Schematics of the mutagenesis and gene fusion versions of SOE-PCR are shown in **Figs. 3** and **4**. The use of SOE-PCR for point mutations and gene fusions prompted the use of this method for assembling syn-

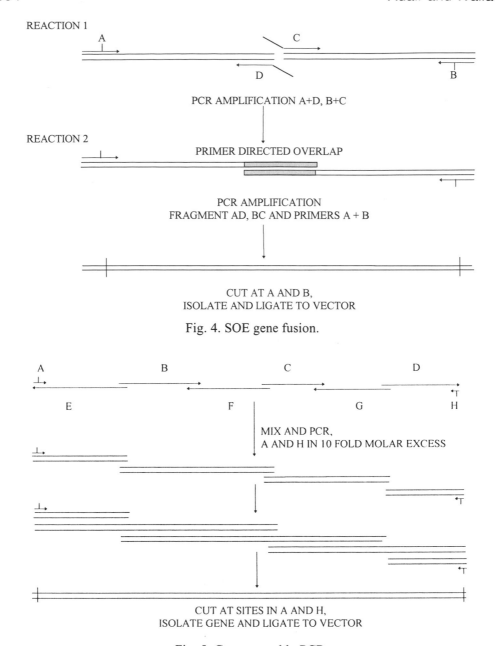

Fig. 4. SOE gene fusion.

Fig. 5. Gene assembly PCR.

thetic genes (e.g., *29,30*). Initially, subsections were extended from partially overlapping oligonucleotides; these subsections were then added together in further PCR reactions, eventually to build the whole gene or domain (*see*, for example, *30,31*). It was also found that adding all of the oligonucleotides together in the PCR reaction could lead to full-length assembly, and sufficient product could be obtained if the full-length fragment was amplified using an excess of the 5' oligonucleotides for each strand (e.g., *32*). However, it was also found that the reactions could be done together as a "one-pot" assembly (e.g., *33,34*; *see* **Fig. 5**). More recently, ligation PCR (LCR) has been

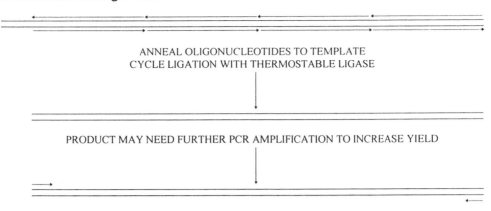

Fig. 6. Ligation chain reaction (LCR) gene assembly/mutagenesis.

used for gene assembly (*35*; *see* **Fig. 6**). This method has the advantage in that template-directed DNA synthesis is not required and offers an improvement over classical methods of oligonucleotide gene assembly. Key technical points to note are the design of the mutagenic oligonucleotides, the choice of polymerases, annealing temperature and cycle number, purification of first-round products, and the need to sequence all of the amplified region.

4.1. Oligonucleotide Design and Annealing Conditions

The oligonucleotides need to be designed with a region of identity to the target sequence of 12 or more bases. Any mismatches to the template sequence should be placed as far from the 3' end of the oligonucleotide as possible and the two oligonucleotides that carry the new sequence (**Fig. 4B,C**) should also have substantial sequence overlap (12–20 bases) so the second round of annealing is efficient; 3' Terminal G:C pairing is also preferred to reduce transient mismatching "breathing" at the 3' end of the annealed oligonucleotide.

4.2. Choice of Polymerases

For some thermostable polymerases, the error rate is quite high (e.g., *36,37*) and the number of errors that accumulate increases with cycle number. Therefore, it is important either to use an enzyme with a proofreading facility and/or to reduce the cycle number to the minimum number that provides acceptable yield after the post-PCR manipulations. In addition, if *Taq* polymerase is used, a nontemplated 3' A is frequently found on the extension product (*38*). As a general procedure, the use of a proofreading enzyme is recommended.

4.3. DNA Handling

In SOE PCR, to avoid amplification of parental sequence in the second-round reactions, it is important to purify the first-round PCR products away from initial template and first-round oligonucleotides. This separation can be quite crude but is worth the effort. In all PCR procedures, it is advisable to sequence all of the DNA that derives from an amplification to ensure that there are no secondary mutations elsewhere in the sequence.

5. Combined Methods

The two approaches described in the previous sections are not mutually exclusive and can be combined to achieve efficient mutagenic yields. If, for instance, the template DNA is particularly large, the non-PCR method can be combined with PCR so pre-existing restriction sites upstream and downstream of the mutation site can be used for the excision of the mutant fragment. If convenient restriction sites do not already exist, they can be introduced by using additional mutagenic oligonucleotides.

6. Applications

The methods described in **Subheading 3.4.** will be discussed by reference to a number of examples. For our examples, we shall be concentrating on modifications to the N-terminal variable domains of antibody heavy- and light-chains, which together form the Fv portion of the antibody that binds to the antigen. Three noncontiguous regions within each of the variable domains are particularly variable in sequence across the antibody population and are usually referred to as the "complementarity determining regions" (CDRs). These regions form a large surface patch and constitute in the main the region of contact with antigen, defining both the specificity and the affinity of the antigen:antibody interaction. Underlying these loops and contributing to their structure is the β-stranded "framework" of the variable domains, which is encoded by four gene segments for each of the VH and VL domains.

6.1. Oligonucleotide Site-Directed Mutagenesis

Mutagenesis of the three CDRs in a human V region DNA sequence has been achieved by three sequential single reactions to substitute the desired sequence in place of a pre-existing CDR sequence in a previously constructed V gene (e.g., *39*), and also by simultaneously annealing all three mutagenic oligonucleotides (e.g., *40*) (**Fig. 1B**). The former method requires three rounds of mutagenesis, but gives the correct product in good yield, whereas the latter method provides the correct product in low yield, but in one step.

More recently, a method combining non-PCR mutagenesis with subsequent, specific PCR amplification of the triple mutant has been used. This basic technique has been used for the successful humanization of several antibodies for the treatment of cancers and infectious diseases (PW, unpublished).

6.2. SOE-PCR

The SOE method has been used very successfully as a rapid method to alter amino acids in antibody variable regions. For example, Tempest and coworkers used the method of Ho et al. (*27, see* **Fig. 3**) to generate a series of point mutations in the heavy-chain variable regions of humanized forms of antibodies recognizing-tumor necrosis factor α (TNFα) and the *Clostridium perfringens* α toxin in order to investigate the effect of individual or group amino-acid sequence alterations on binding affinity *(41,42)*. Similarly, Hsiao et al. *(43)* used the procedure to investigate the role of one particular amino acid in the heavy-chain variable region of a humanized form of antibody 60.3, which recognizes human CD18.

6.3. SOE Gene Fusion

The "gene fusion" format of SOE-PCR has been applied to the reconstruction of the whole variable domain in the humanization process. Lewis and Crowe *(30)* showed that murine CDRs could be attached to human framework sequences in a number of parallel reactions, the products of which could themselves be linked together in a subsequent PCR reaction to assemble a humanized form of the rat antidigoxin monoclonal antibody (MAb) DX48. Similarly, Daugherty et al. *(33)* assembled a humanized form of an antihuman CD18 MAb, whereas Sato et al. *(31)* used the procedure to generate humanized variable regions for an anti-interleukin-6 (IL-6) receptor antibody.

Single-chain Fv (scFv, *44*) are molecules in which the heavy- and light-chain variable-region coding sequences are combined into a single polypeptide chain using a coding sequence for a short-peptide linker. scFvs are designed to prevent the dissociation of Fvs into heavy- and light-variable domains, which often occurs under physiological conditions and thus retains the useful features of Fvs, e.g., their small size and short in vivo half-lives. scFvs have been assembled using the SOE-PCR method (e.g., *45*). The scFv can be assembled from two first-round reactions in which the overlap occurs within the linker-coding sequence (*see* **Fig. 4**), or as a three-fragment reaction with the heavy-chain variable domain, linker, and light-chain variable domain being assembled in a first-round PCR and then mixed in a second round to form the scFv (e.g., *46*).

Hiyashi et al. *(39)* also showed that oligonucleotides of mixed sequence can be utilized. This allows the formation of a number of sequence variants over a given region, for example, over the CDR regions, which could then be screened for differences in affinity or specificity, for example, using phage-display systems *(47)* (*see* Chapter 46).

Clackson et al. *(48)* showed that populations of heavy- and light-chain variable regions could be isolated from cytoplasmic RNA from spleen cells as cDNAs, and linked using the three-fragment PCR reaction to form a library of scFv from immunized mice. The scFvs were then cloned into the genome and subsequently expressed on the surface of a filamentous phage; those phages that specifically adsorbed to the immunogen could be isolated and the scFv sequences determined.

7.4. PCR Gene Assembly

It is reasonably straightforward to assemble novel coding sequences for antibody-variable domains by sequential SOE-PCR, as mentioned in **Subheading 6.3.**, or as a "one pot" method. Daugherty et al., *(33)* used four long, partially overlapping, internal oligonucleotides and two short, external-amplification oligonucleotides (refer to **Fig. 5**, oligonucleotides A and H) to assemble the variable domains of a humanized form of the antihuman CD18 MAb 1B4. The assembled variable-domain coding-sequence fragment was then used in a further three-fragment SOE-PCR to add a secretion-signal sequence and 3' splicing sequences. However, more internal oligonucleotides can be used so the secretion signal and any required 5' and 3' cloning sites can be added in the initial reaction (e.g., *40*).

7.5. LCR Gene Assembly

More recently, the LCR has been adapted for antibody variable-domain assembly (*see* **Fig. 6**). It is also possible to incorporate mixed-sequence oligonucleotides into the reaction to effect random mutagenesis of the CDR regions of the sequence. After molecular cloning, the population of variants can be sorted by screening for differences of phenotype; for example, affinity variants can be identified by presenting the antibody genes, displayed on the surface of filamentous phage, to antigen *(35)*.

8. Conclusions

"Classical" oligonucleotide-directed mutagenesis and the more recent PCR-based methods offer rapid, reliable methods for gene assembly and mutagenesis. Each of the approaches have advantages and disadvantages and each has attracted a following of convinced users. Both methods are flexible and can be used for rapid and reliable gene assembly and mutagenesis for the engineering of protein structure and function.

Acknowledgments

The use of PCR for research is covered by patents owned by Roche Molecular Systems. To avoid infringement of issued patents, a licence is required either directly or by purchase of enzymes from a licensed supplier and the reactions need to be performed in an authorized thermal cycler.

References

1. Smith, M. (1993) *Synthetic DNA and Biology.* The Nobel Foundation 1994, Stockholm, Sweden.
2. Mullis, K. B. (1993) *The Polymerase Chain Reaction.* The Nobel Foundation 1994, Stockholm, Sweden.
3. Carter, P. (1986) Site-directed mutagenesis. *Biochem. J.* **237,** 1–7.
4. Lehman, I. R. (1981) DNA polymerase I of *Escherichia coli*, in *The Enzymes* vol. 14A (Boyer, P. D., ed.), Academic, New York, pp. 16–38.
5. Klenow, H. and Henningsen, I. (1970) Selected elimination of the exonuclease activity of the deoxyribonucleic acid polymerase from *Escherichia coli B* by limited proteolysis. *Proc. Natl. Acad. Sci. USA* **65,** 168–175.
6. Klenow, H., Overgaard-Hansen, K., and Patkar, S. A. (1971) Proteolytic cleavage of native DNA polymerase into two different catalytic fragments. Influence of assay conditions on the change of exonuclease activity and polymerase activity. *Eur. J. Biochem.* **22,** 371–381.
7. Nossal, N. G. (1974) DNA synthesis on a double stranded DNA template by the T4 bacteriophage DNA polymerase and the T4 gene 32 DNA unwinding protein. *J. Biol. Chem.* **249,** 5668–5676.
8. Tabor, S., Huber, H. E., and Richardson, C. C. (1987) *Escherichia coli* thioredoxin confers processivity on the DNA polymerase activity of the gene 5 protein of bacteriophage T7. *J. Biol. Chem.* **262,** 16,212–16,223.
9. Nossal, N. G. (1984) Prokaryotic DNA replication systems. *Annu. Rev. Biochem.* **8,** pp. 581–615.
10. Sayers, J. R., Krekel, C., and Eckstein, F. (1992) Rapid high-efficiency site-directed mutagenesis by the phosphorothioate approach. *BioTechniques* **13,** 592–596.
11. Kunkel, T. A. (1985) Rapid and efficient site specific mutagenesis without phenotypic selection. *Proc. Natl. Acad. Sci. USA* **82,** 488–492.
12. Kramer, W., Schugart, K., and Fritz, H.-J. (1982) Directed mutagenesis of DNA cloned in filamentous phage: influence of hemimethylated GATC sites on marker recovery from restriction fragments. *Nucleic Acids Res.* **10,** 6475–6485.
13. Fritz, H.-J. (1885) The oligonucleotide-directed construction of mutations in recombinant filamentous phage, in *DNA Cloning: A Practical Approach*, vol. I (Glover, D. M., ed.), IRL Press Ltd., Oxford, UK, pp. 151–163.
14. Kramer, B., Kramer, W., and Fritz, H.-J. (1984) Different base/base mismatches are corrected with differing efficiencies by the methyl-directed DNA mismatch-repair system of *E. coli. Cell* **38,** 879–887.

15. Carter, P., Bedouelle, H., and Winter, G. (1985) Improved oligonucleotide site-directed mutagenesis using M13 vectors. *Nucleic Acids Res.* **13,** 4431–4443.

16. Laengle-Rouault, F., Maenhaut-Michel, G., and Radman, M. (1986) GATC sequence and mismatch repair in *Escherichia coli. EMBO J.* **5,** 2009–2013.

17. Deng, W. P. and Nickloff, J. A. (1992) Site directed mutagenesis of virtually any plasmid by eliminating a unique site. *Anal. Biochem.* **200,** 81–88.

18. Kramer, W., Drutsa, V., Jansen, H.-W., Kramer, B., Pflugfelder, M., and Fritz, H.-J. (1984) The gapped duplex DNA approach to oligonucleotide-directed mutation construction. *Nucleic Acids Res.* **12,** 9441–9456.

19. Waye, M. M. Y. (1993) Use of M13 Ping-Pong vectors and T4 DNA polymerase in oligonucleotide-directed mutagenesis. *Meth. Enzymol.* **217,** 259–270.

20. Saiki, R. K., Scharf, S., Faloona, F., Mullis, K. B., Horn, G. T., Erlich, H. A., and Arnheim, N. (1985) Enzymatic amplification of β-globin genomic sequences and restriction site analysis for diagnosis of sickle anemia. *Science* **230,** 1350–1354.

21. Newton, C. R. and Graham, A. (1994) *PCR.* Bios, Oxford, UK.

22. McPherson, M. J., Quirke, P., and Taylor, G. R. (eds.) (1992) *Polymerase Chain Reaction: A Practical Approach.* Oxford University Press, Oxford, UK.

23. Abramson, R. D. and Myers, T. W. (1992) Nucleic acid amplification technologies. *Curr. Opin. Biotech.* **4,** 41–47.

24. Ausubel, F. M., Brent, R., Kingston, R. E., Moore, D. D., Seidman, J. G., Smith, J. A., and Struhl, K. (eds.) (1991) *Current Protocols in Molecular Biology.* Wiley, New York.

25. Sarkar, G. and Sommer, S. S. (1990) The "Megaprimer" method of site-directed mutagenesis. *Biotechniques* **8,** 404–407.

26. Higuchi, R., Krummel, B., and Saiki, R. K. (1988) A general method of *in vitro* preparation and specific mutagenesis of DNA fragments: study of protein and DNA interactions. *Nucleic Acids Res.* **15,** 7351–7367.

27. Ho, S. N., Hunt, H. D., Horton, R. M., Pullen, J. K., and Pease, L. R. (1989) Site-directed mutagenesis by overlap extension using the polymerase chain reaction. *Gene* **77,** 51–59.

28. Horton, R. M., Cai, Z., Ho, S. N., and Pease, L. R. (1990) Gene splicing by overlap extension: tailor made genes using the polymerase chain reaction. *Biotechniques* **8,** 528–535.

29. Jayaraman, K., Fingar, S. A., Shah, J., and Fyles, J. (1991) Polymerase chain reaction-mediated gene synthesis: synthesis of a gene coding for isozyme c of horseradish peroxidase. *Proc. Natl. Acad. Sci. USA* **88,** 4084–4088.

30. Lewis, A. P. and Crowe, J. S. (1991) Immunoglobulin complementarity determining region grafting by recombinant polymerase chain reaction to generate humanised monoclonal antibodies. *Gene* **101,** 297–302.

31. Sato, K., Tsuchiya, M., Saldanha, J., Koishihara, Y., Ohsugi, Y., Kishimoto, T., and Bendig, M. M. (1994) Humanization of a mouse anti-human interleukin-6 receptor antibody comparing two methods for selecting human framework regions. *Mol. Immunol.* **31,** 371–381.

32. Dillon, P. J. and Rosen, C. A. (1990) A rapid method for the construction of synthetic genes using the polymerase chain reaction. *Biotechniques* **9,** 298–300.

33. Daugherty, B. L., DeMartino, J. A., Law, M.-F., Kawka, D. W., Singer, I. I., and Mark, G. E. (1991) Polymerase chain reaction facilitates the cloning, CDR-grafting and rapid expression of a murine monoclonal antibody directed against the CD18 component of leukocyte integrins. *Nucleic Acids Res.* **19,** 2471–2476.

34. Prodromou, C. and Pearl, L. H. (1992) Recursive PCR: a novel technique for total gene synthesis. *Prot. Eng.* **5,** 827–829.

35. Deng, S.-J., MacKenzie, C. R., and Narang, S. A. (1993) Simultaneous randomization of antibody CDRs by a synthetic ligase chain reaction strategy. *Nucleic Acids Res.* **21,** 4418–4419.

36. Eckert, K. A. and Kunkel, T. A. (1991) DNA polymerase fidelity and the polymerase chain reaction. *PCR Methods Appl.* **1,** 17–24.

37. Ling, L. L., Keohavong, P., Dias, C., and Thilly, W. G. (1991) Optimization of the polymerase chain reaction with regard to fidelity: modified T7, Taq, and Vent DNA Polymerases. *PCR Methods Appl.* **1,** 63–69.

38. Clark, J. M. (1988) Novel non-templated nucleotide addition reactions catalyzed by procaryotic and eucaryotic DNA polymerases. *Nucleic Acids Res.* **20,** 9677–9686.

39. Verhoeyen, M., Milstein, C., and Winter, G. (1988) Reshaping human antibodies: grafting an antilysozyme activity. *Science* **239,** 1534–1536.

40. Riechmann, L., Clark, M., Waldmann, H., and Winter, G. (1988) Reshaping human antibodies for therapy. *Nature* **332,** 323–327.

41. Tempest, P. R., Barbanti, E., Bremner, P., Carr, F. J., Ghislieri, M., Rifaldi, B., and Marcucci, M. (1994) A humanized anti-tumour necrosis factor-α monoclonal antibody that acts as a partial, competitive antagonist of the template antibody. *Hybridoma* **13,** 183–190.

42. Tempest, P. R., White, P., Williamson, E. D., Titball, R. W., Kelly, D. C., Kemp, G. J. L., Gray, P. M. D., Forster, S. J., Carr, F. J., and Harris, W. H. (1994) Efficient generation of a reshaped human mAb specific for the α toxin of *Clostridium perfringens. Prot. Eng.* **7,** 1501–1507.

43. Hsiao, K., Bajorath, J., and Harris, L. J. (1994) Humanization of 60.3, an anti-CD18 antibody; importance of the L2 loop. *Prot. Eng.* **7,** 815–822.

44. Huston, J. S., McCartney, J., Tai, M. S., Mottola-Carlshorn, C., Jin, D., Wartren, F., Keck, P., and Oppermann, H. (1993) Medical applications of single-chain antibodies. *Intern. Rev. Immunol.* **10,** 195–217.

45. Hayashi, N., Welschof, M., Zewe, M., Braunagel, M., Dubel, S., Breitling, F., and Little, M. (1994) Simultaneous mutagenesis of antibody CDR regions by overlap extension and PCR. *Biotechniques* **17,** 310–315.

46. Clackson, T., Gussow, D., and Jones, P. (1991) General applications of PCR to gene cloning and manipulation, in *Polymerase Chain Reaction: A Practical Approach* (McPherson, M. J., Quirke, P., and Taylor, G. R., eds.), Oxford University Press, Oxford, UK, pp. 187–214.

47. McCafferty, J., Griffiths, A. D., Winter, G., and Chiswell, D. J. (1990) Phage antibodies: filamentous phage displaying antibody variable domains. *Nature* **348,** 552–554.

48. Clackson, T., Hoogenboom, H. R., Griffiths, A. D., and Winter, G. (1991) Making antibody fragments using phage display libraries. *Nature* **352,** 624–628.

49. Adair, J., Bodmer, M. W., Mountain, A., and Owens, R. J. (1992) *CDR Grafted Anti-CEA Antibodies and Their Production.* WO92/01059.

29

Transgenic Techniques

Roberta M. James and Paul Dickinson

1. Introduction

Advances in molecular biology that have allowed the cloning and transfer of genetic material both within and between organisms have become a key tool in molecular biology. Molecular biological techniques have allowed the role of gene function to be addressed from the cellular level in vitro and to be extended to the whole organism in vivo by the use of transgenic techniques in both plants and animals. The major mammalian species in which transgenic techniques have been applied is the mouse.

There are many advantages to using mice: They are relatively inexpensive, large numbers of early embryos are easily obtained, the generation time is short, and the genetics are well-developed. Additionally, embryonal-stem (ES) cell lines are available that may be grown, modified in vitro, and be used subsequently to generate offspring possessing the modification. Many alternatives have been developed to facilitate the transfer of foreign DNA to the germ cells of mice *(1–3)*. These methods include aggregating ES cells from two genetically distinct animals to produce a chimera or introduction of DNA into totipotent teratocarcinoma cells or ES cells, then incorporation into blastocysts or eight-cell stage embryos. However, the most successful and now routine method is direct microinjection of cloned DNA into the pronuclei of fertilized eggs (**Fig. 1**).

2. Pronuclear Microinjection Transgenics

The first reported production of transgenic mice from microinjected eggs was in 1980 *(4)*, soon followed by several other reports *(5–8)*, some of which reported expression from stably inherited foreign DNA in somatic tissues, as well as transmission through the germ line. It was soon shown that sufficient transgene expression could be achieved to alter the physiology of the animal *(9)*.

2.1. Procedures for Production of Microinjection Transgenics

Extensive experimental details concerning microinjection techniques can be found elsewhere *(10–12)*; therefore, what follows is a brief summary of the various techniques employed and equipment required for successful microinjection followed by a brief description of the main methods of analysis.

The pronuclear injection method of gene transfer results in stable integration of the foreign DNA in 10–40% of the resulting animals. Integration generally occurs at the

From: *Molecular Biomethods Handbook*
Edited by: R. Rapley and J. M. Walker © Humana Press Inc., Totowa, NJ

Generation of transgenic mice by pronuclear microinjection

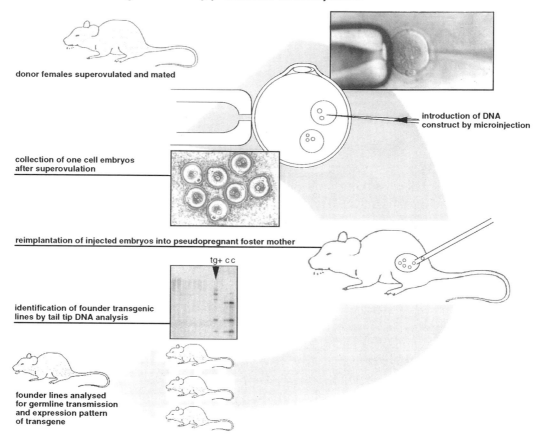

donor females superovulated and mated

introduction of DNA
construct by microinjection

collection of one cell embryos
after superovulation

reimplantation of injected embryos into pseudopregnant foster mother

tg+ c c

identification of founder transgenic
lines by tail tip DNA analysis

founder lines analysed
for germline transmission
and expression pattern
of transgene

Fig. 1. Generation of transgenic mice by pronuclear microinjection. Donor female mice are superovulated and mated with stud males. One-cell embryos are collected the following day and the DNA construct of interest introduced by pronuclear microinjection. Injected embryos are then reintroduced into pseudopregnant foster mothers. The resultant offspring are then analyzed for presence of the transgene and used as founder animals to generate transgenic lines or analyzed directly for expression of the transgene. tg+ve = transgene positive, c = control.

one-cell stage, leading to the presence of the transgene in all cells of the animal, including the germ cells, and allows transmission of the transgene to subsequent generations. However, in approx 20–30% of these cases a mosaic animal results when integration occurs after the first round of replication *(13,14)*. Transgenic lines may vary in copy number from single-copy integrants to hundreds or thousands of copies of the transgene and are usually arranged in tandem arrays at a single integration site, although integration sites on different chromosomes and complex integration sites may also occur *(15)*. The mechanism for integration of the transgene into the host genome is unknown.

2.1.1. Preparation of DNA for Microinjection

The size of DNA construct used to generate lines of transgenic mice is limited by the capacity of the cloning vector and the purification procedure employed. Plasmid, bacteriophage λ, and cosmid inserts *(6)* of up to 60 kb *(16)* have been used to produce transgenic mice. More recently, yeast artificial chromosomes (YACs) (*see* Chapter 24) carrying transgenes several hundred kilobases in size have also been successfully employed to produce transgenic mice *(17)*. Additionally, recombination of overlapping genomic fragments in murine zygotes *(18)* may be used to increase the effective size of the transgene introduced.

Linear DNA molecules give a fivefold greater frequency of integration than supercoiled molecules *(19)* whereas the presence of vector sequences may inhibit transgene expression by up to 1000-fold and are generally removed before microinjection. The purity of DNA used for microinjection is important in order to minimize toxicity to the embryo and prevent blockage of the microinjection needle.

Although optimal integration occurs when the concentration of DNA is 1 ng/µL or higher, to achieve the maximum number of transgenic mice, 1–2 ng/µL (corresponding to 200–400 copies/pL of a ~5 kb fragment) is the optimal balance between integration rate and offspring survival. No correlation has been established between the concentration of injected DNA and the number of copies of the transgene that integrate.

2.1.2. Preparation of Fertilized Eggs Prior to Microinjection

The prepared DNA is injected into the pronuclei of fertilized eggs and collected at the one-cell stage after superovulation and mating of donor females. The choice of donor-mouse strain is important: The females' response to superovulation, the visibility of the pronucleus, the possibility of the cell cycle blocking at the two-cell stage, and litter size are all criteria to consider. Quite commonly, the fertilized eggs are obtained from inbred-hybrid F_1 females, e.g., (CBAxC57BL/6) F_1.

Superovulation involves the administration of the gonadotropins, pregnant mare' serum gonadotropin (PMSG) and human chorionic gonadotropin (hCG) with an approx 45 h interval between injections *(20)*. Immediately after hCG administration, the females are housed with stud males. The following day one-cell stage embryos are collected from the oviducts of mated females into handling medium, the cumulus mass surrounding the eggs removed with the enzyme hyaluronidase and eggs incubated at 37°C, 5% CO_2 in microdrops of culture medium under paraffin oil to await microinjection *(21)*.

2.1.3. Microinjection Procedure

2.1.3.1. EQUIPMENT

To undertake pronuclear injection, a number of specialized pieces of equipment are required, the most essential being a good quality microscope. The most commonly used systems involve the use of an inverted fixed-stage microscope (e.g., Nikon Diaphot, Zeiss Axiovert, LeitzLabovert FS), preferably fitted with Normarski differential-interference optics, which allows easier visualization of the pronucleus, therefore making the microinjection easier. With this system, the microinjection is carried out in an injection chamber consisting of a drop of medium under oil on a tissue-culture dish or siliconized glass depression slide (**Fig. 2**).

During microinjection, DNA is introduced into the pronucleus via a narrow, pointed, glass injection pipet whereas the oocyte is held by suction onto the smoothed end of a glass holding pipet. Both pipets are controlled by micromanipulators, which allow their position in the field of view of the microscope to be controlled by means of control

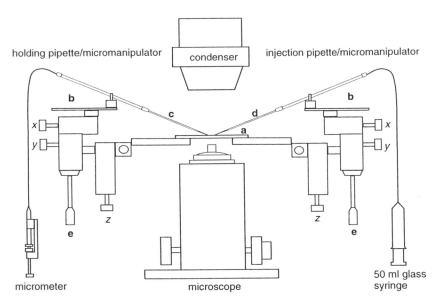

Fig. 2. Schematic diagram of microinjection apparatus. Microinjection of DNA constructs into single-cell embryos and transfer of ES cells into blastocysts may be achieved using an inverted microscope equipped with micromanipulators. Oocytes or blastocysts are transferred to a droplet of medium under oil on a depression slide **(A)** placed on the stage of an inverted microscope. Micromanipulators **(B)** are used to control the position of holding **(C)** and injection **(D)** pipets in *x*, *y*, and *z* dimensions using control knobs or a joystick **(E)**. Single embryos or blastocysts are held in place with pressure from a micrometer syringe or automatic-microinjection apparatus. DNA constructs or ES cells are loaded into the injection pipet, the pipet inserted into the pronucleus or blastocoel cavity by use of the joystick, and injection of DNA or ES cells controlled by the use of a 50-mL glass syringe or automatic-microinjection apparatus.

knobs and joysticks. Suction on the holding pipet is controlled via a micrometer syringe, while pressure is applied to the injection pipet with a 50-mL ground-glass syringe. Alternatively, flow from both pipets can be controlled using an automated injection device.

2.1.3.2. MICROINJECTION PROCEDURE

To commence microinjection, several eggs are placed into the chamber and examined for the presence of pronuclei. The absence of pronuclei may indicate that the embryo is unfertilized, indicated by the absence of the second polar body; fertilization has occurred too recently, or the pronuclei have broken down and the embryo is about to enter division. The injection pipet, filled with DNA for microinjection, is lowered into the chamber and the tip gently broken by touching the holding pipet. The holding pipet is brought into contact with an embryo and the back pressure increased until the embryo is securely held. The focus is adjusted until the pronucleus is visible. Either pronucleus is suitable for microinjection, although the male pronucleus is usually larger and nearer the surface and thus easier to inject. The injection pipet is brought into the same focal plane as the pronucleus and slowly pushed through the zone pellucida toward the nucleus of the embryo, avoiding the nucleolus. Once the needle is inside the pronucleus, pressure is exerted until the pronucleus swells, allowing injection of approx

2 pL of DNA solution. After injection, the pipet is pulled out quickly to minimize loss of cellular contents and the eggs are returned to culture. Fifty to eighty percent of injected eggs should survive and can be transferred to a pseudopregnant foster mother.

2.1.3.4. GENERATION OF TRANSGENIC MICE

Females that have been selected during estrus and housed with vasectomized males are used as foster mothers, because the female prepares for pregnancy although ovulated eggs will not be fertilized. Around 20–30 microinjected embryos can be surgically transferred to the oviduct of a 0.5-d postcoitum pseudopregnant female. The transfer is carried out either on the day of microinjection or on the following day when the embryos are at the two-cell stage (1.5 d p.c.). The asynchronous transfer makes allowance for the slower development of the manipulated embryos. Ten to thirty percent of microinjected embryos transferred to the oviduct will develop to term *(22)*.

2.1.3.5. IDENTIFICATION OF TRANSGENIC MICE

Transgenic mice may be examined for transient expression of the introduced transgene or, alternatively, founder lines may be established. Analysis of transient transgene expression during various stages of embryonic development is possible after sacrifice of the mothers. The phenotype and expression patterns of individual transgenic mice may be studied, often by using a transgene consisting of a promoter of interest coupled to a reporter gene (e.g., *LacZ*) to allow visualization of transgene expression. Alternatively, if permanent transgenic lines are to be established, putative transgenic offspring are screened for the presence of the transgene by DNA analysis shortly after weaning at 21 d post partum. DNA is extracted from tail-tip biopsies for analysis by Southern blotting or polymerase chain reaction (PCR). The mice are marked by ear punching, toe clipping, or tattooing so that future identification is possible. Tail-tip biopsies are generally performed under the anesthetic Halothane, which allows quick postoperative recovery.

2.1.3.6. ESTABLISHMENT OF TRANSGENIC LINES

Once transgenic offspring have been identified, they may be used as founders in the establishment of stably inherited lines. For transgenic lines to be established, the transgene must be transmitted through the germline of the founder to the next generation of mice. Some transgenic constructs may render this impossible and therefore may only be analyzed by the study of transient expression. Once a transgenic line is established, various methods may be used to examine transgene expression. Expression from the transgene may be identified by reverse transcriptase (RT-PCR), RNase-protection (*see* Chapter 5) assay, or Northern blotting of RNA samples (*see* Chapter 9) isolated from a range of tissues. To achieve spatial information regarding transgene expression RNA *in situ* hybridization can be performed on tissue sections. Hybridization to mRNA transcribed from the transgene shows in which tissues, as well as in which cell types the transgene is expressed. Immunohistochemistry may also be performed on tissue sections to reveal whether functional protein is produced and to show its localization within the cell.

2.2. Applications of Microinjection Transgenic Technology

The use of transgenic animals in biomedical research has led to insights into the many mechanisms that control growth and development in an organism. The ability to express genes in a selective manner within an animal has given rise to numerous research possibilities, some of which are discussed in **Subheadings 2.2.1.** and **2.2.2.**

2.2.1. The Use of Transgenic Mice in Research

Although cloned genes introduced into mice integrate at random sites, many show the appropriate tissue-specific and developmental-stage-specific expression determined by the promoter element used to direct expression of the transgene. This phenomenon has proven valuable in the study of tissue-specific and developmental-stage-specific gene regulation. Transgenic mice expressing novel genes normally represent a gain of function that may mimic certain human-dominant conditions. This coupled with the potential to use tissue-specific promoters and enhancers, make transgenic animals suitable models for studying human disorders *(23)*.

Transgenic animals have been used extensively in the study of mammalian development by changing or inactivating existing genes or by the addition of new genetic material. For example, it is possible to disrupt the function of protein complexes by expressing defective subunits from an introduced transgene, which will associate with the normal subunits in a dominant-negative manner *(24)*. Selective ablation of specific cell types can also be performed to address their function and interaction with other cell types. The diphtheria toxin A gene, engineered to prevent export from the cell, when placed under the regulatory-control regions of chosen genes, can selectively destroy cells in which the gene is expressed, leading to insights into the fate of those cells *(25,26)*. In addition, approaches have been developed to allow temporal control of transgene expression by administration of tetracycline to mice expressing a tetracycline-controlled transactivator protein *(27)*. Hybrid or reporter genes have been used in transgenic mice to direct the expression of marker proteins to selected cell types so that the specificity of expression of the regulatory elements involved may be examined. The most commonly used reporter genes include β-galactosidase, chloramphenicol acetyl transferase, and luciferase, which may be assayed biochemically or histochemically. Analysis of reporter-gene expression is usually determined transiently in a large number of transgenic mice. Although patterns of transgene expression largely reflect the elements used to direct expression of the reporter gene, individual transgenic lines are susceptible to position effects, such that nearby controlling elements and genes can affect the transcription of the transgene. It is therefore possible that a transgene in a particular genomic position may be overexpressed, downregulated, or switched off completely. The use of large genomic context YAC-transgenic vectors *(17)* or transgenic vectors possessing locus-control region (LCR) sequences *(28)* have overcome many of these problems to allow copy-number-dependent, position-independent transgene expression in individual transgenic lines. Transgenic mice expressing a normal gene confirm the ability of cloned genes to complement mouse mutations in vivo *(29)*. Studies of this nature allow inference into how much product is required to correct a genetic defect and is an essential step in the development of constructs for human-gene therapy.

2.2.2. Nonrodent Transgenic Applications

Although most widely used in mice, important developments have also been made using transgenic technology in other species. These include the production of enhanced agricultural products, production of unique products for human use, and more appropriate models for human disease *(30)*. Transgenic procedures have been developed in chickens, rabbits, goats, pigs, sheep, and cattle *(31)*, although the efficiency of integration is much lower than in mice. The most significant developments have taken place using sheep and pigs *(32)*. Transgenic lines have been produced to improve animal productivity by attempting to produce larger livestock *(33)*, increase muscle development *(34)*, or afford enhanced resistance to disease *(30)*. Direction of transgene expression to the mammary gland *(35)* allows expression of large quantities of foreign protein in the milk of transgenic animals, which may either modify the milk for agricultural purposes or be of pharmaceutical importance to humans. For example, human α_1-antitrypsin (hα_1AT) genomic sequences under the control of sheep β-lactoglobulin regulatory sequences are shown to express hα_1AT highly in the milk of transgenic ewes. The purified protein is indistinguishable from that purified from human plasma. Generation of transgenic pigs expressing a human complement-regulating protein, decay-accelerating factor (DAF), which blocks elements responsible for hyperacute rejection after xenotransplantation, may open up the future possibility of using transgenic-pig organs for human transplantation *(36)*. In the future, it may be possible to produce livestock that have genetically enhanced traits, such as increased-feed conversion and rate of weight gain, reduction of fat, and improved quality of meat, milk, and wool, all of which would be agriculturally and economically important.

3. Embryonal-Stem (ES) Cell-Derived Transgenics

Microinjection of DNA constructs into the pronuclei of fertilized oocytes generally leads to the random integration of the construct. In many instances, a more directed approach including an in vitro selection procedure may be required that can not be contemplated by microinjection of pronuclei with subsequent random integration of DNA. The use of ES cells allows these modifications to be made. ES cells are derived from the inner-cell mass of blastocysts and may be grown and genetically modified in culture before being reintroduced into blastocysts to produce chimeric animals in which the modified cells may contribute to the germ line (**Fig. 3**). In vitro selection of ES clones can lead to mice incorporating a novel gene, carrying a modified endogenous gene or lacking a specific endogenous gene following gene deletion or "knockout" *(37,38)*. (For a detailed description of the methodology used to generate ES cell-derived transgenic mice, *see* **ref. 10**. What follows is a brief description of the techniques used, including examples of the uses to which this technology can be applied.

3.1. Procedures for Production of ES Cell-Derived Transgenics

In order to contemplate producing ES cell-derived transgenic animals, ES cell lines capable of contributing to the germ line of chimeric animals must be available. DNA constructs may be introduced into these ES cells by a number of techniques, including electroporation, microinjection, transfection, retroviral insertion, and spheroplast fusion *(39–44)*. For experiments in which the modification of an endogenous gene by homologous recombination is to be performed, suitable selectable markers and screening strategies are required. Successfully modified ES cell lines may then be reintroduced into blastocysts either by microinjection of ES cells into the blastocoel or by aggregation of clumps of ES cells with eight-cell stage embryos. Coat-color markers and isoenzyme analysis can be used to monitor efficiency of chimera formation, as

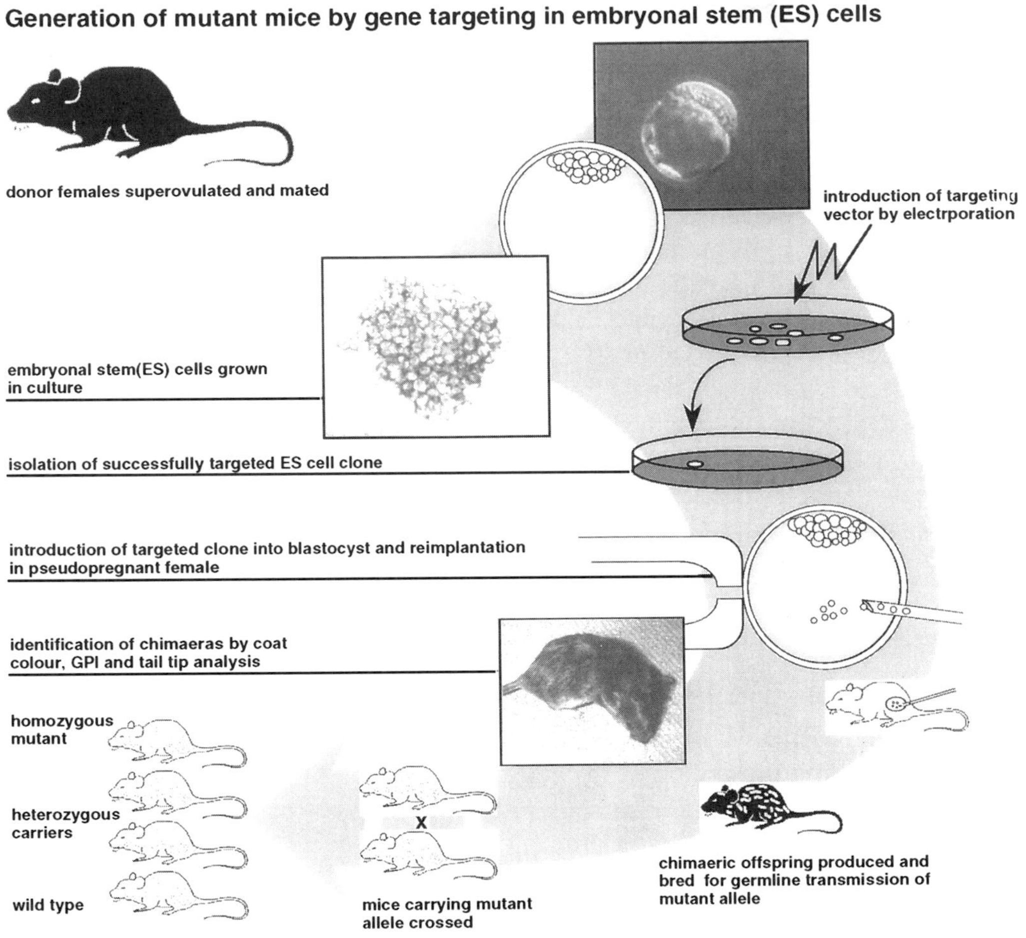

Generation of mutant mice by gene targeting in embryonal stem (ES) cells

donor females superovulated and mated

introduction of targeting
vector by electrporation

embryonal stem(ES) cells grown
in culture

isolation of successfully targeted ES cell clone

introduction of targeted clone into blastocyst and reimplantation
in pseudopregnant female

identification of chimaeras by coat
colour, GPI and tail tip analysis

homozygous
mutant

heterozygous
carriers

wild type

mice carrying mutant
allele crossed

chimaeric offspring produced and
bred for germline transmission of
mutant allele

Fig. 3. Generation of mutant mice by gene targeting in ES cells. ES-cell lines derived from the inner cell mass of mouse blastocysts can be modified in vitro by introduction of a range of DNA constructs. Successfully modified ES clones are selected and introduced into blastocysts collected from superovulated donor females 3.5 d after mating with stud males. Injected blastocysts are then reintroduced into pseudopregnant foster mothers and chimeric offspring identified by coat-color markers. Mating of chimeric offspring may subsequently produce mice heterozygous for the modification introduced into the ES cells and these mice may be intercrossed to generate homozygous mice.

well as subsequent transmission of the genetic modification through the germ line, which may be also monitored by PCR or Southern-blot analysis.

3.1.1. Generation of ES Cell Lines

Murine ES cell lines may be produced by culture of embryos in vitro beyond the point at which they normally implant in the uterus. This may be mimicked in vitro by embryo coculture on fibroblast monolayers or by the use of growth factors (e.g., leukemia inhibitory factor [LIF]). ES-cell lines derived in this manner may retain the totipotent phenotype of the embryonal cells from which they were isolated. However, culture conditions are critical to the maintenance of totipotency, whereas the genetic background of host and donor strains of mice may influence the ability of ES cells to show germ line transmission *(45)*.

Extension of these techniques to other species has been hampered by a general inability to culture embryos in vitro. However, some progress has recently been made toward derivation of ES cells in a number of species that are capable of contributing to chimeric offspring *(46)*, although germ-line transmission has still to be demonstrated.

3.1.2. Genetic Modification of ES Cells

ES cells grown in culture may be transfected using most of the standard techniques developed for transfection of mammalian cells in culture. These include electroporation, microinjection, transfection, retroviral insertion, and spheroplast fusion. However, at all stages of manipulation and especially on implementation of positive- or negative-selection strategies, care must be taken to prevent loss of totipotency.

3.1.2.1. CREATION OF "KNOCKOUT" MICE

The most widespread use of ES cell-derived transgenic mice has been the generation of "knockout" lines by the use of homologous-recombination techniques *(37)*. Electroporation of ES cells with constructs possessing homology to the gene of interest and positive or negative selectable-marker genes (e.g., neomycin phosphotransferase, hygromycin resistance, herpes simplex virus [HSV] thymidine kinase) may result in homologous recombination of the construct with the target locus at frequencies varying from one in <10 to one in several thousand. A number of important factors influencing the frequency of homologous recombination have been identified, including the use of constructs isogenic with the target-DNA sequence *(47)*, the length of homologous sequence used *(48)*, and the nature of the recombination process (e.g., insertional vs replacement targeting [**Fig. 4**]) *(49,50)*. Homologous recombination techniques have been used to abolish the function of many genes, create leaky mutations in which levels of gene expression are reduced but not abolished, and introduce specific-point mutations into genes to produce protein products of altered function.

Replacement gene-targeting strategies use constructs possessing two regions of homology to the target sequence flanking a positive selectable marker. In the positive-negative selection (PNS) strategy *(39)*, one or two negative selectable markers flanking either end of the targeting sequence are included in an attempt to reduce the frequency of random integration. Double-reciprocal crossover between the regions of homology and target sequence leads to disruption of the targeted gene, which may abolish gene function (**Fig. 4A**).

Insertional gene-targeting strategies *(50)* use constructs possessing a single region of homology to the target sequence as well as positive and possibly negative selectable-marker genes. Linearization of the construct within the region of homology followed by recombination with the target locus generates a duplication of the target disrupted by intervening selectable marker and plasmid sequences (**Fig. 4B**). Splicing in of exonic sequences present in the targeting region may disrupt the coding sequence of the targeted gene to abolish gene function. However, the nature of the introduced mutant exon may influence splicing into this sequence and, because no sequences have been deleted from the targeted region, it is possible to generate a normal message if the mutant exon is overlooked by the splicing machinery *(51)*.

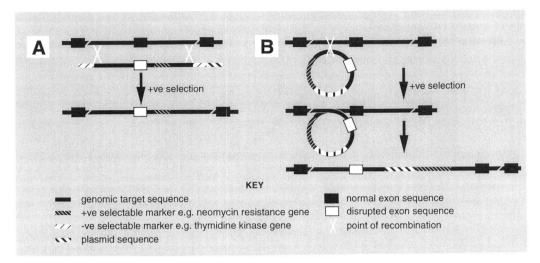

Fig. 4. Comparison of gene targeting by replacement and insertional vectors. Replacement targeting **(A)** involves double reciprocal crossover events (indicated by the cross) to replace the target DNA sequence with a homologous sequence present in the vector and including a positively selectable marker gene, e.g., neo[r]. This strategy can also include a negative selection for sequences present at the ends of the region of homology, e.g., thymidine kinase, to reduce the level of background owing to random integration of the construct. Insertional targeting **(B)** involves a single recombination between the target sequence and homologous sequence present in the vector at the site of linearization. This results in integration of the entire vector, including the positively selectable marker gene, e.g., neo[r], and duplication of the target region of homology.

3.1.2.2. Introduction of Specific Mutations

A number of different strategies have been devised for the introduction of specific mutations into genes by homologous recombination in ES cells (**Fig. 5**). Single-step replacement strategies have been devised, where the targeted region is directly replaced with a homologous sequence bearing a precise mutation, but which also results in the introduction of a positive selectable marker to an intron *(52,53)* (**Fig. 5A**). A two-step replacement targeting strategy *(54)* has also been suggested, in which the gene is first targeted with a back selectable marker (**Fig. 5Bi**), and this is then replaced with a homologous sequence bearing a precise mutation (**Fig. 5Bii**). Additionally, "hit-and-run" or "in-out" strategies have been devised *(55,56)* that use a modification of the insertional gene targeting strategy. In this instance, a duplication of the targeted region that possesses the introduced mutation (hit) is created using an insertional approach (**Fig. 5Ci**), followed by back selection with resolution and loss of the duplicated normal target (run) (**Fig. 5Cii**).

3.1.2.3. Trapping Approaches

In contrast to using techniques of homologous recombination in ES cells to examine the function of particular genes, the ability of an exogenous DNA vector to integrate randomly in the genome when introduced into ES cells can be used to identify new genes in gene-trapping approaches *(57)*. By including in the vector splice- acceptor sequences in frame with a downstream marker/reporter gene, it is possible to detect

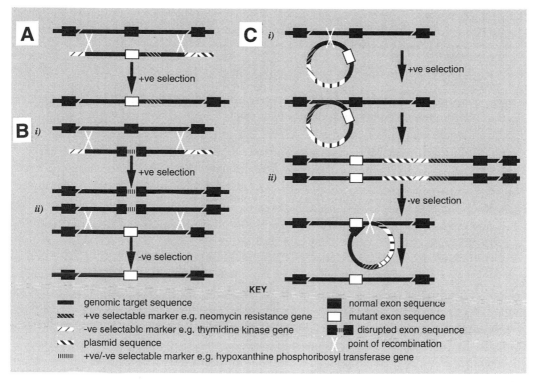

KEY

▬▬	genomic target sequence	■	normal exon sequence
⚡⚡⚡	+ve selectable marker e.g. neomycin resistance gene	□	mutant exon sequence
⁄⁄⁄	-ve selectable marker e.g. thymidine kinase gene	▨▨	disrupted exon sequence
⟍⟍⟍	plasmid sequence	X	point of recombination
⦀⦀	+ve/-ve selectable marker e.g. hypoxanthine phosphoribosyl transferase gene		

Fig. 5. Strategies for the introduction of subtle mutations into ES cells. Single-step replacement **(A)** involves replacement of the targeted region with a homologous sequence bearing a precise mutation, but also results in the introduction of a positive selectable marker, e.g., neo^r, into an intron. Two-step replacement **(B)** involves replacement of the targeted region with a homologous sequence bearing a positively and negatively selectable marker gene, e.g., HPRT **(Bi)**, followed by a second round of replacement targeting in which this is then replaced with a homologous sequence bearing a precise mutation **(Bii)**. Hit-and-run targeting **(C)** involves the use a modification of the insertional gene-targeting strategy. In this instance, a duplication of the targeted region that possesses the introduced mutation (hit) is created using an insertional approach **(Ci)**, followed by back selection with resolution and loss of the duplicated normal target (run) **(Cii)**.

integrations into active transcription units and to isolate clones representing individual insertion events. Reporter activity is dependent on insertion into the intron of an actively transcribed gene, followed by in-frame splicing of the upstream exon to the vector-splice acceptor. As a consequence of splicing, a fusion transcript is produced that both disrupts gene function as an insertional mutation as well as marking the expression pattern of the disrupted gene. Additionally, rapid amplification of cDNA ends (RACE) PCR techniques can be employed to allow cloning of the disrupted gene by virtue of the known vector sequence present in the fusion transcript. Modifications of this technique include the secretory-trap, in which the addition of a transmembrane domain in the gene-trap vector allows selection for integration into genes encoding secreted or membrane-bound proteins *(58)*. Enhancer-trap strategies employ the use of a marker/reporter gene possessing minimal promoter elements in which trapping of the *cis*-regulatory enhancer elements of an endogenous gene are required for expression of the reporter gene *(59)*. This strategy does not require in-frame insertion of the trapping construct and allows identification of novel expression patterns, but does not necessarily lead to gene disruption.

3.1.2.4. GENERATION OF SPECIFIC DELETIONS/REARRANGEMENTS

Homozygous disruption of genes important in early development may produce a phenotype of early embryonic lethality, regardless of other roles the gene may play later in development. To overcome these problems, techniques have been developed to create conditional knockouts of genes in mice. Sequence specific recombination systems present in bacteriophage P1 (Cre-*lox*P) and yeast (FLP-FRT) have been utilized to produce site-specific recombination in mammalian cells in culture by recombinase (Cre or FLP)-mediated recombination between duplicated recognition sequence sites (*lox*P or FRT). Introduction of *lox*P sites by homologous recombination on either side of genic-coding sequences, followed by induction of Cre recombinase activity, can lead to gene inactivation by deletion of the coding sequence *(60)*. In this way, genes can be knocked out after Cre-mediated recombination controlled spatially or temporally during development to help unmask the role the gene may play at specific times or places during development.

Homologous recombination techniques can only be used to make local modifications to specific genomic-target sequences in ES cells. However, for some purposes, such as modeling rearrangements associated with human cancers and developmental abnormalities, the ability to manipulate mammalian chromosomes in vivo to produce translocations, inversions, and deletions is required. The Cre-*lox*P system has been used in ES cells after sequential gene targeting with *lox*P sites and transient expression of Cre recombinase to generate a programmed translocation between the c-*myc* and immunoglobulin heavy-chain genes on chromosomes 15 and 12 *(61)*. The use of inducible recombinase activities (e.g., Cre-human estrogen receptor fusion protein has been shown to exhibit ligand-dependent recombinase activity in embryonal carcinoma cells *[62]*) allows the possibility of chromosome rearrangements or gene knockouts that can be activated in ES cell-derived mice in vivo in a spatiotemporally regulated manner.

3.1.2.5. INTRODUCTION OF LARGE GENOMIC CONSTRUCTS

As mentioned in **Subheading 2.2.1.**, the introduction of large genomic constructs into the mouse germ line may help to overcome many of the problems normally associated with microinjection transgenics *(17)*. Microinjection techniques are limited by both the volume and DNA concentration that can be tolerated by the oocyte. As the size of the vector increases, the number of copies of the vector that can be injected decreases, so that with a vector of 1000 kb injected in a 2-pL volume at an optimal DNA concentration of 2 ng/μL, only 3.6 copies of the vector would be injected. As an alternative to microinjection of large DNA constructs into fertilized oocytes, it is possible to introduce these constructs into ES cells in vitro by lipofection *(41,42)* or spheroplast fusion *(44)*. Selection for the presence of YACs in ES cells can be accomplished by the previous introduction of a mammalian-selectable marker into the YAC-vector arm by homologous recombination in yeast *(42)*, or by colipofection of an unlinked selection cassette *(63)*. YACs purified by preparative pulsed-field gel electrophoresis (PFGE) can be complexed with various lipid reagents and used to transfect ES cells *(41,42)*.

Alternatively, polyethylene glycol-mediated fusion between spheroplasts (produced by enzymatic digestion of the cell wall of YAC containing yeast cells) and ES cells can allow stable transfer of YAC DNA *(44)*. Integrity and expression of the introduced transgene can be examined in vitro in ES cells prior to the production of transgenic mice.

3.1.3. Generation of Chimeras and Germ-Line Transmission

ES-cell clones genetically modified in vitro can subsequently be used to produce ES cell-embryo chimeras to allow transmission of the genetic modification to the offspring of these animals. Two methodologies are currently used for the generation of chimeras; microinjection of ES cells into the blastocoel cavity and aggregation of ES cells with eight-cell embryos.

The microinjection technique *(64)* involves the transfer of 10–15 ES cells into the blastocoel cavity of a recipient embryo using microinjection apparatus similar to that used to produce microinjection-transgenic mice (**Fig. 2**). Recipient blastocysts are collected from females superovulated in the same manner as for production of one-cell stage embryos (*see* **Subheading 2.1.2.**) 3.5 d p.c. by flushing the uterus with handling medium. Blastocysts surrounded by an intact zona pellucida and possessing a visible blastocoel cavity are transferred to the microinjection apparatus, where they are held using suction from a holding pipet. ES cells are drawn into an injection pipet possessing a sharp point or beveled edge of 35–40° angle to allow puncture of the zona pellucida, and 10–15 ES cells are injected into the blastocoel cavity, taking care not to disturb the inner-cell mass. Although this technique has been effectively used by many groups to produce chimeric animals that are capable of transmitting the introduced genetic modification through the germ line, it has a number of limitations. Micromanipulation equipment is both elaborate and expensive, and the manipulations themselves are time-consuming and require substantial training.

In order to overcome these problems, aggregation techniques have been developed that rely on the observation that ES cells readily aggregate with eight-cell embryos if the two cell populations are brought into contact *(65)*. Eight-cell stage embryos are collected from superovulated females 2.5 d p.c. and the zona pellucida removed from the embryos by incubation in acidified Tyrode's solution. As soon as the zona pellucida has disappeared, embryos are washed and transferred to microdrops of medium in bacteriological-grade Petri dishes possessing indentations created by a blunt-ended darning needle. Clumps of 5–10 ES cells are then transferred to each depression and brought into physical contact with an embryo to allow aggregation. Overnight culture at 37°C to the blastocyst stage can result in efficient internalization of the ES cells to produce chimeric embryos containing large numbers of ES cell-derived cells in the inner-cell mass, from which the embryo proper develops.

In both instances, ES cell lines are used that have coat color, eye color, and isoenzyme markers that allow them to be distinguished from the cells derived from the recipient embryo. Chimeric embryos are transferred to the uteri of recipient 2.5 d p.c. pseudo-pregnant females. Six to eight embryos in a small volume of medium can then be transferred to each uterine horn using a transfer pipet. Chimeric offspring may be identified by their patchy coat color owing to the presence of ES cell- derived (129/Rr J-chinchilla) and recipient-derived ([C57BL/6xCBA] F_1 [brown] or CD1 [white]) cells. Additionally, chimeric offspring can be identified by analysis of GPI isoenzyme status

Table 1
Applications of ES Cell Transgenic Technology

Gene modified/knocked out	Phenotype/disease	References
Angiotensinogen	Blood pressure	*76*
Angiotensin II receptor 1a	Blood pressure	*77*
Angiotensin II receptor 2	Blood pressure	*78*
Apolipoprotein E	Hypercholesterolemia, Atherosclerosis	*79–81*
Arginosuccinate synthetase	Citrullinemia	*82*
ATM	Ataxia telangiectasia	*69*
Cystathionine-β synthase	Homocyst(e)inemia	*83*
Cystic fibrosis transmembrane conductance regulator	Cystic fibrosis	*70–72*
Cytochrome b(-245), β polypeptide	X-linked chronic granulomatous disease	*84*
Erythroid Kruppel-like factor	Lethal β-thalassemia	*85*
FMR1	Fragile-X mental-retardation	*86*
Glucocerebrosidase	Gaucher's disease	*73*
Hemoglobin, α adult chains 1 and 2	Lethal α-thalassemia	*87*
Hemoglobin, β chain	Lethal β-thalassemia	*88*
Hemoglobin, β adult chains major and minor	β^0-thalassemia	*89*
Hexosaminidase A	Tay-Sachs disease	*90*
Hox 1.5	DiGeorge's syndrome-like	*91*
Hypoxanthine-guanine phosphoribosyl transferase	Lesch-Nyhan syndrome	*43,74,75*
Interleukin 2	Ulcerative colitis	*92*
Interleukin 10	Chronic enterocolitis	*93*
Low density lipoprotein receptor	Hypercholesterolemia	*94*
RXR-α receptor	Ventricular hypoplasia with fenestrated ventricular septal defects	*95*

of tail-vein blood samples. Mice that show high degrees of chimerism may then be set up in matings to examine the ES cell-derived contribution to the germline. Transmission of coat color and isoenzyme markers as well as any genetic modification introduced to the ES cells in vitro can be analyzed to examine the extent of ES-cell colonization of the germ line. ES cell-derived offspring can then be crossed to generate mice that are homozygous for the introduced genetic modification.

3.2. Applications of ES Cell Transgenic Technology

The ability to precisely modify genes in ES cells and produce animals carrying these alterations has allowed the dissection of gene function in vivo. The application of gene-targeting technology to generate knockouts of gene function has been used to study the role of genes in a vast array of biological processes, including development, cancer, neurobiology, immunology, toxicology, and numerous genetic diseases (*see* **Table 1** or **ref.** *61*).

Knockout mice have proven invaluable in developing an understanding of the mechanisms governing cell division, differentiation, and death and have provided insight into events contributing to cellular dysregulation and loss of growth-control leading to tumorigenesis. In addition, knockout of mouse homologs of genes mutated in human-genetic disorders has given insight into the genetic basis of disease in humans, allowing the generation of numerous mouse models of human-genetic disease. The phenotypes observed in some of these mouse models show a striking similarity with the human condition, e.g., ataxia telangiectasia *(69)*, cystic fibrosis *(70–72)*, and Gaucher's disease *(73)*. However, this is not always the case and it is important to be aware of the differences in physiology that exist between humans and mice. This can be illustrated by the example of Lesch-Nyhan disease. In humans, mutations in the hypoxanthine phosphoribosyltransferase (HPRT) gene lead to Lesch-Nyhan disease, symptoms of which include self-injurious behavior. The mouse HPRT gene was one of the first genes to be knocked out in the mouse *(43,74)*, but no phenotype was initially observed. However, blockage of the alternative route of the purine-salvage pathway, which is of greater importance in the mice than humans, by treatment with inhibitors of adenosine phosphoribosyltransferase (APRT) lead to the self-injurious behavior characteristic of Lesch-Nyhan disease in the HPRT-deficient mice *(75)*.

In addition to the ability to knockout particular genes, technology has been developed to allow the introduction of specific-point mutations, which has allowed the modeling of disease states precisely and aided in the understanding of genotype/phenotype relationships *(54,56,96)*. Techniques have also been developed to allow the creation of specific translocations and rearrangements that are often associated with human cancers and developmental abnormalities *(61)*. In addition to these directed approaches to the understanding of gene function, gene-trap approaches can be used as a genetic screen to select for insertion mutants that display patterns of expression that are of particular interest *(57–59)*. Taken together, these techniques allow unrivaled genetic intervention into a mammalian system to allow the role of gene function to be analyzed.

4. Summary

Techniques now exist that allow the role of gene function to be addressed in mammalian systems in vivo. The major mammalian species in which transgenic techniques have been applied is the mouse. Pronuclear microinjection techniques have allowed the analysis of elements important for correctly regulated gene expression, selective ablation of particular cell types, spatiotemporal control of transgene expression, the effect of overexpression of particular genes to generate models of human disease, and testing of potential gene-therapy constructs for complementation of mutations. These techniques have now been extended to a range of other species to allow the generation of animal bioreactors for the in vivo production of pharmaceutical proteins and may in the future allow the use of animal organs for xenotransplantation. The existence of ES cell lines that may be modified in vitro and subsequently used to generate offspring possessing the modification have created the ability to knockout gene function in the mouse. Homologous recombination can be used to disrupt specific genes and create null alleles. Genes can also be knocked out in a random manner using gene-trap approaches and screened for interesting patterns of expression to generate novel mouse mutants. Site-specific recombinase activities can be used to generate conditional knockouts and induce chromosomal rearrangements or translocations to allow an examination of the effects of these gross rearrangements. Alternatively, techniques now exist that allow the precise introduction of single-codon changes into genes to examine the effect of missense mutations on gene function in vivo. Such a wide range of possibilities can now be contemplated by the use of the transgenic techniques now available that the major limitation to further progress is an understanding of the developmental, biological, and physiological consequences of the intervention.

Acknowledgments

The authors would like to acknowledge funding by the Scottish Hospital Endowments Research Trust (R. M. J.) and Cystic Fibrosis Trust (P. D.) and would like to thank D. J. Porteous and N. D. Hastie for helpful discussion of the manuscript.

References

1. Palmiter, R. D. and Brinster, R. L. (1986) Germ-line transformation of mice. *Ann. Rev. Genet.* **20,** 465–499.
2. Jaenisch, R. (1988) Transgenic animals. *Science* **240,** 1468–1474.
3. Hanahan, D. (1989) Transgenic mice as probes into complex systems. *Science* **246,** 1265–1275.
4. Gordon, J. W., Scangos, G. A., Plotkin, D. J., Barbosa, J. A., and Ruddle, F. H. (1980) Genetic transformation of mouse embryos and bone marrow—a review. *Gene* **33,** 121–136.
5. Brinster, R. L., Chen, H. Y., Trumbauer, M. E., Senear, A. W., Warren. R., and Palmiter, R. D. (1981) Somatic expression of herpes thymidine kinase in mice following injection of a fusion gene into eggs. *Cell* **27,** 223–231.
6. Constantini, F. and Lacy, E. (1981) Introduction of a rabbit β-globin gene into the mouse germline. *Nature* **294,** 92–94.
7. Wagner, E. F., Stewart, T. A., Mintz, B. (1981) The human β-globin gene and a functional thymidine kinase gene in developing mice. *Proc. Natl. Acad. Sci. USA* **78,** 5016–5020.
8. Wagner, T. E., Hoppe, P. C., Jollick, J. D., Scholl, D. R., Hodinka, R. L., and Gault, J. B. (1981) Microinjection of a rabbit β-globin gene in zygotes and its subsequent expression in adult mice and their offspring. *Proc. Natl. Acad. Sci. USA* **78,** 6376–6380.
9. Palmiter, R. D., Brinster, R. L., Hammer, R. E., Trumbauer, M. E., Rosenfeld, M. G., Birnberg, N. C., and Evans, R. M. (1982) Dramatic growth of mice that develop from eggs microinjected with metallothionein-growth hormone fusion genes. *Nature* **300,** 611–615.
10. Hogan, B., Beddington, R., Costantini, F., and Lacy, E. (1994) *Manipulating the Mouse Embryo.* Cold Spring Harbor Laboratory, Cold Spring Harbor, NY.
11. Gordon, J. W. (1993) Production of transgenic mice. *Meth. Enzymol.* **225,** 747–771.
12. Allen, N. D., Barton, S. C., Surani, M. A. H., and Reik, W. (1987) Production of transgenic mice, in *Mammalian Development: A Practical Approach* (Monk, M., ed.), IRL, Oxford, UK, pp. 217–253.
13. Wilkie, T. M., Brinster, R. L., and Palmiter, R. D. (1986) Germline and somatic mosaicism in transgenic mice. *Dev. Biol.* **118,** 9–18.
14. Whitelaw, C. B. A., Springbett, A. J. Webster, J., and Clark, J. (1993) The majority of G_0 transgenic mice are derived from mosaic embryos. *Transgen. Res.* **2,** 29–32.
15. Lacy, E., Roberts, S., Evans, E. P., Burtenshaw, M. D., and Costantini, F. (1983) A foreign β-globin gene in transgenic mice: integration at abnormal chromosomal positions and expression in inappropriate tissues. *Cell* **34,** 343–358.

16. Taylor, L. D., Carmack, C. E., Schramm, S. R., Mashayekh, R., Higgins, K. M., Kuo, C. C., Woodhouse, C., Kay, R. M., and Lonberg, N. (1992) A transgenic mouse that expresses a diversity of human sequence heavy and light chain immunoglobulins. *Nucleic Acids Res.* **20,** 6287–6295.

17. Schedl, A., Montoliu, L., Kelsey, G., and Schutz, G. (1993) A yeast artificial chromosome covering the tyrosinase gene confers copy number-dependent expression in transgenic mice. *Nature* **362,** 258–261.

18. Pieper, F. R., de Wit, I. C. M., Pronk, A. C., Koolman, P. M., Strijker, R., Krimpenfort, P. A. J., Nuyens, J. H., and de Boer, H. A. (1992) Efficient generation of functional transgenes by homologous recombination in murine zygotes. *Nucleic Acids Res.* **20,** 1259–1264.

19. Brinster, H. L., Chen, N. Y., Trumbauer, M. E., Yagle, M. K., and Palmiter, R. D. (1985) Factors affecting the efficiency of introducing foreign DNA into mice by microinjecting eggs. *Proc. Natl. Acad. Sci.* **82,** 4438–4442.

20. Hetherington, C. M. (1987) Mouse husbandry, in *Mammalian Development: A Practical Approach* (Monk, M., ed.), IRL, Oxford, UK, pp. 1–12.

21. Pratt, H. P. M. (1987) Isolation, culture and manipulation of pre-implantation mouse embryos, in *Mammalian Development: A Practical Approach* (Monk, M., ed.), IRL, Oxford, UK, pp. 13–42.

22. Mann, J. R. and McMahon, A. P. (1993) Factors influencing frequency production of transgenic mice. *Meth. Enzymol.* **225,** 771–799.

23. Greaves, D. R., Fraser, P., Vidal, M. A., Hedges, M. J., Ropers, D., Luzzatto, L., and Grosveld, F. (1990) A transgenic mouse model of sickle cell disorder. *Nature* **338,** 352–355.

24. Herskowitz, I. (1987) Functional inactivation of genes by dominant negative mutations. *Nature* **329,** 219–222.

25. Palmiter, R. D., Behringer, R. R., Qualfe, C. J., Maxwell, F., Maxwell, L. H., and Brinster, R. L. (1987) Cell lineage ablation in transgenic mice by cell-specific expression of a toxin gene. *Cell* **50,** 435–443.

26. Behringer, R. R., Mathews, L. S., Palmiter, R. D., and Brinster, R. L. (1988) Dwarf mice produced by genetic ablation of growth hormone-expressing cells. *Genes Dev.* **2,** 453–461.

27. Furth, P. A., St. Onge, L., Boger, H., Gruss, P., Gossen, M., Kistner, A., Bujard, H., and Hennighausen, L. (1994) Temporal control of gene expression in transgenic mice by a tetracycline-responsive promoter. *Proc. Natl. Acad. Sci. USA* **91,** 9302–9306.

28. Talbot, D., Collis, P., Antoniou, M., Vidal, M., Grosveld, F., and Greaves, D. R. (1989) A dominant control region from the human β-globin locus conferring integration site-independent gene expression. *Nature* **338,** 352–355.

29. Schedl, A., Ross, A., Lee, M., Engelkamp, D., Rashbass, P., van Heyningen, V., and Hastie, N. D. (1996) Influence of PAX6 gene dosage on eye development: overexpression causes severe eye abnormalities. *Cell* **86,** 71–82.

30. Pursel, V. G. and Rexroad, C. E., Jr. (1993) Status of research with transgenic farm animals. *J. Anim. Sci.* **71 (Suppl. 3),** 10–19.

31. Ebert, K. M. and Schindler, J. E. S. (1993) Transgenic farm animals: progress report. *Theriogenology* **39,** 121–135.

32. Pursel, V. G. and Rexroad, C. E., Jr. (1993) Recent progress in the transgenic modification of swine and sheep. *Mol. Rep. Dev.* **36,** 251–254.

33. Pursel, V. G., Pinkert, C. A., Miller, K. F., Bolt, D. J., Campbell, R. G., Palmiter, R. D., Brinster, R. L., and Hammer, R. E. (1989) Genetic engineering of livestock. *Science* **244,** 1281–1288.

34. Pursel, V. G., Sutrave, P., Wall, R. J., Kelly, A. M., and Hughes, S. H. (1992) Transfer of c-ski gene into swine to enhance muscle development. *Theriogenology* **37,** 278 (abstr.).

35. Wilmut, I., Archibald, A. L., McClenaghan, M., Simons, J. P., Whitelaw, C. B. A., and Clark, A. J. (1991) Production of pharmaceutical proteins in milk. *Experientia* **47,** 905–912.

36. Storck, M., Abendroth, D., Prestel, R., Pinochavez, G., Pohlein, C., Pascher, A., White, D., and Hammer, C. (1996) Role of human decay-accelerating factor expression on porcine kidneys during xenogeneic *ex-vivo* hemoperfusion. *Transplant. Proc.* **28,** 587,588.

37. Capecchi, M. R. (1989) Altering the genome by homologous recombination. *Science* **244,** 1288–1292.

38. Bradley, A., Hasty, P., Davis, A., and Ramirez-Solis, R. (1992) Modifying the mouse: design and desire. *Biotechnology* **10,** 534–539.

39. Mansour, S. L., Thomas, K. R., and Capecchi, K. R. (1988) Disruption of the proto-oncogene ins-2 in mouse embryo-derived stem cells: a general strategy for targeting mutations to non-selectable genes. *Nature* **336,** 348–352.

40. Zimmer, A. and Gruss, P. (1989) Production of chimaeric mice containing embryonic stem (ES) cells carrying a homeobox *Hox 1. 1* allele mutated by homologous recombination. *Nature* **338,** 150–153.

41. Strauss, W. M., Dausman, J., Beard, C., Johnson, C., Lawrence, J. B., and Jaenisch, R. (1993) Germ line transmission of a yeast artificial chromosome spanning the murine $\alpha_1(l)$ collagen locus. *Science* **259,** 1904–1907.

42. Lamb, B. T., Sisodia, S. S., Lawler, A. M., Slunt, H. H., Kitt, C. A., Kearns, W. G., Pearson, P. L., Price, D. L., and Gearhardt, J. D. (1993) Introduction and expression of the 400 kilobase precursor amyloid protein gene in transgenic mice. *Nature Genet.* **5,** 22–29.

43. Kuehn, M. R., Bradley, A., Robertson, E. J., and Evans, M. J. (1987) A potential animal model for Lesch-Nyhan syndrome through introduction of HPRT mutations into mice. *Nature* **326,** 295–298.

44. Jakobovits, A., Moore, A. L., Green, L. L., Vergara, G. J., Maynard-Currie, C. E., Austin, H. A., and Klapholz, S. (1993) Germ-line transmission and expression of a human- derived yeast artificial chromosome. *Nature* **362,** 255–258.

45. Abbondanzo, S. J., Gadi, I., and Stewart, C. L. (1993) Derivation of embryonic stem cell lines. *Meth. Enzymol.* **225,** 803–823.

46. Campbell, K. H. S., McWhir, J., Ritchie, W. A., and Wilmut, I. (1996) Sheep cloned by transfer from a cultured cell line. *Nature* **380,** 64–66.

47. te Riele, H., Robanus, M., and Berns, A. (1992) Highly efficient gene targeting in embryonic stem cells through homologous recombination with isogenic DNA constructs. *Proc. Natl. Acad. Sci. USA* **89,** 5128–5132.

48. Hasty, P., Rivera-Perez, J., and Bradley, A. (1991) The length of homology required for gene targeting in embryonic stem cells. *Mol. Cell. Biol.* **11,** 5386–5591.

49. Hasty, P., Rivera-Perez, J., Chang, C., and Bradley, A. (1991) Target frequency and integration pattern for insertion and replacement vectors in embryonic stem cells. *Mol. Cell. Biol.* **11,** 4509–4517.

50. Thomas, K. R. and Capecchi, M. R. (1987) Site-directed mutagenesis by gene targeting in mouse embryo-derived stem cells. *Cell* **51,** 503–512.

51. Dorin, J. R., Stevenson, B. J., Fleming, S., Alton, E. W. F. W., Dickinson, P., and Porteous, D. J. (1994) Long-term survival of the exon 10 insertional cystic fibrosis mutant mouse is a consequence of low level residual wild-type *Cftr* gene expression. *Mamm. Genome.* **5,** 465–472.

52. Deng, C., Thomas, K. R., and Capecchi, M. R. (1993) Location of crossovers during gene targeting with insertion and replacement vectors. *Mol. Cell. Biol.* **13,** 2134–2140.

53. Delaney, S., Alton, E. W. F. W., Smith, S. N., Lunn, D. P., Fariey, R., Lovelock, P. K., Thomson, S. A., Hume, D. A., Lamb, D., Porteous, D. J., Dorin, J. R., and Wainwright, B. J. (1996) A cystic fibrosis mouse model carrying the common missense mutation G551 D. *EMBO J.* **15,** 955–963.

54. Wu, H., Liu, X., and Jaenisch, R. (1994) Double replacement: strategy for efficient introduction of subtle mutations into the murine Col 1a-1 gene by homologous recombination in embryonic stem cells. *Proc. Natl. Acad. Sci. USA* **91,** 2819–2823.

55. Hasty, P., Ramirez-Solis, R., Krumlauf, R., and Bradley, A. (1991) Introduction of a subtle mutation into the Hox-2. 6 locus in embryonic stem cells. *Nature* **350,** 243–246.

56. van Doorninck, J. H., French, P. J., Verbeek, E., Peters, R. H. P. C., Morreau, H., Bijman, J., and Scholte, B. J. (1995) A mouse model for the cystic fibrosis ΔF508 mutation. *EMBO J.* **14,** 4403–4411.

57. Friedrich, G. and Soriano, P. (1991) Promoter traps in embryonic stem cells: a genetic screen to identify and mutate developmental genes in mice. *Genes Dev.* **5,** 1513–1523.

58. Skarnes, W. C., Moss, J. E., Hurtley, S. M., and Beddington, R. S. P. (1995) Capturing genes encoding membrane and secreted proteins important for mouse development. *Proc. Natl. Acad. Sci. USA* **92,** 6592–6596.

59. Korn, R., Schoor, M., Neuhaus, H., Henseling, U., Soininen, R., Zachgo, J., and Gossler, A. (1992) Enhancer trap integrations in mouse embryonic stem cells give rise to staining patterns in chimeric embryos with a high-frequency and detect endogenous genes. *Mech. Dev.* **39,** 95–109.

60. Kuhn, R., Schwenk, F., Aguet, M., and Rajewsky, K. (1995) Inducible gene targeting in mice. *Science* **269,** 1427–1429.

61. Smith, A. J. H., De Sousa, M. A., Kwabi-Addo, B., Heppell-Parton, A., Impey, H., and Rabbitts, P. (1995) A site-directed chromosomal translocation induced in embryonic stem cells by Cre-loxP recombination. *Nature Genet.* **9,** 376–385.

62. Metzger, D., Clifford, J., Chiba, H., and Chambon, P. (1995) Conditional site-specific recombination in mammalian cells using a ligand-dependent chimeric Cre recombinase. *Proc. Natl. Acad. Sci. USA* **92,** 6991–6995.

63. Choi, T. K., Hollenbach, P. W., Pearson, B. E., Ueda, R. M., Weddell, G. N., Kurahara, C. G., Woodhouse, C. S., Kay, R. M., and Loring, J. F. (1993) Transgenic mice containing a human heavy chain immunoglobulin gene fragment cloned in a yeast artificial chromosome. *Nature Genet.* **4,** 117–123.

64. Stewart, C. L. (1993) Production of chimeras between embryonic stem cells and embryos. *Meth. Enzymol.* **225,** 823–855.

65. Wood, S. A., Allen, N. D., Rossant, J., Auerbach, A., and Nagy, A. (1993) Non-injection methods for the production of embryonic stem cell-embryo chimaeras. *Nature* **365,** 87–89.

66. Brandon E. P., Idzerda R. L., and McKnight, G. S. (1995) Targeting the mouse genome: a compendium of knockouts (part I). *Curr. Biol.* **5,** 625–634.

67. Brandon E. P., Idzerda R. L., and McKnight, G. S. (1995) Targeting the mouse genome: a compendium of knockouts (part II). *Curr. Biol.* **5,** 758–765.

68. Brandon E. P., Idzerda R. L., and McKnight, G. S. (1995) Targeting the mouse genome: a compendium of knockouts (part III). *Curr. Biol.* **5,** 873–881.

69. Barlow, C., Hirotsune, S., Paylor, R., Liyange, M., Eckhaus, M., Collins, F., Shiloh, Y., Crawley, J. N., Ried, T., Tagle, D., and Wynshaw-Boris, A. (1996) *Atm*-deficient mice: a paradigm of ataxia telangiectasia. *Cell* **86,** 159–171.

70. Snouwaert, J. N., Brigman, K. K., Latour, A. M., Malouf, N. N., Boucher, R. C., Smithies, O., and Koller, B. H. (1992) An animal model for cystic fibrosis made by gene targeting. *Science* **257,** 1083–1088.

71. Dorin, J. R., Dickinson, P., Alton, E. W. F. W., Smith, S. N., Geddes, D. M., Stevenson, B. J., Kimber, W. L., Fleming, S., Clarke, A. R., Hooper, M. L., Anderson, L., Beddington, R. S. P., and Porteous, D. J. (1992) Cystic fibrosis in the mouse by targeted insertional mutagenesis. *Nature* **359,** 211–215.

72. Ratcliff, R., Evans, M. J., Cuthbert, A. W., MacVinish, L. J., Foster, D., Anderson, J. R., and Colledge, W. H. (1993) Production of a severe cystic fibrosis mutation in mice by gene targeting. *Nature Genet.* **4,** 35–41.

73. Tybulewicz, V. L. J., Tremblay, M. L., LaMarca, M. E., Willemsen, R., Stubblefield, B. K., Winfield, S., Zablocka, B., Sidransky, E., Martin, B. M., Huang, S. P., Mintzer, K. A., Westphal, H., Mulligan, R. C., Ginns, E. I. (1992) Animal model of Gaucher's disease from targeted disruption of the mouse glucocerebrosidase gene. *Nature* **357,** 407–410.

74. Hooper, M., Hardy, K., Handyside, A., Hunter, S., and Monk, M. (1987) HPRT-deficient (Lesch-Nyhan) mouse embryos derived from germline colonization by cultured cells. *Nature* **326,** 292–295.

75. Wu, C. L. and Melton, D. W. (1993) Production of a model for Lesch-Nyhan syndrome in hypoxanthine phosphoribosyltransferase mice. *Nature Genet.* **3,** 235–240.

76. Kim, H-S., Krege, J. H., Kluckman, K. D., Hagaman, J. R., Hodgin, J. B., Best, C. F., Jennette, J. C., Coffman, T. M., Maeda, N., and Smithies, O. (1995) Genetic control of blood pressure and the angiotensin locus. *Proc. Natl. Acad. Sci. USA* **92,** 2735–2739.

77. Ito, M., Oliverio, M. I., Mannon, P. J., Best, C. F., Maeda, N., Smithies, O., and Coffman, T. M. (1995) Regulation of blood pressure by the type 1 A angiotensin 11 receptor gene. *Proc. Natl. Acad. Sci. USA* **92,** 3521–3525.

78. Hein, L., Barsh, G. S., Pratt, R. E., Dzau, V. J., and Kobilka, B. K. (1995) Behavioural and cardiovascular effects of disrupting the angiotensin 11 type-2 receptor gene in mice. *Nature* **377,** 744–747.

79. Piedrahita, J. A., Zhang, S. H., Hagaman, J. R., Oliver, P. M., and Maeda, N. (1992) Generation of mice carrying a mutant apolipoprotein E gene inactivated by gene targeting in embryonic stem cells. *Proc. Natl. Acad. Sci. USA* **89,** 4471–4475.

80. Plump, A. S., Smith, J. D., Hayek, T., Aalto-Setala, K., Walsh, A., Verstuyft, J. G., Rubin, E. M., and Breslow, J. L. (1992) Severe hypercholesterolemia and atherosclerosis in apolipoprotein E-deficient mice created by homologous recombination in ES cells. *Cell* **71,** 343–353.

81. Zhang, S. H., Reddick, R. L., Piedrahita, J. A., and Maeda, N. (1992) Spontaneous hyper-cholesterolemia and arterial lesions in mice lacking apolipoprotein E. *Science* **258,** 468–471.

82. Patejunas, G., Bradley, A., Beaudet, A. L., and O'Brien, W. E. (1994) Generation of a mouse model for citrullinemia by targeted disruption of the argininosuccinate synthetase gene. *Somat. Ceil. Mol. Genet.* **20,** 55–60.

83. Watanabe, M., Osada, J., Aratani, Y., Kluckman, K., Reddick, R., Malinow, M. R., and Maeda, N. (1995) Mice deficient in cystathionine beta-synthase: Animal models for mild and severe homocyst(e)inemia. *Proc. Natl. Acad. Sci. USA* **92,** 1585–1589.

84. Pollock, J. D., Williams, D. A., Gifford, M. A. C., Li, L. L., Du, X., Fisherman, J., Orkin, S. H., Doerschuk, C. M., and Dinauer, M. C. (1995) Mouse model of X-linked chronic granulomatous disease, an inherited defect in phagocyte superoxide production. *Nature Gen.* **9,** 202–209.

85. Perkins, A. C., Sharpe, A. H., and Orkin, S. H. (1995) Lethal beta thalassaemia in mice lacking the erythroid CACCC-transcription factor EKLF. *Nature* **375,** 318–322.

86. Bakker, C. E., Verheij, C., Willemsen, R., Vanderheim, R., Oerlmans, F., Vermey, M., Bygrave, A., Hoogeveen, A. T., Oostra, B. A., Reyniers, E., Deboulle, K., Dhooge, R., Cras, P., Vanvelezen, D., Nagels, G., Martin, J. J., Dedeyn, P. P., Darby, J. K., and Willems, P. J. (1994) FMR1 knockout mice—a model to study fragile-X mental retardation. *Cell* **78,** 23–33.

87. Paszty, C., Mohandas, N., Stevens, M. E., Loring, J. F., Liebhaber, S. A., Brion, C. M., and Rubin, E. M. (1995) Lethal alpha-thalassaemia created by gene targeting in mice and its genetic rescue. *Nature Gen.* **11,** 33–39.

88. Shehee, W. R., Oliver, P., and Smithies, O. (1993) Lethal thalassemia after insertional disruption of the mouse major adult beta-globin gene. *Proc. Natl. Acad. Sci. USA* **90**, 3177–3181.

89. Yang, B., Kirby, S., Lewis, J., Detloff, P. J., Maeda, N., and Smithies, O. (1995) A mouse model for β^0-thalassemia. *Proc. Natl. Acad. Sci. USA* **92**, 11,608–11,612.

90. Yamanaka, S., Johnson, M. D., Grinberg, A., Westphal, H., Crawley, J. N., Taniike, M., Suzuki, K., and Proia, R. L. (1994) Targeted disruption of the Hexa gene results in mice with biochemical and pathologic features of Tay-Sachs disease. *Proc. Natl. Acad. Sci. USA* **91**, 9975–9979.

91. Chiaska, O. and Capecchi, M. R. (1991) Regionally restricted developmental defects resulting from targeted disruption of the mouse homeobox gene hox-1.5. *Nature* **350**, 473–479.

92. Sadlack, B., Merz, H., Schorle, H., Schimpl, A., Feller, A. C., and Horak, I. (1993) Ulcerative colitis-like disease in mice with a disrupted interleukin-2 gene. *Cell* **75**, 253–261.

93. Kuhn, R., Lohler, J., Rennick, D., Rajewsky, K., and Muller, W. (1993) Interleukin-10-deficient mice develop chronic enterocolitis. *Cell* **75**, 263–274.

94. Ishibashi, S., Brown, M., Goldstein, J., Gerard, R. D., Hammer, R. E., and Herz, J. (1993) Hypercholesterolemia in LDL receptor knockout mice and its reversal by adenovirus-mediated gene delivery. *J. Clin. Invest.* **92**, 883–893.

95. Sucov, H. M., Dyson, E., Guminger, C. L., Price, J., Chien, K. R., and Evans, R. M. (1994) RXR alpha mutant mice establish a genetic basis for vitamin A signalling in heart morphogenesis. *Genes Dev.* **8**, 1007–1018.

96. Dorin, J. R., Farley, R., Webb, S., Smith, S. N., Farini, E., Delaney, S. J., Wainwright, B. J., Alton, E. W. F. W., and Porteous, D. J. (1996) A demonstration using mouse models that successful gene therapy for cystic fibrosis requires only partial gene correction. *Gene Ther.* **3**, 797–801.

30

DNA Profiling

Theory and Practice

Karen M. Sullivan

1. Introduction

Of the vast range of techniques in use in modern molecular biology, DNA profiling is one of the more difficult to describe, primarily because the term "DNA Profiling" itself means very different things to different people. It is, in fact, a widely applied term that describes a number of technically diverse approaches to the study of hypervariable DNA. This ambiguity is compounded by the fact that the term "hypervariable DNA" describes more than one fraction of genomic DNA. The aim of this chapter is, therefore, to describe the common molecular basis of the varying forms of hypervariable-DNA analyses, and how this may be exploited for genetic identity and relationship testing, to examine the practical approaches available for such studies and their respective strengths and limitations, and finally to look at the range of applications of each DNA-profiling technique.

Over the 10 years since its inception, the application of DNA-profiling technology have grown almost exponentially. It is in use in medical research and diagnostics, veterinary science and animal breeding, plant and animal ecology and conservation, forensic science, epidemiology, and taxonomic studies of plant and animal populations. DNA profiling has become an invaluable research tool in almost every branch of biological and medical science, in no small part owing to the process of technical evolution and refinement that the technology has undergone since it was first reported by Jeffreys et al. in 1985 *(1)*. It is precisely the speed of this evolution that has lead to the generation of the previously mentioned range of techniques under the umbrella term of DNA profiling.

2. Molecular Basis of DNA Profiling

The mean heterozygosity of the human genome is extremely low—around 0.001 per base pair, which at first consideration would make genomic DNA a poor candidate as a means of individual identification. This was, in fact, the case until the publication of Jeffrey's landmark paper in 1985 *(1)*, which described the existence of loci that, unlike the bulk of DNA, exhibited a high degree of individual specificity. Although other groups had previously observed the occurrence of such loci *(2–4)*, they had not explored the use of such loci as tools of genetic individualization. The result of Jeffreys exten-

From: *Molecular Biomethods Handbook*
Edited by: R. Rapley and J. M. Walker © Humana Press Inc., Totowa, NJ

Locus X

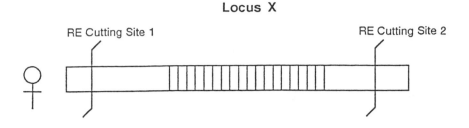

Locus X

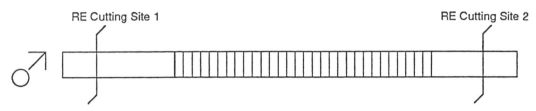

Fig. 1. The maternal and paternal alleles of Locus X are illustrated, showing 18 and 30 tandem repeats, respectively. Note the increase in overall length of the paternal allele with respect to the maternal allele. This will result in a visible RFLP on electrophoretic separation, hybridization, and detection.

sion to this research was the identification of a series of loci that displayed multiple allelism and the simultaneous detection of these loci by Southern blotting (Chapter 8) giving rise to an individual specific banding pattern.

This multiple allelism is the result of the specific sequence structure of these so-called "minisatellite" loci. A sequence motif is tandemly repeated at the minisatellite locus, with the repeat number and hence, the size of the allele varying considerably between individuals. Indeed, the observed range of possible alleles at a given minisatellite was such that they were designated as "hypervariable" regions, to distinguish them from alleles that exhibited a low level of polymorphism. Perhaps the most accurate terminology coined for these regions was Variable Number Tandem Repeats (VNTRs). The accuracy of this desicription is clearly shown in **Fig. 1**, which shows the basic molecular structure of a hypervariable locus.

The quantum leap had been made from the previous perception of DNA as a molecule of low heterozygosity to a molecule that possessed regions that were hypervariable, to a degree that allowed identification of an individual by direct genetic rather than phenotypic means. Allelism at these loci could be detected by alterations in the size of the locus that could easily be visualized by well-established techniques of electrophoresis, blotting, and hybridization (*see* Chapters 3 and 8).

This chapter discusses techniques arising from this original work, which seem technically remote from that first reported, however, the molecular basis of all of the next-generation techniques is essentially the same. The basic principle is that the length of the

Mother   Child 1   Child 2   Alleged Father

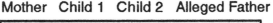
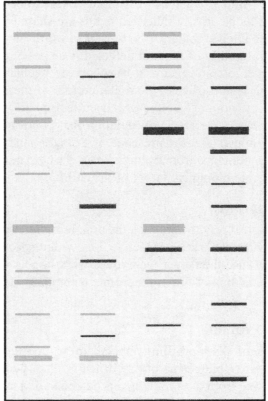

Fig. 2. The schematic of a typical paternity test shows the DNA profiles of the mother, her two children, and the alleged father. The bands in the children that may have been inherited from the mother are those that align with bands in her DNA profile; the remainder must therefore have been inherited paternally. Child 1 cannot be the offspring of the alleged father because the paternal alleles do not find ad match in the male DNA profile. On the other hand, it is very likely that the alleged father is indeed the biological father if Child 2, because all of this child's paternal alleles a find a match in the father's profile, with respect to both size and intensity of hybridization.

allele detected at any given hypervariable locus is likely to be characteristic of the individual in which it was detected, owing to the large number of length variants that may occur at that locus owing to its repetitive structure. Hence, the basis of genetic-identity testing: The size of a series of hypervariable alleles in an individual is very unlikely to be precisely matched by the corresponding allele sizes in any other individual. This led to the technique referred to as DNA fingerprinting, owing to the apparently unique quality of the series of genetic characteristics detected by this methodology. However, this term was soon superceded by the DNA profiling label that is now most commonly used, largely for reasons of political correctness which will be discussed in **Subheading 8.**

Subsequent studies soon indicated that the alleles detected at these loci were largely inherited in a Mendelian fashion *(5)*, so that all of the bands in an individual's DNA profile were directly inherited from his or her biological parents—this is schematized in **Fig. 2.**, which demonstrates how the technology can be used for definitive relationship testing. In the case illustrated, the alleged father in a paternity test can be excluded as a possible father of Child 1, whereas he is highly likely to be the father of Child 2, because all of the child's paternally inherited bands are shared with bands present in the DNA profile of the alleged father. It is, however, worth introducing a cautionary note here: The hypervariability of the loci detected is a consequence of the ability of these sequences to undergo rearrangement, by whatever mechanism, to generate alleles of new and varying length. Hence, the mutation rate of these alleles is extremely high compared to that of "normal" DNA *(6)*. It is quite possible, therefore, that a child may have a "new" allele as the result of mutation that is not present in either of its biological parents. It follows that the presence of a single band that could not be assigned definitively to inheritance from a putative father would not by necessity exclude that person as the possible biological father of the child.

3. The Practical Approach

Perhaps the most logical way to approach the practical strategies that currently exist for DNA profiling is via a historical pespective, i.e., with respect to the chronology of their development, because the newer techniques were largely developed to circumvent technical or financial limitations of the former technique, and are thus best considered in this context.

4. Multilocus DNA Profiling

This is the first type of DNA-profiling analysis to be reported *(1)*, and it uses conventional hybridization techniques as outlined in **Fig. 3** to generate a bar-code-like banding pattern, resulting from the simultaneous detection of a large number of related hypervariable regions scattered throughout the genome. It is imperative that the DNA be digested to completion and it is usual to check that restriction is complete before electrophoresis is carried out. The only other critical part of the procedure is the stringency of hybridization and washing, which varies depending on the precise nature of the probe being used *(1,7,8)* (*see* Chapter 6). This protocol may be varied, however, if oligonucleotide (MLPs) are being used, because the small size of the DNA probe facilitates *in situ* gel hybridization *(9)*.

The genomic location and precise sequence of the alleles detected are not known. They simply have a degree of homology to the conserved-core sequence of the probe used; the varying extent of complementary to the probe is reflected in the variation in intensity of the bands observed.

The advantages of this type of DNA profiling is that it gives a large amount of information in a single test with as many as 20 resolvable alleles being detected with a single probe. It is a relatively quick method, giving results within a few days if sufficient DNA is available for testing, which begs the question regarding how much DNA is required for such a test. To generate a clear multilocus DNA profile, a minimum of 1–2 µg of high-mol-wt DNA is required, which represents a considerable obstacle in some fields of application. The large amount of information generated by the test can, in itself, be a problem, in that if a DNA sample originates from a mixed source, i.e., more than one individual, the resulting DNA-banding pattern is too complex for analysis of the mixed-DNA profiles. The necessity for high-mol-wt DNA may also be prohibitive in some types of application, where DNA samples available for testing may be

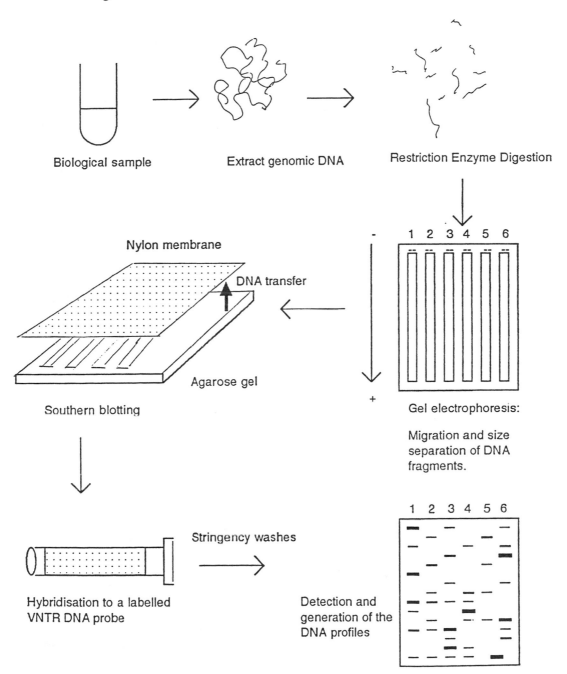

Fig. 3. An outline of the practical approach to MPL DNA profiling.

partially degraded. As mentioned in **Subheading 2.**, the nonuniformity of hybridization of the probe to its target loci results in marked variation in band intensity, from
strong signals to bands that are only just visible to the human eye. The result is that
these profiles are extremely difficult to analysze by automated scanners, owing to the
difficulty in setting parameters that can detect across the whole range of band intensi-

ties. As a result, not only must the profiles be scored manually, but there is also the problem that this introduces an element of subjectivity into the analysis. A consequence of this is that profiles may require independent scoring by two analysts, which increases the costs of the technology considerably.

A further consequence of the varying complementarity of the probe is that stringency is absolutely critical; fluctuations of stringency of hybridization or washing between tests may result in the loss or gain of the weaker bands, so that contemporaneous testing of samples to be compared is essential. Given sufficient DNA, however, this technique has an almost unparalleled discriminating power between individuals, which makes it an exceptionally powerful tool for both identity *(10)* and relationship testing *(11)*.

5. Single-Locus DNA Profiling

The practical approach to this technique is effectively identical to that employed for multilocus DNA profiling, with the exception only of the stringency of the hybridization and washing conditions. The latter arises because the single-locus probe (SLP) and its target hypervariable locus have a higher degree of complementarily than do the equivalent MLPs and target sequence. A SLP is generated by actually subcloning a single hypervariable locus *(12,13)*. At high stringency, this probe will hybridize only to the two alleles of the locus from which it was cloned, giving a two band pattern in heterozygotes, the precise sizes of which represent the variable genetic characteristic. Specific band sizes may be scored for frequency of occurence in a given population, in exactly the same way as the frequencies of phenotypic characteristics may be scored, so that examination of a series of single hypervariable loci may, like its multilocus counterpart, build up a genetic profile of a number of variable genetic characteristics. Together, these characteristics may be used to identify an individual with a high degree of certainty *(14)*.

The use of SLPs overcomes a number of the limitations of MLP, in that the simple two-band pattern per probe facilitates easy analysis of mixed DNA profiles. Mixed samples from two people would show a four-band pattern, which would still be amenable to unambiguous interpretation. The greater degree of sensitivity of the SLPs also enables a DNA profile to be obtained from about 10-fold less DNA, and allows a reasonable profile to be obtained from even partially degraded DNA. The presence of only two bands of approximately equal intensity also makes this technology amenable to automated scanning, which greatly decreases the labor intensity of the process. The SLP also has the advantage that it is amenable to polymerase chain reaction (PCR) amplification *(15)*, because the sequence of the hypervariable locus and its flanking regions can be determined from the cloned-probe fragment, thus allowing the generation of PCR primers to specifically amplify the hypervariable locus of choice.

6. PCR-Based Single-Locus DNA Profiling

The drive toward a PCR-based technique for the analysis of hypervariable DNA was primarily led by the forensic scientists *(16)* in an attempt to obtain evidential value from samples where the DNA content was too small for even SLP analysis; this may frequently arise in cases where the sample is degraded owing to exposure to heat and moisture prior to processing, or in sexual crimes where the perpetrator was oligospermic, for example. The underlying principle is identical to conventional SLP analysis by hybridization, except that PCR amplification of the allele is used to amplify up initially from a minute amount of starting material.

After initial problems with the faithful amplification of larger alleles, the PCR was successfully used, and demonstrated the time-based advantages of PCR testing as compared to conventional hybridization technology. This, in conjunction with parallel developments in MLP technology, which had identified a new type of hypervariable DNA family in genomic DNA, led to the development of a totally new genre of DNA profiling—microsatellite analysis.

Microsatellites are in effect, just smaller versions—with respect to repeat size—of their minisatellite counterparts, commonly comprising di-, tri-, or tetranucleotide tandemly repeated sequence motifs *(17,18)*. The smaller size of these hypervariable elements makes them easier to amplify reproducibly on PCR amplification, and also allows higher-resolution sizing analysis, which obviates many of the problems of accurate band sizing associated with conventional SLP analysis for forensic purposes. The combined development of primers for microsatellite loci with nonoverlapping size ranges, and the use of distinctively colored fluorescent labels for the primers, has allowed coamplification and analysis of up to seven such loci simultaneously. As a result, the whole process of DNA extraction, amplification, and analysis can be achieved in approx 24 h with a huge reduction in the labor intensity originally associated with DNA profiling technology.

Although this type of multiplex-PCR analysis is most definitely the DNA-profiling technology of the future for forensic casework, its application for identity and relationship testing in other areas of science may be limited by the cost of the automated-DNA sequencer and analyzer that is used for detection and analysis of the data obtained from these analyses.

An unforeseen advantage of microsatellite analysis compared to minisatellite-based analysis is that microsatellites are distributed more frequently and randomly throughout the genome, whereas minisatellites tend to be more concentrated around the telomeres of chromosomes *(19)*. Their variability, ubiquity, and ease of detection via PCR analysis have combined to make microsatellites a new and powerful tool in linkage analysis *(20)*, for the construction of genetic maps *(21)*, and the establishment of genomic markers in genetically inherited disease *(22,23)* and in cancers *(24,25)*.

7. Applications of DNA-Profiling Technology

The field of application of these combined technologies is so vast that it is difficult to define a rationale from which to approach this topic. Perhaps the simplest way is to look at the type of information that the technique can generate, and then to show how this type of information may be exploited under various circumstances to answer a variety of biological and medical questions.

If we consider the description of hypervariable-DNA loci—that they are highly individual specific and inherited in a largely Mendelian fashion—then it follows that their analysis forms the basis of two fundamental types of test, namely identity arnd relationship testing. It is under these two headings that the majority of applications may be discussed.

8. Genetic Identity Testing

8.1. Forensic Identification

Forensic science is, without doubt, the area of application where DNA profiling has attracted the greatest degree of publicity. Prior to the initial use of this technology for forensic identification of suspects from biological material left at the scene of a crime, identity testing was carried out primarily by means of blood grouping. Owing to the relatively small number of blood groups and the prevalence of certain of these groups in society, the techniques were only really suitable for exclusion rather than identification purposes. DNA profiling therefore made the quantum leap from providing exclusory evidence to generating evidence of positive identification, so that individuals could be identified to a high degree of certainty, by matching the genetic characteristics present in a forensic stain to those present in the blood sample of a suspect *(26–28)*. The exclusory powers of the technique have also been shown to be of considerable forensic value *(29)*.

This is not to say that the passage of DNA profiling technology into the standard repertoire of forensic techniques was without challenge. Early cases used multilocus DNA profiling analysis to match up samples by a side-by-side analysis, as illustrated in **Fig. 4.** Once visual inspection had identified two profiles as matching, i.e., there were the same number of bands of similar intensities in identical positions in each sample, then the number of bands in the best-resolved part of the profile were counted (normally taken to be above 4 kb in size). It was observed that on average, two unrelated people matched one in four of their bands taken over the whole size range of the scored profile, and so it was adopted that the probability of matching any two bands by chance was, at a conservative estimate 0.25. Hence, to obtain a likelihood of any two profiles having matched by chance, rather than having come from the same person, 0.25 is raised to the power of the number of matching bands. When one considers that as many as 20 or more bands can be detected by a single probe, and further, that two different MLPs can be used sequentially to increase the degree of certainty of a match, it is easy to see that the probability of two unrelated people having matching DNA profiles by chance is so low as to tend to zero. For this reason, the technology was originally referred to as DNA fingerprinting, echoing the uniqueness that we associate with our digital fingerprints. Although some researchers had reservations about treating DNA bands detected by different MLPs as statistically independent, on the basis that linked loci may be detected, and others queried the accuracy of the band share value of 0.25, the statistical evaluation of the weight of DNA profiling evidence was largely accepted.

Eager for the impressive statistical weight of DNA evidence as compared to traditional serological-identification methods, the inability of MLP technology to cope with mixed or denatured DNA samples, or those containing insufficient DNA was seen as a severe limitation of the technology. The development of SLP technology remedied a number of these limitations, as previously discussed, but the statistical evaluation of SLP DNA evidence brought a new set of problems, perhaps even more difficult to overcome than the preceeding technical limitations.

A typical set of SLP data, generated by the serial application of four SLPs is schematized in **Fig. 5.** It is apparent that the simple band-share method used in multilocus analysis cannot be employed here. In this case, it is necessary to calculate the precise size of the bands observed by reference to a set of size markers run contem-

V   CS   Su1   Su2

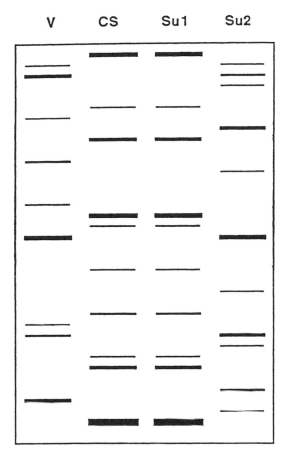

Fig. 4. The schematic shows a typical set of forensic multilocus DNA tests, with the DNA samples of the victim (V), crime sample (CS), and suspects 1 and 2 (Su1 and Su2) being tested contemporaneously. All bands in the crime sample are matched in size and intensity by the bands present in the DNA profile of suspect 1, who has a high probability of being the source of the crime sample DNA. Suspect 2 can clearly be eliminated as a potential source of the DNA in the forensic sample.

poraneously with the test samples, and then to refer to a prevously generated database of frequency of occurrence of each allele size for each locus. Although this method offers the advantage that the crime and suspect samples do not have to be run side by side, because they are each compared to a reference marker track rather than to each other directly, it introduces a whole range of questions on which the scientific community have found it difficult to agree. This method also generates statistical weights of the significance of the match, which are orders of magnitude smaller than those generated by MLP DNA-profiling analysis. For this reason, it was deemed inappropriate to use such a terminology as "fingerprinting," which conveyed uniqueness, under circumstances where there may be several people that possess any given set of SLP characteristics—hence the change from DNA fingerprinting to profiling.

The primary areas of dispute reagrding SLP statistics were the degree of variation that can be exhibited by samples originating from a single source, and hence the win-

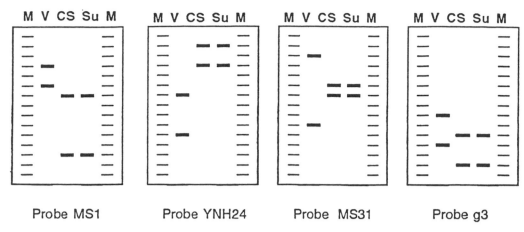

Probe MS1   Probe YNH24   Probe MS31   Probe g3

Fig. 5. The schematic illustrates a set of forensic DNA samples, showing the size marker fragments (M), the victim's DNA sample (V), the DNA from the crime sample (CS), and the DNA sample of the suspect (Su). These samples have been hybridized consecutively to four different SLPs, thus generating four sets of data. In all cases the bands detects in the crime sample are matched by bands present in the DNA sample of the suspect, which means that there is a high probability that the suspect is the source of the DNA found at the scene of the crime.

dows of variation that must be imposed on the calculated band size in order to establish band-matching criteria *(30,31)*, the statistical means of weighting the DNA evidence when bands were not perfectly matched, and the size, applicability and constitution of the databases used *(32–34)*. Although subsequent research has resolved some of these issues satisfactorily *(35)*, others have rumbled on and conventions have been adopted rather than precise scientific answers having been found.

Although certain aspects of the database composition are still contentious, the problems with band matching have been largely obviated by a move to PCR-based microsatellite analysis *(36)*. The high resolution of the electrophoresis system used mean: that the size of any observed band can be determined to the nearest base pair, and because tri- or tetranucleotide repeats are normally analyzed, the alleles size can be assigned unambiguously.

Although some multiplex PCR-based systems currently in use do not provide quite the same level of discrimination as do SLP or MLP analyses, the number of loci that can be coamplified and hence the degree of certainty of the match of any two samples is rapidly increasing. The speed and accuracy of the technique compared to SLP technology will almost certainly supercede completely SLP technology for forensic purposes in the near future.

In summary, then, despite early controversy and setbacks, DNA profiling technology, most recently in the form of mutiplex PCR analysis of microsatellites, is an invaluable tool for forensic-identity testing using biological samples in cases of rape, sexual assault, assault, murder, and even burglary, where a burglar is cut gaining entry to a property.

8.2. Cell-Line Verification

Happily, the forensic application of DNA profiling is almost unique in the difficulties it encounters, because in few, if any, of its other applications is the statistical significance of the match of any relevance whatsoever. In cell-line verification, as in most other applications, it is simply a question of comparing DNA profiles to see whether they match.

Cell lines are used extensively in modern biological and medical science, and it is frequently the case that a tissue culture facility within a company or research institute will have numerous related cell lines in simultaneous use. These are frequently chosen for their specific genetic background, which is essential to the success of the experiment or to the interpretation of the data. If cell lines become contaminated by an invading line that, although phenotypically similar, has a different genetic profile, then there can be a huge loss of time and resources on meaningless research or erroneous production. Previous methods of screening for contamination using biochemical markers were both laborious and in some instances inadequate. DNA fingerprinting, therefore, offers a relatively rapid and totally unambiguous method of determining the identity of a cell line *(37–39)*.

8.3. Zygosity Testing

The majority of applications of DNA profiling in identity testing exploit the fact that each individual's DNA profiles are different. the exception to this rule is identical twins. Monozygotic twins originate from the same fertilized egg and therefore possess the same genetic information; hence identical DNA profiles. Nonidentical or dizygotic twins are essentially normal siblings and share only the same amount of genetic information as do ordinary brothers and sisters, i.e., they exhibit an average band share of 62.5% when their multilocus-DNA profiles are compared. This can be useful to clinicians when the zygosity of newborn twins is in doubt, because identical twins are often less hardy than nonidentical twins, which can affect the prognosis when neonate twins are ill. It may also have implications when one of the twins requires a transplant from the other *(40)*.

8.4. Cancer Research and Diagnostics

In the healthy individual, DNA profiles obtained from any tissue will be identical and will remain so for the whole lifespan of the individual concerned—hence, its application in forensic science. The exception to this rule is the observation that in tumor tissues there are often alterations in the tumor-specific DNA profile with respect to that of the normal tissue, referred to as constitutional tissue. These alterations may be the loss, gain, or amplification of a given locus *(41–45)*.

Multilocus DNA profiling is therefore a useful method of genomic scanning to identify loci that have undergone somatic mutation associated with the neoplastic process. There is also a good correlation of DNA-profile changes with karyotpe changes in tumor progression, in such states as myeloid leukemia, so that the technology can also be viewed as a diagnostic aid *(46)*.

This latter application has been extended since the advent of microsatellite technology, because it has been observed that in a large and growing range of cancers, the onset of neoplasia is characterized by expansion or contraction of certain microsatellite arrays *(24,25)*. These alterations appear to be stable within the clonal populations of cells deriving from the original tumor, so this provides a rapid PCR-based screen for the presence of malignant cells in body fluids and around surgical margins *(47)*.

8.5. Epidemiology

This is not an area where one would intuitively expect DNA profiling to be able to play a role, but some interesting work on mosquito-feeding patterns showed how SLP DNA-profiling technology could be used to determine the number of individuals a mosquito had bitten on a routine feeding foray. Using a human-specific SLP against the mosquito blood meal, counting the number of bands observed in the mixed DNA profile can allow a fairly accurate estimate of the number of distinct individuals on whom the mosquito had fed. This kind of information has obvious ramifications for the study of the mosquito as a disease vector.

8.6. Monitoring Bone-Marrow Transplants

One final medical application of genetic-identity testing is the use of DNA profiling to identify the source of circulating blood cells following bone-marrow transplantation. If the graft has been successful, circulating blood cells should have a different DNA profile from that of the patient, obtained, for example, from hair roots or cheek epithelial cells, and should match that of the donor *(48)*. PCR analysis of a single SLP locus should be sufficient to discriminate between the two sources of DNA. The technique can also be used to provide a genetic marker for relapse in childhood acute lymphoblastic leukemia *(49)*.

9. Relationship Testing

9.1. Human Paternity Testing

Prior to the advent of DNA-profiling technology, paternity testing, like identity testing, relied largely on blood-grouping analysis. Again, these tests were fine for exclusionary purposes, but were of little use in definitively ascribing paternity. As previously described, a side-by-side MLP analysis of a mother's, child's and alleged father's DNA profiles may be used to determine paternity. The bands in the child's DNA profile are first compared to those in the maternal DNA profile and all bands that could possibly have been inherited from the mother are removed from consideration. The remaining bands must, by necessity, have been paternally inherited, and so the next task is to compare these bands to the DNA profile of the alleged father. All bands in the child's DNA profile that are paternally inherited must find a match with the bands in the true biological father's DNA profile. Once a match of the paternal bands has been made, the probability of paternity is determined by comparing the probability of the bands matching if he is the father (in which case they must match, so the probability is 1), to the probability of their matching if he is not the father, i.e., he simply happens to match them by chance. The latter is calculated by using the aforementioned band share value of 0.25 per band. The only difference between identity testing and relationship testing is that in the former, entire DNA profiles are compared, whereas in the latter, only the paternal contribution to the DNA profile is assessed.

The ability to definitively assign paternity has been of use in civil disputes *(50)*, criminal paternity cases *(51)*, e.g., where paternity testing of a child or fetus conceived as the result of a rape constitutes obvious proof of intercourse, and in immigration casework, where entry to the United Kingdom is sought on the basis of biological kinship to British residents *(11)*.

Multilocus-DNA profiling remains the method of choice for relationship testing, owing to the large amount of information provided by even a single MLP test. Its use is not normally problematic, because there is generally more than sufficient DNA available to generate a good profile. In exceptional cases, perhaps when fetal or chorion villus DNA is being used for paternity testing, it may be necessary to use SLP analysis, although the statistical weight of the data will be signficantly reduced.

DNA-profiling analysis has also been used for a variant of paternity testng in humans—namely, genetic origin testing of molar pregancies (52–54). These arise when an egg is abnormally fertilized. In a partial mole, a normal egg is fertilized by two sperm, but in a complete mole, an anucleate egg is fertilized by either two separate sperm or by a single sperm that subsequently divides. Both will result in the formation of an undifferentiated mass of tissue, but the important distinction is that following a complete molar pregnancy, particularly a homozygous mole, the woman's risk of subsequently developing choriocarcinoma is significantly increased. Hence, the identification of the genetic origin of a molar pregnancy is central to the adequate provision of aftercare and screening for the patient.

9.2. Relationship Testing in Animals

The extent to which DNA profiling has been exploited in the definition of precise biological relationships in animals is quite surpising, the technology being used in a number of diverse applications. Most conventionally, DNA profiling can be used for pedigree testing of animals to confirm alleged parentage, which may be important in the case of valuable animals (55,56).

In a similar context, it can be used to define the biological relationships of individuals within a defined social group of animals. This has been invaluable to behavioral scientists in their interpretations of the social roles of animals, particularly with repect to collective rearing of offspring (57–59).

A less conventional approach to relationship testing has been adopted in the relationship testing of animals in captive-breeding programs, in order to determine which pairs of animals are the least related so that breeding pairs will retain the greatest genetic diversity, thus maintaining the maximum gene pool among the captive populations. This same principle has been applied in rerelease programs to ensure that the least genetically similar animals are released back to the wild and also in policing the legitimacy of supposed captive-breeding programs (60).

The determination of genetic relatedness of animal populations, rather than specific individuals, has also been exploited in demographic studies in a number of species, and this principle has been extended further to generate a tool to assist in the control of big-game poaching (61). Elephants from different populations in Africa, owing to their physical separation over a long period, have distinct groups of DNA profiles. Consequently, the origin of ivory can be ascribed to a specific regional group of elephants, enabling game wardens or the police to determine whether the ivory has been legally culled or illegally poached.

DNA-profiling technology is also playing a role in population biology by providing a means of measuring the genetic similarity of populations of animals at the genomic level (62,63). Such information can be of considerable value in demographic studies of wild-animal populations (64,65).

10. Marker-Assisted Selection

Multilocus DNA profiling is also an extremely useful tool in the development of markers for physically desirable traits. The construction and verification of an extensive pedigree by DNA profiling can allow the identification of specific alleles that are coinherited with commercially important traits, such as disease resistance in agricultural livestock. MLP analysis allows a large number of genetic markers to be screened simultaneously, compared to biochemical markers which are usually far more laborious to assay *(66,67)*.

11. Microsatlellites in Linkage Analysis

An alternative way of looking at the use of hypervariable loci is to exploit their level of heterozygosity not to identify individuals, or to assess their relationships, but to discriminate among them. In the past, a great deal of effort has gone into finding such discriminatory markers in the form of RFLPs for linkage anaysis. The latter are rare and show only a low level of polymorphism, whereas hypervariable loci, in particular microsatellites, are common and highly polymorphic. One could therefore consider the use of hypervariable DNA to discriminate among individuals and among loci within an individual as a third type of test.

There is, as previously mentioned, an extensive and rapidly expanding bulk of literature on the use of these sequences for general genome mapping, disease locus localization, and also as diagnostic markers in identifying cells in which these arrays have undergone contraction and expansion in rearrangements associated with neoplasia *(68)*. One recent development is the determination of the genetic loci responsible for the appearance of genetic instability in microsatellites *(69)*. The role of this phenomenon in the neoplastic process itself is also being investigated. All of this is a far cry from the original roots of DNA-profiling technology.

12. Summary

Multilocus DNA profiling is a powerful tool for identity and relationship testing and genomic analysis that has supported significant advances in forensic, medical, and biological sciences. Although microsatellite technology has superceded MLP analysis for identity testing in forensic science, MLP analysis is still the method of choice for relationship testing and many other applications where the power of the technology to scan numerous genomic loci, and hence many genetic characteristics, simultaneously is required. DNA profiling, in one form or another, is therefore a valuable addition to the technical repertoire of a large and growing number of life scientists.

References

1. Jeffreys, A. J., Wilson, V., and Thein, S. L. (1985) Hypervariable 'minisatellite' regions in human DNA. *Nature* **314,** 67–73.
2. Higgs, D. R., Goodbourn, S. K. Y., Wainscoat, J. S., Clegg, J. B., and Weatherall, D. J. (1981) Highly variable regions flank the human alpha-globin genes. *Nucleic Acids Res.* **9,** 4213,4214.
3. Goodbourn, S. K. Y., Higgs, D. R., Clegg, J. B., and Weatherall, D. J. (1983) Molecular basis of length polymorphism in the human zeta globin gene complex. *Proc. Natl. Acad. Sci. USA* **80,** 5022–5026.

4. Goodbourn, S. K. Y., Higgs, D. R., Clegg, J. B., and Weatherall, D. J. (1984) Allelic variation and linkage properties of a highly polymorphic restriction fragment in humans. *Mol. Biol. Med.* **2,** 223–238.

5. Jeffreys, A. J., Wilson, V., Thein, S. L., Weatherall, D. J., and Ponder, B. A. J. (1986) DNA "fingerprints" and segregation analysis of multiple markers in human pedigrees. *Am. J. Hum. Genet.* **39,** 11–34.

6. Jeffreys, A. J., Royle, N. J., Wilson, V., and Wong, Z. (1988) Spontaneous mutation rates to new length alleles at tandem-repetitive hypervariable loci in human DNA. *Nature* **332,** 278–281.

7. Fowler, S. J., Gill, P., Werret, D. J., and Higgs, D. R. (1988) Individual specific DNA fingerprints from a hypervariable region probe: alpha-globin 3'HVR. *Hum. Genet.* **79,** 142–146.

8. Chen, P., Hayward, N. K., Kidson, C., and Ellem, K. A. O. (1990) Conditions for generating well-resolved human DNA fingerprints using M13 phage DNA. *Nucleic Acids Res.* **18,** 1065.

9. Neitzel, H., Digweed, M., Nurnberg, P., Popperl, A., Schmidt, C. A., Tinschert, S., and Sperling, K. (1991) Routine applications of DNA fingerprinting with the oligonucleotide probe (CAC)5/(GTG)5. *Clin. Genet.* **39,** 97–103.

10. Gill, P., Jeffreys, A. J., and Werrett, D. J. (1985) Forensic applications of DNA fingerprints. *Nature* **318,** 577–579.

11. Jeffreys, A. J., Brookfield, J. F. Y., and Semeonoff, R. (1985) Positive identification of an immigration test case using human DNA fingerprints. *Nature* **317,** 818,819.

12. Wong, Z., Wilson, V., Jeffreys, A. J., and Thein, S. L. (1986) Cloning a selected fragment from a human DNA 'fingerprint': isolation of an extremely polymorphic minisatellite. *Nucleic Acids Res.* **14,** 4605–4616.

13. Wong, Z., Wilson, V., Patel, I., Povey, S., and Jeffreys, A. J. (1987) Characterisation of a panel of highly variable minisatellites cloned from human DNA. *Ann. Human Genet.* **51,** 269–288.

14. Allard, J. E. (1992) Murder in South London: a novel use of DNA profiling. *J. Forensic Sci. Soc.* **32,** 49 58.

15. Li, H., Gyllensten, U. B., Cui, X., Saiki, R. K., Erlich, H. A., and Arnheim, N. (1988) Amplification and analysis of DNA sequences in single human sperm and diploid cells. *Nature* **335,** 414–417.

16. Jeffreys, A. J., Royle, N. J., Patel, I., Armour, A. L., MacLeod, A., Collick, A., Gray, I. C., Neumann, R., Gibbs, M., Crosier, M., Hill, M., Signer, E., and Monckton, D. (1991) Principles and recent advances in human DNA fingerprinting, in *DNA Fingerprinting; Approaches and Applications* (Burke, T., Dolf, G., Jeffreys, A. J., and Wolff, R., eds.), Birkhauser Verlag, Basel, Switzerland, pp. 1–19.

17. Litt, M., and Luty, J. A. (1989) A hypervariable microsatellite revealed by in vitro amplification of a dinucleotide repeat within the caridaic muscle actin gene. *Am. J. Hum. Genet.* **44,** 397–401.

18. Tautz, D. (1989) Hypervariability of simple sequences as a general source for polymorphic DNA markers. *Nucleic Acids Res.* **17,** 6463–6471.

19. Royle, N. J., Clarkson, R. E., Wong, Z., and Jeffreys, A. J. (1988) Clustering of hypervariable minisatellites in the proterminal regions of human autosomes. *Genomics* **3,** 352–360.

20. Mansfield, D. C., Brown, A. F., Green, D. K., Carothers, A. D., Morris, S. W., Evans, H. J., and Wright, A. F. (1994) Automation of genetic linkage analysis using fluorescent microsatellite markers. *Genomics* **24,** 225–233.

21. Sander, A., Murray, J. C., Cherpbier-Heddema, T., Buetow, K. H., Weissenbach, J., Zingg, M., Ludwig, K., and Schmelzle, R. (1995) Microsatellite based fine mapping of the Van der Woude Syndrome locus to an interval of 4.1 cM between DIS245 and DIS414. *Am. J. Hum. Genet.* **56,** 310–318.

22. Carinci, F., Pezzetti, F., Scapoli, L., Padula, E., Baciliero, U., Curioni, C., and Tognon, M. (1995) Non-symdromic cleft lip and palate: evidence of linkage to a microsatellite marker on 6p23. *Am. J. Hum. Genet.* **56,** 337–339.

23. Loudianos, G., Figus, A. L., Loi, A., Angius, A., Dessi, V., Deiana, M., De Virgiliis, S., Monni, G., Cao, A., and Pirastu, M. (1994) Improvement of prenatal diagnosis of Wilson Disease using microsatellite markers. *Prenatal Diag.* **14,** 999–1002.

24. Mao, L., Lee, D. J., Tockman, M. S., Erozan, Y. S., Askin, F., and Sidransky, D. (1994) Microsatellite alterations as clonal markers for the detection of human cancer. *Proc. Natl. Acad. Sci. USA* **91,** 9871–9875.

25. Huff, V., Jafe, N., Saunders, G. F., Strong, L. C., Villalba, F., and Routeshouser, E. C. (1995) WT1 exon 1 deletion/insertion mutations in Wilms Tumour patients, associated with di- and trinucleotide repeats and deletion hotspot concensus sequences. *Am. J. Genet.* **56,** 84–90.

26. Gill, P., Lygo, J. E., Fowler, S. J., and Werret, D. J. (1987) An evaluation of DNA finger-printing for forensic purposes. *Electrophoresis* **8,** 38–44.

27. Honma, M., Yoshii, T., Ishiyama, I., Mitani, K., Kominami, R., and Muramatsu, M. (1989) Individual identification from semen by the deoxyribonucleic acid (DNA) finger-print technique. *J. Forensic Sci.* **34,** 222–227.

28. Bar, W. and Hummel, K. (1991) DNA Fingerprinting: its applications in forensic case-work, in *DNA Fingerprinting: Approaches and Applications,* (Burke, T., Dolf, G., Jeffreys, A. J., and Wolff, R., eds.), Birkhauser Verlag, Basel, Switzerland, pp. 349–355.

29. Gill, P. and Werrett, D. J. (1987) Exclusion of a man charged with murder by DNA fin-gerprinting. *Forensic Sci. Intl.* **35,** 145–148.

30. Syndercombe Court, D., Fedor, T., Gouldstone, M., Lincoln, P. J., Phillips, C. P., Tate, V., Thomson, J. A., and Watts, P. W. (1992) Investigation of the between-gel and within-gel variation in fragment size determinations found when using single locus DNA probes. *Forensic Sci. Intl.* **53,** 173–191.

31. Pascali, V. L., Moscetti, A., Dobosz, M., Pescarmona, M., and d'Aloja, E. (1992) Errors in sizing bands of hypervariable DNA profiles on autoradiograms: are they Gaussian? *Electrophoresis* **13,** 341–345.

32. Nichols, R. A. and Balding, D. J. (1993) Effects of population structure on DNA finger-print analysis in forensic science. *Heredity* **66,** 297–302.

33. Chakraborty, R. and Jin, L. (1992) Heterozygote deficiency, poluation substructure and their implications in DNA fingerprinting. *Hum. Genet.* **88,** 267–272.

34. Walsh, J. J. (1992) The population genetics of DNA typing: "Could it have been someone else?" *Crim. Law Quar.* **34,** 469–497.

35. Devlin, B., Risch, N., and Roeder, K. (1990) No excess of homozygosity at loci used for DNA fingerprinting. *Science* **249,** 1416–1419.

36. Gill, P., Kimpton, C., D'Aloja, E., Andersen, J. F., Bar, W., Brinkman, B., Holgersson, S., Johnsson, V., Kloosterman, A. D., Lareu, M. V., Nelleman, L., Pfitzinger, H., Phillips, C. P., Scmitter, H., Schneider, P. M., and Stenersen, M. (1994) Report of the European DNA profiling group (EDNAP)—towards standardisation of short tandem repeat (STR) loci. *Forensic Sci. Int.* **65,** 51–59.

37. van Helden, P. D. and Wiid, I. J. F. (1987) Application of DNA fingerprinting to the investigation of cell line genotype. *SA J. Sci.* **83,** 244.

38. van Helden, P. D., Wiid, I. J. F., Albrecht, C. F., Theron, E., Thornley, A. L., and Hoal-van Helden, K. G. (1988) Cross-contamination of human esophageal squamous carcimona cell lines detected by DNA fingerprint analysis. *Cancer Res.* **48,** 5660–5662.

39. Thacker, J., Webb, M. B. T., and Debenham, P. G. (1988) Fingerprinting cell lines: use of human hypervariable DNA probes to characterise mammalian cell cultures. *Somatic Cell Mol. Genet.* **14,** 519–525.

40. Jones, L., Thein, S. L., Apperley, J. F., Catovsky, D., and Goldman, J. M. (1987) Identical twin marrow transplantation for 5 patients with chronic myeloid leukaemia: role of DNA fingerprinting to confirm monozygosity in 3 cases. *Eur. J. Haematol.* **39,** 144–147.

41. Fey, M. F., Wells, R. A., Wainscoat, J. S., and Thein, S. L. (1988) JAssesment of clonality in gastrointestinal cancer by DNA fingerprinting. *Clin. Invest.* 1532–1537.

42. Boltz, E. M., Harnett, P., Leary, J., Houghton, R., Kefford, R. F., and Friedlander, M. L. (1990) Demonstration of somatic rearrangements and genomic heterogeneity in human ovarian cancer by DNA fingerprinting. *Br. J. Cancer* **62,** 23–37.

43. Hayward, N., Chen, P., Nancarrow, D., Kearsley, J., Smith, P., Kidson, C., and Ellem, K. (1990) Detection of somatic mutations in tumours of diverse types by DNA fingerprinting with M13 phage DNA. *Int. J. Cancer* **45,** 687–690.

44. Lee, J. H., Kavanagh, J. J., Wildrick, D. M., Wharton, J. T., and Thick, M. (1990) Frequent loss of heterozygosity on chromosomes 6q, 11 and 17 in human ovarian cancer. *Cancer Res.* **50,** 2724–2728.

45. Agurell, E., Li, R., Rannug, U., Norming, U., Tribukait, B., and Ramel, C. (1992) Detection of DNA alterations in human bladder tumours by DNA fingerprint analysis. *Cancer Genet. Cytogenet.* **60,** 53–60.

46. Pakkala, S., Knuutila, S., Helminen, P., Ruutu, T., Saarincn, U. M., and Peltonen, L. (1990) DNA-fingerprint changes compared to karyotypes in acute leukemia. *Leukemia* **4,** 866–870.

47. Mao, L., Lee, D. J., Tockman, M. S., Erozan, Y. S., Askin, F., and Sidransky, D. (1994) Microsatellite alterations as clonal markers for the detection of human cancer. *Proc. Natl. Acad. Sci. USA* **91,** 9871–9875.

48. Gineburg, D., Antin, J. H., Smith, B. R., Orkin, S. H., and Rappeport, J. M. (1985) Origin of cell populations after bone marrow transplantation. *J. Clin. Invest.* **75,** 596–603.

49. Pakkala, S., Helminen, P., Saarinen, U., Alitalo, R., and Peltonen, L. (1988) Differences in DNA fingerprints between remission and relapse in childhood acute lymphoblastic leukaemia. *Leukemia Res.* **9,** 757–762.

50. Helminen, P., Ehnholm, C., Lokki, M. L., Jeffreys, A., and Peltonen, L. (1988) Application of DNA "fingerprints" to paternity determinations. *Lancet* **12,** 574–576.

51. Rittner, C., Schaker, U., Rithner, G., and Schneider, P. M. (1988) Application of DNA polymorphisms in paternity testing in Germany: solution of an incest case using bacteriophage M13 hybridisation with hypervariable minisatellite DNA. *Adv. Forensic Haemogenet.* **2,** 388–391.

52. Fisher, R. A., Lawler, S. D., Povey, S., and Bagshawe, K. D. (1988) Genetically homozygous choriocarcinoma following pregnancy with hydatidiform mole. *Br. J. Cancer* **58,** 788–792.

53. Azuma, C., Saji, F., Nobunaga, T., Kamiura, S., Kimura, T., Tokugawa, Y., Koyama, M., and Tanizawa, O. (1990) Studies on the pathogenesis of choriocarcinoma by analysis of restriction fragment length polymorphisms. *Cancer Res.* **50,** 488–491.

54. Nobunaga, T., Azuma, C., Kimura, T., Tokugawa, Y., Takemura, M., Kamiura, S., Saji, F., and Tanizawa, O. (1990) Differential diagnosis between complete mole and hydropic abortus by deoxyribonucleic acid fingerprints. *Am. J. Obstet. Gynecol.* **163,** 634–638.

55. Georges, M., Lequarre, A. S., Castelli, M., Hanset, R., and Vassart, G. (1988) DNA fingerprinting in domestic animals using four different minisatellite probes. *Cytogenet. Cell Genet.* **47,** 127–131.

56. Binns, M. M., Holmes, N. G., Holliman, A., and Scott, A. M. (1995) The indentification of polymorphic microsatellite loci in the hoere and their use in thoroughbred parentage testing. *Br. Vet. J.* **151,** 9–15.

57. Burke, T. and Bruford, M. W. (1987) DNA fingerprinting in birds. *Nature* **327,** 149–152.

58. Jones, C. S., Lessells, C. M., and Krebs, J. R. (1991) Helpers-at-the-nest in european bee-eaters (Merops apiaster): a genetic analysis, in *DNA Fingerprinting: Approaches and Applications* (Burke, T., Dolf, G., Jeffreys, A. J., and Wolff, R., eds.), Birkhauser Verlag, Basel, Switzerland, pp. 169–192.

59. Graves, J., Hay, R. T., Scallan, M., and Rowe, S. (1992) Extra-pair paternity in the shag, Phalacrocorax aristotelis as determined by DNA fingerprinting. *J. Zool. London* **226,** 399–408.

60. Wolfes, R., Mathe, J., and Seitz, A. (1991) Forensics of birds of prey by DNA fingerprinting with 32P-labelled oligonucleotide probes. *Electrophoresis* **12,** 175–180.

61. Blackett, R. S. and Keim, P. (1992) Big game species identification by deoxyribonucleic acid (DNA) probes. *J. Forensic Sci.* **37,** 590–596.

62. Meng, A., Carter, R. E., and Parkin, D. T. (1990) The variability of DNA fingerprints in three species of swan. *Heredity* **64,** 73–80.

63. Scribner, K. T., Arntzen, J. W., and Burke, T. (1994) Comparative analysis of intra- and interpopulation genetic diversity in Bufo bufo, using allozyme, single-locus microsatellite, minisatellite and multilocus minisatellite data. *Mol. Biol. Evol.* **11,** 737–748.

64. Wetton, J. H., Carter, R. E., Parkin, D. T., and Walters, D. (1987) Demographic study of a wild house sparrow population by DNA fingerprinting. *Nature* **327,** 147–149.

65. Gilbert, D. A., Lehman, N., O'Brien, S. J., and Wayne, R. K. (1990) Genetic fingerprinting reflects population differentiation in the California Channel Island fox. *Nature* **344,** 764–766.

66. Georges, M., Lathrop, M., Hilbert, P., Marcotte, A., Scwers, A., Swillens, S., Vassart, G., and Hanset, R. (1990) On the use of DNA fingerprinting for linkage studies in cattle. *Genomics* **6,** 461–474.

67. Kuhnlein, U., Zadworny, D., Gavora, J. S., and Fairfull, R. W. (1991) Identification of markers associated with quantitative trait loci in chickens by DNA fingerprinting, in *DNA Fingerprinting: Approaches and Applications* (Burke, T., Dolf, G., Jeffreys, A. J., and Wolff, R., eds.), Birkhauser Verlag, Basel, Switzerland, pp. 274–282.

68. Trapman, J., Sleddens, H. F. B. M., van der Weiden, M. M., Dinjems, W. M. N., Konig, J. J., Schroder, F. H., Faber, P. W., and Bosman, F. T. (1994) Loss of heterozygosity of chomosome 8 microsatellite loci implicates a candidate tumour suppressor gene between the loci D8S87 and D8S133 in human prostate cancer. *Cancer Res.* **54,** 6061–6064.

69. Pykett, M. J., Murphy, M., Harnisch, P. R., and George, D. L. (1994) Identification of a microsatellite instabiity phenotype in meningiomas. *Cancer Res.* **54,** 6340–6343.

31

Radiolabeling of Peptides and Proteins

Arvind C. Patel and Stewart R. Matthewson

1. Introduction

Radioactive labeled peptides and proteins are used extensively in many areas of biochemistry, pharmacology, and medicine. For example, they are frequently employed as tracer molecules in quantitative determinations, such as measurement of hormone and hormone-receptor concentrations, and kinetic and equilibrium studies of both agonist and antagonist binding to receptors. These studies require the accurate determination of very low amounts of the labeled peptides and protein. Such small amounts can be accurately measured by using a tracer molecule labeled to high-specific radioactivity. The most commonly used radionuclides for peptide and protein studies are tritium and ^{125}I, followed by ^{14}C, ^{35}S, and ^{32}P.

In most cases where $[^{125}I]$iodine is used to label a molecule, it is a foreign label, i.e., it does not normally occur in the molecule. The replacement of nonradioactive carbon or hydrogen by $[^{14}C]$carbon or tritium will have virtually no effect on the biological properties of the molecule. However, the replacement of a proton with a large iodine atom can have a considerable effect on the properties of the protein; this can usually be overcome if the label is at a position that is some distance from the site of biological activity.

There are several major advantages in using $[^{125}I]$iodine over $[^{14}C]$carbon or tritium. The first is the specific activity available (**Table 1**).

There is an inverse relationship between the half life of an isotope and its theoretical specific activity. In some isotopes this maximum is never obtainable. $[^{125}I]$Iodine has a maximum theoretical specific activity of 2175 Ci/mmol and is usually obtainable at ~2000 Ci/mmol. The maximum specific activities of $[^{14}C]$carbon and tritium are 62.4 mCi/mmol and 28.8 Ci/mmol, respectively. Several atoms of $[^{14}C]$carbon or tritium can be substituted in a molecule, but the specific activity obtained is still very much lower than with $[^{125}I]$iodine. Very small amounts of radioiodinated material can be used while maintaining sensitive assays. The count rate obtained from $[^{125}I]$iodine can be 100 times greater than for tritium and 35,000 times greater than $[^{14}C]$carbon. Another major advantage is in the case of detection. $[^{125}I]$iodine decays by electron capture followed by X-ray emission which can be counted directly in a γ counter. $[^{125}I]$iodine is used in vivo for imaging owing to the nonparticulate emission which reduces radiation damage to the biological

From: *Molecular Biomethods Handbook*
Edited by: R. Rapley and J. M. Walker © Humana Press Inc., Totowa, NJ

Table 1
Half-Life and Available Specific Activity of Commonly Used Isotopes

Radionuclide	Type of emission	Half-life	Specific activity (per millimole) at 100% isotopic abundance
^{125}I	$\gamma(EC)$	60.00 d	2175.0 Ci
^{131}I	γ and β^-	8.04 d	16,235.0 Ci
^{14}C	β^-	5730.00 yr	62.4 mCi
^3H	β^-	12.43 yr	28.8 Ci
^{32}P	β^-	14.30 d	6000.0 Ci
^{35}S	β^-	87.40 d	1493.0 Ci

material. Both [^{14}C]carbon and tritium are pure β^- emitters resulting in particulate emission in the form of electrons. To count these, scintillants and a scintillation counter are required, which involves extra sample preparation and counting time, extra cost of scintillant and increased volumes of radioactive material for disposal. The high specific activity and count rate of iodinated compounds are advantageous in autoradiography, especially when very small amounts of receptor are to be localized. In contrast, the time required to autograph tritiated and [^{14}C]carbon ligands can stretch to months.

Complex organic chemistry may be required to label a molecule with [^{14}C]carbon. This can mean starting from [^{14}C]-labeled CO_2, methanol, $BaCO_3$, benzene, and so forth. It is also an expensive radionuclide to obtain, and multistage preparations inevitably decrease overall yields. [^{14}C]-peptides (unless using reductive methylation) must be built from labeled amino acids. Tritium labeling requires synthesis of specific precursors. Tritiation of samples often involves catalytic hydrogenation to add to a double bond or to replace a halogen in a molecule. Radioiodinations are comparatively easy.

There are also disadvantages in using iodine. As previously stated, iodine is usually a foreign label and labeling with [^{125}I]iodine can therefore alter the properties of many molecules. This can be a particular problem in receptor studies. Even a small change in structure, such as oxidation of one amino acid in an iodination, can completely block binding to the receptor. Reaction rates can also be altered. The advantage of having a high specific activity is countered by the disadvantage of a shorter half-life. There is also the possibility of faster decomposition, especially radiation decomposition and also the loss of iodine. A shorter half-life is advantageous for waste disposal. [^{14}C]Labeled materials can remain pure for many years.

2. Labeling Reactions: ^{125}I

2.1. General Consideration

A number of methods are available for the iodination of proteins. These methods can be varied to produce a satisfactory ligand. Radioiodination is the process of chemically modifying a molecule to contain one or more atoms of radioactive iodine. Protein iodinations are broadly divided into two groups: Direct labeling, in which the [^{125}I]iodine is directly incorporated into tyrosines (and/or) histidine residues of proteins, usually in the presence of an oxidizing agent, and conjugation methods in which the radioactive moiety is conjugated via amino or sulfhydryl groups.

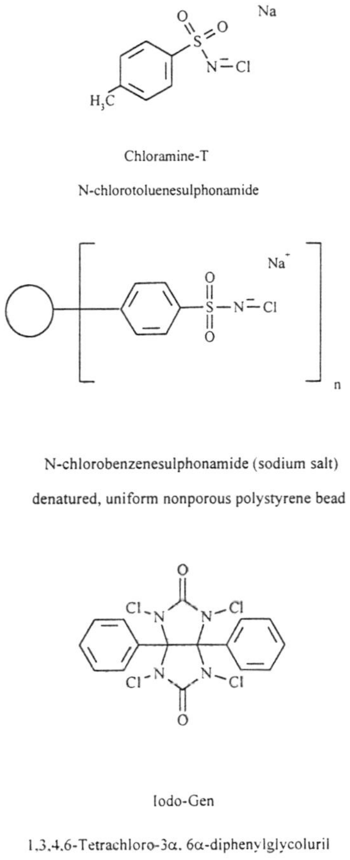

Chloramine-T

N-chlorotoluenesulphonamide

N-chlorobenzenesulphonamide (sodium salt)

denatured, uniform nonporous polystyrene bead

Iodo-Gen

1,3,4,6-Tetrachloro-3α, 6α-diphenylglycoluril

Fig. 1. Oxidizing reagents.

Generally, the direct iodination methods are investigated first, because they have the advantage of involving a one-step reaction stage in which the protein reacts with the [^{125}I]iodine. The reaction involves the oxidation of ^{125}I⁻ to (["^{125}I⁺"]; Iodonium), followed by electrophilic substitution of ^{125}I⁺ (**Fig. 1**). This oxidation is effected by several reagents such as chloramine-T, lactoperoxidase/hydrogen peroxide, Iodogen™, and so forth. The methods are generally rapid, convenient, easy to perform, and result in high yields and specific activities of product. Direct iodinations often have associated with the reactions the unwanted oxidation of susceptible groups. Methionine residues can be oxidized to sulfoxides and sulfones *(1)*; these are often biologically inactive. Methods that use strong oxidizing agents, such as chloramine-T, are more likely to lead to oxidation problems, whereas the milder procedures are less likely to oxidize the protein.

2.2. Radioiodination Using Chloramine-T

This is the most commonly used method for iodinating peptides and proteins to high-specific activity *(2)*. The advantages of the chloramine-T method are its reproducibility, rapidity, efficiency of the iodination in the absence of carrier nonradioactive iodine, and avoidance of extremes of pH or organic solvents, which may denature the protein. Chloramine-T is the sodium salt of the *N*-monochloro derivative of p-toluenesulfonamide. This breaks down slowly in aqueous solution producing hypochlorous acid, which oxidizes the sodium [^{125}I]iodide to "[^{125}I$^+$]" which is incorporated into aromatic rings. The reaction is terminated, usually with an agent, such as sodium metabisulfite, which reduces excess chloramine-T and free iodine. During purification, carrier potassium iodide or a protein containing buffer is often added to prevent losses of labeled material during purification. The degree of iodination is concentration-dependent, and the reaction needs to be carried out at high ligand concentration. This method normally yields higher levels of incorporation compared with other direct methods of radioiodination.

Oxidation damage can occur during the reaction, especially to methionine residues, which can be converted to the sulfoxide. Therefore, the minimum amount of chloramine-T required to produce the degree of iodination and specific activity required should be used. Commercially obtainable [^{125}I]iodide is supplied in sodium hydroxide solution at pH 8.0–10.0. The optimum pH for the iodination of tyrosine is pH 7.2–7.4, with reduced incorporation being obtained at pH 6.5 or 8.5. It is therefore necessary to buffer the reaction to pH 7.2. To iodinate the imidazole ring of histidine, a pH of 8.1 is required.

Reagents in the chloramine-T iodination are added in rapid succession into a polypropylene or Sarstedt microcentrifuge tube, while continuously agitating to ensure thorough mixing of reagents. Poor mixing is probably the most common cause of a low yield of labeled protein by this method. Glass should be avoided if the protein is known to adhere nonspecifically. Chloramine-T, cysteine, and sodium metabisulfite solutions should be freshly prepared prior to use.

The iodination is performed by incubating protein (50 µL; 25 µg) with chloramine-T (50 µL, 50 µg) and sodium [^{125}I]iodide (1.0 mCi; 10 µL) for 30 s. The reaction is terminated by addition of sodium metabisulfite (50 µL, 50 µg) or cysteine (20 µL, 10 mM) and then applied to the separation column. The protein, chloramine-T, and terminating solutions should be prepared in phosphate buffer, pH 7.2–7.4.

The use of cysteine is not advisable in peptides and proteins that contain disulfide bridges. Tyrosine solution (0.2 mg/mL in phosphate buffer [0.2M; pH 7.4]) can be used as a substitute. However, in cases when the specific activity is to be determined prior to termination of the reaction, a small sample of the reaction mixture should be terminated in sodium metabisulfite solution.

2.3. Radioiodination Using Lactoperoxidase

Marchalonis first described this method for radioiodination of immunoglobulin (3,4). Lactoperoxidase was used to catalyze the oxidation of iodide in the presence of a very small amount of hydrogen peroxide. The iodine may react with tyrosine and histidine residues within the protein (see **Fig. 2**). Reaction is terminated by dilution with buffer or quenching the enzyme action with cysteine. Reaction times are longer (~20 min), but, because no strong oxidizing or reducing agents are used, immunological damage in kept to a minimum. In general, this method gives low yields and incorporation at low-specific activities (200–400 Ci/mmol). The optimum pH for iodination is dependent on the nature of the protein; hence, a number of experiments are often required to establish the best protocol. If the presence of small amount of hydrogen peroxide is a problem, this can be produced *in situ* using glucose/glucose oxidase or by using solid-phase lactoperoxidase. The latter was available commercially as Enzymobeads™ (Bio-Rad cat. no. 170-6001 or 170-6003) in which the lactoperoxidase has been immobilized to agarose beads (5,6).

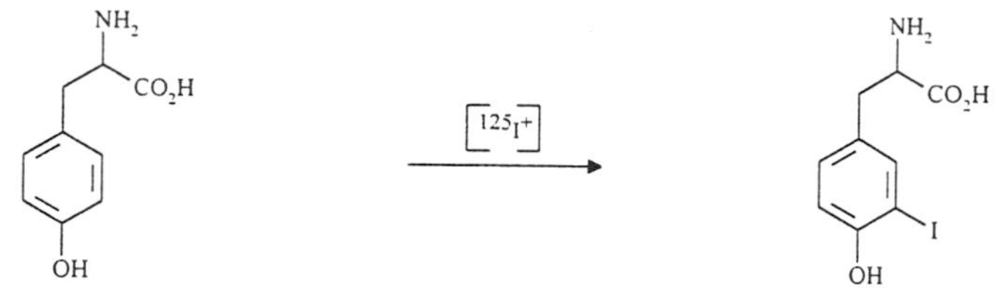

Fig. 2. Iodination of tyrosine and histidine. Reagents in the chloramine-T iodination are added in rapid succession into a polypropylene or Sarstedt microcentrifuge tube, while continuously agitated to ensure thorough mixing of reagents. Poor mixing is probably the most common cause of a low yield of labeled protein by this method. Glass should be avoided if the protein is known to adhere nonspecifically. Chloramine-T, cysteine, and sodium metabisulfite solutions should be freshly prepared prior to use.

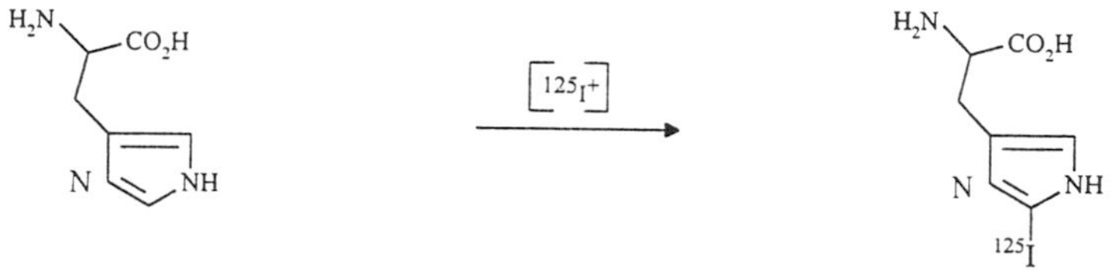

The iodination if performed by incubating protein (50 μL; 25 μg) with chloramine-T (50 μL, 50 μL) and sodium [^{125}I]iodide (1.0 mCi; 20 μL) for 30 s. The reaction is terminated by addition of sodium metabisulfite (50 μL, 50 μg) or cysteine (20 μL, 10 mM) and then applied.

Buffers incorporating sodium azide as a preservative must be avoided, because this is a potent inhibitor of lactoperoxidase. During the iodination, lactoperoxidase is self-iodinated, thus increasing the iodide loss and complicating the separation of labeled protein from labeled enzyme and determination of specific activity. Separation can be achieved typically by reverse-phase high-performance liquid chromatography (HPLC); Sephadex G-25 is generally unsuitable because the self iodinated [^{125}I]lactoperoxidase will coelute in the void volume with the protein.

The lactoperoxidase reaction is carried out by mixing the protein (10 μg; 20 μL), lactoperoxidase (10 μL; 25 U/mL), sodium [^{125}I]iodide (0.5–1.0 mCi) and sodium phosphate buffer (0.2 M; pH 7.4; 100 μL). The reaction is initiated by addition of hydrogen peroxide (10 μL; 0.003%) in deionized water and incubation of the reaction mixture at room temperature for 20 min. Reaction is terminated by addition of cysteine (or tyrosine) solution (10 μL; 1 mg/mL) and then purified by the method of choice.

2.4. Radioiodination Using Iodogen™

One method of avoiding oxidation problems is to use an insoluble oxidizing agent which avoids direct contact between the oxidizing agent and protein. One such agent is 1,3,4,6-tetrachloro-3α,6α-diphenylglycoluril, which is available commercially as Iodogen (Pierce Chemical Company) *(7)*. The oxidant is dissolved in a suitable organic solvent and then coated onto the walls of the reaction vessel by evaporating the solvent in a stream of dry nitrogen. The coated vials can be stored in dry conditions to allow batch iodinations. Once the vessel is coated with the iodogen, it may be stored at –20°C in the dark for up to 1 mo. Buffer, sodium [^{125}I]iodide, and protein are added to start the reaction. Owing to the heterogenous nature of the reaction, incubation times are typically in the 10–20-min range. The reaction is terminated by removal from the reaction vessel and loading straight onto the gel filtration column or by addition of a reducing agent. Although Iodogen is almost completely insoluble in water, the iodination does not immediately stop after removal of solution from the Iodogen vial; therefore, a small amount of reducing agent should be added after removal or by addition of an excess of low-mol-wt iodinatable species, such as tyrosine. The coating of the reaction vial needs a great deal of care and attention, otherwise small parts of the coated Iodogen will fall off and contaminate the reaction mixture. Some proteins may bind to the surface of the reaction vessel; this needs to be taken into account when calculating the specific activity and yield.

Iodogen (0.5 mg) is dissolved in dichloromethane (1.0 mL). Iodogen solution (10 μL) is dispensed into the reaction tube and the solution evaporated to dryness by blowing a gentle stream of nitrogen onto the surface. The Iodogen vial is washed by dispensing sodium phosphate buffer (0.2 M; pH 7.4; 50 μL) into it followed by withdrawal of the solution to remove any nonadherent dry iodogen flakes. Sodium phosphate buffer (0.2 M; pH 7.4; 50 μL), protein (10–20 μg; 10–20 μL), and sodium [^{125}I]iodide (1.0 mCi) are added to the Iodogen vial, which is incubated for 15 min at room temperature with occasional gentle shaking. The reaction mixture is transferred to another vial containing tyrosine solution (50 μg/mL; 250 μL) in sodium phosphate buffer (0.2 M; pH 7.4) to terminate the reaction. The [^{125}I]protein is then purified by the method of choice.

2.5. Radioiodination Using Iodo-Beads™

Markwell introduced a new oxidizing reagent, N-chloro-benzenesulfonamide coupled covalently to nonporous polystyrene spheres (3.2 mm), referred to as Iodo-beads, to facilitate smooth iodinations (8). Iodo-beads are commercially available from Pierce Chemical Company. The oxidizing capacity is about 0.05 mmol/bead and is limited to the outer surface of the bead, which allows reaction times to be extended to about 15 min rather than the 30 s with chloramine-T. The rate of reaction may be changed depending on the number of beads and the concentration of sodium iodide. The reagent functions well over a broad range of pH, as well as temperature conditions. The reaction is terminated by removing the Iodo-beads from the reaction, hence avoiding the use of reducing agents. The performance of Iodo-beads is not inhibited by azide, detergents, urea, or high salt concentration. The Iodo-beads are prone to inactivation by reducing agents, moisture on storage, and certain organic solvents that dissolve the surface of the beads.

2.6. Indirect Methods: Radioiodination Using Bolton and Hunter Reagent

This is by far the most widely used indirect method of radiolabeling. Bolton and Hunter reagent is N-succinimidyl-3(4-hydroxy-5-[^{125}I]iodophenyl)propionate, which is an acylating reagent (9). The reagent itself is prepared by chloramine-T iodination

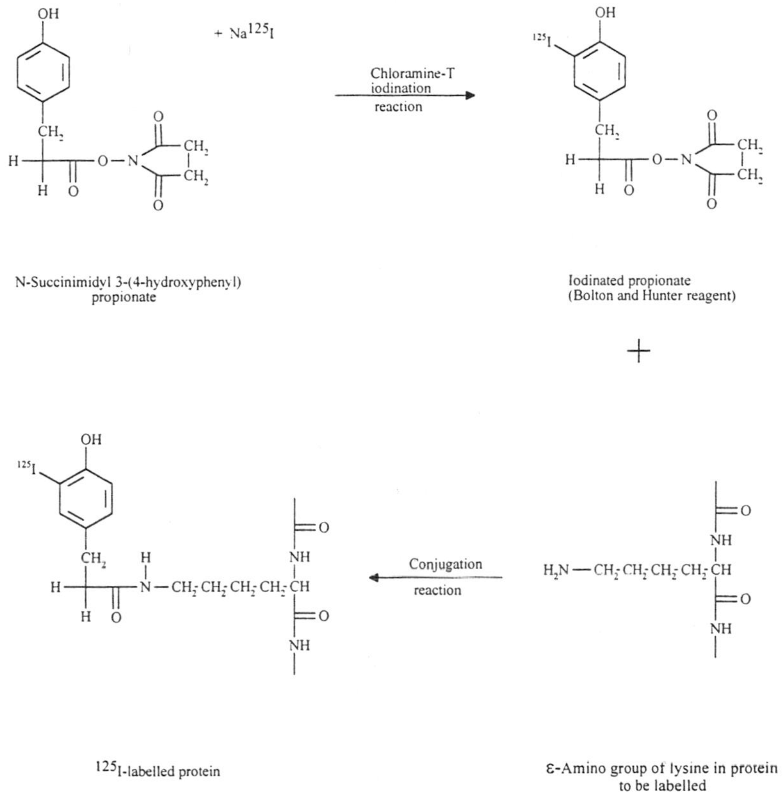

Fig. 3. Bolton and Hunter conjugation reaction.

(*see* **Fig. 3**). Conjugation occurs mainly with ε-amino groups of lysine residues and N-terminal amino groups. Rapid conjugation is essential because the acylating agent in aqueous medium is quickly degraded by hydrolysis to 3-(4-hydroxy-5-[^{125}I]iodo-phenylpropionic acid, hence the reaction is concentration-dependent. The conjugation reaction takes place under mildly alkaline conditions (pH 8.0–8.5). This method also overcomes any problems of contact with oxidizing or reducing agents and can be used when tyrosine residues occur in the biologically active regions of the protein or are not readily accessible for iodination unless some amount of unfolding of the protein molecule occurs. Because the reagent is itself hydrolyzed in aqueous conditions to 3-(4-hydrox-phenyl)propionic acid, incorporation of iodine tends to be lower than by the direct methods, although larger amounts of reagent can be used in the reaction. The reaction needs to be carried out as rapidly as possible to minimize hydrolysis. The conjugation involves three steps: evaporation of the benzene, addition of the protein, and purification.

This operation must be performed in a well-ventilated fume cupboard or similar facility. Additional safety precautions may be taken by using a charcoal trap to absorb volatilized [^{125}I]iodine; this is usually available on request from the manufacturer of the reagent. The method is technically more complex and requires more manipulation of radioactive material; these can be reduced by using commercial sources of the reagent. This reagent is available commercially from a number of suppliers such as Amersham International plc (Cat. code IM.5861), NEN and ICN Biomedicals. It is supplied in dry benzene at 2000 Ci/mmol and has been purified by HPLC. The activity of the product can be increased by using the di-iodo [[125]]Bolton and Hunter reagent (4000 Ci/mmol), however, this is at the expense of reduced stability.

The Bolton and Hunter reagent can be used to conjugate to amine-containing proteins. The derivative can then be iodinated using one of the aforementioned methods. This is useful if a stable Bolton and Hunter derivative can be stored under stable conditions until required for iodination.

The conjugation is carried out by evaporating the [^{125}I]Bolton and Hunter reagent (0.2 mL commercial reagent (185 MBq/mL; 5 mCi/mL) to dryness by blowing a gentle stream of dry nitrogen onto the surface of the solvent. This is done by inserting two hypodermic needles through the seal in the top of the vial and attaching a tube leading from the nitrogen source to one of the needles. The other needle acts as a outlet to which a charcoal trap is attached for the gaseous solvent. The protein (10 μL; 0.5 mg/mL) in sodium borate buffer (0.1 M; pH 8.5) is added to the dry reagent and mixed. The vial is incubated for 15–20 min at room temperature. Glycine solution (0.2 M; 50 μL) in sodium borate buffer (0.1 M; pH 8.5) is added to quench the conjugation and the vial incubated for an additional 5 min at room temperature. The reaction mixture can now be purified by the method of choice.

The low-mol-wt products may bind to serum albumin; therefore, buffers containing it should be avoided during the separation stage. Avoid thiol reagents, free amine-containing buffers, and sodium azide in the reaction mixture because these will react with the Bolton and Hunter reagent.

2.7. Purification of Radioiodinated Peptide and Proteins

Purification entails the isolation of the radiolabeled peptide or protein from the iodination mixture which contains low mol-wt products, e.g., unreacted iodide, oxidizing and reducing agents. This can be achieved rapidly and simply by a desalting procedure using gel filtration. Most commonly this is carried out on a prepacked column of Sephadex G-25 (Pharmacia, PD-10, cat. no. 17-0851-01) which has been equilibrated with sodium phosphate buffer (0.2 M; pH 7.4; 10 mL) containing bovine serum albumin (0.2%) and sodium azide (0.02%) or gelatin (0.2%) and sodium azide (0.02%) as preservative. The BSA or gelatin is added to reduce nonspecific binding, which may result in significant losses, especially when working with very small amounts of starting material. The labeled protein recovered will be contaminated with damaged protein (i.e., multi-iodinated protein, aggregates, and oxidized protein), thus requiring additional purification. The methods used for additional purification will depend upon the individual cases. Powerful separation techniques such as affinity chromatography, ion-exchange, and reverse-phase HPLC are commonly used in most laboratories.

Affinity chromatography is particularly useful when labeling immunologically active materials and receptor ligands. The conditions required to remove the labeled ligand from the affinity material are usually fairly harsh and can damage the products. Affinity chromatography does not remove the unlabeled material. Ion-exchange chromatography, which separates on charge, can be used to remove di-iodinated products and unlabeled material as well as lactoperoxidase, but will not separate various multi-iodinated species or oxidized material. Reverse-phase HPLC can be used to separate various monoiodinated materials from all the other reaction products and yield a pure product at maximum specific activity. The latter technique is ideally suited to small peptides.

2.8. Specific Activity Determination

The specific activity is defined as the unit of radioactivity per mole, which in the case of iodine is usually quoted as TBq/mmol or Ci/mmol. The introduction of just one atom of [^{125}I]iodine into the protein molecule causes the minimum alteration to its structure and thus keeps to a minimum any substitution effects. In general, further incorporation leads to products that are more susceptible to increased radiolytic decomposition. However, the higher the specific activity, the lower the mass of ligand that needs to be used in the experiment, and this normally leads to an increase in sensitivity of detection of either low-density or low-affinity receptors in studying ligand-binding interactions.

In the case of receptor studies, the specific activity needs to be known accurately, because it is used to quantify the density and affinity of ligand for receptors. In order to calculate the specific activity, the amounts of protein and radioactive iodide used in the iodination need to be known, as well as the yield of the product. The radioactivity needs to be measured in a calibrated ionization chamber and in the same type of vial to ensure the same geometry on each occasion.

There are several ways of determining the specific activity. In the case where gel filtration is used as a method of purification, the unreacted [^{125}I]iodide and the labeled protein can both be measured in a calibrated instrument. [^{125}I]Iodide does not absorb onto the surfaces of plastic or gel-filtration columns. Thus, any losses are normally owing to the absorption of [^{125}I]protein on the gel-filtration column. If the amount of radioactivity originally taken is known and the amount of radioactivity in the salt peak (free iodide) is measured, then the amount of radioactivity in the protein can be represented by the equation:

$$\text{Total protein radioactivity} = \text{original activity} - \text{activity in salt peak}$$

The method assumes that the absorption of the unlabeled protein and labeled protein remain the same. The method is suitable for most methods of iodination. In the case of lactoperoxidase methods, self-iodination of the enzyme gives rise to complications. In the case of insoluble-oxidizing media, such as Iodogen, a certain proportion of the [^{125}I]iodide remains attached to the plastic in the absence of protein.

Ascending thin-layer chromatography can also be used to determine specific activity using a suitable medium and solvent. In such cases, using the appropriate mobile phase, the labeled protein remains at the origin, whereas the [^{125}I]iodide moves up the plate. By determining the percentage incorporation and knowing the amount of activity and protein used, the specific activity can be determined. It should be stressed that, in the case of proteins, the radioligand is, in reality, a heterogenous mixture of labeled and unlabeled protein.

3. Labeling of Proteins with ^{3}H and ^{14}C

Tritiated proteins can be prepared by a mild method similar to the Bolton and Hunter reagent to produce a product with high retention of biological activity. [^{3}H]NSP (*N*-succinimidyl [2,3-^{3}H]propionate) is a general-purpose labeling reagent that labels free-amino groups. It reacts in a similar manner to [^{125}I]Bolton and Hunter reagent and, although offering a lower specific activity, the propionyl group introduced is small, and hence there is less alteration to the protein structure. The relative rate of reaction of [^{3}H]NSP with the ε-amino lysine and other groups, such as the α-amino groups of N-terminal amino acids, and with the thiol group of cysteine have been compared to establish the general conditions for use of this reagent (10). [^{3}H]NSP also offers the convenience of a one-step reaction.

The [^{3}H]NSP conjugation reaction must be performed in a well-ventilated fume cupboard. The toluene containing the [^{3}H]NSP is blown to dryness with a stream of dry nitrogen on the surface of the solvent. Addition of the protein to the vial in 25 mM sodium phosphate, pH 8.0, to give molar equivalence of protein to [^{3}H]NSP should be used. Incubate at room temperature for 2–4 h then purify by the method of choice as discussed previously.

[^{14}C]-Formaldehyde reacts with the amino group to form a Schiff's base that is then reduced with borohydride. Potassium borohydride is a milder reducing agent than sodium borohydride. It has the advantage of being stable under neutral aqueous solution for short periods, unlike sodium borohydride, which is quickly hydrolyzed in the absence of base. Sodium cyanhydride has fewer side reactions (such as hydrolysis) and a neutral pH can be used. In the range pH 6.0–7.0, it is completely inert to ketone and aldehyde groups (11).

4. Radioiodination Safety

This section is not intended to be a comprehensive guide to the handling of [^{125}I]iodine, rather, it is a collection of important considerations. For a more comprehensive guide the reader is referred to Amersham (12).

1. It is very strongly emphasized that any iodination-reaction vial must not be handled directly; vials should be handled with forceps in a suitable rack.
2. Use lead shielding or bricks to reduce dose.
3. It is advisable to designate certain adjustable pipets solely for iodinations. This reduces the risk of radioactive contamination elsewhere in the laboratory.
4. Wear disposable plastic or rubber gloves at all times and dispose used gloves within the iodination facility in a waste can.
5. The iodination facility should be a well-ventilated fume cupboard and not a laminar-flow cabinet, which pushes air outward toward the user.
6. Take particular care when carrying out thin-layer chromatography.
7. Plan ahead to minimize time spent handling radioactivity.
8. Use radioactivity hand monitors to provide constant surveillance over the working area and to detect the quantity and extent of radiation contamination.
9. Minimize accumulation of waste.

References

1. Houghten, R. A. and Li, C. H. (1979) Reduction of sulfoxides in peptides and proteins. *Anal. Biochem.* **98,** 36–46.
2. Hunter, W. M. and Greenwood, F. C. (1962) Preparation of iodine 131-labelled growth hormone of high specific activity. *Nature* **194,** 495–496.
3. Marchalonis, J. J. (1969) An enzymic method for the trace iodination of immunoglobulins and other proteins. *Biochem. J.* **113,** 299–305.

4. Huber, R. E., Edwards, L. A., and Carne, T. J. (1989) Studies on the mechanism of the iodination of tyrosine by lactoperoxidase. *J. Biol. Chem.* **264(3),** 1381–1388.

5. Murphy, M. J. (1976) 125I labelling of erythropolition without loss off of biological. *Biochem. J.* **159,** 287–289.

6. Karonen, S. L., Morsby, P., Siren M., and Seuderlug, U. (1975) An enzymatic solid phase method for trace iodination of proteins and peptides with [125]Iodine. *Anal. Biochem.* **67,** 1–10.

7. Fraker, P. J. and Speck, J. C. (1978) Protein and cell membrane iodinations with sparingly soluble chloramide, 1,3,4,6-tetrachloro-3α,6α-diphenylglycoluril. *BBRC* **80(4),** 849–857.

8. Markwell, M. A. K. (1982) A new solid-state reagent to iodinate proteins. *Anal. Biochem.* **125,** 427–432.

9. Bolton, A. E. and Hunter, W. M. (1973) The labelling of proteins to high specific radioactivities by conjugation to ^{125}I-containing acylating agent. *Biochem. J.* **133,** 529–539.

10. Tang, Y. S., et. al. (1982) N-succinimidyl propienate: characterisation and optimum conditions for use as a tritium labelling agent for proteins. *Lab. Comp. Radiopharm.* **20,** 277–284.

11. Dottavio-Martin, D. and Ravel, J. M. (1978) Radiolabeling of proteins by reductive alkylation with carbon-14 formaldehyde and sodium cyanoborohydride. *Anal. Biochem.* **87,** 562–565.

12. *Guide to Radioiodination Techniques.* (1993) Amersham International plc publication, Little Chalfont, UK.

32

Protein Electrophoresis

Paul Richards

1. Theory and Background

There are an estimated 100,000 genes in the entire human genome, 10% of which may be expressed in a particular cell-type at a given time. Purifying and analyzing the products of genes (proteins), from humans and other organisms, offers a unique challenge to the biochemist and one that is essential to our understanding of biological function. The diversity in size, ionic charge, solubility, stability, cellular expression, and posttranslational modification of proteins requires protein-analysis techniques that have high-resolving capacities. Electrophoretic separation techniques provide a high-resolution, adaptable and simple range of methodologies used in nearly all forms of protein analysis *(1)*.

Electrophoresis is defined as the movement of charge particles under the influence of an electric field. The velocity of the particle (electrophoretic mobility) is dependent on several factors, some intrinsic to the molecule itself and others dependent on the particular electrophoretic technique employed in the separation. For example, electrophoretic mobility is critically dependent on the electric-field strength, pH, and ionic strength of the buffers employed; buffers of low-ionic strength generally increase the rate of migration compared with those of high ionic strength. In many forms of protein electrophoresis the charge density of the molecule is critical to the electrophoretic mobility. Charge density is the ratio of net charge on the protein to its molecular mass, the larger the value, the greater the electrophoretic mobility. Thus, differences in charge densities between proteins can be exploited to aid in their separation. Other electrophoretic techniques can be employed that separate proteins according to their net charge or polypeptide molecular mass independently of each other. Combination of electrophoretic techniques helps the analysis of complex biological samples by examination of several different physicochemical parameters of proteins.

If electrophoresis was performed in a free solution of an appropriate buffer, the net effect of diffusion would decrease the resolving capacity of the technique. Furthermore, detection of proteins in a free solution would be quite difficult, requiring the use of UV absorbance or fluorescence, adding cost and complexity to the method. However, to circumvent these problems, most electrophoretic techniques utilize a supporting medium (a gel) containing small pores through which the proteins electrophorese.

From: *Molecular Biomethods Handbook*
Edited by: R. Rapley and J. M. Walker © Humana Press Inc., Totowa, NJ

As the proteins migrate through the gel matrix the pores impede their flow. The molecular sieving effect can be exploited to effect protein separation based on molecular mass. Other advantages of using a gel matrix are the decrease diffusion and the ability to fix (precipitate) the proteins in the gel following electrophoresis. Proteins thus immobilized within the gel matrix can be analyzed by selective staining, allowing the detection of subnanogram amounts of protein. Furthermore, glycoproteins *(2)*, phophoroproteins *(3)*, and lipoproteins *(4)* can be selectively detected to further increase the analytical possibilities of electrophoresis.

1.1. Electrophoresis Formats

Electrophoresis can be carried out using several different types of gel-format. Gels can be formed and run in glass tubes. However, tube-gel electrophoresis has been largely superseded by the slab-gel format. Slab-gels are cast and run between two glass plates, usually held vertically in the case of protein gels. The glass plates (typically about 16×16 cm) are held apart by narrow spacer strips at either side of the apparatus. The gel is then be sealed at the base with waterproof tape or a rubber gasket and the gel mixture poured into the mold and allowed to set. It is usual to insert a square-toothed comb in the top of the mold when the gel is poured. After the gel has set, the comb can be removed, leaving wells cast in the gel into which the samples can be applied. Slab-gels allow several samples to be run in parallel, an obvious advantage over running individual samples in glass tubes. Proteins are usually applied to the gel in a special buffer (sample buffer). The sample buffer contains a marker dye, which runs ahead of the protein samples and is used to monitor the progress of the electrophoretic separation. After electrophoresis, the distances migrated by the proteins can be compared to the distance migrated by the marker dye, allowing the Rf value to be obtained (distance moved by protein/distance moved by the marker dye).

The assembled gel apparatus contains an upper- and a lower-buffer chamber to provide electrical contact with the gel. Two platinum electrodes, one in the upper buffer chamber and one in the lower, allow connection with a power-pack. Electrophoresis can then be conducted for the desired length of time, at either a constant current or constant voltage. After electrophoresis the apparatus can be disassembled the gel-slab removed and proteins detected by a suitable method.

1.2. Support Media

Agarose and polyacrylamide are the main support media used for protein electrophoresis. Agarose is mainly reserved for use with very high molecular mass proteins and protein complexes, immunoelectrophoresis *(5)*, and in certain types of isoelectric focusing *(see* **Subheading 1.6.***)*. Polyacrylamide is the media of choice for nearly all forms of protein electrophoresis because of its mechanical strength, low electrical conductivity, low-UV absorbance, and the controllability of gel pore-size *(6)*. However, acrylamide is a neurotoxin and caution should be exercised when handling the monomer. Many manufacturers supply ready-made aqueous acrylamide solutions so that acrylamide powder is not handled, the dust from which can be inhaled if proper precautions are not taken. When handling the liquid, gloves should always be worn and any spillage immediately mopped up; acrylamide solutions are odorless, colorless, and tasteless.

Polyacrylamide is formed *in situ* by the copolymerization of acrylamide and a suitable crosslinker, typically *N,N'*-methylene bisacrylaide (Bis). Other crosslinkers can be employed to give a particular property to the gel. For example, *N,N'*-(1,2-dihydroxyethylene)bisacrylamide (DHEBA) allows the gel to be readily solubilized in periodic acid after electrophoresis *(7)*, releasing proteins of interest from the matrix. This is particularly useful for detecting radioactively labeled proteins *(see* **Subheading 2.3.**).

The pore-size of polyacrylamide gels can readily be controlled. Pore size is dependent on two parameters; the total percentage (w/v) of monomer (acrylamide plus Bis), %T, and the percentage (w/w) of crosslinker, %C. For example, a 10%T, 2.67%C gel would contain 0.27 g of Bis and 9.73 g of acrylamide per 100 mL of buffer.

Many general gel compositions exist for profiling complex mixture of proteins. However, purification or analysis of a particular protein requires optimization of the gel in terms of %T and %C. Also, by manipulating the aforementioned parameters, we can extract information about the size and charge of the protein. To obtain information on the physical properties of a series of different gel concentrations %T could be run using at a constant crosslinker (%C) concentration and the results for the separation of the protein(s) of interest analyzed. The data consequently generated can be analyzed using a Ferguson plot *(8)*. A Ferguson plot is a graph of $\log_{10}$ mobility (Rf) vs gel composition, %T. The slope of the plot, K_R, and Y-axis intercept, Y_0, can be defined. K_R and Y_0 can be related to the protein's molecular size and charge. K_R is a retardation factor that can be related to molecular size. The variable Y_0 is related to the free solution mobility of the protein.

1.3. Gel Formation

Polyacrylamide is formed from the free-radical polymerization of an aqueous solution of acrylamide and a crosslinking agent to connect the polymer chains. Polymerization of acrylamide is usually initiated by the addition of persulfate ions (usually as ammonium persulfate), which decompose to produce free radicals. *N,N,N',N'*,-tetramethylethylenediamine (TEMED) is added to catalyze the breakdown of persulfate. Gel formation is also dependent on temperature; the higher the temperature the faster the gel formation. Frequently, polymerization is apparent within 10–30 min, at room temperature; gels should, however, be left for at least 2 h to allow for complete polymerization and ensure reproducibility between gels.

1.4. Continuous-Buffer Systems

A gel system, such as the one described, is suitable for use with a continuous-buffer system, i.e., the same buffer ions are present throughout the electrophoresis system. Continuous gels are often employed for use with native PAGE techniques, where subunit interaction and biological function of the proteins being analyzed are preserved. Proteins separated on native PAGE gels are resolved according to their charge-density ratio. Furthermore, because the proteins separated are still active, the gels can be stained to detect enzymatic activity to aid in protein profiling and identification. Such activity stains are commonly employed with proteases *(9)*, but many other enzyme systems can be detected *(10)*.

In running a continuous PAGE-gel, the protein sample needs to be applied in as small a volume of sample buffer as possible and to be applied to the bottom of a sample well. It is usual to include glycerol or sucrose in the sample buffer, to increase its density such that it will sink to the bottom of the well and form a thin layer.

After sample loading, the gel apparatus is assembled and a current applied to induce the proteins to migrate through the gel at a rate dependent on their charge density. The aforementioned phenomenon is termed zone electrophoresis, wherein a narrow layer of proteins, placed at the origin of the gel, separate out into discrete zones according to their electrophoretic mobility.

Several buffer systems available for native PAGE, covering a wide pH range *(11)*. Typically, for acidic proteins, Tris-acetate, or Tris-glycine buffers can be used (covering a pH range 7.2–9.5). Basic proteins can be resolved with a β-alanine-acetate system (covering a pH range 4.0–5.0) or 1 *M* acetic acid can be used for acid stable proteins *(12)*.

1.5. Discontinuous-Buffer Systems

Because of the problems associated with adding small amounts of a concentrated sample of proteins to the origin of the gel, discontinuous-buffer systems were developed. With discontinuous systems, the buffers in the electrode chambers are different to those within the gel. Also, a stacking gel is employed to electrophoretically concentrate the proteins into a narrow zone, allowing larger amounts of more dilute protein to be added to the sample well. The narrow zone contains (hopefully) all the proteins of interest, which are subsequently separated in the resolving gel. The method is illustrated with reference to the Ornstein-Davies buffer system *(11)*.

The rate of electrophoretic mobility of a particular particle is dependent on the length of time it remains ionized at a given pH. Near its isoelectric points (pI) a molecule is ionized for a small percentage of time, thus only carries a partial charge. With the Ornstein-Davies buffer system the stacking gel contains chloride ions in the buffer, whereas the upper-cathode buffer contains only glycine ions. At the pH of the stacking gel (6.7) glycine carries only a partial negative charge. However, chloride ions are fully ionized and consequently have greater electrophoretic mobility than the glycine ions. Proteins have an electrophoretic mobility between the two. When a current is applied, the chloride ions migrate rapidly through the stacking gel, creating a narrow zone of high electrical-field strength. Running behind the chloride ions are the glycine ions from the upper buffer chamber and sandwiched between are the proteins. Ahead of this zone the field strength is lower, so any proteins ahead of it are rapidly overtaken and incorporated into the stacking zone. The net result of this effect is to concentrate the proteins into a narrow zone or stack between the chloride and glycine ions. The stacking gel also has a large pore size to prevent molecular sieving of the proteins. When the protein stack reaches the resolving gel the higher pH value (8.9) allows almost complete ionization of the glycine, increasing its electrophoretic mobility to greater than that of the proteins. The proteins can now move in a zone of uniform voltage and are sieved through the smaller pore-size resolving gel.

1.6. Isoelectric Focusing

Isoelectric focusing (IEF) is a technique used for the separation of proteins according to their pI, the pH at which they carry no net charge *(13)*. An IEF gel is composed of a polyacrylamide or agarose slab-gel matrix containing 2–4% carrier ampholyte. Carrier ampholytes are a mixture of amphoteric molecules with narrowly spaced pI values. After the gel has been cast, it is removed from the mold, laid flat on a cooling plate, and a strip electrode (usually absorbent paper) placed at either end of the gel. An anode strip is wetted in an acid solution (e.g., 1 *M* phosphoric acid) and the cathode in an alkali (e.g., 1 *M* sodium hydroxide). Application of electric field causes the ampholytes to migrate through the gel in a direction dependent on their net charge. Thus, the most acidic migrate to the anode and the most basic to the cathode. This results in the formation of pH gradient through the gel. The pH of surrounding solution is the same as that of the ampholyte in that particular region that of the gel. If buffers, or other charged small molecules, are introduced into the gel, the pH gradient will be disturbed. Disruptions in the pH gradient will cause poor quality separations. When a protein is applied to the gel, it will travel until it reaches the band corresponding to its own pI. If it moves out of this region it will gain a net electrical charge and move back into the zone corresponding to its own pI. The protein is then focused into a narrow region; differences in pI by as little as 0.02 pH units can be resolved by this technique. Low percentage monomer gels are used with IEF to minimize molecular sieving.

An alternative approach is to "cast" the pH gradient into the gel. Immobilized pH gradients are formed using a series of acrylamide derivatives termed Immobilines (Pharmacia). These are made from derivatized acrylamide monomers containing either a carboxyl or tertiary amine group, which buffer in the pH range 3.0–10.0. Thus, a pH gradient can be cast into the gel by selecting the appropriate Immobiline reagent. Immobilized pH gradients allow the selection of a pH range from very wide to ultra-narrow. Proteins can be resolved that differ in pI by as little as 0.001 pH units by this method.

2. Applications of Protein Electrophoresis
2.1. SDS-PAGE

Sodium dodecylsulfate polyacrylamide gel electrophoresis (SDS-PAGE) is a commonly employed technique in protein electrophoresis. It allows estimation of polypeptide-molecular mass and high-resolution analysis of complex mixtures of proteins. However, the technique is a destructive one, requiring the proteins being analyzed to be denatured for an effective separation. To dissociate the proteins into single-polypeptide chains, the detergent SDS is used in combination with a reducing agent. Proteins bind 1.4 g SDS/g protein, the detergent molecules embedding into the protein inducing unfolding. The protein/SDS complex effectively masks the protein charge, sulfate groups on the detergent imparting an overall negative charge. Electrophoresis in the presence of SDS is effectively independent of protein charge and is based on the relative molecular masses of the polypeptide chains. This relationship holds true for many proteins *(14)*. SDS is usually used with Ornstein-Davies buffer system in a method developed by Laemlli *(15)*. To help denaturation of the proteins, the sample buffer contain a reducing agent (dithiotheitol or 2-mercaptoethanol) to break disulfide bonds and allow the proteins to unfold. The sample is usually boiled in SDS-sample buffer for 3 min before loading to ensure complete denaturation of the protein. Typically, for SDS-PAGE, the sample buffer contains 2% (w/v) SDS, 10% glycerol (to make the solution dense so that it sinks to the bottom of the well), and a small amount of bromophenol blue to act a tracking dye (this should migrate ahead of the proteins). It is important to include SDS in the upper-buffer chamber (0.1%), because the SDS ions will migrate toward the anode during electrophoresis and deplete the protein/SDS complexes. A series of known molecular-mass proteins can be run in a lane alongside the samples being analyzed. A plot of $\log_{10}$ molecular mass against Rf should give a straight line from which the molecular mass of any unknown protein can be calculated (**Fig. 1**).

With the Laemlli-gel system, it is usual to employ constant concentration of Bis (2.67%C). Different pore-sizes are formed by varying the concentration of total mono-mer (%T). For example, a 10%T, 2.67%C gel will separate proteins in the range 20–200 kDa. However, the Tris/glycine buffer system of the Laemmli system can only resolve proteins down to a molecular-mass of about 15 kDa. This results from small peptides travelling with the SDS micelles and not resolving as they travel through the resolving gel. In order to resolve smaller proteins urea (usually 8 M) can be incorporated into the gel to reduced the SDS micelle size. Alternatively, modified PAGE systems can be employed. One of the most useful of these uses Tricine in place of Tris. Tricine gels have been reported to resolve proteins with a molecular mass of 1 kDa *(16)*.

To separate complex mixtures of proteins, a concentration gradient of polyacryla-mide can be used. In this case, a plot of $\log_{10}$ molecular mass vs $\log_{10}$ %T will give a straight line *(17)*. A 6–20% Laemlli-gradient gel is linear in the range 14–200 kDa. If the proteins being analyzed fall within a narrow molecular-mass range then it is prob-ably better to use a homogeneous gels.

Glycoproteins can run anomalously on SDS-PAGE gels. The sugar groups on the glycoprotein cannot saturate with SDS, thus, when they are resolved on SDS gels, they give a erroneously high molecular mass. Alternative buffer systems, such a Tris/borate/ EDT, or pore-gradient gels can give a more accurate estimation of glycoprotein-molecular mass.

2.2. Detection Methods

After electrophoresis, the gel can be stained to detect protein bands. There are many types of stain available, varying in their ease of use and sensitivity in detecting protein bands. The most simple method involves soaking the gel in a solution of Coomassie blue containing methanol and acetic acid *(18)*. The cationic dye binds to the protein in the acidic solution and any unbound dye can be washed out with a methanol/acetic acid solution. If the proteins have been resolved by IEF, the carrier ampholytes need to be removed before staining can be achieved. This is done by first fixing the proteins in the gel with a 12% trichloroacetic acid solution, which also removes the ampholytes. The gel can then be stained in a Coomassie blue solution to reveal the protein bands. Because of the low concentration of polyacrylamide used with IEF, the gels tend to be quite "tacky" and so should be handled with care. Staining with Coomassie blue can take from 30 min to several hours depending on the amount of protein loaded on the gel; the limit of sensitivity is 200 ng/protein band.

The sensitivity of Coomassie blue can be increased by employing the dye in a colloi-dal form; here the limit of detection is 1 ng/protein band *(19)*. Several rapid-staining techniques have been described, these can detect protein bands within 1 h (e.g., **ref. 20**).

Finally, the most sensitive general-staining procedure involves the use of silver. Silver cations bind to protein amino and sulfur groups (lysine, cysteine, and methion-ine residues); the silver/protein complexes can subsequently be reduced to produce grains of metallic silver in the gel. Over 100 different methods employing silver-stain-

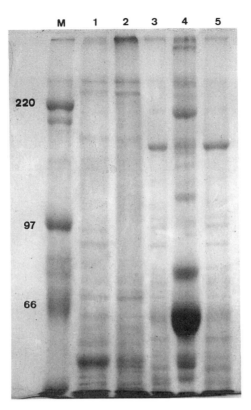

Fig. 1. Separation of rat proteins extracted from various tissues. (1) Total-brain homogenate, (2) brain-microsomal proteins, (3) liver, (4) plasma, (5) spleen. Proteins were run on a 10%T, 2.67%C Laemlli gel. Lane M contains marker proteins, 220, 97, and 66 kDa.

ing have been described *(21)*. The high sensitivity of silver-staining (0.21 ng/protein band) makes it the method of choice for detection of trace impurities or very dilute protein solutions. However, the technique is prone to produce high-background staining and often is quite laborious to perform.

2.3. Detection of Radiolabeled Proteins

Labeling proteins with radioisotopes can be useful in the detection of specific proteins and metabolic studies. Proteins separated by PAGE can be examined using several different radioactive detection methods

First, proteins can be subjected directly to autoradiography. The gel is dried and placed in direct contact with X-ray film and exposed for a suitable length of time. However, the low energy β-emitter tritium, commonly used in biological studies, is hardly detected by autoradiography.

An alternative detection method is fluorography. A gel can be soaked in a scintillation fluid (similar to those used for liquid-scintillation counting) dissolved in a suitable solvent. The gel can then be exposed to X-ray film, where the flashes of light produced from the radioactive decay of the radioactive particles are detected. This method works well for tritium-labeled proteins *(22)*.

Table 1
Specific Protease Useful in Peptide Mapping

Protease	Cut sites
Staphlococcus aureus V$_8$	C-terminal of glutamate
Trypsin	C-termini of lysine and arginine
Lysobacter enzymogenes Lys-C	C-terminal of lysine
Clostrdium histolyticum Arg-C	C-terminal of arginine
Pseudomonas fragi Asp-N	N-terminal of asparatate
Papaya proteinase IV	C-terminal of glycine

2.4. Peptide Mapping

The use of specific proteolytic enzymes allow the digestion of proteins into reproducible peptide fragments. The fragments can be resolved on SDS-PAGE gels to generate patterns that facilitate the analysis of proteins or mixtures of proteins. Peptide-mapping can be useful when comparing families of proteins or newly discovered proteins to existing ones. Specific proteases useful in peptide-mapping are given in **Table 1**.

Proteins can be digested in vitro prior to loading on a gel. Alternatively, the peptide map can be generated *in situ (23)*. A mixture of proteins can be separated on an SDS gel and stained with Coomassie blue. Bands of interest can be excised from the gel, equilibrated in an appropriate buffer, and placed in the well of second gel. A protease solution can be added to the well (V$_8$ protease works well) and electrophoresis commenced. The protease will digest the protein in the stacking gel as electrophoresis proceeds and the peptides will separate in the resolving gel.

2.5. Protein Blotting (see *Chapter 35*)

The transfer of proteins onto membrane from a slab gel opens up the analytical possibilities of PAGE. Transfer is usually accomplished electrophoretically using specially designed apparatus *(24)*. Proteins immobilized on a suitable membrane may be subjected to gas-phase sequencing *(25)*, stained for enzyme activity, or blotted for reaction with antibodies *(26)* (Western blotting) and other ligands

2.6. 2-D PAGE

About 100 bands can be resolved by a single-dimension electrophoretic technique; the implication of this for complex protein mixtures is a single band on a gel may represent several different types of protein. Although single dimension techniques may be useful for profiling a particular sample, it is obvious that resolution of individual proteins would offer an enormous advantage as a tool for protein analysis. The use of two-dimensional (2-D) techniques can increase the resolution of electrophoresis several fold, in-fact the most powerful 2-D techniques can resolve several thousand proteins *(27)*.

In essence the concept of 2-D PAGE is quite simple (**Fig. 2**). A protein sample is first separated on an IEF gel. After suitable equilibration, the IEF strip is placed onto of an SDS gel and the sample run in the second dimension. Thus, the proteins are resolved by two physicochemical properties; in the first dimension, by their pIs and in the second, by their molecular masses. However, the technique is not experimentally simple

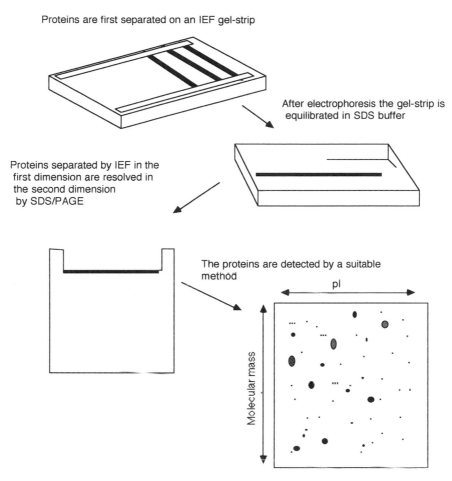

Fig. 2. Procedure for producing two-dimensional maps of proteins.

and many problems can be encountered. First, IEF requires that the protein sample are in a soluble form in a buffer of low-ionic strength and not containing large amounts of charged detergent. For most soluble proteins these requirements can be easily met, however for many proteins, especially membrane proteins, this is not easy. Urea (8 *M*) can be included in the sample preparation buffer along with neutral (Triton X-100, octyl glycoside) or zwitterionic (CHAPS, amidosulfobetaines) detergents *(28)*. The aforementioned detergents, along with urea, can be included in the gel to keep the protein soluble during electrophoresis. Proteins can also be solubilized by including a small amount of SDS in the sample buffer, but this detergent cannot be included in the actual gel. Another problem is the resolving capacity and pH range of IEF. Although carrier ampholytes can be obtained covering a pH range of 2.5–11.0, in practice the pH range 4.5–8.5 is effective, with values outside this giving inconsistent results. Stability of the gradient during long runs is also questionable. Immobilized pH gradients *(see* **Subheading 1.6.**) are more commonly employed with no problems of pH stability and reproducible gradients *(29)*. Also, nonlinear pH gradients are commercially available. These have a flattened pH range of 5.0–7.0 increasing resolution in this critical region.

ISOELECTRIC FOCUSING

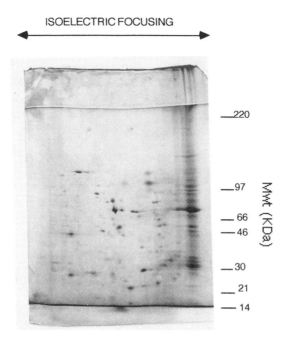

Fig. 3. 2-D gel of rat-brain proteins. The proteins were separated by IEF in the first dimension, followed by SDS-Page on a 10% Laemmli gel. Proteins were detected by silver-staining.

In the second dimension, extra large SDS-gel formats (e.g., 32 × 43 cm) can be used to obtain maximum resolution.

2.7. Detection of Proteins in 2-D Gels

One of the problems associated with 2-D gels is the reduced staining sensitivity resulting from the protein sample being separated over a 2-D surface. For Coomassie blue staining at least 100 μg of protein needs to applied to the first-dimension gel. To increase the sensitivity of detection, radioactive samples are commonly prepared for use with 2-D gel systems. This could be achieved at the translational level by the incorporation of labeled amino acids, or chemically, after synthesis, using radioactive protein-modification reagents. A silver-stained 2-D gel is shown in **Fig. 3**.

Acknowledgment

We thank Tanya Sage for excellent technical support, especially in the production of 2-D gels.

References

1. Dunn, M. J. (1993) *Gel Electrophoresis: Proteins.* Bios, Oxford, UK.
2. Estep, T. N. and Miller, T. J. (1986) Optimisation of erythrocyte membrane glycoprotein fluorescent labelling with dansylhydrazine after polyacrylamide gel electrophoresis. *Anal. Biochem.* **157,** 100–105.
3. Debruyne, I. (1983) Staining of alkali-labile phophoproteins and alkaline phosphates on polyacrylamide gels. *Anal. Biochem.* **133,** 110–115.
4. Tsai, C.-M. and Frasch, C. E. (1982) A sensitive silver stain for detecting lipopolysaccharides in polyacrylamide gels. *Anal. Biochem.* **119,** 115–119.

5. Bøg-Hansen, T. C. (1993) Immunoelectrophoresis, in *Gel Electrophoresis of Proteins* (Hames, B. D. and Rickwood, D., eds.), IRL, Oxford, UK, pp. 273–300.

6. Raymond, S. and Weintraud, L. S. (1959) Acrylamide gel as a supporting medium for zone electrophoresis. *Science* **130**, 711–721.

7. O'Connel, P. B. H. and Brady, C. J. (1976) Polyacrylamide gels with modified cross-linkages. *Anal. Biochem.* **76**, 63–73.

8. Rodbard, D., Chrambach, A., and Weis, G. H. (1974) in *Electrophoresis and Isoelectric Focusing in Polyacrylamide Gel* (Allen, R.C. and Mauvrer, H. R., eds.), Walter de Grugtner, Berlin, p. 62.

9. Heussen, C. and Dowdle, E. B. (1980) Electrophorectic analysis of plasminogen activators in polyacrylamide gel containing sodium dodecyl sulphate and copolymerised substrates. *Anal. Bioche.* **102**, 196–202.

10. Gabriel, O. (1971) Locating enzymes on gels. *Methods Enzymol.* **22**, 578–604.

11. Dunn, M. J. (1993) Electrophoresis under native conditions, in *Gel Electrophoresis: Proteins*, Bios, Oxford, UK, pp. 31–39.

12. Carré-Eusébe, D., Lederer, F., Diêp, K. H., and Elsevier, S. M. (1991) Processing of the precursor of protamine P2 in mouse. *Biochem. J.* **277**, 39–45.

13. Righetti, P. G. (1983) *Isoelectric Focusing: Theory Methodology and Applications.* Elsevier, Amsterdam.

14. Weber, K. and Osborn, M. (1969) The reliability of molecular weight determination by dodecyl sulphate polyacrylamide gel electrophoresis. *J. Biol. Chem.* **244**, 4406–4412.

15. Laemmli, U. K. (1970) Cleavage of structural proteins during the assembly of the head of the bacteriopahge T4. *Nature* **227**, 680–685.

16. Schägger, H. and von Jagow, G. (1987) Tricine-sodium dodecyl sulfate-polyacrylamide gel electrophoresis for the separation of proteins in the range from 1 to 100 KDa. *Anal. Biochem.* **166**, 368–379.

17. Poduslo, J. F. and Rodbard, D. (1982) Molecular weight estimation using sodium dodecyl sulphate-pore gradient electrophoresis. *Anal. Biochem.* **101**, 394–406.

18. Wilson, C. M. (1983) Staining of proteins on gels: comparisons of dyes and procedures. *Methods Enzymol.* **91**, 236–246.

19. Neuhoff, V., Arold, N., Taube, D., and Ehrhardt, W. (1988) Improved staining of proteins in polyacrylamide gels including isoelectric focusing gels with clear background and nanogram sensitivity using Coomassie brilliant blue G–250 and R–250. *Electrophoresis* **9**, 255–262.

20. Ward, A. H. and Allen, W. S. (1972) An improved procedure for protein staining in poly-acrylamide gels with a new type of Coomassie brilliant blue. *Anal. Biochem.* **48**, 617–620.

21. Oakley, B. R., Kirsch, D. R., and Morris, N. R. (1980) A simplified ultrasensitive silver stain for detecting proteins in polyacrylamide gels. *Anal. Biochem.* **105**, 361–363.

22. Bonner, W. M. and Laskey, R. A. (1974) A film detection method for tritium-labelled proteins and nucleic acids in polyacrylamide gels. *Eur. J. Biochem.* **46**, 83–88.

23. Cleveland, D. W., Fischer, S. G., Kirschner, M. W., and Laemmli, U. K. (1977) Peptide mapping by limited proteolysis in sodium dodecyl sulfate and analysis by gel electrophoresis. *J. Biol. Chem.* **252**, 1102–1106.

24. Gültekin, H. and Heermann, K. H. (1988) The use of polyvinylidenedifluoride membranes as a general blotting matrix. *Anal. Biochem.* **172**, 320–329.

25. Kennedy, T. E., Wager-Smith, K., Barzilai, A., Kandel, E. R., and Sweatt, J. D. (1988) Sequencing proteins from acrylamine gels. *Nature* **336**, 499,500.

26. Pluskal, M. G., Przekop, M. B., Kavonian, M. R., Vecoli, C., and Hicks, D. A. (1986) Immobilon™ PVDF transfer membrane: a new membrane substrate for Western blotting of proteins. *BioTechniques* **4**, 272–283.

27. Hochstrasser, D. F., Harrington, M. G., Hochstrasser, A.-C., Miller, M. J., and Merril, C. R. (1988) Methods for increasing the resolution of two-dimensional protein electrophoresis. *Anal. Biochem.* **173,** 424–435.

28. Perdew, G. H., Scaup, H., and Selivonchick, D. P. (1983) The use of detergents in two-dimensional gel electrophoresis of trout-liver microsomes. *Anal. Biochem.* **135,** 453–455.

29. Görg, A. (1993) Two-dimensional electrophoresis with immobilized pH gradients: current state. *Biochem. Soc. Trans.* **21,** 130–133.

33

Free Zone Capillary Electrophoresis

David J. Begley

1. Introduction

Electrophoresis within silica capillaries is a relatively new technique, with the first publications appearing in the 1970s followed by a rapid expansion and establishment of the technique in the 1980s. Commercial machines for the application of capillary electrophoresis appeared in the late 1980s and now a number of machines are manufactured.

All of the commercially available machines function in essentially the same way, although there is, of course, individual variation. This account of capillary electrophoresis is tailored not to any particular machine, but is written in a manner that describes the basic theory and method in a way that can be easily varied to suit a particular application. This chapter contains a detailed description of a capillary electrophoresis apparatus and some theory of the operation, together with protocols and notes for performing separations.

High-performance capillary electrophoresis (HPCE) can be used to separate a wide variety of solutes, both charged and uncharged, and is particularly suited to the separation of small peptides and proteins. Separations can be performed under a variety of conditions with free zone capillary electrophoresis (FZCE), where the solutes being separated are in simple solution in buffer. A variant of the technique is capillary isoelectric focusing (CIF), where the ampholyte and sample are premixed and loaded into the capillary before application of a voltage. A pH gradient is formed, and separation of samples according to their isoelectric point (pI) takes place within the capillary. This chapter will be essentially confined to FZCE separations, because these have found the greatest application in the biological sciences (1–6).

Capillary electrophoresis is normally carried out in fused-silica capillaries with internal diameters between 10 and 100 μm. Capillary length may be varied with a longer capillary producing a greater spatial separation between two solutes of similar mobility. A popular capillary for a variety of separations would be one of 25-μm internal diameter and 20 cm in length. A potential difference is normally applied across the two ends of the capillary with a power source. Electrophoresis is usually conducted under constant voltage conditions with the current finding its own level. Typical operating parameters under differing conditions are shown in **Table 1**. Separation under normal

From: *Molecular Biomethods Handbook*
Edited by: R. Rapley and J. M. Walker © Humana Press Inc., Totowa, NJ

Table 1
Operating Parameters for Capillary Electrophoresis Under Differing Conditions[a]

Buffer	pH	Voltage, kV	Current, µA	Resistance, Meg Ω	Power, watts
Phosphate 100 m*M*	2.5	8.0	12.6	634.9	0.101
Phosphate 100 m*M*	4.4	8.0	25.25	316.8	0.202

[a]Note that the resistance approximately halves with the change in pH, and the current and the power dissipation (heat production) double.

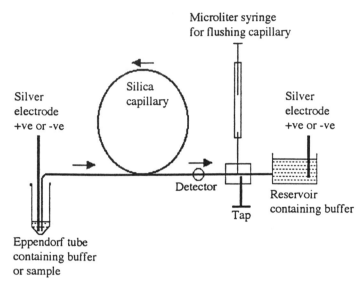

Fig. 1. A typical layout for a HPCE apparatus. For electrophoretic loading, an Eppendorf tube containing the sample is offered up to the left-hand electrode, and the capillary and a loading voltage applied for a set time. With suitable Eppendorf tubes, the sample contained in loading buffer may be as little as 10 µL. The tube containing the sample is then substituted with one containing the running buffer and the run conducted. After the run, the tap can be closed and the capillary back-flushed with running buffer between runs. At pH 2.5 with positively charged analyses, the left-hand electrode will normally be the anode and the right-hand electrode the cathode.

conditions produces significant amounts of heat, but this is not usually a problem that affects the quality of a separation or leads to denaturation of protein, because the large surface area of a capillary in relation to its volume efficiently dissipates the heat produced. Power sources for use in capillary electrophoresis normally supply voltages up to 30 kV or more.

1.1. Apparatus and Theory of Operation

A typical layout for a capillary-electrophoresis apparatus is shown in **Fig. 1**. The apparatus consists of the silica capillary, which is usually coiled to make it more manageable. The ends of the capillary project into two buffer reservoirs that contain silver

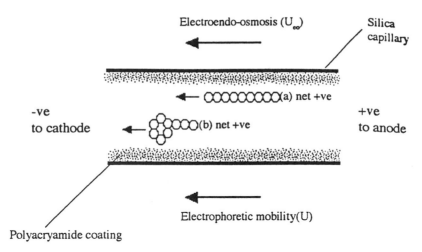

Fig. 2. Migration of solute in the silica capillary. The ζ potential on the wall of the capillary produces an electroendo-osmotic flow (U_{eo}) of water from the anode to cathode. Coating the capillary internally with polyacrylamide reduces this standing charge and reduces the magnitude of U_{eo}. The electrophoretic mobility of a charged solute (U) is dependent on the mass/charge ratio and the applied potential difference. Thus, the actual rate of migration (U_{app}) is equal to $U + U_{eo}$. With coated capillaries, the degree of electroendo-osmosis (U_{eo}) is minimized for two solutes of equal charge and mass, such as (a) and (b). The configuration of the molecule can also influence the rate of migration (U) with (b) exhibiting a greater molecular sieving and/or chromatographic effect than (a). (*See also* **Fig. 4.**)

electrodes. A facility for changing the polarity of these two electrodes usually exists, although for the majority of applications, the left-hand electrode on the figure will be the anode and the right-hand the cathode.

When a potential difference is applied across the ends of the capillary and because the walls of the capillary have a standing charge (the ζ potential), an electroendo-osmotic flow of water is produced from anode to cathode. This is shown diagrammatically in **Fig. 2**. Thus, migration of a positively charged solute from anode to cathode along the capillary is partially produced by the voltage gradient applied and partly the result of electroendo-osmotic flow. This electroendo-osmotic flow makes it possible to achieve the separation of uncharged solutes, such as steroids, in a silica capillary. With uncharged solutes, the electroendo-osmotic flow of water provides the propulsive force, and solutes of similar molecular weight are separated by a combination of differences in their mass and also partly by a chromatographic interaction with the capillary wall. For a charged species, the apparent rate of migration (U_{app}) is the product of the electrophoretic mobility (U), plus the rate at which the solute is propelled by the electroendo-osmotic flow (U_{eo}), and is related to the mass/charge ratio of the solute (**Fig. 2**).

Electroendo-osmosis can be greatly reduced by coating the inner surface of the capillary with a thin layer of polyacrylamide. This layer of polyacrylamide abolishes the ζ potential and reduces the electroendo-osmotic flow to virtually zero. Under these conditions U_{app} and U are virtually identical, and the mobility of solutes in the electrical gradient is a product of the mass charge ratio. Highly charged solutes with a small mass

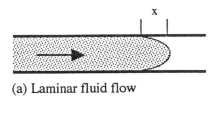

(a) Laminar fluid flow

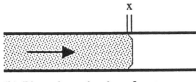

(b) Plug-shaped solute front

Fig. 3. Factors influencing the profile of the solute front. **(A)** When there is solvent movement through the capillary as a result of the electroendo-osmotic flow of water, a laminar flow profile is set up where the solvent near the wall of the capillary moves more slowly as a result of solvent drag. **(B)** In coated capillaries where the solutes are migrating through the solvent down an electrical gradient, this effect is minimized and the solute front is plug-shaped, and thus produces a sharp well-defined zone (x).

exhibit the greatest mobility. The separation of charged solutes is improved in coated capillaries, because when there is fluid flow in the capillary, the solute front is elliptical in shape as a result of laminar flow in the capillary, whereas when the propulsive force is purely the electrical gradient the solute is migrating through the solvent and the solute front is "plug" shaped (**Fig. 3**). This means that solute spread is minimized and peaks produced by the detector are sharper. Coating of the capillaries also reduces any nonspecific adsorption of proteins onto the capillary wall.

The detector usually comprises a beam from a deuterium lamp (190–380 nm; UV) or tungsten lamp (380–800 nm) that is focused through the capillary. The wavelength can be selected via a holographic monochromator to give a bandwidth of approx 5 nm. A typical wavelength setting for the detection of peptides and proteins would be 200 nm. The transmitted light is then detected, and the signal amplified to give typical full-scale deflections of 1.0–0.005 aufs on a chart recorder. Fluorescence, conductivity, and electrochemical methods of detection are all possible with capillary electrophoresis, but UV absorption is most commonly used.

Almost any buffer that does not have an excessive absorption of light at the chosen wavelength for detection can be used. For a buffer of a given pH, if the solute has a net positive charge, migration will be from anode to cathode, the rate of migration for a given molecular mass being proportional to the charge. Thus, if the buffer pH is close to the isoelectric point (pI) of the solute in question, the rate of migration through the capillary will be slower. This effect can be used to advantage to separate similar molecules with similar isoelectric points; for example, isoforms of a protein or enzyme. A buffer with a pH close to the pI of a solute, in this manner, accentuates the relative differences in mobility of solutes with a similar pI with respect to their elution time.

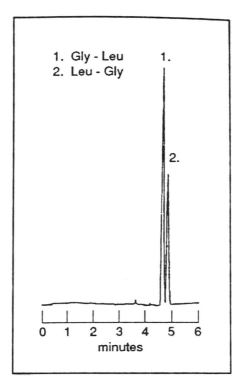

Fig. 4. Separation of two dipeptides, glycyl-L-leucine and leucyl-glygine. These two peptides can be separated with free zone capillary electrophoresis. Their slightly different configurations allow separation to be achieved by a combination of molecular sieving and a chromatographic interaction with the walls of the capillary. Sample: Mixture of glycyl-L-leucine and leucyl-glycine, 25 µg/mL. Buffer: Phosphate/HCl, pH 2.5. Load: 10 m*M* buffer, 4 kV, 4 s, 2.2–2.6 µA. Run: 100 m*M* buffer, 8 kV, 13.8 µA. Detection: 200 nm.

The time taken for the solute to move through the capillary to the detector is called the elution time (T_e), and depends on the mass/charge ratio under the conditions of separation and the length of the capillary and the voltage gradient (i.e., the driving force through the capillary). **Figure 4** illustrates the powerful resolution that can be obtained with FZCE. Typical elution times for some proteins in relation to their physical characteristics are shown in **Table 2**.

The sample can be introduced into the capillary by:

1. Displacement loading (sometimes called vacuum or pressure injection);
2. Electrokinetic loading (sometimes called electroendo-osmotic loading); and
3. Electrophoretic loading.

With displacement loading, the sample is injected directly into the capillary via a suitable manifold or drawn into the capillary by applying a negative pressure at the far end. Both of these methods suffer from problems with reproducibility. Electrokinetic loading is achieved with uncoated capillaries, and the sample is moved into the capillary by applying a potential difference that induces an electrophoretic movement of the solutes and an electroendo-osmotic flow. It is a combination of these two effects that

Table 2
Elution Time, T_e, for Some Proteins at pH 4.4

Protein	T_e	Mol wt, daltons	pI	Mean radius, nm
IgG	8.4	150,000	6.4–8.9	5.34
Transthyretin (prealbumin)	10.2	61,000	4.7	3.25
Albumin	15	69,000	4.8–4.9	3.58

moves sample into the capillary. Only with coated capillaries is pure electrophoretic loading achieved, where the electrophoretic mobility alone moves the solute into the capillary. It is important to appreciate that the amount of solute that moves into the capillary in a given time, and with a given potential difference, is proportional to the electrophoretic mobility of that solute under the loading conditions. Thus, with a mixture of solutes of differing electrophoretic mobilities more of the more mobile solute is loaded relative to the solutes of lesser mobility. With electrokinetic loading, this effect is reduced and, with displacement loading, does not occur. The protocols that follow assume electrophoretic loading with coated capillaries.

2. Practical Procedures

1. Loading the sample: To load the sample electrophoretically, an Eppendorf tube containing the sample, which may be as little as 10 µL in a suitably sized tube, is offered up to the cathode and capillary tip. A potential difference, for example 8 kV, is applied for 8 s, and the sample is loaded. In a mixture, the amount of each solute loaded into the capillary is proportional to the electrophoretic mobility, the loading time, and the voltage applied, and is thus proportionally different for each solute. This is shown graphically for bovine serum albumin (BSA) in **Fig. 5**. Solutes will have a greater mobility in a buffer of lower ionic concentration than the running buffer. Hence, the loading buffer should be 10 m*M,* and the running buffer 100 m*M.* Thus, a greater difference between the ionic concentration of the running and loading buffer will produce a greater loading of the sample. With electrophoretic loading, this phenomenon has the useful effect of also causing the solutes to slow down once they enter the more concentrated buffer at the start of the capillary. This results in a "stacking" of the solutes at the beginning of the capillary (*see* **Fig. 6**) and means that all of the solutes are starting their migration essentially from the start of the capillary. The end result is a better resolution of the individual peaks in a sample, because the zone concentrations may be increased by a factor of ten. This does not occur with electrokinetic or displacement loading, where a substantial length of the capillary will contain the sample, prior to the run. For a particular sample, the experimenter must vary the loading current and loading time until an optimal loading and separation of the solutes is achieved. For electrophoretic loading a good starting point for a peptide or protein is 8 kV for 8 s. Then vary first loading time and then voltage until loading is optimized. A voltage of 8 kV with a 25 µm × 20 cm capillary with pH 2.5 phosphate buffer will give a loading current of approx 11–14 µA. The current should be repeatable from sample to sample, when loaded under the same conditions.

2. Running the separation: The run is initially carried out at the same potential difference as the loading voltage. The run should be carried out with a constant voltage set on the power supply and, at 8 kV, will give approximately the same current as the loading conditions

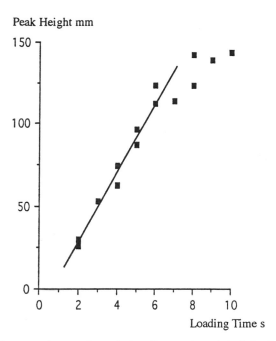

Peak Height mm

Loading Time s

Fig. 5. Effect of time on electrophoretic loading and peak height for BSA. Sample: BSA, 50 µg/mL (Sigma, A7888) Buffer: phosphate/HCl, pH 2.5. Load: 10 mM buffer, 8.0 kV, 11.8–12.6 µA. Run: 100 mM buffer, 7.0 kV, 8.8–11.2 µA. Detection wavelength: 200 nm. Note that the amount of BSA loaded declines with longer loading times.

(a) Load: 10mM phosphate buffer pH 2.5; 8kV,4s

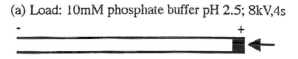

(b) Run: 100mM phosphate buffer pH 2.5; 8kV 15m

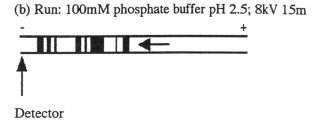

Detector

Fig. 6. Loading and running the separation. **(A)** When the sample is loaded from 10 mM buffer into a capillary containing 100 mM buffer, the effect is to stack the sample at the start of the capillary. **(B)** When the separating voltage is applied, the positively charged solutes in the mixture migrate toward the cathode at a rate predominantly determined by their mass–charge ratio to form discrete zones that are detected as absorbance peaks at the detector.

(see **Table 1**). Elution times for different solutes, of course, vary widely according to mass and charge. However, few peaks yielding useful information will elute after 25 min or so, as a result of zone spreading and distortion of the peaks. The voltage can be increased to speed the migration of slow peaks, or reduced to slow down fast-moving solutes and improve the resolution between peaks with a similar T_e. Small peptides with mol wt of 1 kDa or less usually elute within 15 min, whereas a complex biological sample may still be producing peaks after 20 min or so. Careful experimentation with the separating voltage must be conducted to optimize the separation. In reality, this means a careful balance between the loading conditions and the separating conditions to produce the best separation of the solutes of interest. In a complex mixture, as with the technique of high-performance liquid chromatography (HPLC), these conditions are often a compromise, and it proves impossible to separate all of the contents of a complex mixture with equal resolution.

3. Quantifying the results: The best method for determining the quantity of material represented by a peak on the chart recorder is to measure the peak height. Peak area with HPCE techniques is not reliable as a measure of the quantity of solute, because a rapidly migrating solute will pass the detector quickly and produce a narrow peak, and a slowly moving solute will produce a wider peak. Thus, peak area is related to speed of migration as well as the quantity of material present. This effect is obviously most pronounced with coated capillaries, where electroendo-osmosis is reduced to a minimum. A graph relating peak height to the quantity of albumin loaded is shown in **Fig. 7**. The apparatus must be calibrated with known quantities of the solute being determined to give the relationship between peak height and quantity for that solute.

4. Care of the capillary: The silica capillaries are remarkably robust, and it is in fact quite difficult to damage them. A capillary should perform several hundred separations before it needs replacing. The greatest danger to the capillary is blockage either as the result of particulate material in the sample being drawn into the capillary or because solute has crystallized in the capillary lumen. To guard against the former, all samples should be centrifuged before any attempt is made to load them, and if very fine suspended material is present, the sample should be passed through a 0.45-μ filter. Crystallization of a solute in the capillary is most likely to occur when the solute passes from buffer of one composition to another, i.e., from the loading buffer into the running buffer. This effect should be borne in mind when dealing with solutes of limited solubility. If the capillary blocks, some attempts to unblock it may be successful. Back-flush the capillary with buffer under pressure. This will not damage the capillary and may shift the blockage. Some authorities have been known to use an HPLC pump to do this, producing pressures in excess of 3000 psi to shift the blockage!

    Capillary performance and life can be enhanced by regular flushing with a 0.5% sodium dodecyl sulfate (SDS)/lauryl sulfate solution after use. Capillaries should be stored between use in the refrigerator at 4°C and filled with distilled water. These treatments will greatly minimize the chance of a capillary blocking while being stored.

5. Special techniques: A particular technique, which may be useful if a sample contains a mixture of both charged and noncharged solutes to be separated, is that of micellar electrokinetic capillary chromatography (MECC) *(7–9)*, where separation is achieved by a combination of both electrophoretic migration and micellar partitioning. Micelles are formed by adding SDS 0.05 *M* (approx 1.5%) to the buffer. The SDS will solubilize hydrophobic species and form ion pairs with those of opposite charge. The separation is usually performed at pH 7.0 and with uncoated capillaries to induce electroendo-osmotic flow. Components of the sample have differing degrees of interaction with the micelles formed by the detergent action of the SDS. The micelles formed by SDS are anionic in nature and will therefore migrate in the opposite direction to the induced electroendo-osmotic flow,

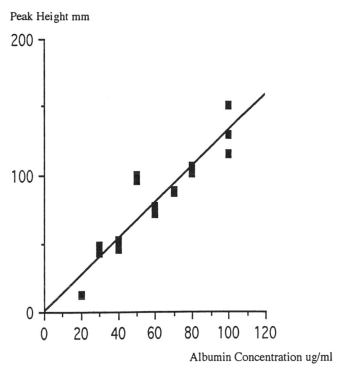

Fig. 7. Relationship between concentration of BSA in sample and peak height. Sample: BSA, 20–100 µg/mL (Sigma, A7888). Buffer: Phosphate/HCl, pH 2.5. Load: 10 mM buffer, 8.0 kV, 8 s, 11.4–12.6 µA. Run: 100 mM buffer, 7 kV, 10.6–12.0 µA. Detection: 200 nm T_e = 4.31 ± 0.03 min (n = 16).

in a countercurrent manner, thus setting up the partitioning effect. The migration of the various solutes in a sample therefore ranges among the rates of electroendo-osmotic flow, electrophoretic migration, and the opposing rate of micelle migration. MECC has the advantage of allowing both charged and noncharged solutes to be separated in a single procedure, and has a far better resolution than an uncoated capillary alone. However, because the electroendo-osmotic flow and micelle movement are in opposite directions, not all of the components of a mixture may be drawn past the detector. To optimize the separation, it may be necessary to modify the composition of the buffer to increase or decrease the rate of electroendo-osmosis. Other buffer additives *(13)* are worth experimenting with, for example, such detergents as Triton X-100 and disaggregating agents, such as urea.

Acknowledgments

The author would like to thank Bio-Rad Laboratories UK Ltd., for the grant of an HPE 100 capillary electrophoresis apparatus and for the donation of research materials to the laboratory.

Further Reading

Horvath, C. and Nikelly, J. G. eds. (1990) Analytical biotechnology: Capillary electrophoresis and chromatography. *Am. Chem. Soc. Symp. Series* **434,** American Chemical Society, Washington, D.C.

Altria K. D. (1998) *Methods in Molecular Biology, vol. 52: Capillary Electrophoresis Guidebook: Principles, Operations, and Applications*, Humana, Totowa, NJ.

References

1. Zhu, M., Hansen, D. L., Burd, S., and Gannon, F. (1989) Factors affecting free zone electrophoresis and isoelectric focusing in capillary electrophoresis. *J. Chromatog.* **480**, 311–319.

2. Hjerten, S., Elenberg, K., Kilar, F., Liao, J-L., Chen, A. J. C., Siebert, C. J., and Zhu, M. (1987) Carrier-free zone electrophoresis, displacement electrophoresis and isoelectric focusing in a high-performance electrophoresis apparatus. *J. Chromatog.* **403**, 47–61.

3. Mazzeo, J. R. and Krull, S. (1991) Capillary isoelectric focusing of proteins in un-coated fused-silica capillaries using polymeric additives. *Anal. Chem.* **63**, 2852–2857.

4. Krull, S. and Mazzeo, J. R. (1992) Capillary electrophoresis: The promise and the practise. *Nature* **357**, 92–94.

5. Karger, B. L., Cohen, A. S., and Guttman, A. (1989) High-performance capillary electrophoresis in the biological sciences. *J. Chromatog.* **492**, 585–614.

6. Higashami, T., Fuchigami, T, Imasak, T., and Ishibashi, N. (1992) Determination of amino acids by capillary zone electrophoresis based on semiconductor laser fluorescence detection. *Anal. Chem.* **64**, 711–714.

7. McDowall, R. D. (1989) Sample preparation for biomedical analysis. *J. Chromatog.* **492**, 3–58.

8. Gorbunoff, M. J. and Timasheff, S. N. (1984) The interaction of proteins with hydroxyapatite: III Mechanism. *Anal. Biochem.* **136**, 440–445.

9. Advis, J. P., Hernandez, L., and Guzman, N. A. (1989) Analysis of brain neuropeptides by capillary electrophoresis: determination of luteinizing hormone-releasing hormone from ovine hypothalamus. *Peptide Res.* **2**, 389–394.

10. Terabe, S. and Isemura, T. (1990) Ion-exchange electrokinetic chromatography with polymer ions for the separation of isomeric ions having identical electrophoretic mobilities. *Anal. Chem.* **62**, 650–652.

11. Ghowsi, K., Foley, J. P., and Gale, R. J. (1990) Micellar electrokinetic capillary chromatography theory based on electrochemical parameters: optimisation for three models of operation. *Anal. Chem.* **62**, 2714–2721.

12. Foley, J. P. (1990) Optimization of micellar electrophoretic chromatography. *Anal. Chem.* **62**, 1302–1308.

13. Burton, D., Sepaniak, M., and Maskarinec, M. (1987) Evaluation of the use of various surfactants in micellar electrokinetic capillary chromatography. *J. Chromatogr. Sci.* **24**, 347–351.

34

Protein Blotting

Principles and Applications

Peter R. Shewry and Roger J. Fido

1. Introduction

Protein blotting was initially introduced in the late 1970s to identify protein antigens that bound to specific antibodies *(1,2)*. In this procedure, a complex protein fraction was separated by electrophoresis and the proteins transferred and bound to a membrane, which was then probed with a radiolabeled antibody. This concept and procedure were based on blotting and hybridization methods developed for DNA by Southern *(3)* (and logically called Southern blotting), which was subsequently modified for RNA *(4)* (less logically called Northern blotting by some authors). It is, therefore, not surprising that the protein/antibody procedure was called Western blotting *(2)*, and that Southwestern (protein/DNA) and Far Western (or West Western) (protein/protein) procedures have since been developed. In addition, it is possible to apply proteins to the membrane in solution (dot-blotting), or to transfer them directly from tissues (squash blots and tissue prints), whereas the development of microscale protein sequencing and amino-acid analysis has allowed the use of blotting to purify and characterize individual components of highly complex mixtures.

It is clear from this brief introduction that protein blotting is a highly versatile procedure with many actual and potential applications. These are briefly discussed in this chapter, with particular reference to plant biochemistry and molecular biology. More detailed protocols for the individual procedures are provided *(5,6)*, or are available in the references and standard texts *(7–9)* as well as equipment suppliers (e.g., Applied Biosystems, Bio-Rad, and so forth).

2. Western Blotting

A range of options are available for Western blotting, in choice of membrane, transfer procedure and protein visualization.

2.1. Choice of Membrane

The most commonly used membranes are based on nitrocellulose (NC), which may be pure or supported. The pore size is important, and we would recommend a 0.45-μm

From: *Molecular Biomethods Handbook*
Edited by: R. Rapley and J. M. Walker © Humana Press Inc., Totowa, NJ

pore membrane for general use but a smaller pore size (0.2 μm) for proteins of M_r below approx 20,000. The main disadvantages of NC membranes are that they tend to be fragile and have low-binding capacity. In contrast, nylon membranes are much stronger than NC membranes and have up to six times the protein binding capacity, although these are not compatible with the most widely used protein stains (coomassie blue and amido black). They also give higher levels of nonspecific binding and hence require more effective blocking before probing (*see* **Subheading 2.3.**).

Perhaps the best available membranes, although also the most expensive, are those based on polyvinylidene difluoride (PVDF; *10)* These combine high-binding capacity with mechanical strength and are particularly suited for microsequencing (*see* **Subheading 2.5.**).

2.2. Protein Transfer

Four methods have been described to transfer protein from the electrophoretic gel, based on simple diffusion, capillary action (similar to Southern's original DNA-blotting procedure), vacuum blotting, and electroblotting. Of these, only electroblotting is widely used today.

Electroblotting uses electroelution to transfer proteins from the gel to the membrane and generally gives more rapid and effective transfer than the other methods. However, the rate and efficiency of transfer are affected by a number of factors, including the protein M_r (large proteins transferring less efficiently) and isoelectric point (pI), the acrylamide concentration in the gel and the transfer buffer. We generally use a buffer containing 20% (v/v) methanol and 0.2% (w/v) sodium dodecyl sulfate (SDS), which gives efficient transfer while preventing swelling of the gel. However, for the transfer of larger size proteins (M_r >75,000), we use a nonmethanol Tris-borate buffer system *(10)*.

A wide range of apparatus is available commercially, which falls into two types: wet vertical tank systems and semidry horizontal systems. Although the latter have been reported to be cheaper (using less buffer) and more rapid to use, we obtain consistently good results with a wet tank system, reusing the buffer up to five times without loss of transfer efficiency.

2.3. Total-Protein Detection

Staining of membrane-bound proteins can be used to confirm visually the efficiency and quality of transfer. It is also used to determine the positions of molecular-weight marker proteins, although prestained markers can also be used.

Filters can be stained with standard stains used for gels, such as Coomassie brilliant blue R250, amido black, India ink, Aurodye, and colloidal iron. In addition, rapid and reversible staining prior to immunodetection can be carried out either with Ponceau S or Amido black. This allows the proteins of interest or standards to be marked lightly with a pencil before destaining with Tris buffered saline containing 0.05% (v/v) Tween 20 (TTBS).

Once proteins have been bound to the filter they can be probed with the antibody to identify those that react. However, because the antibody is also a protein it will bind nonspecifically to the filter unless all unused binding sites are blocked. This can be achieved by using a cheap and widely available protein that will not react with the antibody of interest, such as bovine serum albumin (BSA) used at a concentration of up

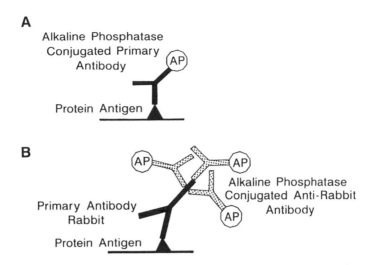

Fig. 1. Detection of bound proteins using labeled primary **(A)** and secondary **(B)** antibodies.

to 5% (w/v). However, even BSA can be expensive when used in large amounts and an even cheaper source of blocking protein is nonfat dried milk powder, of the "Marvel" type. One blocking cocktail, called BLOTTO *(11)*, consists of a 5% solution of low-fat dried milk in Tris-buffered saline containing antifoam and antimicrobial agents. BLOTTO, or 1% BSA, must also be used during the antibody binding, washing, and detection in order to maintain the blocking action.

After blocking, the membrane can be hybridized with the antibody of interest (called the primary antibody). However, in order to visualize the bound antibody, it is necessary to either label it directly or to use a secondary detection system. Labeling of the primary antibody can be achieved by iodination of tyrosine residues, with ^{125}I, allowing detection of the bound antibody by autoradiography. Although highly sensitive, radioiodination is also very hazardous and use of a fluorescent label, such as fluorescein isothiocyanate, is safer.

Although labeling of the primary antibody is still used in some studies, it is more usual to use a secondary detection system, either a second antibody or ligand (**Fig. 1**). A range of second antibodies are available to detect different types of primary antibodies (species and Ig isotypes). Thus, if the primary antibody is an IgG raised in rabbits, it is necessary to use an antirabbit IgG raised in a different species (for example, goat) as the second antibody. This second antibody is itself conjugated to a detection system, allowing its presence to be detected. The most popular detection systems are conjugated enzymes, either horseradish peroxidase (HRP) or alkaline phosphatase (AP). The addition of substrate results in the formation of an insoluble, colored product at the binding site, revealing the bound antibody. HRP was the first enzyme used for this purpose, but is relatively insensitive and the color fades on exposure to light. Alkaline phosphatase is generally preferred because it is more sensitive and produces a stable end color when used with substrates, such as 5-bromo-4-chloro-3-indolyl phosphate and nitro-blue tetrazolium.

Alternative secondary-detection systems are also available, notably the highly sensitive enhanced chemiluminescence (ECL) system marketed by Amersham International plc. In this system, an HRP-labeled second antibody is used to oxidize Luminol, to emit light that is enhanced approx 1000-fold and detected using photographic film. Alternatively, the primary antibody can be reacted with Protein A, Protein G, or gold-labeled second antibody, and enhanced with silver (Bio-Rad).

The advantage of secondary-detection systems is enhancement of the signal, resulting in higher sensitivity. Thus, the use of Protein A, Protein G, or second antibody enhances the reaction of the primary antibody (**Fig. 1**), whereas further enhancement is achieved by the ECL or silver-enhancement procedures.

In most cases, immunoreactive proteins can be readily identified by comparisons of stained gels with blots made from duplicate unstained gels. However, there may be a problem in some cases; for example, with complex patterns of proteins on 2-D gels. In this case, it may be necessary to confirm the identity by staining filters for total protein after immunodetection which can be achieved by using 0.1% (v/v) India ink in TTBS *(12)*, allowing the chromogenically stained immunoreactive proteins to be identified.

2.4. Applications of Western Blotting

The following examples are selected to illustrate the range of applications of Western blotting in plant science.

2.4.1. Analysis of Transgenic Plants

The development of transformation systems for many plants, including major cereal, oilseed, and legume crops, provides an opportunity to manipulate aspects of growth, development, and composition by genetic engineering. Western blotting is an important tool for such studies, in order to confirm the identity of the transgene product and to determine its pattern of expression.

For example, scientists in our institute are transforming wheat and tritordeum (a novel "hybrid" cereal derived from combining the genomes of pasta wheat and a wild-barley species) with genes encoding high-mol-wt subunits of glutenin in order to determine the effects on the elasticity, and hence, breadmaking quality of doughs. This work has been facilitated by the availability of a monoclonal antibody (MAb) (IFRN 1601) that is highly specific for a subgroup of high-mol-wt subunits (called x-type subunits), as shown in **Fig. 2**. There are numerous other examples of this application, for example Higgins and Spencer *(13)* who studied the expression of vicilin protein in transgenic tobacco plants.

2.4.2. Purification of Antibodies by Immunoaffinity Blotting

Polyclonal antisera consist of mixtures of antibodies, including those that are specific for the protein used as antigen. It is possible to purify small amounts of specific antibodies by binding to proteins which have been transferred to diazotized paper *(14)* or nitrocellulose *(15)*. The antibody can then be released by pH shock with 0.2 M glycine HCl buffer at pH 2.8, and the pH adjusted to 7.0 for storage. These methods were initially developed for studies of mammalian cytoskeletal proteins, but we have since used them to purify polyclonal antibodies to plant proteins for use in immunolocalization studies.

2.4.3. Identification of Allergens

A number of plant proteins are major allergens, and can cause severe symptoms in sensitized individuals. The most widely studied are soybean and peanut, both of which contain a number of apparently unrelated food allergens. Other food allergens include

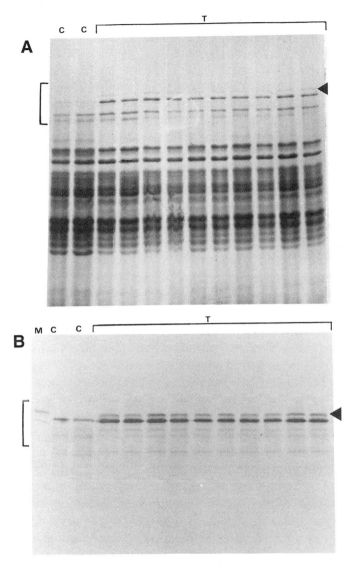

Fig. 2. Expression of wheat high-mol-wt glutenin subunit 1Dx5 in seeds of transformed tritordeum line HT 28. **(A)** Total seed proteins separated by SDS-PAGE and stained with Coomassie BBR250. **(B)** The proteins were transferred to nitrocellulose and detected with the MAb IFRN 1601, followed by second antibody labeled with alkaline phosphatase. M = Seeds of the wheat cv. *Mercia*, containing HMW subunit 1Dx5. T = Single seeds of tritordeum transformed with the wheat 1Dx5 gene. C = seeds from plants of tritordeum that had been regenerated but not transformed. The high-mol-wt subunits are indicated by brackets and subunit 1Dx5 by arrowheads.

2S albumin seed storage proteins of Brazil nut, castor bean, and mustard, and a protease/amylase inhibitor of rice. In addition, inhalation of flours of wheat and barley can result in an allergenic response known as baker's asthma, and the active proteins have been identified as low-M_r inhibitors of the cereal α-amylase/trypsin inhibitor family. Western blotting can be a powerful tool in the analysis of allergenic proteins, using IgE fractions from sera of sensitive individuals to probe proteins extracted from plant materials that elicit a response *(16,17)*.

2.4.4. Characterization of Antibody Specificity

There are two approaches to raising MAbs. The first is to purify the protein of interest, and to use this to raise a library of MAbs. In this case, all of the MAbs should recognize the protein of interest, although they may well recognize different epitopes and differ in their specificity and titer. Alternatively, it is possible to immunogenize with a mixture of proteins and then characterize a larger number of antibodies, in order to identify those that are specific for individual proteins. In both cases, Western blotting can be used to determine antibody specificity.

For example, we have raised a library of MAbs to a wheat glutenin fraction in order to identify clones that were specific for the different protein types present in the mixture. One such MAb, called IFRN067, proved to be of particular interest because its degree of binding to gluten protein extracts from various grain samples was correlated with breadmaking performance ($r = 0.497$). Western blotting of 1-D SDS-polyacrylamide gel electrophoresis (PAGE) separations showed specific binding to a major band of M_r about 45,000, but it was not possible to identify the precise immunoreactive proteins among the complex mixture. This was finally established by double-staining (immunodetection followed by India ink) *(12)* of a Western blot of proteins separated using a 2-D pollectic focusing (IEF)/SDS-PAGE system. This allowed the immunoreactive band to be separated into two major components of similar M_r, which were unequivocally identified in the protein mixture (**Fig. 3**).

2.4.5. Analysis of Plant Development

There are numerous examples of the use of protein blotting to study plant development and to show changes in the amount of individual proteins present in a complex mixture. Blotting can give much greater specificity and sensitivity than conventional protein staining. For example, we have used a wide specificity MAb (IFRN 0610) to study the amount and pattern of gluten-protein synthesis in developing caryopses of wheat (**Fig. 4**). This allows the detection of proteins from only 7 d after anthesis, several days before they can be reliably detected by protein staining.

2.5. Microsequencing of Blotted Proteins

One of the most widely used applications of protein blotting is for microsequencing, using automated pulsed-liquid and gas-phase sequencers that can determine the N-terminal amino sequences of 20 pmols or less of proteins. This eliminates the need to purify proteins by conventional methods and provides protein that is pure and easy to handle. A range of filters based on glass fiber (chemically activated, siliconized or coated with quaternary ammonium polybases) were initially used, but most workers currently prefer PVDF membranes; either Immobilon™ (Millipore Corp.) or Pro-Blot™ (Applied Biosystems, Inc.). The latter has been specially developed for use in the Applied Biosystems Model 477A Pulsed Liquid Sequencer.

2.6. Tissue Prints and Dot-Blots

The presence and location of proteins in plant tissues can be determined using a modified Western-blotting procedure called tissue printing or squash blotting. The

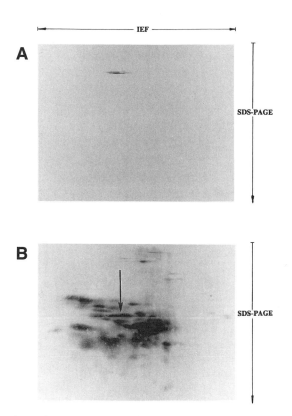

Fig. 3. **(A)** Immunodetection with MAb IFRN067 and **(B)** double-staining procedure using immunodetection followed by total protein staining with India ink, of a Western blot of glutenin proteins separated using a 2 D IEF/SDS-PAGE system. This allowed the immunoreactive band to be separated into two major components of similar M_r, which were identified (observed as purple spots [arrowed]) among the nonreactive spots within the protein mixture. Taken from Brett et al., 1992 (*see* **ref. *18***).

proteins are not extracted from the tissue, but transferred directly to a nitrocellulose membrane and then detected using an antibody and one of the labeling systems discussed previously. A surprisingly high level of resolution can be obtained if the protocol is optimized to suit the experimental material. A good example of tissue printing is provided by a recent study of invertase distribution in leaf tissue *(19)* (**Fig. 5**).

Dot-blotting is a simplified Western-blotting procedure, with the protein extract applied directly to the nitrocellulose membrane as a small spot. It has the advantage of being simple and rapid, and is reported to be approx 10- to 1000-fold more sensitive than electroblotting. It is also possible to load dilute solutions as multiple loadings, with the membrane being allowed to dry in between. The speed and simplicity mean that it is readily adapted to automated or semi-automated analysis of multiple samples, and a number of systems are available commercially. For example, the Bio-Dot system from Bio-Rad has either 96 wells or 48 slots attached to a common manifold base to allow application and washing under vacuum. The membrane is processed and probed as for conventional Western blotting.

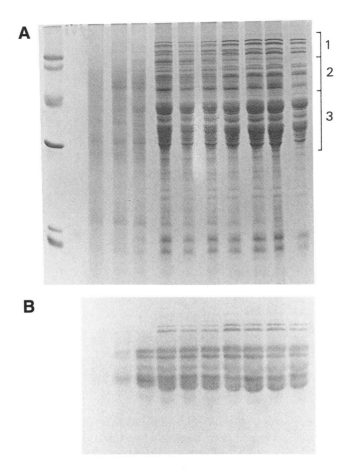

Fig. 4. Use of a wide specificity MAb to study the amount and pattern of gluten-protein synthesis in developing caryopses of Chinese Spring wheat 5–40 d after flowering. **(A)** Total protein extracted on an equal weight basis and SDS-PAGE followed by **(B)** immunodetection of transferred protein by MAb IFRN 0610, as in **Fig. 2**. The groups of gluten proteins are indicated by brackets; 1, high-mol-wt subunits of glutenin; 2, ω-gliadins; 3, low-mol-wt subunits of glutenin, α-type gliadins, and γ-type gliadins.

3. Southwestern and Far-Western Blotting

Southwestern blotting is a combination of protein and DNA blotting that was developed to identify DNA-binding proteins in crude protein extracts from nuclei *(20)*. Such proteins could be involved in DNA replication or, by binding to regulatory sequences present in the 5' upstream regions, the control of gene expression. The concept is simple. The crude-protein extract is initially separated by SDS-PAGE and transferred to a nitrocellulose membrane. It is then probed with a DNA sequence that is labeled either isotopically or using a nonradioactive system. Although Southwestern blotting is most widely used to identify proteins that bind to a characterized DNA sequence *(20,21)*, it

Fig. 5. Tissue printing from peony *(Paeonia officinalis)* leaves. **(A)** Protein transfer from a leaf from which the epidermis has been partially removed, revealed by amido-black staining. **(B)** Immunolocalization of acid-invertase to the vascular regions. Kindly provided by A. Kingston-Smith.

can also be used to select unknown DNA fragments that bind to characterized proteins, as a semipreparative system *(22)*.

Far-Western (or West-Western blotting) is also derived from conventional Western blotting, but the blotted proteins are probed with a nonantibody protein in order to detect protein–protein interactions *(23,24)*. The bound protein can then be detected using a specific antibody that uses a standard labeling system. Neither Southwestern blotting nor Far-Western blotting have so far been widely used in plant science.

Acknowledgments

The authors are grateful to A. Tatham, P. Barcelo, F. Barro, and P. Lazzari (IACR-Long Ashton Research Station) for providing **Fig. 2**. IACR receives grant-aided support from the Biotechnology and Biological Sciences Research Council of the United Kingdom.

References

1. Towbin, S. A., Staehelin, T., and Gordon, J. (1979) Electrophoretic transfer of proteins from polyacrylamide gels to nitrocellulose: procedure and some applications. *Proc. Natl. Acad. Sci. USA* **76,** 4350–4354.
2. Burnette, W. N. (1981) "Western Blotting"; electrophoretic transfer of proteins from sodium dodecylsulphate-polyacrylamide gels to unmodified nitrocellolose and radiographic detection with antibody and radioiodinated Protein A. *Anal. Biochem.* **112,** 195–203.
3. Southern, E. M. (1975) Detection of specific sequences among DNA fragments separated by gel electrophoresis. *J. Mol. Biol.* **98,** 503–517.
4. Alwine, J. C., Kemp, D. J., and Stark, G. R. (1977) Method for detection of specific RNAs in agarose gels by transfer to diazobenzyloxymethyl-paper and hybridisation with DNA probes. *Proc. Natl. Acad. Sci. USA* **74,** 5350–5354.

5. Fido, R. J., Tatham, A. S. and Shewry, P. R. (1993) Applications of protein blotting in plant biochemistry and molecular biology, in *Methods in Plant Biochemistry, vol. 10, Molecular Biology* (Bryant, J., ed.), Academic, London, pp. 101–115.

6. Fido, R. J., Tatham, A. S., and Shewry, P. R. (1995) Western blotting analysis, in *Methods in Molecular Biology, vol. 49: Plant Gene Transfer and Expression Protocols* (Jones, H., ed.), Humana, Totowa, NJ, pp. 423–437.

7. Bjerrun, O. J. and Heegaard, N. H. H., eds. (1988) *Handbook of Immunoblotting of Proteins Volume I, Technical Descriptions* (265 pp.); *Volume II, Experimental and Clinical Applications* (214 pp.), CRC, Boca Raton, FL.

8. Dunbar, B. S., ed. (1994) *Protein Blotting: A Practical Approach.* IRL, Oxford University Press, Oxford, UK, 280 pp.

9. Electrophoresis (1993) Paper symposium—*Protein Blotting: The Second Decade*, vol. 14, pp. 829–960.

10. Baker, C. S., Dunn, M. J., and Yacoub, M. H. (1991) Evaluation of membranes used for electroblotting of proteins for direct automated microsequencing. *Electrophoresis* **12**, 342–348.

11. Johnson, D. A., Gautsch, J. W., Sportsman, J. R., and Elder, J. H. (1984) Improved technique utilizing nonfat dry milk for analysis of proteins and nucleic acids transferred to nitrocellulose. *Gene Anal. Tech.* **1**, 3–8.

12. Ono, T. and Tuan, R. S. (1990) Double staining of immunoblot using enzyme histochemistry and india ink. *Anal. Biochem.* **187**, 324–327.

13. Higgins, T. J. V. and Spencer, D. (1991) The expression of a chimeric cauliflower mosaic virus (CaMV–35S)—pea vicilin gene in tobacco. *Plant Sci.* **74**, 89–98.

14. Olmsted, J. B. (1981) Affinity purification of antibodies from diazotizes paper blots of heterogeneous protein samples. *J. Biol. Chem.* **256**, 11,955–11,957.

15. Tailin, J. C., Olmsted, J. B., and Goldman, R. D. (1983) A rapid procedure for preparing fluorescein–labeled specific antibodies from whole antiserum: its use in analysing cytoskeletal architecture. *J. Cell Biol.* **97**, 1277–1282.

16. Donovan, G. R. and Baldo, B. A. (1993) Immunoaffinity analysis of cross-reacting allergens by protein blotting. *Electrophoresis* **14**, 917–922.

17. Sandiford, C. P., Tee, R. D., and Newman–Taylor, A. J. (1995) Identification of crossreacting wheat, rye, barley and soya flour allergens using sera from individuals with wheat-induced asthma. *Clin. Exp. Allergy* **25**, 340–349.

18. Brett, G. M., Mills, E. N. C., Tatham, A. S., Fido, R. J. Shewry, P. R., and Morgan, M. R. A. (1993) Immunochemical identifiction of LMW subunits of glutenin associated with bread-making quality of wheat flours. *Theor. App. Genet.* **86**, 442–448.

19. Kingston-Smith, A. H., and Pollock, C. J. (1996) Tissue level localization of acid invertase in leaves: an hypothesis for the regulation of carbon export. *New Phytol.* **134**, 423–432.

20. Bowen, B., Steinberg, J., Laemmli, U. K., and Weintraub, H. (1980) The detection of DNA-binding proteins by protein blotting. *Nucleic Acids Res.* **8**, 1–20.

21. Miskimins, W. K., Roberts, M. P., McClelland, A., and Ruddle, F. H. (1985) Use of a protein–blotting procedure and a specific DNA probe to identify nuclear proteins that recognize the promoter region of the transferrin receptor gene. *Proc. Natl. Acad. Sci. USA* **82**, 6741–6744.

22. Keller, A. D. and Maniatis, T. (1991) Selection of sequences recognized by a DNA binding protein using a preparative Southwestern blot. *Nucleic Acids Res.* **19**, 4675–4680.

23. Macgregor, P. F., Abate, C., and Curran, T. (1990) Direct cloning of leucine zipper proteins: Jun binds cooperatively to the CRE with CRE-BP1. *Oncogene* **5**, 451–458.

24. Grasser F. A., Sauder, C., Haiss, P., Hille, A., Konig, S., Gottel, S., Kremmer, E., Leinenbach, H. P., Zeppezauer, M., and Mueller-Lantzsch, N. (1993) Immunological detection of proteins associated with the Epstein-Barr virus nuclear antigen 2A. *Virology* **195**, 550–560.

35

Ion-Exchange Chromatography

David Sheehan and Richard Fitzgerald

1. Introduction

Proteins contain charged groups on their surfaces that enhance their interactions with solvent water and hence their solubility. At physiological pH, some of these charged groups are cationic (positively charged, e.g., lysine), whereas others are anionic (negatively charged, e.g., aspartate). Because proteins differ from each other in their amino-acid sequence, the net charge possessed by a protein at physiological pH is determined ultimately by the balance between these charges (i.e., negatively charged proteins possess more negatively charged groups than positively charged groups). This also underlies the different isoelectric points (pIs) of proteins. Ion-exchange chromatography *(1)* separates proteins first on the basis of their charge type (cationic or anionic) and, second, on the basis of relative charge strength (e.g., strongly anionic from weakly anionic).

The basis of ion-exchange chromatography (**Fig. 1**) is that charged ions can freely exchange with ions of the same type. In this context, the mass of the ion is irrelevant. Therefore, it is possible for a bulky anion like a negatively charged protein to exchange with chloride ions. This process can later be reversed by washing with chloride ions in the form of a NaCl or KCl solution. Such washing removes weakly bound proteins first, followed by more strongly bound proteins with greater net negative charge.

Like most column chromatography techniques, ion-exchange chromatography requires a stationary phase, which is usually composed of insoluble, hydrated polymers, such as cellulose, dextran, and Sephadex *(2)*. The ion-exchange group is immobilized on this stationary phase, and some of the chemical structures of commonly used groups are shown in **Table 1**.

In this chapter, the use of microgranular diethylaminoethyl (DEAE) cellulose manufactured by Whatman (Maidstone, UK) is described. Variations on this method suitable for other resins are summarized in **Subheading 3.** Novel formats for ion-exchange chromatography are provided by the immobilization of DEAE or carboxymethyl (CM) groups on filters packed in a cartridge (MemSep™, PerSeptive Biosystems, Framingham, MA) and by the development of Perfusion Chromatography®) media (Per Septive Biosystems). These may be used either in a low pressure system or with an fast perfor-

From: *Molecular Biomethods Handbook*
Edited by: R. Rapley and J. M. Walker © Humana Press Inc., Totowa, NJ

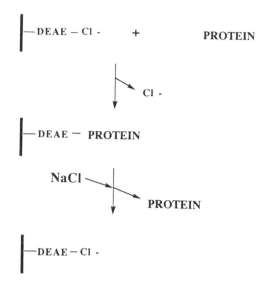

Fig. 1. Ion-exchange chromatography of anionic protein. The protein exchanges with the Cl⁻ ion and binds to the resin. This process is reversed when an NaCl gradient is applied to the column. Each protein has slightly different charge characteristics and, therefore, elutes at a slightly different Cl⁻ concentration. This, therefore, results in separation of proteins.

mance liquid chromatography (FPLC) system (Pharmacia- LKB, Uppsala, Sweden). MemSep cartridges offer high resolution with very low back pressure, owing to the use of thin filters rather than beads as the stationary phase, and can result in separations in as little as 3–5 min. Perfusion Chromatography features media that are rigid and contain large pores with a large surface area. This results in high-column capacity and resolution combined with very fast separations *(3)*.

2. Practical Requirements

1. Buffers: Buffers used will depend on the characteristics of the protein of interest. For proteins not previously purified by DEAE-cellulose chromatography, buffer selection may be helped if the pI of the protein of interest is known. In general, proteins will bind to anion exchangers at pH values above their pIs, while binding to cation exchangers, such as CM-cellulose, at pH values below their pIs. Phosphate, Tris, and other common buffers are used in ion-exchange chromatography in concentrations of 10–50 mM. Urea (desalted on a mixed-bed column of Dowex or else molecular biology grade) may be included in buffers at concentrations of up to 8 M if separation of denatured samples is required (e.g., chromatography of CNBr-generated peptides). Care should be taken that the urea does not precipitate in the column during chromatography (especially at low temperatures). Buffers are prepared fresh before use, and all reagents are of Analar grade. Deionized/distilled water is used. Typically, the following buffers may be employed for a column that requires operation at pH 8.0:
   Buffer A: 200 mM Tris-HCl, pH 8.0.
   Buffer B: 10 mM Tris-HCl, pH 8.0.
   Buffer C: 10 mM Tris-HCl, pH 8.0, 100 mM NaCl.

Table 1

Some Commonly Used Ion-Exchange Groups[a]

Group	pH range	Structure
Anion exchangers		
Q (quaternary ammonium)	2 – 12	$-CH_2-\overset{+}{N}\overset{CH_3}{\underset{CH_3}{\diagup}}CH_3$
DEAE (diethylaminoethyl)	2 – 9	$-O-CH_2-CH_2-\overset{+}{NH}\overset{C_2H_5}{\underset{C_2H_5}{\diagup}}$
QAE (quaternary aminoethyl)	2 – 12	$-O-(CH_2)_2-\overset{+}{N}\overset{C_2H_5}{\underset{C_2H_5}{\diagup}}CH_2-CHOH-CH_3$
Cation exchangers		
SP (sulfopropyl)	2 – 12	$-(CH_2)_2-CH_2-SO_3^-$
S (methyl sulfonate)	2 – 12	$-CH_2-SO_3^-$
CM (carboxymethyl)	6 – 11	$-O-CH_2-COO^-$

[a]These groups may be immobilized on various stationery phases, such as cellulose, dextran, agarose, and glass beads.

2. Resins: The use of microgranular DEAE-cellulose 52 is described in this chapter. Desalting resins, such as Sephadex G-25 (Pharmacia-LKB, Uppsala, Sweden), are also required.

3. Apparatus: A 500-mL sintered-glass funnel (no. 1 sinter) is useful for washing resin during equilibration and regeneration. This is usually connected to water suction. For desalting and ionexchange chromatography (column volumes of 300 mL), sintered glass columns (45 × 3 cm in diameter) are required. Smaller columns could be used for smaller-column volumes. A gradient maker (2 × 500 mL) is used to generate salt gradients. A calibrated conductivity meter is also required.

3. Practical Procedure

1. Sample preparation: Protein samples require desalting before being applied to the ion-exchange column. This may be achieved either by dialysis or by gel filtration on a Sephadex G-25 column *(4)*. Pharmacia produces a PD10 column for rapid desalting of small volumes.

2. Resin equilibration: Before use, 100–200 g resin is washed in 1–2 L water followed by removal of fines. This is achieved by stirring the resin with a glass rod, allowing the bulk of the resin to settle, and then aspirating supernatant, which contains the fines. This process is repeated three times. It is sometimes convenient to carry out this procedure in a graduated cylinder. The defined resin is washed in 500-mL buffer A to achieve equilibration (the use of concentrated buffer allows rapid equilibration with small volumes compared to more dilute buffer concentrations). Again, following settlement of the resin, the supernatant is aspirated off. This process is repeated twice (i.e., a total of three washes). The resin is gently stirred into a slurry, and transferred into a sintered-glass column that contains a small volume of water (this prevents air bubbles becoming trapped in the sinter). Buffer B is passed through the column until it is completely equilibrated. This is determined by measuring the pH and conductivity of buffer B and the column eluate. When these are the same, the column is equilibrated.

3. Chromatography: The desalted sample is applied to the equilibrated resin, and the effluent is collected. This should be assayed in case the protein of interest has not bound. In cases in which a protein has not previously been purified by this method, it may be neccessary to develop a new purification procedure for it. The main variables to investigate are choice of ion-exchange group, pH, salt concentration, and gradient steepness. In general, for example, approx 70% of rat-liver cytosolic proteins will bind to DEAE-cellulose at pH 7.5. If the protein of interest has an alkaline pI, then passage through such a column at this pH might be a useful first step in the purification. Conversely, if the protein of interest binds to DEAE-cellulose at pH 7.5, then it is worth repeating the experiment at progressively higher pH values (e.g., 8.5 and 9.5). Only highly acidic proteins will bind at pH 10.0. The column is developed by applying a suitable salt gradient (2×500 mL), for example, 0–100 mM NaCl in 10 mM Tris buffer, pH 8.0 (buffers B and C, aforementioned). Fractions are collected in a fraction collector and assayed for the protein of interest, protein concentration *(5)*, and conductivity. The conductivity meter should be carefully calibrated (**Note:** This property is temperature-dependent), so that conductivity measurements may be expressed as NaCl concentrations, thus facilitating determination of chromatogram reproducibility. These measurements may also be made continuously using on-line UV and conductivity detectors connected to a PC or chart recorder. Appropriate fractions are pooled and concentrated for further study or purification. This is a major factor in the success of the chromatography. In general, fractions should not be pooled in a new purification until sodium dodecyl sulfate-polyacrylamide gel electrophoresis (SDS-PAGE) analysis *(6,7)* has been carried out on aliquots taken from active fractions. This gives a good indication of fraction purity. It is best to accept some losses of the protein of interest if this means removing significant contaminants. Carrying out the chromatography at different pH values *(see* above) usually gives quite different chromatograms, which is often useful for displacing contaminant peaks.

   Although shallower (e.g., 0–100, 300–400 mM) gradients generally give better resolution, steeper gradients (e.g., 0–1 M) can be used in initial experiments to identify elution positions of proteins of interest. It is also possible to use stepwise washes (e.g., 100 mM NaCl followed by 200 mM NaCl) rather than continuous gradients. The best approach to developing a new method, therefore, is to assess the binding (i.e., binding vs nonbinding) of the protein of interest on a small scale on both CM-cellulose and DEAE- cellulose at a range of pH values in the alkaline and acidic ranges, respectively. Chromatograms where the protein has bound are then developed with a steep gradient. Finally, shallower gradients are assessed at the most promising pH values. It is also possible to use nonlinear gradients for more difficult separations.

4. Resin regeneration: The resin may usually be regenerated by washing with 10 mM Tris-HCl, pH 8.0, buffer containing 2 M NaCl after use. Three to four uses of a batch of resin are possible in most cases before extensive regeneration is neccessary. However, owing to column compaction as a consequence of high-salt concentration gradients, in some cases, only single usage of a packed resin is possible before extensive regeneration is required. The resin is placed in a beaker containing 15 vol of 0.5 M HCl for 30 min, and then the supernatant is decanted. The resin is then washed extensively on a glass sinter until the pH of the wash-through reaches an intermediate value of 4.0. The resin is then placed in a beaker containing 15 vol of 0.5 M NaOH for 30 min, and the supernatant is again decanted. The resin may now be stored in 0.1% Na azide until required. In the case of fibrous DEAE–cellulose, the resin is exposed to 0.5 M HCl for 60 min rather than 30 min. Regeneration of CM–cellulose follows a similar procedure except that the resin is washed first with 0.5 M NaOH for 30 min, followed later by 0.5 M HCl. The intermediate pH is 8.0 whereas the final pH is 5.5–6.5.

References

1. Himmelhoch, S. R. (1971) Ion-exchange chromatography. *Methods Enzymol.* **22,** 273–290.
2. Walsh, G. and Headon, D. R. (1994) Downstream processing of protein products, in *Protein Biotechnology.* Wiley, Chichester, UK, pp. 39–117.
3. Regnier, F. E. (1991) Perfusion chromatography. *Nature* **350,** 634,635.
4. Boyer, R. F. (1993) Gel exclusion chromatography, in *Modern Experimental Biochemistry.* Benjamin Cummings, New York, pp. 81–89.
5. Lowry, O. H., Rosebrough, N. J., Farr, A. L., and Randall, R. J. (1951) Protein measurement with the folin-phenol reagent. *J. Biol. Chem.* **193,** 267–275.
6. Laemmli, U. K. (1970) Cleavage of structural proteins during the assembly of the head of bacteriophage T4. *Nature* **227,** 680–685.
7. Hames, B. D. and Rickwood, D. (1990) *Gel Electrophoresis of Proteins—A Practical Approach.* Oxford University Press, Oxford, UK.

36

Size-Exclusion Chromatography

Paul Cutler

1. Introduction

Size-exclusion chromatography (also known as gel-filtration chromatography) is a technique for separating proteins and other biological macromolecules on the basis of molecular size. Size-exclusion chromatography is a commonly used technique owing to the diversity of the molecular weights of proteins in biological tissues and extracts. It also has the important advantage of being compatible with physiological conditions *(1,2)*.

The solid-phase matrix consists of porous beads (typically 100–250 μm) that are packed into a column with a mobile-liquid phase flowing through the column (**Fig. 1**). The mobile phase has access to both the volume inside the pores and the volume external to the beads. The high porosity typically leads to a total liquid volume of >95% of the packed column.

Separation can be visualized as reversible partitioning into the two liquid volumes. Large molecules remain in the volume external to the beads because they are unable to enter the pores. The resulting shorter flow path means that they pass through the column relatively rapidly, emerging early. Proteins that are excluded from the pores completely elute in what is designated the void volume, V_o. This is often determined experimentally by the use of a high-mol-wt component, such as Blue Dextran or calf thymus DNA. Small molecules that can access the liquid within the pores of the beads are retained longer and therefore pass more slowly through the column. The elution volume for material included in the pores is designated the total volume, V_t. This represents the total liquid volume of the column and is often determined by small molecules, such as vitamin B12.

The elution volume for a given protein will lie between V_o and V_t. and is designated the elution volume, V_e. Intermediate-sized proteins will be fractionally excluded with a characteristic value for V_e. A partition coefficient can be determined for each protein as K_{av} (**Fig. 1**). In size exclusion, the macromolecules are not physically retained, unlike adsorption techniques; therefore, the protein will elute in a defined volume between V_o and V_t. If the protein elutes before the void volume *($V_e < V_o$)* this suggests channeling through the column owing to improper packing or operation of the column. If the protein elutes after the total volume *($V_e > V_t$)*, then some interaction must have occurred between the matrix and the protein of interest. Size exclusion tends to be used at the end of a purification scheme when impurities are low in number and the target protein

From: *Molecular Biomethods Handbook*
Edited by: R. Rapley and J. M. Walker © Humana Press Inc., Totowa, NJ

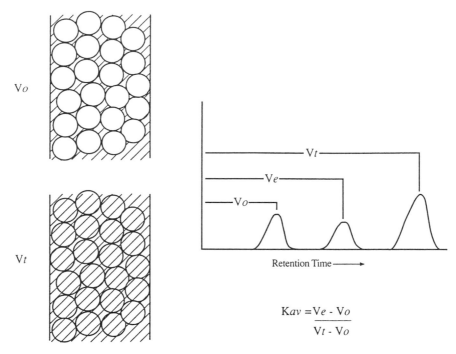

Fig. 1. The basic principle of size exclusion. Solutes are separated according to their molecular size. Large molecules are eluted in the void volume *(V_o)*; small molecules are eluted in the total volume *(V_t)*. Solutes within the separation range of the matrix are fractionally excluded with a characteristic elution volume *(V_e)*.

has been purified and concentrated by earlier chromatography steps. An exception to this is membrane proteins, where gel filtration may be used first because concentration techniques are not readily used and the material will be progressively diluted during the purification scheme *(3)*.

A range of different preparative and analytical matrices are available commercially. High-performance columns are used analytically for studying protein purity, protein folding, protein–protein interactions, and so forth *(4)*. Preparative separations performed on low-pressure matrices are used to resolve proteins from proteins of different molecular weight, proteins from other biological macromolecules, and for the separation of aggregated proteins from monomers *(5)*. Size exclusion is particularly suited for the resolution of protein aggregates from monomers. Aggregates are often formed as a result of the purification procedures used. Size exclusion is often incorporated as a final polishing step to remove aggregates and act as a buffer-exchange mechanism into the final solution.

Several parameters are important in size-exclusion chromatography. The pore diameter controlling the separation is selected for the relative size of proteins to be separated. Many types of matrix are available. Some are used for desalting techniques where proteins are separated from buffer salts *(6)*. Desalting gels are used to rapidly remove low-mol-wt material, such as chemical reagents from proteins and for buffer exchange. Because the molecules to be separated are generally very small, typically <1 kDa, gels

Table 1
Commonly Used Preparative Size Exclusion Matrices

Supplier/matrix	Material type (pH stability)[a]	Separation range[b], kDa
Sephadex		
G25	Dextran (2.0–10.0)[d]	1–5
G50		1.5–30
G75		3–80[c]
G100		4–100[c]
G150		5–150[c]
G200		5–600[c]
Sepharose		
6B	Agarose (3.0–13.0)	10–4000
4B		60–20000
2B		70–40000
Superdex		
30	Agarose/dextran (3.0–12.0)	0–10
75		3–70
200		10–600
Sephacryl		
S100HR	Dextran/bisacrylamide (3.0–11.0)	1–100
S200HR		5–250
S300HR		10–1500
Biogel		
P-2	Polyacrylamide (2.0–10.0)	0.1–1.8
P-4 gel		0.8–4
P-10 gel		1.5–20
P-60 gel		3–60
P-100 gel		5–100

[a]In aqueous buffers.
[b]For globular protein.
[c]Sifferent grades are available that effects performance.
[d]pH stability may vary depending on grade and exclusion limit.

are generally used with an exclusion limit of approx 2–5 kDa. The protein appears in the void volume (V_o) and the reagents and buffer salts are retained. Because of the distinct mol-wt differences, the columns are shorter than other size-exclusion columns and operated at higher flow rates (e.g., 30 cm/h). Some matrices offer a wide range of mol-wt separations and others are high-resolution matrices with a narrow range of operation (**Table 1**).

2. Practical Requirements

The preparative separation of proteins by size exclusion is suited to commercially available standard low-pressure chromatography systems. Systems require a column packed with a matrix offering a suitable fractionation range, a method for mobile-phase delivery, a detector to monitor the eluting proteins, a chart recorder for viewing the

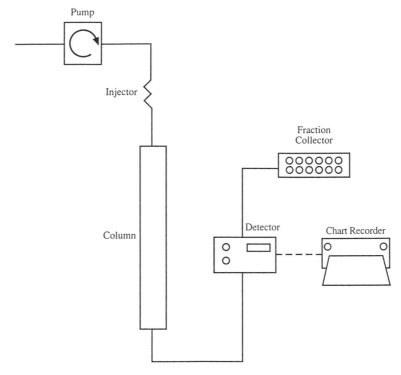

Fig. 2. Schematic diagram of the equipment for preparative size-exclusion chromatography.

detector response, and a fraction collector for recovery of eluted proteins (**Fig. 2**). The system should be plumbed with capillary tubing with a minimum hold up volume.

Early systems were less sophisticated with a gravimetric feed of the mobile phase from a suspended reservoir, whereas the most modern systems now have computers to control operating parameters and to collect and store data. The principle of separation, however, remains the same and high resolution is attainable with relatively simple equipment.

1. Pumps: An important factor in size exclusion is a reproducible and accurate flow rate. The most commonly used pumps are peristaltic pumps, which are relatively effective at low-flow rates, inexpensive, and sanitizable. Peristaltic pumps do, however, create a pulsed flow, and often a bubble trap is incorporated to both prevent air entering the system and to dampen the pulsing effect. More expensive, yet more accurate pumps are syringe pumps, such as those seen on the FPLC® system (Pharmacia).

2. Column: Size exclusion, unlike some commonly used adsorption methods of protein separation, is a true chromatography method based on continuous partitioning; hence, resolution is dependent on column length. Columns tend toward being long and thin, typically 70–100 cm long. In some instances, the length of the column required to obtain a satisfactory separation exceeds that which can be packed into a commercially available column (>1 m). In these cases, columns can be packed in series.

   The tubing connecting the columns should be as narrow and as short as possible to avoid zone spreading. The column must be able to withstand the moderate pressures generated during operation and be resistant to the mobile phase. The use of columns with flow adapters is recommended to allow the packing volume to be varied and provide a finished support with the required minimum of deadspace.

3. Detectors: Protein elution is most often monitored by absorbance in the ultraviolet range, either at 280 nm, which is suitable for proteins with aromatic amino acids, or at 206 nm, which detects the peptide bond. Detection at lower wavelengths may be complicated by the absorbance characteristics of certain mobile phases. The advent of diode-array detectors has enabled continuous detection at multiple wavelengths, enabling characterization of the elutes via analysis of spectral data.

   Fluorescence detection either by direct detection of fluorescent tryptophan and tyrosine residues or after chemical derivitization (e.g., by fluorescein) have been used, as have refractive index, radiochemical, electrochemical, and molecular size (by laser-light scattering). In addition to these nonspecific on-line monitoring systems, it is quite common, particularly when purifying enzymes, to make use of specific assays for individual target molecules.

4. Fraction collectors: A key factor in preparative protein purification is the ability to collect accurate fractions. No matter how efficiently the column may have separated the proteins, the accurate collection of fractions is critical. For the detector to reflect as near as possible in real time the fraction collector, the volume between the detector and the fraction collector should be minimal.

5. Buffers: Size-exclusion matrices tend to be compatible with most aqueous-buffer systems even in the presence of surfactants, reducing agents, or denaturing agents. Size-exclusion matrices are extremely stable, with effective pH ranges of approx 2.0–12.0 (Table 1). An important exception to this are the silica-based matrices, which offer good mechanical rigidity but low chemical stability at alkaline pHs. Some silica matrices have been coated with dextran, and so forth, to increase the chemical stability and increase hydrophilicity.

   The choice of mobile phase is most often dependent on protein stability, necessitating considerations of appropriate pH, and solvent composition as well as the presence or absence of cofactors, protease inhibitors etc. which may be essential to maintain the structural and functional integrity of the target molecule. All buffers used in size exclusion should ideally be filtered through a 0.2-μm filter and degassed by low vacuum or sparging with an inert gas, such as helium.

   The majority of protein separations performed using size exclusion are carried out in the presence of aqueous-phase buffers. Size exclusion of proteins in organic phases (sometimes called gel permeation) is not normally undertaken but is sometimes used for membrane-protein separations. The agarose- and dextran-based matrices are not suitable for separations with organic solvents. Synthetic polymers and, to a certain extent, silica are suitable for separations in organic phases. For separation in acidified organic solvents, polyacrylamide matrices are particularly suitable. The separation of the proteins may be influenced by the denatured state of the protein in the organic phase. The equipment must be compatible with the solvent system, e.g., glass or Teflon.

   Many matrices retain a residual charge owing to, for example, sulfate groups in agarose or carboxyl residues in dextran. The ionic strength of the buffer should be kept at 0.15–2.0 M to avoid electrostatic or Van Der Waals interactions that can lead to nonideal size exclusion (7,8). Crosslinking agents, such as those used in polyacrylamide, may reduce the hydrophilicity of the matrix, leading to the retention of some small proteins, particularly those rich in aromatic amino-acid residues. These interactions have been exploited effectively to enhance purification in some cases, but are generally best avoided. Interactions with the matrix are commonly seen when charged proteins are being resolved. If low-ionic strength is necessary, then the risk of interaction can be reduced by manipulating the charge on the protein via the pH of the buffer. This is best achieved by keeping the mobile phase above or below the pI of the protein as appropriate.

Protein–matrix interaction is a common cause of protein loss during size exclusion. This may be owing to complete retention on the column or retardation sufficient for the material to elute in an extremely broad dilute fraction, thus evading detection above the baseline of the buffer system. Another important consideration is when enzymes are detected off-line by activity assay. The active enzyme may resolve in to inactive subunits. In such cases, a review of the mobile phase is advisable.

6. Selection of matrix: The beads used for size exclusion have a closely controlled pore size, with a high chemical and physical stability. They are hydrophilic and inert to minimize chemical interactions between the solutes (proteins) and the matrix itself. The performance and resolution of the technique has been enhanced by the development of newer matrices with improved properties. Historically, gels were based on starch, although these were superseded by the crosslinked dextran gels (e.g., Sephadex). In addition polystyrene-based matrices were developed for the use of size exclusion in nonaqueous solutions. Polyacrylamide gels (e.g., Biogel P series) are particularly suited to separation at the lower-mol-wt range owing to their microreticular structure. More recently, composite matrices such as the Superdex® gel (Pharmacia) have been developed, where dextran chains have been chemically bonded to a highly crosslinked agarose for high speed size exclusion (**Table 1**).

3. Practical Procedures

1. Flow rates: In chromatography flow rates should be standardized for columns of different dimensions by quoting linear flow rate (cm/h). This is defined as the volumetric flow rate (cm^3/h) per unit cross-sectional area (cm^2) of a given column. Because the principle of size exclusion is based on partitioning, success of the technique is particularly susceptible to variations in flow rates. Conventional low-pressure size exclusion matrices tend to operate at linear flow rates of 5–15 cm/h. Too high a flow rate leads to incomplete partitioning and band spreading. Conversely, very low flow rates may lead to diffusion and band spreading.

2. Preparation of gel matrix: Gel matrices are supplied as either preswollen gels or as dry powder. If the gel is supplied as a dry powder it should be swollen in excess mobile phase as directed by the manufacturers. Swollen gels must be transferred to the appropriate mobile phase. This can be achieved by washing in a sintered-glass funnel under low vacuum. During preparation, the gel should not be allowed to dry. The equilibrated gel should be decanted into a Buchner flask, allowed to settle, and then fines removed from the top by decanting. The equilibrated gel (an approx 75% slurry) is then degassed under low vacuum.

3. Packing the gel: Good column packing is an essential prerequisite for efficient resolution in size-exclusion chromatography. The column should be held vertically in a retort stand, avoiding adverse drafts, direct sunlight, or changes in temperature. The gel should be equilibrated and packed at the final operating temperature. With the bottom frit or flow adaptor in place degassed buffer (5–10% of the bed volume) is poured down the column side to remove any air from the system. In one manipulation, the degassed-gel slurry is poured into the column using a glass rod to direct the gel down the side of the column, avoiding air entrapment. If available, a packing reservoir should be fitted to columns to facilitate easier packing. The column can be packed under gravity, although a more efficient method is to use a pump to push buffer through the packing matrix. The flow rate during packing should be approx 50% higher than the operating flow rate (e.g., 15 cm/h for a 10 cm/h final flow rate). The bed height can be monitored by careful inspection of the column as it is packing. Once packed, a clear layer of buffer will appear above the bed and the level of the gel will remain constant. The flow should then be stopped and excess buffer removed from the top of the column, leaving approx 2 cm buffer above the gel. The

outlet from the column should be closed and the top-flow adaptor carefully placed on top of the gel, avoiding trapped air or disturbing the gel bed.

The packed column should be equilibrated by passing the final buffer through the column at the packing flow rate for at least one column volume. The pump should always be connected to pump the eluent on to the column under positive pressure. Drawing buffer through the column under negative pressure may lead to bubbles forming as a result of the suction. The effluent of the column should be sampled and tested for pH and conductivity in order to establish equilibration in the desired buffer.

4. Sample application: Several methods exist for sample application. It is critical to deliver the sample to the top of the column as a narrow sample zone. This can be achieved by manually loading via a syringe directly on to the column, although this requires skill and practice. The material may be applied through a peristaltic pump, although this will inevitably lead to band spreading owing to sample dilution. The sample should never be loaded through a pump with a large hold-up volume, such as a syringe pump, or upstream of a bubble trap. Arguably, the best method of applying the sample is via a sample loop in conjunction with a switching valve, allowing the sample to be manually or electronically diverted through the loop and directly on to the top of the column.

5. Evaluation of column packing: The partitioning process occurs as the bulk flow of liquid moves down the column. As the sample is loaded, it forms a sample band on the column. In considering the efficiency of the column partitioning can be perceived as occurring in discrete zones along the axis of the columns length. Each zone is referred to as a theoretical plate and the length of the zone termed the theoretical plate height (H). The value of H is a function of the physical properties of the column, the exclusion limit, and the operating conditions, such as flow rate, and so forth. Column efficiency is defined by the number of theoretical plates (N) that can be measured experimentally using a suitable sample, e.g., 1% (v/v) acetone (**Fig. 3**). Resolution and, hence, the number of theoretical plates, is enhanced by increasing column length.

In addition to determination of the number of theoretical plates (N) described, the performance of the column can be assessed qualitatively by the shape of the eluted peaks (**Fig. 4**). The theoretically ideal peak is sharp and triangular with an axis of symmetry around the apex. Deviations from this are seen in practice. Some peak shapes are diagnostic of particular problems that lead to broadening and poor resolution. If the downslope of the peak is significantly shallow, it is possible that the concentration of the load was too high or the material has disturbed the equilibrium between the mobile and stationary phases. If the downslope tends to symmetrical initially, but then becomes shallow, it is common to assume that there is a poorly resolved component; however, it may be suspected that interaction with the matrix is taking place. A shallow upslope may represent insolubility of the loaded material. A valley between two closely eluting peaks may suggest poor resolution, but can also be the result of a faulty sample injection.

6. Standards and calibration: Calibration is obtained by use of standard proteins and plotting retention time (V_e) against log molecular weight. Successful calibration requires accurate flow rates. The resultant plot gives a sigmoidal curve approaching linearity in the effective separation range of the gel (**Fig. 5**).

It should be noted that it is not only the molecular weight that is important in size exclusion, but also the hydrodynamic volume or the Stokes radius of the molecule. Globular proteins appear to have a lower molecular size than proteins with a similar molecular weight, which are in an α-helical form. These, in turn, appear smaller than proteins in random-coil form. It is critical to use the appropriate standards for size exclusion where proteins have a similar shape. This has been useful in some cases for studying protein folding and unfolding. Commonly used mol-wt standards are given in **Table 2**.

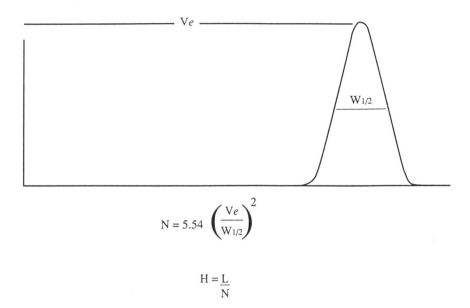

$$N = 5.54 \left(\frac{Ve}{W_{1/2}}\right)^2$$

$$H = \frac{L}{N}$$

Fig. 3. Calculation of the theoretic plate height as a measure of column performance. The number of theoretical plates (N) is related to the peak width at half height ($W_{1/2}$) and the elution volume *(V_e)*. The height of the theoretical plate (H) is related to the column length (L).

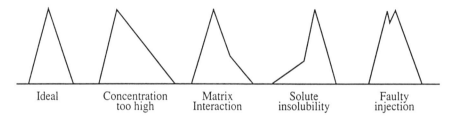

| Ideal | Concentration too high | Matrix Interaction | Solute insolubility | Faulty injection |

Fig. 4. Diagnosis of column performance by consideration of peak shape.

7. Separation of proteins: The optimum load of size-exclusion columns is restricted to <5% (typically 2%) of the column volume in order to maximize resolution. Gel-filtration columns are often loaded at relatively high concentrations of protein, such as 2–20 mg/mL. The concentration is limited by solubility of the protein and the potential for increased viscosity, which begins to have a detrimental effect on resolution. This becomes evident around 50 mg/mL. It is important to remove any insoluble matter prior to loading by either centrifugation or filtration. Owing to the limitation on loading, it is often wise to consider the ability of the method for scale-up when optimizing the operating parameters (*9*).

8. Column cleaning and storage: Size-exclusion matrices can be cleaned *in situ* or as loose gel in a sintered-glass funnel. Suppliers usually offer specific guidelines for cleaning gels. Common general-cleaning agents include nonionic detergents (e.g., 1% [v/v] Triton X-100) for lipids and 0.2–0.5M NaOH for proteins and pyrogens (not recommended for silica-based matrices). In extreme circumstances, contaminating protein can be removed by use of enzymic digestion (pepsin for proteins and nucleases for RNA and DNA). The gel should be stored in a buffer with antimicrobial activity, such as 20% (v/v) ethanol or

Table 2
Molecular Weight Standards
for Size Exclusion Chromatography

Vitamin B-12	1350
Ribonuclease A	13,700
Myoglobin	17,000
Chymotrypsinogen A	25,000
Ovalbumin	43,000
Bovine serum albumin	67,000
Bovine γ-globulin	158,000
Blue dextran 2000	~2,000,000

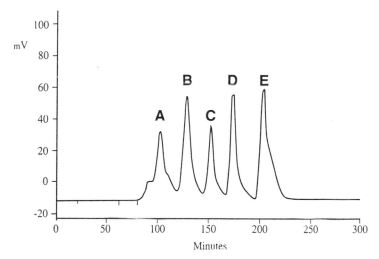

Fig. 5. Size exclusion of five mol-wt markers on Superdex 200 (Pharmacia) column (1.6 × 75 cm) Thyroglobulin, 670,000 **(A)** γ globulin, 158,000 **(B)**; ovalbumin, 44,000 **(C)**; myoglobin, 17,000 **(D)**; and vitamin B12, 1350 **(E)**.

0.02–0.05% (w/v) sodium azide. NaOH is a good storage agent that combines good solubilizing activity with prevention of endotoxin formation. It may, however, lead to chemical breakdown of certain matrices.

References

1. Laurent, T. C. (1993) Chromatography classic: history of a theory. *J. Chrom.* **633,** 1–8.
2. Hagel, L. (1989) Gel filtration, in *Protein Purification* (Jansen, J.-C. and Ryden, L., eds.), VCH, New York, pp. 63–106.
3. Findlay, J. B. C. (1990) Purification of membrane proteins, in *Protein Purification Applications* (Harris, E. L. V. and Angal, S., eds.), IRL, Oxford, UK, pp. 59–82.
4. Hagel, L. (1993) Size exclusion chromatography in an analytical perspective. *J. Chrom.* **648,** 19–25.
5. Stellwagen, E. (1990) Gel filtration. *Methods Enzymol.* **182,** 317–328.
6. Pohl, T. (1990) Concentration of proteins and removal of solutes. *Methods Enzymol.* **182,** 69–83.

7. Kopaciewicz, W. and Regnier, F. E. (1982) Non-ideal size exclusion chromatography of proteins: effects of pH at low ionic strength. *Anal. Biochem.* **126,** 8–16.

8. Dubin, P. L., Edwards, S. L., Mehta, M. S., and Tomalia, D. (1993) Quantitation of non-ideal behavior in protein size exclusion chromatography. *J. Chrom.* **635,** 51–60.

9. Jansen, J.-C. and Hedman, P. (1987) Large scale chromatography of proteins. *Adv. Biochem. Eng.* **25,** 43–97.

37

Hydrophobic Interaction Chromatography

Paul A. O'Farrell

1. Introduction

Hydrophobic interaction chromatography (HIC) is a widely used technique for the separation of proteins and peptides. It is based on a general property of proteins and thus has applicability as wide as the other general chromatographic techniques, ion-exchange chromatography and gel-filtration. However, careful manipulation of the chromatographic conditions allows the technique to have great resolving power.

In the simplest type of procedure, the protein mixture is applied to the column in a buffer containing a high concentration of salt. After nonbinding proteins have been washed off the column, bound proteins are eluted by reducing the salt concentration. This can be done either in a stepwise fashion or via a decreasing salt gradient, with the least-hydrophobic proteins eluting first, and more-hydrophobic proteins being released at lower concentrations of salt.

Proteins contain a variety of amino-acid residues with polar and nonpolar (or hydrophobic) side-chains. Many of the hydrophobic side-chains are buried in the interior of the protein, away from the surrounding (water) solvent ("hydrophobic," after all, means "water-fearing"); however, a sizeable proportion of them can appear on the surface. It has been estimated that as much as 50% of the surface of a soluble protein can consist of hydrophobic residues (1), but this percentage varies over a wide range from one protein to another. It is this difference in the extent and character of surface hydrophobicity that is exploited for protein separation by HIC. However, the technique was not originally developed in an attempt to exploit this characteristic; rather, it was "discovered" from observation of other chromatographic techniques. In affinity chromatography, "spacer" arms consisting of short chains of carbon atoms are often used to keep the affinity ligand at some distance from the insoluble-column matrix, thus helping to avoid steric interference that could result in poor binding. In the development of this technique, experiments were carried out to determine what effect the spacer had on binding of proteins to the column; thus, matrices were produced that had only the spacer attached, with no affinity ligand. It was found that proteins could bind to columns made from such material under certain conditions; this observation led to the first systematic studies of HIC (e.g., **refs.** 2 and 3).

From: *Molecular Biomethods Handbook*
Edited by: R. Rapley and J. M. Walker © Humana Press Inc., Totowa, NJ

Early HIC columns were made from matrices that were charged to a certain extent, thus the interaction of proteins with such columns combines hydrophobic with ion-exchange effects. Such complications made the development of theories about HIC difficult—in fact, the nature of the hydrophobic interaction is still not fully understood. A number of different theories have been put forward, including some based on surface-tension effects, entropy, or van der Waals forces. It is clear, however, that the properties of water, in particular its ability to form structure, are of central importance. Water is a very polar molecule. A particular water molecule will interact with its neighbors by hydrogen bonding. These neighboring molecules will hydrogen bond with molecules lying further out, and so on. In this way, a network of hydrogen bonds forms a fairly ordered structure in liquid water. When a nonpolar solute molecule is dissolved in water, this structure is perturbed—water molecules in contact with the solute molecule cannot hydrogen-bond with it and are forced to form a "shell" around it where the hydrogen-bonding structure between the water molecules is different from that of the surrounding bulk solvent and is, in fact, more ordered. This change to a more ordered structure results in a decrease in entropy, and thus is thermodynamically unfavorable. Consider what happens when two such solvent molecules, each surrounded by a shell of ordered water, come into contact. The two separate, ordered shells of water amalgamate and a single shell is formed around the pair of nonpolar molecules. This amalgamation releases some of the ordered-water molecules, which can then return to their normal, less-ordered state (**Fig. 1**). This results in an increase in entropy and is thermodynamically favorable. Thus, the hydrophobic-solute molecules are not attracted to each other in a conventional sense; their interaction is forced on them by the structure of the solvent, and the entropic advantage gained by their aggregation (*see* **refs. 4–7** for other theoretical aspects). It follows that the addition of modifiers to the solvent that change its ability to form structure will have an effect on hydrophobic interactions. Many salts (called "structure-forming" salts) increase the level of intermolecular structure in water. Addition of these salts increases the strength of hydrophobic interactions. Salts that reduce the amount of structure (chaotropic) have the opposite effect. Addition of nonpolar solvents (e.g., ethanol, ethylene glycol) also has the effect of reducing the strength of hydrophobic interactions.

Proteins are much more complex than the small nonpolar molecules on which much of hydrophobic theory is based, and their interactions are similarly more complex; however, it is not necessary to have a complete understanding of every aspect of the interaction to utilize it for protein isolation. It is always necessary to optimize purification techniques by experimentation with the particular extract of interest, and HIC is no exception.

2. Experimental Considerations

2.1. Column Matrix

A wide variety of hydrophobic resins are available commercially. In addition, a number of products combine hydrophobic properties with others, such as size exclusion. Of the purely hydrophobic matrices, the most commonly used are substituted with phenyl or alkyl ligands. The degree of hydrophobicity of the ligands varies, and the choice between them should be made empirically. Small-scale screening experiments should

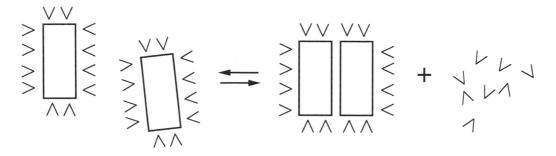

1. Isolated non-polar solute molecules surrounded by order shells of water molecules

2. Aggregation of solute molecules allows some solvent molecules to return to a more thermodynamically favourable state.

Fig. 1. Expulsion of ordered-water molecules in hydrophobic interaction among nonpolar solute molecules.

be carried out with a number of different products to determine which is the most suitable for a particular application. (**Ref. 8** gives a good guide to such screening experiments.) With an uncharacterized protein, it is probably best to begin with a weakly hydrophobic ligand (e.g., phenyl), because a strongly hydrophobic protein may be difficult to elute from a strongly hydrophobic resin. The matrix itself should have sufficient mechanical strength to withstand the pressures and flow-rates used.

2.2. Temperature

Because hydrophobic interactions are, to a large degree, dependent on entropic factors, it is to be expected that temperature will have an effect on strength of binding. It has been shown, for example, that the capacity of a hydrophobic matrix can reduce by as much as 30% on reducing the temperature from 20 to 4°C *(9)*. However, because temperature also affects protein conformation and solubility, it can be difficult to predict the overall effect of temperature changes on a particular separation. Changes in temperature are not generally used as a means of effecting separation in HIC, but their effects should nonetheless be considered. For example, a protein that binds to a particular hydrophobic matrix at room temperature may not bind in the cold room.

2.3. pH

Generally, the strength of the interaction of a protein with a hydrophobic matrix decreases with increasing pH; presumably this is because of a decrease in the hydrophobicity of proteins owing to increasing charge as acidic side-chains are titrated. Because the effect of pH is different for different proteins, it should be possible to improve elution profiles by changing the pH. However, the effect is slight and this is not often done in practice—it is more usual to simply work at a pH where the protein of interest is known to be stable.

2.4. Elution

Binding can be achieved with many different salts, some of which have stronger effects than others. The effect of particular salts on binding to hydrophobic matrices

Table 1
The Hofmeister Series[a]

Anions	Cations
PO_4^{3-}	NH_4^+
SO_4^{2-}	Rb^+
CH_3COO^-	K^+
Cl^-	Na^+
Br^-	Cs^+
NO_3^-	Li^+
ClO_4^-	Mg^{2+}
I^-	Ca^{2+}
SCN^-	Ba^{2+}

[a]Ions are listed in the order from those that favor hydrophobic interactions to those that decrease the strength of the interaction.

generally parallels their effects on salting out. The Hofmeister series lists ions in the order from those that most favor salting-out (**Table 1;** *10*). Various aspects of this series (also called the lyotropic series) have been looked at in the context of HIC *(11)*. It should be noted that some salts can have specific effects on particular proteins so the series may not always be followed precisely by every protein. Often the choice of which salt to use is dictated by other factors than its position in the lyotropic series. HIC will usually be used in combination with other techniques. If it is preceded by, for example, ion-exchange chromatography, the salt that was used for elution in that step will often be used for binding in HIC. Thus, it is not necessary to remove the salt (e.g., by dialysis or gel filtration) before proceeding. Indeed, HIC is often used directly after ion-exchange chromatography or ammonium sulfate precipitation because of this economy of procedure. Elution is achieved by reducing the salt concentration. In some cases, reduction of the salt concentration is not enough to achieve elution and it may be necessary to add a chaotropic salt (e.g., urea), alcohol (e.g., ethylene glycol, up to 80%), or a detergent (e.g., Triton X-100, 1% [w/v]) to recover the protein of interest. Some workers have found that using a decreasing gradient of a structure-forming salt and at the same time increasing the concentration of a chaotrope, can result in higher resolution and better separations *(12)*. However, chaotropic agents have deleterious effects on protein structure, and are normally avoided. In many cases, the requirement for such agents can be escaped by using a less hydrophobic ligand.

3. Applications
3.1. One-Step Purification

Although HIC is generally used in conjunction with other chromatographic techniques, it has been used to achieve purification of a protein in a single step. Pyrrolidone carboxyl peptidase (Pcp) removes pyrrolidone carboxylic acid (PCA) from the N-terminus of some PCA-proteins and peptides. The presence of PCA at the N-terminus results in "blocking," which prevents the Edman-degradation protein sequencing reaction from being carried out. The reaction catalyzed by this enzyme can unblock some proteins, thus allowing N-terminus sequencing. Awade et al. *(13)* have cloned the gene encoding Pcp from *Streptococcus pyrogenes* and have overexpressed the protein in *Escherichia coli*. Using HIC as the sole chromatographic step, they were able to isolate the pure enzyme in a very rapid procedure. The crude extract was first treated with protamine sulfate (to remove DNA) and, after an ammonium sulfate precipitation step (40–55% saturation), the protein pellet was resuspended in buffer containing 1 *M* ammonium sulfate. An aliquot of the preparation was applied to a Progel-TSK phenyl-5PW column, previously equilibrated with 1.2 *M* ammonium sulfate, in a high-performance liquid chromatography (HPLC) system. The column was run at a flow-rate of 0.8 mL/min and was first washed with the equilibration buffer for 20 min to remove unbound protein. Bound proteins were then eluted with a gradient of 1.2–0 *M* ammonium sulfate over 45 min. Two peaks containing Pcp were identified. SDS-PAGE and native PAGE showed a single identical-protein band to be present in both peaks. The entire process was very rapid, taking less than a day to produce the pure enzyme.

3.2. HIC as a Nonspecific Procedure

A protein's affinity for other molecules is dependent on its physicochemical properties. Charge–charge interaction, hydrogen-bonding, and hydrophobic interaction can all be important in binding an enzyme to its substrate, for instance. This type of interaction depends on a specific affinity. However, nonspecific interactions can be just as important to a protein's function. Membrane proteins, for example, need hydrophobic regions on their surfaces to ensure insertion into the nonpolar environment of the cell membrane. HIC has been used in the isolation of proteins whose hydrophobicity is important for function. Onishi and Proudlove *(14)* used the technique to isolate proteins important for the formation of foam on beer; earlier reports had indicated that hydrophobic polypeptides make the most important contribution to the stability of beer foam. HIC was used to separate polypeptides into various groups based on their relative hydrophobicity. Foam was collected from commercial lager beer and allowed to collapse to liquid. This was applied to a column of octyl-sepharose CL-4B that had been pre-equilibrated with water, and unbound proteins were eluted by washing with water. A group of polypeptides ("very strongly hydrophobic") were eluted by washing with 50% ethylene glycol, and a further group ("extremely hydrophobic") with 6 *M* urea. The fraction that had remained unbound in water solution was reapplied in ammonium sulfate solution (8.5%). Washing with this solution released unbound polypeptides. The "moderately hydrophobic "group was eluted with water, and a further group with 50% ethylene glycol. (It should be noted that in this experiment, denaturation in the presence of urea or ethylene glycol was not a problem, because the foam proteins were expected to be in the unfolded state). Polypeptides in each of the groups were further separated by anion-exchange chromatography. The various classes of polypeptide were then tested for their ability to form a stable foam. In general, it was found that the more hydrophobic proteins formed a more stable foam.

3.3. Clinical

HIC also has uses in the clinical field. CD4 is a cell surface marker found on T-lymphocytes and other cells of the immune system. A monoclonal antibody (MAb) directed against CD4 has been purified using HIC and other techniques for use in the treatment of autoimmune disease *(15)*. Throughout the procedure, great care had to be taken to ensure that the final product was extremely pure and satisfied the regulatory requirements for use in human subjects (sterility, freedom from pyrogens, and so forth). HIC was carried out on a phenyl-sepharose column. The extract was applied in 0.5 M ammonium sulfate and elution was achieved by reducing the concentration of the salt via a gradient. Fractions were collected under sterile conditions using a laminar-flow cabinet. Other techniques used in the purification were anion-exchange chromatography, gel filtration, ultrafiltration, and removal of lipids by solvent extraction. In addition to being used in the purification of therapeutic agents, HIC has also been used directly in the treatment of autoimmune disease *(16)*. In some cases, standard immunosupressive therapy is inadequate and more aggressive treatment is necessary. One approach is to reduce the level of antibodies in the patient's blood by extracorporeal elimination. Briefly, blood is removed from the body (not all at once!) and the cells separated from the plasma. Immunoglobulins are removed from the plasma by adsorption to a hydrophobic column. After passing through the column, the plasma is mixed with the cells once again, and the reconstituted, antibody-depleted blood is returned to the patient.

3.4. Analytical

As mentioned in **Subheading 1.**, HIC is capable of great resolving power and can be a powerful analytical technique. Jing et al. *(17)* examined the effect of point mutations in a recombinant protein, staphylococcal nuclease, on its retention in hydrophobic-interaction chromatography. Four different proteins were prepared; the wild-type and three mutants with amino-acid substitutions either on the surface of the protein or in the interior. They found that replacement of a tryptophan in the interior of the protein with phenylalanine (a less-hydrophobic residue) had no effect on retention in HIC, whereas replacement of a surface tyrosine with tryptophan (more hydrophobic) increased the retention time, regardless of which amino-acid residue was at the interior position. In addition, HIC was capable of separating proteins differing by just one surface amino-acid residue from mixtures. They also examined the ability of HIC to differentiate proteins with different conformations. When recombinant proteins are overexpressed in *E. coli*, they often occur as inclusion bodies—insoluble aggregates in which the protein is generally folded incorrectly. To regain the native conformation, the protein must first be denatured (e.g., with guanidinium hydrochloride) and allowed to refold. However, although the refolded molecules may be soluble, they are not necessarily in the correct conformation. HIC on a phenyl-superose HR5/5 column was able to separate the correctly folded protein from another peak which had a longer retention time and only 24% of the activity of the native form. Further denaturation and refolding of this material enabled it to regain the native conformation. It was pointed out that for some proteins, for example, those with large molecular weights and disulfide bonds, even very careful denaturation–renaturation treatment may still result in a mixed population of native and nonnative conformers. They suggest that in such cases HIC may prove to be a very useful method for the analysis of conformational homogeneity.

References

1. Lee, B. and Richards, E. M. (1971) The interpretation of protein structures: estimation of static availability. *J. Mol. Biol.* **55**, 379–400.
2. Er-el, Z., Zaidenzaig, Y., and Shaltiel, S. (1972) Hydrocarbon-coated sepharoses. Use in the purification of glycogen phosphorylase. *Biochem. Biophys. Res. Comm.* **49(2)**, 383–390.
3. Hofstee, B. H. J. (1973) Hydrophobic affinity chromatography of proteins. *Anal. Biochem.* **52**, 430–448.
4. Hjerten, S. (1973) Some general aspects of hydrophobic interaction chromatography. *J. Chromatog.* **87**, 325–331.
5. Melander, W. R., Corradini, D., and Horvath, C. S. (1984) Salt-mediated retention of proteins in hydrophobic-interaction chromatography. Application of solvophobic theory. *J. Chromatogr.* **317**, 67–85.
6. von der Haar, F (1976) Purification of proteins by fractional interfacial salting out on unsubstituted agarose gels. *Biochem. Biophys. Res. Comm.* **70(3)**, 1009–1013.
7. Srinivasan, R. and Ruckenstein, E. (1980) Role of physical forces in hydrophobic interaction chromatography. *Separ. Purific. Methods* **9(2)**, 267–371.
8. Hydrophobic Interaction Chromatography—Principles and Methods (1993) available from Pharmacia. biotech, Bjorkgaten 30, S-751 82 Uppsala, Sweden.
9. Hjerten, J. (1977) Fractionation of proteins by hydrophobic interaction chromatography with reference to serum proteins. *Proceedings of the International Workshop on Technology for Protein Separation and Improvement of Blood Plasma Fraction* (Sandburg, H. A., ed.), US Department of Health Education and Welfare, DHEW publication no. (NIH) 78-1422, pp. 410–421.
10. Hofmeister, F (1888) Zur lohre von der wirkung der salze. Zweite mittheilung. *Arch. Exp. Pathol. Pharmakol.* **24**, 247–260.
11. Melander, W. and Horvath, C. S. (1977) Salt effects on hydrophobic interactions in precipitation and chromatography of proteins: an interpretation of the lyotropic series. *Arch. Biochem. Biophys.* **183**, 200–215.
12. El Rassi, Z., De Ocampo, L. F., and Bacolod, M, D. (1990) Binary and ternary salt gradients in hydrophobic interaction chromatography. *J. Chromatog.* **499**, 141–152.
13. Awade, A., Gonzales, T., Cleuziat, P., and Robert-Baudout, J. (1992) One step purification and characterisation of the pyrrolidone carboxyl peptidase of *Streptococcus pyrogenes* over-expressed in *Escherichia coli*. *FEBS Letts.* **308(1)**, 70–74.
14. Onishi, A. and Proudlove, M. O. (1994) Isolation of beer foam polypeptides by hydrophobic interaction chromatography and their partial characterisation. *J. Sci. Food Agric.* **65**, 233–240.
15. Guse, A. H., Milton, A. D., Schulze-Koops, H., Muller, B., Roth, E., Simmer, B., Wachter, H., Weiss, E., and Emmrich, F. (1994) Purification and analytical characterisation of an anti-CD4 monoclonal antibody for human therapy. *J. Chromatogr. A.* **661**, 13–23.
16. Borberg, H., Jimenez, C., Belak, M., Haupt, W. F., and Spath, P. (1994) Treatment of autoimmune disease through extracorporeal elimination and intravenous immunoglobulin. *Transfus. Sci.* **15(4)**, 409–418.
17. Jing, G. Z., Zhou, B., Liu, L. J., Zhou, J. X., and Liu, Z. (1994) Resolution of proteins on a phenyl-superose HR5/5 column and its application to examining the conformational homogeneity of refolded staphylococcal nuclease. *J. Chromatogr. A.* **685**, 31–37.

38

Affinity Chromatography

George W. Jack

1. Introduction

The concept of affinity as a basis for the purification of proteins has been with us for longer than most of us might realize. Perhaps the oldest example comes from 1953 *(1)*, with the description of a method for the partial purification of tyrosinase using an inhibitor of the enzyme immobilized on cellulose. Given the limited and largely imperfect techniques available at the time for the preparation of proteins it is perhaps surprising that the concept did not gain general acceptance for virtually another 20 years.

For its success, affinity chromatography depends on the ability of biological molecules to recognize complimentary substances, be it an enzyme binding its substrate or inhibitor or an antibody combining with its antigen. The technique may be described in essence as an exploitation of a protein's affinities in its purification rather than its properties as a protein, such as charge, size, or hydrophobicity.

Further early examples of affinity-based purifications are to be found mainly in the field of immunoaffinity chromatography. Currently, this technique is employed to prepare pure antigen on a column of immobilized antibody, but in its early incarnation the reverse method was used to prepare enriched antibody fractions from serum on immobilized-antigen columns *(2)*.

The overall process of affinity chromatography may be split into a number of stages. First the matrix, the particulate insoluble chromatography gel, must be activated chemically to generate reactive groups with which the ligand can react and couple. The ligand, the molecule that will interact with the protein one wishes to purify, can then be coupled to the insoluble matrix either directly as in the case of protein ligands (protein A for antibody purification or a monoclonal antibody [MAb] in immunoaffinity chromatography) or via a spacer arm for small-mol-wt ligands to form the affinity sorbent. Because the binding site of enzymes and proteins is often buried within the structure of the molecule, small ligands coupled directly to the matrix often do not interact owing to steric interference between the matrix and the target protein. The use of a spacer arm to distance the ligand from the matrix can have a dramatic effect in promoting binding of the target protein to the ligand. After blocking any unreacted sites, the gel can be washed free of unreacted compounds and poured as a column ready for the final stage, the chromatographic process. This is essentially an on/off process in which a crude sample

From: *Molecular Biomethods Handbook*
Edited by: R. Rapley and J. M. Walker © Humana Press Inc., Totowa, NJ

containing the protein to be purified is contacted with the affinity matrix under conditions that allow the protein to bind to the ligand. The contaminating proteins can then be washed away before the conditions are changed (change pH, add chaotrope, add cofactor, add substrate) so that the protein no longer binds to the ligand and is eluted from the column.

Despite the appearance of the occasional paper describing affinity methods for enzyme purification *(3)*, the widespread use of affinity chromatography had to await the availability of a better matrix than cellulose on which to immobilize molecules. Such matrices should ideally have low nonspecific interaction with proteins, be mechanically and chemically stable, and consist of a porous infrastructure. The gel itself should be spherical, rigid, and display a narrow-size range. The commercial availability of agarose-derived gels and, to a lesser extent, polyacrylamide gels provided the impetus to develop affinity chromatography as we know it today. Not only were agarose gels mechanically suitable for the task, they allowed a range of chemical reactions to be performed in coupling molecules to the matrix. This was described extensively by Cuatrecasas *(4)* in a paper that became most people's starting point in their investigation of the possibilities of the technique.

Over the years the general methodology has been expanded to encompass a wide range of purifications, some of which are highly specific for a unique molecular species, whereas others cope with groups of proteins. The number of proteins that have now been purified by an affinity method is so large that attempts to list them have become unreasonable owing to the sheer volume of numbers.

2. Matrices

The number of matrices commercially available to the chromatographer is increasingly wide and confusing. In choosing a matrix, there are a number of considerations to be made:

1. Is the matrix commercially available in the quantities you require?
2. Is the matrix readily derivatized and does it contain the appropriate chemical groups for the activation you wish to perform? (Generally an abundance of hydroxyl groups is favored because they will participate in a wide variety of activation chemistries.)
3. Is the matrix macroporous? (It should have a high surface-area-to-volume ratio and contain pores sufficiently large to allow unimpeded passage of protein molecules.)
4. Is the matrix gel uniform and does it have a narrow-size distribution?
5. Is the matrix chemically and physically stable? (It should be able to withstand extremes of pH and high concentrations of chaotropic agents as well as resist compression at high flow rates in a column format.)
6. Is the matrix hydrophilic? (It should exhibit low nonspecific absorption, whereas increasing hydrophobicity encourages nonspecific binding.)
7. Is the matrix nontoxic or can it leach toxic breakdown products?

These are some of the criteria to be borne in mind in selecting a suitable matrix, but also to be considered is whether the gel shrinks or swells with changes in pH or if organic solvents with their attendant problems of handling and disposal are required. The early examples of affinity chromatography used fibrous cellulose as the matrix, although today such material would fail the selection process on **items 3–5**. Some commercially available matrices are described briefly in **Table 1**. Almost all these gels

Table 1
Commercially Available Afffinity Chromatography Matrices

Trade name	Chemical structure	Supplier
Sepharose	Agarose (D-galactose/3-anhydrogalactose)	Pharmacia
Sepse CL	Agarose (react with 2,3-dibromopropanol)	Pharmacia
Bio-Gel A	Crosslinked agarose	BioRad
	Controlled pore glass (silanized)	Sigma
Bio-Gel P	Polyacrylamide	BioRad
Trisacryl	Tris-hydroxymethylacrylamide crosslinked with *N,N'*-diallyltartradiamide	IBF
Ultrogel	Agarose- acrylamide composite	IBF
Sephacryl	Dextran-acrylamide copolymer	Pharmacia
Toyopear	Copolymer of glycidyl methacrylate and polyethylene glycol with pentaerythritol dimethacrylate	TosoHaes
HEMA	Nethacrylate/dimethacrylate	Alltech
Eupergit[a]	Methacrylamide/allyl glycidyl ether	Rohm Pharma
Poros	Poylstyrene/divinylbenzene	PerSeptive
Azlactone[a]	Vinyidimethyl azlactone copolymer with acrylamide	3M

[a]These matrices already contain reactive groups and do not require activation before ligands may be coupled directly.

have appropriate characteristics with the exception of some of the polyacrylamide gels, which display poor flow properties.

3. Activation and Coupling

The range of activation and coupling chemistries at the disposal of the chromatographer is even more varied and bewildering than the available matrices. The aim is to chemically modify the matrix such that the ligand of choice will react with it directly or via an appropriate spacer arm. Care must be exercised in the use of spacer arms because it was shown *(5)* in some instances that the chromatographic properties of the sorbent were owing primarily to the charges present in the spacer arm or to its hydrophobic nature.

Matrix activation is generally effected by highly reactive chemicals, such as cyanogen bromide, and as such requires appropriate containment facilities. Activated matrices are available commercially and are to be recommended because they are a consistent product and avoid the use of hazardous or toxic chemicals. Matrices may be purchased with cyanogen bromide activation, with reactive carboxyl or amino groups, or with spacer arms terminating in epoxide groups, which on reaction yield stable-ether linkages.

To the question, "Which activation and coupling chemistries should I use for a particular application?", there is unfortunately no definitive answer; each problem must be approached *de novo* and solved in a pragmatic manner. This is illustrated by the activities of some immunaffinity ligands produced by a range of chemistries *(6)*, some of which bind the antigen, whereas others showed no binding ability.

The catalogs of firms such as Sigma, Pharmacia, and Bio-Rad show the range of sorbents available for use in a wide selection of separations. For those still driven to

Table 2
Matrix Activation Chemistries

Activation method[a]	Coupling group on protein
Azlactone	NH_2
Benzoquinone	NH_2
Bisoxirane	NH_2, OH, SH
Carbonyidiimidazole	NH_2
Cyanogen bromide	NH_2
Cyanuric chloride	NH_2
Dlvinyl sulfone	NH_2, OH
Epichlorohydrin	NH_2, OH, SH
FMP	NH_2, SH
Hydrazine	NH_2
Tresyl chloride	NH_2, SH

[a]Details of these activation methods are to be found in **ref. 7**.

attempt their own activations, a comprehensive group is described by Hermanson et al. *(7)*. Some of the more commonly used activation methods are listed in **Table 2**.

All supports leak ligand to a certain extent. Work by Pharmacia *(8)* has shown that agarose matrices are stable at all pHs other than extreme acid or alkaline pH, suggesting that the matrix itself will be stable under the conditions employed during chromatography. Some bond solvolysis will occur, but most studies support the view *(9)* that ligand leakage is the product of extreme pH or the presence of high concentrations of chaotropes. An important source of leakage in newly synthesized affinity supports is unreacted reagents remaining in the matrix owing to insufficient washing.

One problem sometimes associated with affinity supports is their nonspecific binding of protein. This can occur with protein ligands and is attributed to the immobilized protein acting as a mixed ion-exchanger. It will, in all probability, possess on its surface both positively and negatively charged groups available to interact nonspecifically. The effect may be minimized by operating the support in a buffer containing 0.5 *M* NaCl. Supports containing low molecular weight ligands can also exhibit nonbiospecific adsorptive effects. In this instance, however, the effect is less likely to be owing to ionic interactions than to hydrophobic interactions with the spacer arm *(5)*. Rather than attempt to control the problem by altering the running conditions, it is better to construct a sorbent with an uncharged and more inherently hydrophilic spacer arm.

4. Chromatography

The practical business of performing the chromatographic process is normally carried out in a column format. As a generality, the sample is applied to the column at approximately neutral pH in a low ionic-strength buffer. Under these conditions, given the biospecificity of the sorbent, only the target molecule should bind to the column.

Further washing with buffer will remove the bulk of the contaminants, whereas washes with water and a moderately high ionic-strength buffer (0.3 *M*) will remove contaminants bound to the column by hydrophobic or ionic forces.

The target molecule may now be eluted from the column, a process always better undertaken using reversal of the direction of liquid flow through the column. The nature of the eluant can only be determined on a case-by-case basis. A successful eluent is one that reduces the affinity of the target molecule for the immobilized ligand. This may be achieved in a number of ways but is commonly done by a change of pH, the introduction of chaotropic ions or denaturing agents, or by use of a substrate or cofactor. The determining factor in all this is finding an eluent that causes dissociation of the protein–ligand complex but retains the activity of the protein and the sorbent for subsequent use.

5. Separations

The purifications achievable by affinity chromatography can be categorized as either specific methods applicable to single proteins or group methods that purify classes of molecules; for example, glycoproteins purified on a lectin column. The specific methods are legion and are scattered throughout the literature although they generally depend on the use of an immobilized substrate or inhibitor of an enzyme or, say, an immobilized receptor protein of a cell effector *(10)*. A number of specific applications are to be found in **ref. 7**.

Immunoaffinity chromatography *(11)* utilizes immobilized antibodies, but because they are generally monoclonals and, hence, specific not just for a given protein but for an individual epitope on the surface of that protein, it is an example of a highly specific method, yet one with a common theme, that of the antibodies, uniting the individual methods.

5.1. Group-Specific Separations

True group methods are perhaps best exemplified by the lectin columns. Lectins are a group of proteins mainly of plant origin that interact with particular sugar residues. Thus, immobilized lectins may be used in the purification of polysaccharides, but are mainly for isolating glycoproteins and glycolipids. The lectin from the common lentil, *Lens culinaris*, is specific for α-D-glucose and α-D-mannose residues and will bind glycoproteins *(12)* containing these sugars even in the presence of low concentrations of detergent. A range of lectins is commercially available with different specificities for the sugar residues recognized so these matrices can be used to study the maturation of carbohydrate-containing molecules. The converse strategy can be employed in that immobilized sugars can be used to purify lectins *(13)*.

A further example of a group separation is antibody purification on immobilized protein A or protein G. Both are cell-wall proteins from staphylococcal and streptococcal strains, respectively, which bind antibodies by nonimmune mechanisms at the Fc region of the molecules *(14)*. Both molecules bind IgG from a variety of species, but in general, protein G binds more strongly, which can be a disadvantage in that harsher conditions may be required to desorb IgG from protein G columns than from protein A columns. Protein G, however, will bind IgG from horse, sheep, and rat, a property not shared by protein A. These sorbents can be used to prepare an IgG fraction from sera or more commonly to purify monoclonal IgG from either culture medium or ascites fluid *(15)*. The purification is generally effected by applying the starting material to the column at slightly alkaline pH and then running a decreasing pH gradient to approx pH 3.0.

The most amazing generalized ligands are the dye ligands, a group of textile dyes known as reactive triazine dyes. These adsorbents bind a huge variety of proteins of seemingly different structure and function but with different affinities *(16)*. The sorts of proteins purified on these dyes include nicotinamide adenine dinucleotide (NAD) and nicotinamide adenine dinucleotide phosphate (NADP)-dependent enzymes, carboxypeptidases, interferon, α-2-macroglobulin, coagulation factors and albumin. The precise nature of the interaction between the dye ligand and protein is not precisely known in all instances but is generally believed to be a mixture of charge and hydrophobic binding. It is not merely an ion-exchange phenomenon, because some proteins bind to the dyes at pH values above their isoelectric point. It has been suggested that the dyes mimic the shape of nucleotides and bind to some enzymes at their $NAD(P)^+$ binding site; a possibility supported by the fact that elusion of dehydrogenases from dye columns may be induced by nucleotide cofactors, such as $NAD(P)^+$ and adenosine-5'-triphosphate (ATP) *(17)*. The use of immobilized Cibacron blue in the purification of human serum albumin is one of the few affinity-chromatographic processes run on a commercial scale.

Heparin, a mucopolysaccharide generally prepared from porcine intestinal mucosa with a molecular-weight range of between 5 and 30 kDa, is an unlikely candidate as a group-separation ligand. This polymer will bind a range of growth factors, coagulation proteins, steroid receptors, lipases, initiation factors for protein synthesis, and many enzymes involved in DNA synthesis, together with bacterial restriction endonucleases *(18)*. Heparin's ability to function in this manner may be owing to its ability to mimic the phosphodiester backbone of nucleic acids.

Next is a collection of small molecule ligands that will bind a diverse population of molecules. Immobilized lysine will bind both double-stranded DNA and ribosomal RNA, but is best known for its ability to bind plasminogen activator and plasminogen from a variety of species *(19)*. Arginine-containing sorbents will bind both serine proteases and enzymes with a more biospecific affinity for arginine *(20)*. *p*-Aminobenzamidine is the preferred ligand for the serine proteases. On immobilization, it will bind urokinase, chymotrypsin and trypsin *(21)*.

Immobilized boronic-acid derivatives have been shown to bind a surprising array of proteins from glycoproteins to enzymes that require diol-containing cofactors, such as $NADP^+$. Enzymes inhibited by boronic acids also bind, which sees such diverse enzymes as serine proteases, lipases, β-lactamases, and urease binding to these sorbents. The most common use of these matrices is in the isolation and quantitation of glycosylated hemoglobin *(22)*.

Calmodulin is a eukaryotic-regulatory protein that will recognize in the presence of Ca^{2+} ions a number of enzymes involved in nucleotide metabolism, Ca^{2+} transport, and contractile processes *(23)*. Immobilized calmodulin is available commercially.

Glutathione has found use both as a ligand for the purification of S-transferases and glutathione-dependent proteins and also recombinant fusion proteins containing a portion of the C-terminus of glutathione S-transferase. Following purification, the unwanted peptide fragment may be cleaved off the protein of interest *(24)*.

Less frequently considered by protein chemists are the affinity matrices containing nucleic acids, nucleotides, or polynucleotide 5'-triphosphates. Immobilization of these compounds is well documented *(7)*. Not surprisingly adenosine monophosphate (AMP) and adenosidne diphosphate (ADP) as ligands will bind NAD^+ and $NADP^+$-dependent dehydrogenases, as do the triazine-dye ligands. As might be expected, numerous enzymes of nucleic-acid synthesis and modification may be purified on immobilized DNA; individual enzymes have been found with preferences for single- or double-stranded DNA and for denatured DNA. Messenger and transfer RNAs may be purified in this manner as well as viral and phage nucleic acids *(25)*. The proteins that may be purified on these sorbents include DNA and RNA polymerases, polynucleotide kineses, exo and endonucleases, mRNA cap-binding protein, interferon, reverse transcriptase and helicases.

5.2. Subtractive Separations

The methods indicated so far use affinity in a direct manner to purify a molecule of interest. Conversely, the technique may employed as a scavenger to remove contaminants from a protein solution. Detergents, frequently used to solubilize membrane proteins, may subsequently be removed from solution by commercially available sorbents, namely Extracti-Gel D from Pierce and Bio-Beads SM-2 from Bio-Rad. Both claim to remove a wide range of detergents while giving high protein recovery.

Endotoxins (pyrogens) are part of the cell wall of gram-negative bacteria and are lipopolysaccharides containing lipid A. They cause a febrile response in animals, which in extreme cases may be fatal. They must be eliminated from biological material used as a therapeutic. This may sometimes be achieved in the preparation of the therapeutic if the process involves chromatography on a diethyl-aminoethyl (DEAE) matrix. Failing that, specific removal of endotoxin may be realized by immobilized polymyxin B *(26)* or immobilized histidine *(27)*.

A final example of scavenging is the use of immobilized protease-inhibitors to remove proteases from protein solutions to improve the stability of the proteins. As in the case of protease purification, *p*-aminobenzamidine may be used as ligand *(21)*, whereas other protease inhibitors, such as soybean-trypsin inhibitor and aprotinin *(28)*, may be used in its stead.

6. Conclusions

From this brief summary of affinity chromatography, it becomes obvious that the possibilities of the technique for the purification of biological molecules are endless. The sheer number and range of molecules purified by the methodology is vast *(29)*. Although this reference is now rather old, it gives some idea of the scope of the purifications already carried out when the technique was only about 10 years old, and the list largely omits proteins prepared by immunoaffinity methods.

The one area in which affinity methods have yet to make a serious impact is in the large-scale manufacture of therapeutic products. In part this is owing to the propensity for sorbents to leak ligand, which may compromise the product for human use. A few are in commercial use, such as blue-sepharose for albumin production, whereas an immunoaffinity method has been a part of the recombinant human interferon-purification protocol. In the laboratory, however, where often only small amounts of highly purified material are required, affinity techniques offer rapid means of providing such material. In general, if one biological molecule interacts with another, then affinity methods can be devised for the purification of both.

References

1. Lerman, L. S. (1953) A biochemically specific method for enzyme isolation. *Proc. Natl. Acad. Sci. USA* **39,** 232–236.
2. Campbell, D. H., Luescher, E., and Lerman, L. S. (1951) Immunologic adsorbants. I. Isolation of antibody by means of a cellulose-protein antigen. *Proc. Natl. Acad. Sci. USA* **37,** 575–578.
3. Arsenis, C. and Mc Cormack, D. B. (1964) Purification of liver flavokinase by column chromatography on flavin-cellulose compounds. *J. Biol. Chem.* **239,** 3093–3097.
4. Cuatrecasas, P. (1970) Protein purification by afffinity chromatography. *J. Biol. Chem.* **245,** 3059–3065.
5. O'Carra, P., Barry, S., and Griffin, T. (1974) Spacer arms in afffinity chromatography: use of hydrophilic arms to control or eliminate nonbiospecific adsorption effects. *FEBS Lett.* **43,** 169–175.
6. Beer, D. J., Yates, A. M., Randles, S. C., and Jack, G. W. (1995) A comparison of the leakage of a monoclonal antibody from various immunoafffinity chromatography matrices. *Bioseparation* **5,** 241–247.
7. Hermanson, G. T., Mallia, A. K., and Smith, P. K. (1992) *Immobilized Affinity Ligand Techniques.* Academic, San Diego, CA.
8. Johansson, B.-L., Hellberg, U., and Wennberg, Q. (1987) Determination of the leakage from Phenyl-Sepharose CL–4B, Phenyl-Sepharose FF and Phenyl-Superose in bulk and column experiments. *J. Chromatogr.* **403,** 85–98.
9. Hulak, I., Nguyen, C., Girot, P., and Boschetti, E. (1991) Immobilised Cibacron Blue-leachables, support stability and toxicity on cultured cells. *J. Chromatogr.* **22,** 355–362.
10. Weber, D. V. and Bailon, P. (1990) Application of receptor-affinity chromatography to bioaffinity purlfication. *J. Chromatogr.* **510,** 59–69.
11. Jack, G. W. (1994) Immunoafffinity chromatography. *Mol. Biotechnol.* **1,** 59–86.
12. Tsujimoto, M., Adachi, H., Kodama, S., Tsuruoka, N., Yamada, Y., Tanaka, S., Mita, S., and Takatsu, K. (1989) Purification and characterization of recombinant human interleukin 5 expressed in Chinese Hamster Ovary cells. *J. Biochem.* **106,** 23–28.
13. Hayes, C. E. and Goldstein, I. J. (1974) A α-D-galactosyl-binding lectin from *Bandeiraea simplicifolia* seeds: isolation by affinity chromatography and characterization. *J. Biol. Chem.* **249,** 1904–1914.
14 Bjork, L. and Kronvall, G. (1984) Purification and some properties of streptococcal Protein G. a novel IgG-binding reagent. *J. Immunol.* **133,** 969–974.
15. Kenney, A. C. and Chase, H. A. (1987) Automated production scale purification of monoclonal antibodies. *J. Chem. Technol. Biotechnol.* **39,** 173–182.
16. Lowe, C. R. and Pearson, J. C. (1984) Afffinity chromatography on immobilised dyes. *Methods Enzymol.* **104,** 97–113.
17. Scopes, R. K. (1986) Strategies for enzyme isolation using dye-ligand and related adsorbants. *J. Chromatogr.* **376,** 131–140.
18. Bhikhabhai, R., Joelson, T., Unge, T., Strandberg, B., Carlsson, T., and Lovgren, S. (1992) Purification, characterization and crystallization of recombinant HIV–1 reverse transcriptase. *J. Chromatogr.* **604,** 157–170.

19. Gonzales-Gronow, M. , Grennet, H. E., Fuller, G. M., and Pizzo, S. V. (1990) The role of carbohydrate in the function of human plasminogen: a comparison of the protein obtained from molecular cloning and expression in *E. coli* and COS cells. *Biochim. Biophys. Acta* **1039,** 269–276.

20. Dendricks, D., Wang, W., Scharpe, S., Lommaert, M.-P., and Van Sande, M. (1990) Purification and characterization of a new arginine carboxypeptidase in human serum. *Biochim. Biophys. Acta* **1034,** 86–92.

21 Winkler, M. E., Blaber, M., Bennet, G. L., Holmes, W., and Vehar, G. A. (1985) Purification and characterization of recombinant urokinase from *E. coli. Bio/Technology* **3,** 990–1000.

22. Hageman, J. H. and Kuehn, G. D. (1992) Boronic acid matrices for the affinity purlfication of glycoproteins and enzymes, in *Methods in Molecular Biology, vol. 11: Practical Protein Chromatography* (Kenney, A. and Fowell, S., eds.), Humana, Totowa, NJ, pp. 45–71.

23. Dedman, J. R., Kaetzel, M. A., Chan, H. C., Nelson, D. J., and Jamieson, G. A. (1993) Selection of targeted biological mod)fiers from a bacteriophage library of random peptides. The identification of novel calmodulin regulatory peptides. *J. Biol. Chem.* **268,** 23,025–23,030.

24. Guan, K. and Dixon, J. E. (1991) Eukaryotic proteins expressed in *Escherichia coli:* an improved thrombin cleavage and purification of fusion proteins with glutathione S transferase. *Anal. Biochem.* **192,** 262–267.

25. Jarret, H. W. (1993) Affinity chromatography with nucleic acid polymers. *J. Chromatogr.* **618,** 315–339.

26. Issekutz, A. C. (1983) Removal of Gram-negative endotoxin from solutions by affinity chromatography. *J. Immunol. Methods* **61,** 275–281.

27. Matsumae, H., Minobe, S., Kindan, K., Watanabe, T., Sato, T., and Tosa, T. (1990) Specific removal of endotoxin from protcin solutions by immobilised histidine. *Biotechnol. Appl. Biochem.* **12,** 129–140.

28. Hewlett, G. (1990) Apropos aprotinin: a review. *Biotechnology* **8,** 565–568.

29. Wilcheck, M., Miron, T., and Kohn, J. (1984) Affinity chromatography. *Methods Enzymol.* **104,** 3–55.

39

Reversed-Phase HPLC

Bill Neville

1. Introduction

Reversed-phase HPLC (RP-HPLC) is now well established as a technique for isolation, analysis, and structural elucidation of peptides and proteins *(1,2)*. Its use in protein isolation and purification may have reached a peak owing to recent developments in high efficiency ion-exchange and hydrophobic-interaction supports, which are now capable of equivalent levels of resolution to RP-HPLC without concomitant risk of denaturation and loss of biological activity. Nevertheless, there are many applications in which denaturation may be unimportant and high concentrations of organic modifier can be tolerated (e.g., purity analyses, structural studies, and micropreparative purification prior to microsequencing). It is a widely used tool in biotechnology for process monitoring, purity studies, and stability determinations.

1.1. Principle of Separation

Application of a solvent gradient is generally superior to isocratic elution because separation is then achievable within a reasonable time frame and peak broadening of later-eluting peaks is reduced, thus increasing sensitivity. In gradient elution RP-HPLC, proteins are retained essentially according to their hydrophobic character. The retention mechanism can be considered either as adsorption of the solute at the hydrophobic stationary surface or as a partition between the mobile and stationary phases *(3)*.

In the first case, retention is related to total interfacial surface of the RP packing and is expressed by the adsorption coefficient K_A. Retention is based on a hydrophobic association between the solute and the hydrophobic ligands of the surface *(4)*. By increasing solvent strength of the mobile phase, attractive forces are weakened and the solute is eluted. The process can be regarded as being entropically driven and endothermic, i.e., both ΔS and ΔH are positive *(5)*. Through the relation between the solute capacity factor k' and the mobile- and stationary-phase properties, solvophobic theory permits prediction of the effect of the organic modifier of the mobile phase, the ionic strength, the ion-pairing reagent, the type of ligand of the RP packing, and other variables on the chromatographic retention of proteins.

The second case assumes a partitioning of the solute between the mobile and stationary phases, the latter being regarded as a hydrophobic-bulk phase. This system resembles

From: *Molecular Biomethods Handbook*
Edited by: R. Rapley and J. M. Walker © Humana Press Inc., Totowa, NJ

an *n*-octanol/water two-phase system where the selectivity is expressed by the partition coefficient P of the solute. Both retention principles have their pros and cons *(3)*. In essence, the effective process is largely dependent on the molecular organization of the bonded *n*-alkyl chains of the RP packing in the solvated state and the size and conformation of the solute.

Several methods have been developed for correlating peptide structures with RP-HPLC retention. One method sums empirically derived retention coefficients, representing the hydrophobic contribution of each amino-acid residue *(6)*. The assumption behind this approach is that each amino-acid residue contacts the adsorbent surface and that the total hydrophobicity of the solute determines RP-HPLC retention. This approach works well for peptides of 20 amino acids or less *(7)*, but generally is not accurate for predicting retention of larger proteins. This is caused by the peculiarities, especially of polar basic moieties, leading to either less retention or irreversible adsorption by residual-matrix silanols. Studies with peptides containing amphiphilic helices indicate that secondary structure plays a large role in determining the surface of a polypeptide, which is exposed to the hydrophobic adsorbent *(8)*. Other studies have shown that proteins are less sensitive to changes in the HPLC support or bonded phase than are corresponding small molecules *(9)*. These results imply that protein retention in RP-HPLC is governed mainly by the protein surface and not by the support surface chemistry. Although some proteins show a loss of biological activity during RP-HPLC, others retain full biological activity. Thus, the hydrophobic forces necessary for RP binding compete with those required to maintain the protein's secondary and tertiary structures. Through steric, hydrophobic, and ionic constraints, these structural features define a chromatographic surface or "footprint" of the protein, which in turn defines its RP binding. This concept of multipoint attachment of a protein to adsorbent surface is consistent with a relatively large chromatographic footprint for a protein compared with that of a small molecule.

1.2. RP-HPLC Columns

Optimization of chromatographic separation is determined by an appropriate choice of the column matrix, the mobile phase, and the temperature of separation. RP-HPLC has rapidly become the most widely used tool for the separation, purification, and analysis of peptides. For proteins, RP-HPLC has suffered from problems associated with denaturation, including loss of activity, poor recoveries, wide and misshapen peaks, and ghost peaks *(10,11)*. Some of these problems may be column related *(12)* and insurmountable, although others can be overcome by selective optimization of extra-column variables, such as sample pretreatment and mobile-phase and hardware considerations *(13)*. As with small-molecule HPLC, there are a large number of RP-HPLC columns for use with peptides and proteins. In trying to select the best column for a new application or for broad applicability, one should consider such column variables as support, bonded phase, pore size, particle size, and column dimensions. These factors are dealt with in the following.

1.2.1. Packing Support

Most commercial RP-HPLC columns for peptides and proteins are silica-based, as a consequence of its use for the separation of small molecules over many years. Silica offers good mechanical stability and allows a wide range of selectivities by virtue of the bonding of various phases. Although it is well known that silica-based columns are not stable at basic pH, recent reports suggest that in the acidic mobile-phases typically used, the bonded phase may also be slowly dissolved from the base silica *(14,15)*. Degradation can thus affect not only reproducibility, stability, and lifetime of a column, but also recovery and, in addition, can modify selectivity as the silica surface becomes uncovered. An added problem could be contamination of recovered products with silica and bonded-phase material. Because of the limitations associated with silica-based columns, polymer-based columns have gained increased popularity for peptide and protein separations by RP-HPLC. Polymer-based columns, such as divinylbenzene-crosslinked polystyrene, are usually stable over the pH range 1.0–14.0, making them more widely applicable with the added advantage that they are easier to clean up after use. Although most polymer-based columns are still inferior to their silica-based counterparts in terms of mechanical strength, selectivity, and efficiency, newer materials, such as PLRP-S (Polymer Laboratories, Church Stretton, Shropshire, UK), show superior performance in complex protein separations *(13)*. Polymer columns unfortunately suffer from a relatively low pressure resistance and, in general, pressure should be kept below 2000 psi. Furthermore, the physicochemical properties are governed by a solvent-dependent swelling and shrinking of the organic matrix, which is associated with concomitant changes in pore diameters, leading to changes in mass-transfer characteristics of the column. This effect may be encountered during gradient elution, starting with an aqueous phase (by which only little or no wetting of the column matrix takes place) and termination with a pure organic solvent (e.g., methanol, acetonitrile, 2-propanol). Pore diameter will continuously change and often may be the cause of poor chromatographic performance. In contrast, silica gel as the starting material for subsequent alkyl silylation offers the advantage that a great number of different alkyl silica substituents can be bound to the silica surface. Recent studies have shown that the use of polymer-based columns at high pH (>8.0) can offer unique selectivities and thus complement separations with classical acidic mobile phases *(16,17)*. These advantages, coupled with superior stability, longer column life, and better reproducibility, make polymer-based columns a good first choice for complex protein separations.

1.2.2. Bonded Phases

A wide variety of bonded phases are available for the separation of peptides and proteins by RP-HPLC, but the most common are *n*-alkyl-bonded phases, namely C_4, C_8, and C_{18}. In general, C_4 and C_8 phases are preferable for more hydrophobic samples and the C_{18} phase for hydrophilic samples. In general, significant differences may be observed in the separations achieved on nominally the same column but from different sources *(12)*. For proteins, a C_8 column is generally a good compromise. If the proteins of interest are too strongly retained, a C_4 column could be substituted.

1.2.3. Particle Size

As with small molecules, theory predicts that column performance should increase with decreasing particle size. This is generally the case for most commercially available RP-HPLC columns used for protein separations. Columns with particle sizes of 5–10 μm show little difference in performance. Columns with particle sizes of 2–3 μm are now available and can give much greater resolution. Any gain in column efficiency can be outweighed quickly by increased column backpressure, susceptibility to plugging, and shorter column lifetime.

1.2.4. Pore Size

The standard pore size for separation of small molecules is 80 Å but for the separation of proteins a pore size of 300 Å has become accepted as the norm. For proteins >100 kDa, pore sizes of >1000 Å may be even better *(12)*. For complex samples, if all other column variables are equal, it would be better to use a larger pore size to ensure that restricted diffusion or exclusion from the pores is not encountered by very large proteins.

1.2.5. Column Dimensions

The two factors to consider when selecting a column size are the efficiency and sample-loading capacity. In the case of proteins, column length contributes little to efficiency. In addition, longer columns may have an adverse effect on protein recovery. When considering gradient rather than isocratic elution, resolution of a mixture of proteins depends directly on the particle diameter of the packing, but less on the column length and flow rate for gradients of equal time. However, peak height, which will determine the detection limit, depends inversely on the particle diameter, flow rate and, to a lesser extent, gradient time. An increase in column length would have little effect on resolution because by the time a protein band reaches the extra column length, it would be overtaken by the gradient and the organic modifier concentration would ensure that the protein spent little time in the adsorbed state. The effect of additional column length is thus largely passive and may even lead to increased band broadening. Long columns might be useful in isocratic protein separations but if extra capacity were needed, a short fat column might be a better choice than a long thin column. Having selected the column length, choice of column diameter should be based on required sample capacity. Columns of 1-mm internal diameter (id) (microbore) to 2.1-mm id (narrowbore) are best suited to submicrogram levels of material, thus minimizing sample loss and also increasing sensitivity of detection. For analytical applications in the microgram to low milligram range, columns of 4.6-mm id (analytical) are best.

1.3. Mobile Phase

The mobile-phase composition is the most readily changed variable in an RP-HPLC separation. In normal usage, the mobile phase consists of a mixture of water, a miscible organic solvent, and dissolved buffers and salts. A buffer, such as a low concentration of a strong acid or salt, is essential for chromatography. If a protein is adsorbed in the absence of a salt or acid, no increase in the proportion of organic component will elute it. In addition, other components, either solvents or solutes, may be added in order to affect the separation. This may be to accelerate or delay elution, improve peak shape, or adjust the elution position of some components with respect to others, thus affecting selectivity. Other important variables are the pH of the mobile phase and the concentration and type of ions present. The ionic strength is a factor in limiting "mixed-mode" retention involving silica hydroxyls, but the choice of ions can also affect the solubility and stability of the protein in the mobile phase and, through ion pairing, the distribution of the protein between stationary and mobile phases. The net effect depends on the combination of column packing and mobile phase.

Numerous studies have examined the effect of mobile phase on the retention and selectivity of proteins in RP-HPLC. Systematic studies by several researchers over the past 10 years have resulted in the publication of a number of fairly standardized sets of elution conditions. The behavior can be altered by the variation of a number of mobile-phase parameters. Altering the nature of the organic component of the mobile phase can alter column behavior, but not usually in a predictable manner. The most successful way to alter column behavior is to manipulate the ionic component of the the solvent system. By altering the nature and strength of the ion-pairing reagent and the pH of the mobile-phase, it is possible to exploit both the acidic and basic character of proteins in a systematic and predictable way. Without prior knowledge of the primary structure of a peptide or protein, one can be certain of some hydrophobic character by virtue of the content of the lipophilic amino acids leucine, isoleucine, methionine, phenylalanine, tyrosine, tryptophan, and valine. It will also have hydrophilic character, largely because of the presence of aspartic acid, glutamic acid, arginine, and lysine. The ratio between hydrophobic and hydrophilic amino acids determines the initial behavior observed for any protein in the most common RP-HPLC solvent system in use today, namely, aqueous acetonitrile containing 0.1% trifluoroacetic acid (TFA). To some extent hydrophilic character of solutes is suppressed at pH 2.0. All carboxylic acid groups are protonated under these conditions and their contribution to hydrophilic character is reduced. This in turn will improve peak shape. Basic amino acids (e.g., arginine, lysine, and histidine) will remain as cations, which in turn will lead to a marked reduction in silanophilic solute–matrix interactions.

1.3.1. Organic Component of Mobile Phase

The most popular organic solvents are acetonitrile, 1-propanol, and 2-propanol. Propanols are much more viscous than acetonitrile, giving higher column back-pressures, but these are not significant when low flow rates and short columns are used. The acetonitrile concentration needed to elute a protein is considerably higher than the equivalent 2-propanol concentration, perhaps half as high again, and thus proteins are likely to be seriously denatured. There is also evidence that acetonitrile is intrinsically a more powerful denaturant than the alcohols. Recoveries are often higher in propanol-containing mobile phases, especially for the more "difficult" model proteins, such as ovalbumin. Acetonitrile and 2-propanol can be frozen in dry-ice/alcohol mixtures and lyophilized. All solvents should be of the highest quality available. Water needs to be free of UV-absorbing organic impurities, which can be adsorbed onto the column during equilibration and the early part of a gradient, and which may be eluted later as "ghost" peaks. Water should be freshly distilled and deionized and stored out of contact with plastic, thus avoiding dissolution of plasticizers.

1.3.2. Minor Mobile-Phase Additives

The most popular minor components of the mobile-phase are the perfluoroalkanoic acids (e.g., TFA and heptafluorobutyric acid). These appear to solubilize proteins in organic solvents. TFA is used in preference to other fully dissociated acids because it is completely volatile and will not corrode stainless steel. TFA also acts as a weak hydrophobic ion-pairing reagent. Most RP columns employed for RP-HPLC are silica-based and apart from the suppression of silanol ionization under acidic conditions (thereby suppressing undesirable ionic interactions with basic residues), silica-based columns are more stable at low pH. As mentioned previously (**Subheading 1.3.**), an acid pH is preferred, although many polypeptides will lose their tertiary structure at pH 2.0–3.0. This may not necessarily be fatal for resultant activity, but if the polypeptide consists of subunits stabilized by noncovalent forces, the individual chains will almost certainly be separated during the chromatography.

Other popular additives are pyridinium acetate buffers and buffers based on phosphoric acid, particularly triethylammonium phosphate (TEAP). TEAP is useful in circumstances where tailing is suspected because of nonspecific interactions with surface silanols. Proteins are eluted at a higher percentage of organic component with an equivalent concentration of phosphoric acid. Sulfate, phosphate, perchlorate, and chloride are all transparent in the far UV, whereas acetate, formate, and fluoroalkanoic acids must be used at concentrations <20 mM if wavelengths below 220 nm are to be monitored. Pyridine-containing eluents cannot be monitored by UV. Ion-pairing reagents, such as heptane and octane sulfonic acids, sodium dodecyl sulfate, or alkyl ammonium salts, may be added to the mobile phase to selectively increase the retention of proteins carrying larger charges of opposite sign. Other additives might be needed depending on the nature of the protein being analyzed. Metal ions, such as calcium, may be included to enhance the stability of a protein and low concentrations of nonionic detergents may improve the behavior of hydrophobic membrane proteins. Finally, guanidinium hydrochloride or urea, at moderate concentrations, may be added to elute very hydrophobic membrane proteins.

1.3.3. Gradient Elution

The theory of gradient elution is now well-established and a good understanding exists concerning the effects of gradient steepness on separation. Geng and Regnier *(18)* developed the stoichiometry factor, Z, which represents the number of solvent molecules displaced during the binding of the protein to the column packing. Thus, the value of Z is a measure of the size of the effective chromatographic footprint of that protein.

For RP-HPLC, **Eq. 1** has been derived, which shows that Z is proportional to the term S:

$$Z = 2.3\phi S \tag{1}$$

where ϕ is the volume fraction of the organic in the mobile-phase *(19)*. Under defined gradient conditions, every protein will have an S value described by **Eq. 2**:

$$\log k' = \log k_w - S\phi \tag{2}$$

where k' is the retention of the solute (capacity factor) and k_w is the value of k' when water is the mobile phase ($\phi = 0$). Values of k_w and S are characteristic of each solute in a sample. Although empirically derived, **Eq. 2** is a good approximation of RP-HPLC retention in the gradient mode. For a series of related proteins of nearly identical composition, the hydrophobic contribution to RP retention will be similar and thus observed changes in Z (and S) should reflect changes in the structure of the protein that contacts the adsorbent surface. Examination of **Eq. 2** reveals why small changes in the organic composition of RP eluents result in large changes in protein retention as opposed to small molecule separations. Proteins will have a much larger chromatographic footprint (Z and S) than small molecules. Thus, changes in ϕ (% organic) will be amplified by large values of S, resulting in even larger changes in retention (log k'). Many examples of the applicability of **Eq. 2** for peptide or protein samples have been reported. Values of S and k_w for each solute can be obtained from two experimental gradient separations of the sample. Retention times in gradient elution then can be predicted as a function of gradient conditions. Two components that elute adjacent to one another in a chromatogram will often show significant changes in band spacing when isocratic solvent strength or gradient steepness is varied. The resolution of such band pairs can usually be accomplished when values of S for the two components differ by 5% or more.

1.4. Temperature

Protein samples will often contain many individual components. A complete separation of such samples poses a real challenge because statistical considerations suggest that one or more peak pairs usually will be resolved poorly. To overcome this dilemma, a systematic variation of separation selectivity would be required. This approach has been widely used for the separation of typical small-molecule samples. In the case of peptide and protein samples, however, the control of selectivity has received little attention. When a change in selectivity is desired, the usual approach is a change of column or mobile phase. The use of elevated temperatures for RP-HPLC of protein samples has been advocated primarily as a means of increasing column efficiency or shortening run time. For samples of this type, a few studies have shown that a change in column temperature can also affect separation selectivity. A combination of low pH and higher temperature can result in a very short life for commonly used alkyl-silica columns, which in turn limits the application of temperature optimization. This problem has recently been overcome by the development of "sterically protected" RP packings that are extremely stable at low pH (<2.0) and high temperature (>90°C). The combined use of temperature and gradient steepness would provide an efficient procedure for the control of peak spacing and optimization of separation. At the same time, this approach to selectivity control is more convenient than alternatives, such as change of column or mobile phases, because temperature and gradient steepness can be varied via the HPLC system controller.

2. Practical Requirements

To illustrate the general points in **Subheading 1.**, the separation of a mixture of standard proteins is described.

1. Equipment: Analysis of proteins by RP-HPLC is suited to most commercially available equipment. Systems require a column packed with a suitable matrix giving adequate retention and resolution, a detector to monitor eluted proteins, a chart recorder or data system to view detector response, and a fraction collector for recovery of eluted proteins, if further analysis is intended. If high sensitivity is a requirement, then narrowbore or microbore chromatography should be considered and specialized equipment may be necessary. In order to obtain good performance from microcolumns, it is very important to have an HPLC system with very low dead volume so that extra-column peak broadening does not destroy the resolution achieved by the column.

   Consideration should be given to the minimization of dead volume at all points in the chromatographic system. If there is a large dead volume between the point at which the gradient is mixed and the injector, then lag will be introduced into the gradient. Also, the gradient system must be capable of delivering the gradient accurately and reproducibly at low flow rates. A suitable mixing device (volume <200 µL) may need to be added to the system to enable a reproducible gradient to be delivered at very low flow rates. Proteins are usually monitored by their absorbance in the UV either at 280 nm, which is suitable for proteins with aromatic amino acids, or at 206 nm, which is characteristic of the peptide bond. During a gradient there is usually an associated change in background absorption. This may be caused directly by the increase in the concentration of organic component or by the effect of this concentration change on other mobile-phase components. In addition, refractive index changes may be detected in the flowcell. Usually not much can be done about flowcell design, but careful choice of wavelength and pH can minimize the change in absorption spectrum of mobile-phase components. For TFA and acetonitrile, the optimum wavelength is 214 nm. Alternatively, the concentration of other components in the mobile phase can be adjusted to compensate or a small amount of a UV absorbing solute can be added to one of the eluent reservoirs. For example, in a TFA/water/acetonitrile system monitored at 214 nm, the rise in background can be balanced by using 0.1% (v/v) aqueous TFA as eluent A and 70% aqueous acetonitrile, containing 0.085% (v/v) TFA as eluent B. Most modern chromatographic systems now have computers to control operating parameters and to collect and store data. Microfraction collectors are also available with low internal-diameter tubing to reduce the volume to a minimum between detector and collection tube in order to more nearly reflect the real time elution situation.

2. Column: A wide variety of suitable C_8 or C_{18} columns are commercially available for protein analysis. The price of modern columns is such that little advantage is gained by self-packing a column. It is generally accepted that Vydac columns are the "industry standard," against which other manufacturers' columns are compared. Packing columns is based on know-how. Packing being commercially important, know-how is not readily imparted. Packing LC columns is still not fully understood. Commercial columns are usually supplied with a test and a guarantee to replace the column if performance does not match that specified. Although packed columns cannot be expected to last indefinitely, a usual lifetime of several months and several hundred injections is a reasonable objective. The column must be able to withstand moderate pressures (250 bar) and be resistant to the mobile phase (pHs typically 1.0–8.0 for silica columns). A guard column (containing the same or a very similar packing material) should be used as a matter of course to extend analytical-column lifetime.

3. Injector: For most purposes a conventional HPLC manual-injection valve (e.g., Valco or Rheodyne) with a fixed volume loop (e.g., 10 µL) is adequate. For microbore analysis, injectors with small internal loops may be better. Many modern chromatographs are now fitted with auto-samplers whereby injection volume can be varied (e.g., 0.1–200 µL) via operating computer.

4. Eluents: All eluents should be filtered through a 0.2-µm filter and degassed under vacuum before use.
   a. Eluent A: 0.1% TFA (v/v). To 1 L of water add 1 mL of TFA.
   b. Eluent B: acetonitrile:water (70:30) + 0.085% TFA (v/v). To 700 mL of acetonitrile add 300 mL of water and 0.85 mL of TFA.

   Eluents should be degassed by a slow stream of helium while on the chromatograph, if at all possible, to counteract the effect of dissolved air. If a ternary system is available, acetonitrile can be programmed into a gradient to fully elute hydrophobic materials that otherwise might be retained by the column packing.

5. Detector: By far the most common detector used in chromatography is the variable wavelength UV detector, operated at 210–220 or 280 nm. Most proteins have an absorption

maximum at about 280 nm because of the presence of aromatic amino acids, which falls to a minimum at 254 nm, a popular wavelength for fixed-wavelength HPLC detectors. Highest sensitivities are obtained by monitoring the strong absorption bands, which peak below 220 nm owing to the "peptide bond" itself.

Proteins are usually eluted as fairly broad peaks (*see* **Fig. 1**, a protein test mixture of insulin, cytochrome *c,* lactalbumin, carbonic anhydrase, and ovalbumin) and thus place fairly modest demands on the dimensions of a normal flowcell (typically 8 µL with pathlength 1 mm). An alternative to UV detection is to monitor intrinsic fluorescence. The usual excitation wavelength is 280 nm with emission monitored at either 320 (for tyrosine) or 340 nm (for tryptophan). The main advantage is a several-fold increase in sensitivity and an improvement in selectivity of detection, particularly in distinguishing protein peaks from "ghost" peaks originating from eluent impurities.

6. Protein sample: A mixture of test proteins, namely insulin (from bovine pancreas), cytochrome *c* (from horse heart), α-lactalbumin (from bovine milk), carbonic anhydrase (from bovine erythrocytes), and ovalbumin (from hen egg whites) can be used as a column test. Two of the proteins, insulin and cytochrome *c,* are difficult to resolve. A partial to complete separation should be achievable when a column with a high enough number of theoretical plates is used. Of the five standard proteins, ovalbumin is not only the most hydrophobic, but also the most difficult to recover. The utility of the column from a recovery perspective will be illustrated by the size of this peak.

3. Outline of the Practical Procedure

1. Set up appropriate pump, temperature, injection, detector, and integration methods, as required by chromatograph being used.
2. Install appropriate column (C_8 or C_{18} bonded silica, 300 Å, 5 µm, 10–25-cm length, 2.1 or 4.6 mm id).
3. Equilibrate the column by pumping eluent A at 200 µL/min (narrowbore, 2.1 mm) or 1 mL/min (normalbore, 4.6 mm) for several column volumes. Monitor detector baseline (214 nm) until flat and then zero the detector. Inject 5 µL of HPLC quality water and run a blank gradient to identify any system peaks and to balance baseline (± TFA), if necessary.
4. Dissolve protein mixture in HPLC quality water at a concentration of approx 2 µg/µL to produce a stock solution. The stock solution should be diluted further with eluent A (1:1) to produce a working solution of 1 µg/µL of each protein.
5. Apply the sample (5 µL) at initial gradient conditions. Proper gradient conditions are critical in optimizing separations of proteins by RP-HPLC. The gradient time, range, and shape are all important and must be optimized for a given sample. A typical starting gradient would be a 2%/min linear increase in acetonitrile over perhaps 70 min. A washing step of 100% eluent B then should be included before returning to initial conditions, allowing sufficient equilibration before the next injection.

   Column re-equilibration time is an extremely important variable. The time must be sufficient to allow the initial mobile phase to equilibrate in the pores of the column packing (at least 3 column volumes) and should be exactly repeatable if precise retention times are required. A linear gradient of the above type (0.5–5%/min increase in acetonitrile) should allow resolution of the proteins in the test mixture. For increased resolution of certain of the proteins, a number of steps may have to be included in the gradient profile.
6. Start gradient to elute proteins.

A typical elution profile is shown in **Fig. 1**.

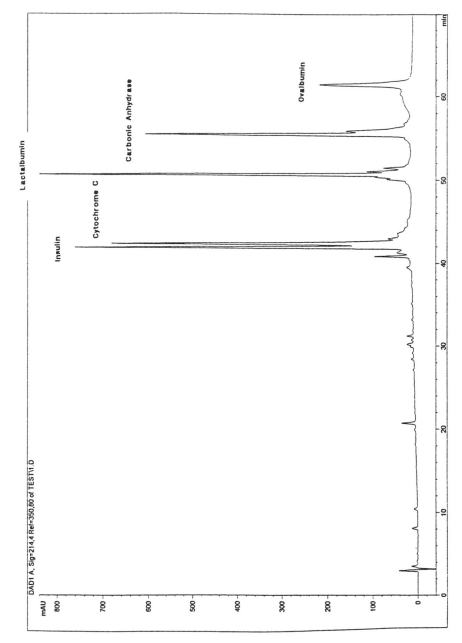

Fig. 1. RP-HPLC of a protein test mixture. The conditions used for chromatography are described in **Subheading 3.**

References

1. Hancock, W. S. and Harding, D. R. K. (1984) Review of separation conditions, in *CRC Handbook of HPLC for the Separation of Amino Acids, Peptides and Proteins,* vol. 2 (Hancock, W. S., ed.), CRC, Boca Raton, FL, pp. 303–312.
2. Regnier, F. E. (1987) Peptide mapping. *LC/GC* **5(5)**, 392–395.
3. Dill, K. A. (1980) The mechanism of solute retention in reversed-phase liquid chromatography. *J. Phys. Chem.* **91**, 1987–1992.
4. Melander, W. and Horvath, C. S. (1980) Reversed-phase chromatography, in *HPLC— Advances and Perspectives,* vol. 2 (Horvath, C. S., ed.), Academic, New York, p. 114.
5. Hearn, M. T. W. (1980) HPLC of peptides, in *HPLC—Advances and Perspectives,* vol. 3 (Horvath, C. S., ed.), Academic, New York, p. 99.
6. Guo, D., Mant, C. T., Parker, J. M. R., and Hodges, R. S. (1986) Prediction of peptide retention times in reversed-phase HPLC: I. Determination of retention coefficients of amino acid residues of model synthetic peptides. *J. Chromatogr.* **359**, 499–517.
7. Meek, J. L. and Rossetti, Z. L. (1981) Factors affecting retention and resolution of peptides in HPLC. *J. Chromatogr.* **211**, 15–28.
8. Heinitz, M. L., Flanigan, E., Orlowski, R. C., and Regnier, F. E. (1988) Correlation of calcitonin structure with chromatographic retention in HPLC. *J. Chromatogr.* **443**, 229–245.
9. O'Hare, M. J., Capp, M. W., Nice, E. C., Cooke, N. H., and Archer, B. G. (1982) Factors influencing chromatography of proteins in short alkylsilane-bonded large pore-size silicas. *Anal. Biochem.* **126**, 17–128.
10. Pearson, J. D., Lin, N. T., and Regnier, F. E. (1982) Separation of proteins, in *HPLC of Peptides and Proteins* (Hearn, M. T. W., Wehr, C. T., and Regnier, F. E., eds.), Academic, New York, p. 81.
11. Hearn, M. T. W. (1986) Reversed-phase chromatography, in *HPLC—Advances and Perspectives,* vol. 3 (Horvath, C. S., ed.), Academic, New York, pp. 87–91.
12. Burton, W. G., Nugent, K. D., Slattery, T. K., and Summers, B. R. (1988) Separation of proteins by reversed-phase HPLC: I. Optimising the column. *J. Chromatogr.* **443**, 363–379.
13. Nugent, K. D., Burton, W. G., Slattery, T. K., and Johnson, B. F. (1988) Separation of proteins by reversed-phase HPLC: II. Optimising sample pre-treatment and mobile phase conditions. *J. Chromatogr.* **443**, 381–397.
14. Glajch, J. L., Kirkland, J. J., and Kohler, J. (1987) Effect of column degradation on the reversed-phase HPLC of peptides and proteins. *J. Chromatogr.* **384**, 81–90.
15. Sagliano, J., Floyd, T. R., Hartwick, R. A., Dibussolo, J. M., and Miller, N. T. (1988) Studies on the stabilisation of reversed-phases for liquid chromatography. *J. Chromatogr.* **443**, 155–172.
16. De Vos, F. L., Robertson, D. M., and Hearn, M. T. W. (1987) Effect of mass loadability, protein concentration and *n*-alkyl chain length on the reversed-phase HPLC behaviour of bovine serum albumin and bovine follicular fluid inhibin. *J. Chromatogr.* **392**, 17–32.
17. Guo, D., Mant, C. T., and Hodges, R. S. (1987) Effects of ion-pairing reagents on the prediction of peptide retention in reversed phase HPLC. *J. Chromatogr.* **386**, 205.
18. Geng, X. and Regnier, F. E. (1984) Retention model for proteins in reversed-phase liquid chromatography. *J. Chromatogr.* **296**, 15–30.
19. Kunitani, M., Johnson, D., and Snyder, L. R. (1986) Model of protein conformation in the reversed-phase separation of interleukin-2 muteins. *J. Chromatogr.* **371**, 313–333.

40

Glycoprotein Analysis

Terry D. Butters

1. Introduction

In eukaryotic cells, one of the most important posttranslational modifications of proteins is the covalent addition of carbohydrate. We can consider two major types of modification to amino-acid residues: *N*-glycosylation of asparagine amine groups and *O*-glycosylation of serine or threonine hydroxyl groups *(1)*. N-linked oligosaccharides can be divided into three major classes; the complex type containing *N*-acetylglucosamine, mannose, galactose, fucose, and sialic acid; the oligomannose type containing *N*-acetylglucosamine and mannose only; and the hybrid type that has features common to both complex and oligomannose chains (**Fig. 1**). All of these structures are synthesized by a common pathway that begins in the endoplasmic reticulum (ER) with the assembly of a lipid-linked donor molecule. The preformed oligosaccharide is transferred to protein cotranslationally in the lumen of the endoplasmic reticulum and by a series of glycosidase (α-glucosidase and α-mannosidase)-trimming reactions is modified as the protein progresses through the ER and Golgi apparatus *(2)*. The diversity of N-linked oligosaccharide structure is dictated by the accessibility of these partially processed chains to Golgi resident glycosyltransferases, a group of enzymes able to add monosaccharides to oligosaccharides directly from nucleotide sugar donors. Glycosyltransferases are specific for nucleotide sugar donor, anomericity, glycosidic linkage between sugars, and acceptor substrates. Consequently, there are a number of different transferases and each cell, tissue, and species has a unique complement of enzymes that control oligosaccharide biosynthesis *(3)*. O-linked oligosaccharides contain similar residues to *N*-glycans, but their synthesis has no requirement for *en bloc* addition of carbohydrate to the polypeptide chain. *O*-glycosylation proceeds by glycosyltransferase-catalyzed, stepwise addition of monosaccharides to generate, as in the case of mucin glycoproteins, a diverse number of branched oligosaccharides *(4,5)* (*see* **Fig. 2**).

The degree of importance of glycosylation is reflected in the amount of energy and genetic information used by cells in providing the machinery that coordinates the correct assembly of protein oligosaccharides. Because many, if not all, of these processes have been conserved throughout the evolution of eukaryotes, there are clearly some vital roles played by the oligosaccharide moieties. These roles are often indirect; in

From: *Molecular Biomethods Handbook*
Edited by: R. Rapley and J. M. Walker © Humana Press Inc., Totowa, NJ

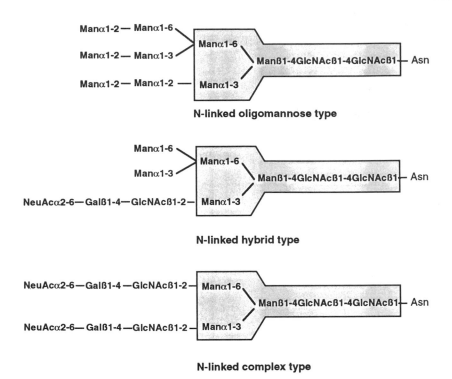

N-linked oligomannose type

N-linked hybrid type

N-linked complex type

Fig. 1. The structure of N-linked oligosaccharides. A core unit (shaded) that is found in all three types is derived from a common lipid donor in the biosynthetic pathway. *See* **ref. 2.**

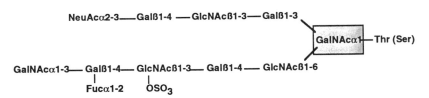

Fig. 2. O-linked oligosaccharide structure. A core unit *N*-acetylgalactosamine residue (shaded) is common to all O-linked oligosaccharides. *See* **ref. 5.**

assisting the folding machinery in the endoplasmic reticulum to ensure the secretion of conformationally correct proteins *(6)*, or in stabilizing the protein to heat or protease digestion. Functions of the oligosaccharide that have a more direct role include carbohydrate–protein interactions, where the specificity is dictated by the expression of an appropriate sequence of carbohydrate residues. For a more thorough review of oligosaccharide functions, *see* Varki *(7)*.

A single glycoprotein may be glycosylated with both *N-* and *O*-glycans and several attachment sites may be present in the polypeptide chain. The heterogeneity introduced during oligosaccharide biosynthesis creates a mixture of glycans at the same attachment site on the polypeptide and presents considerable difficulties in analyzing structure. To obtain a unique "fingerprint" of a single polypeptide species, one therefore needs structural information regarding the oligosaccharide at each attachment point.

Attempts to study the sequence of both N- and O-linked oligosaccharides found on proteins must as a consequence address this enormous diversity of structure *(8)*. This information can then be used to manipulate protein oligosaccharides as a means to understand function. For the biologist/biochemist, two questions are posed when proteins from different experimental systems are to be analyzed. The first is to ask if the protein is glycosylated. Second, if it is a glycoprotein, what is the structural identity, the carbohydrate sequence of the oligosaccharide? An additional question then presents itself: How does the oligosaccharide structure influence the biochemical function of the protein? In this chapter, a review of some of the current and traditional methodologies used to answer these questions will be presented.

2. Detection of Protein-Linked Oligosaccharide After Sodium Dodecyl Sulfate-Polyacrylamide Gel Electrophoresis (SDS-PAGE)

In many biological systems, the target protein may only be available in such small amounts that it precludes gross chemical analysis. However, the separation afforded by SDS-PAGE can be used in conjunction with staining methods that detect the presence of carbohydrate. One of the most simple chemical-detection methods is the periodate oxidation of adjacent hydroxyl groups of monosaccharides to generate aldehydes. These groups are detected *in situ* by Schiff's reagent or on Western blots after reaction with more sensitive reporter molecules such as digoxigenin–hydrazide or biotin-hydrazide *(9)*. Subsequent reaction of these conjugates with antibody or streptavidin-labeled enzymes permits colorimetric detection of glycoproteins at the nanogram (fmol) level.

The contribution of carbohydrate to the mass of the polypeptide chain can be assessed after deglycosylation by chemical means using trifluoromethanesulfonic acid. This procedure efficiently cleaves N- and O-linked oligosaccharide chains, leaving the protein reasonably intact *(10)*. Selective cleavage of N-linked oligosaccharides is provided by endoglycosidase digestion. Endoglycosidases are able to release, nondestructively, most but not all glycoprotein oligosaccharides. The endoglycosidase isolated from *Flavobacerium menigosepticum*, peptide-N^4-(N-acetyl-β-glucosaminyl)asparagine amidase (PNGase), cleaves between the chitobiose core GlcNAc residue and asparagine amino acid of oligomannose and complex type *N*-glycans, but is unable to hydrolyze oligosaccharides from plant proteins substituted with core α-1,3-linked fucose *(11)*. Endo-β-*N*-acetylglucosaminidase from *Streptomyces spp.* cleaves the chitobiose core between GlcNAc residues of oligomannose type *N*-glycans but will not hydrolyze small (<3 mannose units) glycans, especially if these are core fucosylated *(12)*. Estimates of the polypeptide mass by SDS-PAGE before and after deglycosylation can be used to determine the number of N-linked oligosaccharide chains.

With this type of information, proteins can be further probed for particular types of oligosaccharide or carbohydrate sequences using a number of lectins. Lectins are plant and animal proteins that are able to bind sugars specifically either to terminal residues or as part of an extended sequence *(13)*. Using enzymes coupled with lectins, sensitive visualization of oligosaccharide chains of glycoproteins after electrophoretic transfer to Western blots (either nitrocellulose or polyvinylidinefluoride [PDVF] membranes) can identify the type and even portions of the glycan sequence *(14)*. This information aids protein purification or oligosaccharide separation using lectin-affinity chromatography *(15)*. Used in conjunction with specific glycosidase digestion, lectin blot analysis has identified the presence of hybrid-type carbohydrate chains on human monoclonal antibody (MAb) *(16)*. One important consideration when using either chemical or lectin staining methods is the specificity of the reaction. The inclusion of appropriate controls and the additional use of proteins of known glycosylation type can help confirm identification and reduce errors in interpretation owing to erroneous signals *(17)*.

Direct carbohydrate analysis of glycoproteins blotted onto PDVF membranes has been successfully applied to the examination of N-linked oligosaccharides of plant and animal origin. Acid hydrolysis of glycoprotein bands (100 µg protein) and monosaccharide compositional analysis by high-performance liquid chromatography (HPLC) was reproducible and agreed with analyses performed without SDS-PAGE and electroblotting onto PDVF membranes *(18)*. A more sophisticated analysis, using high-pH anion-exchange chromatography (HPAEC) coupled with pulsed-amperometric detection (PAD), allows the unequivocal determination of the presence of carbohydrate, a complete monosaccharide composition, and oligosaccharide mapping to separate size, charge, and isomerity of 10–50 µg electroblotted glycoprotein bands *(19)*.

3. Analysis of Oligosaccharides Isolated from Proteins

3.1. Glycan Release and Labeling

Further quantitation of the types of monosaccharide linkage and number of oligosaccharide structures requires release of the oligosaccharide. Chemical methods, such as hydrazinolysis, involve cleavage of the glycosylamine linkage with anhydrous hydrazine. Glycosidic bonds between monosaccharides are unaffected and the oligosaccharide (N- and O-linked) is released intact *(20)*. Following re-*N*-acetylation, the terminal *N*-acetylglucosamine residue can be labeled by reductive methods using either borotritide to introduce a radioactive group, or fluorescent molecules that provide a more sensitive reporter group, to aid detection. Enzymatic release can be accomplished by endoglycosidases, after first giving some consideration to the potential lack of hydrolysis noted with certain oligosaccharide types *(see* **Subheading 2.***)*. By contrast to hydrazine, endoglycosidase action does not destroy the protein, an important advantage when the contribution of oligosaccharide to biological activity is assessed.

O-linked glycans can be selectively released using anhydrous hydrazine at reduced temperature *(21)*, or by β-elimination in the presence of sodium hydroxide. These reactions must be followed by a reduction step (using $NaBH_4$) to stabilize the glycan to peeling reactions. O-links can also be removed under nonreducing conditions using triethylamine in aqueous hydrazine *(22)*. Endoglycosidase-catalyzed release has been sparingly applied, mostly owing to the lack of well characterized enzymes that allow the release of all O-linked glycans *(23)*.

3.2. Fractionation of Oligosaccharides

The physical techniques described in **Subheading 4.** require that some preliminary fraction of oligosaccharides has been achieved, possibly using one or a combination of methods. These include size separation by low-pressure chromatography using Bio-Gel P-4 *(24)* that can provide some assignment based on the unique hydrodynamic volume of certain uncharged oligosaccharide sequences *(25)*. The application of HPLC technology to carbohydrate analysis provides a greater increase in the resolution of oligosaccharide mixtures. Chromatography times are far less than low-pressure separations and when combined with fluorescent derivatization has also increased sensitivity *(26)*. HPAEC is able to separate isomeric structures nondestructively, despite the use of basic pH eluants *(27,28)*, and electrochemical detection using pulsed amperometry allows picomolar amounts of carbohydrate to be analyzed without the need for derivatization.

The majority of these highly sophisticated technologies have been commercialized but alternative techniques can be performed simply and effectively. The use of immobilized lectins to affinity-fractionate oligosaccharides with similar masses comprised of a number of quite different structures exploits the exquisite specificities of plant lectins *(29,30)*. Polyacrylamide-gel electrophoresis of oligosaccharides labeled with very sensitive fluorophores allows high-resolution separation and detection *(31,32)* and uses equipment that is familiar to most laboratories. Using these methods, coupled with specific glycosidase digestion, oligosaccharide structures have been deduced at the picomolar level.

4. Methods for the Structural Analysis of Oligosaccharides

4.1. Monosaccharide Compositional Analysis

Hydrolysis of the glycosidic bonds of isolated oligosaccharides and intact glycoprotein can be achieved using acidic conditions. Trifluoroacetic acid (2 M TFA) at 100°C for 4–6 h to releases neutral sugars (glucose, galactose, mannose, and fucose) nondestructively, whereas amino sugars (*N*-acetylglucosamine and *N*-acetylglucosamine) are not quantitatively recovered. Amino sugars are quantitatively released by hydrolysis using 6 M HCl under the same conditions, a treatment that destroys neutral sugars. Separation and detection of the monosaccharides at the picomolar level usually require derivatization. Compositional analysis using mass spectrometry of alditol acetates derived from sugars released by acid hydrolysis, separated by gas-liquid chromatography, has been obtained from <5 µg of glycoprotein *(33)*. Derivatization is unnecessary with HPAEC/PAD *(19)* and analyses using similar amounts of protein can accommodated. HPAEC also has the advantage that acid-labile monosaccharides, for example sialic acid, can be quantitated after digestion of oligosaccharides or glycoproteins with mild acid conditions (0.1 N HCl for 1 h at 80°C) or enzymatically using neuraminidases. The availability of specific neuraminidases that discriminate between α2,3- and α2,6-linked sialic-acid residues provides additional information about the glycosidic linkage.

4.2. Glycosidic Linkage Analysis

Information regarding the ways in which monosaccharides are linked to one another in an oligosaccharide can be obtained by permethylation of all the free hydroxyl groups of the monosaccharides. Hydrolysis of the glycosidic bond leaves free hydroxyl groups, indicating where the monosaccharides were linked. These hydroxyl residues are then acetylated and the monosaccharides separated and detected by gas–liquid chromatography–mass spectrometry *(34)*. Very little material is required (low pmol) and the carbon atoms involved in forming the glycosidic linkage can be unambiguously identified.

4.3. Mass Spectrometry

Methods to ionize oligosaccharides and detection by mass spectrometry require relatively little material (low pmol range). Matrix-assisted, laser-desorption techniques involve irradiation of the sample mixed with a UV-absorbing matrix followed by mass separation of the ions. Because only molecular ions are detected using time of flight instruments, this method yields even less inductive information than Bio-Gel P-4 chromatography, but its ease of use and rapid analysis used in combination with other separation technologies and/or glycosidase digestion increases its analytical power. Although fast action bombardment mass spectrometry (FABMS) can give sequence, branching point, and linkage information owing to fragmentation within the ion source of the spectrometer (35), without a compound database or other information none of these parameters can be predictively assigned. To confirm the structure of an unknown glycan, further derivatization techniques, such as permethylation of free-hydroxyl groups with or without periodate oxidation, have to be employed (36). The analysis of oligosaccharides in solution by electrospray mass spectrometry has the advantage that direct coupling to HPLC liquid phase chromatographic separation (37), capillary supercritical-fluid chromatography (38), and capillary electrophoresis (39), is possible.

4.4. Nuclear Magnetic Resonance (NMR)

NMR analysis typically requires much larger amounts of free oligosaccharide to confirm the identity of structure and linkage, but can do so unambiguously and nondestructively (40). For high-resolution, two-dimensional studies at high field (500–600 MHz), 1–5 μmol is required. Much less material is needed for a 1D spectrum (50 nmol) but this may be orders of magnitude higher than is possible to isolate for some biologically active glycoproteins, and the structural identification depends on reference to databases containing the known chemical shifts of anomeric protons. The quantity of primary biological material required for a single analysis using this technique may be beyond the scope of most studies. The NMR machine provides more than an analytical service and has important applications as a research tool in determining oligosaccharide conformation and molecular dynamics in solution.

4.5. Oligosaccharide Sequencing Using Glycosidases

The structural characterization of oligosaccharides can be achieved using glycosidases (41), a group of hydrolytic enzymes abundant in nature. Glycosidases have been purified from many sources, including plant, vertebrate and invertebrate tissues, and microbes. Although most glycosidases are of lysosomal origin or associated with degradative vacuoles involved in the catabolism of glycoconjugates, others, for example α-glucosidases and α-mannosidases, are found in the endomembrane system of eukaryotes, where they participate in N-glycan biosynthesis. Glycosidases are also found as soluble, secreted enzymes or are located in the outer membrane, for example, members of the neuraminidases that are present in bacteria and viruses. Two major groups of enzyme are used in the determination of glycan structure; exoglycosidases that hydrolyze monosaccharides from the nonreducing terminus of glycans and endoglycosidases that cleave between monosaccharide residues located internally to the oligosaccharide.

Structural assignment using sequential exoglycosidase digestion relies on the known specificity and purity of glycosidase-catalyzed hydrolysis. Exoglycosidases are usually named after the monosaccharide that is cleaved and its anomeric configuration. For example, β-galactosidase only hydrolyzes glycans containing terminal β-galactose residues and α-fucosidase, terminal α-fucose residues. The strict observance of both glycon and anomericity by exoglycosidases (an exception being the β-hexosaminidases that hydrolyze both *N*-acetylglucosamine and *N*-acetylgalactosamine) is a feature that has considerable predictive value. The selection of enzyme or even the concentration of enzyme can be used to determine the anomeric linkage among monosaccharides. The α-fucosidase isolated from *Charonia lampas* cleaves all α-fucosyl residues but at markedly different rates. Consequently, concentrations can be used to remove preferentially and predictively α1,6-linked rather than α1,3/4-linked fucose. By contrast, almond-meal α-fucosidase hydrolyses only α1,3/4-linked fucosyl residues and even at very high concentrations will not cleave α1,6 bonds *(42)*.

In addition to the aforementioned substrate specificities, some exoglycosidases have activities that are dependent on the neighboring, or aglycon group. The most complicated example of aglycon specificity is shown by the β-hexosaminidase isolated from *Streptococcus pneumoniae* where hydrolysis of the preferred glycon (GlcNAc linked β1,2 to mannose) is restricted by further substitutions of the mannose residue at C_6, or by the presence of a bisecting GlcNAc *(43)*. Aglycon specificity is also an important property of most endoglycosidases, where an extended sequence of carbohydrate several residues away from the site of catalysis determines hydrolytic rates *(44)*. By paying careful attention to the rules for glycosidic cleavage (including protein and activity concentration, pH and buffer optima, and ion dependence), oligosaccharide sequences are validly assigned. The use of glycosidases can therefore complement physical and chemical analyses (GC-MS, MS, NMR) and in many cases is sufficient alone in providing an appropriate level of information.

Any of the techniques previously described (usually more than one) are appropriate to examine enzyme-cleavage products. Gel filtration (Bio-Gel P-4 and mass spectrometry) relies on the shift in mass after hydrolysis. Usually several rounds of enzyme digestion, recovery of the reaction products, further exoglycosidase digestion, and analysis are required to enable full characterization. The covalently attached reporter group at the reducing terminus is retained during exoglycosidase digestion, which proceeds from the nonreducing terminus allowing the change in mass to be measured or the glycan recovered. Separation using HPLC includes a greater variety of matrices that exploit the physical/chemical properties of the oligosaccharide. Charged glycans can be resolved using weak anion-exchange resins *(45)* and reversed-phase and gel filtration separates carbohydrates according to hydrophilicity or the number of monosaccharide units *(26)*. HPAEC/PAD has the unique advantage that no derivatization or introduction of a reporter group is necessary for detecting separated glycans or monosaccharides and additionally, all the reaction products can be measured. As with other HPLC separations, however, the retention times on these columns are not always predictive of structure and identification must rely on the elution times of known oligosaccharide standards.

Enzyme arrays or the reagent-array analysis method (RAAM), a method where a purified oligosaccharide is subjected to a mixture of glycosidases and the fragmentation or fingerprint obtained is matched to computer databases, allows rapid and repro-ducible analyses to be made *(46)*. Although this technique has only been applied to gel-permeation chromatography to separate the fragments, HPLC, HPAEC, and mass spectrometric methods could be used.

4.6. Strategies for Glycosylation Site Analysis

Important biochemical functions of glycoproteins are mediated by the complement of oligosaccharides at particular attachment points on the polypeptide *(47)*. To identify the oligosaccharide composition at specific sites, a general strategy is to perform a limited proteolytic digestion of the protein. The peptide mixture can be separated by lectin affinity chromatography to isolate the glycopeptides selectively *(48)* that are further fractionated using reversed-phase HPLC *(49)*. Treatment of each glycopeptide fraction with PNGase F results in cleavage of N-linked oligosaccharides and the con-version of the asparagine-linked amino acids to aspartic acid. Automated Edman sequencing is used to confirm amino-acid composition and reveal amino-acid residues substituted by O-linked oligosaccharides. The released oligosaccharide can then be analyzed for carbohydrate composition, reductively labeled if appropriate, and sequenced using any of the aforementioned methods.

5. Glycoprotein Remodeling

Definition of the role(s) played by the oligosaccharide moiety in glycoprotein func-tion can be made by experimental manipulation either during biosynthesis or on the mature protein. Glycosylation inhibitors have been used to great effect in changing the normal *N*-glycosylation pattern *(50)*. The uses of inhibitors that block the addition of N-linked glycans, for example, tunicamycin, have profound effects on protein folding and secretion, and consequently many proteins lack biological activity if cells are grown in the presence of such compounds. More selective effects are gained by subtle modifi-cation of the N-linked glycan after it has been added to the nascent protein. The deoxynojirimycin and deoxymannojirimycin families, castanospermine, and swainsonine are natural products found in plants that are potent glycosidase inhibitors *(51)* and have been used therapeutically to attenuate the infectivity of the HIV virus *(52)*. The addition to tissue-cultured cells of these compounds, inhibit α-glucosidase and α-mannosidase processing reactions and prevent or reduce the complement of com-plex-type oligosaccharides *(53)*. The expression of recombinant glycoproteins in Chinese Hamster Ovary (CHO) cells in the presence of an α-glucosidase inhibitor gen-erates a uniform population of oligosaccharides that are easily cleaved using endoglycosidases. This protocol efficiently produces a properly folded yet deglycosylated protein that can be used to probe function *(54)* or may be used to obtain X-ray crystal structure where the otherwise heterogeneous nature of the oligosaccha-ride precludes such analysis.

Molecular-biology techniques can be applied to glycoprotein remodeling by the substitution of the amino acids (asparagine, serine, and threonine) involved in N- and O-linked glycosylation, at the genetic level. Site-directed mutagenesis has been suc-cessfully applied to the HIV surface glycoprotein, gp 120, to delete *N*-glycosylation attachment points. These experiments provide important information regarding the influence of glycosylation on virus infectivity *(55)*.

Glycoproteins in solution and on the surface of cells can be modified using exoglycosidases and glycosyltransferases to aid detection of terminal carbohydrate residues and to probe function *(56,57)*. Recent efforts to obtain the genes or cDNAs for glycosidases *(43)* and glycosyltransferases *(58)* have been successful in providing a range of specific reagents for biologists to manipulate oligosaccharides.

6. Summary

The many techniques presented here are offered as a guide; the strategy adopted will depend on the quality of information that is most relevant to the needs of the biologist. If the target protein to be analyzed is as complex as the HIV glycoprotein gp120, which contains over 100 glycoforms *(59)*, then only a fully integrated approach will be successful. This includes the release of oligosaccharide, labeling, fractionation, and exoglycosidase, NMR, MS sequence, and compositional analysis. For other studies, the separative power of SDS-PAGE used in conjunction with endoglycosidase release to characterize the type of glycan and its contribution to the mass of the protein is an excellent starting point. Enzyme-based methods to remove the glycan or modify its structure provide important clues to glycoprotein function and should dictate further strategy.

References

1. Allen, H. J. and Kisailus, E. C. (1992) *Glycoconjugates. Composition, Structure and Function.* Marcel Dekker, New York.
2. Kornfeld, R. and Kornfeld, S. (1985) Assembly of asparagine-linked oligosaccharides. *Annu. Rev. Biochem.* **54,** 631–664.
3. Kleene, R. and Berger, E. G. (1993) The molecular and cell biology of glycosyltransferases. *Biochim. Biophys. Acta* **1154,** 283–325.
4. Carraway, K. L. and Hull, S. R. (1991) Cell surface mucin-type glycoproteins and mucin-like domains. *Glycobiology* **1,** 131–138.
5. Corfield, T. (1992) Mucus glycoproteins, super glycoforms: how to solve a sticky problem? *Glycoconjugate J.* **9,** 217–221.
6. Helenius, A. (1994) How N-linked oligosaccharides affect glycoprotein folding in the endoplasmic reticulum. *Mol. Biol. Cell* **5,** 253–265.
7. Varki, A. (1993) Biological roles of oligosaccharides—All of the theories are correct. *Glycobiology* **3,** 97–130.
8. Kobata, A. (1992) Structures and functions of the sugar chains of glycoproteins. *Eur. J. Biochem.* **209,** 483–501.
9. O'Shannessy, D. J. and Quarles, R. H. (1987) Labeling of the oligosaccharide moieties of immunoglobulins. *J. Immunol Methods* **99,** 153–161.
10. Edge, A. S., Faltynek, C. R., Hof, L., Reichert, L. E., Jr., and Weber, P. (1981) Deglycosylation of glycoproteins by trifluoromethanesulfonic acid. *Anal. Biochem.* **118,** 131–137.
11. Tretter, V., Altmann, F., and Marz, L. (1991) Peptide-N4-(N-acetyl-beta-glucosaminyl) asparagine amidase-F cannot release glycans with fucose attached $\alpha-1 \rightarrow 3$ to the asparagine-linked N-acetylglucosamine residue. *Eur. J. Biochem.* **199,** 647–652.
12. Trimble, R. B. and Tarentino, A. L. (1991) Identification of distinct endoglycosidase (Endo) activities in *Flavobacterium-meningosepticum*—Endo-F1, Endo-F2, and Endo-F3—Endo-F1 and Endo-H hydrolyze only high mannose and hybrid glycans. *J. Biol. Chem.* **266,** 1646–1651.

13. Lis, H. and Sharon, N. (1986) Lectins as molecules and as tools. *Annu. Rev. Biochem.* **55,** 35–67.

14. Kijimoto-Ochiai, S., Katagiri, Y. U., and Ochiai, H. (1985) Analysis of N-linked oligosaccharide chains of glycoproteins on nitrocellulose sheets using lectin-peroxidase reagents. *Anal. Biochem.* **147,** 222–229.

15. Cummings, R. D. (1994) Use of lectins in analysis of glycoconjugates. *Methods Enzymol.* **230,** 66–86.

16. Tachibana, H., Seki, K., and Murakami, H. (1993) Identification of hybrid-type carbohydrate chains on the light chain of human monoclonal antibody specific to lung adenocarcinoma. *Biochim. Biophys. Acta* **1182,** 257–263.

17. Leonards, K. S. and Kutchai, H. (1985) Coupling of Ca^{2+} transport to ATP hydrolysis by the Ca^{2+}-ATPase of sarcoplasmic reticulum: potential role of the 53-kilodalton glycoprotein. *Biochemistry* **24,** 4876–4884.

18. Ogawa, H., Ueno, M., Uchibori, H., Matsumoto, I., and Seno, N. (1990) Direct carbohydrate analysis of glycoproteins electroblotted onto polyvinylidene difluoride membrane from sodium dodecyl sulfate-polyacrylamide gel. *Anal. Biochem.* **190,** 165–169.

19. Weitzhandler, M., Kadlecek, D., Avdalovic, N., Forte, J. G., Chow, D., and Townsend, R. R. (1993) Monosaccharide and oligosaccharide analysis of proteins transferred to polyvinylidene fluoride membranes after sodium dodecyl sulfate-polyacrylamide gel electrophoresis. *J. Biol. Chem.* **268,** 5121–5130.

20. Takasaki, S., Mizuochi, T., and Kobata, A. (1982) Hydrazinolysis of asparagine-linked sugar chains to produce free oligosaccharides. *Methods Enzymol.* **83,** 263–268.

21. Patel, T., Bruce, J., Merry, A., Bigge, C., Wormald, M., Jaques, A., and Parekh, R. (1993) Use of hydrazine to release in intact and unreduced form both N-linked and O-linked oligosaccharides from glycoproteins. *Biochemistry* **32,** 679–693.

22. Cooper, C. A., Packer, N. H., and Redmond, J. W. (1994) The elimination of O-linked glycans from glycoproteins under non-reducing conditions. *Glycoconjugate J.* **11,** 163–167.

23. Takahashi, N. and Muramatsu, T. (1992) *Handbook of Endoglycosidases and Glycoamidases.* CRC, Florida.

24. Yamashita, K., Mizuochi, T., and Kobata, A. (1982) Analysis of oligosaccharides by gel filtration. *Methods Enzymol.* **83,** 105–126.

25. Kobata, A., Yamashita, K., and Takasaki, S. (1987) BioGel P-4 column chromatography of oligosaccharides: effective size of oligosaccharides expressed in glucose units. *Methods Enzymol.* **138,** 84–94.

26. Hase, S., Ikenaka, K., Mikoshiba, K., and Ikenaka, T. (1988) Analysis of tissue glycoprotein sugar chains by two dimensional high-performance liquid chromatographic mapping. *J. Chromatog.* **434,** 51–60.

27. Lee, Y. C. (1990) High-performance anion-exchange chromatography for carbohydrate analysis. *Anal. Biochem.* **189,** 151–162.

28. Townsend, R. R. and Hardy, M. R. (1991) Analysis of glycoprotein oligosaccharides using high-pH anion exchange chromatography. *Glycobiology* **1,** 139–147.

29. Merkle, R. K. and Cummings, R. D. (1987) Lectin affinity chromatography of glycopeptides. *Methods Enzymol.* **138,** 232–259.

30. Osawa, T. and Tsuji, T. (1987) Fractionation and structural assessment of oligosaccharides and glycopeptides by use of immobilised lectins. *Annu. Rev. Biochem.* **56,** 21–42.

31. Jackson, P. (1990) The use of polyacrylamide-gel electrophoresis for the high-resolution separation of reducing saccharides labelled with the fluorophore 8-aminonaphthalene-1,3,6-trisulphonic acid. *Biochem. J.* **270,** 705–713.

32. Jackson, P. (1991) Polyacrylamide gel electrophoresis of reducing saccharides labeled with the fluorophore 2-aminoacridone—subpicomolar detection using an imaging system based on a cooled charge-coupled device. *Anal. Biochem.* **196,** 238–244.

33. Kenne, L. and Stromberg, S. (1990) A method for the microanalysis of hexoses in glycoproteins. *Carb. Res.* **198,** 173–179.

34. Montreuil, J., Bouquelet, S., Debray, H., Fournet, B., Spik, G., and Strecker, G. (1986) Glycoproteins, in *Carbohydrate Analysis: A Practical Approach* (Chaplin, M. F. and Kennedy, J. F., eds.), IRL, Oxford, UK, pp. 143–204.

35. Dell, A., Khoo, K.-H., Panico, M., McDowell, R. A., Etienne, A. T., Reason, A. J., and Morris, H. R. (1993) FAB-MS and ES-MS of glycoproteins, in *Glycobiology: A Practical Approach* (Fukuda, M. and Kobata, A., eds.), IRL, Oxford, UK, pp. 187–222.

36. Harvey, D. J. (1992) The role of mass spectrometry in glycobiology. *Glycoconjugate J.* **9,** 1–12.

37. Medzihradszky, K. F., Maltby, D. A., Hall, S. C., Settineri, C. A., and Burlingame, A. L. (1994) Characterisation of protein N-glycosylation by reversed-phase microbore liquid chromatography/electrospray mass spectrometry, complementary mobile phases, and sequential exoglycosidase digestion. *J. Am. Soc. Mass Spectrom.* **5,** 350–358.

38. Leroy, Y., Lemoine, J., Ricart, G., Michalski, J.-C., Montreuil, J., and Fournet, B. (1990) Separation of oligosaccharides by capillary supercritical fluid chromatography and analysis by direct coupling to high-resolution mass spectrometer—application to analysis of oligomannosidic N-glycans. *Anal. Biochem.* **184,** 235–243.

39. Kelly, J. F., Locke, S. J., and Thibault, P. (1993) Analysis of protein glycoforms by capillary electrophoresis-electrospray mass spectrometry. *Discovery Newsletter (Beckman)* **2,** 1–6.

40. Dwek, R. A., Edge, C. J., Harvey, D. J., Wormald, M. R., and Parekh, R. B. (1993) Analysis of glycoprotein-associated oligosaccharides. *Annu. Rev. Biochem.* **62,** 65–100.

41. Jacob, G. S. and Scudder, P. (1994) Glycosidases in structural analysis. *Methods Enzymol.* **230,** 280–299.

42. Butters, T. D., Scudder, P., Rotsaert, J., Petursson, S., Fleet, G. W. J., Willenbrock, F. W., and Jacob, G. S. (1991) Purification to homogeneity of *Charonia lampas* α-fucosidase by using sequential ligand-affinity chromatography. *Biochem. J.* **279,** 189–195.

43. Clarke, V. A., Platt, N., and Butters, T. D. (1995) Cloning and expression of the β-N-acetylglucosaminidase gene from *Streptococcus pneumoniae* —generation of truncated enzymes with modified aglycon specificity. *J. Biol. Chem.* **270,** 8805–8814.

44. Kobata, A. (1979) Use of endo- and exoglycosidases for structural studies of glycoconjugates. *Anal. Biochem.* **100,** 1–14.

45. Guile, G. R., Wong, S. Y. C., and Dwek, R. A. (1994) Analytical and preparative separation of anionic oligosaccharides by weak anion-exchange high-performance liquid chromatography on an inert polymer column. *Anal. Biochem.* **222,** 231–235.

46. Edge, C. J., Rademacher, T. W., Wormald, M. R., Parekh, R. B., Butters, T. D., Wing, D. R., and Dwek, R. A. (1992) Fast sequencing of oligosaccharides—the reagent-array analysis method. *Proc. Natl. Acad. Sci. USA* **89,** 6338–6342.

47. Rademacher, T. W., Parekh, R. B., and Dwek, R. A. (1988) Glycobiology. *Annu. Rev. Biochem.* **57,** 785–838.

48. Yeh, J. C., Seals, J. R., Murphy, C. I., Vanhalbeek, H., and Cummings, R. D. (1993) Site-specific N-glycosylation and oligosaccharide structures of recombinant HIV-1 gp120 derived from a baculovirus expression system. *Biochemistry* **32,** 11,087–11,099.

49. Rohrer, J. S., Cooper, G. A., and Townsend, R. R. (1993) Identification, quantification, and characterization of glycopeptides in reversed-phase HPLC separations of glycoprotein proteolytic digests. *Anal. Biochem.* **212,** 7–16.

50. Elbein, A. D. (1987) Glycosylation inhibitors for N-linked glycoproteins. *Methods Enzymol.* **138,** 661–709.
51. Fleet, G. W. J. (1988) Amino-sugar derivatives and related compounds as glycosidase inhibitors. *Spec. Publ.-R Soc. Chem.* **65,** 149–162.
52. Jacob, G. S., Scudder, P., Butters, T. D., Jones, I., and Tiemeier, D. C. (1992) Aminosugar attenuation of HIV infection, in *Natural Products as Antiviral Agents* (Chu, C. K. and Cutler, H. G., eds.), Plenum, New York, pp. 137–152.
53. Karlsson, G. B., Butters, T. D., Dwek, R. A., and Platt, F. M. (1993) Effects of the imino sugar N-butyldeoxynojirimycin on the N-glycosylation of recombinant gp120. *Biol. Chem.* **268,** 570–576.
54. Davis, S. J., Davies, E. A., Barclay, A. N., Daenke, S., Bodian, D. L., Jones, E. Y., Stuart, D. I., Butters, T. D., Dwek, R. A., and Vandermerwe, P. A. (1995) Ligand binding by the immunoglobulin superfamily recognition molecule CD2 is glycosylation-independent. *J. Biol. Chem.* **270,** 369–375.
55. Lee, W.-R., Syu, W.-J., Du, B., Matsuda, M., Tan, S., Wolf, A., Essex, M., and Lee, T.-H. (1992) Nonrandom distribution of gp120 N-linked glycosylation sites important for infectivity of human immunodeficiency virus type 1. *Proc. Natl. Acad. Sci. USA* **89,** 2213–2217.
56. Paulson, J. C. and Rogers, G. N. (1987) Resialylated erythrocytes for assessment of the specificity of sialyloligosaccharide binding proteins. *Methods Enzymol.* **138,** 162–168.
57. Whiteheart, S. W., Passaniti, A., Reichner, J. S., Holt, G. D., Haltiwanger, R. S., and Hart, G. W. (1989) Glycosyltransferase probes. *Methods Enzymol.* **179,** 82–95.
58. Natsuka, S. and Lowe, J. B. (1994) Enzymes involved in mammalian oligosaccharide biosynthesis. *Curr. Opin. Struct. Biol.* **4,** 683–691.
59. Mizuochi, T., Matthews, T. J., Kato, M., Hamako, J., Titani, K., Solomon, J., and Feizi, T. (1990) Diversity of oligosaccharide structures on the envelope glycoprotein GP 120 of Human Immunodeficiency Virus-1 from the lymphoblastoid cell line H9—presence of complex-type oligosaccharides with bisecting N-acetylglucosamine residues. *J. Biol. Chem.* **265,** 8519–8524.

41

Protein Sequencing

Bryan J. Smith and John R. Chapman

1. Introduction

It is approaching half a century since Edman developed a chemistry for the determination of the N-terminus of a protein or peptide *(1)*, and this chemistry is still being used today. In the intervening decades various ancillary techniques have come (and some have gone) that have allowed the field to develop. Apart from the automation of Edman chemistry itself, the most significant advances have been improvements in sample preparation and the advent of mass spectrometric techniques. In the early days the task of sequencing a protein was a significant one, limited to those proteins that could be prepared in sufficiently large amounts. Then, the aim (other than to develop the technology) was to obtain the full sequence of the protein in question to begin to understand how proteins were structured, which was done by use of Edman chemistry exclusively. Currently, the situation is different, and it is usually necessary to obtain only partial protein sequence from the protein itself. From the partial sequence, suitable oligonucleotides can be designed and used to clone the corresponding gene, which can be rapidly analyzed to yield the full sequence. For identification of a protein the partial sequence may need to be only 3–5 residues long. Such short sequences, known as "sequence tags," are currently of importance in the study of the "proteome," the set of proteins that is expressed by the genome. Studies of the proteome follow the results of genome sequencing, which are giving an explosion of information. The sequence of the genome suggests what proteins might be made, but does not prove which are actually expressed in any given tissue(s), how they are regulated, or what their functions are. For these purposes there is a need to separate and identify individual proteins, then to correlate their presence and modification with function. The large number of proteins in any one organism (possibly as many as 70,000 different proteins in a human) makes this a formidable task. Two-dimensional polyacrylamide gel electrophoresis (2D-PAGE) has the potential to resolve thousands of proteins as spots on a single gel and has become of great use in this area. The task is now one of identification of the many protein spots on gels. Protein sequencing techniques have developed to the point where picomole or femtomole amounts of protein can be analyzed with sufficient sensitivity to partially sequence spots on 2D-PAGE. Partial sequence data derived from a given protein spot can be compared with databases generated by genome sequencing

From: *Molecular Biomethods Handbook*
Edited by: R. Rapley and J. M. Walker © Humana Press Inc., Totowa, NJ

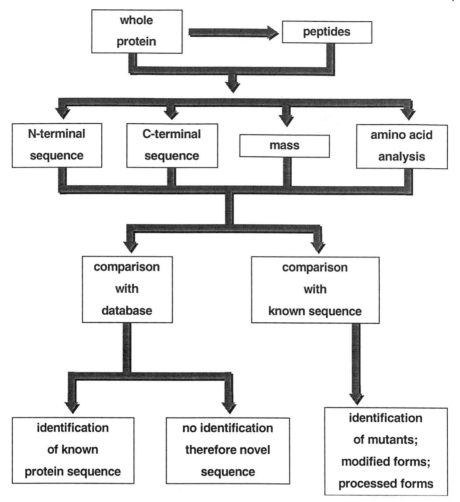

Fig. 1. Protein sequence determination—a basic strategy.

and when combined with other information (e.g., the mass of a protein or of peptides derived from it, amino acid composition, or pI value) can identify a protein with a good degree of confidence. Protein sequencing is also of use in the study of postsynthetic modification—identifying sites and types of modification—something that is not obvious from knowledge of the amino acid sequence alone.

The advent of the biopharmaceutical industry has given protein sequence determination another role, that of quality control. The industry generates products that are proteins, for instance antibodies or hormones, that require rigorous checking before being used. Synthetic peptides may also require analysis to confirm their identities. It is the application of protein sequencing technology to areas such as these that has made the field continually useful and enduring, and that also fuels the drive toward still more sensitive and faster analysis methods.

Analysis of polypeptide primary structure involves a variety of techniques, from sample preparation, through sequence determination, to analysis of data and homology screening of sequence databases. **Figure 1** sketches the sort of strategy that might be employed currently in sequence analysis (dependent on the precise purpose of the study). The steps indicated in **Fig. 1** are described in more detail in **Subheading 2.**

2. Sample Preparation

The preparation of a sample is a critical step in its analysis *(2)*. Current techniques require as little as nanomole or femtomole amounts of sample to produce a result. The handling of such small amounts of protein or peptide presents special problems, however. Some techniques can tolerate the presence of more than one polypeptide in a sample; for instance, mass spectrometry of peptide mixtures, or Edman chemical sequencing, especially if the sequence of some component(s) is known. Generally, however, the aim is for purity (say 10% or less of the sample as other, interfering material). During the purification steps care must be taken to avoid contamination, modification, or loss of the sample protein.

Buffer salts are unavoidable during purification procedures, and detergents are also commonly used to solubilize hydrophobic proteins and to improve recovery from purification steps. To minimize problems, avoid addition to the sample of anything that may spoil subsequent analysis by always using the highest quality water and reagents. For example, trace buffer components may block Edman sequencing by reaction with the polypeptide N-terminal amine. Traces of reactive peroxides in nonionic detergents may do this, so use specially purified solutions stored under nitrogen *(3)*. Again, urea solutions may contain cyanates that react with amines and thiols. Remove these by deionization just prior to use. Amines, such as glycine and Tris, can generate background "noise" in Edman sequencing.

In electrospray ionization (ESI) mass spectrometry, in which surfactants can be used to study higher order protein structure, e.g., protein denaturation and protein–protein interactions, the presence of nonionic saccharide surfactants (e.g. *n*-dodecyl glucoside) affords the strongest analyte ion currents, without serious chemical background, whereas sodium dodecyl sulfate (SDS) and sodium taurocholate are detrimental, with strong signal suppression *(4)*. Nonionic detergents themselves can, however, interfere in other ways; for instance, by contributing UV-absorbing peaks to chromatograms after high-performance liquid chromatography (HPLC) and by inhibiting binding of polypeptides to polyvinylidene difluoride (PVDF) membranes. Some, especially ionic, detergents can also be problematic in matrix-assisted laser desorption/ionization (MALDI) mass spectrometry, leading to signal loss and often to peak broadening. On the other hand, in experiments on direct cellular profiling by MALDI, the addition of a nonionic detergent, Triton X-100, was found to improve the spectrum quality by making the components of the complex mixture available to the ionization process *(5)*. For a similar reason, 2-propanol + water is a generally effective matrix solvent, for use in the recording of MALDI spectra from most proteins *(5)*. A solvent, such as 7:3 (v/v) formic acid + hexafluoroisopropanol, followed by ultrasonication *(6)*, may be used for hydrophobic proteins in MALDI. Ionic buffer components can also give problems in mass spectrometry, especially ESI, by causing suppression of ionization and adduct formation.

Offending buffer components can be diluted or removed by dialysis or chromatographic exchange of buffer, but there is a significant danger of sample loss if the protein is insoluble in the new buffer conditions. Many proteins readily adsorb to dialysis tubing, glass, and plastic, especially when in microgram amounts or less. Polypropylene tubes are generally the best vessels for collection and storage of samples; silylation may reduce the dangers of sample loss in these. Never dry a sample. The act of concentration may adversely affect the buffer conditions, causing adsorption of protein to vessel walls, and once the protein is dry it may never redissolve.

A golden rule in sample preparation is to reduce sample handling to a minimum. This minimizes sample loss and ingress of contaminants. To this end, strategies for purification usually employ high resolution methods to clean up samples by the minimum number of steps. Notable among these are HPLC and capillary electrophoresis (CE): both can be used for microgram or submicrogram amounts of sample, and both can be linked on-line to ESI mass spectrometry. Alternatively, samples can be collected and analyzed off-line by MALDI mass spectrometry. However, perhaps the most generally applicable, quickest, cheapest, and highest-resolving method is PAGE, typically using SDS in 1D-PAGE, or isoelectric focusing combined with SDS-PAGE in 2D-PAGE. A 2D gel of say, 25 by 25 cm can resolve 1000 proteins or more. Individual protein spots can be subject to MALDI mass spectrometry or fragmented *in situ* (and the resulting peptides isolated for analysis), or can be transferred to PVDF (or other) membranes, for sequencing, proteolysis, or modification.

The last-mentioned is a convenient way of trapping a protein and preventing further loss: salts can be washed away, the sample can be applied directly to automated N- and C-terminal sequencers or MALDI mass spectrometry, or it can be fragmented or modified for other analyses. Sample handling, in this case, is reduced to a minimum; from tissue extraction to the gel, then to a membrane.

In many ways the preparation of a sample for structural analysis (arguably) presents more challenges than the analysis itself, since it is so dependent on the characteristics of the particular protein in question, these being different from those of the previous protein!

3. Proteolysis and Modification of Proteins

It may be necessary to cleave a protein into peptides in order to obtain "internal" (i.e., not N- or C-terminal) sequence information from any of them, to obviate the problem of a blocked N-terminus, or to "map" (i.e., locate within the sequence) sites of modification, such as phosphorylation or glycosylation. CE or HPLC chromatograms of peptides ("peptide maps") are characteristic of individual proteins, since peptide generation is dependent on amino acid sequence. They are therefore useful in themselves for checking the quality of proteins, and are also of use for preparation of peptides. The mass spectrometric equivalent is a list of masses of the peptides, and is useful for comparison with the (ever growing) databases of gene sequences to identify a protein as a particular gene's product (*see* **Subheading 4.**).

The various common methods for protein cleavage are summarized in the **Table 1**. Larger amounts of sample allow these methods to be carried out in solution but microgram and submicrogram amounts of proteins on PVDF blots, in gels, or on columns can also be treated by various adaptations of the in-solution methods *(7–10)*. The results of protein cleavage can be monitored, and peptides prepared, by PAGE *(11,12)*; HPLC *(13,14)*; or CE *(15)*. Methods have also been described for the isolation of specific peptides from mixtures with others: glycopeptides *(16)*; a peptide from the C-

Table 1
Methods for Cleavage of Polypeptides

Cleaved bond	Reagent	Comment[a]	Ref[b]
Arg-X[c]	Endo Arg C (protease)	X may be amide or ester group	*85*
	Trypsin (protease)	X may be amide or ester group Lys-X also cleaved (q.v.) unless this is inhibited by succinylation of Lys side-chain.	*21,85*
Asn-Gly	Hydroxylamine, 2 *M*, pH 9.0, 45°C, 4 h	Substrate is denatured by inclusion of 2 *M* guanidine HCl.	*86*
X-Asp	Endo Asp N (protease)	X-Cys may also be cleaved.	*85*
Asp-X	Dilute HCl, pH 2.0, 108°C, 2 h	Frequently, cleavage of all Asp-X bonds is not complete.	*87*
Asp-X	Endo Asp C (protease)	Frequently, cleavage of all Asp-X bonds is not complete.	*91*
Cys-X	2-nitro-5-thio cyanobenzoate, 10-fold molar excess over Cys, 37°C, 20 min, followed by adjustment to pH 9.0, for 16 h at 37°C, then Raney nickel	Disulfides are reduced prior to reaction and the sample denatured in guanidine HCl. Cleavage occurs in the pH 9.0 step.	*88*
Glu-X	Endo Glu C (protease) pH 4.0	If reaction is at neutral pH, Asp-X may also be cleaved.	*85*
Hydrophobic-X	Chymotrypsin (protease)	Hydrophobic = Y; F; W. Other bonds may be cleaved also but at lesser rates.	*85*
Lys-X	Endo Lys C (protease)	X may be amide or ester group.	*85*
	Trypsin (protease)	X may be amide or ester group; Arg-X also cleaved.	
Met-X	Cyanogen bromide in 50% (v/v) trifluoroacetic acid, ambient temp. 24 h	Trifluoroacetic acid is used instead of the traditional formic acid since the latter formylates the polypeptide.	*89*
Trp-X	Cyanogen bromide, 4°C, 30 h, dark	Prior to reaction, Trp residues are oxidised to oxindolyl alanyl residues (in acid–dimethylsulfoxide).	*90*

[a]Proteases listed are commercially available in a form especially purified for sequencing work. The operating pH for the proteases is between 6.0 and 8.0, except for Endo Glu C. Conditions for chemical treatments are more harsh—note that if low pH is used some deamidation of Gln and Asn may occur.

[b]For further details *see* these references and other references quoted therein.

[c]X = Any amino acid; note that susceptibility to proteases may be reduced or lost if X is Pro, or a like residue (e.g., E-E for Endo Glu C).

terminus of a protein *(17)*; or a peptide from the N-terminus of a protein with a blocked α-amino group *(18)*.

As discussed by Tawfik *(19)* and by Speicher *(20)*, the chemical modification of various amino acid side-chains is useful for investigating their role in protein function. Although modification of various residue side-chains is possible (see more detailed reviews in **refs.** *21* and *22*), in the field of protein sequencing the most significant modification is of the cysteine thiol. Reduction of cysteine and derivatization of the resulting thiol causes sample denaturation, allowing more efficient polypeptide cleavage and subsequent separation of otherwise covalently linked peptides. Addition of specific groups to cysteine residues, as well as comparison of nonreduced and reduced-and-alkylated peptide maps can aid assignment of disulfide bonds. Finally, cysteine derivatives are stable to Edman sequencing chemistry, unlike underivatized cysteine, which is unstable and gives a gap in the sequence.

There is a variety of methods for modification of the cysteine thiol *(23)*. Derivatization with 4-vinylpyridine and acrylamide (giving *S*-[2-(4-pyridyl)ethyl]cysteine and cysteinyl-*S*-β-propionamide, respectively) are two such currently popular methods. Derivatization by 4-vinyl pyridine adds a chromophore absorbing at 254 nm that can be useful for the identification of modified peptides. Modification of some cysteine residues by acrylamide can be done in vitro but also occurs adventitiously during PAGE of protein samples. Alkylation of cysteine by iodoacetic acid (to carboxymethyl cysteine) is another useful approach in that radioactive reagent is commercially available and so can radiolabel the protein of interest.

Additional methods for modification of a protein involve enzymes instead of chemicals. For example, attached oligosaccharide chains can be removed by the action of glycosidases *(24)* (or indeed, by chemical methods *[25,26]*), or proteins can be phosphorylated by kinases *(27)*. Again, an N-terminus that has been blocked to Edman chemistry by cyclization of a glutamine to a pyroglutamine residue can be deblocked (usually at <100% yield) by pyroglutaminase enzyme *(28,29)*.

4. Mass Spectrometry and the Analysis of Intact Proteins and Peptide Mixtures

The masses of proteins and peptides are relevant data to peptide sequencing. Crude estimates may be obtained by SDS-PAGE, but more accurate ones are available from mass spectrometry. ESI *(30)* and MALDI *(31)*, have revolutionized the application of mass spectrometry to the analysis of biomolecules, especially in the higher molecular mass range. In ESI, sample solution (a typical solvent for proteins is 1:1 water + acetonitrile containing 0.1% trifluoroacetic acid [TFA] is pumped through a narrow-bore metal capillary held at a few kilovolts potential. The charged liquid sprays into the atmospheric pressure ion source and, as the droplets evaporate further, they release sample ions, contained within the liquid, into the gas phase. These gas-phase ions are introduced into the analyzer for mass analysis. The ESI technique measures ions already present in solution; hence the use of TFA to form, by protonation, ions representative of the protein in solution. ESI is a mild process so that intact molecular ions, from analytes with molecular masses in excess of 100-kDa, may be produced. Sample requirements are in the low- to sub-picomole range.

As a relatively extreme example *(32)*, the molecular mass of the murine B72.3 monoclonal antibody (MAb) has been measured as $146,861 \pm 52$. This figure is approx 3% higher than that determined from the DNA sequence since mass spectrometry measures the whole molecule, including glycosylation. A more routine measurement *(33)* is of αB_2 crystallin protein (calc. mol wt 20,201.0), which confirms the expected component $(20,200 \pm 0.9)$ but also indicates the well-resolved presence of 10% of a component at 20,072.2 (calc. for loss of terminal Lys 20072.8) as well as minor components at 20,241 and 20,280. Normally, only molecular mass information is available, but useful fragmentation can be induced by a voltage change in the ion source–analyzer interface region. Changing solution conditions can affect protein association and can also affect the ESI spectra. For example, concanavalin A, in a pH 8.4, 10 mM NH$_4$OAc solution, produces mainly tetrameric ions, whereas at pH 5.7, ions indicative of a dimer species are recorded. Both observations are consistent with known solution behavior *(34)*.

MALDI is an alternative technique for high-molecular-mass analysis that allows the ready observation of ions from intact proteins with molecular masses in excess of 200 kDa. In MALDI, the sample is prepared as a solid solution in a large excess of a matrix compound (typical matrices for proteins are 2,5-dihydroxybenzoic acid and α-cyano-4-hydroxycinnamic acid). The sample preparation is irradiated with laser light, which the matrix absorbs and transforms into excitation energy for the solid system. The result is the sputtering of surface layers of matrix and analyte as well as ionization of some of the analyte molecules. Overall, the sensitivity of MALDI is equivalent to that of ESI, but an advantage of MALDI, compared with ESI and other techniques, is its relative (but not absolute) tolerance of inorganic or organic contaminants. A disadvantage may be the lower mass resolution and reduced accuracy of mass assignment compared with ESI, although, heterogeneity, e.g., of glycosylation (often present in the samples to which MALDI is applied), can itself diminish mass measurement accuracy *(35)*. A further advantage of ESI is that it can be linked on-line to liquid-phase separation techniques, such as HPLC.

Protein variants having substitutions of one or a few amino acids differ in molecular mass from the normal protein and are, theoretically, detectable by measuring the molecular mass of the intact protein. There is, however, a minimum mass difference that can be measured with sufficient accuracy by mass spectrometry. In addition, this smallest difference in molecular mass increases with increasing size of the proteins. A theoretical analysis suggests that the minimum difference in molecular mass detectable, by ESI mass spectrometry, for an equimolar mixture with normal proteins of M_r over 100,000 is about 20 mass units *(36)*. This means that about half of the possible amino acid substitutions in a protein of this size will escape detection. The usual strategy used to detect these smaller changes in molecular mass or to acquire more detailed structural information is based on mass spectrometric analysis of the products of proteolysis, since peptides are smaller and the errors proportionately smaller, too. The most straightforward use of this strategy is mass spectrometric "fingerprinting." A mass spectrum of the peptide mixture resulting from the proteolytic digestion can provide a fingerprint (list of peptide masses) that is specific enough to identify the starting protein uniquely *(37)*. In practice, digestion usually uses trypsin, which produces fragments that are not too large, and therefore allow easy analysis, but are large enough to provide data that offer discrimination from other proteins. The digest mixture is most often analyzed by MALDI. MALDI is particularly well suited to this application since it more readily tolerates low levels of contaminants (including buffer salts) as well as minimizing discrimination between components of mixtures. Peptide molecular masses, determined from the MALDI spectrum, usually with an accuracy of ± 1 or 2 Daltons for tryptic peptides, are compared with a database of mass values, prepared by applying the enzyme cleavage rules to the entries in collections of sequence data, such as SwissProt or PIR. If the sequence database does not contain the unknown protein, then a search should recover those entries that exhibit the closest sequence homology. Again, the digest mass spectrum contains extensive redundancy so that posttranslational modifications that are not noted in the database should still be identified through related entries.

Peptide mass fingerprinting does not work well with protein mixtures and is therefore often used together with gel electrophoresis, which resolves the mixture. In this case, the most practical method is to electroblot the proteins onto a membrane and then digest the chosen protein on the membrane, using conditions that release the proteolytic peptides into solution *(38)*. It is also generally desirable to cleave disulfide bridges prior to digestion to make the protein more accessible to the protease. An advantage of this method is that any SDS can be rinsed away prior to digestion; otherwise, removal of the SDS by selective precipitation of the protein *(39)* or chromatographic purification of the peptides is necessary. PVDF is preferred to nitrocellulose membrane for *in situ* digests because of the lower chemical background *(37)*. Stains, such as sulforhodamine B for PVDF *(40)*, or zinc imidazole for gels *(41)*, cause less signal suppression in MALDI and do not require destaining. Volatile buffers, such as ammonium bicarbonate, are easily removed by lyophilization. A nonionic detergent, such as octyl glucoside, is better tolerated than SDS in both MALDI and ESI, as has already been discussed in **Subheading 2.** *(42)*. This approach to identification of proteins in 2-D PAGE has found application in studies of the proteome.

This mass spectrometric method is considerably faster and more sensitive than automated Edman sequencing of individual proteins or peptides (performed one at a time), and much more specific (i.e., related to sequence) than PAGE or HPLC. Thus, MALDI analysis takes only a few minutes and, under favorable conditions, requires substantially less than a picomole of sample *(37)*, allowing the remainder to be used for Edman sequencing, if necessary. A typical protein can be uniquely identified using a database, such as SwissProt, with just 4 or 5 peptide mass values. However, peptide mass fingerprinting is limited to the identification of proteins for which sequences are already known. It is not a method of structural elucidation. More recently, the use in this area of mass spectrometer sample tips that carry covalently bound trypsin has been demonstrated *(43)*.

Proteolytic digestion, for example with trypsin, in combination with HPLC separation and on-line ESI, is another form of peptide mapping that separates and provides molecular mass information on the peptides that result from the digestion of a protein *(44)*.

As with fingerprinting, peptide mapping is highly effective in the characterization of protein primary structure. One advantage of coupling ESI with a separation technique is that spectra are simple and suppression (by buffer components) is minimized. With ESI, in addition to molecular mass information, some peptide sequence information can be made available by adjusting voltages in the ion source–analyzer interface region to induce collision-induced dissociation (CID) of the molecular ion into fragment ions (referred to as CID-MS to distinguish it from the CID experiment in tandem mass spectrometry—*see* **Subheading 7.**). The result is a spectrum that also contains signals from any background material in the ion source, although this interference is minimized by the coupling with a separation technique. This extraneous information can complicate interpretation, certainly for full-sequence determination, but partial sequence information can usually be extracted, and this partial data can be a very useful complement in searching a protein database. Recently, several computer methods have been described for matching partial CID data to a sequence data base for the purpose of identifying an unknown *(45–47)*.

Another mapping method that can sometimes be applied involves the use of fast-atom bombardment (FAB) mass spectrometry. In FAB ionization *(48)*, the unseparated proteolysis product is presented, dissolved in a liquid matrix of low volatility (e.g., glycerol), to a bombarding beam of ions or neutrals, and thereby releases sample-related secondary ions. Although, as in all mixture analyses, there is the danger of suppression of ions from certain components, point mutations may often be located in one simple experiment using FAB. As an example *(36)*, β-globin normally gives (among other products) tryptic peptides 13 and 14 at *m/z* 1378.5 and 1149.6, respectively, whereas a FAB spectrum of the variant with Lys-132 (the cleavage point between T13 and T14) replaced by Asn shows a single peptide at *m/z* 2495.3.

Means other than trypsin have been used to fragment proteins for mass spectrometry, for instance cyanogen bromide, which cleaves only at the relatively uncommon methionine (*see* **Table 1**) and therefore generates large peptide fragments of molecular mass 20,000–30,000 kDa, which can, nevertheless, readily be measured to the nearest integer mass *(36)*. Electrospray ionization is used despite the fact that the direct analysis of peptide mixtures by ESI without on-line HPLC separation can present difficulties resulting from varying efficiencies of ionization between peptides. Thus, the use of cyanogen bromide, which produces too few fragments for fingerprinting, can still provide enough data to confine a suspected site of mutation to a small part of the protein molecule by a determination of the shift in molecular mass.

Methods have been devised for analysis of specific types of peptides, e.g., glyco- and phosphopeptides thus, tryptic digest of a glycoprotein will contain glycopeptides. An HPLC/ESI-MS method for specifically locating these materials in a digest *(49,50)*, is based on the fact that all glycopeptides contain isomers of hexose, and/or *N*-acetylhexosamine, and/or neuraminic acid, which, when the voltage in the ion source-analyzer region is increased as described earlier in **Subheading 4.**, produce diagnostic fragment ions (ions of *m/z* 163, *m/z* 204, and *m/z* 292, respectively and *m/z* 366 [hexose-*N*-acetylhexosamine] are used). Peaks in the chromatograms corresponding to these diagnostic ions indicate the elution of a glycopeptide at that point; in addition, full spectra are automatically recorded between each sampling of the diagnostic ions, at a low interface voltage so that the molecular mass of the eluting glycopeptide is immediately available.

A similar method, but for selectively identifying phosphopeptides containing phosphate linked to Ser, Thr, and Tyr in complex mixtures *(51–53)*, uses operation in the negative-ion mode. Diagnostic ions for Ser- and Thr-linked phosphate are *m/z* 97 ($H_2PO_4^-$), *m/z* 79 (PO_3^-), and *m/z* 63 (PO_2^-), whereas only *m/z* 79 and 63 are formed from Tyr-linked phosphate. A peak in the chromatograms for these marker ions indicate the elution of a phosphopeptide and, just as with glycopeptides, full spectra, automatically recorded at a low interface voltage, give the molecular mass of the eluting phosphopeptide.

Mass spectrometry may also be used to rapidly count cysteines, free sulfhydryl groups, and disulfide bonds in proteins *(54)*. The method involves three steps:

1. Reducing all disulfide bonds and alkylating free sulfhydryl groups;
2. Alkylating only the original free sulfhydryl groups without reduction of disulfide bonds; and
3. Determining the molecular mass (by ESI or MALDI) of the native protein and those formed in **steps 1** and **2**.

Thus, the appropriate molecular masses for ribonuclease A (using ESI) were 13,682.3 (MH$^+$, native), 14,156.8 (MH$^+$, reduced and alkylated), and 13,680.8 (MH$^+$ alkylated only). From these measurements, the number of cysteines, for example, is found to be $(14,146.8–13,682.3)/59 = 8$.

Once numbers have been established, the next step is to assign the disulfide linkages in the proteins. The method employed uses digestion, chemical and/or enzymatic, to give a mixture of peptides, each of which contains not more than one disulfide bond. For example, with hen egg-white lysozyme, treatment with cyanogen bromide was followed by tryptic digestion *(55)*. MALDI spectra of the tryptic digest before and after reduction with dithiothreitol (DTT) identify the disulfide-linked peptides since their masses change to those of the corresponding pair of sulfhydryl peptides on reduction. Six such peptides, which account for three of four disulfide bonds in the protein, were identified using only a few picomoles of material. Note that this method requires that proteins are cleaved, between the cysteinyl residues, without disulfide exchange. A high-density core of cysteine residues can make this process difficult, since the cysteine residues are very close to each other; in this case, partial acid hydrolysis can be an alternative where enzymatic cleavage is unsuccessful *(56)*.

In the ESI spectra from the lysozyme tryptic digest referred to above, signals from some of the peptides were very weak or suppressed completely. On-line HPLC/ESI-MS, also in combination with DTT reduction, is, however, a practical method; the HPLC separation provides simple spectra and eliminates suppression. Thus, HPLC peaks that correspond to intermolecular disulfide-bonded peptides are lost after DTT treatment, and new peaks corresponding to their constituent peptides are found in the early eluting fractions The appropriate fractions can be collected by off-line HPLC and further examined by FAB mass spectrometry. For example, on-probe reduction in a glycerol/thioglycerol matrix can be used to verify disulfide-containing peptides *(57)*. If a peptide contains an intramolecular disulfide bond, a new peak, two mass units higher, will appear after reduction. If a peptide contains an intermolecular disulfide bond, the peak corresponding to this peptide will be replaced by peaks corresponding to the two constituent peptides. This reduction approach, using FAB, is simple but cannot identify thiol-containing peptides; it also requires good mass resolution to identify intramolecular disulfide-containing peptides at higher molecular masses. Furthermore, the constituent peptides are not always observed, particularly when the sample is a complex mixture. On-probe oxidation is another simple approach to the identification of both thiol- and disulfide-containing peptides *(58)*. Performic acid converts cysteine to cysteic acid (an increase in the molecular mass of 48 Daltons). Intramolecular disulfide-linked peptides will increase in molecular mass by 98 Daltons, whereas intermolecular disulfide-linked peptides will be replaced by two peptides, each containing cysteic acid.

5. Chemical N-Terminal Sequencing

Chemical sequencing of the N-terminus of a polypeptide utilizes the chemistry described by Edman and illustrated in **Fig. 2**. The terminal residue is derivatized by reaction with phenylisothiocyanate (PITC) and cleaved in acid conditions to give an anilinothiozolinone (ATZ) derivative. This is unstable, but is converted to the stable phenylthiohydrantoin (PTH) derivative. This cycle of chemistry generates a new

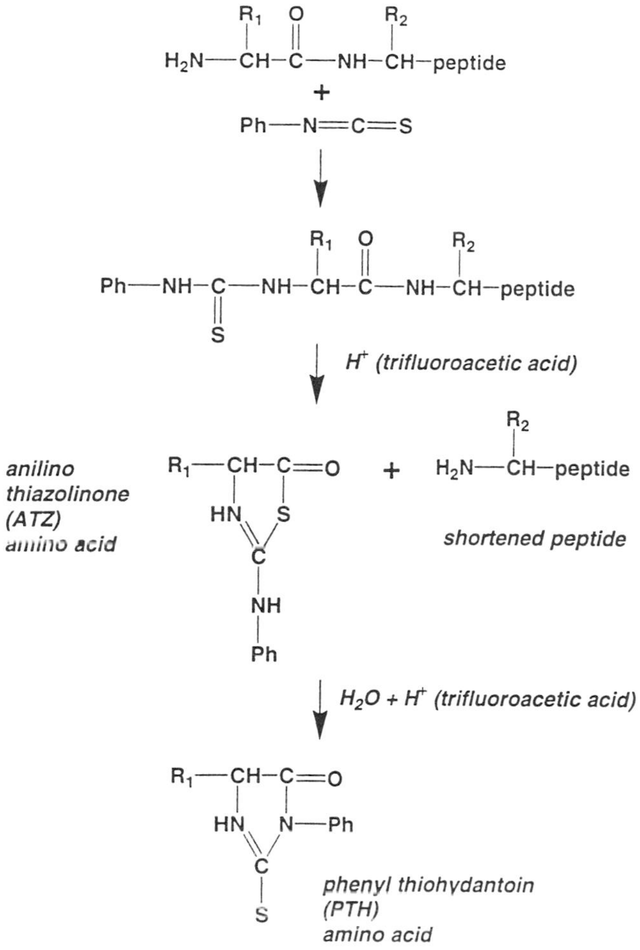

Fig. 2. Edman chemistry for N-terminal sequencing of polypeptides.

N-terminus, the next residue in the sequence. The cycle is repeated numerous times to generate the order of constituent residues. This chemistry may be carried out manually but currently is mostly done in automated protein sequences, in which interference by oxygen and water is minimized. Automated sequencing is also more sensitive, more rapid, and capable of generating longer sequences. Cycles can be carried out with good yields (better than 95% per cycle) to give extended sequences of 50 residues or more if desired. Several designs of sequencer are commercially available utilizing the same basic chemistry and able to analyze samples in solution, bound to PVDF membrane, or linked to a solid phase (for instance synthetic peptide still linked to the resin on which it was synthesized). Solid-phase sequencing can be advantageous (for instance, in studying sites of phosphorylation) because more extreme conditions can be used to extract the cleaved ATZ derivative, without the accompanying danger of washing out the remaining sample.

Edman chemistry represents the first stage of sequencing. The second is identification of the PTH-amino acid derivative produced by each cycle. Manual sequencing achieves this by thin layer chromatography, but automated sequences have on-line HPLC, so PTH-residue analysis occurs automatically. Different designs of sequencer have slightly different HPLC systems but operate basically similar C18 reversed-phase chromatography. The product from each cycle of sequencing is identified by comparing it with standard PTH-amino acids.

As **Fig. 2** indicates, Edman chemistry requires a free amine, but this is frequently not available. It has been estimated that perhaps as many as 80% of proteins are naturally modified on their N-terminal amino group, which is therefore "blocked" to sequencing. In addition, artificial blockage may arise through reaction of the amine with species in buffers and polyacrylamide gels (as mentioned in **Subheading 2.**). Methods have been described for removal of blocking groups, such as acetylserine, acetyl threonine, formylated residues and pyroglutamic acid; for instance, *see* Hirano et al. *(28)* for methods suitable for deblocking of samples on PVDF membrane.

Cysteine residues are unstable in the conditions used in Edman chemistry and so generate a blank at the appropriate position. This can be overcome by modification of the cysteine thiol, as mentioned in **Subheading 3.**, to a stable derivative, such as *S*-[2-(4-pyridyl)ethyl]cysteine. Such derivatives can be positively identified as the PTH derivatives. Tryptophan may undergo some destruction also, thus giving somewhat low yields at its PTH derivative, but yields are still sufficient for sequencing. Serine and threonine likewise give low yields because of their conversion to dehydro-derivatives, but this is generally a problem only in weaker sequences. Glycosylated residues are not soluble in the solvents generally used in automated sequencers for extraction of ATZ derivatives from reaction mixtures. Consequently, glycosylated residues are seen as blanks or as low yields of the unmodified residue (if glycosylation had not occurred on all residues). This result might resemble a "blank" resulting from cysteine unless the latter has been modified as discussed above. Glycosylation may be confirmed by treatment of the sample to remove the oligosaccharide (by enzymatic or chemical means), followed by sequencing, which this time would reveal the unmodified residue. Potential sites of *N*-glycosylation may also be recognized by their consensus sequence: Asn-X-(Ser or Thre) where Asn is glycosylated and X is any amino acid. Sites of *O*-glycosylation are less obvious, having a weaker consensus. Phosphorylated residues present other problems, giving low yields. The efficiency of recovery of the phosphorylated PTH residue depends on the surrounding sequence and on the residue itself (lability in acid increasing in the order phosphotyr > phosphoser > phosphothre). There are various methods for analysis of these residues and the rarer 1- and 3-phosphohistidine and phosphoarginine *(59)*, including detection of incorporated ^{32}P radiolabel and conversion of (nonradiolabeled) phosphoserine to *S*-ethylcysteine (which gives a stable PTH derivative).

With the exception of modified or unusual amino acid residues, such as those mentioned above, sequencing proceeds in a generally predictable way. Low-picomole or femtomole amounts of (unblocked) sample can generate short sequences (sufficient for screening databases). More sample allows longer sequences. The length of sequence possible is finite—each cycle is several percentage points incomplete, and each cycle involves an acid step that may generate cleavages at internal points in the protein, each of these providing a new N-terminus that also becomes sequenced. Thus, yields drop and background increases. How quickly this renders interpretation of the sequencing impossible depends on the sequence of the protein as much as the efficiency of the chemistry itself. Pure samples are best for unequivocal sequencing but on occasion protein mixtures can be analyzed. This can be done if one (or more) protein sequence is known and can be subtracted, and/or if the amounts of the different components are so different that on quantification of data the various sequences can be differentiated. Quantification of chemical sequencing data is an important aspect of the method: the proportions of various polypeptides in a mixture can be estimated and indicate whether a sequence is of the expected protein or a minor (unblocked) contaminant of the blocked sample. The ability to generate quantified data is a significant difference between chemical sequencing and mass spectrometric approaches (which are not quantitative).

Protein ladder sequencing *(60)* is a mass-spectrometry-based variant of Edman sequencing. The method involves two steps: A nested set of peptide fragments from the polypeptide chain is generated by stepwise chemical degradation at the amino terminus. There are several ways in which such fragments can be generated *(60,61)*; the original method *(60)* was based on controlled partial Edman degradation using an automatic gas-phase sequencer. Then the molecular masses of the set of peptides are measured by recording a MALDI mass spectrum of the mixture. The identities of amino acid residues are determined by the mass differences between consecutive molecular ions in the mass spectrum, and their sequence is defined by their order of occurrence. The molecular mass of modified residues provides information on the modification. For example, phosphorylation of serine, threonine, or tyrosine increases the residue mass by 80 Daltons, whereas deamidation of asparagine or glutamine decreases the mass by 1 Dalton.

6. Chemical C-Terminal Sequencing

The sequence of the C-terminus of a polypeptide is of as much interest as that of the N-terminus, but has been more difficult to obtain because an efficient chemical method equivalent to Edman chemistry has not been available. Recent efforts have made advances in this direction, however. A method has been described for sequencing manually that works best on solid phase-linked polypeptide, so that retrieval of the sample for subsequent cycles of chemistry is readily achieved *(62,63)*. The chemistry involves reaction of the C-terminal carboxyl group with acetic anhydride/acetic acid (to give the peptidyl thiohydantoin), then reaction with thiocyanate (the coupling reaction), then treatment with base (to cleave off the C-terminal residue). The released thiohydantoin amino acid is then analyzed by HPLC. Certain residues (such as Asp, Glu, Ser, and Thre) are problematic, and the procedure works best on peptides containing hydrophobic residues. On such a sample a sequence of the order of ten residues might be obtained. Automated N-terminal sequencers have been modified to run this chemistry on samples linked to beads or on PVDF membrane, but recently a different chemistry has been developed together with a suitable commercially available automated sequencer, and it seems likely that most C-terminal sequencing will be carried out this way in the future *(64)*. The chemistry is illustrated in **Fig. 3**. The resultant thiohydantoin amino acids are analyzed by on-line C18 reversed-phase HPLC. The chemistry is suitable for all 20 common amino acids (with cysteine requiring alkylation in order to be distinguishable from serine, which, like unmodified cysteine, generates the product dehydroalanine thiohydantoin). At the present stage of development the method can

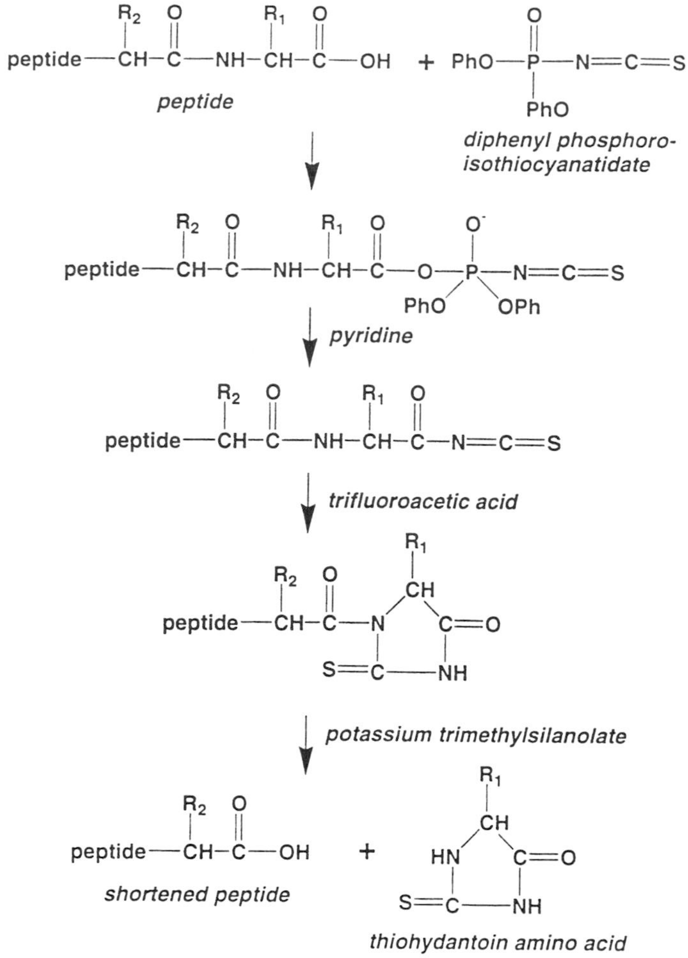

Fig. 3. Automated C-terminal sequencing chemistry.

generate sequences of three to five residues from a sample of a few tens of picomoles or more, at the rate of approx 1 cycle/h. Thus, the method is currently about ten times less sensitive than automated N-terminal sequencing—not sensitive enough for the sequencing of spots on 2D-PAGE. More heavily loaded gels may be blotted to Zitex (porous Teflon) membrane for sequencing, however (PVDF is unsuitable because of its instability toward this chemistry).

An alternative approach to C-terminal sequencing is that using carboxypeptidase *(65)*, an exopeptidase that progressively hydrolyses the C-terminal residue from a polypeptide provided that the C-terminus is accessible (this provision possibly necessitating reduction and denaturation of the substrate). Four carboxypeptidases have been used for this purpose: A, B, C, and Y. They show some degree of substrate specificity, however. As discussed in **ref. 66**, carboxypeptidase A preferentially removes C-terminal residues with aromatic or large side chains, others at lesser rates, and Hydroxypro, Pro, Lys, and Arg not at all, so that digestion and sequencing stops at these residues. On the other hand, carboxypeptidase B shows great preference for the basic residues, Arg and Lys. Carboxypeptidases C and Y have broader specificity. Carboxypeptidase

Y is also a robust enzyme, capable of working in urea and detergents, such as might be used for substrate denaturation. Consequently, carboxypeptidase Y (EC 3.4.16.1) has been the most widely used of these enzymes, either alone or in combination with other carboxypeptidase(s) in order to optimize digestion. Thus, carboxypeptidase Y can be used at an enzyme:substrate ratio of 1:100 in a volatile buffer (pyridine acetate or ammonium acetate, pH 5.5–6.5), with incubation at about 25°C. Samples can then be taken at intervals (minutes or hours) in order to monitor reaction. In the past the reaction was followed by identification and quantification of the liberated amino acids by amino acid analysis. This meant that the method was insensitive—up to 20 nmol or more of sample being required to generate a short sequence (depending on the sensitivity of the analysis method). An improvement in this has come with the development of mass spectrometric techniques that can analyze the shortened peptide products of digestion.

7. Sequencing by Tandem Mass Spectrometry

Tandem mass spectrometry (MS/MS) is an important alternative to the Edman method for obtaining sequence information from peptides. The ion corresponding to the protonated peptide molecule is selected in the first analyzer and is collided with argon gas in a collision cell to generate fragment ions (collision-induced dissociation, CID). The fragment ions generated are then mass-separated in a second analyzer and finally detected. Some mass analyzers, e.g., sector analyzers, provide high collision energies, whereas quadrupole analyzers provide low-energy. More details of these instruments may be found in **ref. 67** and, in particular, the use of time-of-flight instrumentation (used in a conventional manner for most MALDI experiments) for MS/MS is discussed in **ref. 68**.

The ion types produced from CID of peptides (**Fig. 4**) are denoted a_n, b_n, c_n, and d_n when the positive charge is retained on the peptide N-terminal fragment and v_n, w_n, x_n, y_n, and z_n, when the charge is retained on the C-terminal fragment *(69–72)*. Here, the subscript refers to the position of fragmentation in the amino acid chain, whereas the letter (e.g., a, b, v, w) denotes the ion type. The d_n and w_n ions, produced by side-chain cleavage and found only in high-energy CID spectra *(69,70)*, are particularly useful because they allow leucine and isoleucine residues to be distinguished.

The most useful CID spectra are those with one or two complete ion series; in these spectra, the amino acid sequence is derived from the mass difference between successive ions in a series. When more than two ion series are present, they are generally incomplete and interpretation is more difficult, although complete sequencing data are not needed when the aim is to identify, by database searching, the protein from which the peptide was derived. Arginine, which is the most basic amino acid residue, is the preferred site for gas-phase protonation and its location has a strong influence on the appearance of peptide CID spectra. A C-terminal arginine results in predominantly v_n and w_n series, i.e., C-terminal, ions *(69,70)*, whereas N-terminal arginine affords mainly a_n and d_n series, i.e., N-terminal, ions. An arginine residue located at a position other than a terminal one, or no arginine residue, results in a mixture of N-terminal and C-terminal product ions that is more difficult to interpret.

Phosphorylation of serine, threonine, or tyrosine gives a derivative that fragments during CID to form a moderately abundant diagnostic ion, corresponding to the loss of

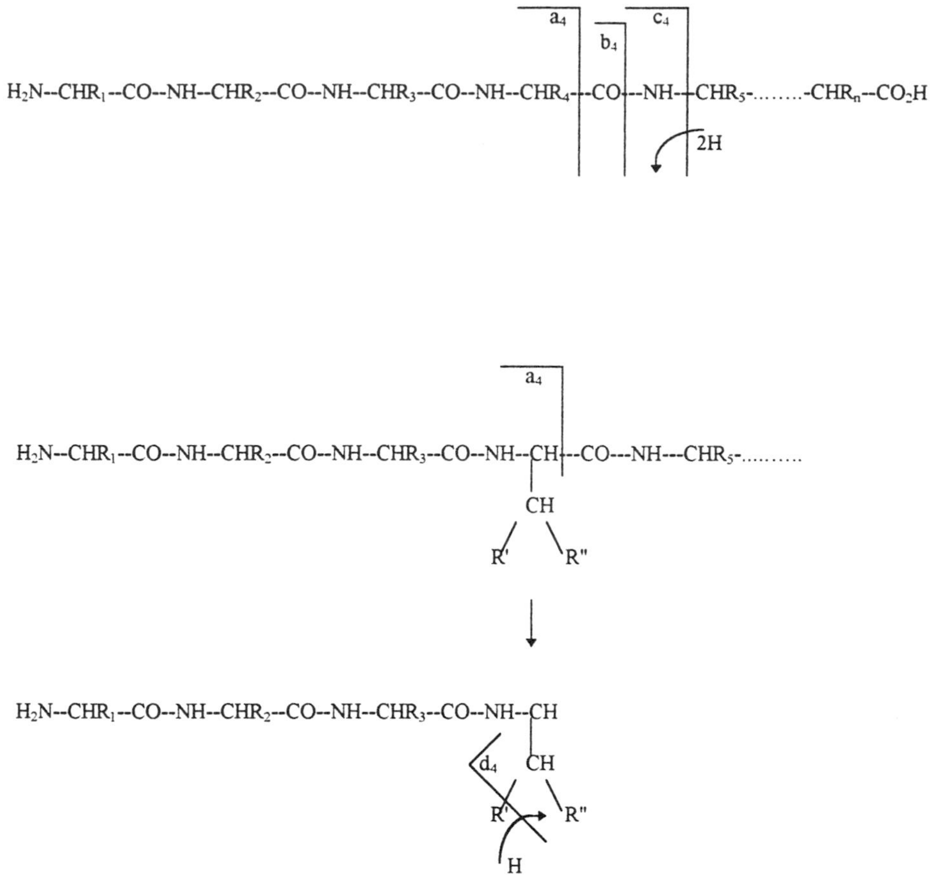

Fig. 4. Ion types produced from dissociation of peptides. The a, b, and c ions are formed by main-chain fragmentation, with the positive charge on the N terminus, whereas the d ion is produced by side-chain fragmentation. The v, w, x, y, and z ions are formed with the positive charge on the C terminus; in this case, the v and w ions are produced by side-chain fragmentation.

H_3PO_4 from the protonated molecule *(73,74)*. Overall, peptide fragmentation is not adversely affected. Again, most -SH modifying groups (methyl, peracetyl, acetamidomethyl, vinylpyridyl) do not adversely affect CID fragmentation of peptides. Suitable modifying groups can be introduced to facilitate the interpretation of high-energy CID spectra. For example, N-terminal derivatization with a fixed-charge bearing group *(69)* results in greatly simplified spectra; the dimethylalkylammonium derivative *(75)* is found to be an effective charge carrier, and these derivatives have no detrimental effect, as well as being easily synthesized.

An ideal case for sequencing occurs when two charges are present and are localized at the N- and C-termini of the peptide, i.e., a doubly charged ion. In this respect, very appropriate peptides are produced by tryptic digestion since these have highly basic functional groups at each terminus (C-terminal arginine or lysine and the amino(N)-terminus itself), on which charges preferentially reside. Electrospray ionization readily produces multiply charged ions *(30)* and, in the case of tryptic digest products, these

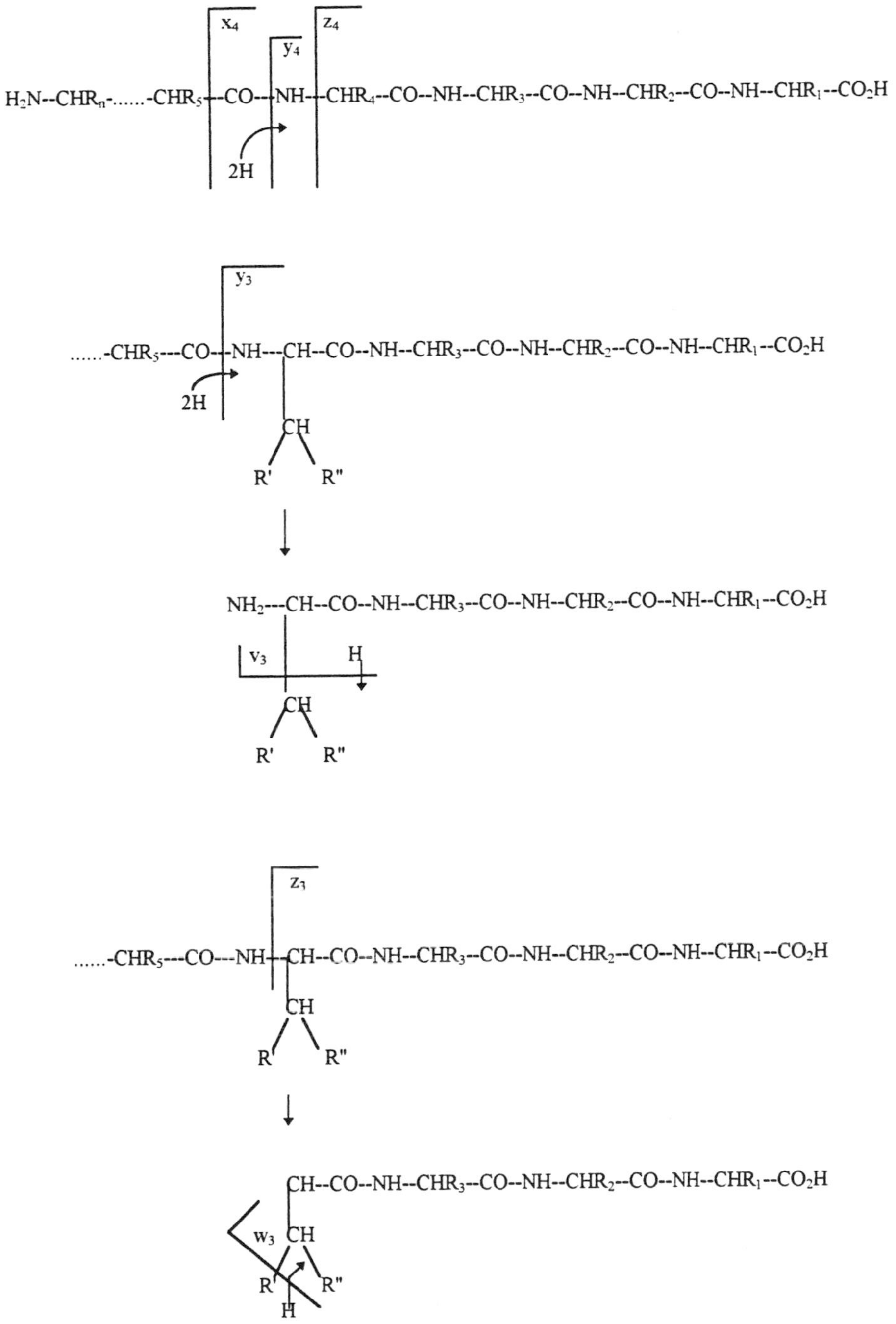

Fig. 4. *(continued)*

are usually doubly charged. With these doubly charged ions, the y_n ion series is easily identified as the predominant ion series with some *m/z* values greater than that of the precursor ion (the doubly charged molecular ion is chosen as the precursor ion for fragmentation) *(76)*.

8. Amino Acid Analysis

This, the quantification of the amino acids present in a protein, is clearly a reflection of the protein's amino acid sequence. Amino acid analysis is still of value in the characterization of synthetic peptides and recombinant proteins and is excellent for quantification of a protein sample, but for most other purposes it has been superceded by sequencing and mass spectrometric techniques. The first stage in amino acid analysis is the conversion of the protein to a mixture of single amino acids by acid hydrolysis (or alkaline hydrolysis if Tryp is to be quantified). This deamidates Asn and Gln to their respective acids so that none of them can be quantified absolutely. The reader is referred to Davidson *(77)* for further discussion of hydrolysis methods. The second stage is the resolution of this mixture (by chromatography), and identification and quantification of each amino acid (by comparison with standards). To gain sensitivity, the amino acids are derivatized to give more highly UV-absorbant or fluorescent forms. For example, amino acids may be derivatized before resolution by HPLC, by reaction with phenylisothiocyanate to give phenylthiocarbamyl amino acids *(78)*. Alternatively, amino acids may be derivatized after resolution, by ion exchange chromatography, by reaction with ninhydrin *(79)* . The precolumn derivatization approach is more sensitive (at about 1 pmol level of sensitivity), but the postcolumn method is affected less by interfering contaminants (which are separated away from the amino acids before derivatization).

Proteins that have been blotted to PVDF membrane may be subjected to hydrolysis and amino acid analysis *(80)*, so the method may be of use in the identification of proteins in polyacrylamide gels. Amino acid analysis data may be combined with other characteristics (i.e., mass or pI) to allow the screening of databases of known proteins *(81)*.

A selective analysis of peptides may be carried out during their purification by HPLC in order to indicate which contain Tryp, Phe, Tyr, and/or phosphotyrosine. This is done by use of an on-line diode array detector and second (or higher order) derivative spectroscopy *(82–84)*. This approach may prove useful in allowing selection of peptides of particular interest (such as phosphorylated ones) for further analysis.

References

1. Edman, P. (1950) Method for determination of the amino acid sequence in peptides. *Acta Chem. Scand.* **4,** 283–293.
2. Smith, B. J. and Tempst, P. (1996) Strategies for handling polypeptides on a microscale, in *Methods in Molecular Biology, vol. 64: Protein Sequencing Protocols* (Smith, B. J., ed.), Humana, Totowa, NJ, pp. 1–16.
3. Chang, H. W. and Bock, E. (1980) Pitfalls in the use of commercial nonionic detergents for the solubilisation of integral membrane proteins: sulfhydryl oxidising contaminants and their elimination. *Anal. Biochem.* **104,** 112–117.
4. Ogorzalek Loo, R. R., Dales, N., and Andrews P. C. (1996) The effect of detergents on proteins analyzed by electrospray ionization, in *Methods in Molecular Biology, vol. 61: Protein and Peptide Analysis by Mass Spectrometry* (Chapman, J. R., ed.), Humana, Totowa, NJ, pp. 141–160.
5. Börnsen, K. O., Gass, M. A., Bruin, G. J., von Adrichem, J. H. A., Biro, M. C., Kresbach, G. M., and Ehrat, M. (1997) Influence of solvents and detergents on matrix-assisted laser desorption/ionization mass spectrometry measurements of proteins and oligonucleotides. *Rapid Commun. Mass Spectrom.* **11,** 603–609.

6. Schey, K. L. (1996) Hydrophobic proteins and peptides analyzed by matrix-assisted laser desorption/ionization, in *Methods in Molecular Biology, vol. 61: Protein and Peptide Analysis by Mass Spectrometry* (Chapman, J. R., ed.), Humana, Totowa, NJ, pp. 227–230.

7. Philp, R. J. (1996) Preparation of peptides for microsequencing from proteins in polyacrylamide gels, in *The Protein Protocols Handbook* (Walker, J. M., ed.), Humana, Totowa, NJ, pp. 393–398.

8. Ward, M. (1996) *In situ* chemical and enzymatic digestions of proteins immobilized on miniature hydrophobic columns, in *The Protein Protocols Handbook* (Walker, J. M., ed.), Humana, Totowa, NJ, pp. 399–403.

9. Fernandez, J. and Mische, S. M. (1996) Enzymatic digestion of membrane-bound proteins for peptide mapping and internal sequence analysis, in *The Protein Protocols Handbook* (Walker, J. M., ed.), Humana, Totowa, NJ, pp. 405–414.

10. Stone, K. L. and Williams, K. R. (1996) Enzymatic digestion of proteins in solution and in SDS polyacrylamide gels, in *The Protein Protocols Handbook* (Walker, J. M., ed.), Humana, Totowa, NJ, pp. 415–425.

11. Judd, R. C. (1996) SDS-polyacrylamide gel electrophoresis of peptides, in *The Protein Protocols Handbook* (Walker, J. M., ed.), Humana, Totowa, NJ, pp. 101–107.

12. Judd, R. C. (1996) Peptide mapping by sodium dodecylsulfate-polyacrylamide gel electrophoresis, in *The Protein Protocols Handbook* (Walker, J. M., ed.), Humana, Totowa, NJ, pp. 447–451.

13. Judd, R. C. (1996) Peptide mapping by high-performance liquid chromatography, in *The Protein Protocols Handbook* (Walker, J. M., ed.), Humana, Totowa, NJ, pp. 453–455.

14. Shaw, C. (1996) Reverse-phase HPLC purification of peptides from natural sources for structural analysis, in *Methods in Molecular Biology, vol. 64: Protein Sequencing Protocols* (Smith, B. J., ed.), Humana, Totowa, NJ, pp. 101–107.

15. Smith, A. J. (1996) Analytical and micropreparative capillary electrophoresis of peptides, in *Methods in Molecular Biology, vol. 64: Protein Sequencing Protocols* (Smith, B. J., ed.), Humana, Totowa, NJ, pp. 91–99.

16. Sutton, C. W. and O'Neill, J. A. (1996) Preparation of glycopeptides, in *Methods in Molecular Biology, vol. 64: Protein Sequencing Protocols* (Smith, B. J., ed.), Humana, Totowa, NJ, pp. 73–79.

17. Hayashi, T. H. and Sasagawa, T. (1996) Selective isolation of the carboxy terminal peptide from a peptide, in *Methods in Molecular Biology, vol. 64: Protein Sequencing Protocols* (Smith, B. J., ed.), Humana, Totowa, NJ, pp. 81–83.

18. Hayashi, T. H. and Sasagawa, T. (1996) Selective isolation of the amino terminal peptide from an α-amino blocked protein, in *Methods in Molecular Biology, vol. 64: Protein Sequencing Protocols* (Smith, B. J., ed.), Humana, Totowa, NJ, pp. 85–89.

19. Tawfik, D. (1996) Side-chain selective chemical modifications of proteins, in *The Protein Protocols Handbook* (Walker, J. M., ed.), Humana, Totowa, NJ, pp. 349–351.

20. Speicher, D. W. (1996) Chemical modification of proteins, in *Current Protocols in Protein Science,* vol. 1 (Coligan, J. E, Dunn, B. M, Ploegh, H. H, Speicher, D. W., and Wingfield, P. T., eds.), Wiley, New York, pp. 15.0.1,15.0.2.

21. Carne, A. F. and U, S. (1996) Chemical modification of proteins for sequence analysis, in *Methods in Molecular Biology, vol. 64: Protein Sequencing Protocols* (Smith, B. J., ed.), Humana, Totowa, NJ, pp. 271–284.

22. Glazer, A. N., Delange, R. J., and Sigman, D. S. (1975) *Chemical Modification of Proteins: Selected Methods and Analytical Procedures.* North Holland Pub. Co. Amsterdam, Oxford, American Elsevier Pub. Co. Inc., NY.

23. Crankshaw, M. W. and Grant, G. A. (1996) Modification of cysteine, in *Current Protocols in Protein Science,* vol. 1 (Coligan, J. E., Dunn, B. M., Ploegh, H. H., Speicher, D. W., and Wingfield, P. T., eds.) Wiley, New York, pp. 15.1.1–15.1.18.

24. Hounsell, E. F, Davies, M. J. and Smith, K. D. (1996) Enzymatic release of O- and N-linked oligosaccharide chains, in *The Protein Protocols Handbook* (Walker, J. M., ed.), Humana, Totowa, NJ, pp. 657,658.

25. Hounsell, E. F., Davies, M. J., and Smith, K. D. (1996) Chemical release of O-linked oligosaccharide chains, in *The Protein Protocols Handbook* (Walker, J. M., ed.), Humana, Totowa, NJ, pp. 647,648.

26. Mizuochi, T. and Hounsell, E. F. (1996) Release of N-linked oligosaccharide chains by hydrazinolysis, in *The Protein Protocols Handbook* (Walker, J. M., ed.), Humana, Totowa, NJ, pp. 653–656.

27. Colyer, J. (1996) Analysing protein phosphorylation, in *The Protein Protocols Handbook* (Walker, J. M., ed.), Humana, Totowa, NJ, pp. 501–506.

28. Hirano, H, Komatsu, S., and Tsunasawa, S. (1996) On-membrane deblocking of proteins, in *Methods in Molecular Biology, vol. 64: Protein Sequencing Protocols* (Smith, B. J., ed.), Humana, Totowa, NJ, pp. 285–292.

29. Walker, J. M. and Sweeney, P. J. (1996) Removal of pyroglutamic acid residues from the N-terminus of peptides and proteins, in *The Protein Protocols Handbook* (Walker, J. M., ed.), Humana, Totowa, NJ, pp. 525–527.

30. Fenn, J. B., Mann, M., Meng, C. K., Wong, S. F., and Whitehouse, C. M. (1990) Electrospray ionization—principles and practice. *Mass Spectrom. Rev.* **9,** 37–70.

31. Hillenkamp, F., Karas, M., Beavis, R. C., and Chait, B. T. (1991) Matrix-assisted laser desorption/ionization of biopolymers. *Anal. Chem.* **63,** 1193A–1203A.

32. Bennett, K. L., Hick, L. A., Truscott, R. J. W., Shiel, M. M., and Smith, S. V. (1995) Optimum conditions for electrospray mass spectrometry of a monoclonal antibody. *J. Mass Spectrom.* **30,** 769–771.

33. He, S., Pan, S., Wu, K., Amster, I. J., and Orlando, R. (1995) Analysis of normal fetal eye lens crystallins by high-performance liquid chromatography/mass spectrometry. *J. Mass Spectrom.* **30,** 424–431.

34. Light-Wahl, K. J., Schwartz, B. L., and Smith, R. D. (1994) Observation of the noncovalent quaternary associations of proteins by electrospray ionization mass spectrometry. *J. Am. Chem. Soc.* **116,** 5271–5278.

35. Jespersen, S., Koedam, J. A., Hoogerbrugge, C. M., Tjaden, U. R., van der Greef, J., and Van den Brande, J. L. (1996) Characterization of *O*-glycosylated precursors of insulin-like growth factor II by matrix-assisted laser desorption/ionization mass spectrometry. *J. Mass Spectrom.* **31,** 893–900.

36. Wada, Y. (1996) Structural analysis of protein variants, in *Methods in Molecular Biology, vol. 61: Protein and Peptide Analysis by Mass Spectrometry* (Chapman, J. R., ed.), Humana, Totowa, NJ, pp. 101–113.

37. Cottrell, J. S. and Sutton, C. W. (1996) The identification of electrophoretically separated proteins by peptide mass fingerprinting, in *Methods in Molecular Biology, vol. 61: Protein and Peptide Analysis by Mass Spectrometry* (Chapman, J. R., ed.), Humana, Totowa, NJ, pp. 67–82.

38. Aebersold, R. H., Leavitt, J., Saavedra, R. A., Hood, L. E., and Kent, S. B. H. (1987) Internal amino acid sequence analysis of proteins separated by one- or two-dimensional gel electrophoresis after *in situ* protease digestion on nitrocellulose. *Proc. Natl. Acad. Sci. USA* **84,** 6970–6974.

39. Henderson, L. E., Oroszlan, S., and Konigsberg, W. (1979) A micromethod for complete removal of dodecyl sulfate from proteins by ion-pair extraction. *Anal. Biochem.* **93,** 153–157.

40. Coull, J. M. and Pappin, D. J. C. (1990) A rapid fluorescent staining procedure for proteins electroblotted onto PVDF membranes. *J. Protein Chem.* **9,** 259,260.

41. Ortiz, M. L., Calero, M., Patron, C. F., Castellanos, L., and Mendez, E. (1992) Imidazole-SDS-zinc reverse staining of proteins in gels containing or not SDS and microsequence of individual unmodified electroblotted proteins. *FEBS Lett.* **296,** 300–304.

42. Vorm, O., Chait, B. T., and Roepstorff, P. (1993) Mass spectrometry of protein samples containing detergents. *Proceedings of the 41st Annual Conference on Mass Spectrometry and Allied Topics,* San Francisco, CA, pp. 621.

43. Nelson, R. W., Dogruel, D., Krone, J. R., and Williams, P. (1995) Peptide characterization using bioreactive mass spectrometer probe tips. *Rapid Commun. Mass Spectrom.* **9,** 1380–1385.

44. Covey, T. (1996) Liquid chromatography/mass spectrometry for the analysis of protein digests, in *Methods in Molecular Biology, vol. 61: Protein and Peptide Analysis by Mass Spectrometry* (Chapman, J. R., ed.), Humana, Totowa, NJ, pp. 83–100.

45. Mann, M. and Wilm, M. (1994) Error-tolerant identification of peptides in sequence databases by peptide sequence tags. *Anal. Chem.* **66,** 4390–4399.

46. Bonner, R. and Shushan, B. (1995) The characterization of proteins and peptides by automated methods. *Rapid Commun. Mass Spectrom.* **9,** 1067–1076.

47. Bonner, R. and Shushan, B. (1995) Error-tolerant protein database searching using peptide product-ion spectra. *Rapid Commun. Mass Spectrom.* **9,** 1077–1080.

48. Barber, M., Bordoli, R. S., Elliott, G. J., Sedgwick, R. D., and Tyler, A. N. (1982) Fast-atom bombardment mass spectrometry. *Anal. Chem.* **54,** 645A–657A.

49. Carr, S. A., Huddleston, M. J., and Bean, M. (1993) Selective identification and differentiation of *N*- and *O*-linked oligosaccharides in glycoproteins by liquid chromatography-mas spectrometry. *Protein Sci.* **2,** 183–196.

50. Huddleston, M. J., Bean, M., and Carr, S. A. (1993) Collisional fragmentation of glycopeptides by electrospray ionization LC/MS and LC/MS/MS: methods for selective detection of glycopeptides in protein digests. *Anal. Chem.* **65,** 877–882.

51. Huddleston, M. J., Annan, R. S., Bean, M., and Carr, S. A. (1993) Selective detection of the thr-, ser-, and tyr-phosphopeptides in complex digests by electrospray LC-MS. *J. Am. Soc. Mass Spectrom.* **4,** 710–717.

52. Ding, J., Burkhary, W., and Kassel, D. B. (1994) Identification of phosphorylated peptides from complex mixtures using negative-ion orifice-potential stepping and capillary liquid chromatography/electrospray ionization mass spectrometry. *Rapid Commun. Mass Spectrom.* **8,** 94–98.

53. Covey, T. R., Shushan, B., Bonner, R., Schroder, W., and Hucho, F. (1991) LC/MS and LC/MS/MS screening for the sites of post-translational modification in proteins, in *Methods in Protein Sequence Analysis* (Jornvall, H., Hoog, J.-O., and Gustavsson, A.-M., eds.), Birkhauser Verlag, Basel, Switzerland, pp. 249–256.

54. Feng, F., Bell, A., Dumas, F., and Konishi, Y. (1990) Reduction/alkylation plus mass spectrometry: a fast and simple method for accurate counting of cysteines, disulfide bridges and free SH groups in proteins. *Proceedings of the 38th Annual Conference on Mass Spectrometry and Allied Topics,* Tuczon, AZ, pp. 273,274.

55. Sun, Y., Bauer, M. D., Keough, T. W., and Lacey, M. P. (1996) Disulfide bond location in proteins, in *Methods in Molecular Biology, vol. 61: Protein and Peptide Analysis by Mass Spectrometry* (Chapman, J. R., ed.), Humana, Totowa, NJ, pp. 185–210.

56. Sun, Y., Zhou, Z., and Smith, D. L. (1989) Location of disulfide bonds in proteins by partial acid hydrolysis and mass spectrometry, in *Techniques in Protein Chemistry* (Hugli T. E., ed.), Academic, New York, pp. 176–185.

57. Yazdanparast, R., Andrews, P. C., Smith, D. L., and Dixon, J. E. (1987) Assignment of disulfide bonds in proteins by fast atom bombardment mass spectrometry. *J. Biol. Chem.* **262,** 2507–2513.

58. Sun, Y. and Smith, D. L. (1988) Identification of disulfide-containing peptides by performic acid oxidation and mass spectrometry. *Anal. Biochem.* **197,** 69–76.

59. Aiken, A. and Learmath, M. (1996) Analysis of sites of protein phosphorylation, in *Methods in Molecular Biology, vol. 64: Protein Sequencing Protocols* (Smith, B. J., ed.), Humana, Totowa, NJ, pp. 293–306.

60. Chait, B. T., Wang, R., Beavis, R. C., and Kent, S. B. H. (1993) Protein ladder sequencing. *Science* **262,** 89–92.

61. Bartlet-Jones, M., Jeffery, W. A., Hansen, H. F., and Pappin, D. J. C. (1994) Peptide ladder sequencing by mass spectrometry using a novel, volatile degradation reagent. *Rapid Commun. Mass Spectrom.* **8,** 737–742.

62. Inglis, A. (1991) Chemical procedures for C-terminal sequencing of peptides and proteins. *Analyt. Biochem.* **195,** 183–196.

63. Casagranda, F. and Wilshire, J. F. K. (1996) Enzymatic and chemical methods for manual C-terminal peptide sequencing. In *Methods in Molecular Biology, vol. 64: Protein Sequencing Protocols* (Smith, B. J., ed.), Humana, Totowa, NJ, pp. 243–257.

64. Bailey, J. M. and Iller, C. G. (1996) Automated methods for C-terminal protein sequencing. In *Methods in Molecular Biology, vol. 64: Protein Sequencing Protocols* (Smith, B. J., ed.), Humana, Totowa, NJ, pp. 259–269.

65. Ambler, R. P. (1972) Enzymatic hydrolysis with carboxypeptidases, in *Methods in Enzymol,* vol. 25. (Hirs, C. H. W. and Timasheff, S. N., eds.), Academic, New York, pp. 143–154.

66. Walker, J. M. and Winder, J. S. (1996) C-terminal sequence analysis with carboxypeptidase Y, in *The Protein Protocols Handbook* (Walker, J. M., ed.), Humana, Totowa, NJ, pp. 569–571.

67. Busch, K. L., Glish, G. L., and McLuckey, S. A. (1988) *Mass Spectrometry/Mass Spectrometry: Techniques and Applications of Tandem Mass Spectrometry* VCH, Weinheim, Germany.

68. Spengler, B. (1996) New instrumental approaches to collision-induced dissociation using a time-of-flight instrument, in *Methods in Molecular Biology, vol. 61: Protein and Peptide Analysis by Mass Spectrometry* (Chapman, J. R., ed.), Humana, Totowa, NJ, pp. 43–56.

69. Johnson, R. S., Martin, S. A., and Biemann, K. (1988) Collision-induced fragmentation of $(M + H)^+$ ions of peptides. Side chain specific sequence ions. *Int. J. Mass Spectrom. Ion Processes* **86,** 137–154.

70. Johnson, R. S., Martin, S. A., Biemann, K., Stults, J. T., and Watson, J. T. (1987) Novel fragmentation processes of peptides by collision-induced decomposition in a tandem mass spectrometer: differentiation of leucine and isoleucine. *Anal. Chem.* **59,** 2621–2625.

71. Biemann, K. (1990) Sequencing of peptides by tandem mass spectrometry and high-energy collision-induced dissociation. *Methods Enzymol.* **193,** 455–479.

72. Biemann, K. (1988) Contributions of mass spectrometry to peptide and protein structure. *Biomed. Environ. Mass Spectrom.* **16,** 99–111.

73. Biemann, K. and Scoble, H. A. (1987) Characterization by tandem mass spectrometry. *Science* **237,** 992–998.

74. Gibson, B. W. and Cohen, P. (1990) Liquid secondary ion mass spectrometry of phosphorylated and sulfated peptides and proteins. *Methods Enzymol.* **193,** 480–501.

75. Zaia, J. and Biemann, K. (1995) Comparison of charged derivatives for high energy collision-induced dissociation mass spectrometry. *J. Am. Soc. Mass Spectrom.* **6,** 428–436.

76. Covey, T. R., Huang, E. C., and Henion, J. D. (1991) Structural characterization of protein tryptic peptides via liquid chromatography/mass spectrometry and collision-induced dissociation of their doubly charged molecular ions. *Anal. Chem.* **63,** 1193–1200.

77. Davidson, I. (1996) Hydrolysis of samples for amino acid analysis, in *Methods in Molecular Biology, vol. 64: Protein Sequencing Protocols* (Smith, B. J., ed.), Humana, Totowa, NJ, pp. 119–129.

78. Irvine, G. B. (1996) Amino acid analysis: precolumn derivatization methods, in *Methods in Molecular Biology, vol. 64: Protein Sequencing Protocols* (Smith, B. J., ed.), Humana, Totowa, NJ, pp. 131–138.

79. Smith, A. J. (1996) Postcolumn amino acid analysis, in *Methods in Molecular Biology, vol. 64: Protein Sequencing Protocols* (Smith, B. J., ed.), Humana, Totowa, NJ, pp. 139–146.

80. Nakagawa, S. and Fukuda, T. (1989) Direct amino acid analysis of proteins electroblotted onto polyvinylidene difluoride membrane from sodium dodecylsulfate-polyacrylamide gel. *Analyt. Biochem.* **181,** 75–78.

81. Hobohm, U., Houthaeve, T., and Sander, C. (1994) Amino acid analysis and protein database composition search as a fast and inexpensive method to identify proteins. *Anal. Biochem.* **222,** 202–209.

82. Turck, C. W. (1996) High performance liquid chromatography on-line derivative spectroscopy for the characterisation of peptides with aromatic amino acid residues, in *Methods in Molecular Biology, vol. 64: Protein Sequencing Protocols* (Smith, B. J., ed.), Humana, Totowa, NJ, pp. 109–117.

83. Bray, M. R., Carriere, A. D., and Clarke, A. J. (1994) Quantitation of tryptophan and tyrosine residues in proteins by fourth derivative spectroscopy. *Analyt. Biochem.* **221,** 278–284.

84. Mach, H. and Middaugh, C. R. (1994) Simultaneous monitoring of the environment of tryptophan, tyrosine and phenylalanine residues in proteins by near-ultraviolet second derivative spectroscopy. *Anal. Biochem.* **222,** 323–331.

85. Smith, B. J. and Wheeler, C. (1996) Enzymatic cleavage of polypeptides, in *Methods in Molecular Biology, vol. 64: Protein Sequencing Protocols* (Smith, B. J., ed.), Humana, Totowa, NJ, pp. 43–55.

86. Bornstein, P. and Balian, G. (1977) Cleavage at Asn-Gly bonds with hydroxylamine. *Methods Enzymol.* **47,** 143–145.

87. Inglis, A. S. (1983) Cleavage at aspartic acid. *Methods Enzymol.* **91,** 324–332.

88. Swenson, C. A. and Fredrickson, R. S. (1992) Interaction of troponin C and troponin C fragments with troponin I and the troponin I inhibitory peptide. *Biochemistry* **31,** 3420–3427.

89. Fontana, A. and Gross, E. (1986) Fragmentation of polypeptides by chemical methods, in *Practical Protein Chemistry. A Handbook* (Darbre, A., ed.), Wiley, Chichester, pp. 67–120.

90. Huang, H. V., Bond, M. W., Hunkapillar, M. W., and Hood, L. E. (1983) Cleavage at tryptophanyl residues with dimethyl sulfoxide-hydrochloric acid and cyanogen bromide. *Methods Enzymol.* **91,** 318–324.

91. Kanda, F., Yoshida, S., Okumura, T., and Takamatsu, T. (1995) Asparaginyl endopeptidase mapping of proteins with subsequent matrix-assisted laser desorption/ionization mass spectrometry. *Rapid Commun. Mass Spectrom.* **9,** 1095–1100.

42

Solid-Phase Peptide Synthesis

Gregg B. Fields

1. Introduction

Methods for synthesizing peptides are divided conveniently into two categories: solution and solid-phase (SPPS). Solution synthesis retains value in large-scale manufacturing and for specialized laboratory applications. However, the need to optimize reaction conditions, yields, and purification procedures for essentially every intermediate (each of which has unpredictable solubility and crystallization characteristics) renders solution methods time-consuming and labor-intensive. Consequently, most workers now requiring peptides for their research opt for the more accessible solid-phase approach. In this chapter, an overview of SPPS is presented. For brevity, only commercially available reagents and derivatives utilized for synthesis will be considered here. The reader is referred to a number of excellent, comprehensive reviews *(1–9)* for further discussion of the solid-phase technique.

2. Background

The solid-phase method was conceived and elaborated by R. B. Merrifield *(10)*. The concept of SPPS (**Fig. 1**) is to retain chemistry proven in solution and to add a covalent attachment step that links the nascent peptide chain to an insoluble polymeric support. Subsequently, the anchored peptide is extended by a series of deprotection and coupling cycles, which are required to proceed with exquisitely high yields and fidelities. An advantage of the solid-phase approach is that reactions are driven to completion by the use of excess soluble reagents, which can be removed by simple filtration and washing without manipulative losses. Because of the speed and simplicity of the repetitive steps, which are carried out in a single reaction vessel at ambient temperature, the major portion of the solid-phase procedure is readily amenable to automation. Once chain elaboration has been accomplished, it is necessary to release (cleave) the crude peptide from the support under conditions that are minimally destructive towards sensitive residues in the sequence. Finally, there must follow prudent purification and scrupulous characterization of the synthetic peptide product, in order to verify that the desired structure is indeed the one obtained.

An appropriate polymeric support (resin) must be chosen that has adequate mechanical stability, as well as desirable physicochemical properties that facilitate solid-phase

From: *Molecular Biomethods Handbook*
Edited by: R. Rapley and J. M. Walker © Humana Press Inc., Totowa, NJ

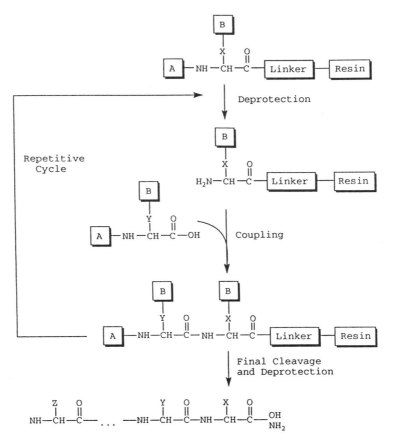

Fig. 1. Stepwise solid-phase synthesis of linear peptides. A is the "temporary" N^α-amino protecting group, B is the "permanent" side-chain protecting group. *See text* for further details.

synthesis. In practice, such supports include those that exhibit significant levels of swelling in useful reaction/wash solvents. Swollen resin beads are reacted and washed batch-wise with agitation, and filtered either with suction or under positive nitrogen pressure. Alternatively, solid-phase synthesis may be carried out in a continuous-flow mode, by pumping reagents and solvents through resins that are packed into columns. The resin support is quite often a polystyrene suspension polymer crosslinked with 1% of 1,3-divinylbenzene. Dry polystyrene beads have an average diameter of about 50 mm, but with the commonly used solvents for peptide synthesis, namely dichloromethane (DCM) and N,N-dimethylformamide (DMF), they swell 2.5–6.2-fold in volume *(11)*. Thus, the chemistry of solid-phase synthesis takes place within a well-solvated gel containing mobile and reagent-accessible chains *(11,12)*. Polymer supports have also been developed based on the concept that the insoluble support and peptide backbone should be of comparable polarities *(7)*. A resin of copolymerized dimethylacrylamide, N,N'-bisacryloylethylenediamine, and acryloylsarcosine methyl ester, commercially known as polyamide or Pepsyn, has been synthesized to satisfy this criteria *(13)*. Increasing popularity has been seen for polyethylene glycol-polystyrene graft supports, which swell in a range of solvents and have excellent physical and mechanical properties for both batch-wise and continuous-flow SPPS *(14–18)*.

Regardless of the structure and nature of the polymeric support chosen, it must contain appropriate functional groups onto which the first amino acid can be anchored. This is achieved by the use of a "linker" (*see* **Fig. 1**), which is a bifunctional spacer that on one end incorporates features of a smoothly cleavable protecting group. The other end of the linker contains a functional group, often a carboxyl, that can be activated to allow coupling to functionalized supports.

The next stage of solid-phase synthesis is the assembly of the peptide chain. Suitably N^α- and side-chain protected amino acids are added stepwise in the C $\rightarrow$ N direction. A particular merit of this strategy is that only negligible levels of racemization are seen. A "temporary" protecting group (A in **Fig. 1**) is removed quantitatively at each step to liberate the N^α-amine of the peptide-resin, following which the next incoming protected amino acid is introduced with its carboxyl group suitably activated.

Once the desired linear sequence has been assembled satisfactorily on the polymeric support, the anchoring linkage must be cleaved. Depending on the chemistry of the original handle and on the cleavage reagent selected, the product from this step can be a *C*-terminal peptide acid or amide. The most widely used approach involves final deprotection carried out essentially concurrent to cleavage; in this way, the released product is directly the free peptide.

3. Protection Schemes

The preceding section outlined the key steps of the solid-phase procedure, but dealt only tangentially with combinations of "temporary" and "permanent" protecting groups (A and B, respectively, in **Fig. 1**) and the corresponding methods for their removal. The choice and optimization of protection chemistry is perhaps the key factor in the success of any synthetic endeavor. Even when a residue has been incorporated safely into the growing resin-bound polypeptide chain, it may still undergo irreversible structural modification or rearrangement during subsequent synthetic steps. The vulnerability to damage is particularly pronounced at the final deprotection/cleavage step, because these are usually the harshest conditions. At least two levels of protecting group stability are required, insofar as the "permanent" groups used to prevent branching or other problems on the side-chains must withstand repeated applications of the conditions for quantitative removal of the "temporary" N^α-amino protecting group. On the other hand, structures of "permanent" groups must be such that conditions can be found to remove them with minimal levels of side reactions that affect the integrity of the desired product. The necessary stability is often approached by kinetic "fine-tuning," which is a reliance on quantitative rate differences whenever the same chemical mechanism (usually acidolysis) serves to remove both "temporary" and "permanent" protecting groups. An often-limiting consequence of such schemes based on graduated lability is that they force adoption of relatively severe final deprotection conditions. Alternatively, orthogonal protection schemes can be used. These involve two or more classes of groups that are removed by differing chemical mechanisms, and therefore can be removed in any order and in the presence of the other classes. Orthogonal schemes offer the possibility of substantially milder overall conditions, because selectivity can be attained on the basis of differences in chemistry rather than in reaction rates.

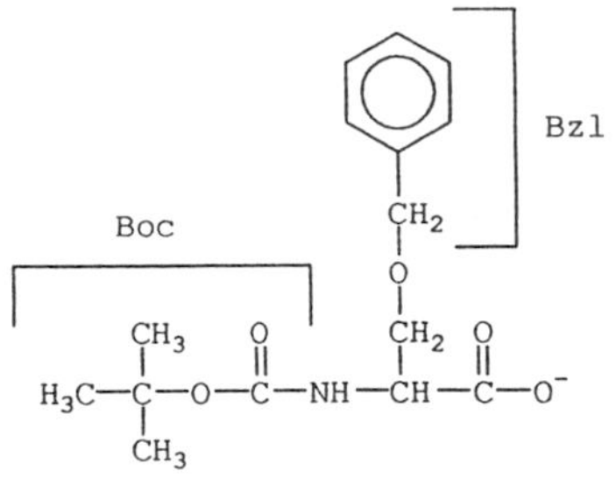

Fig. 2. Boc-protected Ser residue. The "temporary" N^α-amino Boc protecting group is removed by the moderately strong acid TFA, whereas the "permanent" Bzl side-chain protecting group is removed by the strong acid HF (or equivalent).

3.1. Tertiary-*Butyloxycarbonyl (Boc)-Based Chemistry*

The so-called "standard Merrifield" system is based on graduated acid lability (**Fig. 2**). The acidolyzable "temporary" N^α-*tert*-butyloxycarbonyl (Boc) group is stable to alkali and nucleophiles, and removed rapidly by inorganic and organic acids *(2)*. Boc removal is usually carried out with trifluoroacetic acid (TFA) (20–50%) in DCM for 20–30 min, and, for special situations, HCl (4 *N*) in 1,4-dioxane for 35 min. Deprotection with neat (100%) TFA, which offers enhanced peptide-resin solvation compared to TFA-DCM mixtures, proceeds in as little as 4 min *(19,20)*. Following acidolysis, a rapid diffusion-controlled neutralization step with a tertiary amine, usually 5–10% triethylamine (Et₃N) or *N,N*-diisopropylethylamine (DIEA) in DCM for 3–5 min, is interpolated to release the free N^α-amine. Alternatively, Boc-amino acids may be coupled without prior neutralization by using *in situ* neutralization, i.e., coupling in the presence of DIEA or *N*-methylmorpholine (NMM) *(21,22)*. "Permanent" side-chain protecting groups are ether, ester, and urethane derivatives based on benzyl alcohol. Alternatively, ether and ester derivatives based on cyclopentyl or cyclohexyl alcohol are sometimes applied, because their use moderates certain side reactions. These "permanent" groups are sufficiently stable to repeated cycles of Boc removal, yet cleanly cleaved in the presence of appropriate scavengers by use of liquid anhydrous hydrogen fluoride (HF) at 0°C or trimethylsilyl trifluoromethanesulfonate (TMSOTf)/TFA at 25°C. The phenylacetamidomethyl (PAM; for producing peptide acids) or 4-methylbenzhydrylamine (MBHA; for producing peptide amides) anchoring linkages have been designed to be cleaved at the same time.

3.2. 9-Fluorenylmethoxycarbonyl (Fmoc)-Based Chemistry

The electron withdrawing fluorene-ring system of the 9-fluorenylmethyloxycarbonyl (Fmoc) group renders the lone hydrogen on the β-carbon very acidic, and therefore susceptible to removal by weak bases *(23,24)* (**Fig. 3**). Following the abstraction of this acidic proton at the 9-position of the fluorene-ring system, β-elimination proceeds to give a highly reactive dibenzofulvene intermediate *(23–27)*. Dibenzofulvene can be trapped by excess amine cleavage agents to form stable adducts *(23,24)*. The Fmoc group is, in general, rapidly removed by primary (i.e., cyclohexylamine, ethanolamine) and some secondary (i.e., piperidine, piperazine) amines, and slowly removed by tertiary (i.e., Et₃N, DIEA) amines. Removal also occurs more rapidly in a relatively polar

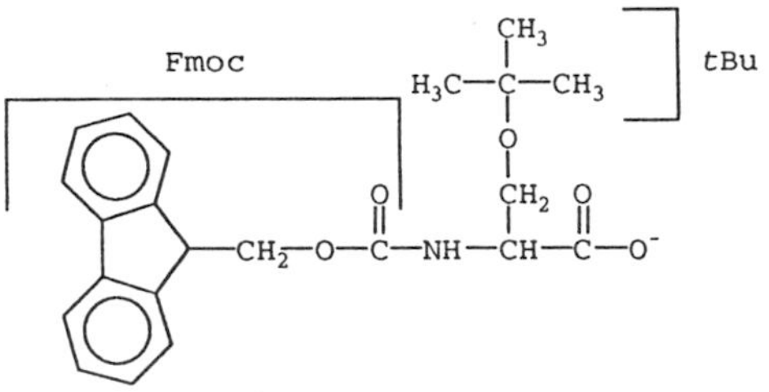

Fig. 3. Fmoc-protected Ser residue. The "temporary" N^α-amino Fmoc protecting group is removed by the moderate base piperidine (or equivalent), while the "permanent" *t*Bu side-chain protecting group is removed by the moderately strong acid TFA.

medium (DMF or *N*-methylpyrrolidone [NMP]) compared to a relatively nonpolar one (DCM). Removal of the Fmoc group is achieved usually with 20–55% piperidine in DMF or NMP for 10–18 min *(28)*; piperidine in DCM is not recommended, because an amine salt precipitates after relatively brief standing. Piperidine scavenges the liberated dibenzofulvene to form a fulvene-piperidine adduct. Two per cent 1,8-diazabicyclo[5.4.0]undec-7-ene (DBU)-DMF can also be used for Fmoc removal *(29)*. This reagent is recommended for continuous-flow syntheses only, because the dibenzofulvene intermediate does not form an adduct with DBU and thus must be washed rapidly from the peptide-resin to avoid reattachment of dibenzofulvene *(29)*. However, a solution of DBU-piperidine-DMF (1:1:48) is effective for batch syntheses, because the piperidine component scavenges the dibenzofulvene *(30)*. After Fmoc removal, the liberated N^α-amine of the peptide-resin is free and ready for immediate acylation without an intervening neutralization step. "Permanent" protection compatible with N^α-Fmoc protection is provided primarily by ether, ester, and urethane derivatives based on *tert*-butanol. These derivatives are cleaved at the same time as appropriate anchoring linkages, by use of TFA at 25°C. Scavengers must be added to the TFA to trap the reactive carbocations that form under the acidolytic cleavage conditions. The TFA-labile 4-hydroxymethylphenoxy (HMP; for producing peptide acids), 2-chlorotrityl (for producing peptide acids), 5-(4-aminomethyl-3,5-dimethoxyphenoxy)valeric acid (PAL; for producing peptide amides), or 4-(2',4'-dimethoxyphenylaminomethyl)phenoxy (Rink amide) anchoring linkages are used in conjunction with Fmoc chemistry.

4. Side-Chain Protection

The side-chain carboxyls of Asp and Glu are protected as benzyl (OBzl) esters for Boc chemistry and as *tert*-butyl (O*t*Bu) esters for Fmoc chemistry. To minimize the imide/$\alpha \rightarrow \beta$ rearrangement side reaction, Fmoc-Asp may be protected with the 1-adamantyl (O-1-Ada) group *(31)*, and Boc-Asp with either the 2-adamantyl (O-2-Ada) *(31)* or cyclohexyl (OcHex) *(32)* groups. The side-chain hydroxyls of Ser, Thr, and Tyr are protected as Bzl and *t*Bu ethers for Boc and Fmoc SPPS, respectively. In strong acid, the Bzl protecting group blocking the Tyr phenol can migrate to the 3-position of the ring *(33)*. This side reaction is decreased greatly when Tyr is protected by the 2,6-dichlorobenzyl (2,6-Cl$_2$Bzl) *(33)* or 2-bromobenzyloxycarbonyl (2-BrZ) *(34)* group; consequently, the latter two derivatives are much preferred for Boc SPPS. The ε-amino group of Lys is best protected by the 2-chlorobenzyloxycarbonyl (2-ClZ) or Fmoc group for Boc chemistry, and reciprocally by the Boc group for Fmoc chemistry. The highly basic trifunctional guanidino side-chain group of Arg may be protected by appropriate benzenesulfonyl derivatives, such as the 4-toluenesulfonyl (Tos) or mesitylene-2-sulfonyl (Mts) groups in conjunction with Boc chemistry, and 4-methoxy-2,3,6-trimethylbenzenesulfonyl (Mtr), 2,2,5,7,8-pentamethylchroman-6-sulfonyl (Pmc), or 2,2,4,6,7-pentamethyldihydro-benzofuran-5-sulfonyl (Pbf) with Fmoc chemistry. These groups most likely block the ω-nitrogen of Arg and their relative acid lability is Pbf > Pmc > Mtr >> Mts > Tos *(35–37)*.

Activated His derivatives are uniquely prone to racemization during stepwise SPPS, owing to an intramolecular abstraction of the proton on the optically active α-carbon by the imidazole π-nitrogen *(38)*. Racemization could be suppressed by either reducing the basicity of the imidazole ring, or by blocking the base directly *(39)*. Consequently, His side-chain protecting groups can be categorized depending on whether the τ- or π-imidazole nitrogen is blocked. The Tos group blocks the N^τ of Boc-His, and is removed by strong acids. However, the Tos group is also lost prematurely during SPPS steps involving 1-hydroxybenzotriazole (HOBt); this allows acylation or acetylation of the imidazole group, followed by chain termination owing to $N^{im} \rightarrow N^\alpha$-amino transfer of the acyl or acetyl group *(40,41)*. Therefore, HOBt should never be used during couplings of amino acids once a His(Tos) residue has been incorporated into the peptide-resin (*see* **Subheading 5.**). An HF-stable, orthogonally removable, N^τ-protecting group for Boc strategies is the 2,4-dinitrophenyl (Dnp) function. Final Dnp deblocking is best carried out at the peptide-resin level prior to the HF cleavage step, by use of thiophenol in DMF. The τ-nitrogen of Fmoc-His can be protected by the Boc and triphenylmethyl (Trt) groups. When His is N^τ-protected by the Boc group, the basicity of the imidazole ring is reduced sufficiently so that acylation by the preformed symmetrical anhydride (PSA) method proceeds with little racemization *(7)*. The Trt group reduces the basicity of the imidazole ring (the pK_a decreases from 6.2 to 4.7), although racemization by the PSA method is not eliminated completely *(42)*. Because Dnp and Trt N^τ-protection do not allow PSA coupling with low racemization, it is recommended that the appropriate derivatives be coupled as preformed esters or *in situ* with carbodiimide in the presence of HOBt *(39,42)*. Boc-His(Tos) is coupled efficiently using benzotriazol-1-yl-oxy-tris(dimethylamino)phosphonium hexafluorophosphate (BOP) (3 Eq) in the presence of DIEA (3 Eq); these conditions minimize racemization and avoid premature side-chain deprotection by HOBt *(43)*.

Although conditions are available for the safe incorporation of Asn and Gln with free side-chains during SPPS, there are compelling reasons for their protection. Side-chain protecting groups such as 9-xanthenyl (Xan) and Trt minimize the occurrence of dehydration *(44–46)* and pyroglutamate formation *(2)*, and may also inhibit hydrogen bonding that otherwise leads to secondary structures that substantially reduce coupling rates. The thioether side-chain of Met survives cycles of Fmoc chemistry, but protection during Boc chemistry is often advisable. The reducible sulfoxide function is applied under these circumstances. Smooth deblocking of Met(O) occurs in 20–25% HF in the presence of dimethylsulfide. The highly sensitive side-chain of Trp is best protected by the N^{in}-formyl (CHO) and N^{in}-Boc groups for Boc and Fmoc chemistry, respectively. Trp(CHO) is deprotected at the peptide-resin level by treatment with piperidine-DMF (9:91), 0°C, 2 h, prior to HF cleavage *(47)*. Boc side-chain protection of Trp is partially reduced to a carboxylate function during the TFA-cleavage procedure; complete deprotection occurs in aqueous solution *(48,49)*.

The most challenging residue to manage in peptide synthesis is Cys. Compatible with Boc chemistry are the 4-methylbenzyl (Meb), acetamidomethyl (Acm), *tert*-butylsulfenyl (S*t*Bu), and 9-fluorenylmethyl (Fm) β-thiol protecting groups; compatible with Fmoc chemistry are the Acm, S*t*Bu, Trt, and Tmob groups. The Meb group is optimized for removal by strong acid *(33)*; Cys(Meb) residues may also be directly converted to the oxidized (cystine) form by thallium (III) trifluoroacetate [Tl(Tfa)$_3$], although some cysteic acid forms at the same time. Cys(Fm) is stable to acid, and cleaved by base. The Trt and Tmob groups are labile in TFA; owing to the tendency of the resultant stable carbonium ions to realkylate Cys *(50,51)*, effective scavengers are needed. The Acm group is acid- and base-stable and removed by mercuric (II) acetate, followed by treatment with H$_2$S or excess mercaptans to free the β-thiol. Mercuric (II) acetate can modify Trp, and thus should be used in the presence of 50% acetic acid for Trp-containing peptides *(52)*. In multiple Cys(Acm)-containing peptides, mercuric (II) acetate may not be a completely effective removal reagent *(53)*. Alternatively, Cys(Acm) residues are converted directly to disulfides by treatment with I$_2$ or Tl(Tfa)$_3$ on-resin *(54,55)*. Finally, the acid-stable S*t*Bu group is removed by reduction with thiols or phosphines.

5. Coupling Reactions

In the solid-phase mode, coupling reagents are used in excess to ensure that reactions reach completion. Recommendations for coupling methods are included in **Tables 1** and **2**.

The classical example of an *in situ* coupling reagent is *N,N'*-dicyclohexyl-carbodiimide (DCC) *(56,57)*. The related *N,N*-diisopropylcarbodiimide (DIPCDI) is more convenient to use under some circumstances, as the resultant urea coproduct is more soluble in DCM. The generality of carbodiimide-mediated couplings is extended significantly by the use of HOBt as an additive, which accelerates carbodiimide-mediated couplings, suppresses racemization, and inhibits dehydration of the carboxamide side-chains of Asn and Gln to the corresponding nitriles *(44,58,59)*. Recently, protocols involving BOP, 2-(1H-benzotriazol-1-yl)-1,1,3,3-tetramethyluronium hexafluoro-phosphate (HBTU), 2-(1H-benzotriazol-1-yl)-1,1,3,3-tetramethyluronium tetra-fluoroborate (TBTU), 2-(2-oxo-1(2H)-pyridyl)-1,1,3,3-bispentamethyleneuronium tetrafluoroborate (TOPPipU), and *O*-(7-azabenzotriazol-1-yl)-1,1,3,3-tetramethyl-uronium hexafluorophosphate (HATU) have deservedly achieved popularity. Interestingly, X-ray crystallographic analysis has shown that the solid-state structures of HBTU and HATU are not tetramethyluronium salts, but guanidinium *N*-oxide isomers *(60)*. BOP, HBTU, TBTU, TOPPipU, and HATU require a tertiary amine, such as NMM or DIEA, for optimal efficiency *(22,61–68)*. HOBt has been reported to accelerate further the rates of BOP- and HBTU-mediated couplings *(66,69)*.

A long-known but steadfast coupling method involves the use of active esters, such as pentafluorophenyl (OPfp), HOBt, and 3-hydroxy-2,3-dihydro-4-oxo-benzotriazine (ODhbt) esters. Boc- and Fmoc-amino acid OPfp esters are prepared from DCC and pentafluorophenol *(70–72)* or pentafluorophenyl trifluoroacetate *(73)*. Although OPfp

Table 1
General Boc Chemistry SPPS

Cycle	Function	Time
1	DCM, DMF, or NMP wash	3×1
2a	TFA-DCM (1:1) deprotection	1×30
or		
2b	TFA deprotection (proceed to cycle 6, then 7*)	2×10
3	DCM wash	3×1
4	DIEA-DCM (1:9) neutralization	2×2
5	DCM wash	3×1
6	DMF or NMP wash	3×1
7a	Boc-amino acid (3 Eq) in DMF	5
7b	DIPCDI (3 Eq) in DMF[a]	60
or		
7a	Boc-amino acid (3 Eq) in DMF	5
7b	BOP (3 Eq):DIEA (5.3 Eq) in DMF[b]	45
or		
7	Boc-amino acid PSA (2 Eq) in DMF or NMP	60
or		
7	Boc-amino acid preformed ester (3 Eq) in DMF or NMP	60
or		
7a*	Boc-amino acid (4 Eq) in DMF or NMP	5
7b*	HBTU (3.9 Eq):DIEA (7.9 Eq):HOBt (4 Eq) in NMP[b]	45
8	DMF or NMP wash	3×1

[a]HOBt (3 Eq) is added for Boc-Asn, -Gln, -Arg(Tos), -Arg(Mts), and -His(Dnp).
[b]Boc-Asn and -Gln are side-chain protected.

Table 2
General Fmoc Chemistry SPPS

Cycle	Function	Time
1	DMF or NMP wash	3×1
2	Piperidine-DMF or NMP (1:4) deprotection	1×20
3	DMF or NMP wash	3×1
4a	Fmoc-amino acid (4 Eq) in DMF or NMP	5
4b	DIPCDI (4 Eq):HOBt (4 Eq) in DMF	60
or		
4a	Fmoc-amino acid (4 Eq) in DMF or NMP	5
4b	BOP (3 Eq):NMM (4.5 Eq):HOBt (3 Eq) in DMF[a]	45
or		
4a	Fmoc-amino acid (4 Eq) in DMF or NMP	5
4b	HBTU (3.9 Eq):DIEA (7.9 Eq):HOBt (4 Eq) in NMP[a]	45
or		
4	Fmoc-amino acid preformed ester (4 Eq) in DMF or NMP	60
5	DMF or NMP wash	3×1

[a]Fmoc-Asn and -Gln are side-chain protected.

esters alone couple slowly, the addition of HOBt (1–2 Eq) increases the reaction rate *(74,75)*. Fmoc-Asn-OPfp allows for efficient incorporation of Asn with little side-chain dehydration *(45)*. HOBt esters of Fmoc-amino acids are rapidly formed (with DIPCDI) and highly reactive *(76,77)*, as are Boc-amino acid HOBt esters *(78)*. N^{α}-protected amino acid ODhbt esters suppress racemization and are highly reactive, in similar fashion to HOBt esters *(79)*.

Preformed symmetrical anhydrides (PSAs) are favored by some workers because of their high reactivity. They are generated *in situ* from the corresponding N^{α}-protected amino acid (2 or 4 Eq) plus DCC (1 or 2 Eq) in DCM; following removal of the urea by filtration, the solvent is exchanged to DMF for optimal couplings. The solubilities of some Fmoc-amino acids make PSAs a less than optimum activated species.

Fmoc-amino acid fluorides react rapidly under SPPS conditions in the presence of DIEA with very low levels of racemization *(80,81)*. Fmoc-amino acid fluorides are an especially effective method for coupling to *N*-alkyl amino acids *(82,83)*.

6. Cleavage and Side-Chain Deprotection

Boc SPPS is designed primarily for simultaneous cleavage of the peptide anchoring linkage and side-chain protecting groups with strong acid (HF or equivalent), whereas Fmoc SPPS is designed primarily to accomplish the same cleavages with moderate-strength acid (TFA or equivalent). In each case, careful attention to cleavage conditions (reagents, scavengers, temperature, and times) is necessary in order to minimize a variety of side reactions.

Treatment with HF simultaneously cleaves PAM and MBHA linkages and removes the side-chain protecting groups commonly applied in Boc chemistry *(84)*. HF cleavages are always carried out in the presence of a carbonium ion scavenger, usually 10% anisole. For cleavages of Cys-containing peptides, further addition of 1.8% 4-thiocresol is recommended. TMSOTf/TFA has also been used for strong acid cleavage and deprotection reactions, which are accelerated by the presence of thioanisole as a "soft" nucleophile *(85,86)*. TMSOTf (1 *M*)-thioanisole (1 *M*) in TFA (also known as DEPRO) efficiently cleaves PAM and MBHA linkages *(86,87)*.

The combination of side-chain protecting groups, e.g., *t*Bu (for Asp, Glu, Ser, Thr, and Tyr), Boc (for His and Lys), Tmob (for Cys), and Trt (for Asn, Cys, Gln, and His), and anchoring linkages, e.g., HMP or PAL, commonly used in Fmoc chemistry are simultaneously deprotected and cleaved by TFA. Such cleavage of *t*Bu and Boc groups results in *tert*-butyl cations and *tert*-butyl trifluoroacetate formation *(88–92)*. These species are responsible for *tert*-butylation of the indole ring of Trp, the thioether group of Met, and, to a very low degree (0.5–1.0%), the 3'-position of Tyr. Modifications can be minimized during TFA cleavage by utilizing effective *tert*-butyl scavengers. The indole ring of Trp can be alkylated irreversibly by Mtr, Pmc, and Pbf groups from Arg *(76,93–96)*, Tmob groups *(45,46)*, and even by some TFA-labile ester and amide linkers *(74,97–100)*. The extent of Pmc modification of Trp is dependent upon the distance between the Arg and Trp residues *(101)*. Cleavage of the Pmc group may also result in *O*-sulfation of Ser, Thr, and Tyr *(94,102)*. Three efficient cleavage "cocktails" for Mtr/Pmc/Pbf/Tmob quenching, and preservation of Trp, Tyr, Ser, Thr, and Met integrity, are TFA-phenol-thioanisole-EDT-H_2O (82.5:5:5:2.5:5) (reagent K) *(95)*, TFA-thioanisole-EDT-anisole (90:5:3:2) (reagent R) *(98)*, and TFA-phenol-H_2O-triisopropylsilane (88:5:5:2) (reagent B) *(103)*. The use of Boc side-chain protection of Trp also significantly reduces alkylation by Pmc/Pbf groups *(96,104)*.

7. Side-Reactions

Side-reactions that occur during SPPS have been reviewed extensively *(2,3,6–9)*. The present discussions focus on new approaches for alleviating well established side-reactions.

7.1. Diketopiperazine Formation

The free N^α-amino group of an anchored dipeptide is poised for a base-catalyzed intramolecular attack of the *C*-terminal carbonyl *(2,105,106)*. Base deprotection (Fmoc) or neutralization (Boc) can thus release a cyclic diketopiperazine while a hydroxy-methyl-handle leaving group remains on the resin. With residues that can form *cis* peptide bonds, e.g., Gly, Pro, *N*-methylamino acids, or D-amino acids, in either the first or second position of the (C → N) synthesis, diketopiperazine formation can be substantial *(106–108)*. For most other sequences, the problem can be adequately controlled. In Boc SPPS, the level of diketopiperazine formation can be suppressed either by removing the Boc group with HCl and coupling the NMM salt of the third Boc-amino acid without neutralization *(21)*, or else by deprotecting the Boc group with TFA and coupling the third Boc-amino acid *in situ* using BOP, DIEA, and HOBt without neutralization *(108)*. For susceptible sequences being addressed by Fmoc chemistry, the use of piperidine-DMF (1:1) deprotection for 5 min *(106)*, or deprotection for 2 min with a 0.1 *M* solution in DMF of tetrabutylammonium fluoride ("quenched" by MeOH) *(109)* has been recommended to minimize cyclization. Alternatively, the second and third amino acids may be coupled as a preformed N^α-protected-dipeptide, avoiding the diketopiperazine-inducing deprotection/neutralization at the second amino acid. The steric hindrance of the 2-chlorotrityl linker may minimize diketopiperazine formation of susceptible sequences during Fmoc chemistry *(110,111)*.

7.2. Aspartimide Formation

A sometimes serious side reaction with protected Asp residues involves an intramolecular elimination to form an aspartimide, which can then partition in water to the desired α-peptide and the undesired by-product with the chain growing from the β-carboxyl *(2,32,112)*. Aspartimide formation is sequence dependent, with Asp(OBzl)-Gly, -Ser, -Thr, -Asn, and -Gln sequences showing the greatest tendency to cyclize under basic conditions *(112–114)*; the same sequences are also quite susceptible in strong acid *(2,32,35)*. For models containing Asp(OBzl)-Gly, the rate and extent of aspartimide formation was substantial both in base (50% after 1–3 h treatment with Et_3N or DIEA) and in strong acid (a typical value is 36% after 1 h treatment with HF at 25°C). Aspartimide formation is minimized during Boc chemistry by using the Asp(OcHex) or Asp(O-2-Ada) derivative.

Sequences containing Asp(OtBu)-Gly are somewhat susceptible to base-catalyzed aspartimide formation (11% after 4 h treatment with 20% piperidine in DMF) *(114)*, but do not rearrange at all in acid *(115)*. Piperidine catalysis of aspartimide formation from side-chain protected Asp residues can be rapid, and is dependent upon the side-chain protecting group. Treatment of Asp(OBzl)-Gly, Asp(OcHex), and Asp(OtBu)-

Gly with 20% piperidine-DMF for 4 h resulted in 100, 67.5, and 11% aspartimide formation, respectively *(114)*, whereas treatment of Asp(OBzl)-Phe with 55% piperidine-DMF for 1 h resulted in 16% aspartimide formation *(116)*. Sequence dependence studies of Asp(O*t*Bu)-X peptides revealed that piperidine could induce aspartimide formation when X = Arg(Pmc), Asn(Trt), Asp(O*t*Bu), Cys(Acm), Gly, Ser, Thr, and Thr(*t*Bu) *(117,118)*. Aspartimide formation can also be conformation dependent *(119)*. This side-reaction can be minimized by including 0.1 *M* HOBt in the piperidine solution *(118)*, or by using an amide backbone protecting group (i.e., 2-hydroxy-4-methoxybenzyl) for the residue in the X position *(120)*.

7.3. Racemization of Cys Residues

C-terminal esterified (but not amidated) Cys residues are racemized by repeated piperidine deprotection treatments during Fmoc SPPS. Following 4 h exposure to piperidine-DMF (1:4), the extent of racemization found was 36% *D*-Cys from Cys(S*t*Bu), 12% D-Cys from Cys(Trt), and 9% D-Cys from Cys(Acm) *(121)*. Racemization of esterified Cys(Trt) was reduced from 11.8% with 20% piperidine-DMF to only 2.6% with 1% DBU-DMF after 4 h treatment *(29,121)*. Additionally, the steric hindrance of the 2-chlorotrityl linker has been shown to minimize racemization of *C*-terminal Cys residues *(122)*.

7.4. Interchain Association

Effective solvation of the peptide-resin is perhaps the most crucial condition for efficient chain assembly *(123)*. Under proper solvent conditions, there is no decrease in synthetic efficiency up to 60 amino acid residues in Boc SPPS *(124)*. The ability of the peptide-resin to swell increases with increasing peptide length owing to a net decrease in free energy from solvation of the linear peptide chains *(11)*. Therefore, there is no theoretical upper limit to efficient amino acid couplings, provided that proper solvation conditions exist *(125)*. In practice, obtaining these conditions is not always straightforward. "Difficult couplings" during SPPS have been attributed to poor solvation of the growing chain by DCM. Infrared and NMR spectroscopies have shown that intermolecular β-sheet aggregates are responsible for lowering coupling efficiencies *(126–128)*. A scale of β-sheet structure-stabilizing potential has been developed for Boc-amino acid derivatives *(129)*. Enhanced coupling efficiencies are seen on the addition of polar solvents, such as DMF, TFE, and NMP *(78,126,130–133)*. Aggregation also occurs in regions of apolar side-chain protecting groups, sometimes resulting in a collapsed gel structure *(134,135)*. In cases where aggregation occurs owing to apolar side-chain protecting groups, increased solvent polarity may not be sufficient to disrupt the aggregate. A problem of Fmoc chemistry is that the lack of polar side-chain protecting groups could, during the course of an extended peptide synthesis, inhibit proper solvation of the peptide-resin *(133,134,136)*. To alleviate this problem, the use of solvent mixtures containing both a polar and nonpolar component, such as THF-NMP (7:13) or TFE-DCM (1:4), is recommended *(133)*. The partial substitution or complete replacement of *t*Bu-based side-chain protecting groups for carboxyl, hydroxyl, and amino side-chains by more polar groups would also aid peptide-resin solvation *(133,134,136)*. The incorporation of reversible, amide backbone protecting groups, such as 2-hydroxy-4-methoxybenzyl (Hmb), has been demonstrated to be an effective method for disrupting interchain aggregates and thus improving solvation and reaction conditions *(137,138)*.

References

1. Erickson, B. W. and Merrifield, R. B. (1976) Solid-phase peptide synthesis, in *The Proteins,* Vol. II, 3rd ed. (Neurath, H. and Hill, R. L., eds.), Academic, New York, pp. 255–527.

2. Barany, G. and Merrifield, R. B. (1979) Solid-phase peptide synthesis, in *The Peptides,* Vol. 2 (Gross, E. and Meienhofer, J., eds.), Academic, New York, pp. 1–284.

3. Stewart, J. M. and Young, J. D. (1984) *Solid Phase Peptide Synthesis,* 2nd ed. Pierce Chemical Co., Rockford, IL.

4. Merrifield, B. (1986) Solid phase synthesis. *Science* **232,** 341–347.

5. Barany, G., Kneib-Cordonier, N., and Mullen, D. G. (1987) Solid-phase peptide synthesis: a silver anniversary report. *Int. J. Peptide Protein Res.* **30,** 705–739.

6. Kent, S. B. H. (1988) Chemical synthesis of peptides and proteins. *Annu. Rev. Biochem.* **57,** 957–989.

7. Atherton, E. and Sheppard, R. C. (1989) *Solid Phase Peptide Synthesis: A Practical Approach.* IRL, Oxford, UK.

8. Fields, G. B. and Noble, R. L. (1990) Solid phase peptide synthesis utilizing 9-fluorenylmethoxycarbonyl amino acids. *Int. J. Peptide Protein Res.* **35,** 161–214.

9. Fields, G. B., Tian, Z., and Barany, G. (1992) Principles and practice of solid-phase peptide synthesis, in *Synthetic Peptides: A User's Guide* (Grant, G. A., ed.), W. H. Freeman & Co., New York, pp. 77–183.

10. Merrifield, R. B. (1963) Solid phase peptide synthesis I: synthesis of a tetrapeptide. *J. Am. Chem. Soc.* **85,** 2149–2154.

11. Sarin, V. K., Kent, S. B. H., and Merrifield, R. B. (1980) Properties of swollen polymer networks: solvation and swelling of peptide-containing resins in solid-phase peptide synthesis. *J. Am. Chem. Soc.* **102,** 5463–5470.

12. Live, D. and Kent, S. B. H. (1982) Fundamental aspects of the chemical applications of cross-linked polymers, in *Elastomers and Rubber Elasticity* (Mark, J. E., ed.), American Chemical Society, Washington, DC, pp. 501–515.

13. Arshady, R., Atherton, E., Clive, D. L. J., and Sheppard, R. C. (1981) Peptide synthesis, part 1: preparation and use of polar supports based on poly(dimethylacrylamide). *J. Chem. Soc. Perkin Trans.* I, 529–537.

14. Hellermann, H., Lucas, H.-W., Maul, J., Pillai, V. N. R., and Mutter, M. (1983) Poly(ethylene glycol)s grafted onto crosslinked polystyrenes, 2: Multidetachably anchored polymer systems for the synthesis of solubilized peptides. *Makromol. Chem.* **184,** 2603–2617.

15. Zalipsky, S., Albericio, F., and Barany, G. (1985) Preparation and use of an aminoethyl polyethylene glycol-crosslinked polystyrene graft resin support for solid-phase peptide synthesis, in *Peptides: Structure and Function* (Deber, C. M., Hruby, V. J., and Kopple, K. D., eds.), Pierce Chemical Co., Rockford, IL, pp. 257–260.

16. Bayer, E. and Rapp, W. (1986) New polymer supports for solid-liquid-phase peptide synthesis, in *Chemistry of Peptides and Proteins, Vol. 3* (Voelter, W., Bayer, E., Ovchinnikov, Y. A., and Ivanov, V. T., eds.), de Gruyter, Berlin, pp. 3–8.

17. Bayer, E., Albert, K., Willisch, H., Rapp, W., and Hemmasi, B. (1990) ^{13}C NMR relaxation times of a tripeptide methyl ester and its polymer-bound analogues. *Macromolecules* **23,** 1937–1940.

18. Zalipsky, S., Chang, J. L., Albericio, F., and Barany, G. (1994) Preparation and applications of polyethylene glycol-polystyrene graft resin supports for solid-phase peptide synthesis. *Reactive Polymers* **22,** 243–258.

19. Kent, S. B. H. and Parker, K. F. (1988) The chemical synthesis of therapeutic peptides and proteins, in *In Banbury Report 29: Therapeutic Peptides and Proteins: Assessing the New Technologies* (Marshak, D. R. and Liu, D. T., eds.), Cold Spring Harbor Laboratory, Cold Spring Harbor, NY, pp. 3–16.

20. Wallace, C. J. A., Mascagni, P., Chait, B. T., Collawn, J. F., Paterson, Y., Proudfoot, A. E. I., and Kent, S. B. H. (1989) Substitutions engineered by chemical synthesis at three conserved sites in mitochondrial cytochrome *c*. *J. Biol. Chem.* **264,** 15,199–15,209.

21. Suzuki, K., Nitta, K., and Endo, N. (1975) Suppression of diketopiperazine formation in solid phase peptide synthesis. *Chem. Pharm. Bull.* **23,** 222–224.

22. Schnölzer, M., Alewood, P., Jones, A., Alewood, D., and Kent, S. B. H. (1992) *In situ* neutralization in Boc-chemistry solid phase peptide synthesis: Rapid, high yield assembly of difficult sequences. *Int. J. Peptide Protein Res.* **40,** 180–193.

23. Carpino, L. A. and Han, G. Y. (1972) The 9-fluorenylmethoxycarbonyl amino-protecting group. *J. Org. Chem.* **37,** 3404–3409.

24. Carpino, L. A. (1987) The 9-fluorenylmethyloxycarbonyl family of base-sensitive amino-protecting groups. *Acc. Chem. Res.* **20,** 401–407.

25. O'Ferrall, R. A. M. and Slae, S. (1970) β-elimination of 9-fluorenylmethanol in aqueous solution: an *E1cB* mechanism. *J. Chem. Soc. (B),* 260–268.

26. O'Ferrall, R. A. M. (1970) β-elimination of 9-fluorenylmethanol in solutions of methanol and t-butyl alcohol. *J. Chem. Soc. (B),* 268–274.

27. O'Ferrall, R. A. M. (1970) Relationships between *E2* and *E1cB* mechanisms of β-elimination. *J. Chem. Soc. (B),* 274–277.

28. Fields, G. B. (1994) Methods for removing the Fmoc group, in *Methods in Molecular Biology, Vol. 35: Peptide Synthesis Protocols* (Pennington, M. W. and Dunn, B. M., eds.), Humana, Totowa, NJ, pp. 17–27.

29. Wade, J. D., Bedford, J., Sheppard, R. C., and Tregear, G. W. (1991) DBU as an N^α-deprotecting reagent for the fluorenylmethoxycarbonyl group in continuous flow solid-phase peptide synthesis. *Peptide Res.* **4,** 194–199.

30. Fields, C. G., Mickelson, D. J., Drake, S. L., McCarthy, J. B., and Fields, G. B. (1993) Melanoma cell adhesion and spreading activities of a synthetic 124-residue triple-helical "mini-collagen." *J. Biol. Chem.* **268,** 14,153–14,160.

31. Okada, Y. and Iguchi, S. (1988) Amino acid and peptides, part 19: synthesis of β 1- and β 2-adamantyl aspartates and their evaluation for peptide synthesis. *J. Chem. Soc. Perkin Trans. I,* 2129–2136.

32. Tam, J. P., Riemen, M. W., and Merrifield, R. B. (1988) Mechanisms of aspartimide formation: The effects of protecting groups, acid, base, temperature and time. *Peptide Res.* **1,** 6–18.

33. Erickson, B. W. and Merrifield, R. B. (1973) Acid stability of several benzylic protecting groups used in solid-phase peptide synthesis: rearrangement of *O*-benzyltyrosine to 3-benzyltyrosine. *J. Am. Chem. Soc.* **95,** 3750–3756.

34. Yamashiro, D. and Li, C. H. (1973) Protection of tyrosine in solid-phase peptide synthesis. *J. Org. Chem.* **38,** 591,592.

35. Fujino, M., Wakimasu, M., and Kitada, C. (1981) Further studies on the use of multi-substituted benzenesulfonyl groups for protection of the guanidino function of arginine. *Chem. Pharm. Bull.* **29,** 2825–2831.

36. Green, J., Ogunjobi, O. M., Ramage, R., Stewart, A. S. J., McCurdy, S., and Noble, R. (1988) Application of the N^G-(*2,2,5,7,8*-pentamethylchroman–6-sulphonyl) derivative of Fmoc-arginine to peptide synthesis. *Tetrahedron Lett.* **29,** 4341–4344.

37. Carpino, L. A., Shroff, H., Triolo, S. A., Mansour, E.-S. M. E., Wenschuh, H., and Albericio, F. (1993) The *2,2,4,6,7*-pentamethyldihydrobenzofuran–5-sulfonyl group (Pbf) as arginine side chain protectant. *Tetrahedron Lett.* **34,** 7829–7832.

38. Jones, J. H., Ramage, W. I., and Witty, M. J. (1980) Mechanism of racemization of histidine derivatives in peptide synthesis. *Int. J. Peptide Protein Res.* **15,** 301–303.

39. Riniker, B. and Sieber, P. (1988) Problems and progress in the synthesis of histidine-containing peptides, in *Peptides: Chemistry, Biology, Interactions with Proteins* (Penke, B. and Torok, A., eds.), de Gruyter, Berlin, pp. 65–74.

40. Ishiguro, T. and Eguchi, C. (1989) Unexpected chain-terminating side reaction caused by histidine and acetic anhydride in solid-phase peptide synthesis. *Chem. Pharm. Bull.* **37,** 506–508.

41. Kusunoki, M., Nakagawa, S., Seo, K., Hamana, T., and Fukuda, T. (1990) A side reaction in solid phase synthesis: Insertion of glycine residues into peptide chains via $N^{im} \rightarrow N^{\alpha}$ transfer. *Int. J. Peptide Protein Res.* **36,** 381–386.

42. Sieber, P. and Riniker, B. (1987) Protection of histidine in peptide synthesis: A reassessment of the trityl group. *Tetrahedron Lett.* **28,** 6031–6034.

43. Forest, M. and Fournier, A. (1990) BOP reagent for the coupling of pGlu and Boc-His(Tos) in solid phase peptide synthesis. *Int. J. Peptide Protein Res.* **35,** 89–94.

44. Mojsov, S., Mitchell, A. R., and Merrifield, R. B. (1980) A quantitative evaluation of methods for coupling asparagine. *J. Org. Chem.* **45,** 555–560.

45. Gausepohl, H., Kraft, M., and Frank, R. W. (1989) Asparagine coupling in Fmoc solid phase peptide synthesis. *Int. J. Peptide Protein Res.* **34,** 287–294.

46. Sieber, P. and Riniker, B. (1990) Side-chain protection of asparagine and glutamine by trityl: application to solid-phase peptide synthesis, in *Innovation and Perspectives in Solid Phase Synthesis* (Epton, R., ed.), Solid Phase Conference Coordination, Ltd., Birmingham, UK, pp. 577–583.

47. Fields, G. B., Carr, S. A., Marshak, D. R., Smith, A. J., Stults, J. T., Williams, L. C., Williams, K. R., and Young, J. D. (1993) Evaluation of peptide synthesis as practiced in 53 different laboratories, in *Techniques In Protein Chemistry IV* (Angeletti, R. H., ed.), Academic, San Diego, CA, pp. 229–238.

48. Franzén, H., Grehn, L., and Ragnarsson, U. (1984) Synthesis, properties, and use of N^{in}-Boc-tryptophan derivatives. *J. Chem. Soc. Chem. Commun.* 1699,1700.

49. White, P. (1992) Fmoc-Trp(Boc)-OH: A new derivative for the synthesis of peptides containing tryptophan, in *Peptides: Chemistry and Biology* (Smith, J. A. and Rivier, J. E., eds.), ESCOM, Leiden, The Netherlands, pp. 537–538.

50. Photaki, I., Taylor-Papadimitriou, J., Sakarellos, C., Mazarakis, P., and Zervas, L. (1970) On cysteine and cystine peptides, part V: *S*-trityl- and *S*-diphenylmethyl-cysteine and-cysteine peptides. *J. Chem. Soc. (C),* 2683–2687.

51. Munson, M. C., García-Echeverría, C., Albericio, F., and Barany, G. (1992) *S*–2,4,6-Trimethoxybenzyl (Tmob): a novel cysteine protecting group for the N^{α}-9-fluorenylmethoxycarbonyl (Fmoc) strategy of peptide synthesis. *J. Org. Chem.* **57,** 3013–3018.

52. Nishio, H., Kimura, T., and Sakakibara, S. (1994) Side reaction in peptide synthesis: modification of tryptophan during treatment with mercury(II) acetate/2-mercaptoethanol in aqueous acetic acid. *Tetrahedron Lett.* **35,** 1239–1242.

53. Kenner, G. W., Galpin, I. J., and Ramage, R. (1979) Synthetic studies directed towards the synthesis of a lysozyme analog, in *Peptides: Structure and Biological Function* (Gross, E. and Meienhofer, J., eds.), Pierce Chemical Co., Rockford, IL, pp. 431–438.

54. Albericio, F., Hammer, R. P., García-Echeverría, C., Molins, M. A., Chang, J. L., Munson, M. C., Pons, M., Giralt, E., and Barany, G. (1991) Cyclization of disulfide-containing peptides in solid-phase synthesis. *Int. J. Peptide Protein Res.* **37,** 402–413.

55. Edwards, W. B., Fields, C. G., Anderson, C. J., Pajeau, T. S., Welch, M. J., and Fields, G. B. (1994) Generally applicable, convenient solid-phase synthesis and receptor affinities of octreotide analogs. *J. Med. Chem.* **37,** 3749–3757.

56. Rich, D. H. and Singh, J. (1979) The carbodiimide method, in *The Peptides,* Vol. 1 (Gross, E. and Meienhofer, J., eds.), Academic, New York, pp. 241–314.

57. Merrifield, R. B., Singer, J., and Chait, B. T. (1988) Mass spectrometric evaluation of synthetic peptides for deletions and insertions. *Anal. Biochem.* **174,** 399–414.

58. König, W. and Geiger, R. (1970) Eine neue methode zur synthese von peptiden: Aktivierung der carboxylgruppe mit dicyclohexylcarbodiimid unter zusatz von 1-hydroxybenzotriazolen. *Chem. Ber.* **103,** 788–798.

59. König, W. and Geiger, R. (1973) *N*-hydroxyverbindungen als katalysatoren für die aminolyse aktivierter ester. *Chem. Ber.* **106,** 3626–3635.

60. Abdelmoty, I., Albericio, F., Carpino, L. A., Foxman, B. M., and Kates, S. A. (1994) Structural studies of reagents for peptide bond formation: crystal and molecular structures of HBTU and HATU. *Lett. Peptide Sci.* **1,** 57–67.

61. Dourtoglou, V., Gross, B., Lambropoulou, V., and Zioudrou, C. (1984) *O*-benzotriazolyl-*N,N,N',N'*-tetramethyluronium hexafluorophosphate as coupling reagent for the synthesis of peptides of biological interest. *Synthesis* 572–574.

62. Fournier, A., Wang, C.-T., and Felix, A. M. (1988) Applications of BOP reagent in solid phase peptide synthesis: Advantages of BOP reagent for difficult couplings exemplified by a synthesis of [Ala[15]]-GRF(1-29)-NH$_2$. *Int. J. Peptide Protein Res.* **31,** 86–97.

63. Ambrosius, D., Casaretto, M., Gerardy-Schahn, R., Saunders, D., Brandenburg, D., and Zahn, H. (1989) Peptide analogues of the anaphylatoxin C3a; synthesis and properties. *Biol. Chem. Hoppe-Seyler* **370,** 217–227.

64. Gausepohl, H., Kraft, M., and Frank, R. (1989) In situ activation of Fmoc-amino acids by BOP in solid phase peptide synthesis, in *Peptides 1988* (Jung, G. and Bayer, E., eds.), de Gruyter, Berlin, pp. 241–243.

65. Seyer, R., Aumelas, A., Caraty, A., Rivaille, P., and Castro, B. (1990) Repetitive BOP coupling (REBOP) in solid phase peptide synthesis: Luliberin synthesis as model. *Int. J. Peptide Protein Res.* **35,** 465–472.

66. Fields, C. G., Lloyd, D. H., Macdonald, R. L., Otteson, K. M., and Noble, R. L. (1991) HBTU activation for automated Fmoc solid-phase peptide synthesis. *Peptide Res.* **4,** 95–101.

67. Knorr, R., Trzeciak, A., Bannwarth, W., and Gillessen, D. (1991) 1,1,3,3-Tetramethyluronium compounds as coupling reagents in peptide and protein chemistry, in *Peptides 1990* (Giralt, E. and Andreu, D., eds.), Escom, Leiden, The Netherlands, pp. 62–64.

68. Carpino, L. A., El-Faham, A., Minor, C. A., and Albericio, F. (1994) Advantageous applications of azabenzotriazole (triazolopyridine)-based coupling reagents to solid-phase peptide synthesis. *J. Chem. Soc. Chem. Commun.* 201–203.

69. Hudson, D. (1988) Methodological implications of simultaneous solid-phase peptide synthesis 1: Comparison of different coupling procedures. *J. Org. Chem.* **53,** 617–624.

70. Kisfaludy, L., Löw, M., Nyéki, O., Szirtes, T., and Schön, I. (1973) Die verwendung von pentafluorophenylestern bei peptid-synthesen. *Justus Liebigs Ann. Chem.* 1421–1429.

71. Penke, B., Baláspiri, L., Pallai, P., and Kovács, K. (1974) Application of pentafluorophenyl esters of Boc-amino acids in solid phase peptide synthesis. *Acta Phys. Chem.* **20,** 471–476.

72. Kisfaludy, L. and Schön, I. (1983) Preparation and applications of pentafluorophenyl esters of 9-fluorenylmethyloxycarbonyl amino acids for peptide synthesis. *Synthesis* 325–327.

73. Green, M. and Berman, J. (1990) Preparation of pentafluorophenyl esters of Fmoc protected amino acids with pentafluorophenyl trifluoroacetate. *Tetrahedron Lett.* **31,** 5851,5852.

74. Atherton, E., Cameron, L. R., and Sheppard, R. C. (1988) Peptide synthesis, part 10: use of pentafluorophenyl esters of fluorenylmethoxycarbonylamino acids in solid phase peptide synthesis. *Tetrahedron* **44,** 843–857.

75. Hudson, D. (1990) Methodological implications of simultaneous solid-phase peptide synthesis: A comparison of active esters. *Peptide Res.* **3**, 51–55.

76. Harrison, J. L., Petrie, G. M., Noble, R. L., Beilan, H. S., McCurdy, S. N., and Culwell, A. R. (1989) Fmoc chemistry: synthesis, kinetics, cleavage, and deprotection of arginine-containing peptides, in *Techniques in Protein Chemistry* (Hugli, T. E., ed.), Academic, San Diego, CA, pp. 506–516.

77. Fields, C. G., Fields, G. B., Noble, R. L., and Cross, T. A. (1989) Solid phase peptide synthesis of [^{15}N]-gramicidins A, B, and C and high performance liquid chromatographic purification. *Int. J. Peptide Protein Res.* **33**, 298–303.

78. Geiser, T., Beilan, H., Bergot, B. J., and Otteson, K. M. (1988) Automation of solid-phase peptide synthesis, in *Macromolecular Sequencing and Synthesis: Selected Methods and Applications* (Schlesinger, D. H., ed.), Liss, New York, pp. 199–218.

79. König, W. and Geiger, R. (1970) Racemisierung bei peptidsynthesen. *Chem. Ber.* **103**, 2024–2033.

80. Carpino, L. A., Sadat-Aalaee, D., Chao, H. G., and DeSelms, R. H. (1990) ((9-Fluorenylmethyl)oxy)carbonyl (Fmoc) amino acid fluorides: convenient new peptide coupling reagents applicable to the Fmoc/*tert*-butyl strategy for solution and solid-phase syntheses. *J. Am. Chem. Soc.* **112**, 9651–9652.

81. Carpino, L. A. and Mansour, E.-S. M. E. (1992) Protected β- and γ-aspartic and-glutamic acid fluorides. *J. Org. Chem.* **57**, 6371–6373.

82. Wenschuh, H., Beyermann, M., Krause, E., Brudel, M., Winter, R., Schümann, M., Carpino, L. A., and Bienert, M. (1994) Fmoc amino acid fluorides: convenient reagents for the solid-phase assembly of peptides incorporating sterically hindered residues. *J. Org. Chem.* **59**, 3275–3280.

83. Wenschuh, H., Beyermann, M., Haber, H., Seydel, J. K., Krause, E., Bienert, M., Carpino, L. A., El-Faham, A., and Albericio, F. (1995) Stepwise automated solid phase synthesis of naturally occurring peptaibols using Fmoc amino acid fluorides. *J. Org. Chem.* **60**, 405–410.

84. Tam, J. P. and Merrifield, R. B. (1987) Strong acid deprotection of synthetic peptides: mechanisms and methods, in *The Peptides,* Vol. 9 (Udenfriend, S. and Meienhofer, J.,eds.), Academic, New York, pp. 185–248.

85. Yajima, H., Fujii, N., Funakoshi, S., Watanabe, T., Murayama, E., and Otaka, A. (1988) New strategy for the chemical synthesis of proteins. *Tetrahedron* **44**, 805–819.

86. Nomizu, M., Inagaki, Y., Yamashita, T., Ohkubo, A., Otaka, A., Fujii, N., Roller, P. P., and Yajima, H. (1991) Two-step hard acid deprotection/cleavage procedure for solid phase peptide synthesis. *Int. J. Peptide Protein Res.* **37**, 145–152.

87. Akaji, K., Fujii, N., Tokunaga, F., Miyata, T., Iwanaga, S., and Yajima, H. (1989) Studies on peptides CLXVIII: Syntheses of three peptides isolated from horseshoe crab hemocytes, tachyplesin I, tachyplesin II, and polyphemusin I. *Chem. Pharm. Bull.* **37**, 2661–2664.

88. Jaeger, E., Thamm, P., Knof, S., Wünsch, E., Löw, M., and Kisfaludy, L. (1978) Nebenreaktionen bei peptidsynthesen III: synthese und charakterisierung von N^{in}-tert-butylierten tryptophan-derivaten. *Hoppe-Seyler's Z. Physiol. Chem.* **359**, 1617–1628.

89. Jaeger, E., Thamm, P., Knof, S., and Wünsch, E. (1978) Nebenreaktionen bei peptidsynthesen IV: Charakterisierung von *C*- und *C,N*-tert-butylierten tryptophan-derivaten. *Hoppe-Seyler's Z. Physiol. Chem.* **359**, 1629–1636.

90. Löw, M., Kisfaludy, L., Jaeger, E., Thamm, P., Knof, S., and Wünsch, E. (1978) Direkte tert-butylierung des tryptophans: Herstellung von 2,5,7-tri-tert-butyltryptophan. *Hoppe-Seyler's Z. Physiol. Chem.* **359**, 1637–1642.

91. Löw, M., Kisfaludy, L., and Sohár, P. (1978) tert-Butylierung des tryptophan-indolringes während der abspaltung der tert-butyloxycarbonyl-gruppe bei peptidsynthesen. *Hoppe-Seyler's Z. Physiol. Chem.* **359**, 1643–1651.

92. Masui, Y., Chino, N., and Sakakibara, S. (1980) The modification of tryptophyl residues during the acidolytic cleavage of Boc-groups I: studies with Boc-tryptophan. *Bull. Chem. Soc. Jpn.* **53**, 464–468.

93. Sieber, P. (1987) Modification of tryptophan residues during acidolysis of 4-methoxy-2,3,6-trimethylbenzenesulfonyl groups: effects of scavengers. *Tetrahedron Lett.* **28**, 1637–1640.

94. Riniker, B. and Hartmann, A. (1990) Deprotection of peptides containing Arg(Pmc) and tryptophan or tyrosine: Elucidation of by-products, in *Peptides: Chemistry, Structure and Biology* (Rivier, J. E. and Marshall, G. R., eds.), Escom, Leiden, The Netherlands, pp. 950–952.

95. King, D. S., Fields, C. G., and Fields, G. B. (1990) A cleavage method for minimizing side reactions following Fmoc solid phase peptide synthesis. *Int. J. Peptide Protein Res.* **36**, 255–266.

96. Fields, C. G. and Fields, G. B. (1993) Minimization of tryptophan alkylation following 9-fluorenylmethoxycarbonyl solid-phase peptide synthesis. *Tetrahedron Lett.* **34**, 6661–6664.

97. Riniker, B. and Kamber, B. (1989) Byproducts of Trp-peptides synthesized on a p-benzyloxybenzyl alcohol polystyrene resin, in *Peptides 1988* (Jung, G. and Bayer, E., eds.), de Gruyter, Berlin, pp. 115–117.

98. Albericio, F., Kneib-Cordonier, N., Biancalana, S., Gera, L., Masada, R. I., Hudson, D., and Barany, G. (1990) Preparation and application of the 5-(4-(9-fluorenylmethyl-oxycarbonyl)aminomethyl-3,5-dimethoxyphenoxy)valeric acid (PAL) handle for the solid-phase synthesis of C-terminal peptide amides under mild conditions. *J. Org. Chem.* **55**, 3730–3743.

99. Gesellchen, P. D., Rothenberger, R. B., Dorman, D. E., Paschal, J. W., Elzey, T. K., and Campbell, C. S. (1990) A new side reaction in solid-phase peptide synthesis: solid support-dependent alkylation of tryptophan, in *Peptides: Chemistry, Structure and Biology* (Rivier, J. E. and Marshall, G. R., eds.), Escom, Leiden, The Netherlands, pp. 957–959.

100. Fields, C. G., VanDrisse, V. L., and Fields, G. B. (1993) Edman degradation sequence analysis of resin-bound peptides synthesized by 9-fluorenylmethoxycarbonyl chemistry. *Peptide Res.* **6**, 39–46.

101. Stierandová, A., Sepetov, N. F., Nikiforovich, G. V., and Lebl, M. (1994) Sequence-dependent modification of Trp by the Pmc protecting group of Arg during TFA deprotection. *Int. J. Peptide Protein Res.* **43**, 31–38.

102. Jaeger, E., Remmer, H. A., Jung, G., Metzger, J., Oberthür, W., Rücknagel, K. P., Schäfer, W., Sonnenbichler, J., and Zetl, I. (1993) Nebenreaktionen bei peptidsynthesen V: O-sulfonierung von serin und threonin während der abspaltung der Pmc- und Mtr-schutzgruppen von argininresten bei Fmoc-festphasen-synthesen. *Biol. Chem. Hoppe-Seyler* **374**, 349–362.

103. Solé, N. A. and Barany, G. (1992) Optimization of solid-phase synthesis of [Ala[8]]-dynorphin A. *J. Org. Chem.* **57**, 5399–5403.

104. Choi, H. and Aldrich, J. V. (1993) Comparison of methods for the Fmoc solid-phase synthesis and cleavage of a peptide containing both tryptophan and arginine. *Int. J. Peptide Protein Res.* **42**, 58–63.

105. Gisin, B. F. and Merrifield, R. B. (1972) Carboxyl-catalyzed intramolecular aminolysis: a side reaction in solid-phase peptide synthesis. *J. Am. Chem. Soc.* **94**, 3102–3106.

106. Pedroso, E., Grandas, A., de las Heras, X., Eritja, R., and Giralt, E. (1986) Diketopiperazine formation in solid phase peptide synthesis using p-alkoxybenzyl ester resins and Fmoc-amino acids. *Tetrahedron Lett.* **27**, 743–746.

107. Albericio, F. and Barany, G. (1985) Improved approach for anchoring N^α–9-fluorenyl-methyloxycarbonylamino acids as *p*-alkoxybenzyl esters in solid-phase peptide synthesis. *Int. J. Peptide Protein Res.* **26**, 92–97.

108. Gairi, M., Lloyd-Williams, P., Albericio, F., and Giralt, E. (1990) Use of BOP reagent for the suppression of diketopiperazine formation in Boc/Bzl solid-phase peptide synthesis. *Tetrahedron Lett.* **31**, 7363–7366.

109. Ueki, M. and Amemiya, M. (1987) Removal of 9-fluorenylmethyloxycarbonyl (Fmoc) group with tetrabutylammonium fluoride. *Tetrahedron Lett.* **28**, 6617–6620.

110. Barlos, K., Gatos, D., Hondrelis, J., Matsoukas, J., Moore, G. J., Schäfer, W., and Sotiriou, P. (1989) Darstellung neuer säureempfindlicker harze vom sek.-alkohol-typ und ihre anwendung zur synthese von peptiden. *Liebigs Ann. Chem.,* 951–955.

111. Barlos, K., Gatos, D., Kallitsis, J., Papaphotiu, G., Sotiriu, P., Wenqing, Y., and Schäfer, W. (1989) Darstellung geschützter peptid-fragmente unter einsatz substituierter triphenylmethyl-harze. *Tetrahedron Lett.* **30**, 3943–3946.

112. Bodanszky, M. and Kwei, J. Z. (1978) Side reactions in peptide synthesis VII: sequence dependence in the formation of aminosuccinyl derivatives from β-benzyl-aspartyl peptides. *Int. J. Peptide Protein Res.* **12**, 69–74.

113. Bodanszky, M., Tolle, J. C., Deshmane, S. S., and Bodanszky, A. (1978) Side reactions in peptide synthesis VI: a reexamination of the benzyl group in the protection of the side chains of tyrosine and aspartic acid. *Int. J. Peptide Protein Res.* **12**, 57–68.

114. Nicolás, E., Pedroso, E., and Giralt, E. (1989) Formation of aspartimide peptides in Asp-Gly sequences. *Tetrahedron Lett.* **30**, 497–500.

115. Kenner, G. W. and Seely, J. H. (1972) Phenyl esters for C-terminal protection in peptide synthesis. *J. Am. Chem. Soc.* **94**, 3259–3260.

116. Schön, I., Colombo, R., and Csehi, A. (1983) Effect of piperidine on benzylaspartyl peptides in solution and in the solid phase. *J. Chem. Soc. Chem. Commun.,* 505–507.

117. Yang, Y., Sweeney, W. V., Scheider, K., Thörnqvist, S., Chait, B. T., and Tam, J. P. (1994) Aspartimide formation in base-driven 9-fluorenylmethoxycarbonyl chemistry. *Tetrahedron Lett.* **35**, 9689–9692.

118. Lauer, J. L., Fields, C. G., and Fields, G. B. (1995) Sequence dependence of aspartimide formation during 9-fluorenylmethoxycarbonyl solid-phase peptide synthesis. *Lett. Peptide Sci.* **1**, 197–205.

119. Dölling, R., Beyermann, M., Haenel, J., Kernchen, F., Krause, E., Franke, P., Brudel, M., and Bienert, M. (1994) Piperidine-mediated side product formation for Asp(OBu^t)-containing peptides. *J. Chem. Soc. Chem. Commun.* 853,854.

120. Quibell, M., Owen, D., Packman, L. C., and Johnson, T. (1994) Suppression of piperidine-mediated side product formation for Asp(OBu^t)-containing peptides by the use of *N*-(2-hydroxy–4-methoxybenzyl) (Hmb) backbone amide protection. *J. Chem. Soc. Chem. Commun.* 2343,2344.

121. Atherton, E., Hardy, P. M., Harris, D. E., and Matthews, B. H. (1991) Racemization of C-terminal cysteine during peptide assembly, in *Peptides 1990* (Giralt, E. and Andreu, D., eds.), Escom, Leiden, The Netherlands, pp. 243,244.

122. Fujiwara, Y., Akaji, K., and Kiso, Y. (1994) Racemization-free synthesis of C-terminal cysteine-peptide using 2-chlorotrityl resin. *Chem. Pharm. Bull.* **42**, 724–726.

123. Fields, C. G. and Fields, G. B. (1994) Solvents for solid-phase peptide synthesis, in *Methods in Molecular Biology, vol. 35: Peptide Synthesis Protocols* (Pennington, M. W. and Dunn, B. M., eds.), Humana, Totowa, NJ, pp. 29–40.

124. Sarin, V. K., Kent, S. B. H., Mitchell, A. R., and Merrifield, R. B. (1984) A general approach to the quantitation of synthetic efficiency in solid-phase peptide synthesis as a function of chain length. *J. Am. Chem. Soc.* **106**, 7845–7850.

125. Pickup, S., Blum, F. D., and Ford, W. T. (1990) Self-diffusion coefficients of Boc-amino acid anhydrides under conditions of solid phase peptide synthesis. *J. Polym. Sci. A: Polym. Chem.* **28,** 931–934.

126. Live, D. H. and Kent, S. B. H. (1983) Correlation of coupling rates with physicochemical properties of resin-bound peptides in solid phase synthesis, in *Peptides: Structure and Function* (Hruby, V. J. and Rich, D. H., eds.), Pierce Chemical Co., Rockford, IL, pp. 65–68.

127. Mutter, M., Altmann, K. H., Bellof, D., Flörsheimer, A., Herbert, J., Huber, M., Klein, B., Strauch, L., Vorherr, T., and Gremlich, H. U. (1985) The impact of secondary structure formation in peptide synthesis, in *Peptides: Structure and Function* (Deber, C. M., Hruby, V. J., and Kopple, K. D., eds.), Pierce Chemical Co., Rockford, IL, pp. 397–405.

128. Ludwick, A. G., Jelinski, L. W., Live, D., Kintanar, A., and Dumais, J. J. (1986) Association of peptide chains during Merrifield solid-phase peptide synthesis: a deuterium NMR study. *J. Am. Chem. Soc.* **108,** 6493–6496.

129. Narita, M. and Kojima, Y. (1989) The β-sheet structure-stabilizing potential of twenty kinds of amino acid residues in protected peptides. *Bull. Chem. Soc. Jpn.* **62,** 3572–3576.

130. Yamashiro, D., Blake, J., and Li, C. H. (1976) The use of trifluoroethanol for improved coupling in solid-phase peptide synthesis. *Tetrahedron Lett.,* 1469–1472.

131. Narita, M., Umeyama, H., and Yoshida, T. (1989) The easy disruption of the β-sheet structure of resin-bound human proinsulin C-peptide fragments by strong electron-donor solvents. *Bull. Chem. Soc. Jpn.* **62,** 3582–3586.

132. Fields, G. B., Otteson, K. M., Fields, C. G., and Noble, R. L. (1990) The versatility of solid phase peptide synthesis, in *Innovation and Perspectives in Solid Phase Synthesis* (Epton, R., ed.), Solid Phase Conference Coordination, Ltd., Birmingham, UK, pp. 241–260.

133. Fields, G. B. and Fields, C. G. (1991) Solvation effects in solid-phase peptide synthesis. *J. Am. Chem. Soc.* **113,** 4202–4207.

134. Atherton E., Woolley, V., and Sheppard, R. C. (1980) Internal association in solid phase peptide synthesis: synthesis of cytochrome C residues 66–104 on polyamide supports. *J. Chem. Soc., Chem. Commun.* 970,971.

135. Atherton, E. and Sheppard, R. C. (1985) Detection of problem sequences in solid phase synthesis, in *Peptides: Structure and Function* (Deber, C. M., Hruby, V. J., and Kopple, K. D., eds.), Pierce Chemical Co., Rockford, IL, pp. 415–418.

136. Bedford, J., Hyde, C., Johnson, T., Jun, W., Owen, D., Quibell, M., and Sheppard, R. C. (1992) Amino acid structure and "difficult sequences" in solid phase peptide synthesis. *Int. J. Peptide Protein Res.* **40,** 300–307.

137. Johnson, T., Quibell, M., Owen, D., and Sheppard, R. C. (1993) A reversible protecting group for the amide bond in peptides: use in the synthesis of 'difficult sequences.' *J. Chem. Soc. Chem. Commun.* 369–372.

138. Hyde, C., Johnson, T., Owen, D., Quibell, M., and Sheppard, R. C. (1994) Some 'difficult sequences' made easy: a study of interchain association in solid-phase peptide synthesis. *Int. J. Peptide Protein Res.* **43,** 431–440.

Protein Engineering

Sudhir Paul

1. Introduction

The three-dimensional arrangement of amino acids in a protein dictates its biological properties. One of the urgent challenges of modern biology is to decipher the relationship between the linear sequence of amino acids and the conformation adopted by proteins. Success is predicting protein conformation from the primary structure can be anticipated to permit rational design of therapeutic formulations to treat intractable human diseases and of probes for research in fundamental biological problems.

The starting point for protein engineering is the determination of the conformational features underlying the biological activity of natural proteins. Modifications of function can then be engineered by introducing structural alterations at the residues or domains responsible for the function while maintaining the backbone structure of the protein. This is the subject of this chapter. The alternative approach, viz. the *de novo* design of new proteins with defined functions is reviewed in **refs.** *1* and *2*. The initial aim of site-directed mutagenesis studies is to dissect the contribution of individual amino acids in protein function. Thereupon, substitution of the wild-type amino acids by residues with varying hydrophobicity, chemical-bonding potential and size can be attempted to achieve improved function. The addition or removal of protein domains is another productive route to protein engineering. Domain manipulations can be done using *de novo* designed polypeptides or fragments of natural enzymes, receptors and antibodies. The occurrence of certain short linear strings of amino acids in peptides proteins is associated with discrete conformations. For example, peptides containing certain arrangements of charged residues interspersed with hydrophobic residues tend to form amphiphilic α-helices with charged residues expressed on one face and uncharged residues on the other face *(3,4)*. The inclusion of these motifs is a useful means to achieve association of heterodimeric proteins (*see* **Subheading 6.**), because they tend to aggregate by formation of stable coiled coil pairs of α-helices.

2. Understanding Natural Protein Structures

Even though proteins are large structures, biological functions like enzymatic catalysis and receptor binding are determined by a restricted set of structural determi-

From: *Molecular Biomethods Handbook*
Edited by: R. Rapley and J. M. Walker © Humana Press Inc., Totowa, NJ

nants on the surface of proteins. Identification of these structural determinants is the prerequisite to engineering of protein biological activities. Intermolecular-binding interactions underlying the biological functions of proteins involve long-range and short-range electrostatic forces, formation of hydrogen bonds, van der Waals forces and the hydrophobic effect. Receptor-ligand and enzyme-substrate binding involves contacts at only a few amino-acid residues. These contacts can occur at the peptide backbone or the side chains of the amino acids. X-ray crystallography has suggested that contacts at 15–22 amino acids are involved in high-affinity antigen-antibody binding *(5)*. Of these contacts, the major contribution toward stabilization of the antigen-antibody complex is contributed by a smaller subset of amino acids *(6)*. This type of information is of obvious importance to the protein engineer, in that enhanced binding interactions can be designed by substitution of energetically neutral or unfavorable contacts with favorable ones.

In addition to X-ray crystallography, various spectroscopic techniques such as nuclear magnetic resonance (NMR), circular dichroism measurement, and fluorescence quenching are available to study the mechanism of protein-ligand binding and the constitution of the active-site microenvironment (reviewed in **refs.** *7* and *8)*. Because the total number of natural and artificial proteins with different primary structures is so vast, it may is impractical to directly analyze active-site conformation by biophysical methods in every case. Homology modeling has proven to be a valid shortcut to deduce conformation, particularly in the case of highly conserved proteins like antibodies. In essence, the backbone conformation of a new antibody is deduced from the data bank X-ray crystallography structure of the most homologous antibody and the model is then refined by energy minimization *(9)*.

Site-directed mutagenesis (*see* Chapter 28) is another powerful means to identify individual amino-acids responsible for protein function. Site-directed mutagenesis is most profitably applied when details of the three-dimensional structure of a protein are available. The technique can be used to manipulate the long-range and short-range electrostatic interactions, hydrogen bonding and hydrophobic-packing characteristics in the binding site of proteins.

Because each of the protein analysis methods has unique drawbacks, final conclusions can be obtained only if the results of different types of experiments are in concordance. For example, protein-ligand complexes in crystals are present at very high concentrations, beyond the range normally necessary for the in vivo biological activity of proteins. Concentration-dependent changes in peptide and protein conformation are well-known, and conformations deduced from crystal studies are therefore subject to this artifact. In site-directed mutagenesis studies, the worry is that substitution of the wild-type amino acid by the mutant amino acid can lead to nonspecific perturbations in conformation. Global-conformation changes can be monitored easily by spectroscopic methods, but local perturbations in conformation owing to backbone readjustments and long range electrostatic forces are difficult to analyze. The conceptual problem is minimized if it is recognized that individual residues in the active site not only interact with the ligand, but are also involved in maintenance of the overall active-site topography. In this sense, local conformational changes arising from mutations represent a means to define factors underlying the overall stability of the active site, rather than "nonspecific" perturbations.

Historically, identification of amino acids responsible for the biological activity of proteins has been done by measurement of activity following derivitization with amino acid-selective reagents. Although the specificity of the available reagents that react with individual amino acids is not absolute, this remains a useful technique when no other information is available on a newly identified protein. Saturation of the active site by the ligand prior to amino-acid modification should protect against loss of activity. This is usually a reliable indicator of the absence of nonspecific effects. Excellent compendiums describing the characteristics of available amino acid derivitizing reagents are available *(10,11)*. Intramolecular activation of certain amino acids is a major factor contributing to the catalytic activity of enzymes. In serine proteases, a hydrogen-bonding network between Asp, His, and Ser residues confers enhanced nucleophilicity to the Ser hydroxyl group, permitting its reaction with the carbonyl group of the scissile-peptide bond and the formation of an acyl-enzyme complex. This unusual activity of the Ser residue can be readily detected by its irreversible reaction with active site reagents like diisopropylfluorophosphate.

3. The Antibody System: A Template-Driven Engineering Route

In the Darwinian model, molecular evolution occurs via adaption of active-site structures to achieve the ability to interact with ligands. This model contains two essential components, a genetic mechanism to achieve variation in active site structure and a mechanism leading to selection of the favored active site structure.

The evolutionary mechanism permitting variation in the structure of protein active sites consists of amino acid sequence diversification. Highly evolved active sites of proteins essential to life display a very high level of genetic stability, in that once a protein has acquired a specific biological function, its amino acid sequence generally displays little variability. In contrast, the antibodies (and certain other immunological molecules, including the T-cell receptor) are naturally designed to undergo sequence diversification as a means to defend the organism against infection and environmental toxins. This feature permits facile "engineering" of binding sites in antibodies, simply by performing immunization of experimental animals with the ligand (or "template") followed by collection of antibodies from the serum of the animal. Techniques have been developed to permit efficient antibody synthesis in experimental animal in response to immunization with small ligands (designated haptens in the immunological jargon) as well as large proteins *(12)*. The repertoire of different binding sites that can be synthesized by the immune system has been estimated to range from 10^{10}–10^{12}. Following immunization of an animal with a polypeptide antigen, various antibodies to individual antigenic epitopes are synthesized (an epitope is usually composed of <30 amino acids). To select unique antibodies with the desired properties, hybridomas secreting monoclonal antibodies (MAbs) can be prepared *(13)*. Alternatively, the cDNA for the antibody domain responsible for ligand binding, the F_v domain can be expressed on the surface of phage particles and phages with the desired binding activity can be selected by chromatography or "panning" on the immobilized ligand *(14,15)*.

The antibody system also offers the possibility of creating variants of naturally occurring receptors and enzymes via the "idiotypic network" *(16)*. When animals are immunized with an antibody capable of binding a ligand, the antibody stimulates the formation of anti-idiotypic antibodies (so designated because they are directed to the idiotope, or the unique arrangement of amino acids present in the binding site of the antiligand antibody). Some of the anti-idiotypic antibodies mimic the structure of the original ligand and thus display biological activities similar to the ligand. Examples of activity mimicry derived by this means include antibodies that express opiate *(17)* and estrogen activities *(18)* and catalytic antibodies, however, with acetylcholinesterase activity *(19)*. The binding site in these antibodies is by no means an accurate replica of the ligand. The screening of libraries of anti-idiotypic antibody libraries may permit isolation of variants that possess biological activities related to, but not identical, to that of the ligand. This possibility is exemplified by observations that some anti-idiotypic antibodies function as antagonists rather than agonists for the ligand receptor, as in the case of thyrotropin-mimicking antibodies *(20)*.

The idea that new catalysts can be derived by immunization with presumed transition state analogs has been heavily promoted in several recent publications *(21–23)*. The premise of this idea is that the transition state analog selectively provokes formation of antibodies that bind the transition structure of the substrate more strongly than the substrate itself, thus lowering the activation energy barrier for the reaction (*see* **ref. 24** for review of the transition-state theory of catalysis). However, difficulties have been encountered in generating antibodies capable of catalyzing energetically demanding reactions by this strategy. The likely reason is that immunization with transition state analogs does not induce intramolecular activation of potential catalytic residues. Enzymes use activated amino acid side chains to catalyze chemical reactions. For example, peptide-bond hydrolysis by serine proteases involves a Ser hydroxyl group that is activated by hydrogen bonding with a His residue to become capable of nucleophilic attack on peptide bonds. Simple shape complementarily provoked by immunization with a transition analog cannot be expected to mimic intramolecular activation reactions responsible for the catalysis in natural enzymes.

The limitations of immunization with transition-state analogs have largely been mitigated by the recent discovery that the ability to catalyze chemical reactions is a physiological property of antibodies produced by the immune system in healthy individuals and patients with autoimmune disease. The serendipitous discovery of neuropeptide-cleaving activity in autoantibodies *(25)* has been followed by observations of autoantibody-mediated DNA *(26)* and thyroglobulin *(27)* cleavage reactions, and by the demonstration of amide-bond cleavage by polyclonal antibodies from normal serum samples *(28)*. Antibody light chains synthesized in response to the substrate ground state (as opposed to the transition-state analogs) have been shown to possess protease *(29,30)* and peroxidase *(31)* activities. The proteolytic light chain mimics certain serine proteases in its cleavage specificity, inhibition by known serine protease inhibitors, and loss of activity by mutagenesis at Ser27a and His93, residues that may serve as components of a catalytic triad *(30,32)*. These considerations are consistent with observations that when comparatively efficient antibody catalysis in response to transition state analog immunization is encountered, the mechanism of catalysis does not appear to be directly related to the elicitation of shape complementarily with the transition-state analog, leading to characterization of the catalytic activity as being "accidental" *(33)*. Thus, immunization with the substrate appears to be a straightforward means to isolate high-affinity, substrate-specific antibodies capable of catalyzing energetically demanding reactions.

The presence of different types of biological activities in the variable domain of antibody molecules is advantageous for several reasons. Amino-acid residues located in the complementarily determining regions (CDRs) of the antibody variable domain are largely responsible for the ligand binding activity. The CDRs are surrounded by comparatively conserved regions, the framework regions. This feature permits rapid cloning of different antibody molecules using PCR primers designed to anneal the conserved regions *(34)*. Second, the isolation of improved antibodies by random mutagenesis is facilitated because the mutations must be performed only in the CDRs, as opposed to the entire length of the V_L and V_H domains *(35)*. Third, the immune system does not respond to the molecular scaffold utilized by antibodies as a "foreign antigen," thus limiting the probability of deleterious immunological reactions in situations where therapeutic use of the engineered protein in humans is contemplated.

4. Imprinting of Proteins In Vitro

Proteins are flexible molecules with the potential to undergo changes in conformation upon contact with the ligand. Refolding of antibody fragments from denaturant solutions in the presence of ligand has been observed to promote their binding *(36)* and catalytic *(37)* activities, presumably because the ligand serves as a template around which the antibody-binding site can fold correctly. Similarly, refolding of the Bowman-Birk trypsin inhibitor from a guanidine hydrochloride solution in the presence of trypsin leads to a hyperactive conformation of the inhibitor that relaxes to its lower activity state with time *(38)*.

It is generally held that proteins in aqueous solution are too flexible to "remember" the imprint of the ligand and will revert to the lowest energy state defined by their primary sequence upon removal of the ligand. In organic solvents, however, proteins display a high level of conformational rigidity. Albumin lyophilized from a tartaric acid containing aqueous solution is reported to display impressive levels of tartaric acid-binding activity, provided the assays are done in anhydrous organic solvents *(39)*. The "imprint" of the ligand detected in organic solvents is lost in aqueous solution. The utility of this procedure is limited, therefore, to reactions conducted in the absence of water. Organic solvents can support reactions such as peptide-bond synthesis catalyzed by proteases *(40,41)*. Useful applications of imprinted proteins can be visualized in these instances.

5. Improvement of Function by Site-Directed Mutagenesis
5.1. Protein Stability

The stability of the folded state of a protein can be increased by introduction of disulfide bridges via placement of Cys residues at key positions known to be intramolecular contact points *(42)*. Another type of stabilizing crosslink consists of a transition metal ion coordinated by two His residues. Introduction of a His residue at position 58 in iso-1-cytochrome *c* from *Saccharomyces cervisiae* by mutagenesis permits formation of a His39-Cu^{2+}-His58 crosslink and increases the conformational stability of this protein by about 2 kcal/mol *(43)*. The "stiffening" of preexisting folding patterns is yet another means to stabilize protein conformations. Comparison of the tertiary structures of several proteins isolated from mesophilic and thermophilic organisms has lead to the suggestion that thermal stability can be engineered in proteins by introducing mutations that decrease flexibility and increase the hydrophobicity in the α-helical segments *(44)*. Conversely, removal of hydrophobic amino acids from poorly soluble recombinant proteins like the single chain T-cell receptor *(45)* and certain antibody fragments *(46)* by site-directed mutagenesis can be applied to increase water solubility.

5.2. Binding Affinity

Once the attribution of binding energy in the three-dimensional structure of a receptor-ligand complex has been solved, energetically neutral or unfavorable contacts can be replaced by favorable ones by site-directed mutagenesis. For example, introduction of residues capable of forming ion pairs or hydrogen bonds with the substrate can be predicted to improve ligand-binding affinity, provided the mutations do not lead to gross perturbations in binding-site topography. Replacement of Asp97 with an apolar residue in an antiphenyloxazolone antibody heavy chain, which was deduced from NMR and molecular-modeling studies to make an unfavorable contact with the antigen, has been observed to result in increased antigen-binding affinity *(47)*. Similarly, the binding affinity of an antibody to a mucin epitope varies by about 10-fold by occupancy of position 99 in CDRs of the heavy chain by different amino acids *(48)*.

5.3. Catalytic Activity

Can a ligand-binding site be converted into a catalytic site by site-directed mutagenesis? Catalysis is dependent on the chemical reactivity of active site residues. Replacement of a Tyr residue in the binding site of an antidinitrophenol antibody by a His residue has been reported to impart a weak esterase activity to the antibody *(49)*. Introduction of an -SH group capable of nucleophilic catalysis in the binding site of another antibody permits ester cleavage by the antibody *(50)*. These examples provide the first indications that rational mutagenesis can be applied to transform a binding site into a catalytic site. It is worth noting, however, that these antibodies do not approach the catalytic power of natural enzymes. As discussed in **Subheading 3.**, efficient catalysis is likely to be achieved only when the overall chemistry of the binding site is optimized, which includes the development of intramolecular mechanisms permitting activation of the catalytic residues.

An interesting variant of the enzyme-engineering problem is described in a site-directed mutagenesis study on tissue-plasminogen activator (tPA) *(51)*. Unlike many other serine proteases, tPA is secreted from cells not as a zymogen, but a catalytically active protease. To reduce the systemic thrombolytic side effects of tPA administered to patients with heart disease, a zymogenic variant of tPA was constructed in which Ser and His were introduced by mutagenesis at positions 292 and 477. The resultant molecule contained a zymogenic, hydrogen-bonded triad, viz., Ser292-His305-Asp477. This triad exerts a regulatory role, in that the mutant is nearly devoid of the catalytic activity of wild-type tPA. Cleavage of the mutant tPA into its two chain form restores the catalytic activity to the level of wild-type two chain tPA. Presumably, the interaction of Asp477 with the mutant Ser/His residues no longer permits participation of the former residue as a component of the normal catalytic triad present in single chain tPA.

Enzymatic catalysis often involves the formation of a short-lived covalent enzyme-substrate complexes. The mutation of Thr26 to Glu in the active-site cleft of phage-T4 lysozyme results in an enzyme that still cleaves its mucopolysaccharide substrate, but the product remains covalently bound to the enzyme in the stable form *(52)*. The mutant Glu residue apparently displaces a water molecule that is normally bound by hydrogen bonding to Thr26 in the wild type enzyme and is responsible for hydrolysis of the covalent glycosyl-enzyme complex. This observation has important implications, in that mutant enzymes capable of forming covalent complexes with specific substrates or products found at high concentrations in a particular organ could be used in biological targeting and imaging applications.

5.4. Substrate Specificity

Several attempts have been made to engineer the substrate specificity of proteases by charge reversal of active-site residues thought to recognize the amino acid at the scissile bond. Asp189 located at the base of the binding pocket in trypsin is not an essential catalytic residue, but is postulated to confer specificity for recognition of Lys-X and Arg-X scissile bonds via formation of an ion pair. Replacement of Asp189 by a Lys residue by mutagenesis leads to loss of cleavage activity at Lys-X bonds but the mutant does not acquire the ability to cleave Asp-X or Glu-X bonds *(53)*. The suggested explanation for this result is the presence of a positively charged Lys residue in the interior of an active site, which is optimized to accommodate the negatively charged Asp189 wild-type residue, leads to an overall destabilization of the essential catalytic machinery of trypsin.

In comparison, attempts to reverse the scissile-bond specificity of neutral endopeptidase by active-site charge reversal have been more encouraging *(54)*. This enzyme has a didpeptidylcarboxypeptidase activity, which is mediated in part by recognition of the C-terminal carboxyl group in the substrate by Arg102. Substitution of a Glu residue in place of Arg102 leads to a 30-fold increase in cleavage of a positively charged substrate. The success in redirecting the cleavage specificity of neutral endopeptidase by active-site charge reversal has been attributed to the fact that the wild type Arg102 is located at the edge of the active site, rather than its interior, with the result that perturbation of active-site geometry is comparatively small.

Mutagenesis at Gly166 located at the bottom of the substrate-binding cleft in subtilisin has suggested that changes in specificity can also be engineered by manipulating hydrophobic packing and steric effects *(55)*. Mutants containing large hydrophobic residues instead of Gly166 displayed increased catalytic activity with hydrophobic substrates. Mutants that produced an increase in volume of the active site beyond the wild-type volume showed sharply decreased activity, emphasizing the importance of maintaining the overall geometry of the active site.

The specificity of a catalyst is determined not only recognition of the substrate structural features at the immediate site of the chemical reaction, but also by remote determinants. This is particularly true in the case of catalysts that contain a large active site, such as catalytic antibodies to polypeptides *(56)*. Thus, manipulation of catalystsubstrate contacts at subsites remote from the chemical reaction center provides yet another means to manipulate substrate specificity. Site-directed mutagenesis at active site residues distant from the catalytic residues in an antibody to vasoactive intestinal peptide (VIP) attenuates the VIP-binding affinity and increases the rate constant for VIP hydrolysis (because decreased substrate-binding energy will lead to a decrease in the energy barrier separating the ground state and the transition state of the substrate) *(32)*. In comparison, the mutants hydrolyze substrates unrelated to VIP with unchanged kinetic constants.

Homology-modeling techniques have enabled some of the most dramatic examples of enzyme-specificity remodeling. Comparison of the active-site regions of the homologous enzymes lactate dehydrogenase and malate dehydrogenase permitted identification of Gln102 as a potential specificity determining residues in the former enzyme. Conversion of Gln102 in lactate dehydrogenase to Arg, which is the potential specificity determining counterpart in malate dehydrogenase, produced a specificity shift of seven orders of magnitude toward the malate dehydrogenase form *(57)*. Similarly, the coenzyme specificity of NAD-dependent dehydrogenase has been converted to an NADP-dependent type by mutations introduced at residues selected by homology modeling of the two wild-type enzymes *(58)*.

The pKa of ionizable groups important in catalysis can be influenced by long-range charge interactions with residues located on the surface of the enzyme *(59)*. Thus, the catalytic characteristics of a protein can be modified even by mutations at charged residues that do not contact the substrate. An example of activity manipulation by this means is the observation that lysozyme is rendered a more efficient catalyst at high ionic strengths and pH compared to the wild-type by mutations producing increased positive charge on the protein surface *(60)*.

The primary challenge in rational enzyme engineering is to design a protein capable of binding the substrate reasonably strongly, and the transition-state structure of the substrate even more strongly. The strength of substrate binding governs the K_m value, and the strength of transition state binding, the reaction-rate constant. Whereas substrate ground-state binding requirements can be analyzed by conventional biophysical methods such as X-ray crystallography and NMR, the optimal enzyme geometry favoring transition state binding is far more difficult to decipher. One useful technique is to study enzyme binding to various substrate analogs containing structural features corresponding to the postulated transition-state geometry. The transition state stabilization reaction is predicted to be favored when the analog binding is strongest, provided that the presumed transition state structural features incorporated in the analog are present in the actual transition state. In the case of proteases, the tetrahedral configuration of the peptide-bond carbon atom is thought to be an important feature influencing transition-state binding, leading to the widespread use of tetrahedral transition state analogs in enzyme-engineering studies. It is important to recognize that further details of the fine structure of the transition state will be necessary to achieve efficient catalysis by engineered proteins. Neighboring chemical groups are as much a part of the transition state structure as the reaction center itself. Using the protease example again, conversion of the trigonal carbon at the peptide bond to a tetrahedral configuration is also accompanied by increased flexibility owing to rotation around the C-N bond. The increased flexibility implies that the catalyst may recognize new subsites formed in the transition state that are remote from the scissile bond. A more accurate understanding of transition state recognition interactions can be anticipated, therefore, to lead to the design of catalysts with superior efficiency and substrate specificity.

6. Domain Expansion, Hybridization, and Combination

Large proteins can often be subdivided into contiguous amino-acid sequence units referred to as domains. Protein domains are essentially smaller structural units with a comparatively well-defined folding pattern. The individual domains in a protein are often also its functional units and are capable of independently mediating ligand binding and catalytic functions. Short-flexible or rigid linker sequences are found between the domains. Gilbert *(61)* proposed that proteins can evolve by varying the assembly of modular units, the domains, in addition to the accumulation of single amino acids mutations. As in nature, artificial combination of different types of domains into a single protein offers the possibility of engineering new or improved biological functions.

6.1. Minimum Domain Size

The initial task for the protein engineer is to dissect out the individual domains capable of the desired function. Historically, this aim has been achieved by assay of activity in fragments of the protein obtained by digestion with proteases. In the case of multimeric proteins, detergents are used to dissociate noncovalent complexes and disulfide bonds linking the subunits are reduced prior to the protease digestion. With the advent of recombinant DNA technology, the individual domains in a protein of known primary structure can be conveniently isolated by amplification of the cDNA corresponding to various segments of the protein by PCR, their cloning in a bacterial or eukaryotic-expression vector, and purification of the expressed polypeptides from culture supernatants, periplasmic extracts, or intracellular inclusion bodies. The size of the expressed protein segment can be varied by selecting PCR primers that anneal to various regions of the gene or the mRNA encoding the protein. These procedures have been applied, for example, to determine that the binding region of antibodies is a complex of the variable (V) domains of the light and heavy chain subunits, each about 110 amino acids in length. The individual V_L and V_H domains can also bind antigens, albeit with reduced affinity *(62,63)*. The antibody amino acids that make contact with antigen are located mainly in the CDRs. Synthetic peptides as small as 12 amino acids corresponding to the individual CDRs are also capable of weak antigen binding *(64,65)*.

The results cited in the preceding paragraph suggest that small peptides derived from large proteins retain, at least in part, the ability to recognize biological ligands. This conclusion is not limited to the antigen–antibody system, because high-affinity receptor–ligand binding in general appears to involve energetically favorable contacts at a limited number of amino acids, as described for the complex of growth hormone with its receptor *(66)*. This conclusion is also evident from the ability of many small peptides with hormonal and neurotransmitter properties to bind their receptors with high affinity.

It can be reasonably anticipated that small peptides may simulate enzymatic catalysis. VIP, a 28-amino acid peptide, possesses a weak autolytic activity *(67)* reminiscent of enzymatic catalysis. Substantial effort is being invested in the design of small peptides that contain active-site mimics of serine proteases *(68)*. The proteolytic activity of recombinant V_L domains of antibodies has been described *(32,69)* and further fragmentation of this domain may permit localization of the activities in individual CDRs.

The large size of natural proteins, then, is not owing to a direct involvement of the entire sequence in ligand binding and chemical catalysis, but rather, to the need to maintain the overall three-dimensional structure that permits selected amino acids to form the correct active-site conformation. Based on this premise, the problem of simulation of protein function by a small peptide reduces to a matter of constraining the peptide into the correct conformation, assuming that all of the chemical groups necessary for function are represented in the primary structure of the peptide. Peptides less than 30–40 amino acids in size generally exist in disordered state in aqueous solution. Once bound to another protein, the peptide conformation becomes comparatively inflexible. Hence, the conformation of a bound peptide is defined not only by its own primary structure, but also the intermolecular interactions, which in the case of cell-surface receptor interactions, includes interactions with the lipid microenvironment. The introduction of conformational constraints that force a small peptide into the putative receptor-bound conformation is a promising design strategy to prepare superactive peptide analogs *(70–72)*. Because a small size confers distinct advantages to an engineered protein related to its stability and biological uptake by various tissues, such peptides are likely to be useful as the building blocks for multifunctional proteins.

6.2. Domain Expansion

The binding or catalytic site of a protein can be expanded to include additional amino-acid residues, oligonucleotides, metals, and other small chemical moieties as a means to modify interactions with the ligand. The additional chemical groups are intended to participate directly in ligand binding and catalysis, and must, therefore, be placed within the active site. In a flexible protein domain, such an expansion might be anticipated to produce large changes in structure, which can potentially lead to loss of the original activity of the active site. Thus, the applicability of engineering via domain expansion may be limited to active sites that are comparatively rigid and are able to accept additional structures without gross perturbation of the original active-site geometry.

The binding site of an antibody is an example of favorable interactions between two comparatively rigid domains (V_H and V_L domains) leading to significantly improved ligand-binding activity. Antibody binding sites can be prepared as single chain F_v constructs by linkage of the cDNA for the VL and VH domains by overlap extension and expression of the protein in bacterial hosts *(73,74)*. A short peptide sequence (10–15 amino acids) links the V_H and V_L domains. An appropriate length and flexibility of the linker sequence is necessary to permit proper interactions between the V_L and V_H domains necessary for high affinity ligand-binding activity *(75)*. Most F_v constructs described in the literature are composed of an N-terminal V_H domain linked to a V_L domain on the C-terminal side, but Fv constructs in the opposite orientation are also fully functional *(76,77)*.

The positioning of negatively charged and positively charged leucine zipper peptides at the C-termini of the recombinant domains is an alternative way to construct the active sites of heterodimeric proteins. The two zipper peptide sequences form coiled-coil complexes, in which heterodimer formation is reportedly favored by five orders of magnitude over homodimer formation. A stable form of recombinant T-lymphocyte receptor composed of the α and β subunits associated via the zipper peptides has been prepared by this strategy *(78)*.

As noted previously, the V_L and V_H domains of antibodies can independently bind ligands, but only the V_L domain is known to possess catalytic activity. A single-chain F_v composed of catalytic anti-VIP V_L domain linked via a flexible 14-residue linker to

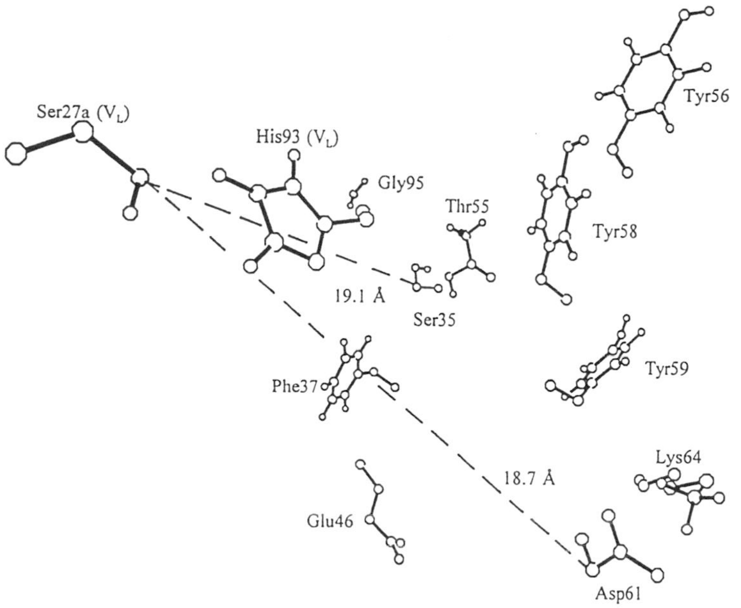

Fig. 1. Spatial position of catalytic residues in the V_L domain (Ser27a, His93) relative to V_H domain residues located within a distance 22Å. Shown are selected residues of a model of the anti-VIP Fv constructed using the computer program AbM (Oxford Molecular Inc.) as described in *(30)*. Distances are between the V_L domain Ser27a Oγ atom and C-β atoms of V_H residues. The relative solvent-accessibilities of the V_H residues are Ser35, 4.4%; Phe37, 12.1%; Thr55, 26.5%; Tyr56, 43.7%; Tyr58, 22.9%; Tyr59, 10.8%; Asp61, 50.1%; Glu46, 21.9%; Lys64, 42.9% and Gly95, 12.2%. Phe57, Tyr59, Glu46, Asp61, and Lys64 are located in framework regions. The remaining residues are located in the CDRs. Three framework-region residues (Trp47, Leu45, Pro60) with low solvent accessibilities (<7.5%) are not shown.

an anti-VIP V_H domain (**Fig. 1**) recognizes VIP with 20-fold improved affinity and hydrolyzes the peptide with fourfold improved catalytic efficiency *(77)*. Expanding the anti-VIP V_L domain with a nonspecific V_H domain leads to dramatic decreases in affinity and catalytic efficiency, suggesting that the beneficial effect of the anti-VIP V_H domain derives from specific substrate binding interactions.

In the case of DNA- and RNA-cleaving enzymes, the catalytic domain can be expanded to include an oligonucleotide that anneals with the substrate and thus renders the enzyme site-specific. The feasibility of this approach has been demonstrated by conjugation of a 15-nucleotide sequence to a Cys residue introduced into the active site of a *Staphylococcus* nuclease *(79)*. Conversely, attachment of the DNA cleaving moiety copper:*o*-phenanthroline to Catabolite Activator Protein (CAP), a DNA binding protein, imparts sequence-specific DNA cleaving activity to the complex *(80)*.

A novel transcription factor capable of binding a DNA sequence that combines the consensus-binding sequences of the natural transcription factors Zif268 and Oct-1 has been generated by linkage of zinc fingers and the homeodomain derived from these proteins *(81)*. Steric conflicts in the expanded bind site of the engineered transcription factor were avoided by using linker peptides designed to correctly span the distance between the individual DNA-binding domains.

A variant of the concept of domain expansion consists of the use of certain proteins as presentation scaffolds to potentiate receptor-binding activity. The tripeptide sequence Arg-Gly-Asp is known to be important in the binding of fibrinogen by platelet receptors. The insertion of a 12-residue peptide containing an Arg-Gly-Asp sequence into CDR3 of the V_L protein REI has been observed to confer potent fibrinogen receptor-antagonist activity to this protein *(82)*. Similarly, the grafting of peptide antigens into CDR3 of the V_H domain has been suggested to result in a superior antigen presentation and immune responsiveness *(83)*.

6.3. Domain Hybridization

In the instance of an active site composed of residues contributed by two domains, substitution of one of the domains with a related but different domain can lead to significant alterations in function. This type of hybridization procedure has been used to construct an F_v domain that recognizes the $V\alpha$ T-cell receptor, by linkage of an antilysozyme V_L domain with V_H domains derived from a mouse immunized with the T-cell receptor *(84)*. Hybridization of V_H and V_L domains from various anti-DNA antibodies has been shown to modulate the fine specificity of the resultant F_v constructs for various conformations of DNA *(85)*. Engineering of the substrate specificity of catalytic antibodies could potentially be achieved by hybridization of a catalytic V_L domain with noncatalytic V_H domains capable of binding the substrate, but this possibility is yet to be realized.

The binding specificity of antibodies is determined mainly by the sequence of the CDRs. Thus, CDR transplantation can be applied to direct the binding specificity. This principle has been demonstrated by conversion of an anti-DNA antibody to an antifluorescein antibody by transplantation of the heavy-chain CDRs *(86,87)*. This result is also relevant to the humanization of murine antibodies, in that transplantation of various specificity-defining CDRs into a single human-antibody scaffold may be a convenient way to minimize untoward immune responses to passively administered antibody.

6.4. Multivalency and Multiple Functions Generated by Domain Combination

Different protein domains often maintain their independent functions when linked together suitably. Steric clashes are usually not a problem because the individual domains tend to adopt their characteristic conformations. The length and flexibility of the interdomain linker sequence are important parameters in ensuing proper folding of the domains. In the enzyme dihydrolipoamide acetyltransferase, variations in the length and flexibility of the linkers connecting the lipoyl, subunit-binding, and catalytic domains have been observed to produce about 25-fold variations in enzyme activity *(88)*.

An interesting application of the linker requirements for domain interactions is the production of bivalent-antibody fragments. The presence of more than one valency in the antibody constructs is beneficial in that it increases the avidity of ligand binding. By using a linker sequence that is too short to permit pairing of the V_L and V_H domains located within a single-chain F_v construct, the domains are forced into intermolecular V_L–V_H pairing, resulting in noncovalent association of two F_v constructs and a resultant bivalent antigen-binding capability *(89)*. Disulfide bonding at Cys residues placed close to the C-terminus of F_v constructs is another means to produce bivalent-antibody constructs *(90,91)*. Tetravalent antibody fragments can be generated by placing a 33-residue amphipathic helical peptide derived from the GCN4 protein at the C-terminus of F_v constructs *(92)*. The C-terminal peptide associates noncovalently into a four-helix bundle, permitting the aggregation of four F_v molecules.

The ability to link two F_v domains permits generation of bispecific antibody constructs in which the individual F_v domains bind different epitopes on the same antigen or different antigens. In the former instance, the antibody construct binds the antigen with a higher avidity owing to the so-called "chelate effect" derived from binding to two adjacent, nonoverlapping epitopes on the same molecule *(93)*. In the latter instance, two different antigens serve as the targets of the antibody constructs. A bispecific dimer with one F_v moiety directed to the transferrin receptor and the second F_v moiety directed to mouse CDϵ has been shown to bind both antigens *(94)*. The anti-CD3/antitransferrin receptor construct is able to direct CD3^+, cytotoxic T- cells to lysis of cells expressing transferrin receptors on their surface. A related strategy that directs cytotoxic T-cells to destroy antigen-coated cells is to induce the expression of an antigen-specific F_v linked to the constant domain of the T-cell receptor on the surface of the T-cells *(95)*.

Receptor-ligand binding interactions are an excellent means of targeting defined molecules and tissues. Ubiquitin-conjugating enzymes tagged with various ligands at their C-termini selectively ubiquitinylate soluble proteins that bind the ligands, thus promoting degradation of the proteins via the ubiquitin-dependent pathway *(96)*. For example, fusion of plant ubiquitin-conjugating enzymes with transforming growth factor α selectively targets the ubiquitinylation reaction to its receptor, the epidermal growth-factor receptor. Antibodies are extensively used in targeting experiments, and, to this end, they have been linked to various types of polypeptides by gene fusion and protein chemistry methods. Potent immunotoxins can be generated by linking toxins like ricin, *Pseudomonas* exotoxin, or tumor-necrosis factor to F_v domains capable of binding tumor-associated antigens like OVB3 and TAG72 *(97–99)*. These constructs deliver the cytotoxic agent selectively to tumor cells. The immunotoxins have been shown to cure tumors in experimental animals *(100)*. Likewise, the radiotherapy of tumors may be facilitated by the availability of antibody fragments containing short peptide motifs that bind metals. An F_v domain directed to the tumor associated protein c-erbB-2 with a C-terminal Gly_4Cys peptide is described to chelate the radiometal 99m-technicium through the Cys sulfur side-chain *(101)*.

Three additional examples of targeting strategies are cited here to illustrate the potential of domain-linkage technologies in developing therapeutic products. First, a bispecific construct consisting of a fibrin-specific F_v domain linked to urokinase for potential use as an anti-thrombolytic agent has been described *(102)*. Second, antibody-fusion proteins offer a means to traverse the blood–brain barrier. Delivery of neuropeptides like VIP into the brain can potentially be used therapeutically to regulate cerebral blood flow, but the blood–brain barrier precludes entrance of peptides greater than five amino acids. A conjugate of an antibody to the transferrin receptor and a VIP analog, however, can be transported into the brain, because the antibody binds the transferrin receptor and crosses brain-capillary endothelial cells by receptor-mediated transcytosis *(103)*. Third, a recombinant fusion protein composed of a β-lactamase linked via a disulfide bond to an F_v moiety directed against the breast tumor associated protein p185^HER2 is reported to efficiently activate a cephalothin doxorubicin prodrug *(104)*. The selective activation of prodrugs at the site of the tumor by such constructs offers the advantage of reduced systemic toxicity.

Domain-fusion techniques can be fruitfully applied to cell-surface receptors because these proteins are generally composed of well-defined ligand binding, transmembrane, and intracellular signal-transducing segments. The aim of receptor engineering is to provide a means to control the metabolism and proliferation of defined cell types, for example, to replenish the supply of essential secretory products. Receptors generally possess hydrophobic surfaces and thus are often insoluble in aqueous solutions. This feature places limits on the type of experiments that can be done using receptors, but the cellular functions of these molecules can be examined by their expression in transfected cell lines. Fusion proteins containing the N-terminal hydrophilic domain of a nicotinic acetylcholine receptor linked to the C-terminal segment of the 5-hydroxytryptamine (5-HT) receptor bind cholinergic ligands, but form membrane spanning ion channels with the distinctive properties of the 5-HT receptor *(105)*. The ion channel is activated by the cholinergic ligands, indicating successful reconstitution of allosteric control of the ion channel domain by the ligand binding domain. Similarly, linkage of the extracellular domain of the growth hormone (GH) receptor to the transmembrane and intracellular domains of the receptor for granulocyte colony-stimulating factor, which belongs to the same receptor superfamily as the GH receptor, imparts GH responsiveness to cells expressing the fusion protein on their surface *(106)*. In this case also, thus, the GH-binding domain is able to activate signal transduction via the transmembrane and intracellular domains of another receptor.

Acknowledgments

The author's work was generously supported by the United Public Health Service, the University of Nebraska Medical Center institutional funds, and IGEN, Inc.

References

1. DeGrado, W. F., Wasserman, Z. R., and Lear, J. D. (1989) Protein design, a minimalist approach. *Science* **243,** 622–628.
2. Sander, C. (1994) Structural design of proteins, in *Concepts in Protein Engineering and Design: An Introduction* (Wrede, P. and Schneider, G., eds), Walter de Gruyter, New York, pp. 209–236.
3. Engel, M., William, S. R., and Erickson, B. (1991) Designed coiled-coil proteins: synthesis and spectroscopy of two 78-residue α-helical dimers. *Biochemistry* **30,** 3161–3169.
4. Regan, L. and DeGrado, W. F. (1988) Characterization of a helical protein designed from first principles. *Science* **241,** 976–978.
5. Davies, D. R., Padlan, E. A., and Sheriff, S. (1990) Antibody–antigen complexes. *Ann. Rev. Biochem.* **59,** 439–474.
6. Novotny, J., Bruccoleri, R. E., and Saul, F. A. (1989) On the attribution of binding energy in antigen–antibody complexes McPc 603, D1. 3 and HyHel–5. *Biochemistry* **28,** 4735–4749.
7. Wdthrich, K. (1989) Protein structure determination in solution by nuclear magnetic resonance spectroscopy. *Science* **243,** 45–50.
8. Hirs, C. H. W. and Timasheff, S. N. (eds.) (1986) *Methods in Enzymology, Vol. 130, Enzyme Structure, Part K.* Academic, New York.

9. Webster, D. M. and Rees, A. R. (1995) Molecular modeling of antibody-combining sites, in *Methods in Molecular Biology, vol. 51: Antibody Engineering Protocols* (Paul, S., ed.), Humana, Totowa, NJ, pp. 17–49.

10. Lundbland, R. and Noyes, C. M. (1984) *Chemical Reagents for Protein Modification,* Vols. I and II. CRC, Boca Raton, FL.

11. Wong, S. S. (1991) *Chemistry of Protein Conjugation and Cross-Linking.* CRC, Boca Raton, FL.

12. Weir, M., Herzenberg, L. A., Blackwell, C., and Herzenberg, L. A. (eds.) (1986) *Handbook of Experimental Immunology.* Blackwell, Oxford, UK.

13. Kohler, G. and Milstein, C. (1975) Continuous cultures of fused cells secreting antibody of predefined specificity. *Nature* **256,** 495–497.

14. Marks, J. D., Hoogenboom, H. R., Bonnert, T. P., McCafferty, J., Griffiths, A. D., and Winter, G. (1991) By-passing immunization. Human antibodies from V-gene libraries displayed on phage. *J. Mol. Biol.* **222,** 581–597.

15. Clackson, T., Hoogenboom, H. R., Griffiths, A. D., and Winter, G. (1991) Making antibody fragments using phage display libraries. *Nature* **352,** 624–628.

16. Jerne, N. K., Roland, J., and Cazenave, P. A. (1982) Recurrent idiotypes and internal images. *EMBO J.* **1,** 243–247.

17. Glasel, J. A. and Agarwal, D. (1995) Anti-idiotypic antibodies that mimic opioids, in *Methods in Molecular Biology, vol. 51: Antibody Engineering Protocols* (Paul, S., ed.), Humana, Totowa, NJ., pp. 183–201.

18. Somjen, D., Amir-Zaltsman, Y., Gayer, B., Mor, G., Jaccard, N., Weisman, Y., Barnard, G., and Kohen, F. (1995) Anti-idiotypic antibody as an oestrogen mimetic: removal of Fc fragment converts agonist to antagonist. *J. Endocrinol.* **145,** 409–416.

19. Izadyar, L., Friboulet, A., Remy, M. H., Roseto, A., and Thomas, D. (1993) Monoclonal anti-idiotypic antibodies as functional internal images of enzymes active sites: production of a catalytic antibody with a cholinesterase activity. *Proc. Natl. Acad. Sci. USA* **90,** 8876–8880.

20. Costagliola, S., Ruf, J., Durand-Gorde, M. J., and Carayon, P. (1991) Monoclonal antiidiotypic antibodies interact with the 93 kilodalton thyrotropin receptor and exhibit heterogeneous biological activities. *Endocrinology* **128,** 1555–1562.

21. Tramontano, A., Janda, K. D., and Lerner, R. A. (1986) Catalytic antibodies. *Science* **234,** 1566–1570.

22. Pollack, S. J., Jacobs, J. W., and Schultz, P. G. (1986) Selective chemical catalysis by an antibody. *Science* **234,** 1570–1573.

23. Stewart, J. D. and Benkovic, S. J. (1995) Transition-state stabilization as a measure of the efficiency of antibody catalysis. *Nature* **375,** 388–391.

24. Kraut, J. (1988) How do enzymes work? *Science* **242,** 533–540.

25. Paul, S., Volle, D. J., Beach, C. M., Johnson, D. R., Powell, M. J., and Massey, R. J. (1989) Catalytic hydrolysis of vasoactive intestinal peptide by human autoantibody. *Science* **244,** 1158–1162.

26. Shuster, A. M., Gololobov, G. V., Kvashuk, O. A., Bogomolova, A. E., Smirnov, I. V., and Gabibov, A. G. (1992) DNA hydrolyzing autoantibodies. *Science* **256,** 665–667.

27. Li, L., Paul, S., Tyutyulkova, S., Kazatchkine, M., and Kaveri, S. (1995) Catalytic activity of anti-thyroglobulin antibodies. *J. Immunol.* **154,** 3328–3332.

28. Kalaga, R., Li, L., O'Dell, J., and Paul, S. (1995) Unexpected presence of polyreactive catalytic antibodies in IgG from unimmunized donors and decreased levels in rheumatoid arthritis. *J. Immunol.* **155,** 2695–2702.

29. Sun, M., Mody, B., Eklund, S. H., and Paul, S. (1991) Vasoactive intestinal peptide hydrolysis by antibody light chains. *J. Biol. Chem.* **266,** 15,571–15,574.

30. Gao, Q. S., Sun, M., Tyutyulkova, S., Webster, D., Rees, A., Tramontano, A., Massey, R., and Paul, S. (1994) Molecular cloning of a proteolytic antibody light chain. *J. Biol. Chem.* **269,** 32,389–32,393.

31. Takagi, M., Kohda, K., Hamuro, T., Harada, A., Yamaguchi, H., Kamachi, M., and Imanaka, T. (1995) Thermostable peroxidase activity with a recombinant antibody L chain-porphyrin Fe(III) complex. *FEBS Lett.* **375,** 273–276.

32. Gao, Q. S., Sun, M., Rees, A., and Paul, S. (1995) Site-directed mutagenesis of proteolytic antibody light chain. *J. Mol. Biol.* **253,** 658–664.

33. Khalaf, A. I., Proctor, G. R., Suckling, C. J., Bence, L. H., Irvine, J. I., and Stimson, W. H. (1992) Remarkably efficient hydrolysis of a 4-nitrophenylester by a catalytic antibody raised to an ammonium hapten. J: *Chem. Soc. (Perkin 1)* **1,** 1475–1481.

34. Marks, J. D., Tristem, M., Karpas, A., and Winter, G. (1991) Oligonucleotide primers for polymerase chain reaction amplification of human immunoglobulin variable genes and design of family-specific oligonucleotide probes. *Eur. J. Immunol.* **21,** 985–991.

35. Deng, S., MacKenzie, C. R., and Narang, S. A. (1995) Synthetic antibody gene libraries for in vitro affinity maturation, in *Methods in Molecular Biology, vol. 51: Antibody Engineering Protocols* (Paul, S., ed.), Humana, Totowa, NJ, pp. 329–342.

36. Glockshuber, R., Schmidt, T., and Pluckthun, A. (1992) The disulfide bonds in antibody variable domains: effects on stability, folding in vitro, and functional expression in *Escherichia coli. Biochemistry* **31,** 1270–1279.

37. Sun, M., Gao, Q. S., Li, L., and Paul, S. (1994) Proteolytic activity of an antibody light chain. *J. Immunol.* **153,** 5121–5126.

38. Flecker, P. (1995) Template-directed protein folding into a metastable state of increased activity. *Eur. J. Biochem.* **232,** 528–535.

39. Braco, L., Dabulis, K., and Klibanov, A. M. (1990) Production of abiotic receptors by molecular imprinting of proteins. *Proc. Natl. Acad. Sci. USA* **87,** 274–277.

40. Chen, S. T., Chen, S. Y., and Wang, K. T. (1992) Kinetically controlled peptide bond formation in anhydrous alcohol catalyzed by the industrial protease alcalase. *J. Org. Chem.* **57,** 6960–6965.

41. Deschrevel, B., Vincent, J. C., and Thellier, M. (1993) Kinetic study of the α-chymotrypsin-catalyzed hydrolysis and synthesis of a peptide bond in a monophasic aqueous/organic reaction medium. *Arch. Biochem. Biophys.* **304,** 45–52.

42. Matsumura, M. and Matthews, B. W. (1991) Stabilization of functional proteins by introduction of multiple disulfide bonds, in *Methods in Enzymology,* Vol. 202 (Langone, J. J., ed.), Academic, New York, pp. 336–356.

43. Arnold, F. H. (1993) Engineering proteins for nonnatural environments. *FASEB J.* **7,** 744–749.

44. Menèndez-Arias, L. and Argos, P. (1989) Engineering protein thermal stability: Sequence statistics point to residue substitutions in cY-helices. *J. Mol. Biol.* **206,** 397–406.

45. Novotny, J., Ganju, R. K., Smiley, S. T., Hussey, R. E., Luther, M. A., Recny, M. A., Siliciano, R. F., and Reinhjerz, E. L. (1991) A soluble, single-chain T-cell receptor fragment endowed with antigen-combining properties. *Proc. Natl. Acad. Sci. USA* **88,** 8646–8650.

46. Bianchi, E., Venturini, S., Pessi, A., Tramontano, A., and Sollazzo, M. (1994) High level expression and rational mutagenesis of a designed protein, the minibody. From an insoluble to a soluble molecule. *J. Mol. Biol.* **236,** 649–659.

47. Reichmann, L., Weill, M., and Cavanagh, J. (1992) Improving the antigen affinity of an antibody Fv-fragment by protein design. *J. Mol. Biol.* **224,** 913–918.

48. Xiang, J., Chen, Z., Delbaere, L. T. J., and Liu, E. (1993) Differences in antigen-binding affinity caused by a single amino acid substitution in the variable region of the heavy chain. *Immunol. Cell Biol.* **71,** 239–247.

49. Baldwin, E. and Schultz, P. G. (1989) Generation of a catalytic antibody by site-directed mutagenesis. *Science* **245,** 1104–1107.

50. Pollack, S. J., Nakayama, G. R., and Schultz, P. G. (1988) Introduction of nucleophiles and spectropscopic probes into antibody combining sites. *Science* **242,** 1038–1040.

51. Kuroki, R., Weaver, L. H., and Matthews, B. W. (1993) A covalent enzyme-substrate intermediate with saccharide distortion in a mutant T4 lysozyme. *Science* **262,** 2030–2034.

52. Madison, E. L., Kobe, A., Gething, M. J., Sambrook, J. F., and Goldsmith, E. J. (1993) Converting tissue plasminogen activator to a zymogen: a regulatory triad of Asp-His-Ser. *Science* **262,** 419–424.

53. Graf, L., Craik, C. S., Patthy, A., Roczniak, S., Fletterick, R. J., and Rutter, W. J. (1987) Selective alteration of substrate specificity by replacement of aspartic acid-189 with lysine in the binding pocket of trypsin. *Biochemistry* **26,** 2616–2623.

54. Beaumont, A., Barbe, B., Le Moual, H., Boileau, G., Crine, P., Fournié-Zaluski, M. C., and Rogues, B. P. (1992) Charge polarity reversal inverses the specificity of neutral endopeptidase–24.11. *J. Biol. Chem.* **267,** 2138–2141.

55. Estell, D. A., Graycar, T. P., Miller, J V., Powers, D. B., Burnier, J. P., Ng, P. G., and Wells, J. A. (1986) Probing steric and hydrophobic effects on enzyme-substrate interactions by protein engineering. *Science* **233,** 659–663.

56. Paul, S., Volle, D. J., Powell, M. J., and Massey, R. J. (1990) Site specificity of a catalytic vasoactive intestinal peptide antibody: An inhibitory vasoactive intestinal peptide subsequence distant from the scissile peptide bond. *J. Biol. Chem.* **265,** 11,910–11,913.

57. Wilks, H. M., Hart, K. W., Feeney, R., Dunn, C. R., Muirhead, H., Chia, W. C., Barsow, D. A., Atkinson, T., Clarke, A. R., and Holbrook. J. J. (1988) A specific, highly active malate dehydrogenase by redesign of a lactate dehydrogenase framework. *Science* **242,** 1541–1544.

58. Scrutton, N. S., Berry, A., and Perham, R. N. (1990) Redesign of the coenzyme specificity of a dehydrogenase by protein engineering. *Nature (London)* **343,** 38–43.

59. Alvaro, G. and Russell, A. J. (1991) Modification of enzyme catalysis by engineering surface charge, in *Methods in Enzymology*, Vol. 202 (Langone, J. J., ed.), New York, Academic, pp. 620–643.

60. Muraki, M., Morikawa, M., Jigami, Y., and Tanaka, H. (1988) Engineering of human lysozyme as a polyelectrolyte by the alteration of molecular surface charge. *Protein Eng.* **21,** 49–54.

61. Gilbert, W. (1978) Why genes in pieces? *Nature (London)* **271,** 501.

62. Sun, M., Li, L., Gao, Q. S., and Paul, S. (1994) Antigen recognition by an antibody light chain. *J. Biol. Chem.* **269,** 734–738.

63. Ward, E. S., Gussow, D., Griffiths, A. D., Jones, P. T., and Winter, G. (1989) Binding activities of a repertoire of single immunoglobulin variable domains secreted for Escherichia coli. *Nature* **341,** 544–546.

64. Kang, C. Y., Brunck, T. K., Kieber-Emmons, T., Blalock, J. E., and Kohler, H. (1988) Inhibition of self-binding antibodies (autoantibodies) by a VH-derived peptide. *Science* **240,** 1034–1036.

65. Igarashi, K., Asai, K., Kaneda, M., Umeda, M., and Inoue, K. (1995) Specific binding of a synthetic peptide derived from an antibody complementarily determining region to phosphatidylserine. *J. Biochem. (Tokyo)* **117,** 452–457.

66. Clackson, T. and Wells, J. A. (1995) A hot spot of binding energy in a hormone-receptor interface. *Science* **267,** 383–386.

67. Mody, R. K., Tramontano, A., and Paul, S. (1994) Spontaneous hydrolysis of vasoactive intestinal peptide in neutral aqueous solution. *Int. J Pept. Protein Res.* **44,** 441–447.

68. Hahn, K. W., Klis, W. A., and Stewart, J. M. (1990) Design and synthesis of a peptide having chymotrypsin-like esterase activity. *Science* **248,** 1544–1546.

69. Paul, S., Li, L., Kalaga, R., Wilkins-Stevens, P., Stevens, F. J., and Solomon, A. (1995) Natural catalytic antibodies: Peptide hydrolyzing activities of Bence Jones proteins and V_L fragment. *J. Biol. Chem.* **270,** 15,257–15,261.

70. Hruby, V. J. and Nikiforovitch, G. V. (1991) The Ramachandran Plot and beyond: conformational and topochemical considerations in the design of peptides and proteins, in *Molecular Conformation and Biological Interactions* (Balaram, P. and Ramasehan, S., eds.), Indian Academy of Sciences, Bangalore, India, pp. 429–445.

71. Marshall, G. (1993) A hierarchical approach to peptidomimetic design. *Tetrahedron* **49,** 3547–3558.

72. Musso, G. F., Patthi, S., Ryskamp, T. C., Provow, S., Kaiser, E. T., and Velicelebi, G. (1988) Development of helix-based vasoactive intestinal peptide analogues: identification of residues required for receptor interaction. *Biochemistry* **27,** 8174–8181.

73. Bird, R. E., Hardman, K. D., Jacobson, J. W., Johnson, S., Kaufman, B. M., Lee, S., Lee, T., Pope, S. H., Riordan, G. S., and Whitlow, M. (1988) Single-chain antigen-binding proteins. *Science* **242,** 423–426.

74. Whitlow, M., Bell, B. A., Feng, S.-L., Filpula, D., Hardman, K. D., Hubert, S. L., Rollence, M. L., Wood, J. F., Schott, M. E., Milenic, D. E., Yokota, T., and Schlom, J. (1993) An improved linker for single-chain Fv with reduced aggregation and enhanced proteolytic stability. *Protein Eng.* **6,** 989–995.

75. Bedzkyk, W. D., Weidner, K. M., Denzin, L. K., Johnson, L. S., Hardman, K. D., Pantoliano, M. W., Asel, E. D., and Voss, E. W., Jr. (1990) Immunological and structural characterization of a high affinity anti-fluorescein single-chain antibody. *J. Biol. Chem.* **265,** 18,615–18,620.

76. Anand, N. N., Mandal, S., MacKenzie, C. R., Sadowska, J., Sigurskjold, B., Young, N. M., Bundle, D. R., and Narang, S. A. (1991) Bacterial expression and secretion of various single-chain Fv genes encoding proteins specific for a *Salmonella* serotype B O-antigen. *Biol. Chem.* **266,** 21,874–21,879.

77. Gao, Q. S. and Paul, S. (1995) Molecular cloning of anti-ground state proteolytic antibody fragments, in *Methods in Molecular Biology, Vol. 51: Antibody Engineering Protocols* (Paul, S., ed.), Humana, Totowa, NJ, pp. 281–296.

78. Chang, H. C., Bao, Z. Z., Yao, Y., Tse, A. G. D., Goyarts, E. C., Madsen, M., Kawasaki, E., Brauer, P. P., Sacchettini, J. C., Nathenson, S. G., and Reinherz, E. L. (1994) A general method for facilitating heterodimeric pairing between two proteins: application to expression of α and β T-cell. *Proc. Natl. Acad. Sci. USA* **91,** 11,408–11,412.

79. Corey, D. R. and Schultz, P. G. (1987) Generation of a hybrid sequence-specific single-stranded deoxyribonuclease. *Science* **238,** 1401–1406.

80. Pendergrast, P. S., Ebright, Y. W., and Ebright, R. H. (1994) High-specificity DNA cleavage agent: design and application to kilobase and megabase DNA substrates. *Science* **265,** 959–962.

81. Pomerantz, J. L., Sharp, P. A., and Pabo, C. O. (1995) Structure-based design of transcription factors. *Science* **267,** 93–96.

82. Lee, G., Chan, W., Hurle, M. R., DesJarlais, R. L., Watson, F., Sathe, G. M., and Wetzel, R. (1993) Strong inhibition of fibrinogen binding to platelet receptor $\alpha II\beta 43$ by RGD sequences installed into a presentation scaffold. *Protein Eng.* **6,** 745–754.

83. Bona, C., Brumeanu, T.-D., and Zaghouani, H. (1994) Immunogenicity of microbial peptides grafted in self immunoglobulin molecules. *Cell. Mol. Biol.* **40(Suppl. I),** 21–30.

84. Ward, E. S. (1995) VH shuffling can be used to convert an Fv fragment of anti-hen egg lysozyme specificity to one that recognizes a T cell receptor Vα. *Mol. Immunol.* **32,** 147–156.

85. Polymenis, M. and Stollar, B. D. (1994) Critical binding site amino acids of anti-Z-DNA single chain Fv molecules: role of heavy and light chain CDR3 and relationship to autoantibody activity. *J. Immunol.* **152,** 5318–5329.

86. Gulliver, G. A., Bedzyk, W. D., Smith, R. G., Bode, S. L., Tetin, S. Y., and Voss, E. W., Jr. (1994) Conversion of an anti-single-stranded DNA active site to an anti-fluorescein active site through heavy chain complementarily determining region transplantation. *J. Biol. Chem.* **269,** 7934–7940.

87. Gulliver, G. A. and Voss, E. W., Jr. (1994) Effect of transplantation of antibody heavy chain complementarily determining regions on ligand binding. *J. Biol. Chem.* **269,** 24,040–24,045.

88. Turner, S. L., Russell, G. C., Williamson, M. P., and Guest, J. R. (1993) Restructuring an interdomain linker in the dihydrolipoamide acetyltransferase component of the pyruvate dehydrolgenase complex of *Escherichia coli. Protein Eng.* **6,** 101–108.

89. Holliger, P., Prospero, T., and Winter, G. (1993) "Diabodies": small bivalent and bispecific antibody fragments. *Proc. Natl. Acad. Sci. USA* **90,** 6444–6448.

90. Reiter, Y., Brinkmann, U., Webber, K. O., Jung, S.-H., Lee, B., and Pastan, I. (1994) Engineering interchain disulfide bonds into conserved framework regions of Fv fragments: Improved biochemical characteristics of recombinant immunotoxins containing disulfide-stabilized F_v. *Protein Eng.* **7,** 697–704.

91. Brinkmann, U., Reiter, Y., Jung, S.-H., Lee, B., and Pastan, I. (1993) A recombinant immunotoxin containing a disulfide-stabilized Fv fragment. *Proc. Natl. Acad. Sci. USA* **90,** 7538–7542.

92. Pack, P., Muller, K., Zahn, R., and Plückthun, A. (1995) Tetravalent miniantibodies with high avidity assembling in *Escherichia coli. J. Mol. Biol.* **246,** 28–34.

93. Neri, D., Momo, M., Prospero, T., and Winter, G. (1995) High-affinity antigen binding by chelating recombinant antibodies (CRAbs). *J. Mol. Biol.* **246,** 367–373.

94. Kurucz, I., Titus, J. A., Jost, C. R., Jacobus, C. M., and Segal, D. M. (1995) Retargeting of CTL by an efficiently refolded bispecific single-chain Fv dimer produced in bacteria. *J. Immunol.* **154,** 4576–4582.

95. Eshhar, Z., Waks, T., Gross, G., and Schindler, D. G. (1993) Specific activation and targeting of cytotoxic lymphocytes through chimeric single chains consisting of antibody-binding domains and the gamma or zeta subunits of the immunoglobulin and T-cell receptors. *Proc. Natl. Acad. Sci. USA* **90,** 720–724.

96. Gosink, M. M. and Vierstra, R. D. (1995) Redirecting the specificity of ubiquitination by modifying ubiquitin-conjugating enzymes. *Proc. Natl. Acad. Sci. USA* **92,** 9117–9121.

97. Chaudhary, V. K., Batra, J. K., Gallo, M. G., Willingham, M. C., Fitzgerald, D. J., and Pastan, I. (1990) A rapid method of cloning functional variable-region antibody genes in *Escherichia cold* as single-chain immunotoxins. *Proc. Natl. Acad. Sci. USA* **87,** 1066–1070.

98. Kreitman, R. J., Bailon, P., Chaudhary, V. K., FitzGerald, D. J. P., and Pastan, I. (1994) Recombinant immunotoxins containing anti-Tac(Fv) and derivatives of *Pseudomonas* exotoxin produce complete regression in mice of an interleukin–2 receptor-expressing human carcinoma. *Blood* **83,** 426–434.

99. Yang, J. B., Moyana, T., and Xiang, J. (1995) A genetically engineered single-chain FV/TNF molecule possesses the anti-tumor immunoreactivity of FV as well as the cytotoxic activity of tumor necrosis factor. *Mol. Immunol.* **32,** 873–881.

100. Pastan, I. H., Archer, G. E., McLendon, R. E., Friedman, H. S., Fuchs, H. E., Wang, Q.-C., Pai, L. H., Herndon, J., and Bigner, D. D. (1995) Intrathecal administration of single-chain immunotoxin, LMB–7 [B3(Fv)-PE38], produces cures of carcinomatous meningitis in a rat model. *Proc. Natl. Acad. Sci. USA* **92,** 2765–2769.

101. George, A. J. T., Jamar, F., Tai, M. S., Heelan, B. T., Adams, G. P., McCartney, J. E., Houston, L. L., Weiner, L. M., Oppermann, H., Peters, A. M., and Huston, J. S. (1995) Radiometal labeling of recombinant proteins by a genetically engineered minimal chelation site: Technetium–99m coordination by single-chain Fv antibody fusion proteins through a C-terminal cysteinyl peptide. *Proc. Natl. Acad. Sci. USA* **92,** 8358–8362.

102. Holvoet, P., Laroche, Y., Lijnen, H. R., Van Hoef, B., Brouwers, E., De Cock, F., Lauwereys, M., Gansemans, Y., and Collen, D. (1992) Biochemical characterization of single-chain chimeric plasminogen activators consisting of a single-chain Fv fragment of a fibrin-specific antibody and single-chain urokinase. *Eur. J. Biochem.* **210,** 945–952.

103. Bickel, U., Yoshikawa, T., Landaw, E. M., Faull, K. F., and Pardridge, W. M. (1993) Pharmacologic effects *in vivo* in brain by vector-mediated peptide drug delivery. *Proc. Natl. Acad. Sci. USA* **90,** 2618–2622.

104. Rodrigues, M. L., Presta, L. G., Kotts, C. E., Wirth, C., Mordenti, J., Osaka, G., Wong, W. L. T., Nuijens, A., Blackburn, B., and Carter, P. (1995) Development of a humanized disulfide-stabilized anti-pi 85[HER2] Fv-β-lactamase fusion protein for activation of a cephalosporin doxorubicin prodrug. *Cancer Res.* **55,** 63–70.

105. Eisele, J. L., Bertrand, S., Galzi, J. L., Devillers-Thiéry, A., Changeux, J. P., and Bertrand, D. (1993) Chimaeric nicotinic-serotonergic receptor combines distinct ligand binding and channel specificitiess. *Nature* **366,** 479–483.

106. Fuh, G., Cunningham, B. C., Fukunaga, R., Nagata, S., Goeddel, D. V., and Wells, J. A. (1992) Rational design of potent antagonists to the human growth hormone receptor. *Science* **256,** 1677–1684.

44

Monoclonal Antibodies

Christopher Dean and Helmout Modjtahedi

1. Introduction

Like all great discoveries, the procedure developed by George Kohler and Cesar Milstein for immortalizing antibody producing B-lymphocytes *(1)* is essentially simple; namely, mix B-lymphocytes from the spleen of an immunized rodent with a continuously proliferating B-lymphoma cell line, then induce their membranes to fuse to give a single antibody-secreting cell (hybridoma) that will proliferate indefinitely. Importantly, each hybridoma will secrete antibodies with a specificity for only one antigenic epitope, i.e., the antibodies are monoclonal. It is this property, together with the ability to produce unlimited amounts of antibody, that has revolutionized the use of antibodies in many areas of biological and medical research, and has found important applications not only in clinical diagnosis and therapy, but also for large-scale purification of biological materials. Indeed, monoclonal antibodies (MAbs) are essential tools for the cell and molecular biologist and their use has unlocked the doors behind which the function of many proteins and other cellular molecules would have lain hidden.

In this chapter we describe the preparation and selection of hybridomas secreting MAbs and discuss those factors that influence the success of cell fusion, together with an outline of the uses to which MAbs have been put. The basic protocol used to generate hybridomas and select those producing specific antibodies is summarized in **Fig. 1.** Essentially, the most important requirements are to:

1. Generate specific immune B-lymphocytes,
2. Successfully fuse them to a continuously growing myeloma cell,
3. Identify the antibodies that are sought in culture supernatants, and
4. Isolate and clone the specific hybridomas.

We discuss in more detail the procedures involved in **Subheading 2.**

2. Preparation of Hybridomas
2.1. Selection of Host for Immunization

The majority of MAbs that have been generated to date have been prepared from immune mice, usually of strain BALB/C, because the fusion partners (myelomas) for hybridoma production have been developed from plasmacytomas induced in this strain

From: *Molecular Biomethods Handbook*
Edited by: R. Rapley and J. M. Walker © Humana Press Inc., Totowa, NJ

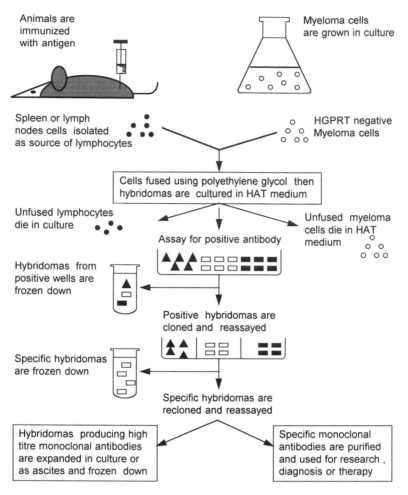

Fig. 1. Production of monoclonal antibodies.

by the intraperitoneal injection of mineral oil. Rats of the LOU/wsl strain develop ileocaecal plasmacytomas with a high incidence, and two of these plasmacytomas have been developed for use in hybridoma production (**Table 1**). An important feature of using lymphoid cells and myelomas from the same inbred strain of rodent is that the hybridomas produced will grow as ascites in these animals. However, because hybridomas can be grown in culture, there is no *a priori* reason why lymphoid cells from donors that are not syngeneic with the myeloma-cell line must be used, and many of the rat MAbs produced to date have been made with lymphocytes taken from rats of many different strains. Indeed, the generation of heteromyelomas by fusing a mouse myeloma with lymphocytes from other animal species has facilitated the generation of hybridomas that secrete equine, bovine, sheep, or rabbit MAbs (**Table 1**). Often, the hybridomas prepared using heteromyelomas are not as stable as their mouse or rat counterparts. However, evidence is accumulating to show that both sheep and rabbit MAbs may have substantially higher affinities for antigen than those of murine origin and this could be an important advantage. In general, the lower cost and ease of handling of the

Table 1
Examples of Myeloma and Hetero-Myeloma Cell Lines

Name	Species/origin	Ig expression	Ref.
Mouse lines			
P3–X63/Ag8	BALB/C mouse	IgG1 (K)	*2*
NSI/1.Ag 4.1	BALB/C mouse	K chain (nonsecreted)	*2*
X63/Ag 8.653	BALB/C mouse	None	*2*
Sp2/0	Sp2 mouse hybridoma	None	*2*
NS0/1	NSI/1.Ag4.1	None	
Rat lines			
Y3–Ag 1.2.3.	Lou/wsl rat myeloma	K chain	*2*
IR984F	Lou/wsl rat myeloma	None	*3*
Human line			
LICR–LON–HMy2	Plasma cell leukemia	IgG1 (K)	*4*
Bovine line			
	Calf–NS0 heterohybridoma	None	*5*
Ovine line			
1C6.3a6T.1D7	Sheep–NS0 heterohybridoma	None	*6*
Equine line			
	Horse–NS0 heterohybridoma		*7*
Rabbit line			
240E 1–1–2	Rabbit plasmacytoma	None	*8*

smaller rodents has made them the first choice of researchers as source of immune lymphocytes. Although the goal of generating human antibodies initially looked achievable, considerable difficulties have been encountered using the hybridoma route. Alternative approaches, such as immortalizing lymphocytes with Epstein-Barr virus or the use of recombinant DNA technologies, are currently being examined.

2.2. Generation of Immune Spleen or Lymph Node Cells

We focus here on the preparation and testing of mouse or rat hybridomas. Detailed laboratory protocols can be found in **refs.** *2* and *9*.

2.2.1. Selection of Antigen for Immunization

It is not essential for purified antigen to be used for immunization, although it can make a difference when poorly immunogenic molecules are involved. Whole cells and partially purified cell extracts can be used, as well as recombinant proteins produced in bacteria, yeast, or insect cells. If antibodies against carbohydrate determinants are required, it should be remembered that glycoproteins produced in bacteria will not be glycosylated, and that recombinant glycoproteins produced in insect or Chinese hamster-ovary cells will be glycosylated according to the host-cell sequences. Peptides based on the predicted coding sequences of cloned cDNAs have also been used for preparing antibodies against sequential determinants on the full-length protein *(10)*.

Unlike rabbits, which are particularly good at making antipeptide antibodies, the smaller rodents rarely respond as well. Also, peptide immunogens are usually ineffective in generating antibodies against conformational determinants, and if these are desired then properly folded protein or protein fragments must be used. Antibodies against some transmembrane glycoproteins, e.g., the receptor (EGFR) for epidermal-growth factor, can be generated easily by immunizing with live cells that overexpress the receptor (*see* **Subheading 3.**). The use of live cells as immunogen has an additional advantage in that the normal conformation of the receptor molecule is retained. Many antibodies have been generated against cell surface determinants on mammalian cells including differentiation antigens on lymphoid and other tissues (discussed in more detail in **Subheading 4.**) as well as specific receptors for other growth factors. Some of the molecules expressed at the cell surface are very antigenic in rodents, e.g., polymorphic epithelial mucin (PEM, **ref. *11***) expressed by both normal epithelia and many tumors of epithelial origin. In consequence, hybridomas secreting MAbs directed against such determinants can predominate in some fusions. When using tumor cells or cell extracts as immunogen it is therefore essential to have suitable screening assays for detecting the specific antibodies that you require.

2.2.2. Preparation of Antigen for Immunizations

Whole cells are generally injected in phosphate-buffered saline or in medium, such as Dulbecco's modified Eagle's medium (DMEM), without addition of an adjuvant. Where soluble materials such as proteins or carbohydrates are to be used, they are mixed 1:1 with Freund's adjuvant to give a stable emulsion. Freund's adjuvant is an oil (croton oil) containing mycobacteria (Freund's complete adjuvant; FCA) that acts as a "slow-release agent" that prevents rapid dispersion of the soluble immunogen and also elicits a strong cellular infiltrate of neutrophils and macrophages at the site of injection. Use of adjuvant substantially improves the antibody response. Subsequent injections are made using an emulsion made with Freund's incomplete adjuvant (FIA) which lacks the mycobacteria. Other adjuvants that have been used include precipitates made with aluminum sulfate or with specific precipitating antibodies. These are less toxic to the recipient animals.

Peptides that contain 20–30 amino acids are often poorly or nonimmunogenic because they do not contain sequences recognized by T-helper cells, which are essential for generating an antibody response. For this reason, it is usual to conjugate them to larger proteins that contain many T-cell reactive epitopes, such as Keyhole lympet hemocyanin or ovalbumin. Conjugation of the peptide to the larger protein is done either with glutaraldehyde, which will link the peptide in random conformations *(10)* or by disulfide linkage, where the peptide has been synthesized with a terminal cysteine. In this case, the peptide will be attached via one end to the carrier protein and this directional orientation will restrict its immunogenicity.

2.2.3. Route of Immunization

To obtain high-titer specific antibodies in serum, it is usual to immunize mice or rats at three or four sites subcutaneously or intramuscularly together with a challenge intraperitoneally. The immunization is repeated at 2–4 wk intervals until bleeds taken either from the tail vein (mice) or jugular (rats) show good titers of specific antibody in the serum. Immunization in this way will stimulate the production of specific antibody-producing cells and their B-cell precursors in the local lymph nodes and the spleen. Usually, splenic lymphocytes are used for fusion and the animals are given a final iv challenge with antigen 2–3 d before the spleens are harvested. Cells from immune lymph nodes are an excellent source of specific B-cells and we use, routinely, cells taken from the mesenteric nodes of rats immunized via the Peyer's patches that lie along the small intestine *(12)*. As with spleen cells, the animals are rechallenged via the Peyer's patches 3 d before removing the mesenteric nodes.

2.2.4. Preparation of Lymphoid Cells for Fusion

The immune rats or mice are killed and the spleens or lymph nodes removed by blunt dissection, taking care to remove any fatty connective tissue. The cells are harvested by disaggregating the spleens or lymph nodes using a sterile spatula to force the spleens or lymph nodes through a stainless-steel mesh (tea strainer) into serum-free DMEM.

The cells are washed once (lymph node) or twice (spleen) in serum-free DMEM and counted. It is not necessary to lyse the red blood cells in the spleen suspension, although it can be done using Alseiver's solution if required. The washed cells are counted in a hemocytometer and 10^8 cells used for each fusion. A spleen from an immune mouse usually yields about 10^8 cells in total, whereas rats may yield up to 3–4×10^8 cells. Surplus lymphocytes can be stored in liquid nitrogen indefinitely if they are first resuspended in a mixture of 95% fetal bovine serum (FBS) and 5% dimethyl sulfoxide (DMSO). The frozen cells will fuse as well, if not better, than fresh lymphocytes, presumably because the DMSO taken up by the plasma membrane aids cell fusion. Specific hybridomas are preserved in the same way.

2.2.5. Preparation of Myeloma Cells for Fusion

Although it is usual to fuse lymphocytes and myelomas from the same species i.e., to give mouse × mouse or rat × rat hybridomas, it is of course possible to make mouse × rat hybridomas when this is desirable. The myelomas used have been selected for loss of the enzyme hypoxanthene–guanine phosphoribosyl transferase (HGPRT) and, in consequence, are unable to utilize the salvage pathway for DNA synthesis in the presence of the folic acid synthesis-inhibitor aminopterin. Normal lymphocytes do not usually proliferate in culture, but because they contain the gene for HGPRT, the hybridomas formed on fusion with the myeloma will grow if the medium containing aminopterin (A) is supplemented with hypoxanthine and thymine (HAT-selection medium).

Exponentially growing myeloma cells are essential and the cells must be passaged regularly to maintain them in this state. Whereas the mouse myelomas usually grow well as single cell suspensions in stationary culture, the rat myelomas Y3 and IR983F are best grown in spinner culture. Indeed, this is essential for the Y3 myeloma, which grows as an adherent cell line if maintained in static culture, and has to be removed by trypsinization for use. We use DMEM for growth of both rat and mouse myelomas supplemented with antibiotics and FBS at 10%. Myeloma cells are harvested just before fusion is carried out, then washed once with serum-free DMEM before use.

2.3. Hybridoma Formation

Freshly prepared spleen or lymph-node cells (10^8 in serum-free DMEM) are mixed with either 10^7 mouse or 5×10^7 rat-myeloma cells and centrifuged at 400*g*. The supernatant is poured off and 1 mL of 50% polyethylene glycol (PEG 1000) in serum free DMEM is stirred into the cell pellet over a period of 1 min. After rocking the fusion tube for a further minute the sample is diluted gradually with serum-free DMEM over a period of 3 min. The cells are pelleted by centrifugation, resuspended in 200 mL DMEM containing HAT, feeder cells, antibiotics, and 20% FBS, and plated into multiwell plates, namely 4 × 26 well or 20 × 96 well. The use of feeder cells is important to provide a suitable environment for the hybridomas to proliferate in the face of the large scale cell death that will take place. We use γ-irradiated rat fibroblasts for both mouse and rat fusions. Many workers making mouse hybridomas use thymocytes obtained from the same animal as the spleen. The plates are incubated at 37°C in an atmosphere of 5% CO_2 and examined daily for possible contamination and after 5–7 d for the presence of hybridoma colonies. In most cases these appear as individually growing spheroids, but the colonies formed with the Y3 myeloma differ with cells scattered over a limited area each cell with the appearance of a small knobbly salad potato. Supernatants from the individual wells are tested for the presence of specific antibodies and individual hybridoma colonies are picked from positive wells and grown up. After rescreening to identify the positive colonies, the cells are cloned by plating about 50 cells in a 96-well plate and rescreening for positive wells that contain only a single colony. Only in this way can we be certain that the hybridoma is truly secreting an antibody which is monoclonal.

2.4. Screening Hybridoma Culture Supernatants for Specific Antibody

Many different procedures have been used to screen hybridoma supernatants from direct-binding assays to cells or tissues using fluorescent, radiolabeled, enzyme-linked, or otherwise tagged second antibodies (antimouse, antirat, antihuman, and so forth) to detect bound MAb, to functional assays, e.g., inhibition of ligand binding to receptor. The importance of the screening assay cannot be overemphasised because it is on the basis of the results that the hybridoma colony is picked, expanded and cloned. Important features of the assay are that it is quick, reliable, and specific. Most hybridomas grow rapidly with generation times of 10–12 h (rat) to 15–24 h (mouse) so that long-term assays cannot be carried out. Assays that determine binding to antigen, whether present on/in cells, to proteins or peptides coated onto plastic multiwell plates or pins or to tissue sections are to be preferred to biological assays that cannot be performed in 1 d. Detailed examples of the assays that can be used are given in *(9,13)*.

A simple assay using multiwell plates coated with soluble antigen or cells is illustrated in **Fig. 2**. When soluble antigens (proteins, peptides, and so forth) are used 50 μL aliquots of phosphate buffer, pH 8.0, containing 10–100 ng of antigen are transferred into each well and the plates incubated overnight at 4°C. The plates are then "blocked" by adding 200 μL of PBS, pH 8.0, containing 3% bovine serum albumin (BSA) or 3% skimmed milk to prevent nonspecific binding of the antibodies present in culture supernatants. Aliquots of the culture hybridoma supernatant are then added and incubated for 1 h at ambient temperature. After thorough washing, the bound antibody can be detected by treatment with specific antirodent antibody, which is either radiolabeled

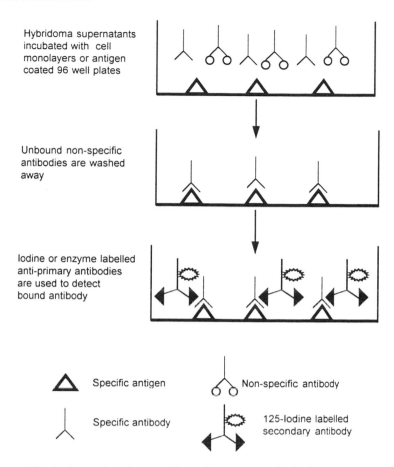

Hybridoma supernatants
incubated with cell
monolayers or antigen
coated 96 well plates

Unbound non-specific
antibodies are washed
away

Iodine or enzyme labelled
anti-primary antibodies
are used to detect
bound antibody

△ Specific antigen

Specific antibody

Non-specific antibody

125-Iodine labelled
secondary antibody

Fig. 2. Screening for specific antibody using the indirect method.

(radioimmunoassay, RIA) or conjugated to an enzyme (enzyme linked immunosorbent assay, ELISA). With the latter, a further incubation with substrate is required which yields either a colored or fluorescent product. Adaptations of these assays can be used for many of the tests required during hybridoma production.

Where more than one hybridoma has been obtained, it is important to assess the relative usefulness of the different antibodies. For example, affinity, i.e., the strength with which it binds antigen, could be important, particularly if the antibody is to be used in radioimmunoassays to estimate antigen concentration. Second, the antibody may have important effector functions when used in vivo, i.e., to activate complement and recruit host effector cells by binding Fc receptors, factors that could affect biological usefulness. This property is critically dependent on the isotype (class or subclass) of the antibody. Third, the antibody may bind to sequential amino acids on the antigen, some antibodies of this type have proved to be ideal for Western blotting of denatured proteins and for binding to cellular determinants in formalin-fixed, paraffin-embedded tissue sections. Some 80% of antibodies generated against biological antigens bind to conformational determinants. These antibodies are of little use to the pathologist who usually works with formalin-fixed sections, but they will bind to frozen sections and to live cells and often they have important biological activities.

For the molecular or cell biologist who is interested in using MAbs to determine the cellular localization and function of novel proteins, all of the criteria previously discussed will apply. Indeed, the use of confocal microscopy plus the ability to inject antibodies or other proteins into individual cells has demonstrated the essential function of a number of cellular proteins. One additional and important application for MAbs is the specific immunoprecipitation of the target antigen from detergent lysates of cells that have been metabolically labeled with radiolabeled precursors such as amino acids (e.g., ^{35}S-methionine) or sugars (e.g., ^{3}H-fucose). The relative ease with which recombinant proteins can be expressed either alone or conjugated to bacterial enzymes (e.g., β-galactosidase) or mammalian enzymes (e.g., glutathione transferase) has simplified their bulk expression. Frequently, however, the antibodies produced are directed against the carrier and not the molecule of interest. Purified protein is usually the solution to the problems of MAb production. There is but one caveat, however, and that concerns the restriction of the mammalian immune system, namely, that it is difficult if not impossible to generate immune responses against a foreign protein which has an identical amino acid sequence to that of the recipient. This restriction may be less stringent with intracellular proteins, e.g., the generation of antibodies to human actin in patients with infectious mononucleosis. Bearing these restrictions in mind it is nevertheless clear that MAbs can be generated to many antigenic determinants which may have been part of the immunological repertoire or not. The fact remains that they have provided us with essential reagents for biomedical research as well as offering some novel procedures for diagnosis and treatment in medicine. Before considering the applications for which MAbs have been used we shall describe the generation of some biologically useful MAbs from our own experience.

3. MAbs to Growth-Factor Receptors

To illustrate how specific MAbs can be generated, we use as the example rat MAbs specific for the extracellular domains of the receptor for EGF and the related product of the *c-erbB-2* proto-oncogene *(14,15)*.

3.1. Production

Although there were a number of ligands that bound specifically to the EGFR, none were available for the *c-erbB-2* receptor. Because of this, a selection protocol for *c-erbB-2* antibodies was used based on binding assays using breast-carcinoma cell lines that were known to express high (BT474), intermediate (MDA–MB 361), or low (MCF7) levels of receptor. Also, in the case of the EGFR, antibodies were required that would prevent the binding of the ligands EGF and TGFα and so block proliferation of tumor cells that were dependent on signals transduced via the receptor. In this case, antibodies were selected on the basis that they inhibited the binding of ^{125}I-EGF to cells expressing the receptor.

As immunogens two breast-cancer cell lines were used namely, MDA–MB 468, which overexpresses the EGFR or BT474, which overexpresses the *c-erbB-2* receptor. CBH/cbi rats were immunized three times at monthly intervals, by injecting each time a total of about 5×10^6 cells into 5–7 of the Peyer's patches associated with the small intestine. This requires the exteriorization of the small gut to enable injection of small quantities (8–10 μL/patch) of the cells under the capsule of the Peyer's patch *(12)*. All procedures using experimental animals are strictly regulated in many countries and only those authorized may perform such surgery. To determine if specific antibodies were elicited following immunization, dilutions of serum were tested for their differential binding to monolayers of cells expressing *c-erbB-2* or for inhibition of the binding of ^{125}I-labeled EGF to monolayers of the bladder-carcinoma cell line EJ grown in 96-well plates. Although sera taken from the rats prior to immunization with MDA–MB 468 did not inhibit the binding of ligand to the EGFR, after the second immunization the sera were found to contain antibodies that actively blocked the binding of ^{125}I-ligand to EJ cells. The rats were rechallenged with either MDA–MB 468 or BT474 cells and 3 d later the mesenteric nodes were removed, the cells fused with the rat myeloma Y3 Ag.1.2.3 using PEG 1500, and the mixture plated into 4×24 well plates containing x-irradiated rat fibroblasts as feeders. Ten days following fusion, aliquots of supernatant from each well were tested for differential binding to monolayers of *c-erbB-2* expressing cells or for inhibition ^{125}I-EGF binding to EJ cells (EGFR). Using a Pasteur pippet individual hybridoma colonies were picked from all of the positive wells and plated into 24-well plates and allowed to grow up. Retesting showed which were the positive colonies and these were cloned twice. At each stage, samples of the hybridoma cells were frozen down in liquid nitrogen for long-term storage.

3.2. Characterization

It was important to determine next if the antibodies bound to the same or different determinants (epitope) on the receptors. The hybridomas were grown up and the individual MAbs purified from culture supernatant by precipitation with $(NH_4)_2SO_4$ followed by ion-exchange chromatography of the redissolved material on a column of Whatman DE52 cellulose *(14,15)*. The purified antibodies against the EGFR or *c-erbB-2* were labeled with ^{125}Iodine and then competed individually with the other unlabeled antibodies (either anti-EGFR or *c-erbB-2*) for binding to monolayers of EJ cells (EGFR) or MDA-MB 361 (*c-erbB-2*). In this way it was found that the antibodies fell into several groups, i.e., they bound to distinct epitopes on the extracellular domains of the receptors *(14,15)*.

Next, the isotype of the MAbs was determined using a capture assay in which wells were coated with mouse MAbs specific for rat antibodies of the IgA or IgM classes or IgG1, IgG2a, or IgG2b subclasses. Binding of the rat MAb to the individual wells was determined by further incubation with ^{125}I-labeled sheep antibody specific for rat F(ab')$_2$. The class or subclass of an antibody is an important indicator of their potential function, e.g., IgA antibodies are usually present in secretory fluids, whereas rat IgG2b antibodies can activate complement and recruit host effector cells by binding to their Fc receptors.

3.3. Biological Activity

Of the antibodies directed against the EGFR, the most effective were found to be ICR64 (Ig1) and ICR16 (IgG2a) or ICR62 (IgG2b), which bound to a different epitope. These antibodies efficiently blocked the binding of growth factors to the EGFR on a number of different cell lines, inhibited the growth in vitro of cells that overexpressed the receptor *(14)*, and induced the regression of established EGFR-overexpressing tumor xenografts grown in nude mice *(16)*. Antibody ICR62 was more effective, because it was an IgG2b and recruited host immune-effector functions more efficiently than the other MAbs. ICR62 has been used in a Phase IB clinical trial in patients with head and neck or lung cancer and the results *(17)* show that the antibody was well-tolerated at doses up to 100 mg and localized well to metastatic lesions. None of the antibodies directed against the *c-erbB-2* product were found to inhibit the growth of cells that overexpressed this receptor. However, some of the antibodies bound to the receptor with high affinity and remained stably expressed at the cell surface. Outstanding among these was antibody ICR12, which has since been used in the clinic to image metastatic disease in patients with breast cancer and has been found, in preclinical testing, to be ideal for use in antibody-directed, enzyme-prodrug therapy *(18)*.

4. MAbs in Biomedical Research

MAbs are essential reagents for the isolation, identification, and cellular localization of specific gene products, and for aiding in the determination of their macromolecular structure. Although the ability to clone and sequence specific genes has revolutionized our understanding of gene structure and function, it is the facility to make MAbs either against the recombinant proteins or peptides based on protein sequences derived from cDNA clones that has revolutionized our understanding of cell biology. Indeed, it is rare to find a scientific paper in the biological sciences that does not employ a MAb at some stage in the investigation.

It is not possible to discuss in detail here the many applications of MAbs in basic research. Instead, we focus briefly on their diagnostic and therapeutic application in medicine. MAbs against cytoplasmic and cell-surface markers have had a major impact not only in basic studies, e.g., in immunology and histopathology, facilitating diagnosis of disease, but also in offering new strategies for therapy. Nowhere is this seen better than in malignancy where these marker antigens are used to stage the disease and provide prognostic indication. For example, the overexpression of the receptor for epidermal-growth factor and the related product of the *c-erbB-2* proto-oncogene have been found to be indicators of poor prognosis in squamous-cell carcinomas or adenomas *(19,20)*. Also, the many surface markers that have been discovered on leukocytes are used both for identifiying malignancies of hemopoetic origin and for determining the type of treatment to be given.

These cell surface molecules are also expresssed by normal leukocytes and constitute receptors, adhesion molecules, enzymes, and so forth. A systematic approach to the naming of leukocyte antigens was made with the "cluster of differentiation" (CD) designation for human leukocyte antigens that were identified with MAbs *(21)*. MAbs were submitted to workshops and were then placed into groups on the basis of their fluorescent-staining patterns for different leukocyte populations. In this classification, the antigens are given a CD number. In most cases, the biochemical nature of the molecule was not known initially, but the antibodies were used subsequently to isolate the antigens from cells and to identify recombinant proteins in expression libraries, and this has enabled the elucidation of the amino-acid sequence and potential structure of these molecules. The very large database on the leukocyte antigens is brilliantly summarized in **ref. 22**.

Although many of the CD antigens are expressed on cells of widely differing origins, the expression of others is confined to leukocytes. For example, some CD antigens delineate particular subsets of leukocytes with important functions in the immune system, such as CD4, expressed by T-helper cells, and CD8, expressed by cytotoxic T-cells. In the last 20 yr, it has been possible to dissect the interactions between the prime movers in the immune response (T-cells and antigen-presenting cells) and to reach an understanding of how T-cells function in the generation of both humoral and cellular-immune responses. The results of these investigations have shown that exogenous antigens, taken up and processed by macrophages, dendritic, or B-cells are presented at the cell surface as peptides usually associated with class II histocompatibility molecules to the CD4-positive T-helper cells, whereas peptides derived from endogenously synthesized proteins are presented at the surface of all cells in association with class I molecules *(23)*. The latter are recognized by the receptors on the CD8 positive cytotoxic T-cells. In this way, normal or abnormal constituents of the cytoplasmic compartment are presented for scrutiny by the CD8 positive T-cells. Elucidation of the moleular events involved in these processes has relied substantially on the use of specific MAbs directed not only against the interacting cell-surface molecules but also against intracellular components of signal transduction and other biochemical pathways that are activated during the immune response.

The therapeutic application of antibodies has a long history. MAbs were hailed as the magic bullets to target cancer and other diseases. One of the success stories is in the area of organ transplantation, where the use of antibodies to T-cell determinants such as the anti-CD3 murine-antibody OKT3 has helped to prevent rejection of cadaveric kidney grafts *(24,25)*. Less successful. however, has been the use of MAbs to treat human cancers. This is owing in part to the fact that all new experimental procedures are tried on patients who have failed conventional treatments, such as surgery, chemotherapy, or radiotherapy, and have a large tumor burden. It is also true to say that some of the earlier MAbs were of relatively low affinity and crossreacted with normal constituents, and this may have limited their ability to localize to tumors in patients. For a recent review of the MAbs that have or are undergoing clinical trial, *see* **ref. 26**.

There have been some notable successes however including the use of [131]Iodine-labeled antibody B1 (anti-CD20) in the treatment of B-cell lymphoma *(27)*, [90]Yttrium-labeled HMFG1 (anti-PEM) for adjuvant therapy of ovarian cancer *(28)* and unconjugated 17-1A for the treatment of colorectal cancer *(29)*. In the last, an extensive and ongoing study involving 189 patients given adjuvant therapy, some promising results have been achieved. Following surgery, the patients were treated over a period of 5 mo with a total of 900 mg of antibody 17-1A, which recognizes a 37–40 kDa cell-surface glycoprotein. Despite HAMA responses in 80% of the patients, the treatment reduced the overall 5-yr death rate by 30% and the recurrance rate by 27%. Interestingly, it appears that this was owing to the prevention of outgrowth of small metastatic deposits and the treatment did not appear to be effective where more extensive tumor deposits were left following surgery.

One problem encountered in the use of rodent antibodies has been the variable development of human antibodies against the injected foreign protein. This may leadto rapid clearance of the therapeutic antibody on repeated challenge. However, such responses could be advantageous and lead to the generation of anti-idiotypic antibodies (Ab2), i.e., antibodies that recognize determinants (complementarity-determining regions) on the rodent antibodies. These, in turn, may initiate a cascade that leads to the induction of anti-anti-idiotypic antibodies (Ab3) some of which may bind the same antigen as the rodent Ab1 but will be of human origin. In this way, long-term immunity may be induced. Considerable effort, however, has been made to reduce the immunogenicity of rodent antibodies by using recombinant-DNA technology either to construct chimeric antibodies consisting of rodent F_v and human F_c, or to insert the rodent CDRs into a human IgG framework *(26,30)*. Certainly, these procedures can reduce the immunogenicity of the parental antibody, but sometimes this is accompanied by a change in affinity for the target antigen.

Currently, there is much interest in the possibility of generating human antibodies from libraries of recombinant VH and VL genes cloned into viral vectors and expressed at the surface of bacteriophage *(31)* (*see* Chapter 46). It remains to be seen how useful such recombinant single-chain Fvs or Fabs will be for the construction of suitable human antibodies and how effective they will be for diagnostic or therapeutic applications. Certainly, immune rodents will continue to be used for the generation of high-affinity antibodies, and recombinant-phage libraries made from lymph node cells may provide an alternative to cell fusion. Indeed, the generation of rabbit or sheep antibodies by these means may lead to the isolation of antibodies with affinities several orders of magnitude greater than those of mouse or rat origin. The therapeutic effectivenes of these super high affinity antibodies may be substantially better than those currently in use today. It is an exciting prospect.

References

1. Kohler, G. and Milstein, C. (1975) Continuous cultures of fused cells secreting antibody of predefined specificity. *Nature (London)* **256,** 495–497.
2. Galfrè, G. and Milstein, C. (1981) Preparation of monoclonal antibodies: strategies and procedures, in *Methods in Enzymology, vol. 73, Immunochemical Techniques* (Langone, J. J. and Van Vunakis, H., eds.), Academic, New York, pp. 3–46.
3. Bazin, H. (1982) Production of rat monoclonal antibodies with the LOU rat non-secreting IR983F myeloma cell line, in *Protides of the Biologcal Fluids, 29th Colloquium* (Peeters, H., ed.), Pergamon, New York, pp. 615–618.
4. Edwards, P. A. W., Smith, C. M., Neville, A. M., and O'Hare, M. J. (1982) A human–human hybridoma system based on a fast-growing mutant of the ARH–77 plasma cell leukaemia-derived line. *Eur. J. Immunol.* **12,** 641–648.
5. Anderson, D. V., Tucker, E. M., Powell, J. R., and Porter, P. (1987) Bovine monoclonal antibodies to the FS (K99) pilus antigen of E. coli produced by murine/bovine hybridomas. *Vet. Immunol. Immunopathol.* **15,** 223–237.
6. Flynn, J. N., Harkiss, G. D., and Hopkins, J. (1989) Generation of a sheep × mouse heterohybridoma cell line (1C6.3a6T.1D7) and evaluation of its use in the production of ovine monoclonal antibodies. *J. Immunol. Meth.* **121,** 237–246.
7. Richards, C. M., Aucken, H. A., Tucker, E. M., Hannant, D., Mumford, J. A., and Powell, J. R. (1992) The production of equine monoclonal antibodies by horse–mouse heterohybridomas. *Vet. Immunol. Immunopathol.* **33,** 129–143.
8. Spieker-Polet, H., Sethupathi, P., Yam, P.-C., and Knight, K. L. (1995) Rabbit monoclonal antibodies: generating a fusion partner to produce rabbit–rabbit hybridomas. *Proc. Natl. Acad. Sci. USA* **92,** 9348–9352
9. Dean, C. J. (1984) Preparation and characterization of monoclonal antibodies to proteins and other cellular components, in *Methods in Molecular Biology, vol. 32: Basic Protein and Peptide Protocols* (Walker, J. M., ed.), Humana, Totowa, NJ, pp. 361–379.

10. Gullick, W. J. (1984) Production of antisera to synthetic peptides, in *Methods in Molecular Biology, vol. 32: Basic Protein and Peptide Protocols* (Walker, J. M., ed.), Humana, Totowa, NJ, pp. 389–399.

11. Taylor-Papadimitriou, J., Stewart, L., Burchell, J., and Beverley, P. (1993) The polymorphic epithelial mucin as a target for immunotherapy. *Ann. NY Acad. Sci.* **690,** 69–79.

12. Dean, C. J., Styles, J. M., Gyure, L. A., Peppard, J., Hobbs, S. M., Jackson, E., and Hall, J. G. (1984) The production of hybridomas from the gut-associated lymphoid tissue of tumour-bearing rats. I. Mesenteric nodes as a source of IgG producing cells. *Clin. Exp. Immunol.* **57,** 358–364.

13. Harlow, E. and Lane, D. P. (1988) *Antibodies, A Laboratory Manual.* Cold Spring Harbor Laboratory, Cold Spring Harbor, NY.

14. Modjtahedi, H., Styles, J. M., and Dean, C. J. (1993) The human EGF receptor as a target for cancer therapy: six new rat antibodies against the receptor on the breast carcinoma MDA-MB 468. *Br. J. Cancer* **67,** 247–253.

15. Styles, J. M., Harrison, S., Gusterson, B. A., and Dean, C. J. (1990) Rat monoclonal antibodies to the extracellular domain of the product of the c-erbB-2 proto-oncogene. *Int. J. Cancer* **45,** 320–324.

16. Modjtahedi, H., Eccles, S., Box, G., Styles, J., and Dean, C. (1993) Immunotherapy of human tumour xenografts overexpressing the EGF receptor with rat antibodies that block gowth factor–receptor interaction. *Br. J. Cancer* **67,** 254–261.

17. Modjtahedi, H., Hickish, T., Nicolson, M., Moore, J., Styles, J., Eccles, S., Jackson, E., Salter, J., Sloane, J., Spencer, L., Priest, K., Smith, I., Dean, C., and Gore, M. (1996) Phase I trial and tumour localisation of the anti-EGFR monoclonal antibody ICR62 in head and neck or lung cancer. *Br. J. Cancer* **73,** 228–235.

18. Eccles, S. A., Court, W. J., Box, G. A., Dean, C. J., Melton, R. J., and Springer, C. J. (1994) Regression of established breast carcinoma xenografts with antibody directed enzyme prodrug therapy (ADEPT) against c-erbB2 p185. *Cancer Res.* **54,** 5171–5177.

19. Modjtahedi, H. and Dean, C. (1994) The receptor for EGF and its ligands: expression, prognostic value and target for therapy in cancer (review). *Int. J. Cancer* **4,** 277–296.

20. Hynes, N. E. and Stern, D. F. (1994) The biology of erbB-2/neu/HER-2 and its role in cancer. *Biochim. Biophys. Acta* **1198,** 165–184.

21. Bernard, A. (1984) *Leukocyte Typing: Human Leucocyte Differentiation Antigens Detected by Monoclonal Antibodies: Specification, Classification, Nomenclature.* Springer-Verlag, Berlin.

22. Barclay, A. N., Brown, M. H., Law, A., McKnight, A. J., Tomlenson, M., Anton vander Mewe, P. (1997) *The Leucocyte Antigen Facts Book.* Academic, London, UK.

23. Rock, K. L. (1996) A new foreign policy: MHC class I molecules monitor the outside world. *Immunol. Today* **17,** 131–137.

24. Goldstein, G., Schindler, J., Tsai, H., et al. (1985) A randomized clinical trial of OKT3 monoclonal antibody for acute rejection of cadaveric renal transplants. *N. Engl. J. Med.* **313,** 337–342.

25. Waldmann, H. and Cobbold, S. (1993) The use of monoclonal antibodies to achieve immunological tolerance. *Immunol. Today* **14,** 247–251.

26. Harris, W. J. and Cunnngham, C. (1995) *Antibody Therapeutics.* R. G. Landes, Austin, TX.

27. Press, O. W., Eary, J. F., Appelbaum, F. R., Martin, P. J., Badger, C. C., Nelp, W. B., Glenn, S., Butchko, G., Fisher, D., Porter, B., Mathews, D. C., Fisher, L. D., and Bernstein, I. D. (1993) Radiolabelled antibody therapy of B cell lymphoma with autologous bone marrow support. *N. Engl. J. Med.* **329,** 1219–1224.

28. Hird, V., Maraveyas, A., Snook, D., Dhokia, B., Soutter, W. P., Meares, C., Stewart, J., Mason, P., Lambert, H., and Epenetos, A. (1993) Adjuvant therapy of ovarian cancer with radioactive monoclonal antibody. *Br. J. Cancer* **68,** 403–406.

29. Reithmüller, G., Schneider-Gadicke, E., Schlimok, G., Schmiegel, W., Raab, R., Pichlmair, H., Hirche, H., Pichlmayr, R., Buggisch, P., Witte, J., and the German Cancer Aid 17-1A Study Group. (1994) Randomised trial of monoclonal antibody for adjuvant therapy of resected Dukes' C colorectal carcinoma. *Lancet* **343**, 1177–1183.

30. Jones, P. T., Dear, P. H., Foote, J., Neuberger, M. S., and Winter, G. (1986) Replacing the complementarity determining regions in a human antibody with those from a mouse. *Nature* **321**, 522–525.

31. Winter, G., Griffiths, A. D., Hawkins R. E., et al. (1994) Making antibodies by phage display technology. *Ann. Rev. Immunol.* **12**, 433–455.

45

Phage-Display Libraries

Julia E. Thompson and Andrew J. Williams

1. Introduction

Foreign proteins can be displayed on the surface of filamentous bacteriophage by fusing them to the phage minor coat protein (**Fig. 1**). Many proteins have been expressed in this way, including human growth factor *(1)*, alkaline phosphatase *(2)*, ricin B *(3)*, pancreatic trypsin inhibitor *(4)*, and a DNA-binding protein *(5)*. Most importantly, the displayed proteins retain their biological activity.

Large libraries of randomly generated antibody fragments and peptides have similarly been constructed. These libraries contain phage with novel biological properties, and selection strategies have been developed to identify clones of interest *(6,7)*. Each clone in the repertoire contains a copy of both the phage genome and the gene encoding a displayed foreign protein. Selection based on the biological activity of a protein (or peptide) can therefore be used to access its nucleotide sequence. Key to this process is the ability of the display phage to replicate and renew itself. During repeated rounds of growth and selection, phage-displaying proteins with specific properties can be enriched within a population until ultimately they predominate (**Fig. 2**). This approach has proved particularly useful in identifying antibodies with improved specificity or affinity. Therefore, the display of antibodies on the surface of phage and the generation of very large antibody libraries has become the most widespread application of phage-display technology.

2. Phage Antibodies
2.1. Background

Since the development of monoclonal antibodies (MAbs) *(8)*, a technology has evolved with diverse research and commercial applications *(9)*. The ability to immortalize an antibody-producing B cell and thereby to derive a cell line secreting an antibody of a single defined specificity has allowed the exploitation of antibodies as both diagnostic and therapeutic reagents. However, although rodent MAbs are relatively easy to produce, they are often recognized as foreign in humans, resulting in an antibody response that blocks any therapeutic effect (HAMA). Attempts have been made to "humanize" rodent antibodies by replacing those sections of the rodent protein not

From: *Molecular Biomethods Handbook*
Edited by: R. Rapley and J. M. Walker © Humana Press Inc., Totowa, NJ

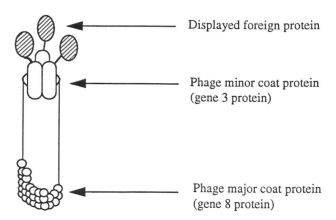

Fig. 1. Display of foreign protein on the surface of phage. A schematic representation of an engineered filamentous phage. A gene encoding a foreign protein in cloned in-frame with the phage minor-coat protein gene, such that the functional-fusion protein is displayed on the surface of the phage particle.

required for its interaction with antigen with the analogous human sequences. Unfortunately, these constructs can still be recognized as foreign *(10)*. It is therefore often preferable to produce fully human antibodies as human therapeutics.

Making human antibodies by the hybridoma route is fraught with difficulties *(11)*. Antibodies to human self-antigens are among the most promising therapeutic targets, but the generation of these in vivo is extremely difficult because the expansion of B-lymphocyte clones is prevented by self-tolerance mechanisms *(12)*. In addition, the immunization of humans with toxic antigens is neither practicable or ethical, and even when an immunized donor can be found, the production of MAbs is still technically demanding.

If the antibody genes, rather than the cells themselves, can be isolated, then the difficulties encountered with B-cell immortalization may be circumvented. The genes encoding the building blocks of human antibodies have been identified, and through the application of PCR technology may be rescued from human lymphocytes *(13–16)*. Although it is not yet possible to efficiently express whole antibodies in *Escherichia coli*, antibody fragments encoding the antigen recognition site are expressed well and can be displayed on phage particles *(17)*. This display of functional antibody fragments on the surface of phage has facilitated the investigation and manipulation of their antigen binding characteristics. Indeed, application of the phage display system to the field of antibody engineering has realized such great potential that the basic technology is now almost routine.

2.2. The Display of Antibodies and the Generation of Libraries

Manipulation and modification of the binding characteristics of antibodies is facilitated by their conserved structure and modular nature. Although whole antibodies are necessary to recruit appropriate effector functions, the variable or "V" domains of the heavy (VH) and light (VL) chains are together all that is necessary for antigen binding (**Fig. 3**). In fact, most of the amino-acid residues involved in an antibody/ antigen interaction reside within three loops at one end of each of the V domains called

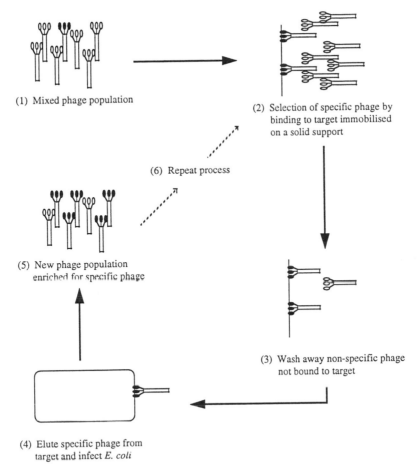

(1) Mixed phage population

(2) Selection of specific phage by binding to target immobilised on a solid support

(6) Repeat process

(5) New phage population enriched for specific phage

(3) Wash away non-specific phage not bound to target

(4) Elute specific phage from target and infect *E. coli*

Fig. 2. Selection of specific phage from a mixed population. A schematic representation of a typical strategy for selecting a novel-binding activity from a phage-display library. (1) The library includes phage that display on their surface proteins with different binding characteristics. (2) Specific phage may be enriched by selection on a target to which only they will bind, e.g., an antigen-coated surface for selection of phage-displaying antibodies of a particular specificity. (3) Extensive washing removes the majority of nonspecifically bound phage from the selection. (4) The eluted phage are used to infect a male *Escherichia coli* strain. Each phage particle contains a copy of the gene encoding its displayed foreign protein; thus, replication will produce more phage with identical binding characteristics. (5) The new phage population is enriched for phage displaying the desired activity. (6) The process can be repeated to enrich further as required.

the complementarity-determining regions (CDRs). McCafferty et al. *(17)* demonstrated that antibody fragments encompassing the CDRs could be displayed on the surface of bacteriophage. These fragments, consisting of the VH and VL domains joined together by a short peptide linker, are known as single-chain Fvs (scFv) (**Fig. 3**). When fused to the N-terminus of the minor coat protein 3 of filamentous bacteriophage, the scFvs were shown both to fold correctly and to bind specifically to antigen *(17)*. The larger Fab-antibody fragment (**Fig. 3**) has also been used for phage display *(18,19)*. In phage

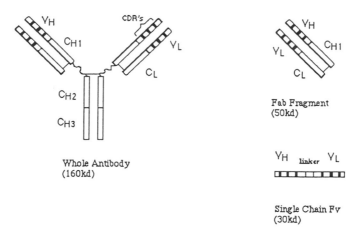

Fig. 3. A typical antibody molecule and the fragments that have been displayed on the surface of phage.

assembly, the coat proteins are exported into the bacterial periplasm; antibody VH and VL chains can be similarly directed into the bacterial periplasm by incorporation of a signal sequence *(20,21)*. For phage display of Fab fragments, one of the chains is exported fused to the capsid protein and the other as a soluble fragment that spontaneously assembles with the fused partner to create a functional heterodimer.

Variability within the VH and VL domains is localized, and there are regions at both the amino and carboxy termini that are well conserved. This has enabled the design of PCR primers for amplification of the different human V gene families *(16,22)*, allowing us to access repertoires of V-genes rearranged in vivo. In the immune system, pairing of different VH and VL sequences generates a large number of different antibody structures. This may be similarly achieved in vitro by the assembly of separate repertoires of VH and VL genes to form scFvs. Cloning the scFv into a suitable vector for display on the surface of phage creates an antibody library from which clones with particular antigen-binding characteristics may be selected (**Fig. 4**).

The random combinatorial approach described can be used to tap the immunoglobulin repertoire of a human donor. Such a repertoire contains antibodies to essentially any antigen, albeit often with only moderate affinity (1–10 μM) *(22–24)*. Repertoires have also been constructed from individuals who have been preimmunized with the antigen of interest *(25–31)*. Although this approach may yield higher-affinity antibodies, there is often not an appropriately immunized human donor, and it is very time consuming to make a new repertoire for each antigen-binding specificity that is selected. Recently, much larger scFv *(32)* and Fab *(33)* libraries, both containing >10^{10} gene pairs, have been generated from nonimmunized sources, providing a universal route for the rapid selection of new antigen-binding specificities. Such libraries have yielded high-affinity antibodies (100 nM–300 pM) to a very wide range of different antigens without the need for immunization.

2.3. Selection of Antibodies from Phage Display Libraries

The antibody fragments within the large phage display libraries potentially represent more than 10^{10} different antigen-binding specificities. It is the ability to quickly

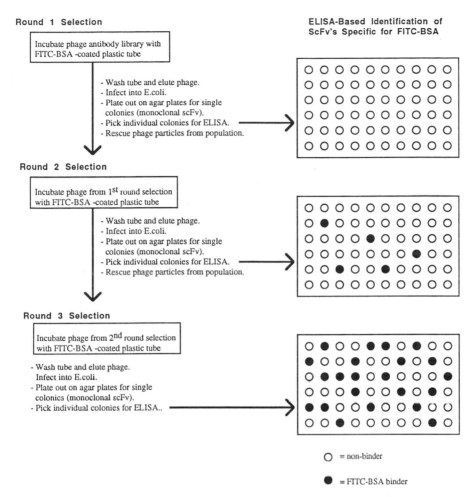

Fig. 4. The isolation and identification of antibodies to FITC-BSA. Phage specific for the hapten fluorescein conjugated to bovine serum albumin (FITC-BSA) were enriched for by incubating a phage antibody library with FITC-BSA immobilized on the surface of a plastic tube, a process known as panning (*see text* for details). Individual colonies (monoclonal scFv) from each round of selection are analyzed on FITC-BSA-coated plates by enzyme-linked immunoassay (ELISA). ScFv produced by each colony are first allowed to bind to the coated ELISA plate. Each scFv has a small tag as its C-terminal that is recognized by an enzyme-conjugated secondary antibody. The addition of this antibody followed by its chromagenic substrate allows the visualization of FITC-BSA-specific binders. The highest affinity FITC-specific scFv isolated using this strategy is approx 300 pM, equivalent to the highest affinity antibodies isolated by traditional hybridoma technology *(32)*.

and efficiently select an individual clone with a particular specificity away from the huge reservoir of irrelevant phage, i.e., to find the proverbial needle in the haystack, that makes phage display technology such a powerful and important tool. In order to achieve efficient selection, it is important that the level of discrimination between specific and nonspecific scFv is as large as possible. There are now a variety of different vector formats for phage display; antibody fragments may be cloned directly into the genome of a filamentous phage, or alternatively, phagemid vectors can be used. Use of a phagemid (a plasmid based vector that can be "rescued" and packaged into phage particles following coinfection with a helper phage *[34]*) leads to fewer copies of the antibody fusion being displayed on each phage particle. This strategy minimizes the effects of avidity and will favor more stringent selection of the highest affinity phage antibody clones *(35,36)*.

Repertoire selection can take many forms; this can be by any method that allows antigen-specific scFv to be partitioned from nonspecific clones. Indeed, because the design of the selection strategy can dramatically affect both the quality of antibodies isolated and the speed with which this is done, often a number of different selection methods are used. Typically, selections have been based on the capture of phage onto antigen that is immobilized either on a column matrix *(17)* or the surface of a plastic dish or tube *(19,22)*. In the method known as panning, the phage library is first incubated in an antigen-coated tube. The tube is then washed extensively to remove nonspecific phage, before elution of antigen-specific phage using either a mild denaturant *(19,22)* or soluble antigen *(37)*. Following infection of *E. coli*, the population of phage antibodies is regrown and the process repeated. Each cycle of enrichment can increase the proportion of phage antibodies within the population by a factor of up to 1000-fold *(17)*. Ultimately, the phage population will become essentially clonal (**Fig. 4**).

Panning does have its drawbacks, however, because the process of immobilizing antigen onto a solid surface or support matrix may cause it to be presented in a nonnative conformation, and the binding of specific phage to the immobilized antigen is also diffusion limited. An alternative selection strategy that avoids these problems uses soluble biotinylated antigen *(38)*. Taking advantage of the very high affinity biotin-streptavidin interaction, phage bound to biotinylated antigen in solution are captured on streptavidin-coated magnetic beads. A magnet is used to separate antigen-bound phage from the nonspecific background. This method allows the experimenter much greater control over the selection conditions, enabling the selection to be tailored to deliver antibodies with particular binding characteristics. Biotin selection has been successfully used to isolate high affinity scFv from a population of lower-affinity variants where panning has failed to discriminate between the same clones *(39)*.

2.4. Affinity Maturation

The affinity of the antibodies derived from the early phage display repertoires was often only in the μM range *(22–24)*, equivalent to an antibody generated in the primary immune response. In vivo, a process called somatic hypermutation produces the high affinity antibodies characteristic of the secondary and tertiary responses. Antibodies of similarly high affinity have now been recovered from the larger phage-display libraries *(32,33)*. However, it is often desirable to build the affinity of a characterized low-affinity clone rather than isolate a new antibody from a larger library. Furthermore, for therapeutic purposes it has been necessary to produce antibodies of even higher affinity than it has been possible to isolate *de novo (39,40)*.

Antibodies isolated from phage-display libraries are readily amenable to further genetic manipulation and thus can be engineered to derive higher-affinity variants. Unlike the immune system, where affinity maturation is limited to mutation of a selected VH and VL pair, random combinatorial repertoires contain an enormous level of untapped diversity that can be readily accessed by a process known as chain shuffling *(37,41)*. In a chain-shuffled library, either the heavy or light chain of an antibody is retained while its partner is replaced with a repertoire of different chains. The shuffled library may be selected for altered binding properties, such as improved affinity or antigen specificity.

Building the affinity of a phage antibody may also use processes that mimic those used in affinity maturation in vivo. Somatic hypermutation can be emulated by reamplifying the antibody sequence using an error-prone polymerase *(42)*. Changes are randomly introduced throughout the sequence and those that are beneficial can be selected (our unpublished observations). Alternatively, sequence-specific oligonucleotides may be used for site-directed mutagenesis, thereby targeting any chosen residue or region of the antibody. Such an approach directed to CDR3 of an antibody heavy chain has given dramatic improvements in affinity *(40),* resulting in antibodies with an affinity higher than is theoretically possible in vivo *(43)*. Pivotal to the success of any mutagenic strategy is the ability to select for the improved antibody variants; this is the main advantage of the phage-display system over other techniques.

3. Uses of Phage-Antibody Libraries

3.1. Antibody Formats

Selecting antibodies in bacterial-expression systems has many practical advantages. One of the more useful features of the phage-display vectors has been the inclusion of an amber codon at the junction between the antibody fragment and the phage capsid protein. This format enables the production of soluble antibody in addition to the phage displayed fusion *(18)*. Production of the soluble antibody fragment in *E. coli* is considerably faster and cheaper than similar utilization of eukaryotic-expression systems, with the added benefit that the stock cultures are easier to establish and maintain.

ScFv fragments have a number of important practical applications (*see*, for example, **ref. 44**). Their small size and rapid rate of clearance in vivo makes them particularly useful agents for imaging purposes, although perhaps limiting their therapeutic potential. However, selection of antibodies on the surface of phage has clear advantages even when the final product is to be a whole antibody molecule. Phage-displayed antibodies are immediately accessible to further genetic manipulation, and thus it is relatively straightforward to reformat them as whole antibodies for expression in eukaryotic cells. Engineering the antibody in this way has an additional bonus in that it is possible to choose the most appropriate antibody isotype for the particular application and the establishment and maintenance of the eukaryotic cell lines is easier than experienced with traditional hybridomas.

3.2. Engineering High-Affinity Therapeutic Antibodies

One important application of phage-antibody libraries has been in the engineering of therapeutic reagents *(39,40)*. Recent work has described the use of phage display as a tool to derive an antibody-based prophylactic for preventing the establishment of human immunodeficiency virus type 1 (HIV-1) infection *(39)*. The principal neutralizing determinant of HIV-1 is the third hypervariable domain (V3 loop) of the viral envelope glycoprotein (gp120), a sequence that is highly variant among viral isolates. Antibodies that recognize this epitope have been shown to be effective in the prevention of viral infection *(45–47)*.

Thompson et al. *(39)* focused on a human anti-V3 loop MAb that was able to bind to a number of different V3 loop sequences and thus had great potential utility for neutralization of diverse viral serotypes. The aim was to increase the affinity of the antibody and thereby its biological potency, while retaining its ability to bind variant V3 loops. Phage display has been used to select antibody variants with improved properties, such as affinity *(38,48)* or specificity *(41)*. The anti-V3 loop antibody, reformatted as an scFv for phage display, was similarly manipulated by sequential replacement of heavy and light chains with repertoires of variable domains derived from unimmunized donors. Higher-affinity variants were then selected using a soluble peptide-based on a V3-loop sequence, and residues where alterations led to improvement were identified. These residues were then used as positional information for the construction of secondary repertoires in which they were simultaneously randomized by site-directed mutagenesis. Following reselection, scFv variants with multiple mutations were found having significantly reduced dissociation rates. Furthermore, these improvements translated to improved recognition of alternative V3 loop sequences. The equilibrium dissociation constant for monomeric antibody fragments on $gp120_{MN}$ was reduced from 2 nM to 280 pM in the final improved candidates, providing an effective demonstration of the use of phage display to tailor the properties of a therapeutic MAb in a predefined fashion.

3.3. Antibodies from Immunized Individuals

Antibody repertoires have been constructed from immunized humans, thereby enabling the study of their response to a particular infection or disease, and potentially allowing direct access to antibodies typical of protective immunity *(25–30)*. Barbas et al. *(31)* used phage display to probe the response of an individual to a viral infection. Antibody fragments reacting with HIV-1 gp120 were generated by selection from a repertoire prepared from a long-term asymptomatic HIV-seropositive individual. This approach allowed the fast and effective isolation of groups of closely related antibody fragments with particular antigen-binding characteristics whose specificity broadly reflected the serum profile of the donor.

Despite the utility of immunized repertoires, sometimes the response to a dominant epitope will prevent the selection of other potentially useful specificities. In one demonstration of how this problem may be circumvented to make more complete use of a phage-antibody repertoire, antibodies against the dominant epitopes from an immunized library have been used to "mask" the antigen. This approach has allowed subsequent selection of novel antibodies directed against more weakly immunogenic epitopes present in a library derived from an asymptomatic HIV-1 seropositive donor *(49)*.

3.4. Isolation of Antibodies Not Possible by Hybridoma Technology

The advent of phage display libraries has provided an answer to some of the intractable problems associated with traditional hybridoma technology. In particular, it is now possible to isolate human antibodies directed against "self" or other nonimmunogenic antigens. Griffiths et al. *(23)* derived antibodies directed against a range of human self-antigens from a nonimmunized phage-antibody repertoire. Unlike natural autoantibodies which tend to be polyspecific, the anti-self scFv fragments selected on the surface of phage are highly antigen-specific. Although the affinity of the antibodies was only moderate, it has subsequently proved possible to isolate very high-affinity anti-self specificities from the larger antibody repertoires now available *(33)*. Thus, nonimmunized phage antibody libraries could be used for the direct isolation of antibodies for use in human therapy.

Antibodies directed against steroid hormones have important applications in the field of immunodiagnostics, but it has been difficult to isolate monoclonals that combine high affinity with the requisite specificity. Because steroids are not themselves immunogenic, they have to be coupled to a larger molecule in order to provoke an immune response in animals. However, the coupling process appears to mask part of the antigen because antibodies raised in this way are often unable to recognize subtle but important differences between structurally related steroid molecules *(50)*. Using phage-display libraries, it has been possible to overcome these problems. A recent report *(51)* describes the selection of estradiol binding scFv from a nonimmunized phage-antibody repertoire. Careful optimization of the selection conditions used in this study reduced the "masking" effect on the coupled steroid antigen and thus enabled the isolation of scFv, which were not only highly specific, but had an affinity in the nanomolar range.

3.5. Humanization by Epitope Imprinting

In some instances it may be desirable to humanize a murine antibody. This is particularly relevant where the murine antibody has already been extensively characterized and shown to have therapeutic value. In this development of the chain-shuffling technique, the VH and VL of the mouse antibody are sequentially replaced with repertoires of human chains and selected on antigen, to derive entirely human antibodies that carry the imprint of the original rodent-antibody specificity. The successful application of this procedure in a number of cases, including antibodies directed against a hapten *(52)* and a human cytokine *(53)*, suggest that it is generally applicable. Unlike alternative methods of humanizing existing rodent antibodies, this technique has the distinct advantage that the end-product is fully human rather than a rodent/human chimera.

4. Alternative Phage-Display Libraries

Although this chapter has concentrated on phage-antibody libraries, the selection of other proteins and peptides from phage display repertoires has widespread application (for recent reviews, *see* **refs.** *7,54*). To date, it would appear that any protein that can be secreted to the bacterial periplasm can be functionally displayed on the surface of filamentous phage. This has enabled the use of phage-display libraries to manipulate other proteins in much the same way as has already been described with reference to antibodies.

The binding affinity of proteins may be increased by the construction and selection of libraries of mutagenized variants. This was the strategy used to increase the affinity of human growth hormone (hGH) for its receptor. Taking advantage of the data available from previous structural and mutagenic studies, hGH phage libraries were made in which carefully chosen regions of the protein were randomized by site-directed mutagenesis. Individual selected mutations with increased affinity for the receptor were then recombined to create a "supermutant" that had 15 changes from wild-type and bound its receptor 400 times more tightly *(55)*. Phage display has also been used to select protein variants with modified binding specificity. Again, the application of this procedure has made use of existing structure–function data to identify residues to be mutated. Bovine pancreatic trypsin inhibitor displayed on the surface of phage has been manipulated in this way to effectively redirect its specificity *(4)*. Selected protein variants bind tightly to human neutrophil elastase, a relatively poor ligand for the wild-type protein.

The first examples of phage-display libraries were repertoires of random linear peptides *(56–58)* providing a source of small binding molecules from which clones with particular properties could be selected. In the simplest case, peptide libraries can be used to select ligands for proteins that recognize linear sequences. For example, the linear epitopes recognized by some MAbs may be identified by selection of a peptide library using immobilized antibody *(59)*. In an extension of this approach, phage display has been used to probe enzyme specificity *(60)*. Libraries of "substrate phage" are constructed in which random peptide sequences are interposed between the phage gene III and a protein that binds specifically to a column matrix. Such phage can be immobilized on the column, and are eluted from the solid phase only when their particular peptide sequence is susceptible to cleavage by a protease of interest. This approach enables the sequence specificity of the protease to be rapidly defined.

Although molecular recognition (including antibody–antigen interaction) is usually effected through discontinuous binding sites (made up of separate regions of the polypeptide chain), large peptide libraries may have sufficient conformational diversity to provide a mimic for the natural epitope using a linear format. Libraries have been selected to derive peptide antagonists or agonists of natural receptor-ligand interactions, including peptides that will compete with nonpeptide ligands for their binding site (for example, **refs. *61,62***). However, peptide mimics of discontinuous binding sites have proved more difficult to isolate. Examinations of the successful mimics of small molecules suggest that this may be owing in part to the general lack of structural integrity of linear peptides. As a result, conformationally constrained peptide libraries are now being constructed in the search for novel binding activities. Such libraries have included repertoires of disulfide-bonded peptide loops *(63)*. Alternatively, random peptides have been constrained by display on a defined protein scaffold, such as a randomized antibody CDR *(64,65)*. This approach, together with the provision of ever larger and more diverse repertoires, should herald the future isolation of peptide mimics for increasingly complex ligands.

5. Future Developments

The new large phage-antibody libraries are already proving their worth for the isolation of high-affinity, fully human antibodies, while they demonstrate further utility in providing a range of different solutions to any one problem. This is of paramount importance where some function other than simple binding is required. For example, a panel of different antibodies needs to be surveyed to find those that are capable of neutralizing the biological activity of their target antigen (*66*, our unpublished observations). Such neutralizing antibodies are likely to prove to be valuable therapeutic agents in the future.

Proteins and peptides selected from phage-display repertoires have had many applications, both commercially as therapeutic and diagnostic reagents, and as research tools. Antibodies directed against a range of antigens have demonstrated utility as immunochemical reagents *(67)*, and thus activities selected from phage-antibody repertoires can, for example, provide a useful tag to follow a target antigen through a drug development program. Phage-display technology has enabled a larger number of protein and peptide variants to be surveyed than has been previously possible. Proteins selected in this way are now becoming increasingly invaluable to new drug-discovery regimes. A binding activity can be selected from a phage-display library with comparative ease and efficiency, and used to provide proof of principle before much greater investment is made into the development of a small molecule therapeutic.

In summary, the phage-display system has rapidly established itself at the forefront of protein and peptide research in both academia and in the pharmaceutical industry, solving problems that have often proved intractable in the past. Within the next few years, products will reach the market place that have either been isolated directly from phage-display libraries or whose identification or development has been aided by the use of this invaluable technology.

References

1. Bass, S., Greene, R., and Wells, J. A. (1990) Hormone phage: an enrichment method for variant proteins with altered binding properties. *Proteins* **8,** 309–314.
2. McCafferty, J., Jackson, R. H., and Chiswell, D. J. (1991) Phage-enzymes: expression and affinity chromatography of functional alkaline phosphatase on the surface of bacteriophage. *Protein Eng.* **4,** 955–961.
3. Swimmer, C., Lehar, S. M., McCafferty, J., Chiswell, D. J., Blattler, W. A., and Guild, B. C. (1992) Phage display of ricin B chain and its single binding domains: system for screening galactose-binding mutants. *Proc. Natl. Acad. Sci. USA* **89,** 3756–3760.
4. Roberts, B. L., Markland, W., Ley, A. C., Kent, R. B., White, D. W., Guterman, S. K., and Ladner, R. C. (1992) Directed evolution of a protein: selection of potent neutrophil elastase inhibitors displayed on M13 fusion phage. *Proc. Natl. Acad. Sci. USA* **89,** 2429–2433.
5. Rebar, E. J. and Pabo, C. O. (1994) Zinc finger phage: affinity selection of fingers with new DNA-binding specificities. *Science* **263,** 671–673.
6. Winter, G., Griffiths, A. D., Hawkins, R. E., and Hoogenboom, II. R. (1994) Making antibodies by phage display technology. *Annu. Rev. Immunol.* **12,** 433–455.
7. Clackson, T. and Wells, J. A. (1994) In vitro selection from protein and peptide libraries. *Trends Biotechnol.* **12,** 173–184.
8. Kohler, G. and Milstein, C. (1975) Continuous cultures of fused cells secreting antibody of predefined specificity. *Nature (Lond.)* **256,** 495–497.
9. Winter, G. and Milstein, C. (1991) Man-made antibodies. *Nature (Lond.)* **349,** 293–299.
10. LoBuglio, A. F., Wheeler, R. H., Trang, J., Haynes, A., Rogers, K., Harvey, E. B., Sun, L., Ghrayeb, J., and Khazeli, M. B. (1989) Mouse/human chimaeric monoclonal antibody in man: kinetics and immune responses. *Proc. Natl. Acad. Sci. USA* **86,** 4220–4224.
11. James, K. and Bell, G. T. (1987) Human monoclonal antibody production: current status and future prospects. *J. Immunol. Meth.* **100,** 5–40.
12. Nossal, G. J. (1989) Immunologic tolerance: collaboration between antigen and lymphokines. *Science* **245,** 147–153.
13. Larrick, J. W., Danielsson, L., Brenner, C. A., Abrahamson, M., Fry, K. E., and Borrebaeck, C. A. (1989) Rapid cloning of rearranged immunoglobulin genes from human hybridoma cells using mixed primers and the polymerase chain reaction. *Biochem. Biophys. Res. Commun.* **160,** 1250–1256.
14. Mullinax, R. L., Gross, E. A., Amberg, J. R., Hay, B. N., Hogrefe, H. H., Kubitz, M. M., Greener, A., Alting, M. M., Ardourel, D., Short, J. M., Sorge, J. A., and Shopes, B. (1990) Identification of a human antibody fragment clones specific for tetanus toxoid in a bacteriophage lambda immunoexpression library. *Proc. Natl. Acad. Sci. USA* **87,** 8095–8099.

15. Persson, M. A. A., Caothien, R. H., and Burton, D. R. (1991) Generation of diverse high-affinity human monoclonal antibodies by repertoire cloning. *Proc. Natl. Acad. Sci. USA* **88,** 2432–2436.

16. Marks, J. D., Tristrem, M., Karpas, A., and Winter, G. (1991) Oligonucleotide primers for polymerase chain reaction amplification of human immunoglobulin variable genes and design of family-specific oligonucleotide probes. *Eur. J. Immunol.* **21,** 985–991.

17. McCafferty, J., Griffiths, A. D., Winter, G., and Chiswell, D. J. (1990) Phage antibodies: filamentous phage displaying antibody variable domains. *Nature (Lond.)* **348,** 552–554.

18. Hoogenboom, H. R., Griffiths, A. D., Johnson, K. S., Chiswell, D. J., Hudson, P., and Winter, G. (1991) Multisubunit proteins on the surface of filamentous phage: methodologies for displaying antibody (Fab) heavy and light chains. *Nucleic Acids Res.* **19,** 4133–4137.

19. Barbas, C. F., Kang, A. S., Lerner, R. A., and Benkovic, S. J. (1991) Assembly of combinatorial antibody libraries on phage surfaces: the gene III site. *Proc. Natl. Acad. Sci. USA* **88,** 7978–7982.

20. Better, M., Chang, C. P., Robinson, R. R., and Horwitz, A. H. (1988) Escherichia coli secretion of an active chimeric antibody fragment. *Science* **240,** 1041–1043.

21. Skera, A. and Pluckthun, A. (1988) Assembly of a functional immunoglobulin Fv fragment in *Escherichia coli. Science* **240,** 1038–1041.

22. Marks, J. D., Hoogenboom, H. R., Bonnert, T. P., McCafferty, J., Griffiths, A. D., and Winter, G. (1991) By-passing immunisation: human antibodies from V-gene libraries displayed on phage. *J. Mol. Biol.* **222,** 581–597.

23. Griffiths, A. D., Malmqvist, M., Marks, J. D., Bye, J. M., Embleton, M. J., McCafferty, J., Baier, M., Holliger, K. P., Gorick, B. D., Hughes-Jones, N. C., Hoogenboom, H. R., and Winter, G. (1993) Human anti-self antibodies with high specificity from phage display libraries. *EMBO J.* **12,** 725–734.

24. de Kruif, J., Boel, E., and Logtenberg, T. (1995) Selection and application of human single chain Fv antibody fragments from a semi-synthetic phage antibody display library with designed CDR3 regions. *J. Mol. Biol.* **248,** 97–105.

25. Caton, A. J. and Koprowski, H. (1990) Influenza virus hemagglutinin-specific antibodies isolated from a combinatorial expression library are closely related to the immune response of the donor. *Proc. Natl. Acad. Sci. USA* **87,** 6450–6454.

26. Lerner, R. A., Barbas, C. R., Kang, A. S., and Burton, D. R. (1991) On the use of combinatorial antibody libraries to clone the "fossil record" of an individual's immune response. *Proc. Natl. Acad. Sci. USA* **88,** 9705,9706.

27. Barbas, C. F., Crowe, J. E., Cababa, D., Jones, T. M., Zebedee, S. L., Murphy, B. R., Chanock, R. M., and Burton, D. R. (1992) Human monoclonal Fab fragments derived from a combinatorial library bind to respiratory syncytial virus F glycoprotein and neutralise infectivity. *Proc. Natl. Acad. Sci. USA* **89,** 10,164–10,168.

28. Williamson, R. A., Burioni, R., Sanna, P. P., and Partridge, L. J. (1993) Human monoclonal antibodies against a plethora of viral pathogens from single combinatorial libraries. *Proc. Natl. Acad. Sci. USA* **90,** 4141–4145.

29. Zebedee, S. L. Barbas, C. F., Hom, Y-L., Caothien, R. H., Graff, R., DeGraw, J., Pyati, J., LaPolla, R., Burton, D. R., Lerner, R. A., and Thornton, G. B. (1992) Human combinatorial antibody libraries to hepatitis B surface antigen. *Proc. Natl. Acad. Sci. USA* **89,** 3175–3179.

30. Rapoport, B., Portolano, S., and McLachlan, S. M. (1995) Combinatorial libraries: new insights into human organ-specific autoantibodies. *Immunol. Today* **16,** 43–49.

31. Barbas, C. F., Collet, T. A., Amberg, W., Roben, P., Binley, J. M., Hoekstra, D., Cababa, D., Jones, T. M., Williamson, R. A., Pilkington, G. R., Haingwood, N. L., Cabezas, E., Satterthwait, A. C., Sanz, I., and Burton, D. R. (1993) Molecular profile of an antibody response to HIV-1 as probed by combinatorial libraries. *J. Mol. Biol.* **230,** 812–823.

32. Vaughan, T. J., Williams, A. J., Pritchard, K., Osbourn, J. K., Pope, A. R., Earnshaw, J. C., McCafferty, J., and Johnson, K. S. (1995) Isolation of human antibodies with sub-nanomolar affinities directly from a large non-immunized phage display library. *Biotechnology*, submitted.

33. Griffiths, A. D., Williams, S. C., Hartley, O., Tomlinson, I. M., Waterhouse, P., Crosby, W. L., Kontermann, R. E., Jones, P. T., Low, N. M., Allison, T. J., Prospero, T. D., Hoogenboom, H. R., Nissim, A., Cox, J. P. L., Harrison, J. L., Zaccolo, M., Gherardi, E., and Winter, G. (1994) Isolation of high affinity human antibodies directly from large synthetic repertoires. *EMBO J.* **13,** 3245–3260.

34. Vieira, J. and Messing, J. (1987) Production of single-stranded plasmid DNA. *Meth. Enzymol.* **153,** 3–11.

35. Garrard, L. J., Yang, M., O'Connell, M. P., Kelley, R. F., and Henner, D. J. (1991) Fab assembly and enrichment in a monovalent phage display system. *Biotechnology* **9,** 1373–1377.

36. Lowman, H. B., Bass, S. H., Simpson, N., and Wells, J. A. (1991) Selecting high-affinity binding proteins by monovalent phage display. *Biochemistry* **30,** 10,832–10,838.

37. Clackson, T., Hoogenboom, H. R., Griffiths, A. D., and Winter, G. (1991) Making antibody fragments using phage display libraries. *Nature (Lond.)* **352,** 624–628.

38. Hawkins, R. E., Russell, S. J., and Winter, G. (1992) Selection of phage antibodies by binding affinity. Mimicking affinity maturation. *J. Mol. Biol.* **226,** 889–896.

39. Thompson, J. E., Pope, A. R., Tung, J-S., Chan, C., Hollis, G., Mark, G., and Johnson, K. S. (1995) Affinity maturation of a high affinity human monoclonal antibody against the third hypervariable loop of Human Immunodeficiency Virus: use of phage display to improve affinity and broaden strain reactivity. *J. Mol. Biol.*, submitted.

40. Barbas, C. F., Hu, D., Dunlop, N., Sawyer, L., Dababa, D., Hendry, R. M., Nara, P. L., and Burton, D. R. (1994) In vitro evolution of a neutralizing human antibody to human immunodeficiency virus type 1 to enhance affinity and broaden strain cross-reactivity. *Proc. Natl. Acad. Sci. USA* **91,** 3809–3813.

41. Kang, A. S., Jones, T. M., and Burton, D. R. (1991) Antibody redesign by chain shuffling from random combinatorial immunoglobulin libraries. *Proc. Natl. Acad. Sci. USA* **88,** 11,120–11,123.

42. Leung, D. W., Chen, E., and Goeddel, D. V. (1989) A method for random mutagenesis of a defined DNA segment using a modified polymerase chain reaction. *Technnique* **1,** 11–15.

43. Foote, J. and Eisen, H. N. (1995) Kinetic and affinity limits on antibodies produced during immune responses. *Proc. Natl. Acad. Sci. USA* **92,** 1254–1256.

44. Huston, J. S., McCartney, J., Tai, M-S., Mottola-Hartshorn, C., Jin, D., Warren, F., Keck, P., and Oppermann, H. (1993) Medical applications of single-chain antibodies. *Intern. Rev. Immunol.* **10,** 195–217.

45. Linsley, P. S., Ledbetter, J. A., Kinney-Thomas, E., and Hu, S. L. (1988) Effects of anti-gp120 monoclonal antibodies on CD4 receptor binding by the env protein of human immunodeficiency virus type I. *J. Virol.* **62,** 3695–3702.

46. Skinner, M. A., Langlois, A. J., McDanal, C. B., McDougal, J. S., Bolognesi, D. P., and Matthews, T. J. (1988) Neutralising antibodies to an immunodominant envelope sequence do not prevent gp120 binding to CD4. *J. Virol.* **62,** 4195–4200.

47. Emini, E. A., Schleif, W. A., Nunberg, J. H., Conley, A. J., Eda, Y., Tokiyoshi, S., Putney, S. D., Matsushita, S., Cobb, K. E., Jett, C. M., Eichberg, J. W., and Murthy, K. K. (1992) Prevention of HIV-1 infection in chimpanzees by gp120 V3 domain-specific monoclonal antibody. *Nature (Lond.)* **355,** 728–730.

48. Marks, J. D., Griffiths, A. D., Malmqvist, M., Clackson, T., Bye, J. M., and Winter, G. (1992) By-passing immunization: building high affinity human antibodies by chain shuffling. *Bio/Technology* **10,** 779–783.

49. Ditzel, H. J., Binley, J. M., Moore, J. P., Sodroski, J., Sullivan, N., Sawyer, L. S. W., Hendry, R. M., Yang, W-P., Barbas, C. F., and Burton, D. R. (1995) Neutralizing recombinant human antibodies to a conformational V-2 and CD4-binding site-sensitive epitope of HIV-1 gp120 isolated by using an epitope masking procedure. *J. Immunol.* **154,** 893–906.

50. Cook, B. and Beastall, G. (1989) Steroid hormones: a practical approach. In *Practical Approach Series* (Rickwood, D. and Hames, B. D., eds.), Oxford University Press, Oxford, UK, pp. 1–65.

51. Pope, A. R., Pritchard, K., Williams, A. J., Hackett, J. R., Roberts, A., Mandecki, W., and Johnson, K. S. (1995) In vitro selection of a high affinity antibody to oestradiol using a phage display human antibody library. *J. Mol. Biol.,* submitted.

52. Figini, M., Marks, J. D., Winter, G., and Griffiths, A. D. (1994) In vitro assembly of repertoires of antibody chains by renaturation on the surface of phage. *J. Mol. Biol.* **239,** 68–78.

53. Jespers, L. S., Roberts, A., Mahler, S. M., Winter, G., and Hoogenboom, H. R. (1994) Guiding the selection of human antibodies from phage display repertoires to a single epitope of an antigen. *Bio/Technology* **12,** 899–903.

54. Barbas, C. F. (1993) Recent advances in phage display. *Curr. Opin. Biotechnol.* **4,** 526–530.

55. Lowman, H. B. and Wells, J. A. (1993) Affinity maturation of human growth hormone by monovalent phage display. *J. Mol. Biol.* **234,** 564–578.

56. Smith, G. P. (1985) Filamentous fusion phage: novel expression vectors that display cloned antigens on the surface of the virion. *Science* **228,** 1315–1317.

57. Scott, J. K. and Smith, G. P. (1990) Searching for peptide ligands with an epitope library. *Science* **249,** 386–390.

58. Devlin, J. J., Panganiban, L. C., and Devlin, P. E. (1990) Random peptide libraries; a source of specific protein binding molecules. *Science* **249,** 386–390.

59. Lane, D. P. and Stephen, C. W. (1993) Epitope mapping using bacteriophage peptide libraries. *Curr. Opin. Immunol.* **5,** 268–271.

60. Matthews, D. J. and Wells, J. A. (1993) Substrate phage, selection of protease substrates by monovalent phage display. *Science* **260,** 1113–1117.

61. Scott, J. K., Loganathan, D., Easley, R. B., Gong, X., and Goldstein, I. J. (1992) A family of concanavalin A-binding peptides from a hexapeptide epitope library. *Proc. Natl. Acad. Sci. USA* **89,** 5398–5402.

62. Oldenburg, K. R., Loganathan, D., Goldstein, I. J., Schultz, P. G., and Gallop, M. A. (1992) Peptide ligands for a sugar-binding protein isolated from a random peptide library. *Proc. Natl. Acad. Sci. USA* **89,** 5393–5397.

63. O'Neil, K. T., Hoess, R. H., Jackson, S. A., Ramachandran, N. S., Mousa, S. A., and DeGrado, W. F. (1992) Identification of novel peptide antagonists for GPIIb/IIa from a conformationally constrained phage peptide library. *Proteins* **14,** 509–515.

64. Barbas, C. F., Bain, J. D., Hoekstra, D. M., and Lerner, R. A. (1992) Semisynthetic combinatorial antibody libraries: a chemical solution to the diversity problem. *Proc. Natl. Acad. Sci. USA* **89,** 4457–4461.

65. Hoogenboom, H. R. and Winter, G. (1992) By-passing immunization: human antibodies from synthetic repertoires of germline VH gene segments rearranged in vitro. *J. Mol. Biol.* **227,** 381–388.

66. Barbas, C. F., Bjorling, E., Chioki, F., Dunlop, N., Cabab, D., Jones, T. M., Zebedee, S. L., Persson, M. A. A., Nara, P. L., Norrby, E., and Burton, D. R. (1992) Recombinant human Fab fragments neutralize human type 1 immunodeficiency virus in vitro. *Proc. Natl. Acad. Sci. USA* **89,** 9339–9343.

67. Nissim, A., Hoogenboom, H. R., Tomlinson, I. M., Flynn, G., Midgley, C., Lane, D., and Winter, G. (1994) Antibody fragments from a 'single pot' phage display library as immunochemical reagents. *EMBO J.* **13,** 692–698.

46

Enzyme-Linked Immunosorbent Assay (ELISA)

John R. Crowther

1. Introduction

Enzyme-Linked Immunosorbent Assays (ELISAs) have had many thousands of applications over the past 25 yr. The technique has been used in all fields of pure and applied aspects of biology, in particular it forms the backbone of diagnostic techniques. The systems used to perform ELISAs make use of antibodies. These are proteins produced in animals in response to antigenic stimuli. Antibodies are specific chemicals that bind to the antigens used for their production thus they can be used to detect the particular antigens if binding can be demonstrated. Conversely, specific antibodies can be measured by the use of defined antigens, and this forms the basis of many assays in diagnostic biology.

This chapter describes methods involved in ELISAs where one of the reagents, usually an antibody, is linked to an enzyme and where one reagent is attached to a solid phase. The systems allow the examination of reactions through the simple addition and incubation of reagents. Bound and free reactants are separated by a simple washing procedure. The end product in an ELISA is the development of color, which can be quantified using a spectrophotometer. These kinds of ELISA are called heterogeneous assays and should be distinguished from homogeneous assays where all reagents are added simultaneously. The latter assays are most suitable for detecting small molecules such as digoxin or gentamicin.

The development of ELISA stemmed from investigations of enzyme-labeled antibodies (1,2), for use in identifying antigens in tissue. The methods of conjugation were exploited to measure serum components in the first "true" ELISAs (3–6).

By far the most exploited ELISAs use plastic microtiter plates in an 8 × 12 well format as the solid phase (7). Such systems benefit from a large selection of specialized commercially available equipment including multichannel pipets for the easy simultaneous dispensing of reagents and multichannel spectrophotometers for rapid data capture. There are many books, manuals, and reviews of ELISA and associated subjects that should be examined for more detailed practical details (8–21).

The key advantages of ELISA over other assays are summarized in **Table 1**.

From: *Molecular Biomethods Handbook*
Edited by: R. Rapley and J. M. Walker © Humana Press Inc., Totowa, NJ

Table 1
Advantages of ELISA

1. Simplicity	a) Reagents added in small volumes.
	b) Separation of bound and free reactants is made by simple washing procedures.
	c) Passive adsorption of proteins to plastic is easy.
2. Reading	a) Colored end-product can be read by eye to assess whether tests have worked (avoiding waiting for results where machine reading essential as in RIA).
	b) Multichannel spectrophotometers quantify results that can be examined statistically.
3. Rapidity	a) Tests can be performed in a few hours.
	b) Spectrophotometric reading of results is rapid (96 wells read in 5 s).
4. Sensitivity	Detection levels of 0.01–1 µg/mL are easily and consistently achievable. These levels are ideal for most diagnostic purposes.
5. Reagent	Commercially available reagents offer great flexibility in ELISA design availability and achievment of specific assays.
6. Adaptability	Different configurations allow different methods to be examined to solve problems. This is useful in research science.
7. Cost	a) Start up costs are low.
	b) Reagent costs are low.
8. Accepability	Fully standardized ELISAs in many fields are now accepted as "gold-standard" assays.
9. Safety	Safe nonmutagenic regents are available. Disposal of waste poses no problems (unlike radioactivity).
10. Availability	ELISAs can be performed anywhere, even in laboratories where facilites are poor.
11. Kits	ELISA kits are widespread and successful. **Table 3** shows the most commonly used enzymes and substrate/chromphore systems used in ELISA and the color changes with relevant stopping agents.

2. What Is ELISA?

Figure 1A illustrates a protein that is adsorbed to a plastic surface. The attachment of proteins and hence the majority of all antigens in nature, to plastic, is the key to most ELISAs performed. This process is passive so that protein solutions, in easy to prepare buffer solutions, can be added to plastic surfaces and will attach after a period of incubation at room temperature. The most commonly used plastic surface is that of small wells in microtiter plates. Such plates contain 96 wells measuring approx 5 mm deep × 8 mm diameter in a 12 × 8 format. The main point here is that there has evolved a whole technology of equipment for rapidly handling materials in association with these plates. After the incubation of antigen excess unbound antigen is washed away by flooding the plastic surface, usually in a buffered solution. The plastic can then be shaken free of excess washing solution and is ready for the addition of a detecting system.

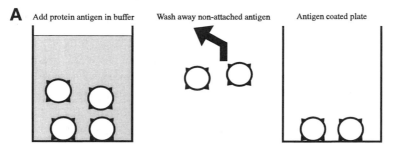

A Add protein antigen in buffer   Wash away non-attached antigen   Antigen coated plate

Antigen is added in buffer. The protein attaches passively to plastic surface of microtitre plates well. After a period of incubation the non-adsorped protein is washed away.

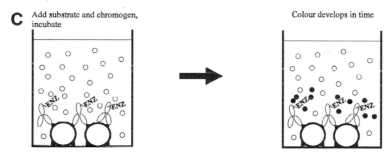

B Add enzyme labelled antibodies in blocking buffer   Wash away non-bound conjugate   Antigen/conjugate complex on solid phase

Antibodies with enzyme co-valently linked (conjugate) is added in a solution containing inert protein and detergent (to prevent non-specific attachment of the antibodies to plastic wells. The antibody binds to the antigen on well surface. After incubation, non-bound antibodies are washed away.

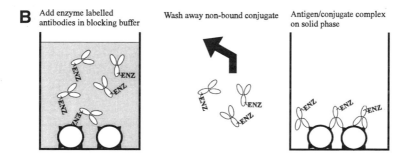

C Add substrate and chromogen, incubate   Colour develops in time

Add substrate and chromogenic dye solution. Substrate interacts with enzyme to affect dye solution to give a colour reaction.

Fig. 1. Illustration of steps in simple direct ELISA.

The simplest application of the ELISA is illustrated in **Fig. 1**. An antibody prepared against the antigen on the plastic is added (**Fig. 1B**). The antibody has been chemically linked to an enzyme, this is usually called a conjugate. The antibody is diluted in a buffer (e.g., phosphate-buffered saline [PBS], pH approx 7.2) containing an excess of protein that has no influence on the possible reaction of the antibody and antigen. A very cheap example is that of approx 5% skimmed-milk powder, Marvel, which is mainly the protein from cows milk, casein. The purpose of this excess protein is to prevent any passive attachment of the conjugate (antibody is protein) to any free sites on the plastic not occupied by the antigen. Such sites will be occupied by the excess milk powder proteins by competition with the low concentration of conjugate. These diluting reagents have been called blocking buffers.

On addition of the conjugate under these conditions the only reaction that occurs is the specific immunological binding of the antigen and antiboby in the conjugate. Thus an antigen-enzyme linked antibody complex is produced, and because the antigen is bound to the plastic, the enzyme is bound also. Such a process requires incubation (15–60 min), which can be at room temperature. After incubation, the plastic surface is washed to remove all unreacted conjugate.

The next stage is illustrated in **Fig. 1C**. Here, a substrate for the enzyme is added in solution with a chromogenic chemical, which is colorless in the absence of enzymic activity on the substrate. Because there is enzyme linked to the antibody that is attached specifically to the antigen on the plastic surface, the substrate is catalyzed, causing a color change. The rates of such color changes are proportional to the amount of enzyme in the complex. Thus, taking the other extreme, if no antigen were attached to the plastic then no antibody would be bound. Therefore no enzyme would be present to catalyze the substrate so that no color change would be observed.The enzymic activity is usually stopped by the addition of a chemical that drastically alters the pH of the reaction or denatures the enzyme e.g., 1 *M* sulfuric acid.

Color can then be assessed by eye or quantified using a multichannel spectrophotometer that is specially designed for use with the microplates and can read a plate (96 samples) in 5 s. Such machines can be interfaced with microcompeters so that a great deal of data can be analyzed in a short time.

There are many systems in ELISA, depending on what initial reagent is attached to the solid phase and what order subsequent reagnts are added. Thus there is great versatility possible for adaption of ELISAs to solve applied and pure problems in science.

The basic principles of ELISA are those be summarized below:

1. Passive attachment of proteins to plastics.
2. Washing away of unattached protein.
3. Addition, at some stage, of a specific antibody linked to an enzyme.
4. Use of competing inert proteins to prevent nonspecific reactions with the plastic.
5. Washing steps to separate reacted (bound) from unreacted (free) reagents.
6. Addition of a specific substrate which changes color on enzymic catalysis or substrate and a colorless chromophore (dye solution) that changes color owing to enzymic catalysis.
7. Incubation steps to allow immunological reactions.
8. Stopping of enzymic catalysis.
9. Reading of the color by spectrophotometer.

Specific details of these stages can be obtained from the references in particular *(8–16)*. The next section illustrates some of the possible variations.

3. Basic Assay Configurations

There are three basic systems used in ELISA: Direct ELISA, Indirect ELISA, and Sandwich ELISA. All these systems can be used to perform competition of inhibition ELISAs. These systems will be described to illustrate the principles involved with the aid of diagrams. The various stages common to ELISAs will then be described in more detail.

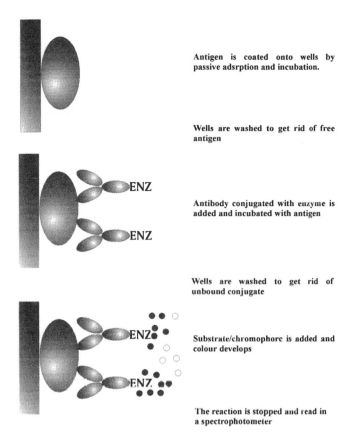

Antigen is coated onto wells by passive adsrption and incubation.

Wells are washed to get rid of free antigen

Antibody conjugated with enzyme is added and incubated with antigen

Wells are washed to get rid of unbound conjugate

Substrate/chromophore is added and colour develops

The reaction is stopped and read in a spectrophotometer

Fig. 2. Direct ELISA. Antigen is attached to the solid phase by passive adsorption. After washing, enzyme-labeled antibodies are added. After an incubation period and washing, a substrate system is added and color allowed to develop.

3.1. Direct ELISA

This is the simplest form of ELISA as shown in **Fig. 2**. Here an antigen is passively attached to a plastic solid phase by a period of incubation. As indicated in **Subheading 2.**, the most useful solid phase is a microtiter plate well. After a simple washing step, antigen is detected by the addition of an antibody that is linked covalently to an enzyme. After incubation and washing the test is developed by the addition of a chromogen/substrate whereby enzymic activity produces a color change. The greater the amount of enzyme in the system, the faster the color develops. Usually color development is read after a defined time or after enzymic activity is stopped by chemical means at a defined time. Color is read in a spectrophotometer.

3.2. Indirect ELISA

Antigen is passively attached to wells by incubation. After washing, antibodies specific for the antigen are incubated with the antigen. Wells are washed and any bound antibodies are detected by the addition of antispecies antibodies covalently linked to an enzyme. Such antibodies are specific for the species in which the first antibody added were produced. After incubation and washing, the test is developed and read as described in **Subheading 3.1.** The scheme is shown in **Fig. 3**.

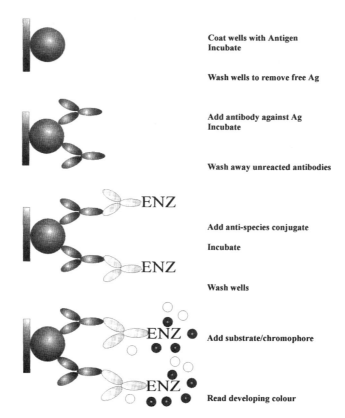

Coat wells with Antigen
Incubate

Wash wells to remove free Ag

Add antibody against Ag
Incubate

Wash away unreacted antibodies

Add anti-species conjugate

Incubate

Wash wells

Add substrate/chromophore

Read developing colour

Fig. 3. Indirect ELISA. Antibodies from a particular species react with antigen attached to the solid phase. Any bound anitbodies are detected by the addition of an antispecies antiserum labeled with enzyme. This is widely used in diagnosis.

3.3. Sandwich ELISA

There are two forms of this ELISA depending on the number of antibodies used. The principle is the same for both whereby instead of adding antigen directly to a solid phase, antibody is added to the solid phase, then acts to capture antigen. These systems are useful where antigens are in a crude form (contaminated with other proteins) or at low concentration. In these cases, the antigen cannot be directly attached to the solid phase at a high enough concentration to allow successful assay based on direct or indirect ELISAs. The sandwich ELISAs depend on antigens having at least two antigenic sites so that at least two antibody populations can bind.

3.3.1. Direct Sandwich ELISA

This is shown in **Fig. 4**. Antibodies are protein in nature and can be passively attached to the solid phase. After washing away excess unbound antibody, antigen is added and is specifically captured. The antigen is then detected by a second enzyme-labeled antibody directed against the antigen. This antibody can be identical to the capture antibody reacting with a repeating antigenic site or an antibody from a different species directed against the same or a different site. Thus a "sandwich" is created. This type of assay is useful where a single-species antiserum is available and where antigen does not attach well to plates.

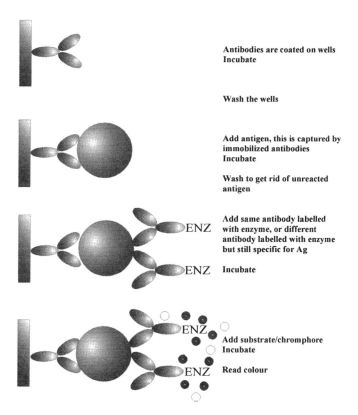

Antibodies are coated on wells
Incubate

Wash the wells

Add antigen, this is captured by
immobilized antibodies
Incubate

Wash to get rid of unreacted
antigen

Add same antibody labelled
with enzyme, or different
antibody labelled with enzyme
but still specific for Ag

Incubate

Add substrate/chromphore
Incubate

Read colour

Fig. 4. Sandwich ELISA-direct. This system exploits antibodies attached to a solid phase to apture antigen. The antigen is then detected using serum specific for the antigen. The detecting antibody is labeled with enzyme. The capture antibody and the detecting antibody can be the same serum or from different animals of the same species or from different species. The antigen must have at least two different antigenic sites.

3.3.2. Indirect Sandwich ELISA

This is similar in principle to the last system but involves three antibodies. **Figure 5** illustrates the scheme. Coating of the solid phase with antibody and capture of antigen are as in **Subheadings 3.1.–3.3.**, however, here the antigen is detected with a second unlabeled antibody. This antibody is in turn detected using an antispecies enzyme-labeled conjugate. It is essential that the antispecies conjugate does not bind to the capture antibody, therefore the species in which the capture antibody is produced must be different. The same considerations about the need for at least two antigenic sites to allow the "sandwich" are relevant. The advantage of this system is that a single antispecies conjugate can be used to evaluate the binding of antibodies from any number of samples. This is not true of the Direct sandwich where each serum tested would have to be labeled with enzyme.

4. Competition/Inhibition ELISAs

The systems described in **Subheadings 3.1.–3.3.** are the basic configurations of ELISA. All of these can be adapted to measure antigens or antibodies using competitive or inhibition conditions. Thus, each the assays described above require pretitration

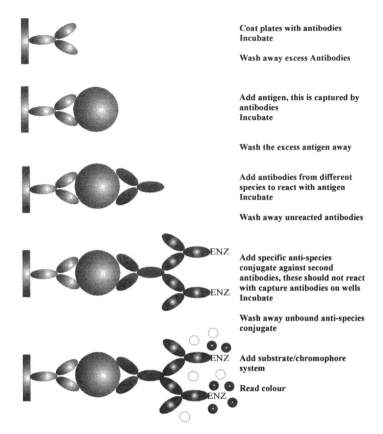

Coat plates with antibodies
Incubate

Wash away excess Antibodies

Add antigen, this is captured by
antibodies
Incubate

Wash the excess antigen away

Add antibodies from different
species to react with antigen
Incubate

Wash away unreacted antibodies

Add specific anti-species
conjugate against second
antibodies, these should not react
with capture antibodies on wells
Incubate

Wash away unbound anti-species
conjugate

Add substrate/chromophore
system

Read colour

Fig. 5. Sandwich ELISA-indirect. The detecting antibody is from a different species to the capture antibody. The antispecies enzyme-labeled antibody binds to the detecting antibody specifically and not to the capture antibody.

of reagents to obtain optimal conditions. These optimal conditions are then challenged either by the addition of antigen or antibody. These will be described.

4.1. Direct ELISA-Antigen Competition

This is shown in **Fig. 6**. The Direct ELISA is optimized whereby a defined amount of antigen coating the plate is bound by an optimal amount of enzyme labeled antibody. Wells coated with the optimal amount of antigen are then set up. This "balanced" situation can then be "'challenged" by the addition of samples that could contain the same (or similar) antigen as that attached to the plate. On addition of the enzyme-labeled conjugate, the test antigen reacts and prevents that antibody binding to the antigen on the solid phase. Thus, the added antigen in the liquid phase and the solid phase antigen, compete for the labeled antibody. The higher the concentration of identical antigen in the test, the greater is the degree of competition. Where the antigen added in the test sample is not the same as the solid-phase antigen, it does not bind to the added conjugate and this can consequently bind without competition to the solid phase antigen.

Pre-titration of labelled antibodies and antigen

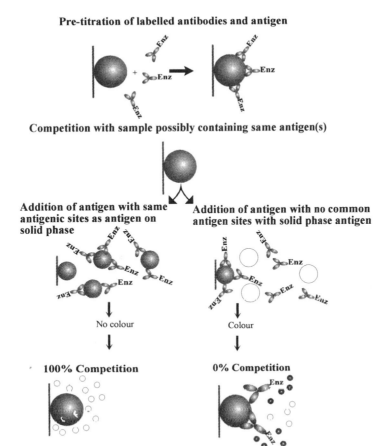

Competition with sample possibly containing same antigen(s)

Addition of antigen with same antigenic sites as antigen on solid phase

No colour

100% Competition

Addition of antigen with no common antigen sites with solid phase antigen

Colour

0% Competition

Fig. 6. Competition ELISA-direct antigen. Reaction of antigen contained in samples with the enzyme-labeled antibody directed against the antigen on the solid phase blocks the label from binding to the solid-phase antigen. If the antigen has no crossreactivity or is absent, then the labeled antibody binds to the solid-phase antigen and a color reaction is observed on developing the test.

4.2. Direct ELISA-Antibody Competition

This is very similar to the assay in **Subheading 4.1.** except that test samples are added containing antibodies possibly directed towards the antigen coated on the solid phase. This is shown in **Fig. 7.** Thus, high concentrations of identical antibody mean that the conjugated pretitrated antibody is inhibited and thus no color reaction is observed (as expected from the pretitration exercise). Such assays are increasing in usefulness with the development of monoclonal antibody (MAb) based tests.

4.3. Indirect ELISA-Antigen Competition

Figure 8 illustrates the principles of this assay. The system relating the antigen, primary antibody and labeled antispecies conjugate is pretitrated. There is inhibition of the binding of primary antibody on addition of test samples containing the same antigen as that coated on the wells. Conversely, where the antigen added does not bind to

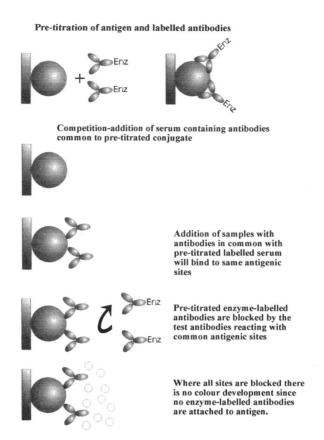

Fig. 7. Competition ELISA-direct antibody. The degree of inhibition by the binding of antibodies in a serum for a pretitrated enzyme-labeled antiserum reaction is determined.

the primary antibody, then no inhibition occurs and on subsequent addition of the conjugate the expected pretitrated level of color is observed.

4.4. Indirect ELISA-Antibody Competition

This is very similar to assay described in **Subheading 2.3.** Here test samples containing antibodies that can bind to the solid phase antigen inhibit the pretitrated primary antibody, as shown in **Fig. 9.** The key problem with this assay is that the test antibody cannot be from the same species as the primary antibody, because this is detected by an antispecies conjugate.

4.5. Sandwich ELISA Direct-Antibody Competition

This is shown in **Fig. 10**. The situation begins to look a little more complex because more reagents are involved. The figure illustrates two methods where a pretitrated Direct sandwich system is competed for by antibody in test samples. The first involves mixing and incubation of the pretitrated antigen with test serum before addition to wells containing the sold-phase antibody. Here, if the antibodies in the test sample bind to the antigen, they stop the antigen binding to the solid-phase antibody. On addition of the pretitrated conjugate, there is no color. The second situation involves the use of

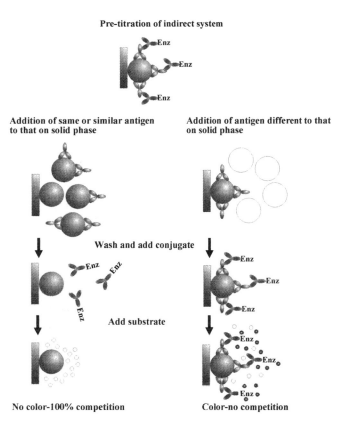

Fig. 8. Competitive ELISA-indirect antigen. The pretitrated indirect ELISA is competed for by antigen. If the antigen shares antigenic determinants with that of the solid-phase antigen, then it binds to the pretitrated antibodies and prevents them binding to the solid-phase antigen. If there is no similarity, then the antibodies are not bound and can react with the solid-phase antigen. Addition of the antispecies enzyme conjugate quantifies the bound antibody.

antigen attached to wells via the capture antibody. After washing the test antibody is added. If this reacts with the captured antigen then it blocks the binding of subsequently added conjugate. In fact, both these forms of assay whereby test antibodies are allowed to bind before the addition of detecting second antibody should be termed inhibition of blocking assays because strictly competition refers to the simultaneous addition of two reagents. This can be illustrated in the second situation where the test antibody and second conjugated antibody could be mixed together before addition to the antigen captured on the wells; this is competitive. Note that because we have an enzyme-labeled detecting serum, then any species serum can be used in the competitive system.

4.6. Sandwich ELISA Indirect-Antibody Competition

Figures 11 and **12** illustrate methods for performing this ELISA. Again we have a more complex situation because five reagents are involved. Basically the Indirect sandwich ELISA is pretitrated. Test antibodies are then added either to the antigen in the

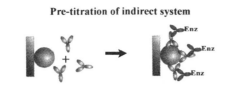

Pre-titration of indirect system

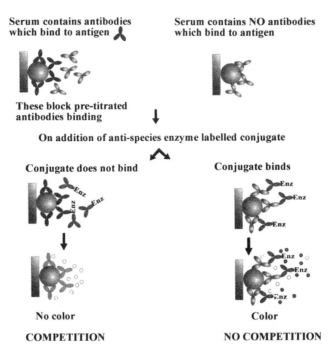

Competition-addition of samples containing antibodies?

Serum contains antibodies
which bind to antigen

Serum contains NO antibodies
which bind to antigen

These block pre-titrated
antibodies binding

On addition of anti-species enzyme labelled conjugate

Conjugate does not bind

Conjugate binds

No color

Color

COMPETITION

NO COMPETITION

Fig. 9. Indirect ELISA-antibody inhibition. The involves the pretitration of an antigen and antiserum in an indirect ELISA. The addition of a serum containing crossreactive antibodies will upset the "balance" of the pretitrated system. Because an antispecies conjugate is used, the species from which the sample of serum is taken cannot be the same as that used for the pretitration (the homologous system).

liquid phase (**Fig. 11**) or to antigen already captured (**Fig. 12**). If test antibodies bind to the antigen in either system, then the subsequent addition of the second antibody and the antispecies conjugate will be negated. Note here that, as with the Indirect ELISA competition, the species from which the test sera came cannot be the same as that used to optimize the assay, i.e., the antispecies conjugate cannot react with the test antibodies.

The assays described are inhibition or blocking assays and can all be further "complicated" with reference to when addition of reagents are made. Thus, in **Fig. 11** (i) the antigen, test antibodies, and detecting antibody could be added together (competition). In **Fig. 11** (ii), the test and detecting antibodies could be added together (competition). Similarly for **Fig. 12**, the test and detecting antibodies could be premixed to offer competitive conditions.

Competition for Ag in liquid phase

Competition by incubation of Ag with test serum followed by addition of conjugated Ab,
OR by incubation of Ag with test serum and labelled serum, simultaneously.

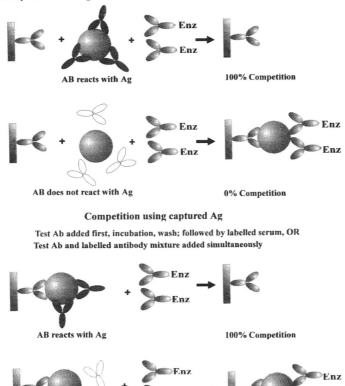

Competition using captured Ag

Test Ab added first, incubation, wash; followed by labelled serum, OR
Test Ab and labelled antibody mixture added simultaneously

4.7. Sandwich ELISAs for Antigen Competition

These have not been illustrated. There is an intrinsic difficulty in that wells are coated
with capture antibody. Addition of competing antigens thus serves to increase the concentration of antigen that can be captured resulting in no competition.

5. Summary of Uses of Various Methods Used in ELISA

This section will consider the reasons how and why different systems need to be
applied, and summarizes the interrelationships of the methods.

An overview of the most commonly used systems is shown in **Fig. 13**. Although not all
applications can be covered here, this will serve to illustrate the versatility of ELISA. Boxes
A, B, C, and D in the figure cover uses of the systems used in noncompetitive ways.

5.1. A: Direct ELISA

The drawback here is that antibodies from each serum have to be labeled. Antispecies
conjugates can be titrated in this method using specific serum proteins from the target

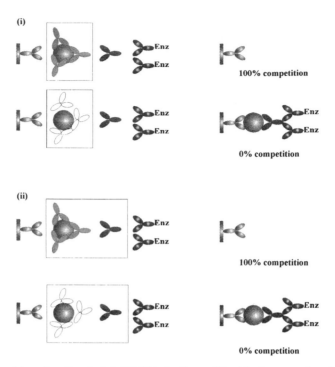

Fig. 11. Competition for Sandwich ELISA-Indirect (liquid-phase antigen). Ag is reacted with competing antibodies followed by addition of second antibody (i) or Ag, competing serum and second antibodies are mixed (ii). Bound second antibody is detected by antispecies against second antibody. This cannot react with species in which test antibody was raised.

species. The Direct ELISA can be used to standardize antispecies conjugates from batch to batch and as a way of estimating the working dilution of conjugate to be used in other ELISA systems.

5.2. B: Indirect ELISA

This is a far more flexible test in that a single antispecies conjugate will detect antibodies. Thus various sera with different specificities can be detected using the same reagent. The Indirect ELISA has ben used widely in diagnosis of diseases through the detection of antibodies. The method depends on the availability of enough specific antigen(s) at a suitable concentration for coating plates. Where antigens are at low concentration or in the presence of other contaminating proteins, the test might be impossible to perform. Thus, antigens have to be relatively pure and be unaffected by adsorption on to the solid phase.

5.3. C: Sandwich ELISA-Direct

The use of antibodies coated to plates as capture reagents is essential where antigens are at low concentration or are contaminated with other proteins, e.g., stool samples. The system also favors presentation of some antigens in a better way than when they are directly coated to plates and also limits changes to conformational epitopes owing to direct coating. The test can be used to detect anti-

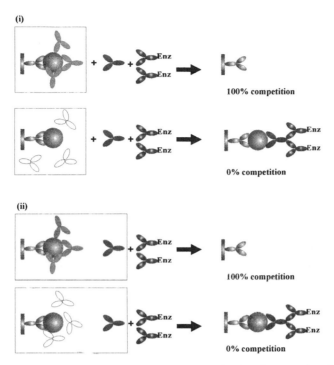

Fig. 12. Competition for Sandwich ELISA-Indirect (solid-phase antigen). Ag is captured first and then either (i) competing antibodies are added, incubated, and then second antibody added, with or without a washing step to remove unbound test antibody and complexes of this antibody and antigen (i) or competing serum and second antibodies are mixed to compete directly (ii). Bound second antibody is detected by antispecies against second antibody. This cannot react with species in which test antibody was raised.

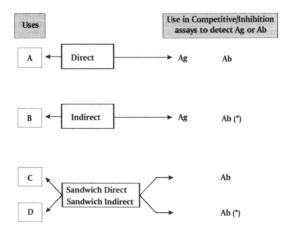

Fig. 13. Relationship of methods used in ELISA. Some uses of assays in noncompetitive ways (A, B, and C) are discussed in text. Ag = antigen; Ab = antibody.

bodies via specifically captured antigens, or antigens through the specificity of the capture antibodies. The Direct Sandwich relies on the conjugation of specific antibodies against the target antigen, which causes some problems with variability where new batches are prepared. The antibodies can be identical to the capture antibodies. The test does rely on there being at least two combining sites (epitopes) on the antigen.

5.4. D: Sandwich ELISA-Indirect

This system offers similar advantages as in **Subheading 3.3.** in that low concentrations of antigen can be specifically captured. This system offers the advantage over the Direct Sandwich ELISA in that a single antispecies conjugate can be used to bind to detecting sera, thus a variety of different species-detecting antibodies can be examined. The antispecies conjugate cannot be allowed to react with the initial-coating antibodies, so that care must be taken to avoid crossreactions and sera must be prepared in at least two species. If only a single species is available, then Fab2 fractions can be prepared from the capture antibodies and a specific anti-Fc conjugate used to develop the test.

5.5. Competition/Inhibition ELISA for all Systems

The advantages examined in the basic ELISAs in **Subheadings 5.1.–5.4.**, all are relevant to the adaptions of the assays in competition/inhibition ELISAs. Thus, the affects of coating on antigen, low concentrations of antigen, contaminating proteins, flexibility of using a single conjugate to detect many sera and orientation of antigens are all pertinent to finding the best system for solving problems. Generally, competitive methods offer advantages over systems where antibodies or antigens are detected directly. The greatest increase in methods has come through the exploitation of MAbs in competition/inhibition systems. **Table 2** briefly defines some elements of ELISA.

6. Uses of ELISA

The purpose of developing ELISAs is to solve problems. These can be divided into pure and applied applications, although the two are interdependent. Thus, a laboratory with a strong research base is essential in providing scientific insight and valuable reagents to allow more routine applications. The methods outlined show the flexibility of the systems. There effective use is up to the ingenuity of scientists. Recent advances in science have given the immunoassayist greater potential for improving the sensitivity and specificity of assays, including ELISA. **Table 3** shows commonly used enzymes and substrate/chromophone systems used in ELISA. In particular the development of MAb technology has given us single chemical reagents (antibodies) of defined specificity that can be standardized in terms of activity as a function of their weight. The development of gene expression systems has also given the possibility of expressing single genes as proteins for use in raising antibodies or acting as pure antigens. This technology goes hand-in-hand with developments in the polymerase chain reaction (PCR) technologies, which enable the very rapid identification of genes and their manipulation. In turn improvements in the fields of rapid sequencing and X-ray

Table 2
Brief Description of Elements Common to ELISAs

Solid phase	This is usually a plastic microtiter plate as well. Specially prepared ELISA plates are commercially available. These have a 8 × 12 well format, the plates can be used with a wide variety of microtiter equipment, such as multichannel pipets, to allow great convience to the rapid manipulation of reagents in small volumes.
Adsorption	This is the process of adding an antigen or antibody, diluted in buffer, so that it attaches passively to the solid phase on incubation. This simple way of immobilization of one of the reactants in ELISA is one of the keys to its success.
Washing	Simply flooding and emptying wells with a buffered solution is enough to separate bound and free reagents. Again, this is a key to the simplicity of the ELISA over methods involving complicated separation methods.
Antigen	These are proteins or carbohydrates, which, when injected into animals or as a result of the disease process, elicit the production of antibodies. Such antibodies usually react specifically with the antigen and therefore can be used to detect that anitgen.
Antibody	Antibodies are produced in response to antigenic stimuli. These are mainly protein in nature. In turn, antibodies are antigenic.
Antispecies antibody	Antibodies obtained when antibodies from one animal are injected into another species. Thus, guinea pig serum injected into a rabbit would elicit a rabbit anti-guinea pig serum.
Enzyme	A substance that can act at low concentration as a catalyst to promote a specific reaction. Several specific enzymes are commonly used in ELISA.
Enzyme conjugate	An enzyme that is attached irreversibily, by chemical means, to a protein, usually an antibody. Thus, an anti-species enzyme conjugate would be guinea pig anti-rabbit conjugated to enzyme.
Substrate	The substrate is the chemical compound on which the enzyme reacts specifically. This reaction is used in some way to produce a signal which is read as a color reaction in ELISA.
Chromophore	This is a chemical that alters color as a result of enzyme interacting with substrate, allowing the ELISA to be quantified.
Stopping	The process of stopping the action of the enzyme and substrate.
Reading	This implies measurement of the color produced in ELISA. This is quantified using special multichannel spectro-photometers reading at the specific wavelength of the color produced. Tests can be read by eye for crude assessment.

Table 3A
Commonly Used Conjugate/Substrate Systems

Enzyme label	Substrate	Dye	Buffer
Horseradish peroxidase	H_2O_2 (0.004%)	OPD (orthophenylene diamine)	Phosphate/citrate, pH 5.0
	H_2O_2 (0.004%)	TMB (tetra methyl benzidine)	Acetate buffer (0.1 M), pH 5.6
	H_2O_2 (0.002%)	ABTS (2,2'-azino di-ethyl thiazoline-sulfonic acid)	Phosphate/citrate, pH 4.2
	H_2O_2 (0.006%)	5AS (5-aminosalicylic acid)	Phosphate (0.2 M), pH 6.8
	H_2O_2 (0.02%)	DAB (diamino benzidine)	Tris or PBS, pH 7.4
Alkaline phosphatase	pnpp	pnpp (paranitrophenyl phosphate)	Diethanolamine (10 mM) plus $MgCl_2$ (0.5 mM), pH 9.5

Table 3B
Common Enzyme Systems, Color Changes, and Stopping Reactions

Enzyme	System	Color change		Reading wavelength		Stopping solution
		Unstopped	Stopped	Unstopped	Stopped	
Horseradish peroxidase	OPD	Light orange	Orange	450 nm	492 nm	1.25 M H_2SO_4
	TMB	Blue	Yellow	650 nm	450 nm	1% SDS
	ABTS	Green	Green	414 nm	414 nm	No stop
	5AS	Black/brown	Black/brown	450 nm	450 nm	No stop
	DAB	Brown	Brown	N/A	N/A	No stop
Alkaline phosphatase	pnpp	Yellow/green	Yellow/green	405 nm	405 nm	2 M sodium carbonate

crystallographic methods has led to a far more intimate understanding of the structure/function relationship of organisms in relation to the immunology of disease. The ELISA fits in rather well in these developments because it is a binding assay requiring defined antibodies and antigens, all of which can be provided. **Table 4** illustrates some applications of ELISA with relevent references.

The ability to develop ELISAs depends on as closer understanding of the immunological/serological/biochemical knowledge of specific biological systems as possible. Such information is already available with reference to literature surveys. Basic skills in immunochemical methods is also a requirement and an excellent manual for this is available *(65)*. References *(66,67)* provide excellent text books on immunology. An invaluable source of commercial immunological reagents is available in **ref. *68***.

Table 4
Applications of ELISA

General	Specific	References
Confirmation of clinical disease.	Titration of specific antibodies	*21, 22, 30, 31, 41, 44, 50, 53, 54, 62–64*
	Single dilution assays.	*35, 41, 53, 54, 62–64*
	Relationship of titer to protection against disease.	*50, 55*
	Kits.	*44, 62, 63*
Analysis of immune response to:	Antibody quantification.	*30, 31, 35, 60, 62, 64*
whole organisms, purified antigens	Antibody class measurement (IgM, IgG, IgA, IgD, IgE).	*25, 33, 34*
extracted from whole organisms, expressed	Antibody subclass measurement (IgG1, IgG2b, IgG3).	*33*
proteins (e.g., vaccinia, baculo, yeast, baceteria), polypeptides, peptides	Antibody G2a, affinity measurement.	*39, 49, 44*
Antigenic comparison	Relative binding antibodies.	*30, 31, 60, 64*
	Affinity differences in binding of antibodies.	*39, 43, 51, 52, 60*
	Measurement of weight of antigens.	*29, 32, 36, 37, 43, 44, 48, 49, 56, 64*
	Examination of treatments to antigen (inactivation for vaccine manufacture, heating, enzyme treatments).	*49*
	Identification of continuous and discontinuous epitopes by examination of binding of polyclonal and MAbs to denatured and nondenatured proteins.	*44, 46, 48, 59*
	Antigenic profiling by MAbs.	*40, 42, 44, 46, 61*
	Comparison of expressed and native problems.	*47, 48, 59*
	Use of MAbs to identify paratopes in polyclonal sera.	*47, 59*
Monoclonal antibodies	Screening during production.	*40, 46*
	Competitive assay-antibody assessment.	*47*
	Comparsion of antigens.	*42, 44, 46, 47, 59, 62*
	Use of MAbs to orientate antigens.	*48*
Novel systems	High-sensitivity assays (Amplified-ELISA)	*58*
	Fluorigenic substrates.	*45*
	Biotin/avidin systems.	*68*

References

1. Avrameas, S. and Uriel, J. (1966) Methode de marquage d'antigenes et d'anticorps avec des enzymes et son application en immunodiffusion. *Comptes Rendus Hendomadaires des Seances de l'Acadamie des Sciences: D: Sciences naturelles (Paris)*, **262**, 2543–2545.

2. Nakane, P. K. and Pierce G. B. (1966) Enzyme-labelled antibodies: preparation and application for the localization of antigens. *J. Histochem. Cytochem.* **14**, 929–931.

3. Avrameas, S. (1969) Coupling of enzymes to proteins with gluteraldehyde. Use of the conjugates for the detection of antgens and antibodies. *Immunochemistry* **6**, 43–52.

4. Avrameas, S. and Guilbert, B. (1971) Dosage enzymo-immunologique de proteines a l'aide d'immunosadorbants et d'antigenes marques aux enzymes. *Comptes Rendus Hendomadaires des Seances de l'Acadamie des Sciences: D: Sciences naturelles (Paris)*, **273**, 2705–2707.

5. Engvall, E. and Perlman, P. (1971) P. Enzyme-linked immunosorbent assay (ELISA). Quantitative assay of immunoglobulin G. *Immunochemistry* **8**, 871–874.

6. Van Weeman, B. K. and Schuurs, A. H. W. M. (1971) Immunoassay using antigen-enzyme conjugates. *FEBS Letters* **15**, 232–236.

7. Voller, A., Bidwell, D. E., Huldt, G., and Engvall, E. (1974) A microplate method of enzyme linked immunosorbent assay and its application to malaria. *Bull. Wld. Hlth. Org.* **51**, 209–213.

8. Burgess, G. W., ed. (1988) *ELISA Technology in Diagnosis and Research.* Graduate School of Tropical Veterinary Science, James Cook University of North Queensland, Townsville, Australia.

9. Collins, W. P. (1985) *Alternative Immunoassays.* Wiley, Chichester, UK.

10. Collins, W. P. (1985) *Complimentary Immunoassays.* Wiley, Chichester, UK.

11. Crowther, J. R. (1995) *ELISA: Theory and Practice.* Humana, Totowa, NJ.

12. Ishikawa, E., Kawia, T., and Miyai, K. (1981) *Enzyme Immunoassay.* Igaku-Shoin, Tokyo, Japan.

13. Kemeny, D. M. and Challacombe, S. J. (1988) *ELISA and Other Solid-Phase Immunoassays. Theoretical and Practical Aspects.* Wiley, Chichester, UK.

14. Maggio, T. (1979) *The Enzyme Immunoassay.* CRC, New York.

15. Ngo, T. T. and Leshoff, H. M. (1985) *Enzyme-Mediated Immunoassay.* Plenum, New York.

16. Voller, A., Bidwell, D. E., and Bartlett, A. (1979) *The Enzyme-Linked Immunosorbent Assay (ELISA).* Dynatech Europe, London, UK.

17. Avrameas, S., Ternynck, T., and Guesdon, J.-L. (1978) Coupling of enzymes to antibodies and antigens. *Scand. J. Immunol.* **8(Suppl. 7),** 7–23.

18. Blake, C. and Gould, B. J. (1984) Use of enzymes in immunoassay techniques. A review. *Analyst.* **109,** 533–542.

19. Guilbault, G. G. (1968) Use of enzymes in analytical chemistry. *Anal. Chem.* **40,** 459.

20. Kemeny, D. M. and Challacombe, S. J. (1986) Advances in ELISA and other solid-phase immunoassays. *Immunol. Today* **7,** 67.

21. Voller A., Bartlett, A., and Bidwell, D. E. (1981) Immunoassays for the 80s. MTP, Lancaster, UK.

22. Kemeny, D. M. (1987) Immunoglobulin and antibody assays, in *Allergy an International Textbook* (Lessoff, M. H., Lee, T. H., and Kemeny, D. M., eds.), Wiley, Chichester, UK, 319.

23. Kemeny, D. M. and Chantler, S. (1988) An introduction to ELISA, in *ELISA and Other Solid Phase Immunoassays. Theoretical and Practical Aspects.* (Kemeny, D. M. and Challacombe, S. J., eds.), Wiley, Chichester, UK.

24. Kemeny D. M. and Challacombe, S. J. (1988) Micrototitre plates and other solid-phase supports, in *ELISA and Other Solid-Phase Immunoassays. Theoretical and Practical Aspects* (Kemeny, D. M. and Challacombe, S. J., eds.), Wiley, Chichester, UK.

25. Kemeny, D. M. (1988) The modified sandwich ELISA (SELISA) for detection of IgE an antibody isotypes, in *ELISA and Other Solid-Phase Immunoassays. Technical and Practical Aspects.* (Kemeny, D. M. and Challacombe, S. J., eds.), Wiley, Chichester, UK.

26. Landon, J. (1977) Enzyme-immunoassay: techniques and uses. *Nature* **268**, 483.

27. Avrameas, S., Nakane, P. K., Papamichail, M., and Pesce, A. J., eds. (1991) 25 Years of immunoenzymatic techniques. *J. Immunol. Methods* International Congress, Athens, Greece, September 9–12.

28. Van Weemen, B. K. (1985) ELISA: highlights of the present state of the art. *J. Virol. Methods* **10**, 371.

29. Yolken, R. H. (1982) Enzyme immunoassays for the detection of infectious antigens in fluids: current limitations and future prospects. *Rev. Infect. Dis.* **4**, 35.

30. Abu Elzein, E. M. E. and Crowther, J. R. (1978) Enzyme-labelled immunosorbent assay technique in FMDV Research. *J. Hyg. Camb.* **80**, 391–399.

31. Abu Elzein, E. M. E. and Crowther, J. R. (1979) Serological comparison of a type SAT2 FMDV isolate from Sudan with other type SAT2 strains. *Bull. Anim. Hlth. Prod. Afr.* **27**, 245–248.

32. Abu Elzein, E. M. E. and Crowther, J. R. (1979) The specific detection of FMDV whole particle antigen (140S) by enzyme labelled immunosorbent assay. *J. Hyg. Camb.* **83**, 127–134.

33. Abu Elzein, E. M. E. and Crowther, J. R. (1981) Detection and quantification of IgM, IgA, IgG1 and IgG2 antibodies against FMDV from bovine sera using an enzyme-linked immunosorbent assay. *J. Hyg. Camb.* **86**, 79–85.

34. Anderson, J., Rowe, L. W., Taylor, W. P., and Crowther, J. R. (1982) An enzyme-linked immunosorbent assay for the detection of IgG, IgA and IgM antibodies to rinderpest virus in experimentally infected cattle. *Res. Vet. Sci.* **32**, 242–247.

35. Armstrong, R. M. A., Crowther, J. R., and Denyer, M. S. (1991) The detection of antibodies against foot-and-mouth disease in filter paper eluates trapping sera or whole blood by ELISA. *J. Immunol. Methods* **34**, 181–192.

36. Crowther, J. R. and Abu Elzein, E. M. E. (1979) Detection and quantification of FMDV by enzyme labelled immunosorbent assay techniques. *J. Gen. Virol.* **42**, 597–602.

37. Crowther, J. R. and Abu Elzein, E. M. E. (1979) Application of the enzyme-linked immunosorbent assay to the detection of FMDVs. *J. Hyg. Camb.* **83**, 513–519.

38. Crowther, J. R. and Abu Elzein, E. M. E. (1980) Detection of antibodies against FMDV using purified Staphylococcus A protein conjugated with alkaline phosphatase. *J. Immunol. Meth.* **34**, 261–267.

39. Abu Elzein, E. M. E. and Crowther, J. R. (1982) Differentiation of FMDV-strains using a competition enzyme-linked immunosorbent assay. *J. Virol. Meth.* **3**, 355–365.

40. Crowther, J. R., McCullough, K. C., Simone E. F. DE., and Brocchi, E. (1984) Monoclonal antibodies against FMDV: applications and potential use. *Rpt. Sess. Res. Gp. Stand. Tech. Comm. Eur. Comm. Cont. FMD,* **7**, pp. 40–45.

41. Crowther, J. R. (1986) Use of enzyme immunoassays in disease diagnosis, with particular reference to rinderpest, In *Nuclear and Related Techniques for Improving Productivity of Indigenous Animals in Harsh Environments.* International Atomic Energy Agency, Vienna, Austria, pp. 197–210.

42. Crowther, J. R. (1986) ELISA, in *FMD Diagnosis and Differentiation and the Use of Monoclonal Antibodies.* Paper pres. 17th Conf. OIE Comm. FMD., Paris, pp. 153–173.

43. Crowther, J. R. (1986) FMDV, in *Methods of Enzymatic Analysis, 3rd ed.* (Bergmeyer, J. and Grassl, M., eds.), VCH Verlagsgesellschaft, Weinheim, Germany, pp. 433–447.

44. Crowther, J. R. (1996) Paper pres. 17th Conf. O. I. E. Comm. FMD., Paris, ELISA in FMD diagnosis and differentiation and the use of monoclonal antibodies. pp. 178–195.

45. Crowther, J. R., and Anguerita, L., and Anderson, J. (1990) Evaluation of the use of chromogenic and fluorigenic substrates in solid phase ELISA. *Biologicals* **18**, 331–336.

46. Crowther, J. R., Rowe, C. A., and Butcher, R. (1993) Characterisation of MAbs against type SAT 2 FMD virus. *Epidemiol., Infect.* **111**, 391–406.

47. Crowther, J. R., Reckziegel, P. O., and Prado, J. A. (1993) The use of MAbs in the molecular typing of animal viruses. *Rev. Sci. Tech. Off. Int. Epiz. 12,* **2**, 369–383.

48. Crowther, J. R. (1995) Quantification of whole virus particles (146S) of foot-and-mouth disease viruses in the presence of virus subunits (12S) using monoclonal antibodies in an ELISA. *Vaccine* **13**, 1064–1075.

49. Curry, S., Abrams, C. C., Fry, E., Crowther, J. R., Belsham, G., Stewart, D., and King, A. Q. (1995) Viral RNA modulates the acid sensitivity of FMDV capsids. *J. Virol.* **69**, 430–438.

50. Denyer, M. S., Crowther, J. R., Wardley, R. C., and Burrows, R. (1984) Development of an enzyme-linked immunosorbent assay (ELISA) for the detection of specific antibodies against H7N7 and an H3N3 equine influenza virus. *J. Hyg. Camb.* **93**, 609–620.

51. Denyer, M. S. and Crowther, J. R. (1986) Use of indirect and competitive ELISAs to compare isolates of equine influenza A virus. *J. Virol. Meth.* **14**, 253–265.

52. Goldberg, M. E. and Djavadi-Ohaniance, L. (1993). Methods for measurement of antibody/antigen affinity based on ELISA and RIA. *Curr. Opin. Immunol.* **5**, 278–281.

53. Hamblin, C. and Crowther, J. R. (1982) Evaluation and use of the enzyme-linked immunosorbent assay in the serology of swine vesicular disease. in *The ELISA: Enzyme-Linked Immunosorbent Assay in Veterinary Research and Diagnosis* (Wardley, R. C. and Crowther, J. R., eds.), Martinus Nijhoff, The Netherlands, pp. 232–241.

54. Hamblin, C. and Crowther, J. R. (1982) A rapid enzyme-linked immunosorbent assay for the serological confirmation of SVD, in *The ELISA: Enzyme-Linked Immunosorbent Assay in Veterinary Research and Diagnosis* (Wardley, R. C. and Crowther, J. R., eds.), Martinus Nijhoff, The Netherlands, pp. 232–241.

55. Hamblin, C., Barnett, I. T. R., and Crowther, J. R. (1986) A new enzyme-linked immunosorbent assay (ELISA) for the detection of antibodies against FMDV. II. *Appl. J. Immun. Meth.* **93**, 123–129.

56. Hamblin, C., Mellor, P., Graham M. S., and Crowther, J. R. (1990) Detection of african horse sickness antibodies by a sandwich competition ELISA. *Epid Infect.* **104**, 303–312.

57. Hamblin, C., Mertens, P. P. C., Mellor, P. S., Burroughs, J. N., and Crowther, J. R. (1991) A serogroup specific enzyme linked immunosorbent assay (ELISA) for the detection and identification of African Horse Sickness Viruses. *J. Vir. Meth.* **31**, 285–292.

58. Johannsson, A., Ellis, D. H., Bates, D. L., Plumb, A. M., and Stanley, C. J. (1986) Enzyme amplification for immunoasays-detection of one hundredth of an attomole. *J. Immunol. Meth.* **87**, 7–11.

59. McCullough, K. C., Crowther, J. R., and Butcher, R. N. (1985) A liquid-phase ELISA and its use in the identification of epitopes on FMDV antigens. *J. Virol. Meth.* **11**, 329–338.

60. Rossiter P. B., Taylor W. P., and Crowther, J. R. (1988) Antigenic variation between three strains of rinderpest virus detected by kinetic neutralisation and competition ELISA using early rabbit antisera. *Vet. Microbiol.* **16**, 195–200.

61. Samuel, A., Knowles, N. J., Samuel, G. D., and Crowther, J. R. (1991) Evaluation of a trapping ELISA for the differentiation of foot-and-mouth disease virus strains using monoclonal antibodies. *Biologicals* **19,** 229–310.

62. Sanchez Vizcaino J. M., Crowther, J. R., and Wardley, R. C. (1981) A collaborative study on the use of the ELISA in the diagnosis of ASF, in *Agriculture. African Swine Fever.* (Wilkinson, P. J., ed.), Luxembourg, pp. 297–325.

63. The sero-monitoring of rinderpest throughout Africa. Phase two. Results for 1993. Proceedings of a Research Coordination Meeting of the FAO/IAEA/SIDA/OAU/IBAR/PARC Coordinated Research Programme organized by the Joint FAO/IAEA Division of Nuclear Techniques in Food and Agriculture, Cairo, Egypt, 7–11, November 1993.

64. Wardley, R. C., Abu Elzein, E. M. E., Crowther, J. R., and Wilkinson, P. J. (1979) A solid-phase enzyme linked immunosorbent assay for the detection of ASFV antigen and antibody. *J. Hyg. Camb.* **83,** 363–369.

65. Harlow, E. and Lane, D. (eds.) (1988) *Antibodies. A Laboratory Manual.* Cold Spring Harbor Laboratory, Cold Spring Harbor, NY.

66. Roitt, I. (1991) *Essential Immunology.* Blackwell Scientific Publications, Oxford, UK.

67. Roitt, I., Brostoff, J., and Male, D. (eds.) (1993) *Immunology.* Mosby.

68. Linscott's Directory of Immunological and Biological Reagents. Linscott's Directory, Santa Rosa, CA.

69. Immunogens, Ag/Ab Purification, Antibodies, Avidin-Bioitn, Protein Modification: PIERCE Immunotechnology Catalog and Handbook, vol 1. Pierce and Warriner, UK. Published yearly.

47

Epitope Mapping

Identification of Antibody-Binding Sites on Protein Antigens

Glenn E. Morris

1. What Is an Epitope?

An epitope can be simply defined as that part of an antigen involved in its recognition by an antibody. In the case of protein antigens, an epitope would consist of a group of individual amino-acid side-chains close together on the protein surface. Epitope mapping, then, becomes the process of locating the epitope, or identifying the individual amino-acids involved. Apart from its intrinsic value for understanding protein structure-function relationships, it also has a practical value in generating antibody probes of defined specificity as research tools and in helping to define the immune response to pathogenic proteins and organisms. Some authors have even extended the epitope concept to the interaction between peptide hormones and their receptors *(1)*; not every immunologist would be happy about this, but it does make the point that in mapping epitopes, we are studying a biological process of fundamental importance, that of protein–protein interaction. Epitope mapping is usually done with monoclonal antibodies (MAbs), though it can be done with polyclonal antisera in a rather less rigorous way, bearing in mind that antisera behave as mixtures of MAbs. Mapping can be done directly by X-ray crystallography of antibody–antigen complexes, but it can also be done by changing individual amino-acids, by using antigen fragments and synthetic peptides or by competition methods in which two or more antibodies compete for the same, or adjacent, epitopes. The term "epitope mapping" has also been used to describe the attempt to determine all the major sites on a protein surface that can elicit an antibody response, at the end of which one might claim to have produced an "epitope map" of the protein immunogen *(2)*. This information might be very useful, for example, to someone wishing to produce antiviral vaccines. However, there is a limit to how far one can go down this road, because the map obtained may be influenced by how MAbs are selected and by the mapping method used. Furthermore, the more strictly correct definition of epitope mapping is based on antigenicity (the ability to recognize a specific antibody), whereas the latter definition is based on immunogenicity (the ability to

From: *Molecular Biomethods Handbook*
Edited by: R. Rapley and J. M. Walker © Humana Press Inc., Totowa, NJ

produce antibodies in a given animal species) and thus depends on the immune system of the recipient animal. Some authors prefer to use the terms "antigenic determinant" and "immunogenic determinant" to make this distinction clear *(3)*. For antigenic determinants, it is possible to think of epitope mapping as a "simple" biochemical problem of finding out how one well-defined protein (a MAb) binds to another (the antigen). Not all antigenic determinants are also immunogenic determinants, however, and Berzofsky *(3)* quotes the example of chicken lysozyme, which is not immunogenic in certain mouse strains although Abs raised against other lysozymes will bind to it. Nevertheless, immunogenicity and antigenicity, although not identical, are sufficiently closely related to make it possible to infer from antibody specificity which regions of a protein are more likely to be immunogenic. It has recently become possible to map T-cell epitopes and this is usually done at present by measuring the ability of peptide fragments of the protein to stimulate cell division of T lymphocytes *(4)*. This approach is clearly based on immunogenicity and the T-cell epitopes identified will depend on the species, strain or individual supplying the T lymphocytes.

B-lymphocytes display immunoglobulin molecules on their surface and are stimulated to divide when these interact with a suitable antigen. They then undergo somatic mutation in a germinal center of the spleen to refine antibody diversity further and there tends to be selection in favor of B-cells, producing higher affinity antibodies *(5)*. The slightly different antibody molecules produced by somatic mutation will generally recognize the same region of protein but with a different affinity or a different tolerance of amino-acid substitutions. These fine specificities can hardly be regarded as defining different epitopes, though it is difficult to decide where exactly to draw the line. A similar problem exists in deciding what point the distinction between two overlapping epitopes should cease to exist. In contrast to the direct stimulation of B-cells by immunogen, protein immunogens have to be processed by antigen-presenting cells (APCs) in order to stimulate T-cells. APCs digest proteins and display short peptides on their surface in association with products of the major histocompatibility complex (MHC); this displayed complex then stimulates T-cells to divide by interacting with specific T-cell receptors (TCRs). Other MHC gene products are involved in proteolytic processing and in transport of peptide fragments to the cell surface *(6)*. At least, T-cell epitopes are simpler that B-cell epitopes in one respect; they can apparently be treated as simple amino-acid sequences without the problems of protein structure and conformation that pervade all aspects of B-cell epitope mapping. The interested reader is referred to reviews of MHC-peptide interactions *(7)* and methods for T-cell epitope mapping *(4)* and prediction *(8)*, because we shall now be leaving immunology behind, to some extent, and be concentrating on the essentially biochemical problem of antibody-antigen interactions; henceforth, epitope means B-cell epitope.

2. Conformational or Linear Epitopes? Structural or Functional Epitopes?

It is essential to distinguish between conformational ("discontinuous," "assembled") epitopes, in which amino-acids far apart in the protein sequence are brought together by protein folding, and linear ("continuous," "sequential") epitopes, which can often be mimicked by simple peptide sequences. Parts of conformational epitopes can sometimes be mimicked by peptides and the term "mimotope" has been coined to describe these peptides. On the other hand, the view that most peptide sequences can produce Abs that recognize native proteins *(3)* is now disputed *(9)*.

Most native proteins are formed of highly convoluted peptide chains, so that residues that lie close together on the protein surface are often far apart in the amino-acid sequence *(10)*. Consequently, most epitopes on native, globular proteins are conformation-dependent or "assembled" and they disappear if the protein is denatured or fragmented. Sometimes, by accident or design, antibodies are produced against "local" (linear, sequential) epitopes that survive denaturation, though such antibodies usually fail to recognize the native protein. Conversely, most antibody molecules in polyclonal antisera raised against native proteins do not recognize unfolded antigens or short peptides *(9)*. There is something of a culture gap between crystallographers who tend to study assembled epitopes exclusively and people who use MAbs as research tools, for whom assembled epitopes can be something of a nuisance if the MAbs do not work on Western blots. Some authors have preferred to emphasize the distinction between epitopes on native proteins and those on denatured proteins by using such terms as "cryptotopes" or "unfoldons" for the latter *(11)*, but they are not commonly used at the present time. The simplest way to find out whether an epitope is conformational is by Western blotting after sodium oecyl sulfate-polyacrylamide gel electrophoresis (SDS-PAGE). If the antibody still binds after the protein has been boiled in SDS and 2-mercaptoethanol, the epitope is unlikely to be highly conformational. It must be remembered, however, that few proteins are completely denatured on Western blots and some epitopes identified by Western blotting may still have a conformational element. Similarly, many, though not all, conformational epitopes are also destroyed when the antigen binds to plastic in an ELISA test.

3. Epitope Mapping Methods

A skeleton outline of the approaches that follow is given in **Table 1**.

3.1. Structural Approach

At their most elaborate, epitope mapping techniques can provide detailed information on the amino-acid residues in a protein antigen that are in direct contact with the antibody binding site ("contact residues"). X-ray crystallography of antibody–antigen complexes can identify contact residues directly and unequivocally *(12)*, though, not surprisingly in view of the effort required, this method is not in routine use. The method is further restricted by the necessity of obtaining good crystals of Ab-Ag complexes and it has usually been to highly-conformational epitopes on the surface of soluble proteins. Van Regenmortel has made the important distinction between "structural" epitopes as defined by X-ray crystallography and related techniques and "functional" epitopes defined by amino-acid residues which are important for binding and cannot be replaced *(9)*. The number of contact residues revealed by X-ray crystallography is usually about 15–20, whereas "functional" mapping methods that depend on Ab binding changes generally find about 4–8 important residues. This difference may be more apparent than real, however, partly because there does not seem to be complete agreement on how close amino-acids in the Ab and Ag must be to constitute a "contact" and also because some residues in the Ag could be "in contact" with the Ab without contributing significantly to the binding. It is equally true, however, that "functional" map-

Table 1
Approaches to Epitope Mapping

Structural
   X-ray diffraction
   Nuclear magnetic resonance
   Electron microscopy
Functional
   Competition
      ELISA
      Ouchterlony plates
      Biosensors
   Antigen modification
      Chemical modification of side-chains
      Protection by Ab from chemical modification
      Site-directed mutagenesis
      PCR-random mutagenesis
      Homolog scanning
      Viral-escape mutants
      Natural variants and isoforms
   Antigen fragmentation
      Chemical fragmentation
      Proteolytic digestion
      Protection by Ab from proteolytic digestion
      Recombinant libraries of random cDNA fragments
      Recombinant subfragments produced by:
         PCR
         Exonuclease III
         Transposon mutagenesis
         Early-translation termination
   Synthetic peptides
      PEPSCAN peptide arrays
      SPOTS synthesis on membranes
      Combinatorial libraries
      Phage-displayed peptide libraries

ping methods may give an incomplete picture; fragmentation methods, for example, may detect only the most important continuous part of a discontinuous epitope and additional amino-acids may contribute significantly to the binding affinity in the intact antigen. Nuclear magnetic resonance (NMR) of Ab-Ag complexes is another "structural" approach that is in many ways complementary to the crystallographic method *(13)*. NMR methods are performed in solution and thus avoid the need for crystals but they are limited by the size of the antigen that can be studied and are usually applied to peptide antigens. NMR is therefore unsuitable for highly assembled epitopes. At the other extreme, electron microscopy of Ab-Ag complexes has also been used to identify Ab-binding sites directly, but this is usually applied to very large antigens, such as whole viruses, and obviously has a rather low resolution *(14)*.

3.2. Functional Approach

The remaining epitope mapping methods are essentially "functional" in approach, because they involve introducing some additional variable into the basic Ab–Ag interaction and then testing for Ab function (i.e., does it still bind?). They can be usefully divided into four groups:

1. Competition methods
2. Ag-modification methods
3. Ag-fragmentation methods
4. The use of synthetic peptides or peptide libraries.

3.2.1. Competition Methods

Competition methods can be very useful when a relatively low degree of mapping resolution is adequate. You may want to establish, for example, that two MAbs recognize different, nonoverlapping epitopes for a two-site immunoassay, or to find MAbs against several different epitopes on the same Ag so that results from cross reactions with other proteins can be rigorously excluded. The principle behind competition methods is to determine whether two different MAbs can bind to a monovalent Ag at the same time (in which case they must recognize different epitopes) or whether they compete with each other for Ag binding. The traditional approaches to competition mapping involve labeling either Ab or Ag with enzymes or radioactivity and immobilizing the Ag (or one of the competing Abs) on a solid support, such as microtiter plates for ELISA or Sepharose beads *(15,16)*. Abs against the same epitope (or one very close) will clearly displace the labeled Ab from immobilized Ag. An even simpler method based on this principle uses Ouchterlony gel-diffusion plates *(17)*, because single MAbs, or mixtures of MAbs that recognize the same epitope, are unable to form precipitin lines. At a more sophisticated and more expensive level, biosensors that follow Ab-binding in real time can be used to determine directly whether two or more unlabeled MAbs will bind to the same unlabeled Ag *(18)*.

3.2.2. Antigen Modification Methods

Chemical modification of amino-acid side-chains is a method that is perhaps less widely used today than previously. In principle, addition of modifying groups specifically to amino acids, such as lysine, should prevent antibody binding to epitopes that contain lysine residues and such an approach should be particularly useful for conformational epitopes that are otherwise difficult to map with simple techniques. Unfortunately, such epitopes are also the most sensitive to indirect disruption by chemicals that cause even small conformational changes and great care is needed to avoid false positives. If the Ag can be expressed from recombinant cDNA and the approximate position of the epitope is known, specific mutations can be introduced by site-directed mutagenesis methods *(19)* (*see* Chapter 29). Alternatively, random mutations can be introduced into part of the antigen by PCR, followed by screening to detect epitope-negative mutants *(20)*. An elegant method for conformational epitopes, homolog scanning *(21)*, requires two forms of the Ag (e.g., from different species) to be expressible from recombinant DNA as native proteins, one of them reactive with the Ab and the other not. Functional chimaeric proteins can then be constructed by genetic engineering and regions responsible for Ab binding identified. Compared with random muta-tion methods, this approach is less likely to disrupt the native conformation because protein function is retained. The "escape mutant" approach for viral-surface epitopes that are recognized by neutralizing antibodies involves selection and sequencing of spontaneous mutants whose infectivity is no longer blocked by the antibody *(22)*. Natu-rally occurring species or isoform differences in amino-acid sequence can also provide very useful information on epitope location, because Abs may or may not crossreact across species or isoforms *(23)*. Protection from chemical modification, as described by Bosshard and coworkers *(24)*, should be more reliable than direct modification because the side-chains in the epitope itself are not altered (protected by Ab) and the modifying groups on the unprotected side-chains are not large (e.g., radioactive acetyl groups). Labeling of individual amino-acids is compared in the presence and absence of the protecting Ab. Protection from proteolytic digestion, also known as "protein footprinting" *(25)*, is similar in principle; antigens are exposed to proteases in the pres-ence or absence of antibody (which is fairly protease-resistant) and differences in digestion are detected by gel electrophoresis. For native proteins, that are often resis-tant to proteases, it does depends on the epitope containing a protease-sensitive site, but assembled epitopes are often found on surface loops which are more likely to be accessible to protease. If the Ab–Ag interaction will survive extensive proteolysis with loss of structure, the Ag fragments remaining attached to the Ab can be identified by mass spectrometry *(26)*; this is really a fragmentation approach rather than a protection or modi-fication method.

3.2.3. Antigen-Fragmentation Methods

A simpler fragmentation approach for epitopes that survive denaturation is partial protease digestion of the Ag alone, followed either by Western blotting for larger frag-ments or by HPLC *(27)*. The fragments that bind Ab can be identified by N-termi-nal microsequencing or by mass spectrometry. Overlapping fragments, produced by different proteases, help to narrow down the epitope location. If the antigen is a recom-binant protein, it can be expressed with affinity tags at each end to enable separate affinity purification of fragments after digestion; the epitope can be localized very sim-ply from the overlap between the shortest N-terminal and the shortest C-terminal frag-ments that bind antibody *(28)*. Chemical fragmentation is an alternative to proteolysis and has the advantage that cleavage sites are less frequent (e.g., for Cys, Trp, and Met residues) so that fragments can often be identified from their size alone *(29,30)*; for this reason, Ag purity is less important than for proteolytic fragmentation. Conditions for chemical cleavage, however, are usually strongly denaturing, so the method is not use-ful for assembled epitopes.

Additional methods of generating and identifying antigen fragments are possible if the Ag can be expressed from recombinant cDNA. Random internal digestion of cDNA with DNaseI, followed by cloning and expression of the cDNA fragments to create "epitope libraries," is a popular way of generating overlapping antigenic fragments *(31,32)*. If a MAb recognizes several different fragments, then the epitope must lie within the region of overlap. Epitope expression from the random colonies is screened using the antibody under study and the precise Ag fragment expressed can be identified by DNA sequencing. The power of this approach can be increased by incorporating phage-display methodology in which the antigen fragments are displayed on the surface of filamentous phage (*see* Chapter 46). This has the important advantage that Ab-positive clones can be obtained by selection rather than screening *(33)*. Another approach is to clone specific, pre-determined (rather than random) fragments that have been generated either by using existing restriction enzyme sites in the cDNA or, more flexibly, by using PCR products that have restriction sites in the primers *(34,35)*. The latter approach is especially useful if you want to know whether an epitope is in a specific domain of the antigen or whether it is encoded by a specific exon in the gene, because other methods may give ambiguous answers to these questions. For PCR products, the necessity to clone may be avoided altogether by including a promoter in the forward primer and transcribing/translating the PCR product in vitro *(36)*. Another major advantage of the PCR approach is that it is not always necessary to have your full-length antigen already cloned. Provided the cDNA sequence is known, RT-PCR (reverse transcriptase-PCR) can be used to clone PCR products directly from mRNA or even total RNA *(35)*. Several methods exist for random shortening of the antigens produced from plasmid vectors. Transposon mutagenesis involves the random insertion of stop codons into plasmid DNA using a bacterial transposon *(37)*. Unlike previous methods, extensive DNA manipulation is not required and the site of introduction of the stop codon can be identified precisely by DNA sequencing. Another method takes advantage of the spontaneous early termination of translation of mRNA, which occurs in in vitro systems *(38)*. When removal of amino-acids abolishes antibody binding, those amino-acids may be contact residues, but they may alternatively be needed to maintain the conformation of the real contact residues in the remaining fragment, so care is needed in the interpretation of any results involving loss of antibody binding. A positive binding result, with a synthetic peptide for example, may be needed to confirm the localization. Exonuclease III digests double stranded DNA nucleotide by nucleotide, but not at 3'-overhangs; because 3' or 5' overhangs can be introduced using restriction enzymes, this method can be used to remove nucleotides progressively from either end *(39)*. This enables production of overlapping recombinant protein fragments that are positive for antibody binding, an approach that determines epitope boundaries reliably.

3.2.4. Peptide Methods

Synthetic peptides have revolutionized our understanding of epitopes to the same extent as X-ray crystallography, though ironically the two approaches are virtually mutually exclusive, because peptides are used for sequential epitopes. In the PEPSCAN method, overlapping peptides (e.g., hexamers) covering the complete Ag sequence are synthesized on pins for repeated screening with different MAbs *(40)*.

Because the synthesis can be done automatically, this popular approach requires very little work by the end-user. The alternative SPOTS technique performs the multiple-peptide synthesis on a cellulose-membrane support *(41)*. Peptides have also been syntheszed on micro-arrays for subsequent detection of antibody binding by fluorescein-labeled second antibody and immunofluorescence microscopy *(42)*. An alternative approach to the synthesis of peptides based on the Ag sequence is the use of combinatorial libraries of random peptide sequences in solution *(43)*. The advent of peptide libraries displayed on the surface of phage took this approach a step further by enabling selection of displayed peptides, as opposed to screening *(44,45)* (*see* Chapter 46). In this case, random oligonucleotides are cloned into an appropriate part of a phag surface protein and the peptide sequence displayed is identified after selection by sequencing the phage DNA. Selection of random peptides is unique in producing a range of sequences that are related, but not identical, to the Ag sequence; this enables inferences to be made about which amino-acids in the epitope are most important for Ab binding. Recently, a method has been developed for displaying peptide libraries directly on the surface of *E. coli* in the major flagellum component, flagellin *(46)* and this may facilitate screening and amplification steps. Another recent development displays random-peptide libraries on polyribosomes and the selected mRNA containing the peptide-encoding sequence is amplified by RT-PCR for reselection or sequencing *(47)*. An advantage shared by all peptide methods is that antigen is not required and this may be important for "rare" Ags that are difficult to purify. Full experimental details of all these epitope mapping methods, together with background and illustrations of their applications, can be found elsewhere *(48)*.

4. Applications and Epitope Prediction

Epitope mapping is to some degree an end in itself, insofar as it provides fundamental information on the way that proteins recognize each other and recognize ligands in general. In other words, it helps us to understand the protein structure-function relationships that underlie all biological processes. Epitope-mapping studies can also suggest regions of viral proteins that are likely to be immunogenic and thus help in the design of potential vaccines. Antibodies that neutralize viral infectivity are of particular interest *(22,49–51)*. Epitope mapping of antibodies present in autoimmune disease may throw light on the cause of these diseases, which are often owing to crossreacting antibodies elicited by unrelated proteins or microorganisms *(52–56)*. Mapping of antibodies that inhibit protein function (e.g., enzyme activity) can be used to determine which parts of the protein are involved in that function *(57–61)*. Similar use can be made of antibodies that recognize more than one state of the antigen (e.g., native and partly unfolded *[62]*, free or complexed with other proteins *[63]*, and so forth). Antibodies with known binding sites can also be used to determine the topology of transmembrane proteins by immunoelectron microscopy *(64)*, the domain structure of proteins *(65)*, and the orientation of proteins in relation to intracellular structures *(66)*, or to detect alternative gene products produced by genetic deletion *(35)* or alternative RNA splicing *(67)*.

Finally, this chapter has dealt with experimental epitope mapping methods only, though many attempts have been made to predict epitopes from the amino-acid sequences of antigens *(68)*. Epitopes show an obvious correlation with antibody accessibility (i.e., they have to be on the surface of the antigen) and possibly with local mobility of the peptide chain, if the Ab–Ag interaction is of the "induced-fit" variety *(9)*. It is also relatively easy to sequence the variable regions of MAb H and L chains, including the hypervariable regions that recognize the antigen, by performing RT-PCR on hybridoma-cell MRNA *(45)*. Such sequences can be used to create 3-D models of the antigen-combining site, or "paratope" on the antibody molecule *(69)*.

References

1. Wells, J. A. (1995) Structural and functional epitopes in the growth hormone receptor complex. *Biotechnology* **13**, 647–651.
2. Atassi, M. Z. (1984) Antigenic structure of proteins. *Eur. J. Biochem.* **145**, 1–20.
3. Berzoksky, J. A. (1985) Intrinsic and extrinsic factors in protein antigenic structure. *Science* **219**, 932–940.
4. Reece, J. C., Geysen, H. M., and Rodda, S. J. (1993) Mapping the major human T helper epitopes of tetanus toxin. The emerging picture. *J. Immunol.* **151**, 6175–6184.
5. Clark, E. A. and Ledbetter, J. A. (1994) How B and T cells talk to each other. *Nature* **367**, 425–428.
6. Howard, J. C. (1993) Restrictions on the use of antigenic peptides by the immune system. *Proc. Natl. Acad. Sci. USA* **90**, 3777–3779.
7. Stern, L. J. and Wiley, D. C. (1994) Antigenic peptide binding by class I and class II histocompatibility proteins. *Structure* **2**, 245–251.
8. Rammensee, H.-G. (1995) Chemistry of peptides associated with MHC class I and class II molecules. *Curr. Opin. Immunol.* **7**, 85–96.
9. van Regenmortel, M. H. V. (1989) Structural and functional approaches to the study of protein antigenicity. *Immunol. Today* **10**, 266–272.
10. Barlow, D. J., Edwards, M. S., and Thornton, J. M. (1986) Continuous and discontinuous protein antigenic determinants. *Nature* **322**, 747–748.
11. Laver, W. G., Air, G. M., Webster, R. G., and Smith-Gill, S. J. (1990) Epitopes on protein antigens: misconceptions and realities. *Cell* **61**, 553–556.
12. Amit, P., Mariuzza, R., Phillips, S., and Poljak, R. (1986) Three-dimensional structure of an antigen-antibody complex. *Science* **233**, 747–753.
13. Zvi, A., Kustanovich, I., Feigelson, D., Levy, R., Eisenstein, M., Matsushita, S., Richalet-Secordel, P., van Regenmortel, M. H. V., and Anglister, J. (1995) NMR mapping of the antigenic determinant recognised by an anti-gp120, human immunodeficiency virus neutralizing antibody. *Eur. J. Biochem.* **229**, 178–187.
14. Dore, I., Weiss, E., Altschuh, D., and van Regenmortel, M. H. V. (1988) Visualization by electron microscopy of the location of tobacco mosaic virus epitopes reacting with monoclonal antibodies. *Virology* **162**, 279–289.
15. Tzartos, S. J., Rand, D. E., Einarson, B. L., and Lindstrom, J. M. (1981) Mapping of surface structures of Electrophorus acetylcholine receptor using monoclonal antibodies. *J. Biol. Chem.* **256**, 8635–8645.
16. Le Thiet Thanh, Nguyen thi Man, Buu Mat, Phan Ngoc Tran, Nguyen thi Vinh Ha, and Morris, G. E. (1991) Structural relationships between hepatitis B surface antigen in human plasma and dimers of recombinant vaccine: a monoclonal antibody study. *Virus Res.* **21**, 141–154.
17. Molinaro, G. A. and Eby, W. C. (1984) One antigen may form two precipitin lines and two spurs when tested with two monoclonal antibodies by gel diffusion assays. *Mol. Immunol.* **21**, 181–184.
18. Johne, B., Gadnell, M., and Hansen, K. (1993) Epitope mapping and binding kinetics of monoclonal antibodies studied by real time biospecific interaction analysis using surface plasmon resonance. *J. Immunol. Methods* **160**, 191–198.
19. Alexander, H., Alexander, S., Getzoff, E. D., Tainer, J. A., Geysen, H. M., and Lerner, R. A. (1992) Altering the antigenicity of proteins. *Proc. Natl. Acad. Sci. USA* **89**, 3352–3356.
20. Ikeda, M., Hamano, K., and Shibata, T. (1992) Epitope mapping of anti-recA protein IgGs by region specified polymerase chain reaction mutagenesis. *J. Biol. Chem.* **267**, 6291–6296.

21. Wang, L. F., Hertzog, P. J., Galanis, M., Overall, M. L., Waine, G. J., and Linnane, A. W. (1994) Structure-function analysis of human IFN-alpha—mapping of a conformational epitope by homologue scanning. *J. Immunol.* **152,** 705–715.

22. Ping, L. H. and Lemon, S. M. (1992) Antigenic structure of human hepatitis-A virus defined by analysis of escape mutants selected against murine monoclonal antibodies. *J. Virol.* **66,** 2208–2216.

23. Nguyen thi Man, Cartwright, A. J., Osborne, M., and Morris, G. E. (1991) Structural changes in the C-terminal region of human brain creatine kinase studied with monoclonal antibodies. *Biochim. Biophys. Acta* **1076,** 245–251.

24. Burnens, A., Demotz, S., Corradin, G., Binz, H., and Bosshard, H. R. (1987) Epitope mapping by differential chemical modification of free and antibody-bound antigen. *Science* **235,** 780–783.

25. Jemmerson, R. and Paterson, Y. (1986) Mapping antigenic sites on a protein antigen by the proteolysis of antigen-antibody complexes. *Science* **232,** 1001–1004.

26. Zhao, Y. and Chait, B. T. (1995) Protein epitope mapping by mass spectrometry. *Anal. Chem.* **66,** 3723–3726.

27. Mazzoni, M. R., Malinski, J. A., and Hamm, H. E. (1991) Structural analysis of rod GTP-binding protein, Gt. Limited proteolytic digestion pattern of Gt with four proteases defines monoclonal antibody epitope. *J. Biol. Chem.* **266,** 14,072–14,081.

28. Ellgaard, L., Holtet, T. L., Moestrup, S. K., Etzerodt, M., and Thogersen, H. C. (1995) Nested sets of protein fragments and their use in epitope mapping—characterization of the epitope for the S4D5 monoclonal antibody binding to receptor-associated protein. *J. Immunol. Meth.* **180,** 53–61.

29. Morris, G. E. (1989) Monoclonal antibody studies of creatine kinase. The ART epitope: evidence for an intermediate in protein folding. *Biochem. J.* **257,** 461–469.

30. Morris, G. E. and Nguyen thi Man. (1992) Changes at the N-terminus of human brain creatine kinase during a transition between inactive folding intermediate and active enzyme. *Biochim. Biophys. Acta* **1120,** 233–238.

31. Stanley, K. K. (1988) Epitope mapping using pEX. *Meth. Mol. Biol.* **4,** 351–361.

32. Nguyen thi Man and Morris, G. E. (1993) Use of epitope libraries to identify exon-specific monoclonal antibodies for characterization of altered dystrophins in muscular dystrophy. *Amer. J. Hum. Genet.* **52,** 1057–1066.

33. Wang, L. F., Du Plessis, D. H., White, J. R., Hyatt, A. R., and Eaton, B. T. (1995) Use of a gene-targeted phage display random epitope library to map an antigenic determinant on the bluetongue virus outer capsid protein VP5. *J. Immunol. Methods* **178,** 1–12.

34. Lenstra, J. A., Kusters, J. G., and van der Zeijst, B. A. M. (1990) Mapping of viral epitopes with procaryotic expression systems (review). *Arch. Virol.* **110,** 1–24.

35. Thanh, L. T., Nguyen thi Man, Hori, S., Sewry, C. A., Dubowitz V., and Morris, G. E. (1995) Characterization of genetic deletions in Becker Muscular Dystrophy using monoclonal antibodies against a deletion-prone region of dystrophin. *Amer. J. Med. Genet.* **58,** 177–186.

36. Burch, H. B., Nagy, E. V., Kain, K. C., Lanar, D. E., Carr, F. E., Wartofsky, L, and Burman, K. D. (1993) Expression polymerase chain reaction for the in vitro synthesis and epitope mapping of autoantigen. Application to the human thyrotropin receptor. *J. Immunol. Methods* **158,** 123–130.

37. Sedgwick, S. G., Nguyen thi Man, Ellis, J. M., Crowne, H., and Morris, G. E. (1991) Rapid mapping by transposon mutagenesis of epitopes on the muscular dystrophy protein, dystrophin. *Nucleic Acids Res.* **19,** 5889–5894.

38. Friguet, B., Fedorov, A. N., and Djavadi-Ohaniance, L. (1993) In vitro gene expression for the localization of antigenic determinants—application to the *E. Coli* tryptophan synthase beta2 subunit. *J. Immunol. Methods* **158,** 243–249.

39. Gross, C. H. and Rohrmann, G. F. (1990) Mapping unprocessed epitopes using deletion mutagenesis of gene fusions. *Biotechniques* **8**, 196–202.

40. Geysen, H. M., Meleon, R. H., and Barteling, S. J. (1984) Use of peptide synthesis to probe viral antigens for epitopes to a resolution of a single amino-acid. *Proc. Natl. Acad. Sci. USA* **81**, 3998–4002.

41. Frank, R., Kiess, M., Lahmann, H., Behn, C. H., and Gausepohl, H. (1995) Combinatorial synthesis on membrane supports by the SPOT technique, in *Peptides 1994* (Maia, H. L. S., ed.), ESCOM, Leiden, pp. 479–480.

42. Holmes, C. P., Adams, C. L., Kochersperger, L. M., Mortensen, R. B., and Aldwin, L. A. (1995) The use of light-directed combinatorial peptide synthesis in epitope mapping. *Biopolymers* **37**, 199–211.

43. Houghten, R. A., Pinilla, C., Blondelle, S. E., Appel, J. R., Dooley, C. T., and Cuervo, J. H. (1991) Generation and use of synthetic peptide combinatorial libraries for basic research and drug discovery. *Nature* **354**, 84–86.

44. Scott, J. K. and Smith, G. P. (1990) Searching for peptide ligands with an epitope library. *Science* **249**, 386–390.

45. Morris, G. E., Nguyen, C., and Nguyen thi Man (1995) Specificity and V_H sequence of two monoclonal antibodies against the N-terminus of dystrophin. *Biochem. J.* **78**, 355–359.

46. Mattheakis, J. C., Bhatt, R. R., and Dower, W. J. (1994) An in vitro display system for identifying ligands from very large peptide libraries. *Proc. Natl. Acad. Sci. USA* **91**, 9022–9026.

47. Lu, Z., Murray, K. S., van Cleave, V., LaVallie, E. R., Stahl, M. L., and McCoy, J. M. (1995) Expression of thioredoxin random peptide libraries on the Escherichia coli cell surface as functional fusions to flagellin. *Bio/technology* **13**, 366–372.

48. Morris, G. E. (ed.) (1996), *Methods in Molecular Biology, vol. 66: Epitope Mapping Protocols*. Humana, Totowa, NJ.

49. Wang, K. S. and Strauss, J. H. (1991) Use of a lambda gt11 expression library to localize a neutralizing antibody-binding site in glycoprotein-E2 of Sindbis virus. *J. Virol.* **65**, 7037–7040.

50. Ho, D. D., Fung, M. S. C., Cao, Y. Z., Li, X. L., Sun, C., Chang, T. W., and Sun, N. C. (1991) Another discontinuous epitope on glycoprotein gp120 that is important in Human Immunodeficiency Virus Type-1 neutralization is identified by a monoclonal antibody. *Proc. Natl. Acad. Sci. USA* **88**, 8949–8952.

51. Rowlands, D. J. (1992) How can peptide vaccines work. *FEMS Microbiol. Lett.* **100**, 479–481.

52. Albani, S. and Roudier, J. (1992) Molecular basis for the association between hla dr4 and rheumatoid arthritis—from the shared epitope hypothesis to a peptidic model of rheumatoid arthritis. *Clin. Biochem.* **25**, 209–212.

53. Butler, M. H., Solimena, M., Dirkx, R., Hayday, A., and Decamilli, P. (1993) Identification of a dominant epitope of glutamic acid decarboxylase (GAD-65) recognized by autoantibodies in Stiff-Man syndrome. *J. Exp. Med.* **178**, 2097–2106.

54. Carson, D. A. (1994) The value of epitope mapping in autoimmune diseases. *J. Clin. Invest.* **94**, 1713.

55. Frank, M. B., Itoh, K., and McCubbin, V. (1994) Epitope mapping of the 52-kD Ro/SSA autoantigen. *Clin. Exp. Immunol.* **95**, 390–396.

56. Palace, J., Vincent, A., Beeson, D., and Newsom-Davis, J. (1994) Immunogenicity of human recombinant acetylcholine receptor alpha subunit: cytoplasmic epitopes dominate the antibody response in four mouse strains. *Autoimmunity* **18**, 113–119.

57. Liu, M. S., Ma, Y. H., Hayden, M. R., and Brunzell, J. D. (1992) Mapping of the epitope on lipoprotein lipase recognized by a monoclonal antibody (5D2) which inhibits lipase activity. *Biochim. Biophys. Acta* **1128**, 113–115.

58. Pietu, G., Ribba, A. S., Cherel, G., and Meyer, D. (1992) Epitope mapping by cDNA expression of a monoclonal antibody which inhibits the binding of von Willebrand factor to platelet glycoprotein-IIb/IIIa. *Biochem. J.* **284,** 711–715.

59. Landis, R. C., Bennett, R. I., and Hogg, N. (1993) A novel LFA-1 activation epitope maps to the I-domain. *J. Cell Biol.* **120,** 1519–1527.

60. Morris, C. A., Underwood, P. A., Bean, P. A., Sheehan, M., and Charlesworth, J. A. (1994) Relative topography of biologically active domains of human vitronectin—evidence from monoclonal antibody epitope and denaturation studies. *J. Biol. Chem.* **269,** 23,845–23,852.

61. Melhus, H., Bavik, C. O., Rask, L., Peterson, P. A., and Eriksson, U. (1995) Epitope mapping of a monoclonal antibody that blocks the binding of retinol-binding protein to its receptor. *Biochem. Biophys. Res. Commun.* **210,** 105–112.

62. Morris, G. E., Frost, L. C., Newport, P. A., and Hudson, N. (1987) Monoclonal antibody studies of creatine kinase. Antibody-binding sites in the N-terminal region of creatine kinase and effects of antibody on enzyme refolding. *Biochem. J.* **248,** 53–59.

63. Syu, W. J. and Kahan, L. (1992) Both ends of Escherichia Coli ribosomal protein-S13 are immunochemically accessible in situ. *J. Protein Chem.* **11,** 225–230.

64. Ning, G., Maunsbach, A. B., Lee, Y. J., and Moller, J. V. (1993) Topology of Na,K-ATPase alpha subunit epitopes analyzed with oligopeptide-specific antibodies and double-labeling immunoelectron microscopy. *FEBS Lett.* **336,** 521–524.

65. Morris, G. E. and Cartwright, A. J. (1990) Monoclonal antibody studies suggest a catalytic site at the interface between domains in creatine kinase. *Biochim. Biophys. Acta* **1039,** 318–322

66. Kruger, M., Wright, J., and Wang, K. (1991) Nebulin as a length regulator of thin filaments of vertebrate skeletal muscles—correlation of thin filament length, nebulin size, and epitopeprofile. *J. Cell Biol.* **115,** 97–107.

67. Morris, G. E., Simmons, C., and Nguyen thi Man (1995) Apo-dystrophins (Dp140 and Dp71) and dystrophin splicing isoforms in developing brain. *Biochem. Biophys. Res. Commun.* **215,** 361–367.

68. Carter, J. M. (1994) Epitope prediction methods, in *Peptide Analysis Protocols* (Dunn, B. M. and Pennington, M. W., eds.), Humana, Totowa, NJ, pp. 193–206.

69. Mandal, C., Kingery, B. D., Anchin, J. M., Subramaniam, S., and Linthicum, D. S. (1996) ABGEN: a knowledge-based automated approach for antibody structure modelling. *Nature Biotechnol.* **14,** 323–328.

48

Immunocytochemistry

Lorette C. Javois

1. Theory and Background

By definition, immunocytochemistry is a biomolecular technique that involves the use of a specific antibody-antigen reaction to identify cellular constituents *in situ*. The technique requires that the antibody be tagged with a label to facilitate its visualization. Immunocytochemistry is a powerful tool for demonstrating not only the presence of an antigen (cellular constituent) but also its location within a cell, tissue, or organism.

Antibodies or immunoglobulins (Igs) are the protein products of cells of the immune system (**Fig. 1**). Polyclonal antibodies recognizing multiple determinants on an antigen may be generated by repeatedly immunizing an animal, such as a rabbit, and then isolating the immunoglobulin-containing fraction from the blood *(1)*. This approach is widely used in institutions equipped with animal-care facilities. In addition, numerous service companies exist that will implement injection and harvesting protocols and provide sera. Alternatively, monoclonal antibodies (MAbs) that recognize a single antigenic determinant may be produced using a spleen cell-hybridoma fusion protocol followed by hybridoma selection, screening, and cloning *(2–4)* (*see* Chapter 45). This widely used technique requires tissue-culture facilities and is initially labor intensive, but it is capable of providing a nearly limitless supply of antibody through the culture of the hybridomas or the production of ascites fluid. Regardless of the source, it is important that antibodies used for immunocytochemical studies are produced in high titer and have high specificity and binding affinity for the antigen of interest, because the power of the technique resides in the specificity of the antibody–antigen interaction.

Whether derived from polyclonal sera, tissue culture medium, or ascites fluid, before antibodies can be labeled they must be purified. Two commonly used techniques for immunoglobulin purification are ammonium-sulfate fractionation and/or gel filtration *(5)*. These techniques may be followed by ion-exchange chromatography and/or affinity chromatography *(6–8)*. Once purified, antibodies may be tagged with a variety of labels, such as fluorescent molecules *(9)*, enzymes *(10)*, biotin *(11)*, colloidal gold *(12)*, or radioactive labels *(13)*. The choice of tag depends on the particular application and instrumentation available for detection of the tag. The use of variously labeled antibodies will be considered in more detail during the discussion in **Subheading 3.**

From: *Molecular Biomethods Handbook*
Edited by: R. Rapley and J. M. Walker © Humana Press Inc., Totowa, NJ

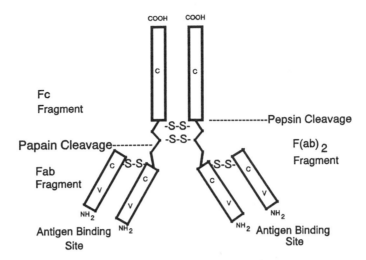

Fig. 1. A diagram representing a typical antibody molecule composed of two identical heavy chains (longer) and two identical light chains (shorter). Each chain is composed of a constant (C) and variable (V) portion. The molecule has two identical antigen-binding sites made up of the amino-terminal (NH_2) variable portions of each heavy- and light-chain combination. The heavy chains and light chains are crosslinked by disulfide bonds (-S-S-). The antibody may be cleaved by enzymes into characteristic fragments. Papain cleavage produces two antibody binding fragments (Fabs) and one crystalizing fragment (Fc). Pepsin cleavage produces one bivalent antibody-binding fragment $F(ab)_2$ and several smaller subfragments of the Fc fragment.

From an historical perspective, immunocytochemistry was first introduced in 1941 by A. H. Coons and associates *(14)* who applied a fluorescently labeled antibody directly to an antigen to localize it within a tissue. This approach is referred to as the direct labeling technique. It required labeling the specific antibody with a fluorescein derivative (fluorescein isothiocyanate; FITC) which was, at the time, a difficult procedure. For this reason, the technique was not widely applied for almost 20 years. The development of the indirect-labeling technique provided a more versatile and sensitive approach to immunocytochemistry *(15)*. In this procedure, the specific primary antibody bound to the antigen is detected by application of a secondary reagent, usually a labeled antibody directed against the primary antibody. This approach is versatile since one labeled secondary antibody may be used to detect numerous different primary antibodies, because binding of the secondary antibody is directed toward determinants common to immunoglobulin molecules of a particular animal species. The technique is more sensitive because the signal is amplified owing to the binding of several labeled secondary antibodies to a single primary antibody. In addition, many commercial sources are available for stable, relatively inexpensive, labeled secondary antibodies, so one does not have to produce, purify and tag antibodies if an indirect labeling technique is being utilized.

2. Practical Steps Involved in Immunocytochemistry
2.1. Direct vs Indirect Labeling

Immunocytochemistry may be performed on living, frozen, or fixed material ranging from single cells to tissues to whole organisms. **Figure 2** provides an overview and comparison of the direct- and indirect-labeling techniques. The direct-labeling tech-

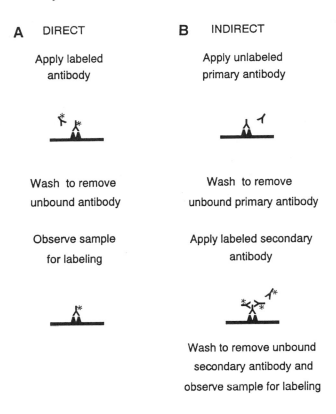

Fig. 2. A comparison of the direct (**A**) and indirect (**B**) immunocytochemical labeling meth-ods. With the direct method, labeled antibody ($\lambda*$) is applied to the sample, allowed to bind, and then unbound labeled antibody is removed by washing with a buffered saline solution. With the indirect method, the sample is incubated with unlabeled primary antibody (λ) first. Unbound primary antibody is removed by washing, and a labeled secondary antibody is then applied. After the final wash to remove unbound secondary antibody, the specimen is examined for labeling. Note that the indirect-labeling method results in an amplified signal as compared with the direct-labeling method.

nique has the advantage of being quick as it involves a single incubation step with the labeled antibody followed by a washing step with buffered-saline solution to remove unbound reagent *(16)*. By comparison, using the indirect-labeling technique, the speci-men is first incubated with the primary antibody directed against the antigen of inter-est. This is followed by a washing step to remove unbound primary reagent. The next step involves incubation of the specimen with the secondary labeled antibody, again followed by a washing step.

In either technique, when working with a fixed specimen, the specimen must be adequately washed to remove residual fixative. The labeling step is often preceded by incubation of the fixed specimen in a blocking solution containing normal serum of the animal in which the labeled antibody was raised. This step serves to mask sites of nonspecific interaction between the labeled antibody and the specimen, thus reducing nonspecific, background labeling (*see* **Subheading 2.2.**). It should be noted that fixation changes the cellular structure and chemical composition of the specimen *(17)*. There is no perfect fixative for immunocytochemistry. Preservation of cellular structure and loss of antigenic determinants are often inversely related, and a fixative must be chosen which will balance the two. Both coagulants (such as alcohols and acetone) and crosslinking reagents (such as formaldehyde/paraformaldehyde and glutaraldehyde) have been used successfully *(18)*. The fixation protocol chosen will greatly depend on the particular antigen being studied and must often be determined empirically.

Mounting and viewing of the specimen will depend on the type of labeled antibody used. Fluorescent molecules (called fluorophores) are not stable and photochemical reactions result in their decomposition and fading of the signal, commonly referred to as bleaching. For this reason, fluorescently-labeled specimens are prepared as wet mounts using glycerol and phosphate-buffered saline containing one of several anti-bleaching compounds *(19)*, such as 1,4-diazo-bicyclo-[2.2.2]-octane (DABCO) *(20)*, *p*-phenylenediamine *(21)*, *n*-propyl-gallate *(22)*, or sodium azide *(23)*. These preparations are not permanent and must be viewed in a darkened room using a special fluorescence microscope. Alternatively, enzyme-labeled specimens can be permanently mounted following dehydration using mounting media such as Euparol or Permount, and they are viewed using conventional light microscopy.

For electron-microscopic immunocytochemistry, the fixation protocol depends on whether the immunogold labeling will be done prior to embedding the sample or postembedding and must be determined empirically for each antigen *(24)*. Colloidal gold conjugated to antibodies, protein A or G, lectins, or enzymes can be used as the probe. Either the direct- or indirect-labeling technique can be followed, and the samples prepared for transmission or scanning-electron microscopy *(25–27)*. The possibility of double-labeling or multiple-labeling of a specimen is considered in **Subheading 3.**

2.2. Artifacts, Pitfalls, and Controls

Generally, immunocytochemical artifacts fall into three categories. The first type is related to immunological specificity resulting from crossreactivity of the antibody. It should be remembered that antibodies recognize determinants consisting of several amino acids or carbohydrate residues which may be present in tissue constituents other than the antigen of interest. The antibody will specifically bind these determinants, and this genuine crossreactivity is impossible to eliminate. If the serum being used is polyclonal and the specific binding is owing to antibody heterogeneity, then the antibody of interest can be isolated by adsorption to a column via the specific antigen (provided it is available). One caveat is that "good" antibodies should have high avidity which makes eluting the bound antibody from the column difficult. There is the risk that eluted antibodies, although pure, will be those of lower avidity. Another option is to switch to a monoclonal serum that recognizes a single antigenic determinant, but even MAbs may specifically bind the same determinant on different antigens. In this case, the only remaining alternative is to screen other antibodies raised to different portions of the antigen molecule for a more suitable reagent, provided these are available.

The second type of artifact results from the nonspecific interaction of the antibody molecule with tissue constituents via hydrophobic or electrostatic interactions. This type of interaction is more easily dealt with. After fixation, potential sites of nonspecific interaction may be "blocked" by incubating the sample in albumin or nonimmune serum from the species in which the labeled antibody has been raised prior to incubation in the primary and secondary antibodies. This preincubation essentially blocks potential sites of nonspecific interaction. Immunoglobulin F_c receptors will bind nonspecifically to unfixed tissue, and F_{ab} fragments should be used in place of whole antibody to avoid this unwanted labeling (**Fig. 1**). Nonspecific interactions may also be disrupted by the addition of detergents such as Triton X-100 to the buffer-wash solutions *(28)*.

The third type of artifact is introduced during tissue preparation, especially fixation. All fixatives cause chemical or conformational changes in the secondary or tertiary structure of proteins, which are often sites recognized by antibodies, and lesser changes in carbohydrate determinants *(29)*. Coagulant fixatives result in tissue extraction and shrinkage while crosslinking fixatives introduce conformational changes by forming chemical bonds between proteins. Both create potential artifacts through the loss of antigens or antigenic sites, steric hindrance of antibody penetration, or even unmasking of nonspecific antigenic determinants. As stated earlier, the preservation of cellular structure and the maintenance of determinants of interest must be balanced when choosing the fixative, and often the "best" fixative can only be determined empirically for any given system *(18)*. Recently, microwave treatments have been developed as a means of retrieving antigens after fixation *(30)* or improving antigenicity during fixation *(31)*.

There are certain controls that should always be performed to demonstrate that a positive immunocytochemical-staining result is genuine and not owing to nonspecific interactions. Substituting nonimmune serum (if available), an inappropriate antibody, or saline in place of the specific primary antibody should result in the absence of staining. Pre-adsorbing the antibody with the specific antigen (if available) prior to using it in the staining procedure should eliminate positive staining. Likewise, pre-adsorbing the antibody with an irrelevant antigen prior to performing the staining procedure should not eliminate positive staining. With every experimental-staining test that is performed a known positive sample should be stained in parallel as a "method control" to show that all solutions are in working order.

It should be noted that false positives in staining procedures involving the use of avidin/biotin or enzyme-conjugated antibodies may be owing to the endogenous presence of biotin or enzyme. This background labeling may be eliminated by introducing a blocking step in which avidin is added to bind endogenous biotin, followed by biotin to block the avidin. Alternatively, streptavidin, which is produced by *Streptomyces avidinni* and is unglycosylated, may be substituted for avidin, and background staining can be virtually eliminated. Endogenous enzymes may be quenched in a blocking step using H_2O_2 in buffer, methanol, or ethanol (for peroxidase), 20% acetic acid or 1 mM levamisole (for alkaline phosphatase) *(32)*, or heat inactivation, providing the antigen withstands this treatment *(33)*.

2.3. Fluorescence Photomicrography

In order to retain a permanent record of fluorescently labeled specimens, photographs are often taken. Fluorescent photomicrography presents certain challenges owing to the low intensity of light emitted by fluorophores relative to the daylight conditions most commercially available films have been designed to work with. Prolonged exposure of these films to low-intensity light results in a phenomenon called reciprocity failure. Long exposures using color film result in incorrect color reproduction owing to the differing reciprocity-failure factors for each of the three primary-color sensitive layers of the film. This effect can be overcome by inserting color-correcting filters in the light path to the camera. A second practical result of prolonged exposures is the fading of the fluorescent image, which in turn necessitates an even longer exposure time. To minimize this problem, emitted fluorescence must be optimized, exposure times must be reduced, and appropriate films must be used *(34)*. Electronic imaging through the use of high-resolution video or solid-state cameras in conjunction with image intensifiers is an alternative to film-based cameras and allows for the capture, storage, and manipulation of dim images through computer processing *(35)*.

3. Imunocytochemistry Detection Systems

The reaction between antibody and isolated antigen can be visualized at the level of electron microscopy where antibodies appear to "decorate" the antigen molecule. However, when the antibody-antigen reaction occurs *in situ*, as with immunocytochemistry, the surrounding tissue components mask this decorating effect. Therefore, it is necessary to tag the antibody in order to visualize it.

3.1. Fluorescent Labeling

Fluorescence is the property of emitting electromagnetic radiation in the form of light as a result of—and only during—the absorption of light from another source. Many molecules fluoresce, but only a few are routinely used in immunocytochemistry. They are used because of the intensity of their fluorescence and the ability to distinguish the fluorescence from background autofluorescence of the specimen arising from excitation of endogenous molecules like the flavin compounds, NADH, elastin, fibronectin, hemoglobin, or chlorophyll. The point source nature of fluorescent light is one of the greatest advantages of this labeling technique because it provides better resolution than systems that depend on refracted light for imaging. Typically, the wavelength of light need to excite the fluorophore is larger and of greater intensity than the wavelength of emitted light. The separation of the excitation wavelength maximum and the emission wavelength maximum is referred to as the Stokes shift and affects both the brightness and detection of the fluorophore. In practice, the configuration of the detection system—the fluorescence microscope, confocal microscope, or fluorescence activated cell sorter/analyzer—also plays a major role in the fluorescence intensity detected. The light source, excitation filter, dichroic mirror, emission filter, and objective all contribute to the intensity of the fluorescence image.

Table 1 provides the spectral properties of fluorochromes commonly used in immunocytochemistry. Coons and colleagues *(14)* pioneered the fluorescent labeling of antibodies, and their direct-labeling technique formed the basis from which all subsequent immunocytochemistry methods have been derived. Today, fluorescent molecules are covalently linked directly to antibody molecules or indirectly coupled to antibodies via an avidin-biotin bridge or conjugation to Protein A or G. Protein A, a cell-wall compo-

Table 1
Spectral Properties of Some Fluorochromes
Employed for Immunofluorescence[a]

Fluorophore	Excitation Maximum (nm)	Emission Maximum (nm)
SITS[b]	336	438
DAMC	354	441
AMCA	355	450
Cascade Blue	396	410
Oregon Green 488	490	514
Alexa 488	491	515
FITC	494	519
BODIPY FL	503	512
B-PE	545	575
TRITC	547	572
CY3.18	554	568
R-PE	565	578
RB-200-SC	568	583
XRITC	582	601
Alexa 594	585	610
Texas Red	589	615
BODIPY TR	592	618

[a]Arranged by ascending order of excitation wavelengths.
[b]Abbreviations used: AMCA = 7-amino-4-methylcoumarin; B-PE = B phycoerythrin; CY3.18 = cyanine 3.18; DAMC = diethyl-amino-coumarin; FITC = fluorescein isothiocyanate; RB-200-SC = lissamine rhodamine sulfonylchloride; R-PE = R phycoerythrin; SITS = 4-acetamido-4'-isothiocyanatostilbene-2,2'-disulfonic acid; TRITC = tetramethyl rhodamine isothiocyanate; XRITC = rhodamine X isothiocyanate. (Reprinted with permission from [36].)

nent of *Staphylococcus aureus*, has a high affinity for the F_c region of human and rabbit IgG. Protein G from *Streptococcal* strain G148 binds to immunoglobulins from a wider range of species including the mouse *(37,38)*.

The green-emitting fluorophore FITC was first introduced by Coons and continues to be one of the most widely used fluorophores. Its small size, hydrophilic nature, and strong fluorescence intensity are all advantageous. The fact that it has a short Stokes shift, readily photobleaches, and fluoresces maximally in the pH 8.0–9.0 range (problematic for living cells) are all drawbacks.

Alternative fluorophores are the red-emitting rhodamine derivatives like tetra-methylrhodamine isothiocyanate (TRITC) and Texas Red. The fluorescence of TRITC is less sensitive to pH and does not photobleach as readily as FITC. Also, tissue-culture dishes, which are used during staining of adherent-cell cultures, do not autofluoresce in the rhodamine spectrum as badly as they do in the fluorescein spectrum. Recently introduced substitutes for the rhedamines, which offer superior photostability, include BODIPY TR and TMR, and Alexa 568 and 594.

Phycobilli proteins of the algal photosynthetic system also fluoresce in the red range. Their extremely high-fluorescence intensity and insensitivity to pH are advantageous. However, they are large molecules. When coupled to antibodies, tissue penetration, stearic interference during antigen binding, and background resulting from unsuccessful washing may be problematic *(39)*.

Multiple-fluorescent labeling of a cell or tissue sample may be performed providing the excitation and emission spectra of the fluorophores have minimal overlap *(40)*. The fluorescein-rhodamine double-labeling technique is most widely used despite some overlap of the excitation and emission spectra. With appropriate optical filters on the microscope, the two fluorescent signals are readily distinguishable. Triple labeling has been performed by using blue-emitting fluorophores such as SITS *(41)*, coumarin derivatives *(42)*, or Cascade Blue *(43)* in addition to fluorescein and rhodamine. Using flow-cytometric analysis, five-color immunofluorescent analysis has been performed *(44)*. Filter sets are now available that allow simultaneous excitation and observation of two to four different fluorochromes.

3.2. Enzyme Labeling

Nakane and Pierce *(45)* pioneered the substitution of the peroxidase enzyme for the fluorescent tag on the labeled immunoglobulin. This direct-labeling technique brought the enzyme into close proximity of the antigen where the enzyme catalyzed the conversion of a colorless solution into a colored precipitate, thus marking the location of the antigen. Subsequently, numerous indirect methods have been developed that use universal secondary reagents and also amplify the signal *(46)*. In addition to substrate solutions containing the widely used chromogen 3,3'-diaminobenzidine tetrachloride (DAB), alternative chromogens that result in insoluble reaction products are available for use with horseradish peroxidase. Likewise, alternative enzymes exist that can be substituted for peroxidase. Examples of commonly used enzymes and chromogens are summarized in **Table 2**.

Increased sensitivity was achieved by Sternberger *(48)* through the use of a peroxidase-antiperoxidase (PAP) complex that relies entirely on immunological interactions rather than conjugation of the enzyme to an antibody (**Fig. 3**). This technique is more sensitive than the standard indirect technique, though it includes an additional incubation step and requires the primary antibody and the PAP complex be generated from the same animal species *(49)*.

To overcome problems associated with conjugating two relatively large molecules—horseradish peroxidase and immunoglobulin—a new detection system was developed in which biotinylated antibody was detected via avidin-labeled enzyme. The avidin-biotin complex (ABC) molecule is an improvement that increases the number of enzyme molecules associated with the primary antibody *(50)*. Essentially, enzyme-labeled biotin is reacted with avidin to build the initial complex and then the complex is reacted with biotin-labeled secondary antibody. This in turn reacts with the primary antibody that detects the antigen of interest (**Fig. 4**) *(51)*. In combination, the ABC- and PAP-labeling systems provide a method for double-labeling a sample with little chance of crossover reactions between the systems, especially if a second enzyme, alkaline phosphatase is used in addition to peroxidase.

Table 2
Enzyme–Chromogen Systems for Immunocytochemistry

Enzyme	Chromogen	Product color
Horseradish peroxidase	3,3'-diaminobenzidine tetrahydrochloride (DAB)	Brown
	p-phenylene diamine HCl and pyrocatechol (PPD/PC)[b]	Black[a]
	4-chloro-1-napthol	Blue/brown
	3-amino-9-ethylcarbazole (AEC)	Dark blue
Alkaline phosphatase	5-bromo-4-chloro-3-indolyl phosphate and nitro blue tetrazolium (BCIP/NBT)	Red
	Fast Red TR and Naphthol AS–MX	Blue
β–galactosidase	5-bromo-4-chloro-3'-indoyl	Red
	β-D-galactopyranoside (X-Gal)	Blue
Glucose oxidase	phenazine methosulfate (PMS)	Blue

[a]With nickel or cobalt enhancement *(47)*.
[b]Hanker Yates reagent.

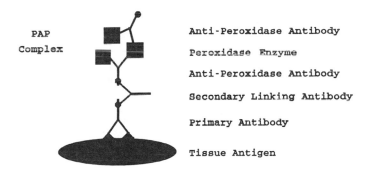

Fig. 3. Diagram illustrating the molecular interactions of the PAP procedure. The PAP complex is comprised of horseradish peroxidase bound to an antiperoxidase antibody generated in the same animal species as the primary antibody that recognizes the antigen of interest. The primary antibody and the PAP complex are linked via a secondary antibody generated in a second animal species against immunoglobulin of the primary animal species. (λ, immunoglobulin; ■, peroxidase enzyme; reprinted with permission from *[49]*).

The most sensitive enzyme-labeling technique is the labeled avidin binding (LAB) method *(52)*. Here the peroxidase is covalently attached to avidin and this complex is linked to the primary antibody via and intermediate biotinylated secondary antibody. The biotin is attached to the secondary antibody via a long carbon arm extension to make it more accessible for binding to the avidin-peroxidase complex (**Fig. 5**) *(53)*.

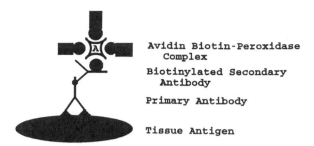

Fig. 4. Diagram illustrating the molecular interactions of the ABC procedure. The primary antibody against the antigen of interest is linked to the avidin-biotinylated peroxidase complex via a biotinylated secondary antibody raised against immunoglobulin of the animal species used to generate the primary antibody (λ, immunoglobulin; ●, biotin; A, avidin; ■, peroxidase; reprinted with permission from *[51]*).

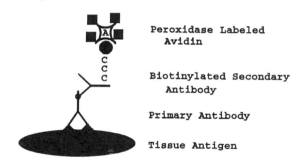

Fig. 5. Diagram illustrating the molecular interactions of the LAB procedure. Horseradish peroxidase is covalently linked to avidin. The primary antibody against the antigen of interest is linked to the enzyme labeled avidin complex (LAB) via a biotinylated secondary antibody raised against immunoglobulin of the animal species used to generate the primary antibody (λ, immunoglobulin; CCC, long carbon arm extension; ●, biotin; A, avidin; ■, peroxidase; reprinted with permission from *[53]*).

When double labeling using enzyme immunocytochemistry is desired, this can be accomplished by altering the enzyme system used for one of the assays or by altering the chromogen used for the enzyme (**Table 2**).

The advantages provided by the enzyme-labeling techniques include the generation of intensely colored reaction products which can be viewed with conventional light microscopy. This is less expensive than investing in a fluorescence microscope, confocal microscope, or fluorescence-activated cell sorter/analyzer required for analysis of fluorescently labeled samples. In addition to the techniques' sensitivity, the resulting labeled specimens are permanent and do not require immediate analysis as do fluorescently labeled samples that fade significantly over the first 24 h. Disadvantages include the possibility of endogenous-enzyme activity and the hazardous nature of the substrate solutions, as most chromogens are carcinogenic or potentially carcinogenic.

3.3. Ultrastructural Labeling

Immunocytochemical-ultrastructural localization of antigens was first achieved by Singer and coworkers *(54)* who used ferritin-labeled antibodies. Ferritin is a large electron dense protein. While immunoferritin methods were important for expanding immunocytochemistry to the ultrastructural level, they are not widely used today. Instead, immunocolloids, especially immunogold, which was introduced by Falk and Taylor in the early 1970s *(55)*, are widely used and extremely useful. More recently, *Staphylococcal* protein A-gold *(56)* and IgG-gold *(57)* techniques have made immunogold-detection methods more universal.

Numerous methods of producing colloidal-gold sols by boiling tetrachloroauric acid with a reducing agent exist *(58)*. As the chloroauric acid is reduced, gold atoms aggregate to form microcrystals. Crystal size, which can range from 2 to 40 nm, depends on the type of reducing agent used and its concentration.

After verifying the size of the gold with the electron microscope, it may be conjugated to antibodies or protein A *(12)*. Samples may be labeled with the colloidal-gold detection system prior to embedding *(26)* or after embedding and sectioning *(27)*. The post-embedding technique is the most widely used immunolabeling technique for electron microscopy and often makes use of gold conjugated to streptavidin and an avidin-biotin detection system. One advantage of the avidin/gold-biotin detection system is the ability to amplify a weak signal *(25)*. Colloidal gold conjugated to protein A or G is also widely used because of its more universal application and greater stability compared with colloidal gold-antibody conjugates. A silver enhancement method may also be applied to colloidal gold label specimens to enlarge the gold particles facilitating their visualization with both the scanning electron and light microscopes *(59,60)*.

Because gold crystals can be made in different sizes, it is possible to double label a sample using direct labeling with antibodies conjugated to differently sized gold particles. The indirect-labeling technique can also be used if staining is carried out sequentially with antibodies that do not cross-react. Alternatively, different faces of the same grid may be stained with separate antibody conjugates *(61)*.

3.3. Radioactive Labeling

Radiolabeled tumor-specific antibodies are used in the imaging of tumors through a direct-labeling technique. The radiolabeled reagent is introduced intravenously, and although there is some unspecific uptake by normal tissues like the thyroid gland, stomach, intestine, and liver, these reagents have proven to be stable in vivo and specifically label the targeted tumors. Iodine, indium, technetium, and rhenium have all been conjugated to antibodies in an effort to develop a simple, effective method for cancer radioimmunotherapy *(62)*.

4. Applications of Immunocytochemistry

Immunocytochemistry is now one of the most widely used tools in research and diagnosis. Provided that antibodies of high avidity and specificity are available, the applications of the technique are unlimited. Using either direct or indirect labeling, the technique can be applied to samples prepared in a variety of ways, with the choice of preparation method determined by such factors as the availability of the material, fixability of the antigen of interest, time constraints on staining and evaluation, and equipment needed for the analysis.

4.1. Paraffin-Embedded and Sectioned-Tissue Preparations

Immunohistochemistry, particularly the immunoperoxidase technique, is widely employed in medical-pathology laboratories that routinely process paraffin-embedded and sectioned specimens for diagnosis. The specificity of the antibody–antigen reaction in combination with an assessment of morphology by a pathologist has revolutionized the differential diagnosis of tumors. Tissue derivation, presence of an infectious agent, or assessment of malignancy are all issues that can be resolved through the use of immunocytochemistry *(63)*. Clinical pathologists and oncologists can even use immunocytochemical methods on archival material for retrospective studies examining recovered nuclei for DNA content by fluorescence activated cell-sorter analysis *(64,65; see* **Subheading 4.3.**). This approach allows one to examine total DNA content and aneuploidy of large numbers of samples with respect to the original clinical diagnosis *(66)*. For laboratories that routinely process hundreds of slides, automated immunostainers have been introduced to reduce the labor-intensive nature of the procedure as well as to minimize variability between specimens and conserve valuable reagents *(67)*.

4.2. Frozen Sections and Touch Preparations

Although fixed, embedded, and sectioned material is the mainstay of the pathology laboratory, often formaldehyde-fixed specimens have high levels of autofluorescence, not to mention decreased accessibility to antigens. For this reason, the use of frozen sections is the preferred approach for examining intact tissues *(68)*. Combining flash freezing in liquid nitrogen, cryosectioning, and immunofluorescence is a rapid way to obtain high-resolution staining, which minimizes autofluorescence. In cases where more immediate results are warranted, whole cells may be removed from a tissue sample via touch preparations or imprints made onto glass slides, which can then be processed for immunocytochemistry *(69)*.

4.3. Cell Suspension and Culture Preparations

In the clinical laboratory, cells collected from suspensions can be subjected to immunocytochemistry. Depending on the viscosity of the medium in which cells are suspended, they may be analyzed as smears (low viscosity), cytospins (medium viscosity), or swabs (high viscosity) *(70)*. In all these preparations, whole cells are applied to a glass slide, fixed, and processed for immunocytochemistry. Cells grown in tissue culture are also routinely used in the research laboratory and are particularly amenable to analysis by immunofluorescent techniques. If grown in suspension, they may be analyzed by cytospin application to slides or by flow cytometric analysis. Flow-cytometric analysis offers the advantages of rapid analysis, quantification of multiple cellular parameters, and sorting of individual cells on the basis of the optical properties of fluorescent probes and the light scattering capabilities of the cells *(71)*. Adherent cell cultures can be examined following growth on coverslips or Petri dishes. The fact that these cells are flat and virtually nonautofluorescent results in exceptional immunofluorescent resolution (**Fig. 6**). Many patterns of intracellular organelles are very characteristic under these conditions and have been documented in an atlas *(73)*.

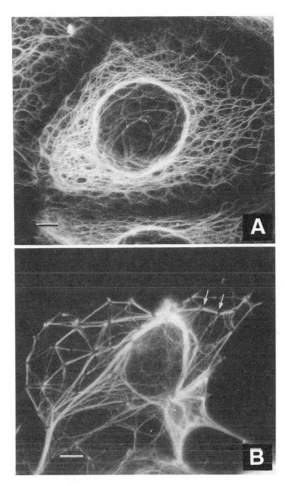

Fig. 6. Indirect immunofluorescence pattern of cytokeratin using monoclonal antibody BG3 in **(A)** control PtK1 cells and **(B)** cytochalasin B-treated PtK$_1$ cells. Arrows indicate focal centers where cytokeratin bundles appear to terminate. Scale bar = 5 μm. (Reprinted with permission from *[72]*.)

Fluorescently labeled adherent cells are also perfect targets for laser-based microscope systems or laser microbeams *(74)*. These workstations combine a laser that can be microscopically focused to excite fluorescent probes in a spot of about 0.5 μm diameter, a system which allows laser- or stage-scanning, fluorescence monitoring with sensitive photomultiplier tubes, and a personal computer to facilitate data acquisition and analysis. Numerous specialized techniques have been developed utilizing this sophisticated instrumentation. Fluorescence resonance energy transfer (FRET) allows the distance between two fluorescently labeled epitopes to be calculated based on the transfer of energy from a donor fluorophore emitted in the absorption spectrum of the second, acceptor fluorophore *(75)*. Applications of FRET include measuring conformational changes in receptors upon ligand binding, as well as the relationship between receptors and nearby accessory molecules. Fluorescence redistribution after photobleaching (FRAP) by the laser microbeam is used to measure the mobility of fluorescently labeled proteins within cell membranes or through gap junctions (GAP-FRAP) *(76)*. Extremely short pulses of high-energy laser light result in photoablation. Antibody-targeted proteins can be selectively ablated through the chromophore-assisted laser inactivation (CALI) technique *(77)*. The ablative technique may be broadened to selectively target and kill whole cells that meet specific fluorescent labeling criteria. Additionally, laser light may be used to optically "trap" antibody-labeled cells or organelles within cells to sort or manipulate them *(78)*.

4.4. Whole-Mount Preparations

Whole mounts of tissue or even entire organisms may be examined using traditional immunocytochemical approaches *(79)*. One of the greatest advantages to this application is the ability to examine large areas of tissue for the pattern and distribution of a relatively small population of cells. In particular, neuronal cell types, their distributions, and innervation patterns have been studied using this approach. The very fine processes associated with neurons are particularly visible (**Fig. 7**). In tissue section, these cells and their processes would be very difficult to observe and the distribution patterns would have to be discerned from tedious three-dimensional reconstructions. Whole-mount technology has also facilitated the study of three-dimensional arrays of cytoskeletal components within tissues relative to changes in morphology or developmental time courses *(80)*. One disadvantage of the whole-mount technique is the increased out-of-focus fluorescence owing to the thickness of the sample, auto-fluorescence of the tissue, and labeled structures outside the focal plane. Some of these effects can be reduced by using low-power objectives with longer depths of focus, but the best solution is to use a confocal microscope. This instrument is designed to focus on one plane of the specimen while eliminating stray fluorescent light arising from other out-of-focus planes *(81)*. The primary advantage is improvement in depth discrimination (**Fig. 8**). Additionally, by optically sectioning through the specimen, a series of images from different focal planes can be collected, saved, and a three-dimensional reconstruction made with the aid of the computer.

4.5. Immunocytochemistry and In Situ Hybridization

The most recent advance in the application of immunocytochemical techniques involves the combination of traditional immunocytochemistry and the nucleic acid hybridization technique *(82)*. This approach allows the detection of DNA or RNA sequences within cells. An antibody-nucleic acid probe is used to target either DNA or RNA endogenous sequences present in the cytoplasm, nucleoplasm, or on chromosomes, or exogenous sequences belonging to infectious agents. The biotinylated nucleic-acid portion of the probe hybridizes to the targeted sequence. The antibody portion is applied in a separate step and is linked via streptavidin to the biotinylated nucleic acid probe. The antibody portion carries with it the means to detect the total probe. The most frequently used detection systems are the enzymes alkaline phosphatase *(83)* and hydrogen peroxidase, or the fluorophores, fluorescein *(84)*, rhodamine, and hydroxycoumarin. The combination of the two traditional techniques results in an approach which is exquisitely specific, sensitive, and widely applicable. The sensitivity of the technique may be enhanced even further through the use of the polymerase chain reaction (PCR) *(85)* (*see* Chapter 25). Amplifying the targeted nucleic-acid

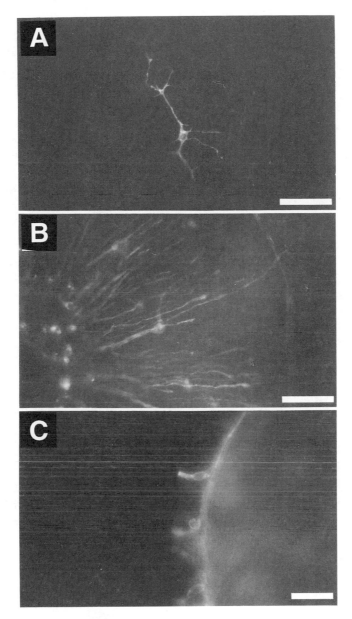

Fig. 7. *Hydra oligactis* whole mount labeled with monoclonal antibody JD1. **(A)** Isolated ganglionic neuron in the body column, scale bar = 50 μm; **(B)** hypostomal nerve net with sensory neurons of the mouth at the left and ganglionic neurons of the perihypostomal ring to the right, scale bar = 50 μm; **(C)** cell bodies of hypostomal sensory neurons extending from the mesoglea (processes) to the surface of the ectoderm, scale bar = 25 μm. (Reprinted with permission from *[78]*.)

sequence through *in situ* PCR prior to detection is an approach that promises to have a great impact on both clinical and research applications, including the detection of viral sequences, single copy genes, gene rearrangements, or gene mutations. In particular, multicolor fluorescence *in situ* hybridization (FISH) can identify as many labeled fea-

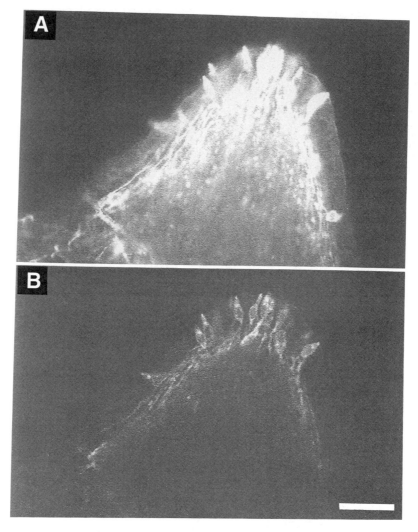

Fig. 8. Hypostome (mouth) of a *Hydra oligactis* whole mount labeled with monoclonal antibody DB5. **(A)** Nonconfocal image of sensory neurons and processes; **(B)** confocal optical section of the same field as (A) illustrating details of neuronal cell bodies and processes. Scale bar = 25 µm. (Reprinted with permission from *[78]*.)

tures as there are different fluorophores used in the hybridization *(86)*. The potential of this technique for both pure research and clinical diagnoses is extraordinary and includes applications in cancer research, pathology, cytogenetics, prenatal diagnosis, and cell and developmental biology.

References

1. DeMey, J. and Moeremans, M. (1986) Raising and testing polyclonal antibodies for immunocytochemistry, in *Immunocytochemistry: Modern Methods and Applications* (Polak, J. M. and VanNoorden, S., eds.), Wright, Bristol, UK, pp. 3–12.
2. Kohler, G. and Milstein, C. (1975) Continuous cultures of fused cells secreting antibody of predefined specificity. *Nature* **256,** 495–497.

3. Ritter, M. A. (1986) Raising and testing monoclonal antibodies for immunocytochemistry, in *Immunocytochemistry: Modern Methods and Applications* (Polak, J. M. and VanNoorden, S., eds.), Wright, Bristol, England, pp. 13–25.

4. Ritter, M. A. and Ladyman, H. M. (eds.) (1995) *Monoclonal Antibodies.* Cambridge University Press, New York.

5. Kent, U. M. (1994) Purification of antibodies using ammonium sulfate fractionation or gel filtration, in *Immunocytochemical Methods and Protocols* (Javois, L. C., ed.), Humana, Totowa, NJ, pp. 13–21.

6. Kent, U. M. (1994) Purification of antibodies using ion-exchange chromatography, in *Immunocytochemical Methods and Protocols* (Javois, L. C., ed.), Humana, Totowa, NJ, pp. 23–27.

7. Kent, U. M. (1994) Purification of antibodies using affinity chromatography, in *Immunocytochemical Methods and Protocols* (Javois, L. C., ed.), Humana, Totowa, NJ, pp. 29–35.

8. Kent, U. M. (1994) Purification of antibodies using protein A-sepharose and FPLC, in *Immunocytochemical Methods and Protocols* (Javois, L. C., ed.), Humana, Totowa, NJ, pp. 37–41.

9. Mao, S.-Y. (1994) Conjugation of Fluorochromes to antibodies, in *Immunocytochemical Methods and Protocols* (Javois, L. C., ed.), Humana, Totowa, NJ, pp. 43–47.

10. Bratthauer, G. L. (1994) Overview of antigen detection through enzymatic activity, in *Immunocytochemical Methods and Protocols* (Javois, L. C., ed.), Humana, Totowa, NJ, pp. 155–164.

11. Mao, S.-Y. (1994) Biotinylation of antibodies, in *Immunocytochemical Methods and Protocols* (Javois, L. C., ed.), Humana, Totowa, NJ, pp. 49–52.

12. Oliver, C. (1994) Conjugation of colloidal gold to proteins, in *Immunocytochemical Methods and Protocols* (Javois, L. C., ed.), Humana, Totowa, NJ, pp. 303–307.

13. Hnatowich, D. J. (1990) Recent developments in the radiolabeling of antibodies with iodine, indium, and technetium. *Sem. Nuclear. Med.* **20,** 80–91.

14. Coons, A. H., Creech, H. J., and Jones, R. N. (1941) Immunological properties of an antibody containing a fluorescent group. *Proc. Soc. Exp. Biol. Med.* **47,** 200–202.

15. Coons, A. H., Leduc, E. H., and Connolly, J. M. (1955) Studies on antibody production. I. A method for the histochemical demonstration of specific antibody and its application to a study of the hyperimmune rabbit. *J. Exp. Med.* **102,** 49–60.

16. Javois, L. C. (1994) Direct immunofluorescent labeling of cells, in *Immunocytochemical Methods and Protocols* (Javois, L. C., ed.), Humana, Totowa, NJ, pp. 117–121.

17. Bowers, B. and Maser, M. (1988) Artifacts in fixation for transmission electron microscopy, in *Artifacts in Biological Electron Microscopy* (Crang, R. F. E. and Klomparens, K. L., eds.), Plenum, New York, pp. 13–42.

18. Melan, M. A. (1994) Overview of cell fixation and permeabilization, in *Immunocytochemical Methods and Protocols* (Javois, L. C., ed.), Humana, Totowa, NJ, pp. 55–66.

19. Florijn, R. J., Slats, J., Tanke, H. J., and Raap, A. K. (1995) Analysis of antifading reagents for fluorescence microscopy. *Cytometry* **19,** 177–182.

20. Langanger, G., DeMay, J., and Adam, H. (1983) 1,4-Diazobizyklo-[2. 2. 2.] oktan (DABCO) verzogest das Ausbleichen von immunofluorenzpreparaten. *Mikroskopie* **40,** 237–241.

21. Johnson, G. D., Davidson, R. S., McNamee, K. C., Russell, G., Goodwin, D., and Holborow, E. J. (1982) A simple method of reducing the fading of immunofluorescence during microscopy. *J. Immunol. Methods* **43,** 349, 350.

22. Giloh, H. and Sadat, J. W. (1982) Fluorescence microscopy: reduced photobleaching of rhodamine and fluorescein protein conjugates by *n*-propyl gallate. *Science* **217,** 1252–1255.

23. Bock, G., Hilchenbach, M., Schauenstein, K., and Wick, G. (1985) Photometric analysis of anti-fading agents for immunofluorescence with laser and conventional illumination sources. *J. Histochem. Cytochem.* **33,** 699–705.

24. Oliver, C. (1994) Fixation and embedding, in *Immunocytochemical Methods and Protocols* (Javois, L. C., ed.), Humana, Totowa, NJ, pp. 291–298.

25. Oliver, C. (1994) Colloidal gold/streptavidin methods, in *Immunocytochemical Methods and Protocols* (Javois, L. C., ed.), Humana, Totowa, NJ, pp. 309–313.

26. Oliver, C. (1994) Pre-embedding labeling methods, in *Immunocytochemical Methods and Protocols* (Javois, L. C., ed.), Humana, Totowa, NJ, pp. 315–319.

27. Oliver, C. (1994) Postembedding labeling methods, in *Immunocytochemical Methods and Protocols* (Javois, L. C., ed.), Humana, Totowa, NJ, pp. 321–328.

28. Grube, D. (1980) Immunoreactivities of gastrin (G) cells. II. Nonspecific binding of immunoglobulins to G-cells by ionic interactions. *Histochemistry* **66,** 149–167.

29. Humason, G. L. (1967) *Animal Tissue Techniques.* W. H. Freeman, San Francisco, CA.

30. Boon, M. E. and Kok, L. P. (1994) Microwaves for immunohistochemistry. *Micron* **25,** 151–170.

31. Jamus, M. C., Faraco, C. D., Lunardi, L. O., Siraganian, R. P., and Oliver, C. (1995) Microwave fixation improves antigenicity of glutaraldehyde-sensitive antigens while preserving ultrastructural detail. *J. Histochem. Cytochem.* **43,** 307–311.

32. Ponder, B. A. J. and Wilkinson, M. M. (1981) Inhibition of endogenous tissue alkaline phosphatase with the use of alkaline phosphatase conjugates in immunohistochemistry. *J. Histochem. Cytochem.* **29,** 981–984.

33. Javois, L. C., Bode, P. M., and Bode, H. R. (1988) Patterning in the head of hydra as visualized by a monoclonal antibody. II. The initiation and localization of head structures in regenerating pieces of tissue. *Develop. Biol.* **129,** 390–399.

34. Javois, L. C. and Mullins, J. M. (1994) Overview of fluorescence photomicrography, in *Immunocytochemical Methods and Protocols* (Javois, L. C., ed.), Humana, Totowa, NJ, pp. 331–334.

35. Inoué, S. (1986) *Video Microscopy.* Plenum, New York.

36. Mullins, J. M. (1999) Overview of fluorochromes, in *Immunocytochemical Methods and Protocols,* Second Edition (Javois, L. C., ed.), Humana, Totowa, NJ, in press.

37. Forsgren, A. and Sjöquist, J. (1966) 'Protein A' from *S. aureus*. I. Pseudoimmune reaction with human gamma-globulin. *J. Immunol.* **97,** 822–827.

38. Björck, L. and Kronvall, G. (1984) Purification and some properties of streptococcal protein G, a novel IgG-binding reagent. *J. Immunol.* **133,** 969–974.

39. Kornick, M. N. (1986) The use of phycobilliproteins as fluorescent labels in immunoassay. *J. Immunol. Methods* **92,** 1–13.

40. Bruins, S., Dejong, M. C. J. M., Heeres, K., Wilkinson, M. H. F., Jonkman, M. F., and Vandermeer, J. B. (1995) Fluorescence overlay antigen mapping of the epidermal basement membrane zone. 3. Topographic staining and effective resolution. *J. Histochem. Cytochem.* **43,** 649–656.

41. Rothbarth, Ph. H., Tanke, H. J., Mul, N. A. J., Ploem, J. S., Vliegenthart, J. F. G., and Ballieux, R. E. (1978) Immunofluorescence studies with 4-acetamido-4'-isothiocyanatostilbene2,2'-disulphonic acid (SITS). *J. Immunol. Methods* **19,** 101–109.

42. Staines, W. A., Meister, B., Melander, T., Nagy, J. I., and Hökfelt, T. (1988) Three-color immunofluorescence histochemistry allowing triple labeling within a single section. *J. Histochem. Cytochem.* **36,** 145–151.

43. Whitaker, J. E., Haughland, R. P., Moore, P. C., Hewitt. P. C., Reese, M., and Haughland, R. P. (1991) Cascade blue derivatives: water soluable, reactive, blue emission dyes evaluated as fluorescent labels and tracers. *Anal. Biochem.* **198,** 119–130.

44. Beavis, A. and Pennline, K. J. (1994) Simultaneous measurement of 5-cell surface antigens by 5-colour immunofluorescence. *Cytometry* **15**, 371–376.

45. Nakane, P. and Pierce, G. (1966) Enzyme-labeled antibodies: preparation and application for the localization of antigens. *J. Histochem. Cytochem.* **14**, 929–931.

46. Farr, A. and Nakane, P. (1981) Immunohistochemistry with enzyme labeled antibodies: a brief review. *J. Immunol. Methods* **47**, 129–144.

47. Hsu, S. and Soban, E. (1982) Color modification of diaminobenzadine (DAB) precipitation by metallic ions and its application for double immunohistochemistry. *J. Histochem. Cytochem.* **30**, 1079.

48. Sternberger, L. (1979) *Immunocytochemistry,* 2nd ed. Wiley, NY.

49. Bratthauer, G. L. (1994) The peroxidase-antiperoxidase (PAP) method, in *Immunocytochemical Methods and Protocols* (Javois, L. C., ed.), Humana, Totowa, NJ, pp. 165–173.

50. Hsu, S., Raine, L., and Fanger, H. (1981) The use of avidin-biotin-peroxidase complex (ABC) in immunoperoxidase technique- A comparison between ABC and unlabeled antibody (PAP) procedures. *J. Histochem. Cytochem.* **29**, 577.

51. Bratthauer, G. L. (1994) The avidin-biotin complex (ABC) method, in *Immunocytochemical Methods and Protocols* (Javois, L. C., ed.), Humana, Totowa, NJ, pp. 175–184.

52. Elias, J., Margiotta, M., and Gabore, D. (1989) Sensitivity and detection efficiency of the peroxidase antiperoxidase (PAP), avidin-biotin complex (ABC), and the peroxidase-labeled avidin-biotin (LAB) methods. *Am. J. Clin. Pathol.* **92**, 62.

53. Bratthauer, G. L. (1994) Labeled avidin binding (LAB) method, in *Immunocytochemical Methods and Protocols* (Javois, L. C., ed.), Humana, Totowa, NJ, pp. 185–193.

54. Singer, S. J. and Schick, A. F. (1961) The properties of specific stains for electron microscopy prepared by conjugation of antibody molecules with ferritin. *J. Biophys. Biochem. Cytol.* **9**, 519–537.

55. Faulk, W. P. and Taylor, G. M. (1971) An immunocolloid method for the electron microscope. *Immunocytochem.* **8**, 1081–1083.

56. Romano, E. L. and Romano, M. (1977) Staphylococcal protein A bound to colloidal gold: a useful reagent to label antigen-antibody sites in electron microscopy. *Immunocytochem.* **14**, 711–715.

57. DeMay, J., Moeremans, M., Geuens, G., Nuyden, R., and DeBrabander, M. (1981) High resolution light and electron microscopic localization of tubulin with IGS (immunogold staining) method. *Cell Biol. Int. Rep.* **5**, 889–893.

58. Oliver, C. (1994) Preparation of colloidal gold, in *Immunocytochemical Methods and Protocols* (Javois, L. C., ed.), Humana, Totowa, NJ, pp. 299–302.

59. Oliver, C. (1994) Use of immunoglold with silver enhancement, in *Immunocytochemical Methods and Protocols* (Javois, L. C., ed.), Humana, Totowa, NJ, pp. 211–216.

60. Humbel, B. M., Sibon, O. C. M., Stierhof, Y. D., and Schwarz, H. (1995) Ultra-small gold particles and silver enhancement as a detection system in immunolabeling and in situ hybridization experiments. *J. Histochem. Cytochem.* **43**, 735–737.

61. Bendayan, M. (1982) Double immunocytochemical labeling applying the protein A-gold technique. *J. Histochem. Cytochem.* **30**, 81–85.

62. Hnatowich, D. J. (1990) Recent developments in the radiolabeling of antibodies with iodine, indium, and technetium. *Sem. Nuclear. Med.* **20**, 80–91.

63. Frisman, D. M. (1994) Overview of immunocytochemical approaches to the differential diagnosis of tumors, in *Immunocytochemical Methods and Protocols* (Javois, L. C., ed.), Humana, Totowa, NJ, pp. 405–427.

64. Cunningham, R. E. (1994) Fluorescent labeling of DNA, in *Immunocytochemical Methods and Protocols* (Javois, L. C., ed.), Humana, Totowa, NJ, pp. 239–241.

65. Cunningham, R. E. (1994) Deparaffinization and processing of pathologic material, in *Immunocytochemical Methods and Protocols* (Javois, L. C., ed.), Humana, Totowa, NJ, pp. 243–247.

66. Coon, J. S., Landay, A. L., and Weinstein, R. S. (1986) Flow cytometric analysis of paraffin-embedded tumors: implications for diagnostic pathology. *Human Pathol.* **17,** 435–437.

67. Herman, G. E., Elfont, E. A., and Floyd, A. D. (1994) Overview of automated immunostainers, in *Immunocytochemical Methods and Protocols* (Javois, L. C., ed.), Humana, Totowa, NJ, pp. 383–403.

68. Bratthauer, G. L. (1994) Preparation of frozen sections for analysis, in *Immunocytochemical Methods and Protocols* (Javois, L. C., ed.), Humana, Totowa, NJ, pp. 67–73.

69. Bratthauer, G. L. (1994) Processing cell-touch preparations, in *Immunocytochemical Methods and Protocols* (Javois, L. C., ed.), Humana, Totowa, NJ, pp. 75–79.

70. Bratthauer, G. L. (1994) Processing of cytological specimens, in *Immunocytochemical Methods and Protocols* (Javois, L. C., ed.), Humana, Totowa, NJ, pp. 81–87.

71. Cunningham, R. E. (1994) Overview of flow cytometry and fluorescent probes for cytometry, in *Immunocytochemical Methods and Protocols* (Javois, L. C., ed.), Humana, Totowa, NJ, pp. 219–224.

72. Wolf, K. M. and Mullins. J. M. (1987) Cytochalasin B-induced redistribution of cytokeratin filaments in PtK_1 cells. *Cell Motility Cytoskel.* **7,** 347–360.

73. Willingham, M. C. and Pastan, I. (1985) *An Atlas of Immunofluorescence in Cultured Cells.* Academic, Orlando, FL.

74. Pine, P. S. (1994) Overview of laser microbeam applications as related to antibody targeting, in *Immunocytochemical Methods and Protocols* (Javois, L. C., ed.), Humana, Totowa, NJ, pp. 349–365.

75. Matyus, L. (1992) Fluorescence resonance energy transfer measurements on cell surfaces. A spectroscopic tool for determining protein interactions. *J. Photochem. Photobiol. B: Biol.* **12,** 323–337.

76. Anders, J. J. and Woolery, S. (1992) Microbeam laser-injured neurons increase in vitro astrocyte gap junctional communication as measured by fluorescence recovery after photobleaching. *Lasers Surg. Med.* **12,** 51–62.

77. Jay, D. G. (1988) Selective destruction of protein function by chromophore-assisted laser inactivation. *Biochemistry* **85,** 5454–5458.

78. Seeger, S., Monjembashi, S., Hutter, K-J., Futterman, G., Wolfrum, J., and Greulich, K. O. (1991) Application of laser optical tweezers in immunology and molecular genetics. *Cytometry* **12,** 497–504.

79. Javois, L. C. (1994) Fluorescent labeling of surface or intracellular antigens in whole-mounts, in *Immunocytochemical Methods and Protocols* (Javois, L. C., ed.), Humana, Totowa, NJ, pp. 143–153.

80. Vielkinc, U. and Swierenga, S. H. (1989) A simple fixation procedure for immunofluorescent detection of different cytoskeletal components within the same cell. *Histochemistry* **91,** 81–88.

81. Harvath, L. (1994) Overview of fluorescence analysis with the confocal microscope, in *Immunocytochemical Methods and Protocols* (Javois, L. C., ed.), Humana, Totowa, NJ, pp. 337–347.

82. Pardue, M. L. and Gall, J. G. (1969) Molecular hybridization of radioactive DNA to the DNA of cytological preparations. *Proc. Natl. Acad. Sci. USA* **64,** 600–604.

83. Nardone, R. M. (1994) Nucleic acid immunocytochemistry, in *Immunocytochemical Methods and Protocols* (Javois, L. C., ed.), Humana, Totowa, NJ, pp. 367–374.

84. Nardone, R. M. (1994) Fluorescence in situ hybridization using whole chromosome library probes, in *Immunocytochemical Methods and Protocols* (Javois, L. C., ed.), Humana, Totowa, NJ, pp. 375–382.

85. Komminoth, P., Long, A. A., Ray, R., and Wolfe, H. J. (1992) In situ polymerase chain reaction detection of viral DNA, single copy genes, and gene rearrangements in cell suspensions and cytospins. *Diag. Mol. Pathol.* **1,** 85–97.

86. Reid, T., Baldini, A., Rand, T. C., and Ward, D. C. (1992) Simultaneous visualization of seven different DNA probes by in situ hybridization using combinatorial fluorescence and digital imaging microscopy. *Proc. Natl. Acad. Sci. USA* **89,** 1388–1392.

Flow Cytometry

Robert E. Cunningham

1. Introduction

Flow cytometry (FCM) is a powerful technique for rapid analysis and sorting (separation) of cells and particles. In theory, it can be used to measure any cell component, provided that a measurable physical property or fluorescent tracer is available that reacts specifically and stoichiometrically with that constituent, i.e., cell-surface receptor, DNA, intracellular receptor, RNA, or chromosomes. The technique provides statistical accuracy, reproducibility, and sensitivity. For approximately the past 20 years FCM has enjoyed increasing utilization in the areas of hematology, immunology, cancer biology, pharmacology and karyotype analysis, mechanisms of multidrug resistance (MDR), cell-cycle analysis, chromosome-ploidy analysis, and analysis for somatic-cell genetics and chromosome sorting (1).

Although sorting is still one of FCM's most powerful aspects, it is not an extensively used procedure. This is changing. In the early days of FCM one of the potential users of sorted cells was in the area of biochemical assays, which necessitated millions of isolated cells. To sort a million cells of a rare subpopulation, <0.01% of the total population, would mean days of cell sorting. This is not to say that this cannot be done. Today there are some very elegant techniques for cell separation, such as velocity sedimentation, isopycnic centrifugation (Percoll, Ficoll, Metrizamide), isoelectrophoresis, (isoelectric focusing) and affinity methods (affinity chromatography, immunoaffinity, magnetic immunologic beads) that can sort with minimal effort and have large returns in the number of cells recovered.

Although FCM was initially used just to count cells, it has developed into a method of measuring the physical and chemical properties of individual cells when they are detected in a fluid stream by electronic sensors. With the application of fluorescence molecules, especially fluorescence-conjugated antibodies, the spectrum of application has expanded and the advent of molecular biology has further extended the use of FCM.

1.1. Principles of Flow Cytometry

Today, more sophisticated flow cytometers combine state-of-the-art advances in computers, laser technology, and optical sensing to obtain an objective and precise

From: *Molecular Biomethods Handbook*
Edited by: R. Rapley and J. M. Walker ©Humana Press Inc., Totowa, NJ

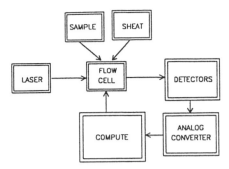

Fig. 1. The necessary components of a flow cytometer.

measurement of multiple-characteristic parameters of an individual cell-by-cell basis. These parameters include size, volume, granularity (refractive index), and fluorescence caused by interactions of specific probes with cell-surface antigens or cytoplasmic molecules *(2,3)*. Flow cytometers now range from instruments with as many as three or four different lasers designed for research applications costing as much as $250,000, to clinical instruments with one argon laser, which are less expensive and versatile but can provide rapid and quantitative clinical information *(4,5)*.

Comparisons can be made between FCM and light microscopy. The microscope slide would equate to the fluidics system to deliver sample for investigation. The lens and light of the microscope are the lasers and beamshaping lenses of the FCM which are quite similar. The most interesting comparison is in the signal-detection mechanisms. Where light microscopy utilizes the human eye and brain to detect morphologic and colorimetric differences, FCM uses a computer, photomultiplier tubes, and signal processor to monitor samples. An advantage of FCM, is that each signal from each parameter can be digitized over a 10-Amp continuum that is unbiased and reproducible. The eye makes measurements on a more subjective and less reproducible scale.

Basically, a flow cytometer has a light source, usually a laser or a mercury arc lamp; a sample chamber; flow cell or jet-in-air nozzle; sheath-fluid stream; a photodetector or photomultiplier tubes (PMTS) that converts the collected light to electronic signals; a signal-processing system that processes digital signals from analog output; and a computer to direct operations, store the collected signals, display data and drive the sorting process *(6)*. The fluid stream, laser, detector system are designed in an orthogonal format, i.e., right angled (**Fig. 1**).

When a cell passes through a sample tube into the center of the nozzle, it creates hyearsodynamic focusing. This centers cells to pass in front of an elliptically shaped narrow laser beam and light is scattered in all directions. The scattered light is first used to evaluate the intrinsic properties of a cell. The narrow angle (2–15%) forward scattered laser light (FALS) is a measure of size, and right angle scattered laser light (RALS) is a measure of general intracellular complexity and granularity. Fluorescence is also gathered at a 90° angle and is further defined with the use of dichroic mirrors, band-pass filters, and cut-off filters, and consequently directed to separate PMTs. Small cells do not scatter as much light as larger cells do. Similarly, light scattered and collected at right angles to the intersection of the sheath-fluid stream and the laser is proportional to the granularity or internal complexity of the cell (**Fig. 2**). Therefore, cells

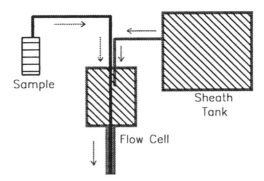

Fig. 2. Optical sensing area and sorting area of a flow cytometer.

with fewer granules or less internal structure do not scatter as much laser light at 90° (RALS) as do cells with more granules *(7)*. Likewise, cells with dim fluorescence do not emit as much fluorescence light as cells with bright fluorescence. Fluorescence emitted by cells is usually attributable to the interaction of cells with dye or fluorescent antibodies and is used to measure extrinsic properties as surface antigen or DNA content *(8,9)*. Fluorescence wil be discussed more fully later in the chapter. So, the amount of scattered laser light and laser induced fluorescence falling on each PMT or detector is converted into a proportional electronic signal that can be observed as an electronic pulse on an oscilloscope. The pulse height, which is proportional to the light scatter and fluorescence, is measured, amplified, and converted by analog-to-digital converters into a digitized value and fed to a computer.

The most commonly used light source is a laser and the most commonly used is an argon laser, which can emit a very narrow elliptical beam of light ranging from 365 to 514 nm. The wavelength most often used is 488 nm, the excitation wavelength for a number of common fluorochromes. Although the earliest flow cytometers used powerful (5 Watt) water-cooled Argon lasers, in recent years smaller (15 milliwatt) air-cooled lasers have proved satisfactory for many applications. Other types of lasers such as Helium-Neon, Krypton, and Helium-Cadmium, dye and can be used in combination with certain dyes for special applications and thus the need for more data parameters *(10)*.

The power of FCM lies not only in real-time acquisition and analysis/sorting but also in reevaluating results of previous samples. The rendering of multiparameters allows for gating and further analysis of collected data in the manner of a "what if" scenario. This can be invaluable in analysis and then re-analysis of data. The multiparameter use in sorting as a means of further limiting the population to select only the desired subpopulation. This could be particularly useful if a small cell with large amount of fluorescent antibody or a large cell with a large amount of antibody A and smaller amount of antibody B *(11)*. The idea of a signal continuum vs a less quantification of merely a presence or absence of antibody is especially critical in the area of sorting cells *(12,13)*.

1.2. Cell Sorting

One of the most powerful aspects of FCM is the ability to physically separate single cells from a mixed population. For sorting, the flow cell or nozzle assembly is vibrated

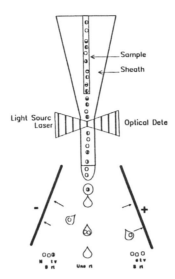

Fig. 3. A more defined area of the flow portion where fluid dynamics are confined.

by an oscillator-driven piezoelectric crystal at a frequency near the natural drop break-up frequency of the flow stream. This stabilizes the drop formation at that frequency, resulting in uniform drop size and a well-defined time between detection of a cell at the laser beam intersection and incorporation of the cell into a free drop. If a cell is to be sorted, a potential in the 100 V range is applied to the fluid inside the nozzle/flow cell. Because the fluid is conductive, drops that break from the jet while the voltage is applied will carry a corresponding electric charge. The timing and duration of the applied voltage are chosen to charge one or more drops which will contain the desired cell. Two populations of cells can be sorted simultaneously by applying a positive charge to drops containing one population and a negative charge to drops containing the other. The procession of drops passes between two deflection plates charged at plus and minus several thousand volts. The electric field between the plates deflects the charged drops away from the uncharged ones. Undetected drops are deposited in a central-waste aspirator, and the deflected drops are collected in appropriate tubes or tissue-culture plates. The aspirator is very effective in preventing formation of aerosols by the undeflected drops. This can be important when handling biohazard materials (**Fig. 3**).

Sorting decisions are usually made by combining independent "windows" on the signals so that a cell selected for sorting if its light scatter signal falls within a selected range and each of its fluorescence signals falls within the range selected on that channel. In two dimensions (e.g., light scatter vs green fluorescence), the selection region can be visualized as a rectangular or a free-form polygon, such that cells whose signals correspond to a point in the selected region are to be sorted, and those outside the box are not sorted.

Once the decision about the preferred disposition of the cell has been made (i.e., left sort, right sort, or no sort) the electronics must produce the drop-charging pulse at the appropriate time. It is also usually desirable to protect the purity of sorted fractions by examining situations in which a cell that fulfill the sorting criteria is close to a cell that does not ("coincidences"), so that sorting of any drop that might contain the wrong cell can be avoided.

As the cell populations under investigation and the questions being asked about them become more complex, it becomes important to increase the array of measurements made on each cell *(14)*. A number of systems have been produced using two or three lasers to excite several dyes *(15)*. Use of such systems may require three or four fluorescence-detection channels, two channels for light scatter, one channel for time tags, and computerized collection and monitoring of the data and sorting. Also light-scatter measurements in angular ranges other than the moderate forward angles aforementioned have proven useful in analyzing complex populations.

FCM and cell sorting have become established technologies in cell biology. New protein and DNA dyes have made it possible to analyze large number of cells individually for multiple properties. These techniques have had a large impact on cellular immunology and the study of cell proliferation. New fluorescent molecules that report on intracellular conditions are used increasingly to study cell physiology.

2. Fluorescent Probes

The basis for a fluorescent signal is a physical paradigm known as a Stoke's shift. This occurs when a molecule absorbs energy as light, it goes from the ground state to an excited state, and then releases excess energy, i.e., fluorescent emission. The time period between the initial absorption and the emission is referred to as the fluorescent lifetime. The shift can be a few nanometers to as many as 400 nm and fluorescence always emits at a longer wavelength than its absorbance wavelength. Fluorescence staining for FCM analysis falls into three categories: methods in which a fluorescent ligand accumulates on or within the cell, methods that require the ligand to interact with a cellular component to release the fluorophore and result in light emission; and methods that rely on fluorophore-coupled antibody binding. In order to identify complex patterns of cell antigens, it is necessary to be able to visualize several antigens at one time. This can be accomplished through the use of fluorochromes, that are excited by the same wavelength of light but have distinct emission spectra. Through the appropriate combination of optical filters, each fluorochrome's emission can be separated and distinguished. The number of different fluorochromes that can be detected on a single cell is limited only by the ability of the optical filters to separate emitted light. Ideally, the fluorochromes should also have high-absorption coefficients. This fluorescent-detection technique provides statistical accuracy, reproducibility, and sensitivity, and allows the simultaneous measurement of several constituents on a cell-by-cell basis. Information is derived from optical responses of the fluorescent probe and the light scattering capability of the labeled, individual cells as they stream single file through a laser beam. Each flash of fluorescence is scattered light is collected, electronically filtered, converted to an electronic signal, amplified, digitized, and stored for later analysis. Cells may be fluorescently labeled with probes that bind directly to cellular constituents or indirectly using an unlabeled first probe followed by a second, labeled probe (e.g., fluorescently labeled antibody. Extrinsic and intrinsic fluorescent probes allow selective examination of both functional and structural components of cells.

It was shown by Creech and Jones *(16)* in 1940 that proteins, including antibodies could be labeled with a fluorescent dye (phenylisocyanate) without biological or immunological effect to the intended target. In theory, fluorescent reporters (tracers, probes, antibodies, stains, and so forth) can be used to measure any cell constituent, provided that the tag reacts specifically and stoichiometrically with the cellular constituent in question *(17)*. Today, the repertoire of fluorescent probes is expanding almost daily (*see* Further Reading). One area that has benefitted from the ever-increasing number of fluorescent probes is FCM and more recently, molecular biology. The use of monoclonal antibodies (MAbs) with flow cytometry for the analysis of hematopoietic cells has greatly advanced the field of diagnostic pathology; new concepts for diagnosis and classification based on quantitative measurements of cellular parameters and the expression of specific differentiation antigens on the surface of cells have been established *(18–20)*.

2.1. Sample Preparation, Disaggregation, and Deparaffinization

Before any FCM, or any molecular techniques, for that matter, can be performed, a single cell or DNA suspension must be obtained. The extraction of material for FCM and molecule techniques is overall very similar except that the FCM technique stops with whole nuclei. The extracellular matrix of mammalian tissue is composed of a complex mix of constitutive proteins. This matrix must be broken down to effectively recover single cells or DNA for culture and/or staining *(21)*. Tissue dissociation and its affiliated problems were described and defined over 70 years ago by Rous and Jones *(22)*. More recent reviews *(23,24)* have revealed newer methods for creating single-cell suspensions. Numerous procedures exist for dissociating solid tumors. They are usually multistep procedures involving one or a combination of mechanical, enzymatic, or chemical manipulations.

Ideally, the dissociation protocol is individualized for the tissue of interest and evaluated relative to both optimal and representative cell yield. Mechanical dissociation of tissue may involve repeated mincing with scissors or sharp blades, scrapping the tissue surface, homogenization, filtration through a nylon or steel mesh, vortexing, repeated aspiration through pipets or small-gauge needles, abnormal osmolality stress, or any combination of these techniques. These methods result in variable cell yields and cell viability. There are various enzymes which can be used, alone or in combination, to digest desmosomes, stromal elements, and extracellular and intercellular adhesions. Enzymes commonly used include: trypsin, pepsin, papain, collagenase, elastase, hyaluronidase, pronase, chymotrypsin, catalase, and dispase. The most routinely used enzymes are collagenase and dispase. DNase is used with these proteolytic enzymes to hydrolyze the DNA-protein complexes, which often entrap cells and can lead to reaggregation of suspended cells. The different specificities of these enzymes for intercellular components allows one to design a dissociation protocol to a specific tumor and for specific purposes. Many enzymes are crude extracts that contain varying amounts of contaminating proteolytic enzymes. The enzymatic method *(25,26)* is probably the method of choice as a starting point for most tissue types because of its ability not only to release a large number of cells but also to preserve cellular integrity and viability *(27)*. Lastly, chemical dissociation is commonly used in conjunction with mechanical or enzymatic procedures. Chemical methods are designed to omit or sequester the Ca^{2+} and Mg^{2+} ions needed for maintenance of the intercellular matrix and cell surface integrity. Ethylene-diamine tetraacetic acid (EDTA) or citrate ion are commonly used to remove these cations, but do not adequately dissociate all types of tissue.

DNA content has become an important diagnostic, as well as prognostic, tool in clinical pathology and investigative oncology *(28)*. Paraffin-embedded tissue (PET) can be examined by FCM methods for total DNA content and aneuploidy with respect to the classification of the original pathologic diagnosis. The relative significance of studies on archival material permits retrospective analysis on a great number of cases, studying different specimen of a tumor for intratumor heterogeneity while comparing results from previous pathologic evaluations *(29,30)*. DNA content as measured in PET is closely related to that obtained from fresh specimens. Still, a major drawback in the procedure is that only nuclei are recovered for analysis of DNA content.

The method that has been used as a template is the Hedley technique *(31,32)* is used to prepare nuclear suspensions from paraffin-embedded tissue samples. Microtome sections are dewaxed, hydrated, and incubated in a proteolytic enzyme (pepsin, pronase, trypsin) with intermittent vortexing and mechanical disruption to release the nuclei. After completion of the tissue digestion, the nuclei are either suspended in 70% EtOH for storage or stained with propidium iodide (PI) for FCM analysis. There are three alternative techniques for preparation of the nuclei: on microslides, in "tea bags," or in test tubes.

2.2. Cellular Staining: Antibody and DNA

The combination of the specificity of the antigen-antibody interaction with the exquisite sensitivity of fluorescence detection and quantitation yields one of the most widely applicable analytical tools in cell biology *(33–35)*. Within the last decade, FCM has become an integral part of basic immunological research. Elaboration of this technology has been intensively stimulated by a rapidly growing sophistication in MAb technology and vice versa *(36)*.

The added specificity of MAbs in immunocytochemical technology provides a consistent and reliable method for exploiting the range of pure antibodies and subclasses of antibodies. These antibodies provide a means of defining cell surface, intracellular, and membrane epitopes for single cells as well as tissue sections. When these antibodies are "tagged" with a fluorescent-reporter antibody, multiple markers are possible. In particular, methods using protein A or the avidin-biotin complexed with alternative fluorescent tags as second steps have added significant latitude to the immunofluorescence technique. An increasing number of clinical laboratories are using flow cytometry to analyze cells stained with fluorescent antibodies, dyes, or receptors *(37–39)*. Other techniques are available as in the area of hematology, where isolation of cell types may be deemed necessary to give good immunocytochemical results. Also, human genes coding cell-surface molecules can be introduced into host cells using a variety of somatic-cell genetic techniques *(40)*, where flow cytometry can then be used to monitor the effectiveness of the genetic techniques.

Another technique, although not as practiced, is the use of antibodies on fixed cells or fixing cells after antibody staining. The detection of intracellular proteins is less well-developed, in large part because antibodies can bind nonspecifically to dying cells and dead-cell components, which leads to considerable biological noise in the fluorescence detectors. There is also signal-to-noise caused by the intra- and/or intermolecular ionic interactions during the process of fixation, which reduces the ability to detect the desired protein(s) within the cell. This is a double edged sword for labeling cells. It is very important to start the staining procedure with a viable cell suspension. If the starting material is viable, at least one of the two problems associated with cell fixation and staining is remedied. The fixation protocols are varied not only for their uses of crosslinking agents, permeabilization agents, and/or precipitating agents, but also for time and temperature *(41)*.

The fixation conditions used to prepare cells for antibody application are assumed to preserve the distributions of the protein(s) being examined *(42)*. Soluble proteins can be redistributed into inappropriate locations and can be differentially extracted from native locations during the permeabilization and fixation of the cells before antibody application *(43,44)*. Further, no cell aggregation or alteration of the intracellular antigenicity should occur in the permeabilization/fixation treatment. The fixation/stain methodology, with and without permeabilization, can be accomplished in various ways depending upon the exact site of the organelle or cell constituent to be stained.

The stain/fixation method is usually used for surface markers, that can withstand fixation, followed with a DNA fluorochrome. The fixation/stain method is used not only for surface markers that can withstand fixation but also for intracellular constituents such as cytoplasmic proteins, membrane and cytoplasmic antigens, nuclear membrane and nuclear protein staining. This is accomplished by using a crosslinking fixative (e.g., Paraformaldehyde (PFA), Formalin) followed by a permeabilization agent, such as Triton X-100, Tween-20, saponin, or lysolecithin. Some of the precipitating agents (e.g., Ethanol, Methanol, Acetone) can also be used for permeabilization after the initial fixation with PFA or formalin, or they can be used alone for both fixation and permeabilization.

Finally, the determination of methodology for cell staining must be evaluated on the type of tissue/cells being examined. It is absolutely critical that the sample be a viable, single-cell suspension. Not only is this important during the staining and data collection, but it is also important in the analysis of the specimen as representative of the pathologic sample.

Cellular DNA content along with gene products *(45)* is one of the most widely measured parameters using FCM, especially in the area of paraffin-embedded tissue where only the nucleus is remaining after release from the tissue. Nuclei can be extracted from archival material for ploidy studies. DNA- or nucleic acid-specific ligands (DAPI, Hoechst, ethidium bromide, propidium iodide) as well as polyanion dyes (acridine orange, acriflavine) are commonly employed. The quantitative cytochemical determination of DNA has been carried out using cytofluorochemical stains and offers a direct measurement of the DNA content of individual cells in a population *(46,47)*. The fluorescence distribution produced by a cell suspension provides a representation of the cell-cycle distribution of this population. Based upon histograms generated from FCM data of the DNA content of individual cells, three groups can be identified in an asynchronous and mitotically active cell population. Most cells are in a resting (G0/G1) phase also known as Gap0 and Gap1. As cells enter the synthesis (S) phase, the amount of cellular DNA increases, resulting in increased fluorescence. After S phase, cells enter the Gap2/Mitosis (G2/M) phase where very little additional DNA is synthesized and cell division occurs. Now the cells contain twice the amount of DNA with approximately twice the staining intensity of the G0/G1 phase. One DNA dye that receives a great deal of use is propidium iodide (PI) *(48)*. It binds to both double-stranded DNA and double-stranded RNA. Therefore, RNase is used to reduce the double-stranded RNA resulting in only DNA staining. Acridine orange has the unusual property of fluorescing green (520 nm) when bound to double-stranded nucleic acids and red (640 nm) when bound to single-stranded nucleic acid . Finally, flow cytometric DNA staining has been regarded as an objective prognostic parameter in several types of human cancer, but it is important to remember that this is a "snapshot" of the cell cycle. From a DNA-content histogram, it cannot be determined if a cell is actively moving through the cell cycle, has slowed, or even stopped its traverse through the cell cycle.

DNA content has become an important diagnostic, as well as prognostic, method for clinical pathology and investigative oncology. PET can be examined by FCM methods for total DNA content and aneuploidy with respect to the classification of the original pathologic diagnosis. The relative significance of studies on archival material permits retrospective analysis on a great number of cases, studying different specimen of a tumor for intratumor heterogeneity while comparing results from previous pathologic evaluations *(49)*. DNA content as measured in PET is closely related to that obtained from fresh specimens. Still, a major drawback in the procedure is that only nuclei are recovered for analysis of DNA content. Novel use of the nuclei isolated from PET is to study a low number of nuclei by the polymerase chain reaction (PCR). PCR has the end result of making DNA from a complementary DNA template *(50,51)* cells can be sorted on the basis of cell cycle, digested to single stranded DNA and the desired segment of DNA amplified. Lastly, the isolated nuclei cannot only be stained for DNA content but also stained for nuclear proteins, proliferation factors, and other nuclear proteins.

2.3. Flow Cytometry in Molecular Biology

The development of nucleic-acid probes, other reporters, and techniques within molecular biology has been of revolutionary significance in the biological and medical sciences *(52)*. These probes offer the promise of disease diagnosis through direct-genomic analysis and the detection of genomic products at an unprecedented level of sensitivity. cDNA probes have been used for the diagnosis of infectious diseases, genetic diseases, and neoplasia. The combination of FCM and fluorescent molecular-biology techniques have unprecedented power in the biological sciences and combined the ability of the flow cytometer to identify, quantify, and separate rare-cell populations with the unique specificity of cDNA probes for genetic sequences in cells or microbial organisms *(53)*. One technique that was exclusively used for slide-based assays has recently been availed to FCM. Fluorescence *in situ* hybridization (FISH) techniques also demonstrate great promise for detecting specific sequences of nucleic acid by FCM, particular in instances where multiple copies of the nucleic-acid sequence are present. Detection of a single-gene copy remains difficult owing to poor signal-to-noise ratios and limits the application of FCM assays for direct examination of individual nucleic-acid sequences, such as point mutations of oncogenes *(54)*. It is quite possible that this technical obstacle will be overcome in the near future

One of the most eloquent techniques in the realm of molecular biology is a micro-technique known as the PCR and combined with FCM has further expanded cell sorting *(55)*. Nuclei can be sorted on the basis of cell cycle, digested to single-stranded DNA and the desired segment of DNA amplified. PCR has the end result of making DNA from a complementary DNA template. Further, sorting does not necessarily require live cells for post-sorting analysis and/or PCR, thereby decreasing the complexity of the sorting process. The requirement for live cells can be replaced by the retrospective analysis of paraffin-embedded tissue. Another molecular technique that utilizes FCM and sorting is cytogentics, which is the study of karyotype anomalies by loss or gain of chromosome material and structural changes. Molecular biology gives a means of recognizing chromosome losses and especially to study oncogenic or antioncogenic mutations *(56)*. Sorting allows for the separation of individual chromosomes. Studying these alterations will allow better prediction of high-risk subjects in cancer families *(57)*. Lastly, the integration of FISH and FCM analysis provide more information on the chromosomal abnormalities of these neoplasms *(58)*. A large number of methods for staining and measuring properties of individual cells are available *(59)*. New protein and DNA dyes have made it possible to analyze large number of cells individually for multiple properties. These techniques have had a large impact on cellular immunology and the study of cell proliferation. New fluorescent molecules that report on intracellular conditions are used increasingly to study cell physiology. Chromosome analysis and sorting by FCM becoming a valuable tool and refinements in the techniques for manipulating small quantities of DNA will increase the application of chromosome sorting in molecular biology. The analysis of rare cell populations is still hampered by shortcomings in the present generation of commercial instruments

Lastly, FCM can be used for the study of multidrug resistance (MDR) which is the study of the capacity of modulating agents to result in overexpression of the P-glycoprotein and the functional aspect of MDR in expulsion of the cytotoxic agents *(60,62)*.

3. Summary

Chromosome analysis and sorting by FCM is becoming a valuable tool and refinements in the techniques for manipulating small quantities of DNA and will increase the application of chromosome sorting in molecular biology. Future technology will make use of alternative effects in fluorescence induction, and improved understanding of nonlinear phenomena will lead to new techniques for single-cell analysis.

The techniques of molecular biology are creating new insight into the subcellular pathology of malignancy. The crucial events of carcinogenesis, tumor progression, and metastatic spread are coming into focus at a molecular level. It is clear that FCM and molecular biology will always have a nexus where the capabilities of both disciplines can be utilized, explored and enhanced. The extrinsic and intrinsic fluorescent probes allow selective examination of both functional and structural components of cells. Many commercial flow cytometers are now equipped with up to six detectors and two lasers for two- or three-color fluorescence analysis. Combined with computer processing and display procedures, up to eight one-dimensional fully cross correlated data parameters may be analyzed. Successful flow-cytometric analysis depends upon adequate sample preparation, appropriate selection of probes or reporter markers, instrumention, data display and analysis. Each of these areas is subject to individual interpretation and requires adequate attention to avoid the introduction of artifacts and misinterpretation of results. Flow cytometers tend to be complicated and require a skilled operator for sample analysis, though manufacturers are introducing more compact, user-friendly systems.

The future is bright and awash in new technologies, such as oncogene amplification and/ or overexpression, growth factors, cellular-proliferation rate and ploidy, gene rearrangement, oncogene products *(63)*, estrogens induced proteins, expression of metastasis related molecules, gene mapping, applications to RNA *(64)*, and gene therapy *(65,66)*.

4. Notes

1. The "Handbook of Fluorescent Probes and Research Chemicals" by R. P. Haugland of Molecular Probes, Inc. is a useful resource for over 3000 fluorescent probes and their applications. It is currently in its sixth edition 1995–1997) and is available from Molecular Probes, Inc. 4849 Pitchford Ave., Eugene, OR 97402.

Further Reading

Flow Cytometry and Sorting, Myron R. Melamed, Paul F. Mullaney, Mortimer L. Mendelsohn. (eds.) New York: Wiley, 1979, 716 p.

Flow Cytometry and Sorting, Myron R. Melamed, Tore Lindmo, Mortimer L. Mendelsohn (eds.), New York: Wiley-Liss, 1990, 824 p.

Practical Flow Cytometry, Howard M. Shapiro: New York: Liss, 1985, 295 p.

Practical Flow Cytometry, Howard M. Shapiro: New York: Liss, 1988, 353 p.

Practical Flow Cytometry, Howard M. Shapiro: New York: Wiley-Liss, 1995, 542 p.

Flow Cytometry IV: Proceedings of the IVth International Symposium on Flow Cytometry (Pulse Cytophotometry), O. D. Laerum, T. Lindmo, E. Thorud (eds.), New York: Columbia University Press, 1980, 535 p.

Flow Cytometry, A Powerful Technology for Measuring Biomarkers, James H. Jett (ed.), Los Alamos, Los Alamos, NM: Los Alamos National Laboratory, 1994.

Flow Cytometry, Zbigniew Darzynkiewicz, J. Paul Robinson, Harry A. Crissman (eds.) San Diego: Academic Press, 1994.

Flow Cytometry, Practical Approach, M. G. Ormerod (ed.), New York: IRL Press, 1994, 282 p.

Flow Cytometry: Clinical Applications, Marion G. Macey (ed.), Boston: Blackwell Scientific Publications, 1994.

Handbook of Flow Cytometry Methods, J. Paul Robinson, (ed.), New York: Wiley-Liss, 1993, 246 p.

Flow Cytometry and Clinical Diagnosis, David F. Keren, Curtis A. Hanson, Paul Hurtubise (eds.), Chicago: ASCP Press, 1993.

Flow Cytometry in Clinical Diagnosis, David F. Keren (ed.), Chicago: ASCP Press, 1989, 343p.

Clinical Applications of Flow Cytometry, Roger S. Riley, Edwin J. Mahin, William Ross (eds.), New York: Igaku-Shoin, 1993, 914 p.

Clinical Flow Cytometry, Alan L. Landay (ed.), New York: New York Academy of Sciences, 1993, 468 p.

Clinical Flow Cytometry: Principles and Application, Kenneth D. Bauer, Ricardo E. Duque, T. Vincent Shankey (eds.), Baltimore: Williams & Wilkins, 1993, 635 p.

Introduction to Flow Cytometry, James V. Watson (ed.), New York: Cambridge University Press, 1991, 443 p.

Flow Cytometry and Cell Sorting, A. Radbruch (ed.), New York: Springer-Verlag, 1992, 223 p.

Flow Cytogenetics, Joe W. Gray (ed.), San Diego: Academic Press, 1989, 312 p.

References

1. Wilson, G. D. (1992) The future of flow cytometry. *BJR Suppl.* **24,** 158–162.
2. Fouchet, P., Jayat, C., Hechard, Y. Ratinaud, M. H., and Frelat, G. (1993) Recent advances of flow cytometry in fundamental and applied microbiology. *Biol. Cell.* **78(1–2),** 95–109.

3. Melamed, M. R., Mullaney, P. F., and Mendelsohn, M. L. (1979) *Flow Cytometry* and *Sorting.* John Wiley, New York.

4. Terstappen, L. W. M. M. and Laken, M. R. (1988) Five-dimensional flow cytometry as a new approach for blood and bone marrow differentiation. *Cytometry* **9,** 548–556.

5. Camplejohn, R. S. (1994)The measurement of intracellular antigens and DNA by multiparametric flow cytometry. *J. Microsci.* **176(Pt 1),** 1–7.

6. Giaretti, W. and Nusse, M. (1994) Light scatter of isolated cell nuclei as a parameter discriminating the cell-cycle subcompartments. *Methods Cell Biol.* **41,** 389–400.

7. Cunningham, R. E. (1994) Overview of flow cytometry and fluorescent probes, in *Methods in Molecular Biology vol. 34 Immunocytochemical Methods and Protocols* (Javois, L. C., ed.), Humana,Totowa, NJ, pp. 219–225.

8. van Dam, P. A., Watson, J. V., Lowe, D. G., and Shepherd J. H. (1992) Flow cytometric measurement of cell components other than DNA: virtues, limitations, and applications in gynecologic oncology. *Obstet. Gynecol.* **79(4),** 616–621.

9. Stewart, C. C. (1990) Multiparameter analysis of leukocytes by flow cytometry. *Methods Cell Biol.* **33,** 427–450.

10. Shapiro, H. M. (1988) *Practical Flow Cytometry* 2nd ed., Liss, New York.

11. Tanke, H. J. and van der Keur, M. (1993) Selection of defined cell types by flow-cytometric cell sorting. *Trends Biotechnol.* **11(2),** 55–62.

12. Assenmacher, M., Manz, R., Miltenyi, S., Scheffold, A., and Radbruch, A. (1995) Fluorescence-activated cytometry cell sorting based on immunological recognition. *Clin. Biochem.* **28(1),** 39–40.

13. Knuchel, R. (1994) Analysis by flow cytometry and cell sorting. Report of current status and perspectives for pathology. *Pathologe* **15(2),** 85–95.

14. van den Engh G. (1993) New applications of flow cytometry. *Curr. Opin. Biotechnol.* **4(1),** 63–68.

15. Noguchi, Y. and Sugishita, T. (1992) A historical review of flow cytometers and characteristics of commercial instrumentsy. *Nippon Rinsho* **50(10),** 2299–2306.

16. Creech, H. J. and Jones, R. N. (1940) The conjugation of horse serum albumin with 1,2-benzanthryl isothiocyanate. *J. Am. Chem. Soc.* **62,** 1970–1975.

17. Maftah, A., Huet, O., Gallet, P. F., and Ratinaud, M. H. (1993) Flow cytometry's contribution to the measurement of cell functions. *Biol. Cell.* **78(1–2),** 85–93.

18. Loken, M. R., Brosnan, J. M., Bach, B. A., and Ault, K. A. (1990) Establishing optimal lymphocyte gates for immunophenotyping by flow Cytometry. **11,** 453–459.

19. Garratty, G. (1990) Flow cytometry: its applications to immunohaematology. *Baillieres Clin. Haematol.* **3,** 267–287.

20. Sasaki, K. and Kurose, A. (1992) Cell staining for flow cytometry. *Nippon Rinsho* **50,** 2307–2311.

21. Cunningham, R. E. (1994) Tissue disaggregation, in *Methods in Molecular Biology, vol. 34 Immunocytochemical Methods and Protocols,* (Javois L. C., ed.), Humana,Totowa, NJ, pp. 225–229.

22. Berwick, L. and Corman, D. R. (1962) Some chemical factors in cellular adhesion and stickiness. *Cancer Res.* **22,** 982–986.

23. Rous, P. and Jones, F. S. (1916) A method for obtaining suspensions of living cells from the fixed tissues and for the plating of individual cells. *J. Exp. Med.* **23,** 549–555.

24. Waymouth, C. (1974) To disaggregate or not to disaggregate. Injury and cell disaggregation, transient or permanent? *In Vitro* **10,** 97–111.

25. Freshney, R. I. (1983) *Culture of Animal Cells. A Manual of Basic Technique.* Liss, New York.

26. Lewin, M. J. M. and Cheret, A. M. (1989) Cell isolation techniques: Use of enzymes and chelators. *Methods Enzymol.* **171,** 444–461.

27. Cerra, R., Zarbo, R. J., and Crissman, J. D. (1990) Dissociation of cells from solid tumors. *Methods Cell Biol.* **33,** 1–12.

28. Costa, A., Silvestrini, R., Del Bino, G., and Motta, R. (1987) Implications of disaggregation procedures on biological representation of human solid tumors. *Cell Tissue Kinet.* **20,** 171–180.

29. Cunningham R. E. (1994) Deparaffinization and processing of pathologic material, in *Methods in Molecular Biology, vol. 34 Immunocytochemical Methods and Protocols:* (Javois, L. C., ed.), Humana, Totowa, NJ, pp. 243–249.

30. Coon, J. S., Landay, A. L., and Weinstein, R. S. (1986) Flow cytometric analysis of paraffin-embedded tumors: implications for diagnostic pathology. *Human Pathol.* **17,** 435–437.

31. Hedley, D. W., Friedlander, M. L., Taylor, I. W., Rugg, C. A., and Musgrove, E. A. (1983) Method for analysis of cellular DNA content of paraffin-embedded pathological material using flow cytometry. *J. Histochem. Cytochem.* **31,** 1333–1335.

32. Hedley, D. W., Friedlander, M. L., and Taylor, I. W. (1985) Application of DNA flow cytometry to paraffin-embedded archival material for the study of aneuploidy and its clinical significance. *Cytometry* **64,** 327–333.

33. Cunningham, R. E. (1994) Indirect Immunofluorescent labeling of viable cells, in *Methods in Molecular Biology, vol. 34: Immunocytochemical Methods and Protocols* (Javois, L. C., ed.), Humana,Totowa, NJ, pp. 229–233.

34. Cunningham, R. E. (1994) Indirect Immunofluorescent labeling of fixed cells, in *Methods in Molecular Biology, vol. 34: Immunocytochemical Methods and Protocols* (Javois, L. C., ed.), Humana, Totowa, NJ, pp. 233–239.

35. Bosman, F. T. (1983) Some recent developments in immunocytochemistry. *Histochem. J.* **15,** 189–200.

36. Kung, P. C., Talle, M. A., DeMarie, M. E., Butler, M. S., Lifter, J., and Goldstein, G. (1980) Strategies for generating monoclonal antibodies defining human T lymphocyte differentiation antigens. *Transplant. Proc. XIII* **3,** 141–146.

37. Othmer, M. and Zepp, F. (1992) Flow cytometric immunophenotyping: principles and pitfalls. *Eur. J. Pediatr.* **151,** 398–406.

38. Haaijman, J. J. (1988) Immunofluorescence: quantitative considerations. *Acta. Histochem.* **35(Suppl.),** 77–83.

39. Zola, H., Flego, L., and Sheldon, A. (1992) Detection of cytokine receptors by high-sensitivity immunofluorescence/flow cytometry. *Immunobiology* **185,** 350–365.

40. Kamarck, M. E., Barbosa, J. A., Kuhn, L., Peters, P. G., Shulman, L., and Ruddle, F. H. (1983) Somatic cell genetics and flow cytometry. *Cytometry* **4,** 99–108.

41. Vyth-Dreese, F. A., Kipp, J. B. A., and DeJohn, T. A. M. (1980) Simultaneous measurement of surface immunoglobulins and cell cycle phase of human lymphocytes, in *Flow Cytometry IV* (Laerum, O. D., Lindo, T., and Thorn, E., eds.), Universitetsforlaget, Oslo, Norway, pp. 207–212.

42. Labalette-Houache, M., Torpier, G., Capron, A., and Dessaint, J. P. (1991). Improved permeabilization procedure for flow cytometric detection of internal antigens. Analysis of interleukin–2 production. *J. Immunol. Methods* **138,** 143–153.

43. Pollice, A. A., McCoy, J. P., Jr., Shackney, S. E., Smith, C. A., Agarwal, J., Burholt, D. R., Janocko, L. E., Hornicek, F. J., Singh, S. G., and Hartsock, R. J. (1992) Sequential paraformaldehyde and methanol fixation for simultaneous flow cytometric analysis of DNA, cell surface proteins, and intracellular proteins. *Cytometry* **13,** 432–444.

44. Schmid, I., Uittenbogaart, C. H., and Giorgi, J. V. (1991) A gentle fixation and permeabilization method for combined cell surface and intracellular staining with improved precision in DNA quantification. *Cytometry* **12,** 279–285.

45. Andreeff, M., Slater, D. E., Bressler, J., and Furth, M. E. (1986) Cellular ras oncogene expression and cell cycle measured by flow cytometry in hematopoietic cell lines. *Blood* **67,** 676–681.

46. Fraker, P. J., King, L. E., Lill-Elghanian, D., and Telford, W. G. (1995) Quantification of apoptotic events in pure and heterogeneous populations of cells using the flow cytometer. *Methods Cell Biol.* **46,** 57–76 .

47. Taylor, I. W. (1980) A rapid single step staining technique for DNA analysis by flow microfluorimetry. *J. Histochem. Cytechem.* **28(11),** 1224–1232.

48. Krishan, A. (1975) Rapid flow cytofluorometric analysis of mammalian cell cycle by propidium iodide staining. *J. Cell Biol.* **66,** 188–193.

49. Joensuu, H. and Kallioniemi, O. P. (1989) Different opinions on classification of DNA histograms produced from paraffin-embedded tissue. *Cytometry* **10,** 711–717.

50. Gibbs, R. A. (1990) DNA amplification by the polymerase chain reaction. *Anal. Chem.* **62,** 1202–1214.

51. Rodu, B. (1990) The polymerase chain reaction: the revolution within. *Am. J. Med. Sci.* **299,** 210–216.

52. Keren, D. F. (1989) Clinical Molecular Cytometry: merging flow cytometry with molecular biology in laboratory medicine, in *Flow Cytometry in Clinical Diagnosis* (Keren, D. F, Hanson, C. A., and Hurtubise, P. E., eds.), ASCP, Chicago, pp. 614–634.

53. Tim, E. A. and Stewart, C. C (1992) Fluorescence in situ hybridization in suspension, (FISHES) using digoxigenin-labeled probes and flow cytometry. *Biotechniques* **12(3),** 362–367.

54. Bauman, J., Bentvelzen, P., and van Bekkum, D. (1987) Fluorescent in situ hybridization of MRNA in bone marrow and leukemic cells measured by flow cytometry. *Cytometry* **4(Suppl. 1),** 94–100.

55. Gyllensten, U. B. (1989) PCR and DNA sequencing. *Biotechniques* **7,** 700–708.

56. Milan, D., Yerle, M., Schmitz, A., Chaput, B., Vaiman, M., Frelat, G., and Gellin, J. (1993) A PCR-based method to amplify DNA with random primers: determining the chromosomal content of porcine flow-karyotype peaks by chromosome painting. *Cytogenet Cell Genet.* **62,** 139–141.

57. Gray, J. W. and Cram, L. S. (1990) Flow karyotyping and chromosome sorting, in *Flow Cytometry and Sorting* (Melamed, M. R., Lindmo, T., and Mendelsohn, M. I., eds.), Wiley-Liss, New York, pp. 503–530.

58. Carrano, A. V., Gray, J. W., Langlois, R. G., Burkhart, S. K., and Van Dilla, D. M. (1979) Measurement and purification of human chromosomes by flow cytometry and sorting. *Proc. Natl. Acad. Sci. USA* **76,** 1382–1384,.

59. Cram, L. S., Bartholdi, M. F., Ray, F. A., Meyne, J., Moyzis, R. K., Schwarzacher-Robinson, T., and Kraemer, P. M. (1988) Overview of flow cytogenetics for clinical applications. *Cytometry* **3(Suppl.),** 94–100.

60. Herzog, C. E. and Bates, S. E. (1994) Molecular diagnosis of multidrug resistance. *Cancer Treat. Res.* **73,** 129–147.

61. Leonce, S. and Burbridge, M. (1993) Flow cytometry: a useful technique in the study of multidrug resistance. *Biol. Cell.* **78(1–2),** 63–68.

62. Morrow, C. S. and Cowan, K. H. (1988) Mechanisms and clinical significance of multidrug resistance. *Oncology* **2,** 55–63.

63. Niman, H. L. (1987) Detection of oncogene-related proteins with site-directed monoclonal antibody probes. *J. Clin. Lab. Analysis* **1,** 28–41.

64. Grunwald, D. (1993) Flow cytometry and RNA studies. *Biol. Cell.* **78(1–2),** 27–30.
65. Rice, G. C. and Pennica, D. (1989) Detection by flow cytometry of protoplast fusion and transient expression of transferred heterologoug CD4 sequence in COS–7 cells. *Cytometry* **10,** 103–107.
66. Scharf, S. J., Hom, G. R., and Erlich, H. A. (1986) Direct cloning and sequence analysis of enzymatically amplified genomic sequences. *Science* **233,** 1076–1078.

50

Mass Spectrometry

John R. Chapman

1. The Mass Spectrometer
1.1. Introduction

Mass spectrometry (MS) *(1)* is one of the most important physical methods in analytical chemistry today. A particular advantage of mass spectrometry, compared with other molecular spectroscopies, is its high sensitivity, so that it provides one of the few methods that is entirely suitable for the identification or quantitative measurement of trace amounts of samples.

A mass spectrometer, in its simplest form, is designed to perform the following basic tasks:

1. Produce gas-phase ions from sample molecules (ionization). This is accomplished in the ion source and may involve sample molecules that are in a vapor or solid state or that are in solution.
2. Separate gas-phase ions according to their mass-to-charge ratio. This takes place in the mass analyzer.
3. Detect and record the separated ions.

Conventionally, the processes of ion formation, analysis and detection all take place in vacuum. Recent developments have, however, led to instruments in which ions are, in fact, produced in a source that operates at atmospheric pressure. Overall, a large number of different instrumental configurations can be used to perform these three tasks, and these will now be introduced in the remainder of this section.

1.2. Ionization

1.2.1. Electron Ionization

Electron ionization (EI) was the first ionization method to be used routinely and is still probably the most widely employed method in mass spectrometry overall. It is suitable for a range of synthetic and naturally occurring compounds, but is unsuitable for compounds of any real lability or with a molecular weight much above 800 Daltons because of the need for sample vaporization prior to ionization.

The EI source is a small enclosure traversed by an electron beam that originates from a heated filament and that has been accelerated through a potential of about 70 V

From: *Molecular Biomethods Handbook*
Edited by: R. Rapley and J. M. Walker © Humana Press Inc., Totowa, NJ

into the source. Gas-phase molecules entering the ion source interact with the electrons. Some of these molecules then lose an electron to form a positively charged ion whose mass corresponds to that of the original neutral molecule—this is the molecular ion (**Eq. 1**). Many molecular ions, have sufficient excess energy to decompose further to form fragment ions, which are characteristic of the structure of the neutral molecule (**Eq. 2**). Thus, the molecular ion gives an immediate measurement of the mol wt of the sample, whereas the mass and abundance values of the fragment ions may be used to elicit specific structural information. Taken together the molecular and fragment ions constitute a fingerprint that is the mass spectrum of the original compound (**Fig. 1**).

$$M \xrightarrow{e^-} M^+ + e^- \tag{1}$$
$$M^+ \rightarrow A^+ + B \tag{2}$$

1.2.2. Chemical Ionization

In chemical ionization (CI) mass spectrometry *(2)*, ions that are characteristic of the analyte are produced by ion-molecule reactions rather than by electron ionization. Chemical ionization requires a high pressure (approx 1 torr) of a so-called reagent gas held in an ion source, which is basically a more gas-tight version of the EI source. Electron ionization of the reagent gas, which is present in at least 10,000-fold excess compared with the sample, produces reagent ions (*see* **Eqs. 3** and **4** for reactions of methane reagent gas) that either are nonreactive or react only very slightly with the reagent gas itself, but that react readily, by an ion-molecule reaction (**Eq. 5**), to ionize the sample.

$$CH_4 \rightarrow CH_4, CH_3, CH_2 \tag{3}$$
$$CH_4^+ + CH_4 \rightarrow CH_5^+ + CH_3 \tag{4}$$
$$M + CH_5+ \rightarrow MH^+ + CH_4 \tag{5}$$

Many compounds that fail to give a molecular ion in electron ionization, afford an intense quasi molecular ion indicative of mol wt, with chemical ionization. For example, reagent ions from isobutane or methane ionize sample molecules by proton transfer (**Eq. 5**) to give a positively charged quasimolecular ion at a mass which is one unit higher than the true molecular weight. In addition, unlike electron ionization, which produces only positive ions, chemical ionization can be used to produce useful ion currents of positive or negative ions representative of different samples.

Just as with electron ionization, a major disadvantage of chemical ionization is the need for sample vaporization prior to ionization, which again rules out any application to higher mol-wt, labile materials. On the other hand, chemical ionization processes are implicated in a number of other ionization techniques, such as fast-atom bombardment (**Subheading 1.2.4.**) and matrix-assisted laser desorption/ionization (**Subheading 1.2.5.**), which are used for the analysis of macromolecules.

1.2.3. Ionization of Labile and Involatile Materials

The need to extend the applicability of mass spectrometry to much less volatile molecules and to molecules of high lability led to the development of new methods of ionization that do not require prior sample volatilization. One approach (**Subheading 1.2.4. and 1.2.5.**) to this problem was the use of energetic particle bombardment as a means of both volatilization and ionization, whereas another approach (**Subheading**

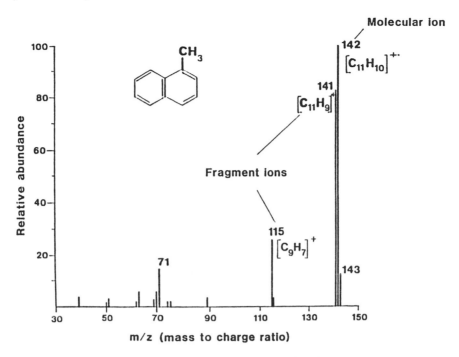

Fig. 1. EI spectrum of methylnaphthalene.

1.2.6.) was the direct field desorption of sample ions from the liquid phase using an intense electrical field.

1.2.4. Ionization by Fast-Atom Bombardment

In fast-atom bombardment (FAB) *(3)* (**Fig. 2**), a beam of fast-moving neutral argon or xenon atoms, directed to strike the sample, which is deposited on a metal-probe tip, produces an intense thermal spike whose energy is dissipated through the outer layers of the sample lattice. Molecules are detached from these surface layers to form a dense gas containing positive and negative ions, as well as neutrals, just above the sample surface. Neutrals may subsequently be ionized by other ions contained within the plasma. Depending on the voltages employed, positive or negative ions may be extracted into the mass analyzer.

With a dry-deposited sample there is a rapid decay in the yield of sample ions owing to surface damage by the incident beam. In fast-atom bombardment, however, the sample is routinely dissolved in a relatively involatile liquid matrix, such as glycerol. The use of a liquid matrix provides continuous surface renewal so that sample ion beams with a useful intensity may be prolonged for periods of several minutes. Subsequently, the use of a neutral primary beam was replaced by the use of a beam of more energetic primary ions, such as Cs^+, and the technique renamed liquid secondary-ion mass spectrometry (LSIMS). Fragment ions are generally of low abundance in FAB spectra and often only mol-wt information is available, e.g., as $(M + H)^+$ ions. Useful fragmentation can, however, be deliberately introduced using tandem mass spectrometry techniques (**Subheading 1.4.**).

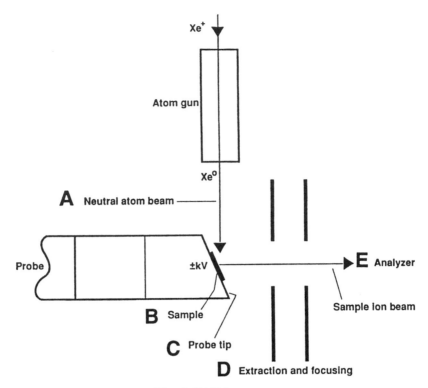

Fig. 2. FAB ion source.

The introduction of fast-atom bombardment as an ionization method marked the first routine use of an energetic-particle bombardment method and the effective entry of routine mass spectrometry into the field of biopolymer analysis. The practical result was the application of mass spectrometry to the analysis of a wide range of thermo-labile and ionic materials as well as biopolymers such as peptides, oligosaccharides, and oligonucleotides.

1.2.5. Matrix-Assisted Laser Desorption/Ionization

Another obvious source of energetic particles for sample bombardment is the laser. Just as with fast-atom bombardment, the successful use of a laser was found to depend on the provision of a suitable matrix material with which the sample was admixed and, in this way, the technique of matrix-assisted laser desorption/ionization (MALDI) (**Fig. 3**) was born. In MALDI *(4)*, the matrix material (generally a solid and present in large excess) absorbs at the laser wavelength and thereby transforms the laser energy into excitation energy for the solid system, which leads to the sputtering of surface molecular layers of matrix and sample.

A typical MALDI-matrix compound combines a number of desirable properties:

1. An ability to absorb energy at the laser wavelength, whereas the analyte generally does not do so.
2. Sufficient volatility to be rapidly vaporized by the laser.

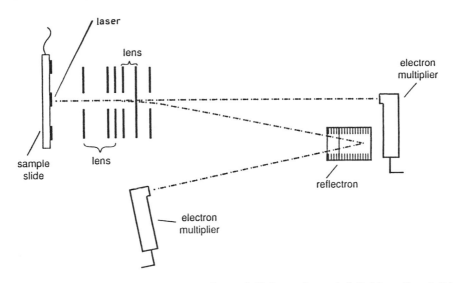

Fig. 3. MALDI source installed on a time-of-flight analyzer (cf. **Subheading 1.3.**).

3. The appropriate chemistry so that matrix molecules excited by the laser can ionize analyte molecules, usually by proton transfer.

Using this technique, ionized proteins with molecular masses in excess of 100 kDa were readily observed—considerably greater than anything previously achieved. MALDI is, in fact, applicable to a wide range of higher mol-wt biopolymers, e.g., proteins, glycoproteins, oligonucleotides, and oligosaccharides. Again, with an appropriate choice of matrix, MALDI is more tolerant than other ionization techniques toward the presence of inorganic or organic contaminants. Just as with FAB, MALDI produces quasimolecular ions with little or no fragmentation.

Overall, compared with FAB, with a practical mol-wt limit of about 2500 Daltons, MALDI has enormously extended the mol wt and polarity range of samples that are amenable to MS analysis, while also providing an analytical technique which is easy to use and is more tolerant of the difficulties encountered in purifying biochemical samples.

Unlike most other ionization techniques, MALDI is a discontinuous technique. Analyte ions are only produced, over a very short period, each time the laser is fired. For this reason, conventional mass analyzers, which present the result of scanning over a wide *m/z* range to the detector, are replaced by a time-of-flight mass analyzer (**Subheading 1.3.**) for MALDI. Alternatively, an integrating detector, used, for example with a magnetic-sector analyzer, is also suitable for MALDI.

1.2.6. Electrospray and Ion-Spray Ionization

Electrospray ionization (ES) *(5)* (**Fig. 4**) is a field desorption process. To initiate electrospray, a flow of sample solution is pumped through a narrow-bore metal capillary held at a potential of a few kilovolts relative to a counter electrode. Charging of the liquid occurs and it emerges from the capillary orifice, into an atmospheric-pressure region, as a mist of very fine, charged droplets.

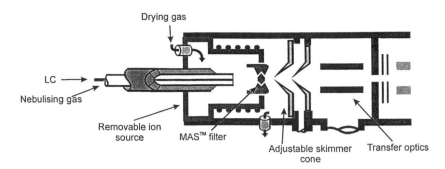

Fig. 4. Electrospray/ion-spray ion source. (Courtesy Micromass Ltd., Manchester, UK.)

The charged droplets, with a flow of a warm drying gas to assist solvent evaporation, decrease in size until they become unstable and explode (Coulomb explosion) to form a number of smaller droplets. Finally, at a still smaller size, the field owing to the excess charge is large enough to cause the desorption of ionized sample molecules from the droplet. These ions, which are formed at atmospheric pressure, are then sampled through a system of small orifices with differential pumping, so that they pass through successive regions of decreasing pressure, into the vacuum system for mass analysis.

ES is a very mild process, with little thermal input, that provides ions from analyte molecules with mol wts in excess of 100 kDa. Fragmentation is virtually absent and usually only mol-wt information is available from the spectra. Useful fragmentation can, however, be deliberately induced either by tandem mass spectrometry techniques (**Subheading 1.4.**) or by an increase in the voltage (ΔV) between the plates which define the first differential pumping region. Thus, with a lower-voltage difference, ions will pass undisturbed through the intermediate pressure region between these plates. On the other hand, an increase in the voltage difference causes sample ions to undergo more energetic collisions with gas molecules and, as a consequence, the ions dissociate into structurally significant fragment ions. Unlike the use of tandem mass spectrometry techniques, however, this in-source collision-induced decomposition is not selective and has some effect, at a given voltage difference, on a significant proportion of all the different ion types passing through the region.

A particular feature of ES spectra is that the molecular ions recorded are multiply charged, so that, for example, an ion of mass 10,000 that carries 10 charges will actually be recorded at m/z 1000. Thus, ES is able to generate, from relatively massive molecules, ions that can readily be analyzed using a simpler "low-mass" analyzer, such as a quadrupole mass filter (**Subheading 1.3.**).

In the ion-spray technique, a flow of nebulizing gas in an annular sheath, which surrounds the spraying needle, is used to input extra energy to the process of droplet formation. Using this technique, the practical upper limit for the liquid flow that can be sprayed is increased. Ion-spray systems also provide increased tolerance towards higher water levels in the solvent flow, as well as being less affected by electrolytes and, as a result, provide a more robust system that is often more suitable than electrospray for routine analysis and especially for liquid chromatography (LC)/MS.

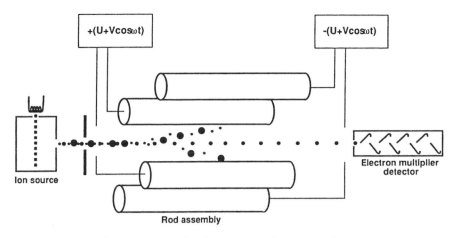

+(U+Vcosωt)   -(U+Vcosωt)

Ion source

Electron multiplier detector

Rod assembly

Fig. 5. Schematic of a quadrupole mass analyzer.

1.3. Mass Analyzers

The function of the mass analyzer *(1)* is to separate the ions according to their mass-to-charge ratio (m/z). **Figure 5** shows one of the most commonly used analyzers—the quadrupole mass filter *(6)*. In this device, a voltage made up of a DC component U and an RF component Vcosωt is applied between adjacent rods of the quadrupole assembly and opposite rods are connected electrically. With a correct choice of voltages only ions of a given mass-to-charge value traverse the analyzer to the detector, whereas ions having other mass-to-charge values collide with the rods and are lost. By scanning the DC and RF voltages and keeping their ratio constant, ions with different mass-to- charge ratios will pass successively through the analyzer. In this way, the whole-mass range may be scanned and a complete mass spectrum recorded.

In a magnetic-sector analyzer *(7)* (**Fig. 6**), accelerated ions are constrained to follow circular paths by the magnetic field. For any one magnetic-field strength, only ions with a given mass-to-charge ratio will follow a path with the correct radius to arrive at the detector. Other ions will be deflected either too much or too little. Thus, by scanning the magnetic field, a complete mass spectrum may be recorded, just as with a quadrupole analyzer.

When the magnetic analyzer is operated in conjunction with an electrostatic analyzer (**Fig. 6**), the instrument then provides energy as well as direction focusing and can attain much higher mass resolution. This double-focusing magnetic sector instrument *(7)* is also much more suitable than the quadrupole analyzer for the analysis of ions with higher mass-to-charge ratio. Detection of ions is accomplished in all the instruments discussed here by such a device as an electron multiplier placed at the end of the analyzer. The output from the electron multiplier is then recorded by means of a data system.

For routine analyzes, the foregoing analyzers can be operated in one of two modes. The first of these is the scanning mode where the mass analyzer is scanned over a complete mass range, usually repetitively, in order to record successive full spectra throughout an analysis. This type of analysis is a survey analysis where the spectra provide information on every component that enters the ion source. The other mode is called

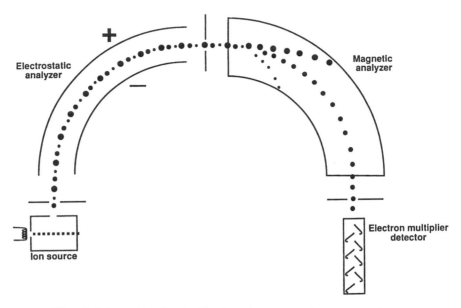

Fig. 6. Schematic of a double-focusing magnetic sector analyzer.

selected-ion monitoring. In this case, the instrument is set to successively monitor specific *m/z* values chosen to be representative of the compounds sought. This type of analysis detects only targeted compounds but does so with a much higher sensitivity because of the longer monitoring time at the selected *m/z* values.

A third analyzer system, which is increasingly common, notably with MALDI, is the time-of-flight analyzer *(8)* (**Fig. 3**). In this simple device, ions are accelerated down a long, field-free tube to a detector. The mass-to-charge ratio of each ion is calculated from a measurement of the time from their start, e.g., the ion-formation laser pulse in MALDI, to the time at which they reach the detector. Unlike the quadrupole and magnetic-sector analyzers, ions with a mass-to-charge ratio other than that which is being recorded are not rejected; all ions that leave the ion source can, in principle, reach the detector. With this type of instrument therefore, there is no real distinction between scanning and selected-ion monitoring modes. A time-of-flight analyzer of improved mass resolution, the so-called reflectron instrument, uses an electrostatic mirror to compensate for energy differences among the ions.

Another type of analyzer that does not distinguish between these two modes is the trapping analyzer. Examples of this type of device are the quadrupole-ion trap (ITMS) *(9)* and the Fourier transform ion cyclotron-resonance instrument (FTMS) *(10)*. In either case, ions are either made by ionization within the trap, or injected into the trap from an external source. These ions are maintained in stable orbits within the trap by means of electrostatic (ITMS) or magnetic fields (FTMS). Ions formed within the ITMS can then be ejected and detected, in order of mass-to-charge ratio, so that the spectrum, which potentially contains a contribution from every ion in the trap, results. With the FTMS instrument, a detector located within the cell is again, in principle, able to record all the ions formed. Ion-trapping devices also have some interesting potential advantages in their ability to carry out tandem mass spectrometry experiments (**Subheading 1.4.**).

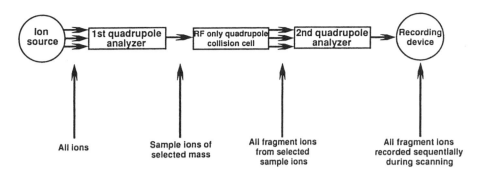

Fig. 7. Functional schematic of a tandem mass spectrometer based on a triple-quadru-pole instrument.

1.4. Tandem Mass Spectrometry

Most tandem mass spectrometry (MS/MS, sometimes written MS^2) *(11)* instruments consist of two mass analyzers arranged in tandem but separated by a collision cell (**Fig. 7**). In MS/MS, sample ions with a specific m/z value are selected by the first analyzer and then accelerated into the collision cell that contains a relatively high pressure of neutral gas molecules. The result of these interactions is ion fragmentation (collision-induced dissociation; CID) that has been induced deliberately.

In a triple-quadrupole MS/MS instrument (**Fig. 7**), for example, the first analyzer is a conventional quadrupole analyzer that transmits ions of the required m/z value, whereas the collision cell is an RF-only quadrupole analyzer that contains collision gas and that transmits fragmentation products of whatever m/z value. Scanning the second analytical quadrupole, which follows the collision cell, records all the fragment ions that originate from the precursor ion selected by the first analyzer. This is the so-called product-ion scan.

Other types of mass analyzers may be used in tandem to give alternative forms of MS/MS instrumentation. For example, a collision cell may be interposed between a double-focusing magnetic sector analyzer and a quadrupole analyzer (hybrid instrument) or between two double-focusing magnetic sector analyzers (four-sector instrument). Ion-trap instruments are suitable for MS/MS experiments without being combined with other instrumentation because they provide facilities for MS/MS experiments in which an ion population is manipulated in a single physical location. Thus, the steps of an ion-trap MS/MS experiment are separated in time but not in space. Typically, all ions except those of a selected m/z value are ejected from the trap and then a spectrum of the product ions from collisions of the selected ions is recorded in the usual manner. In particular, ion trapping is very well suited to the implementation of sequential MS/MS[$(MS)^n$] experiments.

MS/MS instrumentation has found increasing use in conjunction with soft ionization techniques such as fast-atom bombardment and electrospray. For example, a conventional spectrum recorded by these techniques will often show a quasimolecular (QM) ion and no fragment ions. If, however, this QM ion is now selected for fragmentation by CID, product ions, which are fragment ions and are indicative of structural detail, may be recorded using MS/MS techniques. Another application of MS/MS techniques is to remove interfering ions caused by unresolved sample components. Thus, if the first analyzer is set to transmit the sample ion of interest, then none of the interfering ions will reach the collision cell and the collision-induced spectrum will consequently contain only information that relates to the component of interest.

2. Applications of Mass Spectrometry

2.1. Ionspray LC/MS: Molecular Weights of Lens Crystallin Proteins (12)

The characteristics of the lens that enable vision are thought to result from the tight, stable packing of the lens proteins, called crystalline, which account for most of the water-soluble protein in lens cells. Understanding of the role of crystallins in cataract formation requires a comparison of changes in their primary structure associated with both normal aging and cataract formation. A convenient baseline for such comparisons is provided by fetal-lens proteins, because these materials should not have accumulated age- or cataract-related modifications. The use of LC/MS for this type of analysis permits the separation of complex protein mixtures with simultaneous determination of the molecular weight of each component.

2.1.1. Materials and Methods

Lens crystallins were isolated from normal 21–24-wk-old human fetuses by direct lyophilization of the lenses followed by centrifugation of a suspension of the crude protein powder in Tris buffer. The supernatant was defined as the water-soluble protein fraction while the insoluble pellet was homogenized in urea buffer and centrifuged. The urea-soluble fraction was defined as the water-insoluble fraction.

Microbore LC separations used a 1 mm i.d. C18 column with a flow rate of 50 µL/min. A splitting ratio of 1:9 in the LC effluent directed toward the MS ionspray source and UV detector respectively was accomplished by adjusting the lengths of fused-silica tubing that ran to each of these detectors. The quadrupole mass analyzer was scanned continuously from m/z 500 to 1500. Samples passing through the UV detector could be collected for further investigation.

2.1.2. Results and Discussion

The total-ion chromatogram (TIC) from the LC/MS analysis of cat 30 µg of the water-soluble fraction is shown in **Fig. 8**. In TIC traces, the intensities of all ions detected by the mass spectrometer in each scan are summed and the summed values plotted vs time to permit the nonselective detection of proteins and other materials. The mass spectra obtained from each of the peaks in **Fig. 8** indicate that many of the fractions contain multiple-protein components although these can be separated. For example, the mol-wt plot for peak 7 (**Fig. 9**), reconstructed from the mass spectrum, shows two components with molecular weights of 20 200.2 + 1.7 and 20 072.2 + 2.4. These molecular weights identify the components as crystallin αB_2 itself as well as crystallin αB_2 following the loss of the C-terminal lysine residue (theoretical molecular weights are 20, 201.0, and 20, 072.8 respectively).

Overall, it was possible to identify two α-crystallins, two β-crystallins, and a λ-crystallin among the approx 24 proteins detected. The excellent agreement between experimental and theoretical mol wts in these cases implies that the actual primary

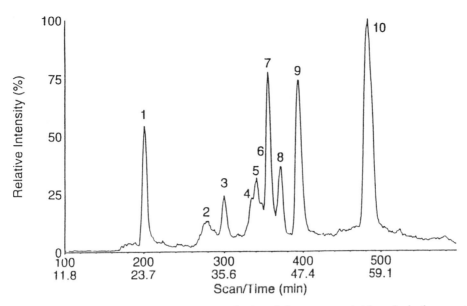

Fig. 8. TIC trace from the HPLC/MS analysis of the water-soluble whole-lens extract. (Reproduced with permission from **ref. *12*.**)

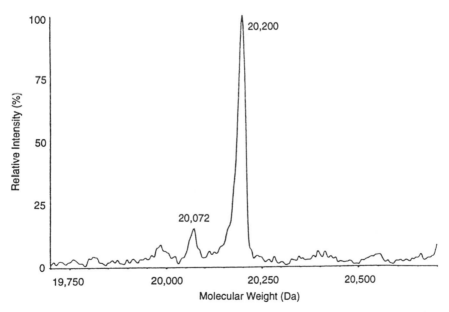

Fig. 9. Reconstructed MW plot acquired from peak 7 (αB_2) in **Fig. 8**. (Reproduced with permission from **ref. *12*.**)

structure corresponds to the published structure and that significant posttranslational modifications are absent. Other crystallins with published primary structures seem to contain some modification or error, which leads to a discrepancy between the experimental and theoretical molecular weights. Further use of LC/MS will be as a screening technique to identify modified crystallins from normal aged and cataract lenses by comparison with fetal-lens crystalline.

2.2. LSIMS-MS/MS: Structures of the Oligosaccharides of Interleukin-3 (13)

Interleukin-3 (IL-3) is a T-cell derived factor that supports the division and differentiation of a broad spectrum of hemopoietic cells. Its ability to promote the proliferation of multipotential stem cells makes it a primary regulator in the early stages of hemopoiesis and a major factor in the inflammatory process. Mouse IL-3 purified from WEHI-3 cell medium is a single polypeptide chain of 140 amino acids with four N-glycosylation sites and a mol wt of 28–32.5 kDa. Mouse IL-3 has also been expressed in *Bombyx mori* using a baculovirus expression vector and contains at least three species that are glycosylated with N-linked oligosaccharides and have mol wts of 18, 20, and 22 kDa. LSIMS-MS/MS was used for the full-structure elucidation of the predominant species in baculovirus-expressed mouse IL-3.

2.2.1. Materials and Methods

Mouse IL-3, dissolved in buffer, was hydrolysed with PNGase F to release N-linked oligosaccharides. Protein was removed by ultrafiltration and the filtrate concentrated to dryness. Separation of released oligosaccharides, by high-pH anion-exchange chromatography, was followed by reaction of the eluted fractions with *n*-hexylamine and sodium cyanoborohydride to give a reductively aminated oligosaccharide. This product was then permethylated in dimethyl sulfoxide using iodomethane in the presence of sodium hydroxide.

The permethylated products were dried (under N_2) before being taken up in acetonitrile and mixed with 1–2 μL thioglycerol + glycerol (2:1 v/v) for LSIMS analysis using a 7–8 keV primary ion beam. Product-ion scans used CID with helium collision gas.

2.2.2. Results and Discussion

The chromatogram of the N-linked oligosaccharides, released from the recombinant mouse IL-3, shows a single major component and several minor ones. The spectrum (**Fig. 10**) obtained from the major component, after derivatization, shows a molecular ion at *m/z* 1233, which cleaves at *N*-acetylglucosarnine to an ion at *m/z* 668. This is consistent with the monosaccharide composition Man₂GlcNAc-(Fuc)GlcNAc with fucose bound to the reducing-end *N*-acetylglucosamine. The elimination of methanol from *m/z* 668 to give an ion at *m/z* 636 suggests that mannose is linked to *N*-acetylglucosamine through α1,4-linkage.

Sequence information for this oligosaccharide was obtained from the CID spectrum recorded with *m/z* 1233 as the precursor ion (**Fig. 11**). Ions at *m/z* 1073, 1043, and 839 result from cleavage of the fucose and the two mannose residues respectively. The ion with *m/z* 793 comes from cleavage between a mannose and an *N*-acetylglucosamine. Cleavage between two *N*-acetylglucosamine residues gives the ion at *m/z* 668 and then at *m/z* 636 after the elimination of methanol. These data establish that the oligosaccharide is linear with two mannose residues linked successively to a chitobiose-type disaccharide. The fragment ions at *m/z* 793 and *m/z* 839 show that the two mannoses are

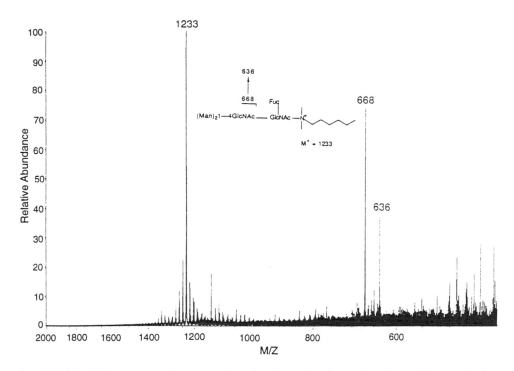

Fig. 10. LSIMS spectrum of oligosaccharide after reductive amination and permethylation. (Reproduced with permission from **ref. 13**.)

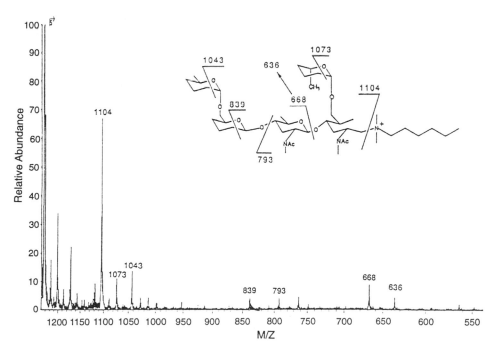

Fig. 11. CID spectrum of the ion at *m/z* 1233 showing fragmentation (methyl esters has been omitted in the formula for clarity). (Reproduced with permission from **ref. 13**.)

linked to each other and do not have glycosidic linkages to *N*-acetylglucosamine. Fucose can only be linked to reducing-end *N*-acetylglucosamine as indicated by the fragment ion at *m/z* 668. The ion at *m/z* 1073 indicates that this can only be a terminal residue.

The production of baculovirus-expressed mouse IL-3 by infected *Bombyx mori* larvae gives material that is heterogeneously glycosylated with N-linked oligosaccharides. These oligosaccharides are core-type structures that are mostly fucosylated and that contain two or three mannose residues. The complex-type oligosaccharides found on vertebrate glycoproteins are not produced. The predominant glycoside from *Bombyx mori* is Manα1-6 Manβ1-4GlcNAc1-4(Fucα1-6)GlcNAc.

2.3. Electrospray-MS and H/D Exchange: An Indicator of Protein Structure (14)

Protein structure may be characterized by the determination of charge-state distributions and H/D exchange rates using ESI-MS. The exchange studies are based on the fact that slowly exchanging H-atoms are often involved in H-bonds found in secondary structures such as β-sheets or α-helices or are shielded from solvent. The advantages of MS for the characterization of higher-order protein structure are speed and the ability to operate at the nanomole level. In this application, conformational perturbations in *Rhodobacter capsulatus* cytochrome c_2 wild-type and two site-directed mutants (G34S and P35A) were investigated. An additional method, LC/ESI-MS analysis of the peptic digest of a deuterated protein, can be used to locate the sites of perturbations more precisely.

2.3.1. Materials and Methods

For H/D exchange experiments, cytochrome c_2 samples were dissolved in deuterated water (pD 5.8) or in deuterated water containing 0.5% d-acetic acid (pD 3.2) at a concentration of 30 pmol/μL. Exchange reactions were performed at 25°C for 3 h. Aliquots were taken at intervals and measured by ESI-MS, without further treatment, by infusion into an electrospray source in deuterated water containing 0.5% d-acetic acid at a flow rate of 5 μL/min. Nitrogen, used as a curtain gas in the source, helps to avoid back-exchange with atmospheric water during the spraying period.

For the analysis of charge-state distribution, samples were dissolved in pure water and their ESI spectra recorded. This experiment was then repeated in the presence of 0.5% acetic acid. The mass spectrometer was scanned from *m/z* 400 to 1400 during these analyses.

2.3.2. Results and Discussion

When dissolved in pure water, no differences were observed between the charge-state distributions of the wild-type and of mutants G34S and P35A. In each case, the maximum number of basic sites observed was 9, indicating that 11 of the 20 basic sites were shielded from solvent or otherwise not protonated. In the presence of 0.5% acetic acid, however, the charge state distribution of the P35A mutant was slightly different, and that of the G34S significantly different, from that of the wild-type.

The total number of exchangeable hydrogens is calculated to be 189 in the wild-type and the G34S mutant, and 190 in the P35A mutant. At pD = 3.2, the charge-state distributions for the G34S mutant, centered at 8+ (A form) and 13+ (B form), represented

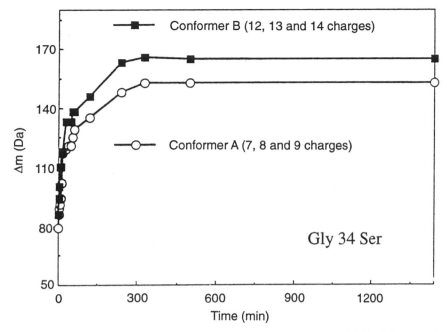

Fig. 12. H/D exchange curves for *Rb. Capsulatus* cytochrome c_2 G34S. The reactions were carried out in D_2O containing 0.5% of d-acetic acid (pD 3.2) at 25°C. (Reproduced with permission from **ref. *14*.**)

Table 1
ESI-Determined Mass of Wild-Type
and Mutant Cytochromes c_2
After 3 h in Pure Deuterated Water (pD 5.8)

Sample	Observed mass (Daltons)	Calculated mass[a] (Daltons)	Δm
Wild-type	12953	12883	70
Pro 35 Ala	12947	12857	90
Gly 34 Ser	13022	12913	109

[a]Without deuterium incorporation.
(Data reproduced with permission from **ref. *14*.**)

25% and 75% of the total ionization, respectively. These two forms exhibited different kinetic behavior (**Fig. 12**), suggesting the presence of two conformers. The difference in mass owing to deuterium incorporation in the A and B forms of G34S (147 and 165 H/D exchanges respectively) was still detectable after 24 h, indicating that the conformers are stable and do not exchange during this time period.

There are significant differences in the H/D exchange properties of the wild-type and mutants at pH 5.8, despite equivalent charge-state distributions. Thus, after 3 h, the total number of H/D exchanges was 70 for the wild-type protein, 90 for P35A, and 109 for G34S (**Table 1**). Compared with the wild-type, 39 and 20 additional exchanges take place in G34S and P35A respectively. Thus, both mutations appear to either alter H-bonds present in the wild-type and/or increase the solvent accessibility to exchangeable hydrogens with respect to the wild-type. Overall, the H/D exchange properties and charge-state distributions of the G34S and P35A mutants are in agreement with previous Gdn-HCl denaturation studies.

2.4. Electrospray-MS: The Observation of Noncovalent Associations (15)

Molecules that form structurally specific noncovalent associations in solution are of fundamental biological importance. ESI-MS is rapidly becoming a useful technique for the detection of such complexes, and potentially for probing their relative stability. Through careful control of instrumental variables in the atmosphere-vacuum interface region of the ion source, several types of noncovalent association, including multimeric proteins, enzyme-substrate, protein-nucleic acid, and oligonucleotide interactions have been transferred into the gas phase from solution and detected intact by ESI.

2.4.1. Materials and Methods

Placing a large, multimeric protein in a conventional ESI-MS solvent (e.g., 5% acetic acid) results in the observation of ions corresponding to the individual subunits, consistent with predicted solution behavior. On the other hand, a 10 mM NH$_4$OAc solution can be used to preserve the tetrameric associations of proteins such as adult human hemoglobin (Hb A$_0$, mol wt ~ 64.5 kDa), concanavalin A(Con A, mol wt ~ 102 kDa), and avidin (mol wt ~ 64 kDa). For the positive-ion analysis of avidin, for example, the protein was placed, at 1 μg/μL, in a pH 6.7 10 mM NH$_4$OAc solution and spectra acquired on a quadrupole instrument with an extended m/z range.

The most important variables in the electrospray atmosphere-vacuum interface which affect the observation of ions from intact noncovalent species are the heat applied to the metal inlet capillary and the capillary-skimmer offset voltage (ΔV). Heating/activation from these areas must be adjusted to provide sufficient desolvation for resolution of ions without destroying the intact complex.

2.4.2. Results and Discussion

A capillary temperature of approx. 174°C and a ΔV value of 80 V were suitable for the observation of the ions Q^{18+} to Q^{16+} (m/z 3500–4500), which are indicative of the intact tetrameric form of avidin (**Fig. 13**). The relatively narrow charge distribution and limited amount of charging demonstrates the retention of higher-order structure in the gas phase. The ESI-MS spectrum also indicates a small octamer contribution at higher m/z (approx 5000) from aggregation of the tetramer. The absence of any nonspecific species (e.g., trimer, pentamer) indicates that the ESI-MS spectrum reflects only stoichiometrically specific protein subunit association into a tetramer, consistent with solution behavior.

Further evidence that the noncovalent species observed in the ESI spectrum are structurally specific, with the correct stoichiometry, and do not arise from random aggregation in the electrospray process can be provided as follows:

1. Changing the solution conditions to affect the associations present in solution should alter the appearance of the spectra. ESI-MS can be used to probe the known solution pH depen-

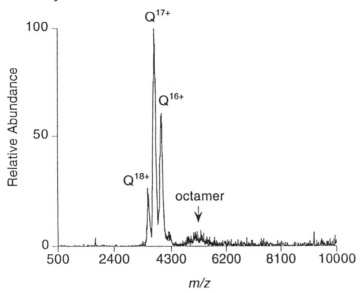

Fig. 13. Positive-ion ESI spectrum of egg-white avidin in 10 m*M* NH$_4$OAc (pH 6.7). The spectrum was acquired with 15 W capillary heating ($\approx$174°C) and ΔV = 80 V in order to observe the active form of the protein. (Reproduced with permission from **ref.** *15*.)

dence for the dimer-tetramer equilibrium of certain proteins. For example, placing the tetrameric protein Con A in a pH 8.4 solution produces mainly tetrameric ions in the mass spectrum, which is expected from solution behavior. However, placing the protein in a pH 5.7 solution and utilizing the identical mild ESI-interface conditions, produces mainly ions indicative of a dimer species, also consistent with solution behavior.

2. Creating harsher conditions in the interface region by increasing either the metal inlet-capillary temperature or the ΔV voltage will generally dissociate multimers into monomer species if the associations are noncovalent. For example, increasing the capillary heating to approx 276°C causes dissociation of the avidin tetramer into predominantly monomer species (M^{13+} to M^{7+}), of relatively low charge state, with some additional lower charge state ions, indicative of a trimer species (T^{9+} to T^{6+}), at high *m/z* (5325–8000) (**Fig. 14**).

2.5. MALDI-MS: Protein Fingerprinting from Tryptic Digests (16)

A mass spectrum of the peptide mixture resulting from the digestion of a protein by an enzyme can provide a fingerprint that is specific enough to identify the protein uniquely. The general approach is to take a small sample of the protein of interest, digest it with a proteolytic enzyme, such as trypsin, and analyze the resulting digest using MALDI-MS. The experimental mol wts are then compared with a database of peptide mol wts, calculated by applying enzyme cleavage rules to the entries in one of the major collections of sequence data, such as SwissProt or PIR.

2.5.1. Materials and Methods

The starting point was an SDS-PAGE gel on which 1 mg of a protein mixture had been separated by 2D-electrophoresis. The gel was equilibrated in the blotting buffer and then electroblotted onto FluoroTrans membrane. The membrane was rinsed thoroughly with deionized water and the proteins visualized with either Coomassie bril-

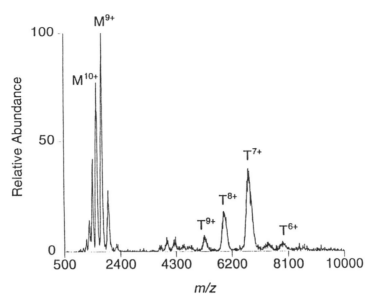

Fig. 14. Positive-ion ESI spectrum of egg-white avidin acquired under harsher interface conditions by increasing the capillary heating to 28 W (≈276°C). Other experimental conditions were as in **Fig. 13**. (Reproduced with permission from **ref. *15*.**)

liant blue or acidic sulforhodamine B followed by thorough rinsing. The rinsed blots, from which selected protein spots were excised with a scalpel for subsequent fingerprinting, were stored in sealed plastic bags.

As a preliminary to analysis, Coomassie-stained spots only were destained with aqueous acetonitrile. Each spot was then dried and cut into pieces no more than 2 mm square. An aliquot of trypsin solution was added and the mixture left overnight. The resultant peptides were solubilized using two extractions with warm formic acid + ethanol (1: 1). These extracts were then lyophilized and finally resuspended in 10% methanol. An aliquot of this digest solution was mixed with an equal volume of matrix solution (10 mg/mL α-cyanohydroxycinnamic acid in 70% acetonitrile) on a stainless-steel target and allowed to dry before being introduced into the mass spectrometer.

2.5.2. Results and Discussion

Figure 15 shows a typical spectrum obtained from the digestion of a single spot. It had not been possible to identify this protein by Edman sequencing because it appeared to have a blocked N-terminus. The peak at *m/z* 1348.8 is a spike of 100 fmol Substance P, which was added to provide an internal standard for mass calibration. The mass values selected for database searching were 1037.6, 1199.9, 1502.6, 1792.8, 1957.9, and 2247.7. The error in these experimental peptide mass values was estimated to be less than ± 3 Da based on experimental observations. From its migration distance on the gel, the molecular weights of the intact protein was thought to be around 43 kDa.

The result of matching the peptide masses against a database built from the complete Release 15.0 of Entrez Sequences (>217,000 sequences) is shown in **Fig. 16**. It can be seen that the list of highest scoring proteins is dominated by actins. (The first entry, 1ATN, is a complex of actin with deoxyribonuclease I from the Brookhaven structural

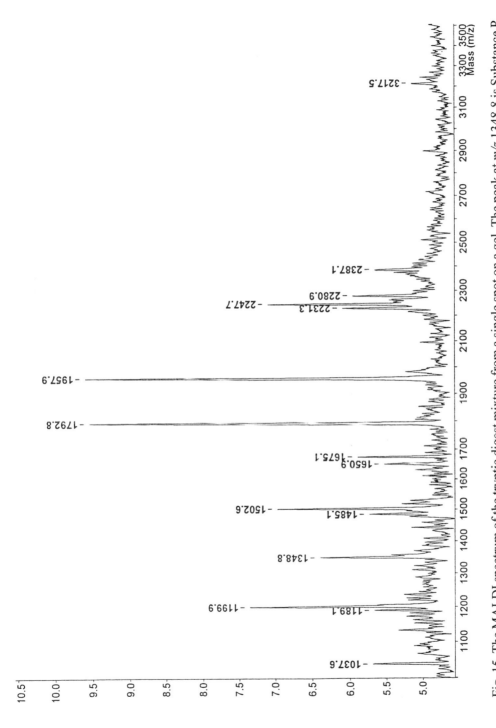

Fig. 15. The MALDI spectrum of the tryptic digest mixture from a single spot on a gel. The peak at *m/z* 1348.8 is Substance P, added as an internal mass calibrant. (Reproduced with permission from **ref. 16**.)

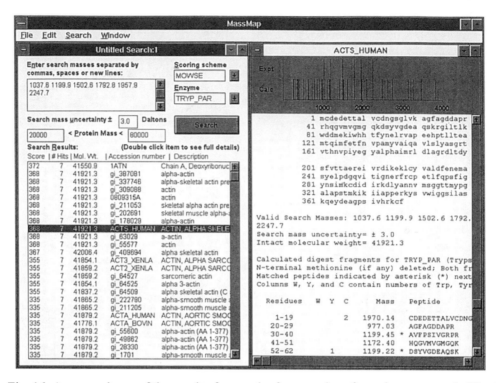

Fig. 16. A screen dump of the result of a search of mass values from the spectrum in **Fig. 15** against a database built from Entrez Sequences release 15.0 using the MassMap software. The protein was clearly identified as skeletal α actin. (Reproduced with permission from **ref. *16*.)

database). The next ten entries are all the same sequence, human-skeletal alpha actin. The sequence annotations for ACTS_HUMAN state that this protein is known to be acylated at the N-terminus, which explains why Edman sequencing failed.

2.6. Electrospray-MS: Molecular Weight Determination of PCR-Amplification Products (17)

Polymerase chain reaction (PCR) products, amplified from human DNA, can be analyzed using EI-MS. In this instance, human-genomic DNA was prepared from blood samples and double-stranded DNA of 57-bp length from the human APC gene was amplified by PCR. APC gene is one of the cancer-suppressor genes and causes familial adenomatous polyposis coli, which is an autosomal dominant inherited disease, conferring a high risk of colon cancer.

It is difficult to measure PCR products themselves without further treatment because *Taq* polymerase, a thermostable DNA polymerase used in PCR, adds nontemplated nucleotide to the 3'-end of the PCR products. Therefore, the PCR products were amplified by a pair of specific synthetic oligonucleotides, which included restriction enzyme *Eco*RI recognition sites at the 5'-end. The PCR products were then digested by *Eco*RI to cut off the nontemplated ends.

2.6.1. Materials and Methods

Owing to the efficiency of PCR as an amplification method, only ≈1 μL of blood, which contains 100 ng of genomic DNA, is sufficient for this measurement. In this instance, a two-stage PCR was used to obtain the required amount of PCR products. To avoid problems with *Taq* polymerase, oligonucleotide PCR primers, which included an *Eco*RI recognition site (GAATTC) on the 5'-end were designed and synthesized. The PCR products amplified with these primers were then treated with *Eco*RI and, after this treatment, either end, including any nontemplated nucleotide adduct, was completely cut off.

Samples were dissolved in aqueous 80% acetonitrile (v/v), containing 5% TEA, at a final concentration of 1–5 pmol/µL and 50 µL aliquots introduced into the ESI source at a flow rate of 2 µL/min. There was no voltage difference between the extraction cone and the first skimmer element in the electrospray ion source. A particular difficulty with the mass-spectrometric analysis of nucleotides is the affinity of the negatively charged backbone for nonvolatile cations, especially Na^+ and K^+. The use of three overnight ethanol precipitations to reduce the level of these adducts made it possible to assign masses to the PCR products with ≈0.005% accuracy and ≈0.02% precision.

2.6.2. Results and Discussion

Double-stranded DNA is denatured under normal ES conditions. Thus, the mass spectrum (**Fig. 17**) of the PCR products corresponds to the two complementary single strands with the measured mol wt values of 15,275.2 ± 1.6 Daltons and 15,029.0 ± 1.3 Daltons being very close to the expected values (15,275.8 and 15,029.8 Daltons, respectively). The mol wt values of the nuclotides A, G, C, T are 313.2, 329.2, 289.2, and 304.2 respectively with a minimum difference among these of 9.0 Da (between A and T). On this basis, ≈0.005% accuracy and ≈0.02% precision for an approx 50 bp oligonucleotide should be sufficient for the recognition of a single substitution. Thus, it is clear that this technique can provide the basis for an accurate, sensitive, and rapid method for gene diagnosis.

2.7. MALDI-MS: Gene Polymorphism Detected in PCR-Amplification Products (18)

Another approach to the rapid and cost-effective detection of human-genetic polymorphisms employs PCR amplification of extracted DNA, followed by direct digestion with restriction enzymes and analysis by MALDI-MS. In this work, two disease-related human genes, carbonic anhydrase (CA) and cystic fibrosis transmembrane-conductance regulator (CFTR) were studied as model genes using DNA collected from buccal cells. the primers used for PCR amplification of the CA II gene and of the exon 6a and exon 12 genes of CFTR are listed in **Table 2**.

2.7.1. Materials and Methods

PCR amplification used *Taq* polymerase with 5 µL of the buccal cell DNA solution and 0.1 µg of each of the primers. Direct restriction-enzyme digestion of these PCR-reaction products was used to simplify the digestion process. A suitable restriction enzyme was added directly to the PCR-reaction products and the mixture incubated to provide appropriate conditions for digestion. The PCR products and their digests were purified using a commercial purification kit and then dried. Finally, the DNA contents were resuspended in deionized water. An aliquot of this solution, taken for MALDI-MS analysis, was deposited on a nitrocellulose substrate, previously applied to a stainless-

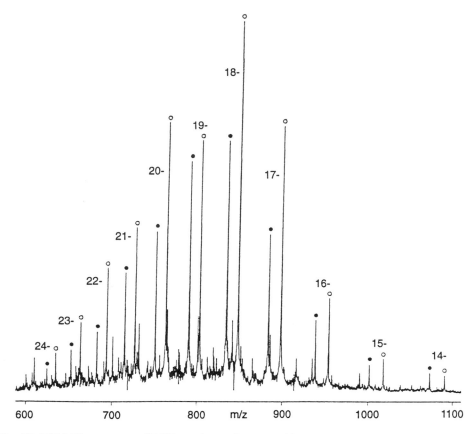

Fig. 17. ESI-MS spectrum of PCR products. One strand is 5'-AAT TCC CTG CAA ATA GCA GAA ATA AAA GAA AAG ATT GGA ACT AGG TCA G-3' (○), expected mass 15 275.8 Daltons; the other is 5'-GGG ACG TTT ATC GTC TTT ATT TTC TTT TCT AAC CTT GAT CCA GTC TTA A-3' (●), expected mass 15 029.8 Daltons. (Reproduced with permission from **ref. 17.**)

Table 2
Primers Used for PCR Amplification of the CA II Gene
and of Exon 6a and 12 Genes of CTFR

Genes or exons	Primers (5' → 3')	Product (bp)
CA II	TGT GTC TGC TGC TCT CCT ACC (1)	212
	GGG TTC CTT GAG CAC AAT (2)	
CA III	GGT CCT GAC CAC TGG CAT GA (1)	86
	TGC CTG ATG TCT TTA GTA TGC AGG[a] TCA A (2)	
	CGG CAG GTC TTC CCA TTA TT[2] (3)	167
CFTR exon 6a	TTA GTG TGC TCA GAA CCA CG (1)	385
	CTA TGC ATA GAG CAG TCC TG (2)	
CFTR exon 12	GTG AAT CGA TGT GGT GAC CA (1)	426
	CTG GTT TAG CAT GAG GCG GT (2)	

[a]This base G in primer 2 is deliberately mismatched
[b]This primer is used for hemi-nesting
(Data reproduced with permission from **ref. 18.**)

steel probe tip, and allowed to air dry. A mixed matrix solution of 3-hydroxypicolinic acid and picolinic acid (4:1 molar ratio) in ≈36% aqueous acetonitrile was then applied over the dried sample deposit.

2.7.2. Results and Discussion

Figure 18 shows the MALDI spectra of restriction enzyme digests of the PCR products obtained from the CA II gene. Data from three representative individuals, two of whom are homozygous and one heterozygous, are readily distinguished. **Figure 19** shows the MALDI spectra of digests from CFTR genes at exon 6a and 12, an example in which PCR products of mutated alleles are not cleaved by the restriction enzymes. In this case, only samples from homozygous individuals, i.e., with both alleles normal (unmutated), were analyzed and are exemplified both by the presence of fragments of 94 and 291 bp and the absence of the 385 bp fragment for the exon 6a gene as well as by the presence of fragments of 125 and 301 bp and the absence of the 426 bp fragment for the exon 12 gene. PCR amplification of these particular genes was chosen in preference to other forms of the CFTR gene because larger PCR fragments, which are more suitable for evaluation of the MALDI technique, were generated. Overall, the results from MALDI-MS are comparable with those obtained from gel electrophoresis but the MALDI method is several orders of magnitude faster than gel electrophoresis techniques.

2.8. MALDI-MS: Identification of Protein Glycosylation Sites (19)

The carbohydrate side-chains of glycoproteins have a crucial role in numerous biological systems. Structurally, these side chains facilitate proper folding and glycoprotein stability and, in general, they mediate an immense range of biological-recognition events. Techniques based on MS play an integral role in characterizing the carbohydrate portion of a glycoprotein, and are particularly useful for identifying the sites of glycosylation. The following method, for identifying both N- and O-linked glycosylation sites, uses specific digest procedures followed by MALDI-MS.

2.8.1. Materials and Methods

Glycoprotein (10 pmol) was reduced with dithiothreitol and then carboxymethylated using iodoacetic acid. The reaction product was then purified by dialysis against water and the retentate dried. The dried material was redissolved in buffer and digested with trypsin. The digestion was stopped with 0.05% trifluoroacetic acid (TFA).

An aliquot of the peptide/glycopeptide mixture from trypsin digestion was removed for MALDI-MS analysis while the remainder was divided into two portions and dried. One portion was resuspended in the buffer and treated with PNGase F to selectively release any N-linked carbohydrate side-chains. The second portion was resuspended in the buffer and then treated with both PNGase F and *O*-glycanase to release both the N- and O-linked carbohydrate chains. Both digestions were stopped with 0.05% TFA and aliquots removed for MALDI-MS analysis in each case.

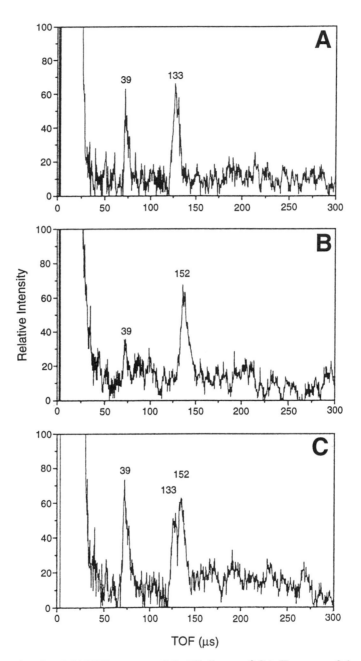

Fig. 18. Negative-ion MALDI spectra of *Bst*NI digest of CA II genes of three individuals:
(A) homozygous, 133 bp, **(B)** homozygous, 152 bp, and **(C)** heterozygous, 133 and 152 bp.
(Reproduced with permission from **ref. *18*.**)

2.8.2. Results and Discussion

The spectrum of the tryptic digest of bovine asialofetuin, which is cleaved by trypsin
into 25 peptide fragments may be compared with the spectrum recorded following

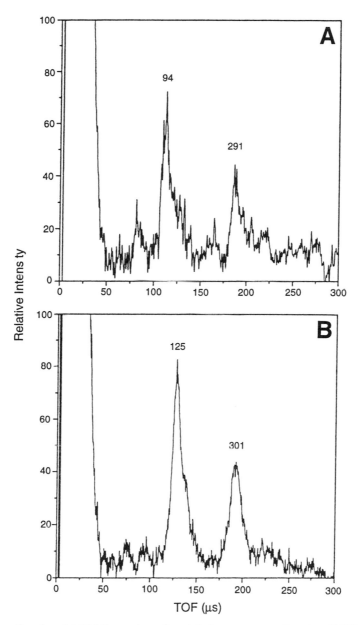

Fig. 19. Negative-ion MALDI spectra of restriction-enzyme digests of CFTR genes. **(A)** *Acc*I digest at exon 6a and **(B)** *Rsa* I digest at exon 12. (Reproduced with permission from **ref. *18*.**)

PNGase treatment (**Fig. 20A,B**). PNGase causes the loss of four [M + H]⁺ ions (*m/z* 3861.0, 5007.2, 5299.0, and 5665.0) and the appearance of two new species (*m/z* 1873.0 and 3677.0). These latter masses correspond to two of the known tryptic-peptide sequences, which contain the N-linked glycosylation-consensus sequence (Asn-Xxx-Ser/ Thr). In order to identify the relationship between a specific side-chain and the corresponding deglycosylated peptide, all of the possible mol-wt differences were calcu-

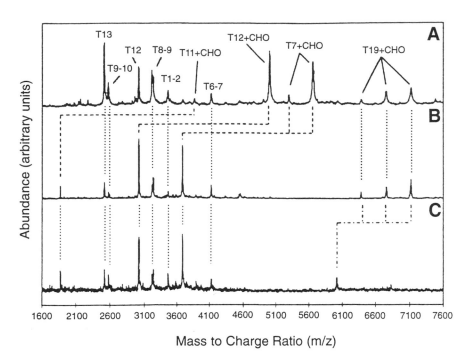

Fig. 20. MALDI-MS analysis of **(A)** trypsin digest of bovine asialofetuin, **(B)** the digest after release of N-linked carbohydrate side-chains and, **(C)** the digest after release of both N- and O-linked carbohydrate side-chains; ...... indicates no change, ---- indicates loss of an N-linked carbohydrate side-chain, and -•-•-•- indicates loss of O-linked carbohydrate side-chains. (Reproduced with permission from **ref. *19*.**)

lated and these putative carbohydrate masses used to search CarbBank. By this means, *m/z* 3861.0 was identified as T11 with a nonfucosylated complex triantennary carbohydrate side-chain, whereas *m/z* 5299.0 and 5665.0 could only correspond to T7 with nonfucosylated complex bi- and triantennary carbohydrate side-chains respectively (**Table 3**).

A third peptide (T12) in bovine asialofetuin is also known to contain the N-linked glycosylation consensus sequence. In this case, an $[M + H]^+$ ion (**Fig. 20A**) indicates that the T12 peptide does occur in unconjugated form before digestion. In addition, however, the difference between this mol wt and that of the fourth N-linked glycopeptide (*m/z* 5007.2) corresponds to a nonfucosylated complex triantennary carbohydrate side-chain (**Table 3**), indicating partial glycosylation of the Asn residue in T12.

The O-linked glycopeptide (T 19) in bovine asialofetuin can be identified by a similar comparison of the MALDI-MS spectrum from the third aliquot (**Fig. 20C**: *N-* and O-linked chains removed) with that from the second (**Fig. 20B**: only N-linked chains removed). Only Gal (β1-3) GalNAc (α1) disaccharide can be released from a peptide by *O*-glycanase so that these spectra can be interpreted to show that up to three disaccharides may be simultaneously present on the glycopeptide. Thus, T19 has at least three separate O-linked glycosylation sites, although the specific attachment sites cannot be identified by this method, because there are more possible O-linked

Table 3
Molecular Masses and Proposed Oligosaccharide Compositions of the Glycopeptides from Asialofetuin

Glycopeptides	Observed mass (Daltons)	Calculated mass (Daltons)	Proposed oligosaccharide composition
T12 + CHO (N-linked)	5007.0	5007.2	$Gal_3Man_3GlcNAc_5$ (Triantennary)
T11 + CHO (N-linked)	3861.0	3861.0	$Gal_3Man_3GlcNAc_5$ (Triantennary)
T7 + CHO (N-linked)	5665.0	5664.8	$Gal_3Man_3GlcNAc_5$ (Triantennary)
	5299.0	5299.5	$Gal_2Man_3GlcNAc_4$ (Biantennary)
T19 + CHO (O-linked)	6384.8	6385.2	GalGalNAc
	6750.1	6750.5	$(GalGalNAc)_2$
	7114.8	7115.8	$(GalGalNAc)_3$

(Data reproduced with permission from **ref. *19*.**)

glycosylation sites (Ser or Thr) than there are carbohydrate side-chains attached to the glycopeptide.

References

1. Chapman, J. R. (1993) *Practical Organic Mass Spectrometry: A Guide for Chemical and Biochemical Analysis.* Wiley, Chichester, UK.
2. Harrison, A. G. (1983) *Chemical Ionization Mass Spectrometry.* CRC, Boca Raton, FL.
3. Barber, M., Bordoli, R. S., Elliott, G. J., Sedgwick, R. D., and Tyler A. N. (1982) Fast atom bombardment mass spectrometry. *Anal. Chem.* **54,** 645A–657A.
4. Hillenkamp, F., Karas, M., Beavis, R. C., and Chait, B. T. (1991) Matrix-assisted laser desorption/ionization of biopolymers. *Anal. Chem.* **63,** 1193A–1203A.
5. Fenn, J. B., Mann, M., Meng, C. K., Wong, S. F., and Whitehouse, C. M. (1990) Electrospray ionization—principles and practice. *Mass Spectrom. Rev.* **9,** 37–70.
6. Dawson, P. H. (1976) *Quadrupole Mass Spectrometry and its Applications.* Elsevier, NY.
7. Duckworth, H. E., Barber, R. C., and Venkatasubramanian, V. S. (1988) *Mass Spectroscopy, 2nd. ed.* Cambridge University Press, Cambridge. UK.
8. Cotter, R. J. (1992) Time-of-flight mass spectrometry for the structural analysis of biological molecules. *Anal. Chem.* **64,** 1027A–1039A.
9. March, R. E. and Hughes, R. J. (1989) *Quadrupole Storage Mass Spectrometry.* Wiley, New York.
10. Asamoto, B. (1991) *FT-ICR MS: Analytical Applications of Fourier Transform Ion Cyclotron Resonance Mass Spectrometry.* VCH, Weinheim.
11. Busch, K. L., Glish, G. L., and McLuckey, S. A. (1988) *Mass Spectrometry/Mass Spectrometry: Techniques and Applications of Tandem Mass Spectrometry.* VCH, New York.
12. He, S., Pan, S., Wu, K., Amster, I. J., and Orlando, R. (1995) Analysis of normal human fetal eye lens crystalline by high-performance liquid chromatography/mass spectrometry. *J. Mass Spectrom.* **30,** 424–431.
13. Hogeland, K. E., Jr. and Deinzer, M. L. (1994) Mass spectrometric studies on the N-linked oligosaccharides of baculovirus-expressed mouse interleukin-3. *Biomed. Environ. Mass Spectrom.* **23,** 218–224.

14. Jacquinod, M., Halgand, F., Caffrey, M., Saint-Pierre, C., Gagnon, J., Fitch, J., Cusanovich, M., and Forest, E. (1995) Conformational properties of Rhodobacter capsulatus cytochrome c2 wild-type and site-directed mutants using hydrogen/deuterium exchange monitored by electrospray ionization mass spectrometry. *Rapid Commun. Mass Spectrom.* **9,** 1135–1140.

15. Schwartz, B. L., Gale, D. C., and Smith, R. D. (1996) Noncovalent interactions observed using electrospray ionization, in *Methods in Molecular Biology: Protein and Peptide Analysis by Mass Spectromeny* (Chapman, J. R., ed.), Humana, Totowa, NJ.

16. Cottrell, J. S. and Sutton, C. W. (1996) The identification of electrophoretically separated proteins by peptide mass fingerprinting, in *Methods in Molecular Biology: Protein and Peptide Analysis by Mass Spectrometry* (Chapman, J. R., ed.), Humana, Totowa, NJ.

17. Naito, Y., Ishikawa, K., Koga, Y., Tsuneyoshi, T., Terunuma, H., and Arakawa, R. (1995) Molecular mass measurement of polymerase chain reaction products amplified from human blood DNA by electrospray ionization mass spectrometry. *Rapid Commun. Mass Spectrom.* **9,** 1484–1486.

18. Liu, Y.-H., Bai, J., Zhu, Y., Liang, X., Siemieniak, D., yenta, P. J., and Lubman, D. M. (1995) Rapid screening of genetic polymorphisms using buccal cell DNA with detection by matrix-assisted laser desorption/ionization mass spectrometry. *Rapid Commun. Mass Spectrom.* **9,** 735–743.

19. Yang, Y. and Orlando, R. (1996) Identifying the glycosylation sites and site-specific carbohydrate heterogeneity of glycoproteins by matrix-assisted laser desorption/ionization mass spectrometry. *Rapid Commun. Mass Spectrom.* **10,** 932–936.

51

The Technique of *In Situ* Hybridization

Principles and Applications

Desirée du Sart and K. H. Andy Choo

1. Introduction

The aim of this chapter is to give a general overview of the technique of *in situ* hybridization. By way of introducing the technique, the history and background of *in situ* hybridization will first be discussed, followed by the principles and basic steps involved and lastly, the various applications of *in situ* hybridization. Detailed descriptions of the individual procedures and methods are outside the scope of this chapter, but readers are referred to In Situ *Hybridization Protocols*, edited by K. H. Andy Choo, 1994, and *Nonradioactive* In Situ *Hybridization Application Manual*, Boehringer Mannheim, 2nd Edition, 1996.

2. History and Background

The technique of *in situ* hybridization was originally developed by two independent groups in 1969 *(1,2)*. At that time, incorporation of radioactive isotope was the only method available for probe labeling, autoradiography was the only means of detecting hybridized sequences, and probes used for *in situ* hybridization were restricted to sequences that could be purified and isolated by conventional biochemical methods. Since then, the technique has undergone a tremendous amount of change. As can be seen, real advancement came in the 1980s with accelerated improvement in molecular cloning and nonradioactive labeling techniques. Because of problems associated with radioactive probes, namely its safety, limited shelf-life, and extensive autoradiography time, nonradioactive *in situ* hybridization is now the preferred method.

The improved nonradioactive technique was essentially a 2–3 d procedure. It involved the stable labeling of the probe nucleic acid (with no theoretical shelf-half-life), generally an overnight hybridization of probe onto target on d 1, simple post-hybridization washes, followed by fluorescent- or enzyme-immunochemistry for hybrid molecule detection, and fluorescent or light microscopy on d 2. *In situ* hybridization, now free of radioactivity, became a safer and faster technique with higher resolution. The first single-copy gene to be mapped using nonisotopic *in situ* hybridization was achieved using fluorescence contrast microscopy in 1985.

From: *Molecular Biomethods Handbook*
Edited by: R. Rapley and J. M. Walker © Humana Press Inc., Totowa, NJ

The current fluorescence *in situ* hybridization (FISH) technology was developed in 1986. In the late 1980s phase, there was a great deal of excitement over this new technology, with FISH being used in both research and diagnostic laboratories and applied to many biological questions. However, this enthusiasm was soon diminished as the pitfalls and limitations of the technique became a hindrance.

In the early 1990s, breakthroughs in technology soon started removing some of the pitfalls and limitations. Image capture and computer software removed the problem of signal fading when using fluorochromes. One of the limitations of earlier FISH methods was the inability to distinguish more than one target sequence simultaneously. The introduction of multicolor FISH overcame this, but was still hampered by the limited number of fluorochromes available. Then, combinatorial or ratio labeling was introduced, where a probe can be labeled with more than one fluorochrome- or hapten-conjugated nucleotide in varying ratios and detected with a color combination mix of more than one fluorochrome. This was demonstrated in 1992, when three fluorochromes along with digital imaging microscopy was used to detect seven DNA probes simultaneously in metaphase cells. Now in 1996, the goal of visualizing all 22 different human autosomes and the two sex chromosomes using 24 different colors has been achieved by the development of epifluorescence filter sets and computer software to detect and discriminate 27 different DNA probes hybridized simultaneously. This major advancement will have a marked impact on the various applications of FISH and will allow the operators to be mused by the world of multicolor!

3. Theory and Principle of *In Situ* Hybridization

The principle behind *in situ* hybridization is the specific annealing of a labeled nucleic acid probe to its complementary sequences in fixed tissue or cells, followed by visualization of the hybridization with radioactive decay, fluorescent signals, or enzyme histochemistry. A critical aspect of the procedure is that the target nucleic acid be preserved *in situ* and be accessible for hybridization to the probe. Unlike the hybridization of nucleic acids in solution to target sequences on membrane filters, the target in this case is crosslinked and embedded in a complex matrix that hinders access of the probe and decreases stability of the hybrids.

There are a number of different techniques in *in situ* hybridization, each dependent on the application of the technique, the type of probe and target nucleic acid to be used, and the type of visualization system available to display the hybridization signals. For example, the probe could be double-stranded DNA, cDNA, single-stranded DNA oligonucleotides, or RNA; the target could be metaphase chromosomes, interphase cells, chromatin fibers, or tissue sections; the hybridization could be displayed by radioactive decay signals or by fluorescent dyes—each part requires a variation of the basic series of procedures for an optimal result. The basic steps for *in situ* hybridization and their underlying principles are discussed below.

3.1. Preparation of Probe

Preparation of the probe involves incorporating labeled nucleotides into the nucleic acid structure of the probe. The application of the *in situ* hybridization, the type of nucleic acid, and the type of label used to tag the probe will determine the method of probe preparation (discussed below). Once tagged with a reporter molecule, the probe is denatured (if double-stranded), diluted in probe-mix (a solution containing the right concentration of sodium ions, formamide, and dextran sulfate; *see* **Subheading 3.3.**) and reannealed with unlabeled competitor DNA for competition hybridization if required, before adding to the target nucleic acid.

3.1.1. Type of Nucleic Acid Used as Probe

3.1.1.1. DOUBLE-STRANDED DNA PROBES

Double-stranded DNA probes can be labeled by nick-translation, random priming, or polymerase chain reaction (PCR) in the presence of labeled nucleotide. In the nick translation procedure the DNA template can be supercoiled or linear; after nicking the DNA with DNase I, the 5' to 3' exonuclease activity of DNA polymerase I extends the nicks to gaps and then the polymerase replaces the excised nucleotides with labeled ones; the size of the probe after the reaction is in the 200–500 bp range. In random primed labeling the template DNA is linearized and denatured; Klenow enzyme is used to synthesize new DNA along the single-stranded substrate starting from the 3' hydroxyl end of the annealed primer; the size of the probe is in the 100–800 bp range. In both these methods, a heterogeneous population of probe strands that have overlapping complementary regions are produced, which could lead to amplification of the signal in the hybridization experiment. In PCR, two oligonucleotide primers hybridize to opposite DNA strands and flank a specific target sequence; a thermostable polymerase (Taq DNA polymerase) then elongates the two PCR primers. A repetitive series of cycles (template denaturation, primer annealing, and extension of primers) result in an exponential accumulation of copies of target sequence. Incorporation of labeled nucleotide during PCR can produce large amounts of labeled probe from minimal amounts of template DNA, or from only a specific region of template DNA. This method characteristically produces sequence-specific and size-specific fragments, depending on the specificity of the primers used. Double-stranded DNA probes have to be denatured before use to allow binding with the target, which can be achieved with heat. These probes are less sensitive than single-stranded probes because the two strands reanneal, thus reducing the concentration of probe available for hybridizing to the target. Nevertheless, the sensitivity obtained with double-stranded probes is sufficient for many purposes and they have found widespread application.

3.1.1.2. SINGLE-STRANDED DNA PROBES

Single-stranded DNA probes can be prepared by primer extension on single-stranded templates by PCR or by the chemical synthesis of oligonucleotides. The PCR method is the easiest and it allows probe to be generated from very small amounts of starting material. Oligonucleotides can be synthesized for use as probes (typically 20–30 bases in length) and labeled either by incorporating a labeled nucleotide in the chemical synthesis or by end labeling the oligonucleotide after synthesis. However, neither labeling method allows for labeling with a high specific activity and so oligonucleotides are less sensitive than longer nucleic acid probes for single copy sequences, though not a problem when probing for repetitive sequences. This can be overcome to a certain extent by using a mixture of labeled oligonucleotides that are complementary to different regions of the target tissue. The small size of oligonucleotide probes allows good penetration during the hybridization procedure, which is especially important when using interphase nuclei as target DNA. Single-stranded probes do not self-anneal during hybridization, so high concentrations of probe can be used without significant increase in background signals. When using synthetic oligonucleotides of repetitive sequences as probe, hybridization time can be reduced to 3–5 h because they are very short, controlled sequence probes and the target DNA is abundant so they renature very rapidly during hybridization. On the other hand, hybridization for the more standard 16 h does not cause a detectable increase in nonspecific background. Also, the probe DNA contains only the sequence of interest so there is no need for blocking with competitor DNA during hybridization.

3.1.1.3. RNA Probes

RNA probes are single-stranded and are labeled by the use of a purified RNA polymerase, namely SP6, T7, or T3 RNA polymerase, to transcribe sequences downstream of the polymerase initiation site in the presence of one or more labeled nucleotides. The probe sequence is usually first cloned into a plasmid vector in such a way that it is flanked by two different RNA polymerase initiation sites. This would then enable either the sense-strand RNA (used as a control) or the antisense RNA (used as the probe) to be synthesized. The plasmid is linearized with a restriction enzyme before synthesis, so that plasmid sequences are not transcribed, since these will contribute to high background in the preparation. Alternatively, RNA probes could be labeled directly without cloning by using PCR, achieved by including an RNA polymerase initiation site in one of the PCR primers.

3.1.2. Label for Tagging Probes

For radioactive *in situ* hybridization, several different isotopes have been used to radiolabel probe nucleotides, differing with regard to the resolution of the signal, the speed of obtaining the results, and the stability of the labeled probe. Three isotopes will be discussed here, namely, ^{3}H, ^{35}S, and ^{32}P. ^{3}H probes offer a subcellular resolution of signal but since only relatively low specific activity probes can be generated, they require long autoradiographic exposures (in the order of several weeks) for signal visualization. ^{3}H has a half-life of 12 yr, so the labeled probes are relatively stable for many years. ^{35}S-labeled probes give less resolution (about one cell diameter) but more rapid results than ^{3}H probes (~ 1 wk). However, because ^{35}S has a half-life of 87 d, probes need to be used within 1 mo of being prepared. Further, ^{35}S-labeled probes require protection from oxidation. ^{32}P-labeled probes give an even poorer resolution of signal than ^{35}S and because of the low efficiency of autoradiography offers no advantage of speed in obtaining results. It also has a half-life of 14 d, so probes have to be used within 1 wk of being made.

The nonradioactive, hapten-labeled nucleotides for probe labeling in fluorescence *in situ* hybridization (FISH) can be digoxygenin- (or DIG-), biotin-, or fluorescein-conjugated nucleotides *(3,4)*. Digoxygenin is a steroid isolated from digitalis plants, namely *Digitalis purpurea* and *Digitalis lanata*, and is linked to the fifth position of the pyrimidine ring of uridine nucleotides via a spacer arm made up of 11 carbon atoms (DIG-11-dUTP). Biotin is a member of the vitamin B complex (also called vitamin H), and is linked to the 5th position of the pyrimidine ring of uridine nucleotides via a spacer arm made up of 16 carbon atoms (Biotin-16-dUTP). Both of these reporter molecules require immunocytochemistry for visualizing the hybridization (an indirect hybridization experiment). Fluorescein, a fluorescent dye, is linked to the pyrimidine ring of uridine nucleotides via a spacer arm of 12 carbons (fluorescein-12-dUTP) and can be used in a direct or an indirect hybridization experiment. Other fluorochrome-labeled nucleotides, Tetramethylrhodamine-6-dUTP (red fluorescence dye) and Amino-methylcoumarin-6-dUTP (blue fluorescence dye), can also be used for direct hybridization, and so allow multicolor direct FISH.

Radioactive labeling has been superseded by the faster, safer, and higher resolution of nonradioactive hapten-labeling. However, the radioactive procedure still holds a significant place in various applications because of its unique features. For example, it offers an acceptable way of quantitating signal distribution on different tissue or chromosomal sites; in addition, if the more elaborate digital imaging system is unavailable for use with the fluorescent visualization, then extended autoradiography offered by the radioactive procedure provides greater sensitivity, especially for small or heterologous probes *(5)*.

3.2. Preparation of Target Tissue

The method used to prepare tissue for *in situ* hybridization is dependent on the nature of the target tissue. This may be cultured cells that have been grown directly onto coverslips or slides, or cultured cells that have been made into a cell suspension, or it may be tissue sections or whole mounts of tissues or embryos. Preservation of the cellular and tissue structure by fixation is essential in this part of the procedure, but it must be balanced with accessibility for probe binding to the target tissue. Stronger fixation yields a better preservation of cellular morphology, but increased crosslinking lowers the accessibility of the target to the probe.

3.2.1. Fixation

Fixation can be done by precipitation when using alcohol-based fixatives, or by crosslinkage when using formaldehyde-based fixatives. The crosslinking fixatives give greater accessibility and/or stable retention of cellular RNA than precipitating fixatives by introducing chemical bonds between nucleic acids and proteins. To date there is no one fixation protocol that can be used for all target tissues, so the fixation and pretreatment protocols must be modified for different applications. It should be noted that the DNA and RNA target sequences are surrounded by proteins and that extensive crosslinking of these proteins masks the target nucleic acid. Permeabilization procedures are therefore often required.

3.2.1.1. Cultured Cells

Cells may be cultured using procedures and culture medium specific for the particular cell type. For example, fibroblast cells are cultured in DMM with 10% fetal calf serum; blood cells are cultured in RPMI 1640 with 20% FCS and a mitogen to stimulate lymphocyte proliferation. For *in situ* hybridization, the cells could be grown directly onto coverslips or microscope slides within Petri dishes and then fixed onto the coverslip or slide. Cultured cells could also be harvested into a cell suspension and then transferred onto slides either by dropping suspension with a pipet or by cytospinning onto slides and then fixed, or the cell suspension may be fixed prior to transferring onto slides, as is the case with normal chromosome preparations. For cultured cell preparations, being either in cells suspension or attached to a coverslip or slide (e.g., chromosome spreads, interphase nuclei, DNA halo preparations, or extended chromatin fiber preparations), methanol/acetic acid fixation is sufficient for good preservation of cellular morphology and allows good accessibility for probe hybridization. It is also not necessary to coat slides to prevent loss of material, as is the case with tissue sections; alcohol-cleaned slides are adequate.

3.2.1.2. Tissue Sections

For paraffin-embedded tissue sections, formalin fixation is performed prior to tissue sectioning. Cryostat sections can be fixed with formaldehyde or Bouin's fixative or paraformaldehyde vapor after tissue sectioning, allowing for a variety of immunochemistry and hybridization techniques on adjacent tissue sections. Embedded tissue does have the advantage of better preservation of tissue morphology and it allows for thinner sections to be cut, increasing the clarity of the preparations. Embedding tissue in plastic allows for very thin sections to be cut, as is required for electron microscopy, but it masks much of the target nucleic acid and so can only be used for high abundance DNA or RNA target hybridizations. For tissue sections, it is advised to coat slides with gelatin or polylysine to help sections adhere to the slide and be retained during subsequent steps of the *in situ* procedure. Although FISH is widely used for mapping chromosomes or in cell culture, its application on tissue sections is limited. Most investigators applying nonradioactive *in situ* hybridization on paraffin or frozen sections prefer using enzyme histochemistry in the detection step because it offers the possibility of conventional counterstaining, which allows for better histological orientation. This is especially important for *in situ* hybridization studies on calcified tissue characterized by a complex histology. However, recently propidium iodide, a fluorescent dye for DNA counterstaining, was used to allow histological orientation in combination with FISH on calcified tissue *(6)*. This would now allow the wide application of FISH on tissue sections as well.

3.2.2. Pretreatment of Material on the Slide

Depending on the application of the *in situ* technique, it may be necessary to perform one or more of the following steps, intended to increase the efficiency of hybridization and/or decrease nonspecific background.

1. For wax-embedded tissue, the first step is to thoroughly remove the wax by dissolving it in an organic solvent.
2. Treatments to prevent background staining: When an enzyme is used as the label, e.g., peroxidase or alkaline phosphatase, the endogenous enzyme activity may have to be inactivated before hybridization. It may also be necessary to RNase-treat the slides to remove endogenous RNA to improve the signal-to-noise ratio in hybridizations to DNA targets. This treatment may also be used as a control in hybridizations with mRNA as target. If a tissue contains pigment granules that can obscure the signal, a bleaching step with hydrogen peroxide is required.
3. Treatment to increase permeabilization: This treatment serves to increase accessibility by either extraction or digestion of protein or lipid membrane components that surround the target nucleic acid. It can be done by treating slides with dilute acid, e.g., $0.2\ M$ HCl, or with detergent/alcohol, e.g., Triton X-100, sodium dodecyl sulfate, or with proteases, e.g.,

Proteinase K and pepsin. It is essential that protease treatment be followed by a refixation step, otherwise disintegration of the tissue will occur. Protease treatment substantially improves the signal obtained with long (>100 bases) probes, but is not required when oligonucleotide probes are being used.

4. Prehybridization: A prehybridization incubation is often necessary to prevent background staining in hybridizations with RNA as target. It is also recommended for whole mount hybridizations. The prehybridization mixture contains all components of a hybridization mixture except for the probe and it is performed at the same temperature as the hybridization.

3.2.3. Denaturation of Target Tissue

For *in situ* hybridization to chromosomal, nuclear, or extended chromatin fiber DNA, the DNA target must be denatured. This can be achieved by the use of a dilution of deionized formamide in a salt solution or by heat, or a combination of the two, for example, 70% formamide in 2X SSC at 72°C is used for denaturing chromosome target DNA. Variations in time and temperature should be evaluated to find the best conditions for denaturation. After denaturation, in most instances, the target tissue is dehydrated in an ethanol series (at −20°C to keep target denatured) before the addition of the probe solution to prevent dilution of the probe.

3.3. Hybridization

Hybridization depends on the ability of the probe nucleic acid to reanneal with its complementary strand of target nucleic acid in an environment just below their melting point (T_m). The T_m is the temperature at which half of the nucleic acid is present in a single-stranded form. It can be calculated by measuring the absorption of ultraviolet light at a wavelength of 260 nm. The stability of the nucleic acid is directly dependent on its base composition, or rather, its GC content. The greater the GC content, the higher the melting point T_m. The melting point is influenced by a number of factors:

1. The nature of the probe and the target nucleic acid: RNA/RNA hybrids are more stable than RNA/DNA hybrids, which are more stable than DNA/DNA hybrids;
2. The length of the probe: longer probes form more stable hybrids; however, short probes are required for *in situ* hybridization because the probe has to diffuse into the dense matrix of cells or chromosomes.
3. The extent of sequence identity between the probe and target: probes that are more divergent from the target sequence form less stable hybrids than homologous probes. On average, the T_m decreases about 1°C per percent of bases mismatched for large probes. The effect on the T_m is greater for smaller probes.
4. The composition of the hybridization or wash solution. Four parameters influence the T_m and renaturation of nucleic acid in the hybridization or wash solution:
   a. Temperature: The maximum rate of renaturation (hybridization) of DNA ranges from a temperature of 16–32°C below T_m and depends on the length of the probe. RNA/RNA hybrids are 10–15°C more stable than DNA/DNA hybrids in standard hybridization buffer (5X SSC, 50% formamide).
   b. pH: In the pH range 5.0–9.0 the rate of renaturation is fairly independent of pH. For buffers containing 20–50 mM phosphate, pH 6.5–7.5 are frequently used. Higher pH can be used to produce more stringent hybridization conditions.
   c. Monovalent cations: Monovalent cations (e.g., sodium ions) interact electrostatically with especially the phosphate groups of nucleic acids, so that electrostatic repulsion

between the two strands of the duplex decreases with increasing salt concentration. Therefore, higher salt concentrations increase the stability of the hybrid; low sodium ion concentrations (below 0.4 *M*), on the other hand, drastically affect the T_m, as well as the renaturation rate.

d. Presence of organic solvents—Formamide: DNA denatures at 90–100°C in 0.1–0.2 *M* sodium ions. For *in situ* hybridization, this implies that preparations must be hybridized at 65–75°C for prolonged periods, which may lead to deterioration of morphology. Organic solvents reduce the thermal stability of double-stranded polynucleotides so that hybridization can be performed at lower temperatures. Formamide reduces the melting temperature of DNA/DNA and DNA/RNA duplexes in a linear fashion by 0.72°C for each percent formamide. Thus, hybridization can be performed at 30–45°C with 50% formamide present in the hybridization mixture.

The melting temperature of hybrids taking the above parameters into consideration can be calculated according to the formulas listed below. These formulas apply to nucleic acids hybridizing in solution, but can be used as a rough guide to hybrids formed with the crosslinked targets present in fixed tissue, which are less stable because of steric hindrance preventing the annealing of probe along its full length. RNA/RNA hybrids formed during *in situ* hybridization have a melting temperature about 5°C lower than hybrids formed in solution. For 0.01–0.2 *M* sodium ions,

$$T_m = 16.6 \log M + 0.41(GC) + 81.5 - 0.72(\%formamide) \tag{1}$$

For sodium ion concentration > 0.4 *M*,

$$T_m = 81.5 + 0.41(GC) - 0.72(\%formamide), \tag{2}$$

where *M* is the molar salt concentration (in the range 0.01–0.20 *M*) and GC the molar percentage of guanine plus cytosine.

Dextran sulfate is included in the probe mix solution because it is regarded as a volume exclusion agent. In aqueous solutions it is strongly hydrated and as a result, other macromolecules in the solution have no access to the hydrating water, which causes an apparent increase in probe concentration and consequently higher hybridization rates.

The prepared probe nucleic acid is added to the prepared target tissue and mounted under a coverslip, sealed with rubber cement, and allowed to hybridize at a specific temperature for the required period of time.

3.4. Post-Hybridization Steps

Labeled probes can hybridize nonspecifically to sequences that bear homology but are not entirely homologous to the probe sequence. The extent to which this occurs can be manipulated to some extent by varying the stringency of the hybridization reaction. Such hybrids are less stable than the perfectly matched hybrids. They can be dissociated by performing washes of various stringencies. The stringency of the washes can be manipulated by varying the formamide concentration, the salt concentration, and the temperature of the wash solution. The higher the formamide concentration, the lower the salt concentration and the higher the temperature; each contribute to a higher wash stringency. It is also possible to reduce the nonspecific signals by treating the slides with single-strand specific nucleases.

3.5. Visualization of the Probe on Target Tissue

The method for visualizing the probe signal depends on the type of label that was incorporated into the probe. Radioactive probes require either an X-ray film placed adjacent to the slide (for ^{35}S) to obtain a quick, low resolution signal, or a much greater resolution is obtained by dipping the slide into a liquid nuclear track emulsion that is then dried, exposed, and developed. The quality of the results obtained is dependent on the grade of the emulsion (coarse grades are more sensitive but give poorer resolution), the thickness of the emulsion (a thicker layer gives greater signal but poorer resolution), the storage and handling of the emulsion (mechanical stress and penetrating radiation can produce high background), exposure time (overexposure leads to reduction of signal-to-noise ratio) and conditions for developing (higher temperature, more agitation of slides, or longer time of developing yields larger silver grains but a lower resolution). After developing, the tissue is stained with a nuclear stain, mounted under a coverslip, and examined under bright-field illumination on a light microscope.

In FISH, the type of hapten labeling used to tag the probe will determine the procedure for detecting the hybridization. In direct hybridization, using fluorochrome-conjugated nucleotides, there is no requirement for immunocytochemistry, and the hybridization can be viewed directly after post-hybridization washes using a fluorescence microscope. In indirect hybridization, immunochemistry is required to indicate the location of the hybridization.

Hybridized DIG-labeled probes may be detected with high affinity antidigoxygenin (anti-DIG) antibodies that are conjugated to alkaline phosphatase or peroxidase (for brightfield or fluorescence microscopy), or fluorescein (or FITC—fluorescein isothiocyanate), rhodamine, Texas Red and Amino-methylcoumarin acetic acid (AMCA) (for fluorescence microscopy), or colloidal gold (for electron microscopy). Alternatively, unconjugated anti-DIG antibodies and conjugated secondary antibodies may be used to increase signal amplification.

In principle, biotin can be used in the same way as digoxygenin, and be detected by antibiotin antibodies. However, streptavidin or avidin, conjugated to a fluorochrome, is more frequently used because these molecules have a high binding capacity for biotin. In order to amplify the signal, a secondary layer of antiavidin antibodies could be applied, followed by the first layer of conjugated avidin.

When fluorescein is used as a direct label, no immunocytochemical visualization procedure is necessary and there is the advantage of very low background. The disadvantage of the direct method is the loss of sensitivity, a problem when using unique sequence or low-copy number probes, but overcome when using repetitive or high-copy number probes. Alternatively, fluorescein-labeled nucleotides can be detected with antifluorescein antibody–enzyme conjugate or with an unconjugated antibody and fluorescein-labeled secondary antibody. The sensitivity of this detection corresponds to other indirect methods *(7)*.

Fluorochrome-labeled nucleotides or antibodies require the availability of a fluorescent microscope and specific filters that allow visualization of the wavelength emitted by the fluorescent dye. Because with exposure to UV light emitted wavelengths become exhausted so signals obtained with the fluorochromes fade quickly usually antifading reagents have to be added before analysis. Advancement in technology in the form of image capture and computer software solved the problem of signal fading when using fluorochromes. For optimal sensitivity and permanence to be achieved, a digital imaging system is required. Digital imaging microscopes detect signals that cannot be seen with conventional microscopes. Also, image processing technology provides enhancement of signal-to-noise ratio as well as measurement of quantitative data. The most sensitive system to date, the cooled CCD (charged coupled device) camera, counts emitted photons with a high efficiency over a broad spectral range of wavelengths, and is the instrument of choice for two-dimensional analysis. For three-dimensional analysis, confocal laser scanning microscopy is used. However, if the probes being used are large (e.g., cosmids or YACs) or contain repeat sequences, thus producing easy to see signals, then ordinary fluorescence microscopy with a camera attachment is sufficient to capture the resulting image.

Multicolor FISH is possible and allows for the detection of more than one target sequence simultaneously in one hybridization, by labeling the different probe nucleic acids with a different tag. In 1990 Nederlof *(8)* described a method for detecting more than three target DNA sequences using only three fluorochromes, by combinatorial labeling or ratio labeling—a probe can be labeled with more than one fluorochrome or hapten-conjugated nucleotide in varying ratios and detected with a color combination mix of more than one fluorochrome. For example, by ratio labeling with three fluorochromes, it is possible to detect seven DNA probes simultaneously in metaphase cells *(9)*, using digital imaging microscopy. The number of useful combinations of N fluorochromes is $2^N - 1$; thus increasing the number of fluorochromes to four would allow the detection of 15 different probe sequences simultaneously, and so on. With the development of epifluorescence filter sets *(10)* and accompanying computer software, multicolor FISH can now detect and discriminate 27 different probes hybridized simultaneously in different colors.

4. Application of *In Situ* Hybridization

The following applications of *in situ* hybridization will be discussed:

1. Chromosome painting;
2. Visualization of chromatin higher order structure;
3. Positional cloning;
4. RNA *in situ* hybridization;
5. Primed *in situ* hybridization (PRINS);
6. Immunochemistry in combination with *in situ* hybridization; and
7. Detection of viruses by *in situ* hybridization.

4.1. Chromosome Painting

Chromosome painting is a technique of FISH using a chromosome-specific DNA library as a probe pool combined with chromosome *in situ* suppression (CISS) hybridization, to depict an entire chromosome or chromosomal region. The probes used in chromosome painting are usually a complex mixture of both repeat and unique sequences derived from the amplification of DNA from a number of different sources, namely, chromosome-specific somatic cell hybrids, microdissected chromosomes or chromosomal regions, flow-sorted chromosome libraries, and chromosome-specific libraries. To ensure specific or at least preferential labeling of homologous chromosomes or regions, dispersed repetitive sequences have to be excluded from hybridization by prehybridization with total unlabeled genomic DNA in excess *(11,12)*, i.e., CISS hybridization. Synthetic oligonucleotides made from sequences of repetitive DNA can also be used in chromosome painting without the need for blocking with competitor DNA during hybridization because they are sequence-specific.

4.1.1. DNA Probes from Somatic Cell Hybrids

Genomic DNA from hybrid cell lines retaining specific chromosomes has been used directly as a source material for painting *in situ (13,14)*, but because the chromosomes of interest represent only a small component of the total DNA, the sensitivity of this technique is not always satisfactory. It is possible to increase the sensitivity of this method when using human chromosomes by selectively amplifying the human DNA present. This can be achieved using an Alu-derived primer or L1 primer in polymerase chain reaction (PCR) to amplify the inter-Alu or inter-L1 regions of the human DNA, i.e., interspersed repetitive sequence-PCR (IRS-PCR) *(15–18)*. Labeled IRS-PCR products have been shown to be efficient for painting specific human chromosomes. Alu and LINE sequences are interspersed in the human genome; their primary distribution correlates with G-negative and G-positive bands, respectively. Thus, FISH with Alu-PCR products will generate a banding pattern corresponding to R-banding. PCR products from IRS-PCR using L1 primer can be used in combination with the Alu-PCR products to increase the amount of chromosome painting obtained to a less banded appearance. It should be noted, however, that there would be "holes" or nonpainted regions produced by using these products as probes because most inter-L1 DNA segments are out of the amplification range of Taq polymerase and, there is extreme paucity of Alu or LINE sequences in the centromeric regions of chromosomes. If centromeric regions are involved in rearrangements, centromeric alphoid probes, or cloned or synthesized oligonucleotides can be successfully used instead. The painting efficiency of the IRS-PCR can be increased to a level comparable to the painting performances using flow sorted chromosome libraries by the use of accurately designed dual-Alu primers in the PCR amplification *(18)*. This technique, though it has limited applications in diagnostic and research laboratories, because it only offers unequivocal identification of rearranged chromosomes that have already been partly or completely indentified by conventional cytogenetic methods, it is useful when using a hybrid panel to investigate chromosomal distribution and organization of DNA sequences between species *(19)*.

4.1.2. DNA Probes from Microdissected Chromosomes

Chromosome microdissection has become a very powerful approach to generate chromosome band-specific paints and whole chromosome paints for physical mapping or cytogenetic analysis. It involves microdissecting part or all of a trypsin-Giemsa banded metaphase chromosome (or a rearranged or marker chromosome), prepared by conventional cytogenetic techniques, using a fine glass needle with an electronic micromanipulator, on the stage of an inverted microscope *(20–22)*, or laser microdissection of unbanded chromosomes *(23)* using focused UV laser microbeams. The microdissected material is then amplified by degenerate oligonucleotide-primed PCR (DOP-PCR) *(20,22)*. The resulting PCR-product probe pool is biotin-labeled and used in FISH onto normal metaphase chromosomes. The signals obtained on the normal metaphase chromosomes will reflect the origin of the dissected chromosome. It is possible to label the PCR-product probe pool with other haptens, e.g., Digoxygenin, spectrum orange, or a combination of these, in order to produce multicolor FISH, where a number of probe pools can be investigated simultaneously. This is a rapid technique; FISH probes can be generated in 1–2 d *(20,22)* and it is now possible to identify unequivocally the chromosome constitution of virtually any cytologically visible chromosome rearrangement. Chromosome abnormalities are an important cause of congenital malformation and are responsible for at least half of all spontaneous miscarriages. The analysis of nonrandom chromosome abnormalities in malignant cells has become an integral part of diagnosis and prognosis of many human cancers, and their molecular analysis has facilitated the identification of genes (including oncogenes and tumor suppressor genes) related to the pathogenesis of both hereditary diseases and cancer. Thus, this technique to identify virtually any chromosome rearrangement without prior knowledge of its origin has a wide application in both diagnostic and research laboratories.

4.1.3. DNA Probes from Flow-Sorted Chromosome Libraries

Rearranged or marker chromosomes can be isolated by fluorescence-activated cell sorting (FACS). In this technique, a chromosome suspension is stained with fluorescent dyes (Hoechst 33258 and chromomycin A3) and passed through a dual-laser cell sorter, in which the two lasers are set to excite the dyes separately, allowing a bivariate analysis of the chromosomes according to their size and nucleotide composition *(24,25)*. If the chromosome of interest can be resolved in the normal flow karyotype, then flow sorting can be achieved with a high degree of purity and can be considered as essentially pure DNA of interest. The flow-sorted material is then amplified by DOP-PCR to generate a complex probe mixture *(26)*. The PCR product is then labeled and used as a probe in FISH to normal metaphase preparations. The composition of the flow-sorted chromosome is directly reflected in the distribution of fluorescence signals on the normal chromosomes. The advantage of this approach is that no prior knowledge of the origin or the intrinsic components of the chromosome is required. This technique also has wide applications in both diagnostic and research laboratories for investigating structural chromosome abnormalities, provided that the aberrant chromosome(s) can be resolved in a flow karyotype.

4.1.4. DNA Probes from Chromosome-Specific Libraries

Chromosome-specific plasmid DNA libraries can be made from flow-sorted chromosomes *(11,27)*, microdissected chromosomes/parts of chromosomes *(23,28)*, and hybrids containing particular chromosomes *(15)*. The genomic inserts contain specific sequences for individual chromosomes as well as interspersed repetitive elements, so hybridization using these DNA as probes requires suppression of repetitive DNA without blocking the single-copy chromosome-specific sequences, i.e., CISS hybridization. These libraries exist for all human chromosomes and the plasmids contain inserts ranging in size from approx 0.1 to 6 kb. There is limited application of this technique in prenatal and postnatal cytogenetics to assist in identifying rearranged/marker chromosomes, for which conventional techniques have indicated a possible origin, and it can also be used in genome comparison between different species to reflect evolutionary changes.

4.1.5. DNA Probes from Synthetic Oligonucleotides

Probe generation using synthetic oligonucleotides allows *in situ* hybridization to be used for mapping repetitive sequences to chromosomal location, for visual analysis of polymorphisms of these repeats, and in the clinical study of aneuploidy, particularly for quantitation of chromosomes in interphase nuclei and analysis of marker chromosomes. The advantage of using synthetic oligonucleotides is that the exact sequence of the probe is known and controlled. They can be constructed to specifications that reproduce all or a segment of the sequence from any cloned repetitive DNA probe, reproduce all or a segment of consensus sequence of a repeat family, or reproduce any modification of any sequence. More detail is given in the single-strand DNA probe section. Examples of application of this method are in the study of telomeric and centromeric repeat sequences.

Chromosome painting is a tool that can be used to analyze small structural chromosomal aberrations that are not readily detected by standard high resolution banding techniques; it can assist in the identification of marker chromosomes, clarify balanced or unbalanced translocations, define chromosomal deletions or duplications, and analyze complex chromosomal rearrangements *(15,29,30)*. This technique can detect an individual chromosome distinctly from other chromosomes in both metaphase cells and interphase nuclei, leading to the diagnosis of numerical chromosome aberrations as well.

Chromosome painting can assist in the establishment of karyotypes for animals or plants in which the conventional cytogenetic chromosome banding is unable to clearly distinguish between the different chromosomes. For example, flow cytometry combined with chromosome painting was used to establish the pig karyotype *(31)* and the same technique was used to establish the dog karyotype *(32)*.

Chromosome painting can be used in evolutionary studies to identify homologous chromosomes or regions between species. Chromosomal rearrangements that occurred during speciation may be identified by interspecific chromosome paints, and chromosome painting across species is also useful for the recognition and chromosomal assignment of linkage groups of less well investigated species. This technique was used to compare primate genomes, to compare homologous segments in distantly related mammals, and to define the distribution and reorganization of repeat DNA in various species. Genomic *in situ* hybridization has also been established for use in studying plant genomes, allowing the origin of chromosomes in hybrids to be distinguished by the labeling of mainly or exclusively homologous sequences with parental DNA probes. However, at this stage, this technique can only be successfully used for lower plants with small, noncomplex genomes. Higher plant species with large complex genomes have dispersed repetitive sequences too complex for efficient blocking by conventional CISS hybridization; the presence of only one or a few such sequences that are not suppressed completely by annealing to unlabeled homologous sequences of the competitor DNA, and therefore persist within a labeled probe, may cause a dense dispersed labeling pattern and so affect interpretation of the results *(33)*.

A very powerful recent development in chromosome painting is comparative genome hybridization (CGH) *(34,35)*. This involves the combined hybridization to a normal metaphase spread, of equal amounts of test and reference DNA, each labeled with a different hapten to be tagged with a different fluorochrome (red or green fluorescence). Regions of equimolar DNA appear yellow, whereas regions with over or underrepresented DNA appear red or green. CGH applications lie mainly in the cancer field, to detect and map deletions and amplifications in malignant cells including primary and metastatic tumor biopsies, cancer cell lines, and tumor cells microdissected from paraffin-embedded tissue sections *(36)*. CGH is a considerable breakthrough for the analysis of solid tumor and other cancer tissues from which only a very few or no high-quality metaphases can be obtained. With future improvements in resolution, the use of CGH will be extended to include small deletions in contiguous gene syndromes and malformations. It has already been used as an adjunct to conventional cytogenetic techniques to examine unbalanced rearrangements and to detect the origin of duplicated or deleted chromosomal material in a single hybridization *(37)*. Thus, this method has future widespread applications in clinical cytogenetics and research laboratories.

4.2. Visualization of Chromatin High Order Structure

Advances in molecular biology combined with high-resolution FISH have made it possible to visualize individual genes, selected chromosome domains, and entire single chromosomes in interphase nuclei, demonstrating that chromatin can be highly folded and confined to discrete, spatially limited nuclear domains, even in genetically active regions. Chromosomes are not randomly dispersed in the nucleus, but occupy distinct and well-defined domains in the interphase nucleus; and at least some chromosomal domains are nonrandomly arranged in a cell type-specific manner *(38,39)*. Thus, the existence of different levels of order within the mammalian cell nucleus is beyond any doubt. The spatial rearrangement of definite chromosome regions (i.e., centromeres, telomeres, constitutive heterochromatin, and nucleolus organizer) is thought to provide a structural framework for the higher order nuclear organization, which may be involved in the regulation of gene expression above the level of single genes *(40)*. The movement and arrangement of these regions can be visualized by applying painting FISH to cells at different stages of the cell cycle *(41)*.

Defined spacial arrangements of chromosomes may facilitate homologous interactions during meiosis. In Drosophila, it has been shown genetically and cytologically that the homologs are associated in somatic cells *(42)*. In yeast, it has been shown that homologous chromosomes share joint territory even before entry into meiosis. During meiotic prophase of some organisms, all telomeres become clustered in an area near the centrosome, leading to a bouquet arrangement of meiotic chromosomes *(43)*. The centromeres of all the yeast chromosomes are clustered in vegetative and meiotic prophase cells, whereas the telomeres cluster near the nucleolus early in meiosis and maintain this configuration throughout meiotic prophase. Several reports suggest that the telomeres represent important initiation sites for meiotic chromosome pairing *(40,43)*. FISH with differentially labeled whole chromosome paints has also been applied to examining meiotic chromosome behavior, by analyzing meiotic segregation products of reciprocal translocations *(44)*.

4.3. Positional Cloning

Positional cloning is defined as the pursuit of genes involved in genetic disorders without foreknowledge of function, i.e., a disease phenotype is first mapped to a subchromosomal region, then the gene is isolated, and then the function of the gene is studied. To do this, genetic segregation of the disorder in disease-prone families is used to pin down a probable chromosomal location. This is then followed by the isolation and testing of genes for mutations segregating with the disorder. High-resolution physical mapping, ordering of probes, and construction of contigs of YAC or cosmid clones are essential steps in positional cloning projects. The rapid developments in FISH techniques have paved the way for the application of new visual physical mapping methods. By proper selection of the degree of condensation of the target nucleic acid on the slides, FISH can be applied at several levels of resolution. The simultaneous use of multiple colors in one hybridization also greatly enhances the resolving power. It permits multiclone orientation and the rapid positioning and ordering of new clones relative to known loci or previously mapped ones.

The use of conventional metaphase chromosomes provides a resolution of 3–10 Mb in fractional length determinations relative to the telomeres and centromere, and 1–3 Mb resolution in multicolor applications (45). This can be used to localize a disease phenotype to a chromosomal subband. By using mechanically stretched chromosomes as hybridization targets, mapping resolution is increased by 10–20-fold (46). FISH on interphase or extended pronuclear chromatin yields a mapping resolution of ~50–1000 kb (47,48). By the preparation of "DNA halos," in which most of the DNA is expelled from the nucleus in loops of 200 μm or more, single cosmid clones can be shown to hybridize over 10–15 μm (49). FISH on extended chromatin fibers provides an unprecedented resolution. It allows a visual inspection of specific DNA sequences along a stretched single DNA fiber and provides high resolution mapping in the range 1–300 kb (50–57).

In situ hybridization can expedite positional cloning by characterizing variable size subchromosomal rearrangements that result in a disease phenotype. Because FISH can visibly detect numerical and structural chromosome rearrangements like deletions and duplications, it accurately reflects the unique locus order of the rearrangement, whereas abnormal fragments resolved by pulsed field gel electrophoresis do not distinguish tandem and inverted duplications without complicated restriction maps.

The use of in situ hybridization in positional cloning allows the detection and mapping of homologous genes and pseudogenes as well as low copy homologous sequences, speeding up fine mapping and linkage studies by characterizing the number of sites from which homologous clones may be isolated.

In situ hybridization allows characterization of rearranged cloned DNA; this is especially important when analyzing YAC clones, since these are notorious for forming chimeric sequences. Hybridizing labeled cloned DNA to normal metaphase chromosome complements is the most direct way to detect and characterize clone rearrangements and hybrid clones.

4.4. RNA In Situ and Its Application

The RNA in situ technique is used for examining mRNA expression in tissues and to localize specific RNA expression at the individual cell level. Conventional molecular biology techniques for RNA analysis, such as Northern hybridizations, RNase protection assays, and reverse transcription polymerase chain reactions (RT-PCR), require the homogenization of the tissue or cell sample, and as a result, it is impossible to relate RNA expression, or changes therein, to histology and cell morphology of the tissue.

With the expanding field of recombinant DNA technology, a variety of cDNA probes is available to detect mRNAs. Originally, radiolabeled cDNA or RNA probes were used *(58)*. These radiolabeled probes have also allowed quantitative *in situ* hybridization studies to analyze gene expression within complex heterocellular systems, such as the central nervous system, and their variations under different experimental or physiological conditions *(59)*. Recently, nonisotopic detection systems have been introduced to overcome the limitations of radioactivity, namely, the high level of background, the long exposure time, and the biohazards normally associated with the use of radioisotopes *(60–62)*. RNA *in situ* hybridization has been further extended to multicolor-hybridization to analyze intron and exon mRNA sequences *(63)*. The nonradioactive *in situ* hybridization on tissue sections is usually followed by enzyme histochemistry for signal detection, to allow conventional counterstaining for histological examination. FISH has just recently been applied to RNA *in situ* hybridization on tissue sections with the introduction of the fluorescent dye, propidium iodide, as a counter stain to allow histological orientation *(6)*. This now opens the wide application of FISH and multicolor-FISH to RNA *in situ* hybridization on tissue sections. Already RNA *in situ* hybridization is a powerful tool in examining tissue distribution of gene expression in development, allowing serial reconstructions for the precise structural identification of the rapidly changing, complex embryo anatomy with the use of computer image analysis *(64)*. It is also being applied to the expression of transcriptionally active transgenes and, in particular situations, to gain further insight into the phenomenon of imprinting *(65)*.

4.5. Primed In Situ Labeling Technique (PRINS)

Primed *in situ* labeling is a technique combining the extreme sensitivity of the PCR and the cell-localization ability of *in situ* hybridization. It has become an alternative to the conventional *in situ* hybridization for the localization of nucleic acid sequences in cells and tissues *(66)*. The procedure is based on the rapid annealing of an unlabeled DNA primer, which could be a restriction fragment, a PCR product, or a synthetic oligonucleotide, to its complementary target sequence *in situ*. The primer serves as an initiation site for *in situ* chain elongation using a thermostable DNA polymerase (*Taq* DNA polymerase) and fluorochrom-, biotin-, or digoxygenin-conjugated nucleotides. The labeled DNA chain can be detected directly by fluorescence microscopy or indirectly using fluorochrome-conjugated avidin or antibody molecules *(67)*. The PRINS technique labels the site of hybridization rather than the probe nucleic acid and the amplified signal remains localized to the site of synthesis with very little diffusion of the PCR product into surrounding regions. Chromatin spatial organization may affect primer accessibility and influence the success of the reaction. The use of short, highly specific primers can overcome the problem of crosshybridization, which may also occur.

Compared with traditional *in situ* hybridization, PRINS has a number of advantages. The signal intensity is unaffected by probe (primer) size; because it is the site rather than the probe that is labeled, very high concentrations of probe can be used without significant increase in background staining, thus obtaining very short reaction times and the short incubation times also allows better preservation of cellular structures *(68)*. The procedure is very fast, allowing good signals to be obtained from repeated DNA sequences in <1 h and from unique DNA sequences in <3 h.

Recently, the detection of multiple DNA targets using different fluorochromes in sequential PRINS reactions has been developed *(69,70)*. However, for the simultaneous identification of more than two DNA sequences, DNA counterstaining or chromosome banding is, in principle, not possible because the available fluorescence colors are used for specific target detection. A more recently developed method is a triple-color, brightfield procedure that allows for the detection of PRINS-labeled nucleic acid sequences by enzyme precipitation reactions, inconjunction with visualizing both chromosomes and nuclei with hematoxylin counterstaining *(71)*. This brightfield approach does not require a fluorescence microscope with a confocal system or charge-coupled device (CCD) camera for image analysis and processing, and has the advantage that no fading of the PRINS signal occurs since they are permanently localized and can be examined in relation to cell morphology. The same brightfield detection approach can be used in applying the PRINS technique on frozen tissue sections.

Since its introduction, the PRINS technique has improved considerably for the detection of specific DNA sequences in cell preparations. It has been used as a rapid and reliable method for chromosome screening, in the assessment of aneuploidy and assisting the clarification of complex structural chromosome rearrangements *(72)*. This technique has not been restricted to animal cells; it has been applied to the study of DNA repeat sequences in different plant species *(73)*.

PRINS has been applied to the detection and quantitation of specific mRNA sequences, thereby determining the frequency of gene expression in cells. The principle of PRINS as applied to mRNA analysis is the *in situ* PCR amplification of cDNA using reverse transcriptase: short, unlabeled oligonucleotides are annealed to the target RNA and used as primers for chain elongation in a reaction catalyzed by AMV-reverse-transcriptase in the presence of radiolabeled or hapten-labeled nucleotides (directly with fluorescein or indirectly with biotin or digoxygenin). This offers a powerful research and clinical tool for examining genes with low expression, and still allows identification of the cellular origin of the signal *(74,75)*, and in particular, it allows for the analysis of closely related RNA populations, by detecting small differences, such as single base mutations or minor splicing-errors, not possible with normal RNA *in situ*.

4.6. Immunocytochemistry Combined with FISH

Fluorescence *in situ* hybridization can be combined with immunocytochemistry to visualize specific genomic DNA or RNA sequences and protein components simultaneously in interphase or metaphase cells. This technique enables the simultaneous demonstration of mRNA and its protein product in the same cell *(76)*, immunotyping cells containing a specific chromosomal aberration or viral infection *(77)*, characterizing cytokinetic parameters of tumor cell populations that are genetically or phenotypically aberrant *(78)*, and analysis of cellular structures that have both DNA and protein components, e.g., centromeres. The factors that determine the success and sensitivity of a combined technique include: preservation of cell morphology and protein epitopes, accessibility of nucleic acid targets, lack of crossreaction between the different detection procedures, good color contrast, and stability of enzyme cytochemical precipitates and fluorochromes. Since several steps in the *in situ* hybridization procedure may destroy the antigenic determinants, immunocytochemistry is usually performed first. Two examples of application this combined technique are discussed below.

For studying centromeres, the centromere protein components can be localized with anticentromere protein antisera and then visualized with a secondary fluorescence-conjugated antibody (e.g., rabbit antihuman conjugated Texas Red). The proteins are first antibody-annealed and crosslinked to the nuclear or chromosome structure, then the FISH technique is performed to localize the probe DNA in relation to the protein. This technique has been applied to the study of candidate centromere DNA sequences *(46,79)*, because it provides a visual tag for the centromeric region, which can otherwise only be indicated by the cytogenetic primary constriction.

Haaf and Ward *(76)* have applied this technique to analyze the effect of transcriptional inhibitors on chromatin structure and chromosomal domains. They were able to visually show nuclear proteins binding onto highly extended linear arrays of ribosomal genes because of the unraveling effect of a transcriptional inhibitor on the normally compact nucleolus.

These are just two examples of the application of immunochemistry-combined FISH. With many questions to be answered concerning DNA, RNA, and protein interactions, the visual, *in situ* aspect of this technique would have many future applications.

4.7. Detection of Viruses with In Situ *Hybridization*

Detecting viral infections depends largely on serological methods to demonstrate the presence of circulating viral antibodies and molecular techniques to identify viral particles in tissue extracts. However, in many situations, only a small fraction of a population of cells in a tissue actually harbors virus genes, and the extent of gene expression may be so curtailed that these usual indicators of viral infection may not be evident, and although sensitive, none of these methods provides the simultaneous morphological information that is unique to techniques related to *in situ* hybridization (ISH) for locating viral genomes or immunocytochemistry for locating viral antigens *(80)*.

Unlike other application of *in situ* hybridization, where the target is likely to be chromosomal DNA or RNA transcripts, the target in studies to detect viral nucleic acids may represent double- or single-stranded DNA or RNA, nucleic acid replicative intermediates, or mRNA. As a result, the sample may require an optional denaturation step to distinguish double- or single-stranded nucleic acids, and/or a prior nuclease digestion step. Because different viruses appear to replicate with different efficiency and because the replication efficiency may differ in acute and chronic infection, the target for ISH may vary between several hundred to several thousand genome copies. These factors may influence the fixative chosen to preserve the tissue and consequently, even though formalin fixation can be used, fresh, frozen tissue will result in higher sensitivity. There is no one particular protocol that would cover each of the possibilities; minor variations to the standard protocol is required to optimize the technique for viral studies.

ISH was developed initially using radiolabeled probes (^{35}S and ^{25}I). Using cloned restriction endonuclease fragments of viruses (or other sources of viral genome nuclei acids) as probes, target viral DNA (if there are many copies in individual cells) or its amplified mRNA can be detected *(81)*. The use of radiolabeled probes can provide good sensitivity and quantitative information about the concentrations of viral genes in single cells *(81)*, and also insight into virus dissemination and mechanisms of tissue injury by the analysis of cellular distribution of viral genomes and cytopathological changes in chronic infections *(80)*.

In recent years, there has been a move toward the use of nonisotopic ISH methods. Biotin, digoxygenin, and fluorochrome-labeled probes provide a safe, rapid means to detect virus nucleic acids at high resolution, without many of the problems associated with the use of radioisotopes *(82,83)*.

As discussed in the previous section, combining immunohistochemistry and *in situ* technique has made it possible to visualize the specific host cells infected with the virus, by using specific antisera conjugated to a color tag to indicate the host cells and autoradiograph silver grain production to localize the virus *(84)*. ISH to detect virus nucleic acids can be combined with other ISH applications (e.g., chromosome painting, analysis of chromatin structure, effect on DNA/protein interactions) to further analyze virus life cycles and their effects on host cells and tissues. One such combined application was investigated by Besse and Puvion-Dutilleul in 1993, who demonstrated the complete compartmentalization of viral DNA without any intrusion into the surrounding cellular condensed chromatin and conversely an Alu DNA probe only localized to the cellular condensed chromatin and not to any virus-induced structures at any stage of the infectious cycle.

Rapid advances in the methods of viral probe preparation and probe labeling have allowed the routine use of FISH for diagnosis of viral infections in laboratories. The technique is further facilitated by the commercial availability of nonradioisotopic labeling kits and prelabeled viral probes. This poses a wide application in both diagnostic and research laboratories.

5. Conclusion

When it was first developed in 1969, *in situ* hybridization was regarded as a specialized technique useable only by some workers for the first 15 years. Nonisotopic technology removed the stopgap in the mid 1980s and *in situ* hybridization was being applied to many exciting biological questions in both research and diagnostic laboratories. The big boom occurred with the development of multicolor fluorescence *in situ* hybridization and ratio-labeling of the probe nucleic acid. The advancement of fluorescence microscopy combined with computer technology, the development of chromosome flowsorting, and chromosome microdissection technology to create a probe from almost any fragment of nucleic acid, and the continued development of improved target tissue morphology preservation has virtually removed all the earlier major limitations of the *in situ* hybridization technique. It is now a routine technique in some diagnostic laboratories, especially in clinical cytogenetics to complement standard cytogenetic procedures in identifying marker chromosomes and assisting in the detection of numerical and structural chromosome abnormalities. With the introduction of comparative genome hybridization, FISH has a major application in both diagnosis and research of cancer. It can enable the monitoring of the effects of therapy and the detection of minimal residual disease, and the examination of the karyotypic pattern of nondividing or interphase cells.

References

1. Pardue, M. L. and Gall, J. G. (1969) Molecular hybridization of radioactive DNA to the DNA of cytological preparations. *Proc. Natl. Acad. Sci. USA* **64,** 600–604.
2. John, H., Birnstiel, M., and Jones, K. (1969) RNA-DNA hybrids at the cytological level. *Nature* **223,** 582–587.

3. Kessler, C. (1990) The digoxygenin system: principle and applications of the novel non-radioactive DNA labeling and detection system. *BioTechnology Int.* **1990**, 183–194.

4. Wiegant, J., Ried, T., Nederlof, P. M., van der Ploeg, M., Tanke, H. J., and Raap, A. K. (1991) In situ hybridization with fluoresceinated DNA. *Nucleic Acids Res.* **19**, 3237–3241.

5. Earle, E. and Choo, K. (1994) in *Methods in Molecular Biology, vol. 33:* In Situ *Hybridization Protocols* (Choo, K., ed.), vol. 33, pp. 147–158, Humana, Totowa, NJ, pp. 147–158.

6. Wulf, M., Bosse, A., Voss, B., and Muller, K. (1995) Improvement of fluorescence in situ hybridization (RNA-FISH) on human paraffin sections by Propidium Iodide counterstaining. *Biotechniques* **19**, 168–172.

7. Pinkel, D., Gray, J. W., Trask, B., van den Engh, G., Fuscoe, J., and van Dekken, H. (1986) Cytogenetic analysis by in situ hybridization with fluorescently labeled nucleic acid probes. *Cold Spring Harb Symp Quant Biol* **51(Pt 1)**, 151–157.

8. Nederlof, P. M., van der Flier, S., Wiegant, J., Raap, A. K., Tanke, H. J., Ploem, J. S., and van der Ploeg, M. (1990) Multiple fluorescence in situ hybridization. *Cytometry* **11**, 126–131.

9. Ried, T., Baldini, A., Rand, T. C., and Ward, D. C. (1992) Simultaneous visualization of seven different DNA probes by in situ hybridization using combinatorial fluorescence and digital imaging microscopy. *Proc. Natl. Acad. Sci. USA* **89**, 1388–1392.

10. Speicher, M., Ballard, S., and Ward, D. (1996) Karyotyping human chromosomes by combinatorial multi-fluor FISH. *Nature Genet.* **12**, 368–375.

11. Lichter, P., Cremer, T., Tang, C. J., Watkins, P. C., Manuelidis, L., and Ward, D. C. (1988) Rapid detection of human chromosome 21 aberrations by in situ hybridization. *Proc. Natl. Acad. Sci. USA* **85**, 9664–9668.

12. Jauch, A., Daumer, C., Lichter, P., Murken, J., Schroeder Kurth, T., and Cremer, T. (1990) Chromosomal in situ suppression hybridization of human gonosomes and autosomes and its use in clinical cytogenetics. *Hum. Genet.* **85**, 145–150.

13. Kievits, T., Dauwerse, J. G., Wiegant, J., Devilee, P., Breuning, M. H., Cornelisse, C. J., van Ommen, G. J., and Pearson, P. L. (1990) Rapid subchromosomal localization of cosmids by nonradioactive in situ hybridization. *Cytogenet Cell Genet* **53**, 134–136.

14. Boyle, A., Lichter, P., and Ward, D. (1990) Rapid analysis of mouse-hamster hybrid cell lines by in situ hybridization. *Genomics* **7**.

15. Lengauer, C., Riethman, H., and Cremer, T. (1990) Painting of human chromosomes with probes generated from hybrid cell lines by PCR with Alu and L1 primers. *Hum. Genet.* **86**, 1–6.

16. Lichter, P., Ledbetter, S., Ledbetter, D., and Ward, D. (1990) Fluorescence in situ hybridization with Alu and L1 polymerase chain reaction probes for rapid characterization of human chromosomes in hybrid cell lines. *Proc. Natl. Acad. Sci. USA* **87**, 6634–6638.

17. Baldini, A., Ross, M., Nizetic, D., Vatcheva, R., Lindsay, E., Lehrach, H., and Siniscalco, M. (1992) Chromosomal assignment of human YAC clones by fluorescence in situ hybridization: use of single colony PCR and multiple labeling. *Genomics* **14**, 181–184.

18. Liu, P., Siciliano, J., Seong, D., Craig, J., Zhao, Y., Jong, P. D., and Siciliano, M. (1993) Dual PCR primers and conditions for isolation of human chromosome painting probes from hybrid cell. *Cancer Genet. Cytogenet.* **65**, 93–99.

19. Muller, S., Koehler, U., Wienberg, J., Marsella, R., Finelli, P., Antonacci, R., Rocchi, M., and Archidiacono, N. (1996) Comparative fluorescence in situ hybridization mapping of primate chromosomes with Alu polymerase chain reaction generated probes from human/rodent somatic cell hybrids. *Chromosome Res.* **4**, 38–42.

20. Bohlander, S. K., Espinosa, R. d., Le Beau, M. M., Rowley, J. D., and Diaz, M. O. (1992) A method for the rapid sequence-independent amplification of microdissected chromosomal material. *Genomics* **13**, 1322–1324.

21. Meltzer, P. S., Guan, X. Y., Burgess, A., and Trent, J. M. (1992) Rapid generation of region specific probes by chromosome microdissection and their application. *Nat. Genet.* **1,** 24–28.

22. Guan, X. Y., Meltzer, P. S., and Trent, J. M. (1994) Rapid generation of whole chromosome painting probes (WCPs) by chromosome microdissection. *Genomics* **22,** 101–107.

23. Lengauer, C., Eckelt, A., Weith, A., Endlich, N., Ponelies, N., Lichter, P., Greulich, K. O., and Cremer, T. (1991) Painting of defined chromosomal regions by in situ suppression hybridization of libraries from laser-microdissected chromosomes. *Cytogenet. Cell Genet.* **56,** 27–30.

24. Gray, J., Langlois, G., Carrano, A., Burkhart-Schultz, K., and Dilla, M. V. (1979) High resolution chromosome analysis: One and two parameter flow cytometry. *Chromosoma* **73,** 9–27.

25. Telenius, H., Pelmear, A. H., Tunnacliffe, A., Carter, N. P., Behmel, A., Ferguson Smith, M. A., Nordenskjold, M., Pfragner, R., and Ponder, B. A. (1992) Cytogenetic analysis by chromosome painting using DOP-PCR amplified flow-sorted chromosomes. *Genes Chromosom. Cancer* **4,** 257–263.

26. Telenius, H., Carter, N. P., Bebb, C. E., Nordenskjold, M., Ponder, B. A., and Tunnacliffe, A. (1992) Degenerate oligonucleotide-primed PCR: general amplification of target DNA by a single degenerate primer. *Genomics* **13,** 718–725.

27. Vooijs, M., Yu, L. C., Tkachuk, D., Pinkel, D., Johnson, D., and Gray, J. W. (1993) Libraries for each human chromosome, constructed from sorter-enriched chromosomes by using linker-adaptor PCR. *Am. J. Hum. Genet.* **52,** 586–597.

28. Saltman, D. L., Dolganov, G. M., Pearce, B. S., Kuo, S. S., Callahan, P. J., Cleary, M. L., and Lovett, M. (1992) Isolation of region-specific cosmids from chromosome 5 by hybridization with microdissection clones. *Nucleic Acids Res.* **20,** 1401–1404.

29. Pinkel, D., Landegent, J., Collins, C., Fuscoe, J., Segraves, R., Lucas, J., and Gray, J. (1988) Fluorescence in situ hybridization with human chromosome-specific libraries: detection of trisomy 21 and translocations of chromosome 4. *Proc. Natl. Acad. Sci. USA* **85,** 9138–9142.

30. Suijkerbuijk, R. F., Matthopoulos, D., Kearney, L., Monard, S., Uhut, S., Cotter, F. E., Herbergs, J., van Kessel, A. G., and Young, B. D. (1992) Fluorescence in situ identification of human marker chromosomes using flow sorting and Alu element-mediated PCR. *Genomics* **13,** 355–362.

31. Langford, C. F., Telenius, H., Carter, N. P., Miller, N. G., and Tucker, E. M. (1992) Chromosome painting using chromosome-specific probes from flow-sorted pig chromosomes. *Cytogenet. Cell Genet.* **61,** 221–223.

32. Langford, C., Fischer, P., Binns, M., Holmes, N., and Carter, N. (1996) Chromosome-specific paints from a high resolution flow karyotype of the dog. *Chromosome Res.* **4,** 115–123.

33. Fuchs, J., Houben, A., Brandes, A., and Schubert, I. (1996) Chromosome 'painting' in plants—a feasible technique? *Chromosoma* **104,** 315–320.

34. Kallioniemi, A., Kallioniemi, O. P., Sudar, D., Rutovitz, D., Gray, J. W., Waldman, F., and Pinkel, D. (1992) Comparative genomic hybridization for molecular cytogenetic analysis of solid tumors. *Science* **258,** 818–821.

35. du Manoir, S., Speicher, M. R., Joos, S., Schrock, E., Popp, S., Dohner, H., Kovacs, G., Robert Nicoud, M., Lichter, P., and Cremer, T. (1993) Detection of complete and partial chromosome gains and losses by comparative genomic in situ hybridization. *Hum. Genet.* **90,** 590–610.

36. James, L. and Varley, J. (1996) Preparation, labeling and detection of DNA from archival tissue sections suitable for comparative genomic hybridization. *Chromosome Res.* **4,** 163,164.

37. Bryndorf, T., Kirchhoff, M., Rose, H., Maahr, J., Gerdes, T., Karhu, R., Kallioniemi, A., Christensen, B., Lundsteen, C., and Philip, J. (1995) Comparative genomic hybridization in clinical cytogenetics. *Am. J. Hum. Genet.* **57,** 1211–1220.

38. Manuelidis, L. (1990) A view of interphase chromosomes. *Science* **250,** 1533–1540.

39. Haaf, T. and Schmid, M. (1991) Chromosome topology in mammalian interphase nuclei. *Exp. Cell. Res.* **192,** 325–332.

40. Gilson, E., Laroche, T., and Gasser, S. (1993) Telomeres and the functional architecture of the nucleus. *Trends Cell Biol.* **3,** 128–134.

41. Nagele, R., Freeman, T., McMorrow, L., and Lee, H. Y. (1995) Precise spatial positioning of chromosomes during prometaphase: evidence for chromosomal order. *Science* **270,** 1831–1835.

42. Henikoff, S. and Dreesen, T. D. (1989) Trans-inactivation of the Drosophila brown gene: evidence for transcriptional repression and somatic pairing dependence. *Proc. Natl. Acad. Sci. USA* **86,** 6704–6708.

43. Loidl, J. (1990) The initiation of meiotic chromosome pairing: the cytological view. *Genome* **33,** 759–778.

44. Tease, C. (1996) Analysis using dual-color fluorescence in situ hybridization of meiotic chromosome segregation in male mice heterozygous for a reciprocal translocation. *Chromosome Res.* **4,** 61–68.

45. Trask, B. J. (1991) DNA sequence localization in metaphase and interphase cells by fluorescence in situ hybridization. *Methods Cell Biol.* **35,** 3–35.

46. Haaf, T. and Ward, D. C. (1994) Structural analysis of alpha-satellite DNA and centromere proteins using extended chromatin and chromosomes. *Hum. Mol. Genet.* **3,** 697–709.

47. Trask, B., Pinkel, D., and van den Engh, G. (1989) The proximity of DNA sequences in interphase cell nuclei correlated to genomic distance and permits ordering of cosmids spanning 250 kilobase pairs. *Genomics* **5,** 710–717.

48. Lawrence, J. B., Carter, K. C., and Gerdes, M. J. (1992) Extending the capabilities of interphase chromatin mapping [news]. *Nat. Genet.* **2,** 171,172.

49. Wiegant, J., Kalle, W., Mullenders, L., Brookes, S., Hoovers, J. M., Dauwerse, J. G., van Ommen, G. J., and Raap, A. K. (1992) High-resolution in situ hybridization using DNA halo preparations. *Hum. Mol. Genet.* **1,** 587–591.

50. Lawrence, J. B., Singer, R. H., and McNeil, J. A. (1990) Interphase and metaphase resolution of different distances within the human dystrophin gene. *Science* **249,** 928–932.

51. Heng, H. H., Squire, J., and Tsui, L. C. (1992) High-resolution mapping of mammalian genes by in situ hybridization to free chromatin. *Proc. Natl. Acad. Sci. USA* **89,** 9509–9513.

52. Parra, I. and Windle, B. (1993) High resolution visual mapping of stretched DNA by fluorescent hybridization [see comments]. *Nat. Genet.* **5,** 17–21.

53. Haaf, T. and Ward, D. C. (1994) High resolution ordering of YAC contigs using extended chromatin and chromosomes. *Hum. Mol. Genet.* **3,** 629–633.

54. Fidlerova, H., Senger, G., Kost, M., Sanseau, P., and Sheer, D. (1994) Two simple procedures for releasing chromatin from routinely fixed cells for fluorescence in situ hybridization. *Cytogenet. Cell Genet.* **65,** 203–205.

55. Heiskanen, M., Karhu, R., Hellsten, E., Peltonen, L., Kallioniemi, O. P., and Palotie, A. (1994) High resolution mapping using fluorescence in situ hybridization to extended DNA fibers prepared from agarose-embedded cells. *Biotechniques* **17,** 928,929, 932,923.

56. Heiskanen, M., Hellsten, E., Kallioniemi, O. P., Makela, T. P., Alitalo, K., Peltonen, L., and Palotie, A. (1995) Visual mapping by fiber-FISH. *Genomics* **30,** 31–36.

57. Senger, G., Jones, T. A., Fidlerova, H., Sanseau, P., Trowsdale, J., Duff, M., and Sheer, D. (1994) Released chromatin: linearized DNA for high resolution fluorescence in situ hybridization. *Hum. Mol. Genet.* **3,** 1275–1280.

58. Hudson, P., Penschow, J., Shine, J., Ryan, G., Niall, H., and Coghlan, J. (1981) Hybridization histochemistry: Use of recombinant DNA as a 'homing probe' for tissue localization of specific mRNA populations. *Endocrinology* **108**, 353–356.

59. Gerfen, C. (1989) Quantification of in situ hybridization histochemistry for analysis of brain function. *Methods Neurosci.* **1**, 79–97.

60. Pringle, J. H., Ruprai, A. K., Primrose, L., Keyte, J., Potter, L., Close, P., and Lauder, I. (1990) In situ hybridization of immunoglobulin light chain mRNA in paraffin sections using biotinylated or hapten-labeled oligonucleotide probes. *J. Pathol.* **162**, 197–207.

61. McNeil, J. A., Johnson, C. V., Carter, K. C., Singer, R. H., and Lawrence, J. B. (1991) Localizing DNA and RNA within nuclei and chromosomes by fluorescence in situ hybridization. *Genet. Anal. Tech. Appl.* **8**, 41–58.

62. Birk, P. and Grimm, P. (1994) Rapid nonradioactive in situ hybridization for interleukin-2 mRNA with riboprobes generated using the polymerase chain reaction. *J. Immunol. Methods* **167**, 83.

63. Dirks, R., Van de Rijke, F., Fujishita, S., Van der Ploeg, M., and Raap, A. (1993) Methodology for specific intron and exon RNA detection in cultured cells by haptenized and fluorochromized probes. *J. Cell Sci.* **104**, 1187–1197.

64. Wilkinson, D. and Green, J. (1990) *In Situ Hybridization and the Three Dimensional Reconstruction of Serial Sections.* IRL, Oxford, UK.

65. De Chiara, T. M., Robertson, E. J., and Efstratadis, A. (1991) Parental imprinting of the mouse insulin-like growth factor II gene. *Cell* **64**, 849–859.

66. Koch, J., Kolvraa, S., Petersen, K., Gregersen, N., and Bolund, L. (1989) Oligonucleotide-priming methods for the chromosome specific labeling of alpha satellite DNA in situ. *Chromosoma* **98**, 259–265.

67. Hindkjaer, J., Koch, J., Mogensen, J., Kolvraa, S., and Bolund, L. (1994) Primed in situ (PRINS) labeling of DNA. *Methods Mol. Biol.* **33**, 95–107.

68. Hindkjaer, J., Koch, J., and Mogensen, J. (1991) In situ labeling of nucleic acids for gene mapping, diagnostics and functional cytogenetics. *Adv. Mol. Genet.* **4**, 45–59.

69. Volpi, E. V. and Baldini, A. (1993) MULTIPRINS: a method for multicolor primed in situ labeling. *Chromosome Res.* **1**, 257–260.

70. Speel, E. J., Lawson, D., Hopman, A. H., and Gosden, J. (1995) Multi-PRINS: multiple sequential oligonucleotide primed in situ DNA synthesis reactions label specific chromosomes and produce bands. *Hum. Genet.* **95**, 29–33.

71. Speel, E. J., Jansen, M. P., Ramaekers, F. C., and Hopman, A. H. (1994) A novel triple-color detection procedure for brightfield microscopy, combining in situ hybridization with immunocytochemistry. *J. Histochem. Cytochem.* **42**, 1299–1307.

72. Pellestor, F., Girardet, A., Lefort, G., Andreo, B., and Charlieu, J. P. (1995) Use of the primed in situ labeling (PRINS) technique for a rapid detection of chromosomes 13, 16, 18, 21, X and Y. *Hum. Genet.* **95**, 12–17.

73. Macas, J., Dolezel, J., Pich, U., Schubert, I., and Lucretti, S. (1995) Primer-induced labeling of pea and field bean chromosomes in situ and in suspension. *Biotechniques* **19**, 402–408.

74. Mogensen, J., Kolvraa, S., Hindkjaer, J., Petersen, S., Koch, J., Nygard, M., Jensen, T., Gregersen, N., Junker, S., and Bolund, L. (1991) Nonradioactive, sequence-specific detection of RNA in situ by primed in situ labeling (PRINS) *Exp. Cell. Res.* **196**, 92–98.

75. Chen, R. H. and Fuggle, S. V. (1993) In situ cDNA polymerase chain reaction. A novel technique for detecting mRNA expression. *Am. J. Pathol.* **143**, 1527–1534.

76. Haaf, T. and Ward, D. (1996) Inhibition of RNA polymerase transcription causes chromatin decondensation, loss of nucleolar structure, and dispersion of chromosomal domains. *Exp. Cell Res.* **224**, 163–173.

77. Weber-Matthiesen, K., Deerberg, J., Muller-Hermelink, A., Schlegelberger, B., and Grote, W. (1993) Rapid immunophenotypic characterization of chromosomally aberrant cells by the new fiction method. *Cytogenet. Cell. Genet.* **63,** 123–125.

78. Speel, E. J., Herbergs, J., Ramaekers, F. C., and Hopman, A. H. (1994) Combined immunocytochemistry and fluorescence in situ hybridization for simultaneous tricolor detection of cell cycle, genomic, and phenotypic parameters of tumor cells. *J. Histochem. Cytochem.* **42,** 961–966.

79. Farr, C. J., Bayne, R. A. L., Kipling, D., Mills, W., Critcher, R., and Cooke, H. (1995) Generation of a human X-derived minichromosome using telomere-associated chromosome fragmentation. *EMBO J.* **14,** 5444–5454.

80. Haase, A., Brahic, M., Stowring, L., and Blum, H. (1984) Detection of viral nucleic acids by in situ hybridization. *Meth. Virol.* **7,** 189–226.

81. Gowans, E., Jilbert, A., and Burrell, C. (1989) in *Nucleic Acid Probes* (Symons, R., ed.), CRC, Boca Raton, FL, pp. 130–158.

82. Brigati, D. J., Myerson, D., Leary, J. J., Spalholz, B. N., Travis, S. Z., Fong, C. K. Y., Hsiung, D. G., and Ward, D. C. (1983) Detection of viral genomes in cultured cells and paraffin embedded tissue sections using biotin labeled hybridization probes. *Virology* **126**.

83. McDougall, J. K., Myerson, D., and Beckmann, A. M. (1986) Detection of viral DNA and RNA by insitu hybridization. *J. Histochem. Cytochem.* **34,** 33–38.

84. Brahic, M., Haase, A., and Cash, E. (1984) Simultaneous in situ detection of viral RNA and antigens. *Proc. Natl. Acad. Sci. USA* **81,** 5445.

Index

A

Acrylamide, *see* Polyacrylamide gels
Affinity chromatography, 469–476
Affinity maturation, 586, 587
Agarose gels, 17–20, 414
Agrobacterium tumefaciens, 215–254
Alkaline phosphatase, 74, 437, 639
Alpha particles, 109
Amido black, 436
Amino acid analysis, 520
Ampholytes, 416, 417
Anion exchangers, 447
Antibodies, 549–551, 567–578, 581–591,
    631, 632
    detection of proteins on blots, 437
    enzyme labeling of, 638–640
    epitope mapping of, 619–626
    structure, 583, 632
    use in immunocytochemistry, 631–646
Antisense, 329
Autoradiography, 26, 109–118, 419, 420
Avidin–biotin, 638–640
Azlactone, 471

B

Bacterial artificial chromosomes (BACs), 148
Bacteriophage vectors, 133–135
Baculovirus vectors, 219–229
Beta-galactosidase, 245–246, 639
Beta particles, 110
Biogel, 453, 471
Biolistic process, 258–260
BLOTTO, 437
Bolton and Hunter reagent, 406–408
Broad-host range vectors, 170, 171
Bromophenol blue, 19

C

Capillary electrophoresis, *see* Free zone
    capillary electrophoresis
Capillary isoelectric focusing, 425
Carboxypeptidases, 516
Cation exchanges, 447

Catrimix, 1, 14
cDNA libraries, 131–141
Cell cycle, 660
Cell sorting, 655–657
Chaotropic agents, 462
Chemical ionization, 670
Chloramine T, 403, 404
Chromatography
    affinity, 469–476
    hydrophobic interaction, 461–466
    ion exchange, 445–449
    reversed phase HPLC, 479–488
    size-exclusion, 451–459
Chromosome walking, 294
Cibacron blue, 474
Competition ELISA, 601–607, 613
Complementarity determining regions
    (CDR), 583
Constant region (antibody), 584
Continuous buffer systems, 415, 416
Coomassie blue, 418
Copy number estimation, 82–84
Cosmids, 147
C-terminal sequence determination, 515–517
Cyanogen bromide, 507
Cycle sequencing, 101, 102
Cysteine
    modification of, 508
Cytochrome P450 enzymes, 227, 228

D

2-D PAGE, *see* Two-dimensional
    polyacrylamide gel electrophoresis
Diagnostic antibodies, 576–578
Didoxigenin, 75
Diethylpyrocarbonate, 3
Digoxigenin, 74
Direct ELISA, 599, 607
Direct PCR sequencing, 313, 314
Direct selection vectors, 169, 170
Discontinuous buffer systems, 407
DNA
    binding proteins, 121–128

cDNA libraries, 131–141
    fluorescent labeling of, 105, 106, 660
    gel electrophoresis of, 17–31
        buffers for, 21
        preparation, 23, 24
    genomic libraries, 145–151
    in vitro transcription of, 327–333
    isolation from cultured cells, 9–15
    isolation from blood, 13
    isolation from tissues, 9–15
    non-radioactive labels, 62, 63, 10–106
    profiling, 383–396
    radiolabeling of, 62–64
    sequencing gels, 28, 102, 103
    sequencing of, 95–107, 184, 185
DNA, 251–254
DNases, 11
DNA polymerase, 99, 100, 308, 309
DNA size markers, 22, 26, 27
Domains, 555–560
Dot-blot, 440, 441

E

ECL, *see* Enhanced chemiluminescence
Edman degradation, 512–515
Electroblotting, 436
Electroendosmosis, 427
Electron ionization, 669, 670
Electroporation, 244, 245, 260, 261
Electrospray mass spectrometry, 508, 509,
        673, 674, 682–685, 688, 689
Electrostatic analyzer, 675
ELISA, *see* Enzyme linked immunosorbent
        assay
Embryonal stem (ES) cells, 362–375
Endoglycosidases, 493, 496–498
Endotoxins, 475
Enhanced chemiluminescence, 437, 438
Enzyme labeling of antibodies, 638–640
Enzyme-linked immunosorbent assay, 595–613
    competition, 601–607, 613
    direct, 599, 607
    indirect 599, 609
    sandwich, 600–607, 613
Enzymobeads, 404
Epitope mapping, 619–626
Ethidium bromide, 20–22
Eupergit, 471
Exoglycosidases, 496–498
Exon–intron mapping, 47
Expression vectors, 173–176

F

FAB, *see* Fast atom bombardment mass
        spectrometry
Fab antibody fragment, 583, 632
$F_{(ab)2}$ antibody fragment, 632
Far Western blotting, 442, 443
Fast atom bombardment mass spectrometry,
        511, 671, 672
Fast performance liquid chromatography,
        445, 446, 454
F_c antibody fragment, 632
FISH, *see* Fluorescence *in situ* hybridization
Flow cytometry, 653–663
Fluorescein isothiocyanate, 74
Fluorescence *in situ* hybridization, 283, 284
Fluorescent labeling
    of DNA, 105, 106
    for immunocytochemistry, 636–638
    for low cytometry, 657, 658
Fluorochromes, 634, 637
Fluorography, 109–118
Forensic identification, 390–392
FPLC, *see* Fast performance liquid
        chromatography
Free zone capillary electrophoresis, 425–433
Freund's adjuvant, 570
Functional screening of libraries, 159, 160

G

Gamma ray, 110
Gel filtration chromatography, *see* Size
        exclusion chromatography
Gene mapping, 84, 85, 274–277
Gene probes, 59–70
Gene therapy, 202–209
Genetic mapping, 281, 282
Genomic library, 145–151
Glucose oxidase, 404, 639
Glycoproteins, 514, 591–599
    glycosylation sites by ms, 691–695
    oligosaccharide structure by ms, 680–682
Glycosidases, 496–498
Glycosidic linkage, 495
Gram positive bacteria
    vectors for, 171, 172

H

HAT selection medium, 571
Heparin, 474
HIC, *see* Hydrophobic interaction
        chromatography

Histidine,
  radioiodination of, 404–406
Hofmeister series, 464
HRP, *see* Horseradish peroxidase
Horseradish peroxidase (HRP), 75, 437, 639
Humanization of antibodies, 578, 581, 582, 589
Hybridization, 68–70, 80–82
Hybridomas, 567–574
Hydrophobic interaction chromatography
      (HIC), 461–466
Hydroxylamine, 507
Hypervariable regions, 277, 278, 383

I

IEF, *see* Isoelectric focusing,
Immunoaffinity blotting, 438
Immunodetection of protein blots, 437
Immunoscreening, 159
Indirect ELISA, 599, 608
In situ hybridization, 283, 284, 644–646
Intensifying screen, 114
In vitro transcription, 327–333
  coupled with translation, 336, 337
In vitro translation, 335–343
  coupled with transcription, 336, 337
  eukaryotic systems, 337
Iodination of proteins, 402–406
Iodine 125, 402, 403
Iodo Beads, 406
Iodogen, 405, 406
Ion exchange chromatography, 445–449
Ion-spray ionization, 673, 674, 678–680
Isoelectric focusing, 416, 417, 420, 421

K

Knockouts, 369, 370, 374
Kunkel method, 350

L

Lactoperoxidase
  radiolabeling with, 404, 405
Ladder sequencing, 515
Lambda gt10, 135
Lambda gt11, 135
Lambda phage, 136, 153–161
Lambda ZAP, 135, 136
LCR, *see* Ligase chain reaction
Lectins, 473
Libraries
  cDNA, 131–141
  genomic, 145–151

  screening of, 139, 140, 149, 150
  YACs, 290, 291
Ligase chain reaction, 322, 354, 355, 357
Linkage analysis, 281, 282
Lipofection, 236, 242, 243
Luminol, 437
Lyotropic series, 464

M

M13 vectors, 181–185
MAbs, *see* Monoclonal antibodies
Magnetic sector analyser, 675
MALDI, *see* Matrix assisted laser
      desorption/ionization
Mass analysers, 675, 676
Mass spectrometry, 508–512, 669–695
  of oligosaccharides, 496
  protein sequence determination, 508–
      512, 517–520
  tandem, 517–520, 677, 678
Matrix assisted laser desorption/ionisation,
      50–509, 672, 673, 685–688, 689–695
Megaprimer technique, 317, 318
Microinjection, 261, 361–367
Microprojectiles, *see* Biolistic process
Microtiter plate, 596
Mobility shift assay, 121–128
Molecular weight markers
  DNA, 22, 26, 27
  proteins, 459
Monoclonal antibodies, 567–578
MS/MS, *see* Tandem mass spectrometry
MS^2, *see* Tandem mass spectrometry
Multiallelic markers, 278
Multigene families, 86, 87
Multilocus DNA profiling, 386–388
Mutant analysis, 86
Mutation
  analysis, 86
  detection by PCR, 319, 320
  screening, 340, 341, 349, 350
  site-directed, 347, 358
Myelomas, 567, 568

N

Nick translation, 63, 64
Nitrocellulose membranes, 435, 436
N-linked oligosaccharides, 492
Non-radioactive labeling
  of DNA, 62–64, 73–76

Northern blotting, 89–93
N-Terminal sequence determination, 512–515

O

Oligonucleotide probes, *see* Gene probes
Oligonucleotide site-directed mutagenesis, 356
Oligosaccharides, *see* Glycoproteins
O-linked oligosaccharides, 492
Orange G, 19

P

P1-derived artificial chromosomes (PACs), 148
Paternity testing, 394, 395
PCR labeling of DNA , 64
PCR, *see* Polymerase chain reaction
Peptide mapping, 420
PEPSCAN, method, 625
Peptides
    electrophoresis of, 425–433
    generation of, 506, 507
    radiolabeling of, 401–410
    synthesis of, 537, 538
Periodate-Schiffs stain, 493
Peroxidase-antiperoxidase (PAP) method, 638
PFGE, *see* Pulsed field gel electrophoresis
Phage display libraries, 581–591
Phagemid vectors, 186–190
Phenol, 10
Photobiotin labeling, 64
Photographic process, 110, 111
Physical mapping, 282, 294–296
Plants,
    transformation of, 251–265
Plasmid vectors, 133–135, 165–177
Polyacrylamide gels, 24, 25, 414, 415
Polybrene, 243, 244
Polyethyleneimine, 236
Polymerase chain reaction, 305–323
    mutagenesis , 317, 318, 352–355
    single overlap extension (SOE), 353–357
Polymorphisms
    detection by PCR, 318, 319
Polyvinylidene difluoride membrane, 436,
    440, 510, 513
Ponceau S, 436
Poros, 471
Pre-flashing, 114
Primers
    for PCR, 307, 308
Proteases, *see* Proteolytic enzymes

Protein A, 438, 473
Protein G, 438, 473
Proteins
    blotting of, 435–443
    ^{14}C labeling of, 409, 410
    cleavage of, 506, 507
    C-terminal sequencing, 515–517
    electrophoresis of, 413–422, 425–433
    engineering of, 547–560
    fingerprinting of by MS, 685–688
    glycoproteins, 491–499
    iodination of, 402–406
    ladder sequencing, 515
    molecular weight determination by ms,
        678, 679
    molecular weight standards, 459
    radiolabeling of, 401–410, 641
    sequence determination, 440, 503–520
    tritium labeling of, 409, 410
Protein–DNA interactions, 121–128, 342
Protein engineering, 547–560
Protein–protein interactions, 341, 342
Protein synthesis in vitro, 335–343
Proteome, 503
Proteolytic enzymes, 420, 506, 507
PTH amino acid, 514
pUC vectors, 167, 168
Pulsed field gel electrophoresis, 29–31, 282,
    283, 292–294
PVDF, *see* Polyvinylidene difluoride
Pyrogens, 475

Q

Quadruple mass analyzer, 675
Quantitation of RNA, 47, 48

R

Radioactivity safety, 117, 410
Radiolabeling
    of DNA, 62–64, 73–76
    of proteins and peptides, 401–410, 641
Reagent array analysis method (RAAM),
    497, 498
Reporter genes, 237, 238, 245, 246
Restriction endonucleases, 272
Restriction fragment length polymorphisms,
    271–279
Retroviral vectors, 193–212
Retroviruses, 193–197
Reversed phase high performance liquid
    chromatography, 479–488

Reverse transcriptase, 131
RFLP, *see* Restriction fragment length polymorphisms
Ribonuclease protection assay, 51–56
Ribosome binding site, 335
RNA
    analysis by PCR, 311, 312
    detection by RPA, 51–56
    extraction from cultured cells, 1–6
    extraction from tissues, 1–6
    northern blotting, 89–93
    processing, 329
    quantitation, 47, 48
    S1 nuclease, 35–48
    S1 nuclease digests, 43
    S1 nuclease mapping, 37–43
RNase, 3
RPA, *see* Ribonuclease protection assay
RP-HPLC, *see* Reversed phase high performance liquid chromatography
Runaway vectors, 170

S

S1 nuclease, 35–48
S1 nuclease mapping, 37–43
Sandwich ELISA, 600–601, 608–613
Screening
    libraries, 140
SDS PAGE, *see* SDS Polyacrylamide gel electrophoresis
SDS Polyacrylamide gel electrophoresis 417, 418 (*see also* 2D-PAGE)
Selection markers for transfected cells, 241
Sephacryl, 453, 471
Sephadex, 453
Sepharose, 453, 471
Sepse CL, 471
Sequence determination of proteins, 503–520
Sequencing gels (DNA), 28
Serine protease, 549
Shuttle vectors, 172
Silicon carbide whisker, 261
Silver staining, 418
Single chain Fv, 583
Single locus DNA profiling, 388
Single overlap extension (SOE) PCR, 353–357
Site-directed mutagenesis, 347–358, 551, 552
Size exclusion chromatography, 451–459
Southern blotting, 23, 77–86

Southwestern blotting, 442, 443
Specific activity (radioactive), 409
Stutter band, 278
Superdex, 453

T

T-DNA, 251–254
Tandem mass spectrometry, 517–520, 677, 678
Taq polymerase, 308, 309
Therapeutic antibodies, 576–578, 587, 588
Thiol group, *see* Cysteine
Ti plasmid, 251, 252
Toyopear, 471
Transcription
    in vitro, 327–331
Transfection, 235–237
Transformation
    of plants, 251–265
    of tissue culture cells, 235–247
Transgenesis, 361–376
Translation
    in vitro, 335–343
Triazine dyes, 474
Trisacryl, 471
Two dimensional (2D) polyacrylamide gel electrophoresis 420–422, 503–506
Tyrosine,
    iodination of, 403–406

U

Ultrogel, 471

V

Variable region (antibody), 584
Void volume, 451

W

Western blotting, *see* Proteins, blotting of

X

X-ray, 110
Xylene cyanol, 19

Y

YACs, *see* Yeast artificial chromosomes
Yeast artificial chromosomes, 148, 282, 287–300
Yeast two hybrid system, 140, 141

Z

Zygosity testing, 393